Student Solutions Manual

Mark A. McKibben
Goucher College

to accompany

College Algebra

Third Edition

Cynthia Y. Young
University of Central Florida

WILEY

PUBLISHER	Laurie Rosatone
ACQUISITIONS EDITOR	Joanna Dingle
PROJECT EDITOR	Jennifer Brady
ASSOCIATE CONTENT EDITOR	Beth Pearson
EDITORIAL ASSISTANT	Elizabeth Baird
SENIOR CONTENT MANAGER	Karoline Luciano
SENIOR PRODUCTION EDITOR	Kerry Weinstein

Founded in 1807, John Wiley & Sons, Inc. has been a valued source of knowledge and understanding for more than 200 years, helping people around the world meet their needs and fulfill their aspirations. Our company is built on a foundation of principles that include responsibility to the communities we serve and where we live and work. In 2008, we launched a Corporate Citizenship Initiative, a global effort to address the environmental, social, economic, and ethical challenges we face in our business. Among the issues we are addressing are carbon impact, paper specifications and procurement, ethical conduct within our business and among our vendors, and community and charitable support. For more information, please visit our website: www.wiley.com/go/citizenship.

ISBN 978-1-118-13757-4

10 9 8 7 6 5 4 3 2 1

Table of Contents

CHAPTER 0

Section 0.1 Solutions ---

1. rational (integer/integer)

3. irrational (doesn't repeat)

5. rational (repeats)

7. $\sqrt{5} = 2.2360\ldots$ irrational (doesn't repeat)

9. a. Rounding: 1 is less than 5, so 7 stays. $\boxed{7.347}$ **b.** Truncating: $7.347\|1$ $\boxed{7.347}$

11. a. Rounding: 9 is greater than 5, so 4 rounds up to 5. $\boxed{2.995}$ **b.** Truncating: $2.994\|9$ $\boxed{2.994}$

13. a. Rounding: 4 is less than 5, so 4 stays: $\boxed{0.234}$
b. Truncating: $0.234\|492$ $\boxed{0.234}$

15. a. Rounding: 4 is less than 5, so 4 stays: $\boxed{5.238}$
b. Truncating: $5.238\|473$ $\boxed{5.238}$

17.

$$5+\underbrace{2\cdot 3}_{6}-7=\underbrace{5+6}_{11}-7=11-7=\boxed{4}$$

19.

$$2\cdot\left(5+\underbrace{7\cdot 4}_{28}-20\right)=2\cdot\left(\underbrace{5+28}_{33}-20\right)=2\cdot(33-20)=2\cdot 13=\boxed{26}$$

21.

$$2-3\left[4(2\cdot 3+5)\right]=2-3\left[4(11)\right]=2-3(44)=\boxed{-130}$$

23. $8-(-2)+7=10+7=\boxed{17}$

25. $-3-(-6)=-3+6=\boxed{3}$

27. $x-(-y)-z=\boxed{x+y-z}$

29. $-(3x+y)=\boxed{-3x-y}$

31. $\dfrac{-3}{(5)(-1)}=\dfrac{-3}{-5}=\boxed{\dfrac{3}{5}}$

33.

$$\begin{aligned}-4-6\left[(5-8)(4)\right]&=-4-6\left[(-3)(4)\right]\\&=-4-6(-12)\\&=-4+72\\&=\boxed{68}\end{aligned}$$

35.

$$\begin{aligned}-(6x-4y)-(3x+5y)&=-6x+4y-3x-5y\\&=\boxed{-9x-y}\end{aligned}$$

37.

$$\begin{aligned}-(3-4x)-(4x+7)&=-3+4x-4x-7\\&=\boxed{-10}\end{aligned}$$

39.

$$\frac{-4(5)-5}{-5}=\frac{-20-5}{-5}=\boxed{5}$$

41. $\dfrac{1}{3}+\dfrac{5}{4}=\dfrac{1(4)+5(3)}{12}=\dfrac{4+15}{12}=\boxed{\dfrac{19}{12}}$

43. $\dfrac{5}{6}-\dfrac{1}{3}=\dfrac{5-1\cdot(2)}{6}=\dfrac{5-2}{6}=\dfrac{3}{6}=\boxed{\dfrac{1}{2}}$

45. $\frac{3}{2}+\frac{5}{12}=\frac{18}{12}+\frac{5}{12}=\boxed{\frac{23}{12}}$

47. $\frac{1}{9}-\frac{2}{27}=\frac{3}{27}-\frac{2}{27}=\boxed{\frac{1}{27}}$

49. $\dfrac{x}{5}+\dfrac{x\cdot(2)}{15}=\dfrac{3\cdot x+2\cdot x}{15}=\dfrac{5\cdot x}{15}=\boxed{\dfrac{x}{3}}$

51. $\frac{x}{3}-\frac{2x}{7}=\frac{7x-6x}{21}=\boxed{\frac{x}{21}}$

53. $\frac{4y}{15}-\frac{(-3y)}{4}=\frac{16y+(3y)(15)}{60}=\boxed{\frac{61y}{60}}$

55. $\frac{3}{40}+\frac{7}{24}=\frac{3(3)+5(7)}{120}=\boxed{\frac{11}{30}}$

57. $\frac{2}{7}\cdot\frac{14}{3}=\frac{2}{1}\cdot\frac{2}{3}=\boxed{\frac{4}{3}}$

59. $\frac{2}{7}\div\frac{10}{3}=\frac{2}{7}\cdot\frac{3}{10}=\frac{1}{7}\cdot\frac{3}{5}=\boxed{\frac{3}{35}}$

61. $\frac{4b}{9}\div\frac{a}{27}=\frac{4b}{9}\cdot\frac{27}{a}=\boxed{\frac{12b}{a}}$

63. $\frac{3x}{10}\cdot\frac{15}{6x}=\frac{3}{2\cdot 2}=\boxed{\frac{3}{4}}$

65. $\frac{3x}{4}\div\frac{9}{16y}=\frac{3x}{4}\cdot\frac{16y}{9}=\boxed{\frac{4xy}{3}}$

67. $\frac{6x}{7}\cdot\frac{28}{3y}=\boxed{\frac{8x}{y}}$

69. $-\frac{(-4)}{2(3)}=\frac{4}{6}=\boxed{\frac{2}{3}}$

71. $\frac{3(4)}{10^2}=\frac{3\cdot 2\cdot 2}{10\cdot 10}=\boxed{\frac{3}{25}}$

73. \$9,176,366,000,000

75. $\frac{9{,}176{,}366{,}494{,}947}{303{,}818{,}361}\approx \$30{,}203$

77. Only look to the right of the digit to round. 13.2$\underline{7}$49: the 4 is less than 5, so 7 remains the same. Don't round the 9 first. $\boxed{13.27}$

79.

$$\begin{aligned}3\cdot(x+5)-2\cdot(4+y)&=3\cdot x+15-8-2\cdot y\\&=3\cdot x-2\cdot y+7.\end{aligned}$$

Don't forget to distribute the -2 to both terms inside the parenthesis.

81. False. Not all student-athletes are honors students.

83. True. Every integer can be written as $\frac{\text{integer}}{1}$.

85. $x\neq 0$

87.

$$-2[3(x-2y)+7]+[3(2-5x)+10]-7[-2(x-3)+5]=$$
$$-2[3x-6y+7]+[6-15x+10]-7[-2x+6+5]=$$
$$-6x+12y-14+6-15x+10+14x-42-35=$$
$$\boxed{-7x+12y-75}$$

89. $\sqrt{1260}=35.4964787\ldots$ irrational

91. 67, so rational

Section 0.2 Solutions --

1. 256	**3.** $(-1)^5 3^5 = -243$
5. -25	**7.** $-4\cdot 4 = -16$
9. 1	**11.** $\frac{1}{10} = 0.1$
13. $\frac{1}{8^2} = \frac{1}{64}$	**15.** $-6\cdot 25 = -150$
17. $\cancel{8}\cdot\frac{1}{\cancel{2^3}}\cdot 5 = 5$	**19.** $-6\cdot\frac{1}{3^2}\cdot 81 = -6\cdot\frac{1}{9}\cdot 81 = -54$
21. $x^2\cdot x^3 = x^{2+3} = \boxed{x^5}$	**23.** $x^2\cdot x^{-3} = x^{2-3} = x^{-1} = \boxed{\frac{1}{x}}$
25. $\left(x^2\right)^3 = x^{2\cdot 3} = \boxed{x^6}$	**27.** $(4a)^3 = (4)^3\cdot(a)^3 = \boxed{64a^3}$
29. $(-2t)^3 = (-2)^3(t)^3 = \boxed{-8t^3}$	**31.** $\left(5xy^2\right)^2\left(3x^3y\right) = \left(25x^2y^4\right)\left(3x^3y\right) = 75x^5y^5$
33. $\frac{x^5y^3}{x^7y} = \left(x^{5-7}\right)\left(y^{3-1}\right) = x^{-2}\cdot y^2 = \boxed{\frac{y^2}{x^2}}$	**35.** $\frac{(2xy)^2}{(-2xy)^3} = \frac{4x^2y^2}{-8x^3y^3} = \left(\frac{-4}{8}\right)x^{2-3}y^{2-3} = \boxed{-\frac{1}{2xy}}$
37. $\left(\frac{b}{2}\right)^{-4} = \left(\frac{2}{b}\right)^4 = \frac{2^4}{b^4} = \boxed{\frac{16}{b^4}}$	**39.** $\left(9a^{-2}b^3\right)^{-2} = (9)^{-2}\left(a^{-2}\right)^{-2}\left(b^3\right)^{-2}$ $= \left(\frac{1}{9^2}\right)a^4b^{-6} = \boxed{\frac{a^4}{81b^6}}$
41. $\frac{a^{-2}b^3}{a^4b^5} = \left(\frac{a^{-2}}{a^4}\right)\left(\frac{b^3}{b^5}\right)$ $= \left(\frac{1}{a^6}\right)\left(\frac{1}{b^2}\right)$ $= \boxed{\frac{1}{a^6b^2}}$	**43.** $\frac{\left(x^3y^{-1}\right)^2}{\left(xy^2\right)^{-2}} = \frac{x^6y^{-2}}{x^{-2}y^{-4}}$ $= x^{6-(-2)}y^{-2-(-4)}$ $= \boxed{x^8y^2}$
45. $\frac{3\left(x^2y\right)^3}{12\left(x^{-2}y\right)^4} = \frac{3x^6y^3}{12x^{-8}y^4} = \frac{x^{14}}{4y}$	**47.** $\frac{\left(x^{-4}y^5\right)^{-2}}{\left[-2\left(x^3\right)^2y^{-4}\right]^5} = \frac{x^8y^{-10}}{\left(-2x^6y^{-4}\right)^5} = \frac{x^8y^{-10}}{-32x^{30}y^{-20}}$ $= \frac{y^{10}}{-32x^{22}}$
49. $\left[\frac{a^2(-1)x^3y^{12}}{a^6x^4y^4}\right]^3 = \frac{-a^6x^9y^{36}}{a^{18}x^{12}y^{12}} = -\frac{y^{24}}{a^{12}x^3}$	**51.** $2^8\cdot 16^3\cdot 64 = 2^8\cdot\left(2^4\right)^3\cdot 2^6 = 2^{26}$

53. $2.76\cdot 10^7$	**55.** 9.3×10^7	**57.** $5.67\cdot 10^{-8}$	**59.** 1.23×10^{-7}

61. 47,000,000	**63.** 23,000
65. 0.000041	
67. a. $(5.0\times10^9)(5\text{in.})=(5.0\times10^9)(\tfrac{5}{12}\text{ft.})\approx2.08\times10^9$ ft. **b.** First, convert miles to feet: $25{,}000\,\text{mi.}\times\frac{5{,}280\text{ ft.}}{1\text{ mi.}}=1.32\times10^8$ ft. This is the circumference of the earth. Now, $\frac{2.08\times10^9\text{ ft.}}{1.32\times10^8\text{ ft.}}\approx15.76$ So, yes, the line of cell phones would wrap around Earth nearly 16 times.	
69. 2.0×10^8 miles	
71. 0.00000155 meter	**73.** Should be adding exponents here: y^5
75. Computed $\left(y^3\right)^2$ incorrectly. Should be y^6.	**77.** False. For instance, if $n=2$, then $-2^2=-4$, while $(-2)^2=(-2)(-2)=4$.
79. False. $x\neq0$	**81.** Multiply all exponents here to get a^{mnk}.
83. $-(-2)^2+2(-2)(3)=-4-12=-16$	**85.** $-16(3)^2+100(3)=-144+300=156$
87. $\frac{(1.5\times10^8)(247)}{6.6\times10^9}=\frac{370.5\times10^8}{6.6\times10^9}\approx5.6$ acres per person.	**89.** $\frac{\left(4\times10^{-23}\right)\left(3\times10^{12}\right)}{6\times10^{-10}}=\frac{12\times10^{-11}}{6\times10^{-10}}=0.2$
91. Obtained the same answer using the calculator.	**93.** 5.11×10^{14}

Section 0.3 Solutions ---

1. $-7x^4-2x^3+5x^2+16$ $\boxed{\text{Degree 4}}$	**3.** $-6x^3+4x+3$ $\boxed{\text{Degree 3}}$
5. $15=15$ $\boxed{\text{Degree 0}}$	**7.** $y-2=y^1-2$ $\boxed{\text{Degree 1}}$
9. $2x^2-x+7-3x^2+6x-2$ $=\boxed{-x^2+5x+5}$	**11.** $-7x^2-5x-8+4x+9x^2-10$ $=2x^2-x-18$
13. $2x^4-7x^2+8-3x^2+2x^4-9$ $=4x^4-10x^2-1$	**15.** $7z^2-2-5z^2+2z-1=\boxed{2z^2+2z-3}$
17. $3y^3-7y^2+8y-4-14y^3+8y-9y^2$ $=-11y^3-16y^2+16y-4$	**19.** $6x-2y-10x+14y=-4x+12y$

21. $\underline{2x^2}-2-x-1-\underline{x^2}+5=\boxed{x^2-x+2}$	**23.** $4t-\underline{\underline{t^2}}-\underline{t^3}-\underline{\underline{3t^2}}+2t-\underline{2t^3}+\underline{3t^3}-1$ $=\boxed{-4t^2+6t-1}$
25. $(5\cdot 7)x^{1+1}y^{2+1}=\boxed{35x^2y^3}$	**27.** $2x^3-2x^4+2x^5=\boxed{2x^5-2x^4+2x^3}$
29. Distribute and arrange terms in decreasing order (according to degree): $10x^4-2x^3-10x^2$	**31.** Distribute and arrange terms in decreasing order (according to degree): $2x^5+2x^4-4x^3$
33. $2a^3b^2+4a^2b^3-6ab^4$	**35.** $6x^2-8x+3x-4=\boxed{6x^2-5x-4}$
37. $x^2+2x-2x-4=\boxed{x^2-4}$	**39.** $4x^2-6x+6x-9=\boxed{4x^2-9}$
41. $2x-4x^2-1+2x=\boxed{-4x^2+4x-1}$	**43.** $4x^4-9$
45. $7y^2-7y^3+7y-2y^3+2y^4-2y^2=$ $\boxed{2y^4-9y^3+5y^2+7y}$	**47.** $x^3-2x^2+3x+x^2-2x+3=$ $\boxed{x^3-x^2+x+3}$
49. $(t-2)(t-2)=t^2-2t-2t+4$ $=\boxed{t^2-4t+4}$	**51.** $(z+2)(z+2)=z^2+2z+2z+4$ $=\boxed{z^2+4z+4}$
53. $(x+y)^2-6(x+y)+9=$ $\boxed{x^2+2xy+y^2-6x-6y+9}$	**55.** $25x^2-20x+4$
57. $y\left(6y^2-3y+8y-4\right)=y\left(6y^2+5y-4\right)$ $=\boxed{6y^3+5y^2-4y}$	**59.** $x^4-x^2+x^2-1=\boxed{x^4-1}$
61. $(a+2b)(b-3a)(b+3a)=(a+2b)\left(b^2-9a^2\right)=$ $\boxed{ab^2+2b^3-9a^3-18a^2b}$	**63.** $2x^2-3xy+5xz+2xy-3y^2+5yz-2xz+3yz-5z^2$ $=\boxed{2x^2-3y^2-xy+3xz+8yz-5z^2}$

65. Revenue $=20x$
Cost $=100+9x$
Since Profit $=P=$ Revenue – Cost, we have
$P=20x-(100+9x)=20x-9x-100=\boxed{11x-100}$

67. Revenue $=-x^2+100x$
Cost $=-100x+7500$
Since Profit $=P=$ Revenue – Cost, we have
$P=-x^2+100x+100x-7500=\boxed{-x^2+200x-7500}$

69. Cutting a square with dimensions x from the four corners of the material create sides with lengths $15-2x$ and $8-2x$, and height x. The resulting volume is $V=(15-2x)(8-2x)x=4x^3-46x^2+120x$.

71. a. The perimeters of the semi-circular pieces are each $\frac{1}{2}(2\pi x) = \pi x$ feet.
The perimeter of the exposed rectangular piece is $2(2x+5)$ feet.
Thus, the perimeter of the track is $P = (2\pi x + 4x + 10)$ feet.

b. The areas of the semi-circular pieces are each $\frac{1}{2}(\pi x^2) = \frac{\pi}{2}x^2$ feet^2.
The area of the rectangular piece is $(2x)(2x+5)$ feet^2.
Thus, the area of the track is $A = (\pi x^2 + 4x^2 + 10x)$ feet^2.

73. $F = \frac{k(x)(3x)}{(10x)^2} = \frac{3kx^2}{100x^2} = \frac{3k}{100}$

75. Forgot to distribute the negative sign through the entire second polynomial.
$2x^2 - 5 - 3x + x^2 - 1 = \boxed{3x^2 - 3x - 6}$

77. True.

79. False.
$(x+y)^3 = (x+y)(x^2 + 2xy + y^2)$
$= x^3 + 3xy^2 + 3x^2y + y^3$

81. Add the degrees, so $m + n$.

83.
$\left[(7x-4y^2)(7x+4y^2)\right]^2 = \left[49x^2 - 16y^4\right]^2$
$= (49x^2 - 16y^4)(49x^2 - 16y^4)$
$= 2401x^4 - 784x^2y^4 - 784x^2y^4 + 256y^8$
$= \boxed{2401x^4 - 1568x^2y^4 + 256y^8}$

85.
$(x-a)(x^2 + ax + a^2) =$
$x^3 + ax^2 + a^2x - ax^2 - a^2x - a^3 =$
$x^3 - a^3$

87.

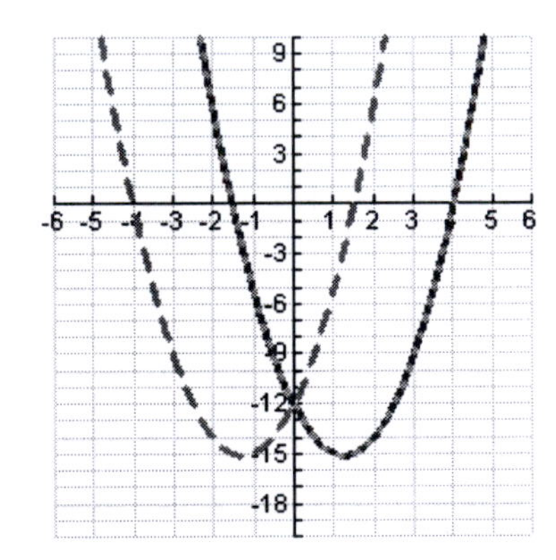

The solid curve represents the graph of both $y = (2x+3)(x-4)$ and $y = 2x^2 - 5x - 12$.

Section 0.4 Solutions ---

1. $5(x+5)$	**3.** $2(2t^2-1)$
5. $2x(x^2-25)=2x(x-5)(x+5)$	**7.** $3x(x^2-3x+4)$
9. $x(x^2-3x-40)=x(x-8)(x+5)$	**11.** $2xy(2xy^2+3)$
13. $(x-3)(x+3)$	**15.** $(2x-3)(2x+3)$
17. $2(x^2-49)=2(x-7)(x+7)$	**19.** $(15x-13y)(15x+13y)$
21. $(x+4)^2$	**23.** $(x^2-2)^2$
25. $(2x+3y)^2$	**27.** $(x-3)^2$
29. $(x^2+1)^2$	**31.** $(p+q)^2$
33. $(t+3)(t^2-3t+9)$	**35.** $(y-4)(y^2+4y+16)$
37. $(2-x)(4+2x+x^2)$	**39.** $(y+5)(y^2-5y+25)$
41. $(3+x)(9-3x+x^2)$	**43.** $(x-5)(x-1)$
45. $(y-3)(y+1)$	**47.** $(2y+1)(y-3)$
49. $(3t+1)(t+2)$	**51.** $(-3t+2)(2t+1)$
53. $(x^3-3x^2)+(2x-6)=x^2(x-3)+2(x-3)$ $=\boxed{(x^2+2)(x-3)}$	**55.** $a^3(a+2)-8(a+2)=(a^3-8)(a+2)$ $=(a+2)(a-2)(a^2+2a+4)$
57. $3xy+6sy-5rx-10rs=$ $3y(x+2s)-5r(x+2s)=$ $(3y-5r)(x+2s)$	**59.** $4x(5x+2y)-y(5x+2y)=$ $(4x-y)(5x+2y)$
61. $(x-2y)(x+2y)$	**63.** $(3a+7)(a-2)$
65. prime	**67.** prime
69. $2(3x^2+5x+2)=2(3x+2)(x+1)$	**71.** $(3x-y)(2x+5y)$
73. $9(4s^2-t^2)=9(2s-t)(2s+t)$	**75.** $(ab)^2-(5c)^2=(ab-5c)(ab+5c)$
77. $(x-2)(4x+5)$	**79.** $x(3x^2-5x-2)=\boxed{x(3x+1)(x-2)}$
81. $x(x^2-9)=\boxed{x(x-3)(x+3)}$	**83.** $x(y-1)-(y-1)=(x-1)(y-1)$
85. $(x^2+3)(x^2+2)$	**87.** $(x-6)(x+4)$

89. $x\left(x^3+125\right)=x(x+5)\left(x^2-5x+25\right)$	**91.** $\left(x^2-9\right)\left(x^2+9\right)=(x-3)(x+3)\left(x^2+9\right)$
93. $p=2(2x+4)+2x=6x+8=2(3x+4)$	**95.** $p=x(2x-15)+2(2x-15)$ $=(x+2)(2x-15)$
97. $-2\left(8t^2+39t-5\right)=-2(8t-1)(t+5)$	**99.** Last step. $(x-1)\left(x^2-9\right)=(x-1)(x-3)(x+3)$
101. False	**103.** True
105. $a^{2n}-b^{2n}=\boxed{\left(a^n-b^n\right)\left(a^n+b^n\right)}$	

107.

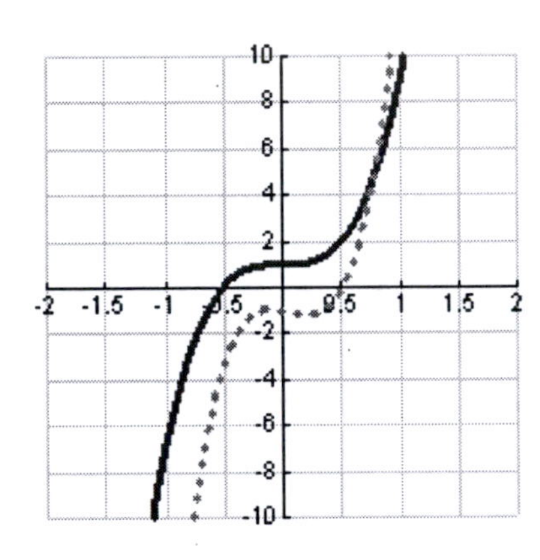

The solid curve represents the graph of both $y=8x^3+1$ and $y=(2x+1)(4x^2-2x+1)$.

Section 0.5 Solutions ---

1. $\boxed{x\neq 0}$	**3.** $x-1\neq 0 \Rightarrow \boxed{x\neq 1}$
5. $x+1\neq 0 \Rightarrow \boxed{x\neq -1}$	**7.** $p^2-1\neq 0 \Rightarrow \boxed{p\neq \pm 1}$
9. p^2+1 is never equal to zero for real values of p. So, no restrictions.	
11. $\dfrac{(x-9)\cancel{(x+3)}}{2(x+9)\cancel{(x+3)}}=\dfrac{(x-9)}{2(x+9)}$ Note: $x\neq -9,-3$	**13.** $\dfrac{(x-3)\cancel{(x+1)}}{2\cancel{(x+1)}}=\boxed{\dfrac{x-3}{2}}$ Note: $x\neq -1$

15. $\frac{2(3y+1)\cancel{(2y-1)}}{3(3y)\cancel{(2y-1)}}=\frac{2(3y+1)}{9y}$ Note: $y\neq 0,\frac{1}{2}$	**17.** $\frac{\cancel{(5y-1)}(y+1)}{5\cancel{(5y-1)}}=\boxed{\frac{y+1}{5}}$ Note: $y\neq\frac{1}{5}$
19. $\frac{(3x+7)\cancel{(x-4)}}{4\cancel{(x-4)}}=\frac{(3x+7)}{4}$ Note: $x\neq 4$	**21.** $\frac{\cancel{(x-2)}(x+2)}{\cancel{(x-2)}}=\boxed{x+2}$ Note: $x\neq 2$
23. $\boxed{1}$ Domain: Note $x\neq -7$	**25.** $\boxed{\frac{x^2+9}{2x+9}}$ Note: $x\neq -9/2$
27. $\frac{(x+3)\cancel{(x+2)}}{(x-5)\cancel{(x+2)}}=\frac{(x+3)}{(x-5)}$ Note: $x\neq -2,5$	**29.** $\frac{(3x+1)\cancel{(2x-1)}}{(x+5)\cancel{(2x-1)}}=\frac{(3x+1)}{(x+5)}$ Note: $x\neq -5,\frac{1}{2}$
31. $\frac{3x+5}{x+1}$ Note: $x\neq -1,2$	**33.** $\frac{2(5x+6)}{5x-6}$ Note: $x\neq 0,\frac{6}{5}$
35. $\frac{2\cancel{(x-1)}}{3x}\cdot\frac{x\cancel{(x+1)}}{\cancel{(x+1)}\cancel{(x-1)}}=\boxed{\frac{2}{3}}$ Note: $x\neq -1,0,1$	**37.** $\frac{3(x^2-4)}{\cancel{x}}\cdot\frac{\cancel{x}\cancel{(x+5)}}{\cancel{(x+5)}(x-2)}=\frac{3(x+2)\cancel{(x-2)}}{\cancel{(x-2)}}=\boxed{3(x+2)}$ Note: $x\neq -5,0,2$
39. $\frac{\cancel{t+2}}{3\cancel{(t-3)}}\cdot\frac{\cancel{(t-3)}(t-3)}{\cancel{(t+2)}(t+2)}=\boxed{\frac{t-3}{3(t+2)}}$ Note: $t\neq -2,3$	**41.** $\frac{3t(t^2+4)}{(t-3)(t+2)}$ Cannot be simplified. Note: $t\neq -2,3$
43. $\frac{(y-2)\cancel{(y+2)}}{y-3}\cdot\frac{3y}{\cancel{y+2}}=\frac{3y(y-2)}{y-3}$ Note: $y\neq -2,3$	**45.** $\frac{\cancel{3x}\cancel{(x-5)}}{2\cancel{x}(x+5)\cancel{(x-5)}}\cdot\frac{(2x+3)(x-5)}{\cancel{3}x(x+5)}=\frac{(2x+3)(x-5)}{2x(x+5)^2}$ Note: $x\neq 0,\pm 5$
47. $\frac{\cancel{(3x+5)}\cancel{(2x-7)}}{(4x+3)\cancel{(2x-7)}}\cdot\frac{(2x-7)(2x+7)}{(3x-5)\cancel{(3x+5)}}=\frac{(2x-7)(2x+7)}{(3x-5)(4x+3)}$ Note: $x\neq -\frac{3}{4},\frac{7}{2},\pm\frac{5}{3}$	**49.** $\frac{3}{x}\cdot\frac{x^2}{12}=\boxed{\frac{x}{4}}$ Note: $x\neq 0$

51. $$\frac{\cancel{6}}{\cancel{x-2}}\cdot\frac{\cancel{(x-2)}(x+2)}{\cancel{12}^{2}}=\frac{x+2}{2}$$

Note: $x \neq \pm 2$

53. $$\frac{1}{x-1}\div\frac{5}{(x-1)(x+1)}=\frac{1}{x-1}\cdot\frac{(x-1)(x+1)}{5}$$
$$=\boxed{\frac{x+1}{5}}$$

Note: $x \neq -1,1$

55. $$\frac{-\cancel{(p-2)}}{(p-1)\cancel{(p+1)}}\cdot\frac{\cancel{(p+1)}}{2\cancel{(p-2)}}=\frac{-1}{2(p-1)}$$
$$=\boxed{\frac{1}{2(1-p)}}$$

Note: $p \neq 2,1,-1$

57. $$\frac{(6-n)\cancel{(6+n)}}{(n-3)\cancel{(n+3)}}\cdot\frac{\cancel{(n+3)}}{\cancel{(n+6)}}=\boxed{\frac{6-n}{n-3}}$$

Note: $n \neq -6,-3,3$

59. $$\frac{3t(t-3)(t+1)}{5(t-2)}\div\frac{6(t+1)}{4(t-2)}=$$
$$\frac{\cancel{3}t(t-3)\cancel{(t+1)}}{5\cancel{(t-2)}}\cdot\frac{\cancel{4}^{2}\cancel{(t-2)}}{\cancel{6}\cancel{(t+1)}}=\boxed{\frac{2t(t-3)}{5}}$$

Note: $t \neq -1,2$

61. $$\frac{\cancel{w}(w-1)}{\cancel{w}}\div\frac{\cancel{w}(w^2-1)}{5w^{\cancel{3}2}}=\frac{(w-1)}{1}\div\frac{(w-1)(w+1)}{5w^2}$$
$$=\frac{\cancel{(w-1)}5w^2}{\cancel{(w-1)}(w+1)}=\boxed{\frac{5w^2}{w+1}}$$

Note: $w \neq -1,1,0$

63. $$\frac{(x-3)\cancel{(x+7)}}{(x-2)\cancel{(x+5)}}\cdot\frac{(x-4)\cancel{(x+5)}}{(x-9)\cancel{(x+7)}}=$$
$$\frac{(x-3)(x-4)}{(x-2)(x-9)}$$

Note: $x \neq -7,-5,2,4,9$

65. $$\frac{\cancel{(5x-2)}\cancel{(4x+1)}}{(5x+2)\cancel{(5x-2)}}\cdot\frac{x\cancel{(3x+5)}}{\cancel{(3x+5)}\cancel{(4x+1)}}=$$
$$\frac{x}{5x+2}$$

Note: $x \neq -\frac{5}{3},-\frac{1}{4},0,\pm\frac{2}{5}$

67. $$\frac{3(5)-2}{5x}=\frac{15-2}{5x}=\boxed{\frac{13}{5x}}$$

Note: $x \neq 0$

69. $$\frac{3(p+1)+5p(p-2)}{(p-2)(p+1)}=\frac{3p+3+5p^2-10p}{(p-2)(p+1)}$$
$$=\boxed{\frac{5p^2-7p+3}{(p-2)(p+1)}}$$

Note: $p \neq -1,2$

71. $$\frac{2x+1}{5x-1}+\frac{3-2x}{5x-1}=\frac{2x+1+3-2x}{5x-1}=\boxed{\frac{4}{5x-1}}$$

Note: $x \neq \frac{1}{5}$

73. $$\frac{3y^2(y-1)+(1-2y)(y+1)}{(y+1)(y-1)}=\frac{3y^3-3y^2+y+1-2y^2-2y}{(y+1)(y-1)}=\boxed{\frac{3y^3-5y^2-y+1}{y^2-1}}$$

Note: $y \neq \pm 1$

75. $$\frac{3x}{(x-2)(x+2)}+\frac{(3+x)}{(x+2)}=\frac{3x+(3+x)(x-2)}{(x-2)(x+2)}$$
$$=\frac{3x+3x-6+x^2-2x}{(x-2)(x+2)}=\boxed{\frac{x^2+4x-6}{(x-2)(x+2)}}$$
$$=\frac{x^2+4x-6}{x^2-4}$$

Note: $x \neq \pm 2$

77. $$\frac{(x-1)(x+2)+(x-6)}{(x-2)(x+2)}=\frac{x^2+x-2+x-6}{(x-2)(x+2)}$$
$$=\frac{x^2+2x-8}{(x-2)(x+2)}=\frac{(x+4)\cancel{(x-2)}}{\cancel{(x-2)}(x+2)}$$
$$=\frac{x+4}{x+2}$$

Note: $x \neq \pm 2$

79. $\dfrac{5a}{(a-b)(a+b)}+\dfrac{7}{a-b}=\dfrac{5a+7(a+b)}{(a-b)(a+b)}$

$=\dfrac{5a+7a+7b}{(a-b)(a+b)}=\dfrac{12a+7b}{a^2-b^2}$

Note: $a \neq \pm b$

81. $\dfrac{7(x-3)+1}{x-3}=\dfrac{7x-20}{x-3}$

Note: $x \neq 3$

83. $\dfrac{\frac{1}{x}-1}{1-\frac{2}{x}}\cdot\dfrac{x}{x}=\boxed{\dfrac{1-x}{x-2}}$ Note: $x \neq 0,2$

85. $\dfrac{\frac{3x+1}{x}}{\frac{9x^2-1}{x^2}}=\dfrac{\cancel{3x+1}}{\cancel{x}}\cdot\dfrac{x^{\cancel{2}}}{\cancel{(3x+1)}(3x-1)}=\dfrac{x}{3x-1}$

Note: $x \neq 0, \pm\frac{1}{3}$

87. $\dfrac{\frac{1}{x-1}+1}{1-\frac{1}{x+1}}\cdot\dfrac{(x-1)(x+1)}{(x-1)(x+1)}=\dfrac{x+1+x^2-1}{x^2-1-(x-1)}$

$=\dfrac{x^2+x}{x^2-x}=\dfrac{\cancel{x}(x+1)}{\cancel{x}(x-1)}=\boxed{\dfrac{x+1}{x-1}}$

Note: $x \neq 0, \pm 1$

89. $\dfrac{\frac{1+x-1}{x-1}}{\frac{1+x+1}{x+1}}=\dfrac{x}{x-1}\cdot\dfrac{x+1}{x+2}=\dfrac{x(x+1)}{(x-1)(x+2)}$

Note: $x \neq \pm 1, -2$

91. $A=\dfrac{pi}{\frac{(1+i)^5-1}{(1+i)^5}}=\dfrac{pi(1+i)^5}{(1+i)^5-1}$

93. $\dfrac{1}{\frac{1}{R_1}+\frac{1}{R_2}}=\dfrac{1}{\frac{R_2+R_1}{R_1R_2}}=\dfrac{R_1R_2}{R_2+R_1}$

95. Initially, $\dfrac{x^2+2x+1}{x+1}$ has the restriction $x \neq -1$. $\dfrac{x^2+2x+1}{x+1}=\boxed{x+1}$

97. False. $\dfrac{x^2-81}{x-9}=x+9\ \boxed{x \neq 9}$

99. False. $\dfrac{1}{3x}-\dfrac{1}{6x}$ LCD $=6x\left(\text{not } 18x^2\right)$

101. $\dfrac{(x+a)}{(x+b)}\div\dfrac{(x+c)}{(x+d)}=\dfrac{(x+a)(x+d)}{(x+b)(x+c)}$

Note: $x \neq -b, -c, -d$

103. 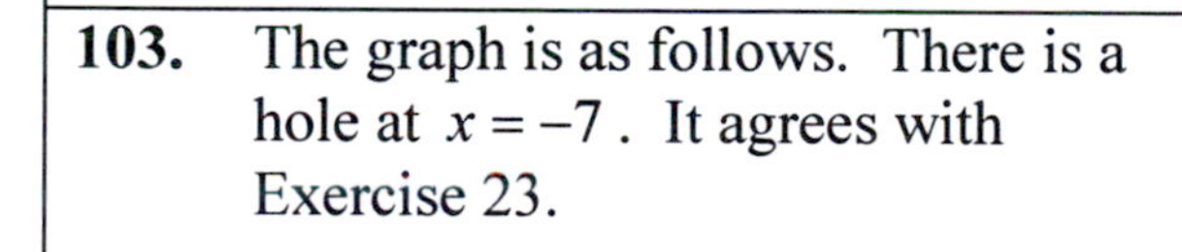
The graph is as follows. There is a hole at $x=-7$. It agrees with Exercise 23.

105. a. $\dfrac{1+\frac{1}{x-2}}{1-\frac{1}{x+2}}=\dfrac{\frac{x-1}{x-2}}{\frac{x+1}{x+2}}=\dfrac{(x-1)(x+2)}{(x+1)(x-2)}$

b.

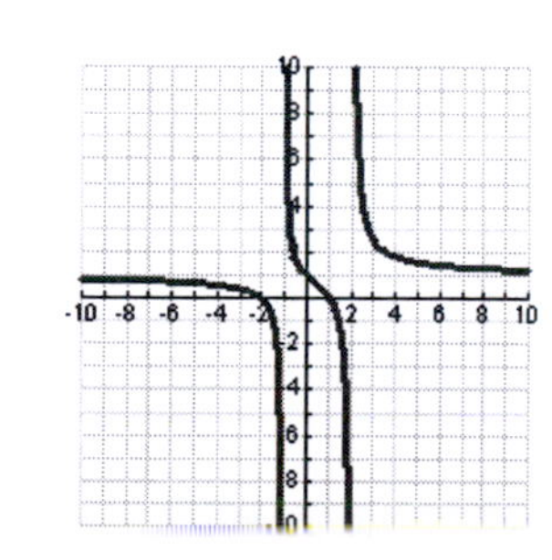

c. The graphs agree as long as $x \neq -1, \pm 2$.

Section 0.6 Solutions --

1. 10	**3.** -12
5. -6	**7.** 7
9. 1	**11.** 0
13. not real	**15.** -3
17. $\left(8^{1/3}\right)^2 = 2^2 = 4$	**19.** -2
21. -1	**23.** $\left(9^{1/2}\right)^3 = 3^3 = 27$
25. $-4\sqrt{2}$	**27.** $8\sqrt{5}$
29. $2\sqrt{3}\cdot\sqrt{2} = 2\sqrt{6}$	**31.** $\sqrt[3]{12}\cdot\sqrt[3]{4} = \sqrt[3]{48} = \sqrt[3]{8\cdot 6} = 2\sqrt[3]{6}$
33. $\sqrt{21}$	**35.** $8\sqrt{25x^2} = 40\lvert x\rvert$
37. $2\lvert x\rvert\sqrt{y}$	**39.** $-3x^2y^2\sqrt[3]{3y^2}$
41. $\frac{1}{\sqrt{3}}\cdot\frac{\sqrt{3}}{\sqrt{3}} = \frac{\sqrt{3}}{3}$	**43.** $\frac{2}{3\sqrt{11}}\cdot\frac{\sqrt{11}}{\sqrt{11}} = \frac{2\sqrt{11}}{33}$
45. $\frac{3}{1-\sqrt{5}}\cdot\frac{1+\sqrt{5}}{1+\sqrt{5}} = \frac{3\left(1+\sqrt{5}\right)}{1-5} = \frac{3+3\sqrt{5}}{-4}$	**47.** $\frac{1+\sqrt{2}}{1-\sqrt{2}}\cdot\frac{1+\sqrt{2}}{1+\sqrt{2}} = \frac{1+2\sqrt{2}+2}{1-2} = -3-2\sqrt{2}$
49. $\frac{3}{\sqrt{2}-\sqrt{3}}\cdot\frac{\sqrt{2}+\sqrt{3}}{\sqrt{2}+\sqrt{3}} = \frac{3\left(\sqrt{2}+\sqrt{3}\right)}{2-3} = -3\left(\sqrt{2}+\sqrt{3}\right)$	**51.** $\frac{4}{3\sqrt{2}+2\sqrt{3}}\cdot\frac{3\sqrt{2}-2\sqrt{3}}{3\sqrt{2}-2\sqrt{3}} = \frac{4\left(3\sqrt{2}-2\sqrt{3}\right)}{18-12}$ $= \frac{2\left(3\sqrt{2}-2\sqrt{3}\right)}{3}$
53. $\frac{4+\sqrt{5}}{3+2\sqrt{5}}\cdot\frac{3-2\sqrt{5}}{3-2\sqrt{5}} = \frac{12+3\sqrt{5}-8\sqrt{5}-10}{9-20}$ $= \frac{-5\sqrt{5}+2}{-11} = \frac{5\sqrt{5}-2}{11}$	**55.** $\frac{\sqrt{7}+\sqrt{3}}{\sqrt{2}-\sqrt{5}}\cdot\frac{\sqrt{2}+\sqrt{5}}{\sqrt{2}+\sqrt{5}} = \frac{\left(\sqrt{7}+\sqrt{3}\right)\left(\sqrt{2}+\sqrt{5}\right)}{-3}$
57. x^3y^4	**59.** $\frac{x^{-1}y^{-3/2}}{x^{-1}y^{1/2}} = \frac{1}{y^2}$
61. $x^{7/6}y^2$	**63.** $\frac{8x^2}{16x^{-2/3}} = \frac{x^{8/3}}{2}$
65. $x^{1/3}\left(x^2-x-2\right) = x^{1/3}(x-2)(x+1)$	**67.** $7x^{3/7}\left(1-2x^{3/7}+3x\right)$

69. $\sqrt{\frac{1280}{16}} = \sqrt{80} \approx 9$ sec.	**71.** $d = (29.46)^{2/3} \approx 9.54$ astronomical units
73. Forgot to square the 4	**75.** False. Only give principal root
77. False. If $a = 3, b = 4$, then $\sqrt{a^2 + b^2} = \sqrt{25} = 5$, while $\sqrt{a} + \sqrt{b} = \sqrt{3} + 2 \neq 5$.	**79.** Multiply the exponents to get a^{mnk}.
81. $\frac{\sqrt{7^2 - 4(1)(12)}}{2(1)} = \frac{1}{2}$	
83. $\frac{1}{\left(\sqrt{a}+\sqrt{b}\right)^2} = \left(\frac{1}{\sqrt{a}+\sqrt{b}}\right)^2 = \left(\frac{1}{\sqrt{a}+\sqrt{b}} \cdot \frac{\sqrt{a}-\sqrt{b}}{\sqrt{a}-\sqrt{b}}\right)^2 = \left(\frac{\sqrt{a}-\sqrt{b}}{a-b}\right)^2 = \frac{\left(\sqrt{a}-\sqrt{b}\right)^2}{(a-b)^2} = \frac{a - 2\sqrt{a}\sqrt{b} + b}{(a-b)^2} = \frac{a+b-2\sqrt{ab}}{(a-b)^2}$	**85.** 3.317 **87.** **a.** $\frac{4}{5\sqrt{2}+4\sqrt{3}} \cdot \frac{5\sqrt{2}-4\sqrt{3}}{5\sqrt{2}-4\sqrt{3}} = \frac{4\left(5\sqrt{2}-4\sqrt{3}\right)}{50-48} = 10\sqrt{2} - 8\sqrt{3}$ **b.** 0.2857291632 **c.** Yes

Section 0.7 Solutions ---

1. $\sqrt{-16} = \underbrace{\sqrt{16}}_{4} \cdot \underbrace{\sqrt{-1}}_{i} = \boxed{4i}$	**3.** $\sqrt{-20} = \underbrace{\sqrt{20}}_{2\sqrt{5}} \underbrace{\sqrt{-1}}_{i} = \boxed{2i\sqrt{5}}$
5. -4	**7.** $8i$
9. $3 - 10i$	**11.** $-10 - 12i$
13. $3 - 7i - 1 - 2i = \boxed{2 - 9i}$	**15.** $10 - 14i$
17. $(4 - 5i) - (2 - 3i) = 4 - 5i - 2 + 3i = \boxed{2 - 2i}$	**19.** $-1 + 2i$
21. $12 - 6i$	**23.** $96 - 60i$
25. $-48 + 27i$	**27.** $-102 + 30i$
29. $(1 - i)(3 + 2i) = 3 + 2i - 3i - 2i^2 = 3 - i + 2 = \boxed{5 - i}$	
31. $-15 + 21i + 20i + 28 = 13 + 41i$	**33.** $42 - 30i + 63i + 45 = 87 + 33i$
35. $-24 + 36i + 12i + 18 = -6 + 48i$	**37.** $\frac{2}{9} + \frac{8}{9}i - \frac{3}{2}i + 6 = \frac{56}{9} - \frac{11}{18}i$
39. $-2i + 34 + 51i + 3 = 37 + 49i$	**41.** $\bar{z} = 4 - 7i$ $z\bar{z} = 4^2 + 7^2 = 65$

43. $\bar{z}=2+3i$

$z\bar{z}=2^2+3^2=13$

45. $\bar{z}=6-4i$

$z\bar{z}=6^2+4^2=52$

47. $\bar{z}=-2+6i$

$z\bar{z}=2^2+6^2=40$

49. $\frac{2}{i}\cdot\frac{i}{i}=-2i$

51. $\frac{1}{3-i}\cdot\frac{(3+i)}{(3+i)}=\frac{3+i}{9-i^2}=\frac{3+i}{10}=\boxed{\frac{3}{10}+\frac{1}{10}i}$

53. $\frac{1}{3+2i}\cdot\frac{(3-2i)}{(3-2i)}=\frac{3-2i}{9-4i^2}=\frac{3-2i}{9+4}=\frac{3-2i}{13}=\boxed{\frac{3}{13}-\frac{2}{13}i}$

55. $\frac{2}{7+2i}\cdot\frac{7-2i}{7-2i}=\frac{14-4i}{49+4}=\frac{14}{53}-\frac{4}{53}i$

57.

$$\frac{1-i}{1+i}\cdot\frac{(1-i)}{(1-i)}=\frac{1-2i+i^2}{1-i^2}$$

$$=\frac{1-2i-1}{1-(-1)}=\frac{-2i}{2}=\boxed{-i}$$

59.

$$\frac{2+3i}{3-5i}\cdot\frac{3+5i}{3+5i}=\frac{6+9i+10i-15}{9+25}$$

$$=-\frac{9}{34}+\frac{19}{34}i$$

61. $\frac{4-5i}{7+2i}\cdot\frac{(7-2i)}{(7-2i)}=\frac{28-8i-35i+10i^2}{49-4i^2}$

$$=\frac{28-43i-10}{49+4}$$

$$=\frac{18-43i}{53}=\boxed{\frac{18}{53}-\frac{43}{53}i}$$

63. $\frac{8+3i}{9-2i}\cdot\frac{9+2i}{9+2i}=\frac{72+27i+16i-6}{81+4}$

$$=\frac{66}{85}+\frac{43}{85}i$$

65.

$$i^{15}=i^{12}\cdot i^3=\underbrace{\left(i^4\right)^3}_{1}\cdot i^3=i\left(i^2\right)=i(-1)=\boxed{-i}$$

67. $i^{40}=\left(i^4\right)^{10}=(1)^{10}=\boxed{1}$

69. $25-20i-4=21-20i$

71. $4+12i-9=-5+12i$

73.

$$(3+i)^2(3+i)=(9+6i-1)(3+i)$$
$$=(8+6i)(3+i)$$
$$=24+18i+8i-6$$
$$=18+26i$$

75.

$$(1-i)^2(1-i)=(1-2i-1)(1-i)$$
$$=-2i(1-i)$$
$$=-2i-2$$
$$=-2-2i$$

77. $z=(3-6i)+(5+4i)=8-2i$ ohms

79. Should have multiplied by the conjugate $4+i$ (not $4-i$).

$$\frac{2}{4-i}\cdot\frac{(4+i)}{(4+i)}=\frac{8+2i}{16-i^2}=\frac{8+2i}{17}=\boxed{\frac{8}{17}+\frac{2}{17}i}$$

81. True.

83. True.

85. $x^4+2x^2+1=\left(x^2+1\right)^2=(x+i)^2(x-i)^2$

87. $41-38i$

89. $\frac{2}{125}+\frac{11}{125}i$

Chapter 0 Review Solutions --

1. a. 5.22 **b.** 5.21	**3.** $2-10+12=\boxed{4}$
5. -2	**7.** $\frac{3x}{12}-\frac{4x}{12}=\boxed{-\frac{x}{12}}$
9. $\frac{3}{1}\cdot\frac{3}{1}=\boxed{9}$	**11.** $-8z^3$
13. $\frac{9x^6y^4}{2x^8y^4}=\frac{9}{2x^2}$	**15.** 2.15×10^{-6}
17. $14z^2+3z-2$	**19.** $36x^2-4x-5-6x+9x^2-10=45x^2-10x-15$
21. $15x^2y^2-20xy^3$	**23.** $x^2+2x-63$
25. $4x^2-12x+9$	**27.** x^4+2x^2+1
29. $2xy^2(7x-5y)$	**31.** $(x+5)(2x-1)$
33. $(4x+5)(4x-5)$	**35.** $(x+5)(x^2-5x+25)$
37. $2x(x+5)(x-3)$	**39.** $(x^2-2)(x+1)$
41. $x\neq\pm3$	**43.** $x+2,\ x\neq2$
45. $\frac{(t+3)\cancel{(t-2)}}{\cancel{(t-2)}(t+1)}=\frac{(t+3)}{(t+1)},\quad t\neq-1,2$	**47.** $\frac{(x+5)\cancel{(x-2)}}{(x+3)\cancel{(x-1)}}\cdot\frac{(x+2)\cancel{(x-1)}}{(x+3)\cancel{(x-2)}}=\boxed{\frac{(x+5)(x+2)}{(x+3)^2},x\neq-3,1,2}$
49. $\frac{x+3}{(x+1)(x+3)}-\frac{x+1}{(x+1)(x+3)}=\boxed{\frac{2}{(x+1)(x+3)}\ \ x\neq-1,-3}$	**51.** $\frac{\frac{2(x-3)+1}{x-3}}{\frac{1+4(5x-15)}{5x-15}}=\frac{2x-5}{\cancel{x-3}}\cdot\frac{5\cancel{(x-3)}}{20x-59}=\frac{10x-25}{20x-59}\qquad x\neq3,\frac{59}{20}$
53. $\sqrt{4\cdot5}=2\sqrt{5}$	**55.** $-5xy\sqrt[3]{x^2y}$
57. $3(2\sqrt{5})+5(4\sqrt{5})=26\sqrt{5}$	**59.** $2+\sqrt{5}-2\sqrt{5}-5=-3-\sqrt{5}$
61. $\frac{1}{2-\sqrt{3}}\cdot\frac{2+\sqrt{3}}{2+\sqrt{3}}=\frac{2+\sqrt{3}}{4-3}=2+\sqrt{3}$	**63.** $\frac{9x^{4/3}}{16x^{2/3}}=\frac{9x^{2/3}}{16}$
65. $5^{\frac{1}{2}-\frac{1}{3}}=5^{\frac{1}{6}}$	**67.** $13i$
69. $(i^4)^5\cdot i^{-1}=\boxed{-i}$	**71.** $(3-2i)+(5-4i)=8-6i$
73. $(12-i)-(-2-5i)=(12-i)+(2+5i)$ $=14+4i$	**75.** $(2i+2)(3-3i)=\boxed{12}$

77. $16+56i-49=-33+56i$

79. $\dfrac{1}{2-i}\cdot\dfrac{2+i}{2+i}=\dfrac{2+i}{4+1}=\boxed{\dfrac{2}{5}+\dfrac{1}{5}i}$

81. $\dfrac{7+2i}{4+5i}\cdot\dfrac{4-5i}{4-5i}=\dfrac{28-27i-10i^2}{16+25}=\boxed{\dfrac{38}{41}-\dfrac{27}{41}i}$

83. $\dfrac{10}{3i}\cdot\dfrac{i}{i}=-\dfrac{10}{3}i$

85. Simplifying the radical using the calculator gives exactly 16.5. So, it is rational.

87. 1.945×10^{-6}

89.

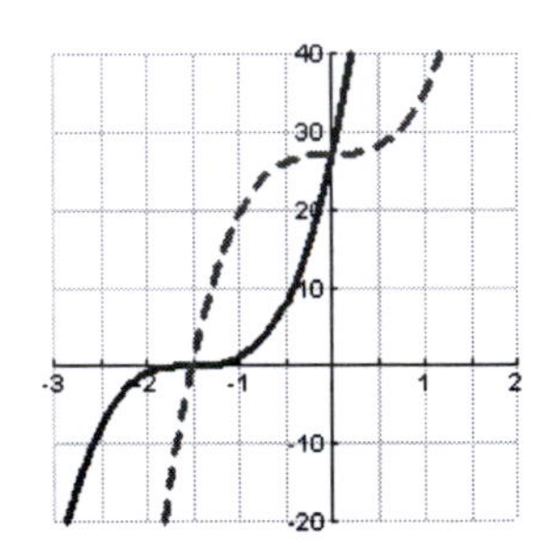

The solid curve represents the graph of both $y=(2x+3)^3$ and $y=8x^3+36x^2+54x+27$.

91.

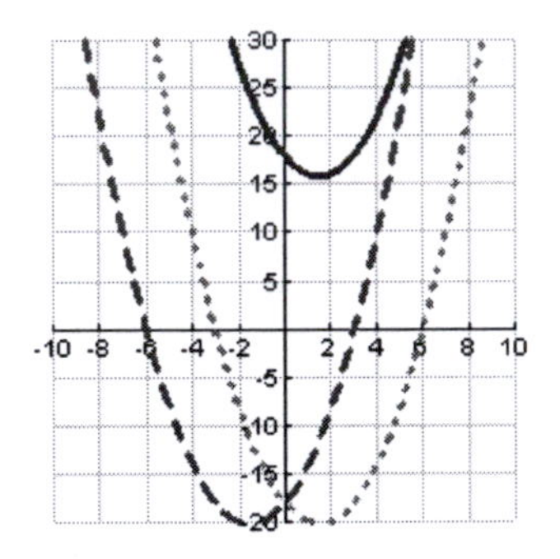

All three graphs are different.

93. a. $\dfrac{1-\frac{4}{x}}{1-\left(\frac{4}{x}\right)^2}=\dfrac{\frac{x-4}{x}}{\frac{x^2-16}{x^2}}=\dfrac{x^2(x-4)}{x\left(x^2-16\right)}$

$=\dfrac{x(x-4)}{(x+4)(x-4)}=\dfrac{x}{x+4}$

b.

c. The graphs agree as long as $x\neq0,\pm4$.

95.

a. $\dfrac{6}{\sqrt{5}-\sqrt{2}}\cdot\dfrac{\sqrt{5}+\sqrt{2}}{\sqrt{5}+\sqrt{2}}=\dfrac{6\left(\sqrt{5}+\sqrt{2}\right)}{5-2}$

$=2\sqrt{5}+2\sqrt{2}$

b. 7.30056308

c. Yes

97. $2868-6100i$

99. 1.6×10^{14}

Chapter 0 Practice Test Solutions

1. $\sqrt{16}=\sqrt{4^2}=\boxed{4}$

3. $-3(27)+2(-4)-(8)=-97$

5. $\sqrt{-12x^2}=\sqrt{(-1)\cdot 2^2\cdot 3\cdot x^2}=\boxed{2i\lvert x\rvert\sqrt{3}}$	**7.** $\dfrac{\left(x^2y^{-3}z^{-1}\right)^{-2}}{\left(x^{-1}y^2z^3\right)^{1/2}}=\dfrac{x^{-4}y^6z^2}{x^{-1/2}yz^{3/2}}=\boxed{\dfrac{y^5z^{1/2}}{x^{7/2}}}$
9. $3\left(3\sqrt{2}\right)-4\left(4\sqrt{2}\right)=-7\sqrt{2}$	**11.** $2y^2-12y+20$
13. $(x-4)(x+4)$	**15.** $(2x+3y)^2$
17. $(2x+1)(x-1)$	**19.** $t(t+1)(2t-3)$
21. $x(x-3y)+4y(x-3y)=(x+4y)(x-3y)$	**23.** $3\left(27+x^3\right)=3(3+x)\left(9-3x+x^2\right)$

25. $\dfrac{2(x-1)+3x}{x(x-1)}=\dfrac{5x-2}{x(x-1)},\ x\neq 0,1$

27.

$$\frac{\cancel{x-1}}{(x-1)(x+1)}\cdot\frac{\cancel{x^2+x+1}}{\cancel{(x-1)}\cancel{\left(x^2+x+1\right)}}=\frac{1}{(x-1)(x+1)}\quad \text{Note: } x\neq\pm 1$$

29.

$$\frac{x-3}{2x-5}\div\frac{x^2-9}{5-2x}=\frac{\cancel{x-3}}{\cancel{2x-5}}\cdot\frac{-\cancel{(2x-5)}}{\cancel{(x-3)}(x+3)}=-\frac{1}{x+3}\quad \text{Note: } x\neq\tfrac{5}{2},\pm 3$$

31. $(1-3i)(7-5i)=7-26i+15i^2=\boxed{-8-26i}$

33.

$$\frac{7-2\sqrt{3}}{4-5\sqrt{3}}\cdot\frac{4+5\sqrt{3}}{4+5\sqrt{3}}=\frac{28+27\sqrt{3}-10\cdot 3}{16-25\cdot 3}=\boxed{\frac{2-27\sqrt{3}}{59}}$$

35.

$$\frac{\dfrac{x+1-2x}{x(x+1)}}{x-1}=\frac{1-x}{x(x+1)(x-1)}=-\frac{1}{x(x+1)}$$

Note: $x\neq 0,\pm 1$

37. 2.330

CHAPTER 1

Section 1.1 Solutions --

1.

$$5x = 35$$
$$\frac{1}{5}\cdot 5x = \frac{1}{5}\cdot 35$$
$$\boxed{x = 7}$$

3.

$$-3 + n = 12$$
$$3 + -3 + n = 3 + 12$$
$$\boxed{n = 15}$$

5.

$$24 = -3x$$
$$-\frac{1}{3}\cdot 24 = -\frac{1}{3}\cdot(-3x)$$
$$\boxed{-8 = x}$$

7.

$$\frac{1}{5}n = 3$$
$$5\cdot\frac{1}{5}n = 5\cdot 3$$
$$\boxed{n = 15}$$

9.

$$3x - 5 = 7$$
$$3x = 12$$
$$\boxed{x = 4}$$

11.

$$9m - 7 = 11$$
$$9m = 18$$
$$\boxed{m = 2}$$

13.

$$5t + 11 = 18$$
$$5t = 7$$
$$\boxed{t = 7/5}$$

15.

$$3x - 5 = 25 + 6x$$
$$3x = 30 + 6x$$
$$-3x = 30$$
$$\boxed{x = -10}$$

17.

$$20n - 30 = 20 - 5n$$
$$20n = 50 - 5n$$
$$25n = 50$$
$$\boxed{n = 2}$$

19.

$$4(x-3) = 2(x+6)$$
$$4x - 12 = 2x + 12$$
$$2x = 24$$
$$\boxed{x = 12}$$

21.

$$-3(4t-5) = 5(6-2t)$$
$$-12t + 15 = 30 - 10t$$
$$-15 = 2t$$
$$\boxed{-\tfrac{15}{2} = t}$$

23.

$$2(x-1) + 3 = x - 3(x+1)$$
$$2x - 2 + 3 = x - 3x - 3$$
$$2x + 1 = -2x - 3$$
$$4x = -4$$
$$\boxed{x = -1}$$

25.

$$5p + 6(p+7) = 3(p+2)$$
$$5p + 6p + 42 = 3p + 6$$
$$11p + 42 = 3p + 6$$
$$8p = -36$$
$$\boxed{p = -\frac{36}{8} = -\frac{9}{2}}$$

27.

$$7x-(2x+3)=x-2$$
$$7x-2x-3=x-2$$
$$5x-3=x-2$$
$$4x=1$$
$$\boxed{x=\tfrac{1}{4}}$$

29.

$$2-(4x+1)=3-(2x-1)$$
$$2-4x-1=3-2x+1$$
$$1-4x=4-2x$$
$$-3=2x$$
$$\boxed{-\tfrac{3}{2}=x}$$

31.

$$2a-9(a+6)=6(a+3)-4a$$
$$-7a-54=6a+18-4a$$
$$-7a-54=2a+18$$
$$-9a=72$$
$$\boxed{a=-8}$$

33.

$$32-\left[4+6x-5(x+4)\right]=4(3x+4)-\left[6(3x-4)+7-4x\right]$$
$$32-[4+6x-5x-20]=12x+16-[18x-24+7-4x]$$
$$32-4-6x+5x+20=12x+16-18x+24-7+4x$$
$$48-x=-2x+33$$
$$\boxed{x=-15}$$

35.

$$20-4\left[c-3-6(2c+3)\right]=5(3c-2)-\left[2(7c-8)-4c+7\right]$$
$$20-4[c-3-12c-18]=15c-10-[14c-16-4c+7]$$
$$20-4c+12+48c+72=15c-10-14c+16+4c-7$$
$$44c+104=5c-1$$
$$39c=-105$$
$$\boxed{c=\frac{-105}{39}=\frac{-35}{13}}$$

37.

$$60\left(\frac{1}{5}m\right)=60\left(\frac{1}{60}m+1\right)$$
$$12m=m+60$$
$$11m=60$$
$$\boxed{m=\frac{60}{11}}$$

39.

$$63\left(\frac{x}{7}\right)=63\left(\frac{2x}{63}+4\right)$$
$$9x=2x+252$$
$$7x=252$$
$$\boxed{x=36}$$

41. $24\left(\frac{1}{3}p\right)=24\left(3-\frac{1}{24}p\right)$ $8p=72-p$ $9p=72$ $\boxed{p=8}$	**43.** $84\left(\frac{5y}{3}-2y\right)=84\left(\frac{2y}{84}+\frac{5}{7}\right)$ $140y-168y=2y+60$ $-30y=60$ $\boxed{y=\frac{60}{-30}=-2}$
45. $8\left(p+\frac{p}{4}\right)=8\left(\frac{5}{2}\right)$ $8p+2p=20$ $10p=20$ $\boxed{p=2}$	**47.** $\frac{x-3}{3}-\frac{x-4}{2}=1-\frac{x-6}{6}$ $6\cdot\left[\frac{x-3}{3}-\frac{x-4}{2}\right]=6\cdot\left[1-\frac{x-6}{6}\right]$ $2(x-3)-3(x-4)=6-(x-6)$ $2x-6-3x+12=6-x+6$ $-x+6=-x+12$ $6=12$, which is false. Hence, $\boxed{\text{no solution}}$.
49. $2y\left(\frac{4}{y}-5\right)=2y\left(\frac{5}{2y}\right)$ $\boxed{y\neq 0}$ $8-10y=5$ $-10y=-3$ $\boxed{y=\frac{3}{10}}$	**51.** $6x\left(7-\frac{1}{6x}\right)=6x\left(\frac{10}{3x}\right)$ $\boxed{x\neq 0}$ $42x-1=20$ $42x=21$ $\boxed{x=\frac{1}{2}}$
53. $3a\left(\frac{2}{a}-4\right)=3a\left(\frac{4}{3a}\right)$ $\boxed{a\neq 0}$ $6-12a=4$ $-12a=-2$ $\boxed{a=\frac{1}{6}}$	**55.** $(x-2)\left(\frac{x}{x-2}+5\right)=(x-2)\left(\frac{2}{x-2}\right)$ $\boxed{x\neq 2}$ $x+5(x-2)=2$ $x+5x-10=2$ $6x=12$ $x=2$ $\boxed{\text{No solution}}$ since 2 was excluded from the solution set.
57. $(p-1)\left(\frac{2p}{p-1}\right)=(p-1)\left(3+\frac{2}{p-1}\right)$ $\boxed{p\neq 1}$ $2p=3(p-1)+2$ $2p=3p-3+2$ $2p=3p-1$ $p=1$ $\boxed{\text{No solution}}$ since 1 was excluded from the solution set.	**59.** $(x+2)\left(\frac{3x}{x+2}-4\right)=(x+2)\left(\frac{2}{x+2}\right)$ $\boxed{x\neq -2}$ $3x-4(x+2)=2$ $-x-8=2$ $\boxed{x=-10}$

61. $\frac{1}{n}+\frac{1}{n+1}=\frac{-1}{n(n+1)}$ $\boxed{n\neq -1,0}$

LCD is $n(n+1)$. So,

$$(n+1)+n=-1$$
$$n+1+n=-1$$
$$2n=-2$$
$$n=-1$$

But since we have already stipulated that $n\neq -1$, there is $\boxed{\text{no solution.}}$

63. $\frac{3}{a}-\frac{2}{a+3}=\frac{9}{a(a+3)}$ $\boxed{a\neq 0,-3}$

LCD is $a(a+3)$. So,

$$3(a+3)-2a=9$$
$$3a+9-2a=9$$
$$a=0$$

But since we have already stipulated that $a\neq 0$, there is $\boxed{\text{no solution.}}$

65.

$$\frac{n-5}{6(n-1)}=\frac{1}{9}-\frac{n-3}{4(n-1)} \quad \boxed{n\neq 1}$$

LCD is $36(n-1)$. So,

$$\frac{(n-5)(36)(n-1)}{6(n-1)}=\frac{36(n-1)}{9}-\frac{(n-3)(36)(n-1)}{4(n-1)}$$
$$6(n-5)=4(n-1)-9(n-3)$$
$$6n-30=4n-4-9n+27$$
$$6n-30=-5n+23$$
$$11n=53$$

So, the final solution is: $\boxed{n=\frac{53}{11}}$

67.

$$\frac{2}{5x+1}=\frac{1}{2x-1} \quad \boxed{x\neq -\tfrac{1}{5},\tfrac{1}{2}}$$
$$2(2x-1)=1(5x+1)$$
$$4x-2=5x+1$$
$$\boxed{x=-3}$$

69.

$$\frac{t-1}{1-t}=\frac{3}{2} \quad \boxed{t\neq 1}$$
$$3(1-t)=2(t-1)$$
$$3-3t=2t-2$$
$$-5t=-5$$
$$t=1$$

$\boxed{\text{No solution}}$ since 1 was excluded from the solution set.

71. $F=\frac{9}{5}C+32$ $F-32=\frac{9}{5}C$ $\frac{5}{9}(F-32)=C$ $\boxed{C=\frac{5}{9}F-\frac{160}{9}}$	**73.** Let x = number of minutes you use the cell phone. Solve: $25.08=15+0.12x$ $10.08=0.12x$ $84=\frac{10.08}{0.12}=x$ So, you used your cell phone for $\boxed{\text{84 min.}}$
75. Let x = number of minutes logged on Solve: $2+0.10x=3.70$ $0.10x=1.70$ $x=17$ So, logged on for $\boxed{\text{17 min.}}$	**77.** **a.** $C(x)=15{,}000+2{,}500x$ **b.** Solve for x: $15{,}000+2{,}500x=5{,}515{,}000$ $2{,}500x=5{,}500{,}000$ $x=2{,}200$
79. Using $a=\frac{d}{c}$ with d = 600mg and c = 125mg/ 5mL = 25mg/mL, we see that $a=\frac{600mg}{25mg/mL}=24mL$.	**81.** $f=\frac{c}{\lambda}$ $\boxed{\lambda\neq 0}$
83. Should have subtracted $4x$ and added 7 to both sides. The correct answer is $x=5$.	**85.** Cannot cross multiply- must multiply by LCD first. The correct answer is $p=6/5$.
87. False $\boxed{x\neq 0}$	**89.** True

91.

$$ax+b=c \quad \boxed{a\neq 0}$$

$$ax=c-b$$

$$\boxed{x=\frac{c-b}{a}}$$

93.

$$\frac{b+c}{x+a} = \frac{b-c}{x-a} \quad \boxed{x \neq \pm a}$$

$$(b+c)(x-a) = (b-c)(x+a)$$

$$bx - ba + cx - ca = bx + ba - cx - ca$$

$$2cx = 2ba \qquad \boxed{x = \frac{ba}{c}}$$

95.

$$\frac{1 - \frac{1}{x}}{1 + \frac{1}{x}} = 1 \qquad \boxed{x \neq -1, 0}$$

$$1 - \frac{1}{x} = 1 + \frac{1}{x} \quad \Rightarrow \quad \frac{2}{x} = 0$$

$\boxed{\text{no solution}}$

97.

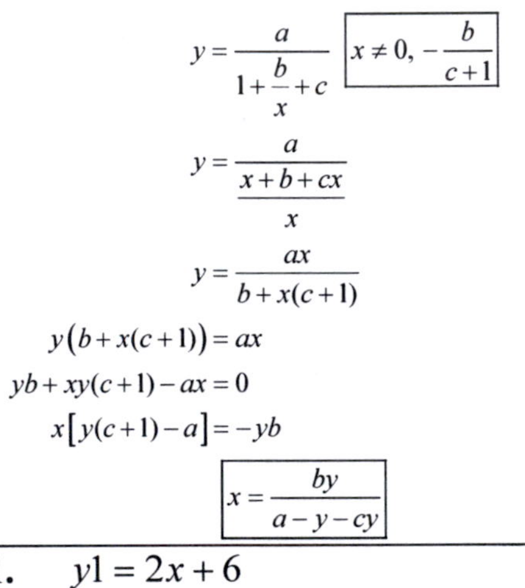

$$y = \frac{a}{1 + \frac{b}{x} + c} \qquad \boxed{x \neq 0, -\frac{b}{c+1}}$$

$$y = \frac{a}{\frac{x+b+cx}{x}}$$

$$y = \frac{ax}{b + x(c+1)}$$

$$y(b + x(c+1)) = ax$$

$$yb + xy(c+1) - ax = 0$$

$$x[y(c+1) - a] = -yb$$

$$\boxed{x = \frac{by}{a - y - cy}}$$

99. $y1 = 3(x+2) - 5x$

$y2 = 3x - 4$

$\boxed{x = 2}$

101. $y1 = 2x + 6$

$y2 = 4x - 2x + 8 - 2$

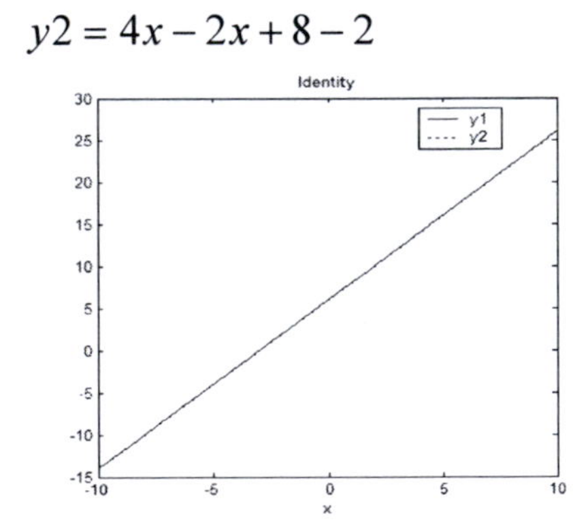

All real numbers

103. $y1 = \dfrac{x(x-1)}{x^2} \quad y2 = 1$

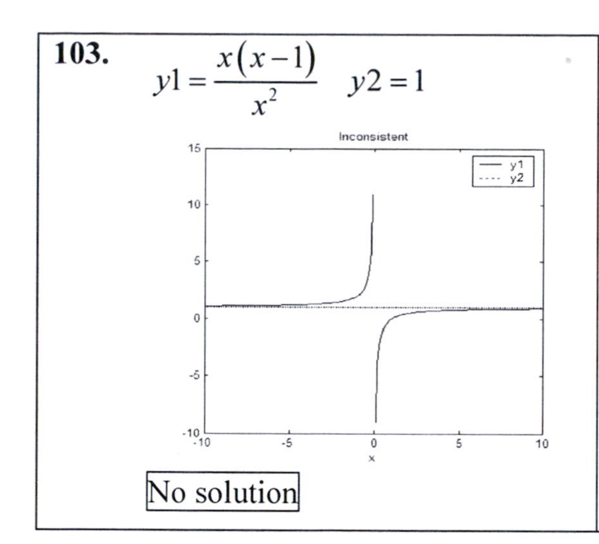

$\boxed{\text{No solution}}$

105. $y1 = 0.035x + 0.029(8706 - x) \quad y2 = 285.03$

(5426, 285.03)

$\boxed{x = 5426}$

Section 1.2 Solutions

1. Let x = price without coupon

$0.9x = 217.95$

$\boxed{x = \$242.17}$

3. Let x = cost of pizza

Tom: 5.16

Chelsea: $1/8\ x$

Jeff: $1/2\ x$

$5.16 + \frac{1}{8}x + \frac{1}{2}x = x$

$41.28 + x + 4x = 8x$

$3x = 41.28$

$\boxed{x = \$13.76}$

5. Let x = original price

$0.85x = 125{,}000$

$x = 147{,}058.82$

$\boxed{\text{Original price} \cong \$147{,}058.82}$

Model price = \$125,000

$\boxed{\text{Savings} = \$22{,}058.82}$

7. Let x = distance from Angela's home to the restaurant.

Home $\rightarrow$ Train station = 1 mile

On train $\rightarrow \frac{3}{4}x$ In taxi $\rightarrow \frac{1}{6}x$

$1 + \frac{3}{4}x + \frac{1}{6}x = x$

LCD = 12

$12 + 9x + 2x = 12x$

$12 + 11x = 12x$

$x = 12$

Angela travels $\boxed{\text{12 miles}}$ to the restaurant.

9. x = hours awake

Class: $\frac{1}{3}x$

Eating: $\frac{1}{5}x$

Working out: $\frac{1}{10}x$

Studying: 3

Other things: 2.5

$\frac{1}{3}x+\frac{1}{5}x+\frac{1}{10}x+3+2.5=x$

$10x+6x+3x+165=30x$

$19x+165=30x$

$11x=165$

$x=15$ awake

$\boxed{\text{9 hours of sleep}}$

11. Fixed costs = 15,000

Variable costs = $18.50x$

Total costs = 20,000

$18.50x+15,000=20,000$

$18.50x=5000$

$x=270.27$

Approximately $\boxed{\text{270 units}}$ can be produced.

13. $\frac{2}{3}x-10=\frac{1}{4}x$

$\frac{5}{12}x=10$

$\boxed{x=10\left(\frac{12}{5}\right)=24}$

15. Let the numbers be x, $x+2$

$4(x)=2+3(x+2)$

$4x=2+3x+6$

$x=8$

The numbers are $\boxed{8,10}$.

17. Let p = perimeter.

First side = 11

Second side = $\frac{1}{5}p$

Third side = $\frac{1}{4}p$

$11+\frac{1}{5}p+\frac{1}{4}p=p$

LCD = 20

$220+4p+5p=20p$

$220=11p$

$p=20$

The perimeter is $\boxed{\text{20 inches}}$.

19. w = width

l = length = $2w+40$

$p=2l+2w$

$260=2(2w+40)+2w$

$260=4w+80+2w$

$180=6w$

$w=30$

$\boxed{\begin{array}{l}\text{width} = 30 \text{ yards}\\ \text{length} = 100 \text{ yards}\end{array}}$

21. r_1 = radius of smaller circle

r_2 = radius of larger circle

$r_2 = r_1 + 3$

Circumference of smaller circle $= 2\pi r_1$

Circumference of larger circle $= 2\pi r_2$

Ratio of circumferences $= \dfrac{2\pi r_2}{2\pi r_1} = \dfrac{r_2}{r_1} = \dfrac{2}{1}$

$r_2 = 2r_1$

$2r_1 = r_1 + 3$

$r_1 = 3$

$\boxed{r_1 = 3 \text{ feet} \quad r_2 = 6 \text{ feet}}$

23. $\dfrac{x}{225} = \dfrac{4}{3}$

$3x = 900$

$x = 300$

The tree is $\boxed{300 \text{ feet}}$ tall.

25. Let $x =$ length of alligator in feet.

Solve:

$\dfrac{3.5}{0.5} = \dfrac{x}{0.75}$

$0.5x = 2.625$

$x = 5.25$

The alligator is about $\boxed{5.25 \text{ feet}}$.

27. Let x = amount invested at 4%.

$120{,}000 - x$ = amount invested at 7%

Solve:

$0.04x + 0.07(120{,}000 - x) = 7{,}800$

$0.04x + 8400 - 0.07x = 7{,}800$

$-0.03x = -600$

$x = 20{,}000$

$\boxed{\$20{,}000 \text{ at } 4\% \text{ and } \$100{,}000 \text{ at } 7\%}$

29. Let x = amount invested at 10%.

$\dfrac{14{,}000 - x}{2}$ = amount invested at 2%

$\dfrac{14{,}000 - x}{2}$ = amount invested at 40%

Interest earned $= 16{,}610 - 14{,}000 = 2{,}610$

Solve:

$0.1x + 0.02\left(\dfrac{14{,}000 - x}{2}\right) + 0.4\left(\dfrac{14{,}000 - x}{2}\right) = 2610$

$0.1x + 140 - 0.01x + 2800 - 0.2x = 2610$

$-0.11x = -330$

$x = 3{,}000$

$\boxed{\begin{array}{l}\$3{,}000 \text{ at } 10\% \\ \$5{,}500 \text{ at } 2\% \\ \$5{,}500 \text{ at } 40\%\end{array}}$

31. Money for plants $= 4200 - 2400 - 1500 = 300$

Let x be the number of trees (\$32 each).

Let $33 - x$ be the number of shrubs (\$4 each).

Solve:

$32x + 4(33 - x) = 300$

$32x + 132 - 4x = 300$

$28x = 168$

$x = 6$

$\boxed{6 \text{ trees and } 27 \text{ shrubs}}$

33. Let $x = $ ml of 5% HCl

Solve:

$100 - x = $ ml of 15% HCl

$$0.05x + 0.15(100 - x) = 0.08(100)$$
$$0.05x + 15 - 0.15x = 8$$
$$-0.1x = -7$$
$$x = 70$$

$\boxed{\text{70ml of 5\% HCl} \quad \text{30ml of 15\% HCl}}$

35. Let $x = $ number of gallons to be drained.

Solve:

$$0.40(5 - x) + 1.00x = 0.80(5)$$
$$2 - 0.40x + x = 4$$
$$2 + 0.60x = 4$$
$$0.60x = 2$$
$$x \approx 3.3$$

About $\boxed{\text{3.3 gallons}}$.

37. $x = $ lbs of caramels ($1.50/lb)

$1.25 - x = $ lbs of gummy bears ($2/lb)

Solve:

$$1.5x + 2(1.25 - x) = 2.50$$
$$1.5x + 2.5 - 2x = 2.50$$
$$-0.5x = 0$$
$$x = 0$$

$\boxed{\text{No caramels, 1.25lb of gummy bears}}$

39. distance = rate · time

distance = 100,000,000 miles

rate = 670,616,629 mph

$$\text{time} = \frac{\text{distance}}{\text{rate}}$$

$= 0.15$ hours $\cong$ $\boxed{\text{9 minutes}}$

41.

$$x + 0.047x = 3.21$$
$$1.047x = 3.21$$
$$x = 3.065$$

So, at the beginning of February, gas was $3.07 per gallon.

43. Let $x = $ number of mL of distilled water (which has 0% salt).

Solve for x:

$$0.03(100\,mL) + 0.00(x\,mL) = 0.009(100 + x)mL$$
$$3mL + 0mL = (0.9mL + 0.009x)$$
$$2.1mL = 0.009x$$
$$x \approx 233mL$$

45. rate (r) = boat speed $(s) \pm$ current speed (c)

boat speed: $s = 16$ mph

upstream: $r = s - c$, $t = 1/3$ hours

downstream: $r = s + c$, $t = 1/4$ hours

Distance is the same both ways $(\text{rate} \cdot \text{time})$

Solve:

$$(16-c)\left(\frac{1}{3}\right) = (16+c)\left(\frac{1}{4}\right)$$

$$4(16-c) = 3(16+c)$$

$$64 - 4c = 48 + 3c$$

$$7c = 16$$

$$\boxed{c = \frac{16}{7} \cong 2.3 \text{ mph}}$$

47. rate of walker $= r_w$

rate of jogger $= r_w + 2$

time of walker = 1 hour

time of jogger $= \frac{2}{3}$ hour

$$r_w(1) = (r_w + 2)(2/3)$$

$$r_w = \frac{2}{3}r_w + 4/3$$

$$\frac{1}{3}r_w = \frac{4}{3}$$

$$r_w = 4$$

$$\boxed{\begin{array}{l}\text{walker: 4 mph}\\ \text{jogger: 6 mph}\end{array}}$$

49. Let x = number of minutes it takes a rider to get to class

Using Distance = Rate × Time, and the fact that since they use the same path, their distances are the same, we must solve the equation:

$$2(12 + x) = 6(x)$$

$$24 + 2x = 6x$$

$$24 = 4x$$

$$x = 6$$

So, it takes the $\boxed{\text{bicyclist 6 minutes to get to class, and the walker 18 minutes}}$.

51. Let x = hours it takes Cynthia to paint house alone. Christopher can paint 1/15 house per hour. Cynthia can paint $1/x$ house per hour.

Together they paint $\left(\frac{1}{15} + \frac{1}{x}\right)$ house per hour.

$$\frac{1}{15} + \frac{1}{x} = \frac{1}{9}$$

$$3x + 45 = 5x$$

$$2x = 45$$

$$x = 22.5$$

$$\boxed{\begin{array}{l}\text{Cynthia can paint the house}\\ \text{alone in 22.5 hours.}\end{array}}$$

53. Tracey can do 1/4 of a delivery per hour, and Robin can do 1/6 of a delivery per hour. Together, they complete 1/4 + 1/6 = 1/(12/5) of the delivery in an hour. So, together, they complete the job in $\boxed{\text{2.4 hours}}$.

55.

$$\frac{4}{5}=\frac{264}{x_1}$$
$$4x_1 = 264(5)$$
$$4x_1 = 1320$$
$$\boxed{x_1 = 330 \text{ hertz}}$$

$$\frac{4}{6}=\frac{264}{x_2}$$
$$4x_2 = 264(6)$$
$$4x_2 = 1584$$
$$\boxed{x_2 = 396 \text{ hertz}}$$

57. Let x = exam grade needed

Test average $= \dfrac{86+80+84+90}{4} = 85$

To earn a "B":

$$\frac{1}{3}(85)+\frac{2}{3}x = 80$$
LCD = 3
$$85+2x = 240$$
$$2x = 155$$
$$\boxed{x = 77.5}$$

To earn an "A":

$$\frac{1}{3}(85)+\frac{2}{3}x = 90$$
LCD = 3
$$85+2x = 270$$
$$2x = 185$$
$$\boxed{x = 92.5}$$

59. Let x = # field goals

$8-x$ = # touchdowns

$$3x+7(8-x) = 48$$
$$3x+56-7x = 48$$
$$-4x = -8$$
$$x = 2$$

2 field goals, 6 touchdowns (boxed)

61.

$$(42)(5) = (60)(x)$$
$$210 = 60x$$
$$x = 3.5$$

Maria should sit 3.5 feet from the center. (boxed)

63. Let the board be 1 unit long.

Let x = distance from Maria to fulcrum.

$1-x$ = distance from Max to fulcrum.

$$60x = 42(1-x)$$
$$60x = 42-42x$$
$$102x = 42$$
$$x \cong 0.4$$

Fulcrum is 0.4 units from Maria and 0.6 units from Max. (boxed)

65.

$$\frac{1}{f}=\frac{1}{d_0}+\frac{1}{d_i}$$
$$f = 3,\ d_i = 5$$
$$\frac{1}{3}=\frac{1}{d_0}+\frac{1}{5}$$
$$\text{LCD} = 15d_0$$
$$5d_0 = 15+3d_0$$
$$2d_0 = 15$$

Object is $\boxed{d_0 = 7.5}$ cm from lens.

67.

$$\frac{1}{f}=\frac{1}{d_0}+\frac{1}{d_i}$$

$$f=2,\ d_i=\tfrac{1}{2}d_0$$

$$\frac{1}{2}=\frac{1}{d_0}+\frac{1}{\frac{1}{2}d_0}$$

Since $\frac{1}{\frac{1}{2}d_0}=\frac{2}{d_0}$,

$$\frac{1}{2}=\frac{1}{d_0}+\frac{2}{d_0}=\frac{3}{d_0}\ \Rightarrow\ d_0=6$$

$\boxed{\text{Object distance} = 6\,\text{cm}}$

69.

$$P=2l+2w$$

$$P-2l=2w$$

$$\frac{P-2l}{2}=w$$

71.

$$A=\tfrac{1}{2}bh$$

$$2A=bh$$

$$\frac{2A}{b}=h$$

73.

$$A=lw$$

$$\frac{A}{l}=w$$

75.

$$V=lwh$$

$$\frac{V}{lw}=h$$

77. Let x = Janine's average speed (in mph).
Then, Tricia's speed = $(12 + x)$ mph. We must solve the equation:

$$2.5(12+x)+2.5x=320$$

$$30+2.5x+2.5x=320$$

$$5x=290$$

$$x=58$$

So, $\boxed{\text{Janine's average speed is 58 mph and Tricia's average speed is 70 mph.}}$

79. $y = 11896.67x + 132500$

$\boxed{\$191,983.35}$

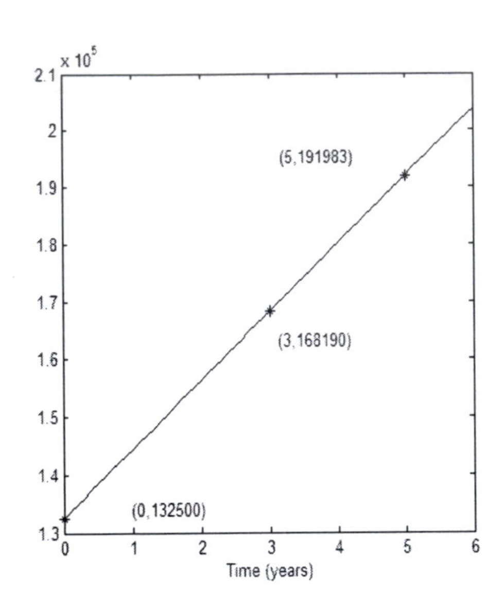

81. Let $x =$ number of times you play.

Option A: $y_1 = 300 + 15x$

Option B: $y_2 = 150 + 42x$

Option B is better if you play about 5 times or less per month. Option A is better if you play 6 times or more per month.

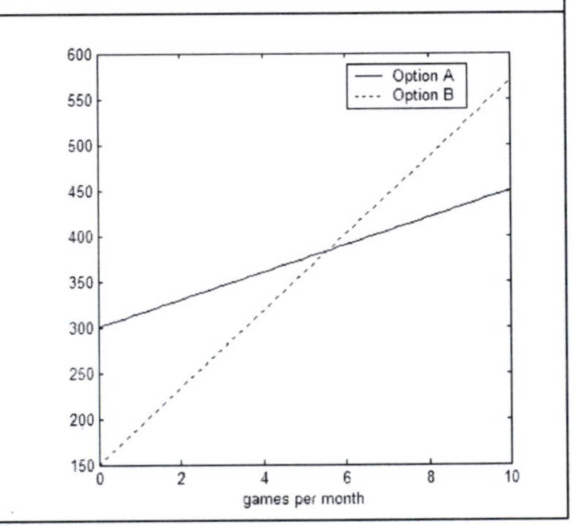

Section 1.3 Solutions ---

1.

$$x^2 - 5x + 6 = 0$$

$$(x-3)(x-2) = 0$$

$$x - 3 = 0 \text{ or } x - 2 = 0$$

$$\boxed{x = 3 \text{ or } x = 2}$$

3.

$$p^2 - 8p + 15 = 0$$

$$(p-5)(p-3) = 0$$

$$\boxed{p = 5 \text{ or } p = 3}$$

5.	$x^2 = 12 - x$ $x^2 + x - 12 = 0$ $(x+4)(x-3) = 0$ $x+4=0$ or $x-3=0$ $\boxed{x=-4 \text{ or } x=3}$	**7.**	$16x^2 + 8x = -1$ $16x^2 + 8x + 1 = 0$ $(4x+1)(4x+1) = 0$ $4x+1=0$ $\boxed{x=-1/4}$
9.	$9y^2 + 1 = 6y$ $9y^2 - 6y + 1 = 0$ $(3y-1)(3y-1) = 0$ $\boxed{y = \frac{1}{3}}$	**11.**	$8y^2 - 16y = 0$ $8y(y-2) = 0$ $8y = 0$ or $y-2=0$ $\boxed{y=0 \text{ or } y=2}$
13.	$9p^2 = 12p - 4$ $9p^2 - 12p + 4 = 0$ $(3p-2)(3p-2) = 0$ $3p-2=0$ $\boxed{p = \frac{2}{3}}$	**15.**	$x^2 - 9 = 0$ $(x+3)(x-3) = 0$ $x+3=0$ or $x-3=0$ $\boxed{x=-3 \text{ or } x=3}$
17.	$x(x+4) = 12$ $x^2 + 4x = 12$ $x^2 + 4x - 12 = 0$ $(x+6)(x-2) = 0$ $x+6=0$ or $x-2=0$ $\boxed{x=-6 \text{ or } x=2}$	**19.**	$2p^2 - 50 = 0$ $2(p^2 - 25) = 0$ $2(p-5)(p+5) = 0$ $\boxed{p=-5 \text{ or } p=5}$
21.	$3x^2 = 12$ $3x^2 - 12 = 0$ $3(x^2 - 4) = 0$ $3(x-2)(x+2) = 0$ $\boxed{x=-2 \text{ or } x=2}$	**23.**	$p^2 - 8 = 0$ $p^2 = 8$ $p = \pm\sqrt{8}$ $\boxed{p = \pm 2\sqrt{2}}$
25.	$x^2 + 9 = 0$ $x^2 = -9$ $\boxed{x = \pm 3i}$	**27.**	$(x-3)^2 = 36$ $x-3 = \pm 6$ $x = 3 \pm 6$ $\boxed{x = -3, 9}$

29.

$$(2x+3)^2 = -4$$
$$2x+3 = \pm 2i$$
$$2x = -3 \pm 2i$$
$$\boxed{x = \frac{-3 \pm 2i}{2}}$$

31.

$$(5x-2)^2 = 27$$
$$5x-2 = \pm\sqrt{27}$$
$$5x = 2 \pm 3\sqrt{3}$$
$$\boxed{x = \frac{2 \pm 3\sqrt{3}}{5}}$$

33.

$$(1-x)^2 = 9$$
$$1-x = \pm 3$$
$$-x = -1 \pm 3$$
$$\boxed{x = 1 \pm 3 = -2,\ 4}$$

35.

$$x^2 + 6x$$
$$\left(\frac{1}{2}\cdot 6\right)^2 = 3^2 = 9$$
$$x^2 + 6x + \boxed{9}$$

37.

$$x^2 - 12x$$
$$\left(\frac{1}{2}\cdot 12\right)^2 = 6^2 = 36$$
$$x^2 - 12x + \boxed{36}$$

39.

$$x^2 - \frac{1}{2}x$$
$$\left(\frac{1}{2}\cdot\frac{1}{2}\right)^2 = \left(\frac{1}{4}\right)^2 = \frac{1}{16}$$
$$x^2 - \frac{1}{2}x + \boxed{\frac{1}{16}}$$

41.

$$x^2 + \frac{2}{5}x$$
$$\left(\frac{1}{2}\cdot\frac{2}{5}\right)^2 = \left(\frac{1}{5}\right)^2 = \frac{1}{25}$$
$$x^2 + \frac{2}{5}x + \boxed{\frac{1}{25}}$$

43.

$$x^2 - 2.4x$$
$$\left(\frac{1}{2}\cdot 2.4\right)^2 = 1.2^2 = 1.44$$
$$x^2 - 2.4 + \boxed{1.44}$$

45.

$$x^2 + 2x = 3$$
$$x^2 + 2x + 1 = 3 + 1$$
$$(x+1)^2 = 4$$
$$x+1 = \pm 2$$
$$x = -1 \pm 2$$
$$\boxed{x = -3,\ 1}$$

47.

$$t^2 - 6t = -5$$
$$t^2 - 6t + 9 = -5 + 9$$
$$(t-3)^2 = 4$$
$$t-3 = \pm 2$$
$$\boxed{t = 3 \pm 2 = 1,\ 5}$$

49.

$$y^2 - 4y = -3$$
$$y^2 - 4y + 4 = -3 + 4$$
$$(y-2)^2 = 1$$
$$y - 2 = \pm 1$$
$$\boxed{y = \pm 1 + 2 = 1, 3}$$

51.

$$2p^2 + 8p = -3$$
$$2\left(p^2 + 4p\right) = -3$$
$$2\left(p^2 + 4p + 4\right) = -3 + 8$$
$$2(p+2)^2 = 5$$
$$(p+2)^2 = \frac{5}{2}$$
$$p + 2 = \pm\sqrt{\frac{5}{2}}$$
$$\boxed{p = -2 \pm \sqrt{\frac{5}{2}} = \frac{-4 \pm \sqrt{10}}{2}}$$

53.

$$2x^2 - 7x = -3$$
$$2\left(x^2 - \frac{7}{2}x\right) = -3$$
$$2\left(x^2 - \frac{7}{2}x + \left(\frac{7}{4}\right)^2\right) = -3 + 2\left(\frac{7}{4}\right)^2$$
$$2\left(x - \frac{7}{4}\right)^2 = -3 + 2\left(\frac{49}{16}\right)$$
$$\left(x - \frac{7}{4}\right)^2 = \frac{-3}{2} + \frac{49}{16} = \frac{25}{16}$$
$$x - \frac{7}{4} = \pm\frac{5}{4}$$
$$\boxed{x = \frac{7}{4} \pm \frac{5}{4} = \frac{1}{2}, 3}$$

55.

$$\frac{x^2}{2} - 2x = \frac{1}{4}$$
$$x^2 - 4x = \frac{1}{2}$$
$$x^2 - 4x + 4 = \frac{1}{2} + 4$$
$$(x-2)^2 = \frac{9}{2}$$
$$x - 2 = \pm\frac{3}{\sqrt{2}}$$
$$\boxed{x = 2 \pm \frac{3}{\sqrt{2}} = \frac{4 \pm 3\sqrt{2}}{2}}$$

57.

$$t^2 + 3t - 1 = 0$$
$$t = \frac{-3 \pm \sqrt{9+4}}{2}$$
$$\boxed{t = \frac{-3 \pm \sqrt{13}}{2}}$$

59.

$$s^2 + s + 1 = 0$$
$$s = \frac{-1 \pm \sqrt{1-4}}{2} = \frac{-1 \pm \sqrt{-3}}{2}$$
$$\boxed{s = \frac{-1 \pm i\sqrt{3}}{2}}$$

61. $3x^2-3x-4=0$ $x=\frac{3\pm\sqrt{9+48}}{6}=\frac{1}{2}\pm\frac{\sqrt{57}}{6}$ $\boxed{x=\frac{3\pm\sqrt{57}}{6}}$	**63.** $x^2-2x+17=0$ $x=\frac{2\pm\sqrt{4-4\cdot17}}{2}=\frac{2\pm\sqrt{-64}}{2}$ $\boxed{x=\frac{2\pm 8i}{2}=1\pm 4i}$
65. $5x^2+7x-3=0$ $x=\frac{-7\pm\sqrt{49+60}}{10}$ $\boxed{x=\frac{-7\pm\sqrt{109}}{10}}$	**67.** $\frac{1}{4}x^2+\frac{2}{3}x-\frac{1}{2}=0$ $3x^2+8x-6=0$ $x=\frac{-8\pm\sqrt{64-4(3)(-6)}}{2(3)}=\frac{-8\pm2\sqrt{34}}{2(3)}$ $\boxed{x=\frac{-4\pm\sqrt{34}}{3}}$
69. $(-22)^2-4(1)(121)=484-484=\boxed{0}$ $\boxed{\text{1 real solution}}$ (repeated root)	**71.** $(-30)^2-4(2)(68)=900-544=\boxed{356}$ $\boxed{\text{2 real solutions}}$ (distinct)
73. $(-7)^2-4(9)(8)=49-288=\boxed{-239}$ $\boxed{\text{2 complex solutions}}$ (complex conjugate)	**75.** $v^2-8v-20=0$ $(v-10)(v+2)=0$ $\boxed{v=-2,10}$
77. $t^2+5t-6=0$ $(t+6)(t-1)=0$ $\boxed{t=-6,1}$	**79.** $(x+3)^2=16$ $x+3=\pm4$ $\boxed{x=-3\pm4=-7,1}$
81. $(p-2)^2=4p$ $p^2-4p+4=4p$ $p^2-8p+4=0$ $p=\frac{8\pm\sqrt{64-4(1)(4)}}{2(1)}=\frac{8\pm4\sqrt{3}}{2}$ $\boxed{p=4\pm2\sqrt{3}}$	**83.** $8w^2+2w+21=0$ $w=\frac{-2\pm\sqrt{4-4\cdot8\cdot21}}{16}$ $w=\frac{-2\pm\sqrt{-668}}{16}=\frac{-2\pm2i\sqrt{167}}{16}$ $\boxed{w=\frac{-1\pm i\sqrt{167}}{8}}$

85.

$$3p^2 - 9p + 1 = 0$$

$$p = \frac{9 \pm \sqrt{81 - 12}}{6}$$

$$\boxed{p = \frac{9 \pm \sqrt{69}}{6}}$$

87.

$$\frac{2}{3}t^2 - \frac{4}{3}t - \frac{1}{5} = 0$$

$$\text{LCD} = 15$$

$$10t^2 - 20t - 3 = 0$$

$$t = \frac{20 \pm \sqrt{400 + 120}}{20}$$

$$t = \frac{20 \pm \sqrt{520}}{20} = \frac{20 \pm 2\sqrt{130}}{20}$$

$$\boxed{t = \frac{10 \pm \sqrt{130}}{10}}$$

89.

$$x + \frac{12}{x} = 7 \quad \boxed{x \neq 0}$$

$$x^2 + 12 = 7x$$

$$x^2 - 7x + 12 = 0$$

$$(x - 3)(x - 4) = 0$$

$$x - 3 = 0 \text{ or } x - 4 = 0$$

$$\boxed{x = 3 \text{ or } x = 4}$$

91.

$$\frac{4(x-2)}{x-3} + \frac{3}{x} = \frac{-3}{x(x-3)} \quad \boxed{x \neq 0, 3}$$

$$\text{LCD} = x(x-3)$$

$$4x(x-2) + 3(x-3) = -3$$

$$4x^2 - 8x + 3x - 9 = -3$$

$$4x^2 - 5x - 6 = 0$$

$$(4x + 3)(x - 2) = 0$$

$$4x + 3 = 0 \text{ or } x - 2 = 0$$

$$\boxed{x = -3/4 \text{ or } x = 2}$$

93.

$$x^2 - 0.1x - 0.12 = 0$$

$$(x - 0.4)(x + 0.3) = 0$$

$$\boxed{x = -0.3, 0.4}$$

95.

$$-4t^2 + 80t - 360 = 24$$

$$t^2 - 20t + 90 = -6$$

$$t^2 - 20t + 96 = 0$$

$$(t - 8)(t - 12) = 0$$

$$\boxed{t = 8 \text{ (August 2003) and } 12 \text{ (Dec. 2003)}}$$

97. Solve $P(q)=0$:

$$-100+(0.2q-3)q=0$$
$$-100+0.2q^2-3q=0$$
$$0.2q^2-3q-100=0$$
$$q^2-15q-500=0$$
$$q=\frac{15\pm\sqrt{(-15)^2-4(1)(-500)}}{2(1)}$$
$$=\frac{15\pm\sqrt{2,225}}{2}=\frac{15\pm 47.17}{2}$$
$$=31.085,\ \cancel{-16.09}$$

So, approximately 31,000 units must be sold to break even.

99. Solve $P(x)=460$:

$$-5(x+3)(x-24)=460$$
$$-5x^2+105x+360=460$$
$$-5x^2+105x-100=0$$
$$x^2-21x+20=0$$
$$(x-20)(x-1)=0$$
$$x=1,\ 20$$

So, the smallest price increase that will produce a weekly profit of \$460 is \$1 per bottle.

101. Solve $P(t)=160,\ 1\le t\le 6$:

$$-t^2+13t+130=160$$
$$-t^2+13t-30=0$$
$$t^2-13t+30=0$$
$$(t-10)(t-3)=0$$
$$t=3,\ \cancel{10}$$

So, 160 people would have contracted the flu after 3 days.

103. a.

The width of useable space = (8.5 – 2(1)) inches = 6.5 inches
The length of useable space = (11 – 2(1.25)) inches = 8.5 inches
So, the amount of useable space is the area, namely (6.5 in)(8.5 in) = 55.25 in^2.

b. Let x = amount of margin reduction (in inches)
Width of useable space = 8.5 – 2(1) + $2x$ = 6.5 + $2x$
Length of useable space = 11 – 2(1.25) + $2x$ = 8.5 + $2x$
So, the useable area is $(6.5+2x)(8.5+2x)=55.25+30x+4x^2$.

Continued onto next page.

c. $55.25+30x+4x^2 - 55.25 = 4x^2+30x$
This represents the increase in useable area of the paper.

d. Find x such that $10\left(55.25+30x+4x^2\right)=11(55.25)$.
Solving for x yields:

$$552.5+300x+40x^2=607.75$$
$$40x^2+300x-55.25=0$$
$$8x^2+60x-11.05=0$$
$$x=\frac{-60\pm\sqrt{60^2-4(8)(-11.05)}}{2(8)}$$
$$=\frac{-60\pm\sqrt{3{,}953.6}}{16}\approx\frac{2.877}{16}\approx 0.2$$

So, about 0.2 inches.

105. Form a right triangle with legs of length x and 25in. and hypotenuse of length 32in. Then, by the Pythagorean Theorem, we solve:

$$x^2+25^2=32^2$$
$$x^2=399$$
$$x=\pm\sqrt{399}\approx\pm 20$$

So, the TV is approximately 20 inches high.

107. Let the numbers be $x, x+1$.

$$x+(x+1)=35$$
$$2x=34 \Rightarrow x=17$$
$$x(x+1)=306$$
$$x^2+x=306$$
$$x^2+x-306=0$$
$$(x+18)(x-17)=0$$
$$x=\cancel{-18},17$$

So, the numbers are 17 and 18.

109. Let l = length of the rectangle (in ft.) Then, the width $w=l-6$ (in ft.) We must solve:

$$135=lw$$
$$135=l(l-6)$$
$$l^2-6l-135=0$$
$$(l-15)(l+9)=0$$
$$l=15, \cancel{-9}$$

So, the rectangle has:
length 15ft. and width 9ft.

111.

$$\text{Area} = \frac{1}{2}b \cdot h = 60$$
$$h = 3b + 2$$
$$\frac{1}{2}b(3b+2) = 60$$
$$\frac{3}{2}b^2 + b = 60$$
$$3b^2 + 2b - 120 = 0$$
$$(3b+20)(b-6) = 0$$
$$\boxed{b = \cancel{\frac{-20}{3}}, 6;\ h = 20}$$

113.

$$h = -16t^2 + 100$$
$$\text{Ground} \to h = 0$$
$$-16t^2 + 100 = 0$$
$$t^2 = \frac{100}{16}$$
$$t = \pm\frac{10}{4}\ (\text{Time must be} \geq 0)$$
$$\boxed{\text{Impact with ground in 2.5 sec}}$$

115.

$$15^2 + 15^2 = r^2$$
$$r^2 = 450$$
$$r = \pm\sqrt{450} = \pm 15\sqrt{2}$$
$$\boxed{r \approx 21.2\,\text{feet}}$$

117.

$$\text{volume} = l \cdot w \cdot h$$
$$v = (x-2)(x-2)(1)$$
$$9 = (x-2)^2$$
$$x - 2 = \pm 3$$
$$x = 2 \pm 3 = -1, 5$$
$$x = 5$$
$$\boxed{\text{Original square was 5ft} \times \text{5ft}}$$

119.

Let w = width of border

Total area of garden + border $= (8+2w)(5+2w) = 4w^2 + 26w + 40$

Area of garden $= 8 \cdot 5 = 40$

Area of border $= \underbrace{(4w^2 + 26w + 40)}_{total} - \underbrace{40}_{garden} = 4w^2 + 26w$

Volume of border = Area $\cdot$ depth (depth = 4 in. = 1/3 ft)

$$= (4w^2 + 26w)(1/3)$$

Volume = 27 ft^3

$$\frac{1}{3}(4w^2 + 26w) = 27$$
$$4w^2 + 26w = 81$$
$$4w^2 + 26w - 81 = 0$$
$$w = \frac{-26 \pm \sqrt{26^2 + 4 \cdot 4 \cdot 81}}{2 \cdot 4} = \frac{-26 \pm \sqrt{1972}}{8}$$
$$w \cong \cancel{-8.8}, 2.3$$

$$\boxed{\text{Width of border is 2.3 feet.}}$$

121. Let x = days for Kimmie to complete job herself.

$x-5$ = days for Lindsey to complete job herself.

$\frac{1}{x}$ = % of job Kimmie can do per day.

$\frac{1}{x-5}$ = % of job Lindsey can do per day.

$\frac{1}{\text{x}}+\frac{1}{x-5}=\frac{1}{6}$ (Together they can do it in 6 days.)

LCD $= x(x-5)6$ $\boxed{x \neq 0,5}$

$6(x-5)+6x = x(x-5)$

$6x-30+6x = x^2-5x$

$x^2-17x+30=0$

$(x-15)(x-2)=0$

$x = \cancel{2}, 15$

Kimmie alone: 15 days

$\boxed{\text{Lindsey alone: 10 days}}$

123. Factored incorrectly

$t^2-5t-6=0$

$(t+1)(t-6)=0$

$\boxed{t=-1, 6}$

125. $\sqrt{-a}$ is imaginary for positive a

$a^2=-\frac{9}{16}$, so $\boxed{a=\pm\sqrt{\frac{9}{16}}=\pm\frac{3}{4}i}$

127. False

$x=-5/3$ satisfies 1st equation but not 2nd

129. True

131. If $x=a$ is a repeated root for a quadratic equation, then $(x-a)^2=0$. Simplifying yields:

$\boxed{x^2-2ax+a^2=0}$

133. $(x-2)(x-5)=0$

$\boxed{x^2-7x+10=0}$

135.

$s=\frac{1}{2}gt^2 \Rightarrow t^2=\frac{2s}{g} \Rightarrow \boxed{t=\pm\sqrt{\frac{2s}{g}}}$

137.

$a^2+b^2=c^2$

$\boxed{c=\pm\sqrt{a^2+b^2}}$

139.

$$x^4 - 4x^2 = 0$$
$$x^2\left(x^2 - 4\right) = 0$$
$$x^2(x-2)(x+2) = 0$$
$$\boxed{x = 0, \pm 2}$$

141.

$$x^3 + x^2 - 4x - 4 = 0$$
$$\left(x^3 + x^2\right) - 4(x+1) = 0$$
$$x^2(x+1) - 4(x+1) = 0$$
$$\left(x^2 - 4\right)(x+1) = 0$$
$$(x-2)(x+2)(x+1) = 0$$
$$\boxed{x = -1, \pm 2}$$

143.

$$x_1 = \frac{-b + \sqrt{b^2 - 4ac}}{2a} \qquad x_2 = \frac{-b - \sqrt{b^2 - 4ac}}{2a}$$

$$x_1 + x_2 = \frac{-b}{2a} + \frac{\sqrt{b^2 - 4ac}}{2a} - \frac{b}{2a} - \frac{\sqrt{b^2 - 4ac}}{2a}$$
$$= \frac{-2b}{2a} = \boxed{\frac{-b}{a}}$$

145.

$$\left[x - \left(3 + \sqrt{5}\right)\right]\left[x - \left(3 - \sqrt{5}\right)\right] = 0$$
$$\left[(x-3) - \sqrt{5}\right]\left[(x-3) + \sqrt{5}\right] = 0$$
$$(x-3)^2 - 5 = 0$$
$$x^2 - 6x + 9 - 5 = 0$$
$$\boxed{x^2 - 6x + 4 = 0}$$

147.

Let x = speed in still air and y = time to make the trip with a tail wind.
Using Distance = Rate × Time, we obtain the following two equations:
With tail wind: $(x+50)y = 600$ **(1)**
Against head wind: $(x-50)(y+1) = 600$ **(2)**
Solve **(1)** for y: $y = \dfrac{600}{x+50}$
Substitute this into **(2)** and solve for x:

$$(x-50)\left(\frac{600}{x+50}+1\right)=600$$

$$(x-50)\left(\frac{600+x+50}{x+50}\right)=600$$

$$(x-50)(650+x)=600(x+50)$$

$$650x-32,500-50x+x^2=600x+30,000$$

$$x^2-62,500=0$$

$$(x-250)(x+250)=0$$

$$x=250,\ \cancel{-250}$$

So, the plane in still air travels at $\boxed{250\text{mph}}$.

149.

2 distinct real roots of $ax^2+bx+c=0$ are: $x_1=\frac{-b+\sqrt{b^2-4ac}}{2a}$ $\quad x_2=\frac{-b-\sqrt{b^2-4ac}}{2a}$

If real roots are negatives of x_1, x_2, then $x_1^*=\frac{b-\sqrt{b^2-4ac}}{2a}$ $\quad x_2^*=\frac{b+\sqrt{b^2-4ac}}{2a}$

Replace b with $-b$. So, $\boxed{ax^2-bx+c=0}$.

151. Let x = speed of small jet (in mph). Then, the speed of the 757-jet = x+100 (mph) Form a right triangle depicting the relative position of the jets after two hours of flight. Using Distance = Rate × time, this triangle will have legs of length $2x$ and $2(x+100)$, and hypotenuse of length 1000 miles. Using the Pythagorean Theorem then yields

$$(2x)^2+(2(x+100))^2=1000^2$$

$$4x^2+4x^2+800x+40,000=1,000,000$$

$$x^2+100x-120,000=0$$

$$x=\frac{-100\pm\sqrt{100^2+4(120,000)}}{2}=\frac{-100\pm 700}{2}=\cancel{-400},\ 300$$

So, $\boxed{\text{the speed of the small jet is 300mph and the speed of the 757-jet is 400mph}}$.

153. $x^2-x-2=0$

$$(x-2)(x+1)=0$$

$$\boxed{x=-1,\ 2}$$

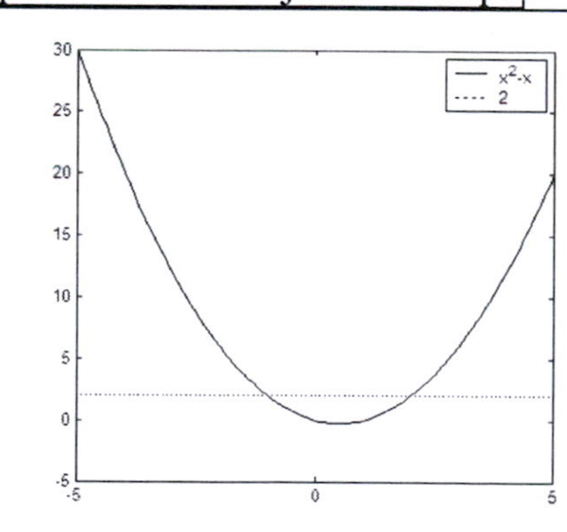

155. (a) Consider $x^2 - 2x - b = 0$. **(1)**
For $b = 8$, **(1)** factors as $(x-4)(x+2) = 0$, so that $x = -2, 4$.
Graphically, we let $y1 = x^2 - 2x,\ y2 = 8$ and look for the intersection points of the graphs:

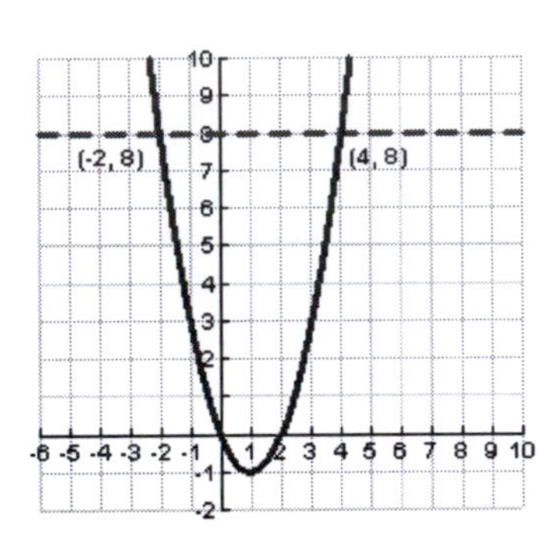

Note that they intersect at precisely the x-values obtained algebraically. So, yes, these values agree with the points of intersections.

(b) We do the same thing now for different values of b.

$\underline{b = -3}$:

$$x^2 - 2x + 3 = 0$$

$$x = \frac{2 \pm \sqrt{4 - 4(3)}}{2} = 1 \pm i\sqrt{2}$$

So, we don't expect the graphs to intersect. Indeed, we have:

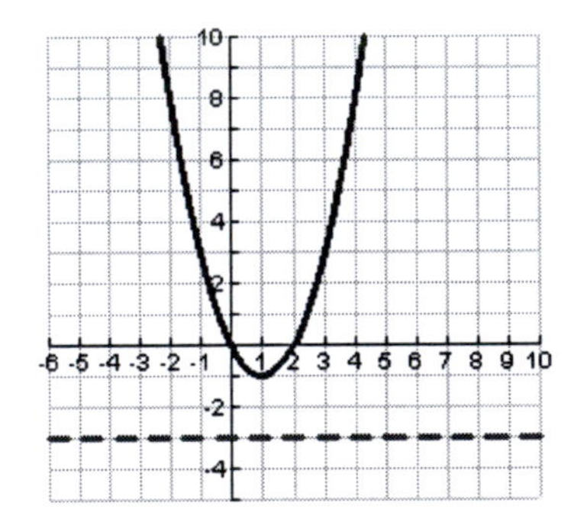

$\underline{b = -1}$:

$$x^2 - 2x + 1 = 0$$

$$(x-1)^2 = 0$$

$$x = 1$$

So, we expect the graphs to intersect once. Indeed, we have:

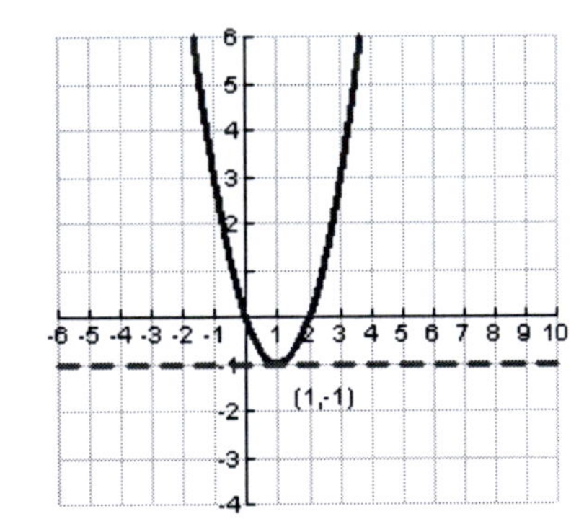

(Continued onto next page)

$\underline{b=0}$:

$$x^2-2x=0$$
$$x(x-2)=0$$
$$x=0,2$$

So, we expect the graphs to intersect twice as in part **(a)**. Indeed, we have:

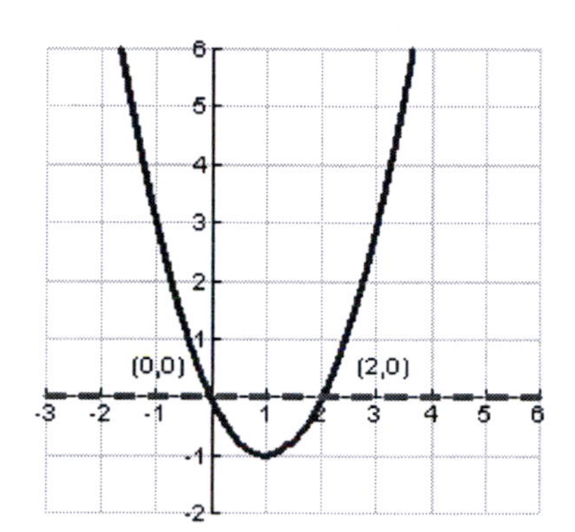

$\underline{b=5}$:

$$x^2-2x-5=0$$
$$x=\frac{2\pm\sqrt{4+4(5)}}{2}=1\pm\sqrt{6}$$

So, we expect the graphs to intersect twice as in part **(a)**. Indeed, we have:

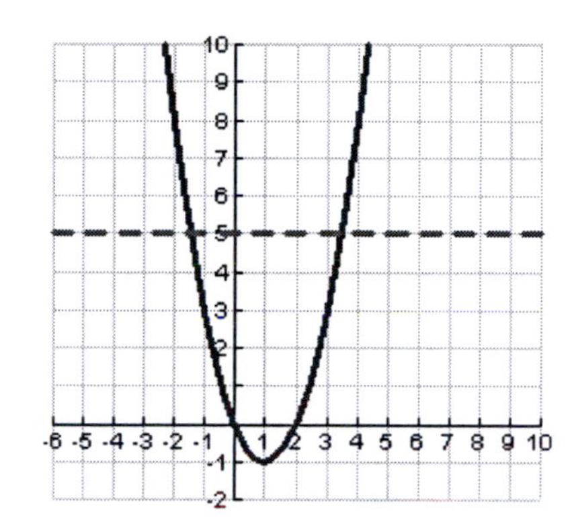

Section 1.4 Solutions ---

1. $\sqrt{t-5}=2$ $t-5=4$ $\boxed{t=9}$	**3.** $(4p-7)^{1/2}=5$ $4p-7=25$ $4p=32$ $\boxed{p=8}$	**5.** $\sqrt{u+1}=-4$ $\boxed{\text{no solution}}$ $u+1=16$ $u=15$ Check: $\sqrt{15+1}$ $=\sqrt{16}=4$	**7.** $\sqrt[3]{5x+2}=3$ $5x+2=3^3=27$ $5x=25$ $\boxed{x=5}$
9. $(4y+1)^{\frac{1}{3}}=-1$ $4y+1=-1$ $4y=-2$ $\boxed{y=-\frac{1}{2}}$	**11.** $\sqrt{12+x}=x$ $12+x=x^2$ $x^2-x-12=0$ $(x+3)(x-4)=0$ $x=-3,\boxed{4}$ Check -3: $\sqrt{12-3}=\sqrt{9}\neq-3$ Check 4: $\sqrt{12+4}=\sqrt{16}=4$	**13.** $y=5\sqrt{y}$ $y^2=25y$ $y^2-25y=0$ $y(y-25)=0$ $\boxed{y=0,25}$ Check 0: $0=5\sqrt{0}$ Check 25: $25=5\sqrt{25}$	**15.** $s=3\sqrt{s-2}$ $s^2=9(s-2)$ $s^2=9s-18$ $s^2-9s+18=0$ $(s-3)(s-6)=0$ $\boxed{s=3,6}$ Check 3: $3=3\sqrt{3-2}=3\sqrt{1}$ Check 6: $6=3\sqrt{6-2}=3\sqrt{4}$

17.

$$\sqrt{2x+6} = x+3$$
$$2x+6 = (x+3)^2$$
$$x^2+4x+3=0$$
$$(x+3)(x+1)=0$$
$$x = \boxed{-3, -1}$$

Check -3:

$$\sqrt{2(-3)+6} = -3+3$$
$$\sqrt{0}=0$$

Check -1:

$$\sqrt{2(-1)+6} = -1+3$$
$$\sqrt{4}=2$$

19.

$$\sqrt{1-3x} = x+1$$
$$1-3x = x^2+2x+1$$
$$x^2+5x=0$$
$$x(x+5)=0$$
$$x=-5, \boxed{0}$$

Check -5:

$$\sqrt{1+15} \neq -4$$

Check 0:

$$\sqrt{1}=1$$

21.

$$3x-6\sqrt{x-1}=3$$
$$3x-3=6\sqrt{x-1}$$
$$x-1=2\sqrt{x-1}$$
$$(x-1)^2=\left(2\sqrt{x-1}\right)^2$$
$$(x-1)^2-4(x-1)=0$$
$$(x-1)(x-1-4)=0$$
$$(x-1)(x-5)=0$$
$$x=1,5$$

23.

$$3x-6\sqrt{x+2}=3$$
$$x-2\sqrt{x+2}=1$$
$$x-1=2\sqrt{x+2}$$
$$(x-1)^2=\left(2\sqrt{x+2}\right)^2$$
$$(x-1)^2=4(x+2)$$
$$x^2-2x+1=4x+8$$
$$x^2-6x-7=0$$
$$(x-7)(x+1)=0$$
$$x=\cancel{-1},7$$

25.

$$3\sqrt{x+4}-2x=9$$
$$3\sqrt{x+4}=2x+9$$
$$\left(3\sqrt{x+4}\right)^2=(2x+9)^2$$
$$9(x+4)=4x^2+36x+81$$
$$9x+36=4x^2+36x+81$$
$$4x^2+27x+45=0$$
$$(4x+15)(x+3)=0$$
$$x=-\tfrac{15}{4},-3$$

27.

$$\sqrt{x^2-4}=x-1$$
$$x^2-4=(x-1)^2$$
$$x^2-4=x^2-2x+1$$
$$2x=5$$
$$\boxed{x=\frac{5}{2}}$$

29.

$$\sqrt{x^2-2x-5}=x+1$$
$$x^2-2x-5=(x+1)^2$$
$$x^2-2x-5=x^2+2x+1$$
$$-6=4x$$
$$\cancel{-\tfrac{3}{2}}=x$$

$\boxed{\text{No solution}}$.

31.

$$\sqrt{3x+1}-\sqrt{6x-5}=1$$
$$\sqrt{3x+1}=\sqrt{6x-5}+1$$
$$\left(\sqrt{3x+1}\right)^2=\left(\sqrt{6x-5}+1\right)^2$$
$$3x+1=6x-5+2\sqrt{6x-5}+1$$
$$3x+1=6x-4+2\sqrt{6x-5}$$
$$(-3x+5)^2=\left(2\sqrt{6x-5}\right)^2$$
$$9x^2-30x+25=4(6x-5)$$
$$9x^2-30x+25=24x-20$$
$$9x^2-54x+45=0$$
$$(9x-9)(x-5)=0$$
$$x=1,\cancel{5}$$

33.

$$\sqrt{x+12}+\sqrt{8-x}=6$$
$$\sqrt{x+12}=6-\sqrt{8-x}$$
$$\left(\sqrt{x+12}\right)^2=\left(6-\sqrt{8-x}\right)^2$$
$$x+12=36-12\sqrt{8-x}+(8-x)$$
$$2x-32=-12\sqrt{8-x}$$
$$x-16=-6\sqrt{8-x}$$
$$(x-16)^2=\left(-6\sqrt{8-x}\right)^2$$
$$x^2-32x+256=36(8-x)$$
$$x^2-32x+256=288-36x$$
$$x^2+4x-32=0$$
$$(x-4)(x+8)=0$$
$$x=4,-8$$

35.

$$\sqrt{2x-1}=1+\sqrt{x-1}$$
$$2x-1=1+2\sqrt{x-1}+x-1$$
$$x-1=2\sqrt{x-1}$$
$$x^2-2x+1=4(x-1)$$
$$x^2-2x+1=4x-4$$
$$x^2-6x+5=0$$
$$(x-5)(x-1)=0$$
$$\boxed{x=1,5}$$

37.

$$\sqrt{3x-5}=7-\sqrt{x+2}$$
$$3x-5=49-14\sqrt{x+2}+x+2$$
$$2x-56=-14\sqrt{x+2}$$
$$x-28=-7\sqrt{x+2}$$
$$x^2-56x+784=49(x+2)$$
$$x^2-56x+784=49x+98$$
$$x^2-105x+686=0$$
$$(x-98)(x-7)=0$$
$$x=\boxed{7},\ \cancel{98}$$

39.

$$\sqrt{2+\sqrt{x}}=\sqrt{x}$$
$$2+\sqrt{x}=x$$
$$\sqrt{x}=x-2$$
$$x=x^2-4x+4$$
$$x^2-5x+4=0$$
$$(x-4)(x-1)=0$$
$$x=\cancel{1},\ \boxed{4}$$

41. Let $u=x^{1/3}$

$$u^2+2u=0$$
$$u(u+2)=0$$
$$u=-2,0$$
$$x^{1/3}=0\rightarrow\boxed{x=0}$$
$$x^{1/3}=-2\rightarrow\boxed{x=-8}$$

43. Let $u=x^2$

$$u^2-3u+2=0$$
$$(u-1)(u-2)=0$$
$$u=1,2$$
$$x^2=1\rightarrow\boxed{x=\pm 1}$$
$$x^2=2\rightarrow\boxed{x=\pm\sqrt{2}}$$

45. Let $u = x^2$

$2u^2 + 7u + 6 = 0$

$(2u+3)(u+2) = 0$

$u = -3/2$ $\quad u = -2$

$x^2 = -3/2$ $\quad x^2 = -2$

$x = \pm i\sqrt{3/2}$ $\quad x = \pm i\sqrt{2}$

$\boxed{x = \frac{\pm i\sqrt{6}}{2}}$ $\quad \boxed{x = \pm i\sqrt{2}}$

47. Let $u = 2x+1$

$u^2 + 5u + 4 = 0$

$(u+4)(u+1) = 0$

$u = -4$ $\quad u = -1$

$2x+1 = -4$ $\quad 2x+1 = -1$

$2x = -5$ $\quad 2x = -2$

$\boxed{x = -5/2}$ $\quad \boxed{x = -1}$

49. Let $u = t-1$

$4u^2 - 9u + 2 = 0$

$(4u-1)(u-2) = 0$

$u = 1/4$ $\quad u = 2$

$t-1 = 1/4$ $\quad t-1 = 2$

$\boxed{t = 5/4}$ $\quad \boxed{t = 3}$

51. Let $u = x^{-4}$

$u^2 - 17u + 16 = 0$

$(u-16)(u-1) = 0$

$u = 1$ $\quad u = 16$

$x^{-4} = 1$ $\quad x^{-4} = 16$

$x^2 = \pm 1$ $\quad x^2 = \pm 1/4$

$\boxed{x = \pm 1, \pm i}$ $\quad \boxed{x = \pm\frac{1}{2}, \pm\frac{1}{2}i}$

53. Let $u = y^{-1}$

$3u^2 + u - 4 = 0$

$(3u+4)(u-1) = 0$

$u = -4/3$ $\quad u = 1$

$y^{-1} = -4/3$ $\quad y^{-1} = 1$

$\boxed{y = -3/4}$ $\quad \boxed{y = 1}$

55. Let $u = z^{1/5}$

$u^2 - 2u + 1 = 0$

$(u-1)^2 = 0$

$u = 1$

$z^{1/5} = 1$

$\boxed{z = 1}$

57.

$(x+3)^{5/3} = 32$

$x+3 = 32^{3/5}$

$x = -3 + \left(32^{1/5}\right)^3 = -3 + 2^3 = -3 + 8 = 5$

59.

$(x+1)^{2/3} = 4$

$x+1 = \pm 4^{3/2}$

$x = -1 \pm 4^{3/2} = -1 \pm 8$

$x = -9$ or $x = 7$

61 Let $u = t^{-1/3}$

$6u^2 - u - 1 = 0$

$(3u+1)(2u-1) = 0$

$u = -1/3$ $\quad u = 1/2$

$t^{-1/3} = -1/3$ $\quad t^{-1/3} = 1/2$

$t = (-1/3)^{-3}$ $\quad t = (1/2)^{-3}$

$\boxed{t = -27}$ $\quad \boxed{t = 8}$

63.

$3 = \frac{1}{(x+1)^2} + \frac{2}{(x+1)}$ $\quad \boxed{x \neq -1}$

$3(x+1)^2 = 1 + 2(x+1)$

$3(x+1)^2 - 2(x+1) - 1 = 0$

Let $u = x+1$

$3u^2 - 2u - 1 = 0$

$(3u+1)(u-1) = 0$

$u = -1/3$ $\quad u = 1$

$x+1 = -1/3$ $\quad x+1 = 1$

$\boxed{x = -4/3}$ $\quad \boxed{x = 0}$

65.

$$\left(\frac{1}{2x-1}\right)^2+\frac{1}{2x-1}-12=0$$

$$\boxed{x\neq 1/2}$$

Let $u=\dfrac{1}{2x-1}$

$$u^2+u-12=0$$

$$(u+4)(u-3)=0$$

Then, we have:

$u=-4$

$$\frac{1}{2x-1}=-4$$

$$-4(2x-1)=1$$

$$-8x+4=1$$

$$-8x=-3$$

$$\boxed{x=3/8}$$

$u=3$

$$\frac{1}{2x-1}=3$$

$$3(2x-1)=1$$

$$6x-3=1$$

$$6x=4$$

$$\boxed{x=2/3}$$

67. Let $x=u^{2/3}$

$$x^2-5x+4=0$$

$$(x-4)(x-1)=0$$

$x=4$ $\quad$ $x=1$

$u^{2/3}=4$ $\quad$ $u^{2/3}=1$

$u=\pm 4^{3/2}$ $\quad$ $u=\pm 1^{3/2}$

$\boxed{u=\pm 8}$ $\quad$ $\boxed{u=\pm 1}$

69. $t^4-t^2-6=0$

Let $u=t^2$

$$u^2-u-6=0$$

$$(u-3)(u+2)=0$$

$u=-2$ $\quad$ $u=3$

$t^2=-2$ $\quad$ $t^2=3$

$\cancel{t=\pm i\sqrt{2}}$ $\quad$ $\boxed{t=\sqrt{3},\ \cancel{-\sqrt{3}}}$

71.

$$x^3-x^2-12x=0$$

$$x\left(x^2-x-12\right)=0$$

$$x(x-4)(x+3)=0$$

$$\boxed{x=0,-3,4}$$

73.

$$4p^3-9p=0$$

$$p\left(4p^2-9\right)=0$$

$$p(2p-3)(2p+3)=0$$

$$\boxed{p=0,\ \pm\tfrac{3}{2}}$$

75.

$$u^5-16u=0$$

$$u\left(u^4-16\right)=0$$

$$u\left(u^2-4\right)\left(u^2+4\right)=0$$

$$u(u-2)(u+2)(u-2i)(u+2i)=0$$

$$\boxed{u=0,\ \pm 2,\ \pm 2i}$$

77.

$$x^3 - 5x^2 - 9x + 45 = 0$$
$$\left(x^3 - 5x^2\right) - \left(9x - 45\right) = 0$$
$$x^2(x-5) - 9(x-5) = 0$$
$$\left(x^2 - 9\right)(x-5) = 0$$
$$(x-3)(x+3)(x-5) = 0$$
$$\boxed{x = \pm 3,\ 5}$$

79.

$$y(y-5)^3 - 14(y-5)^2 = 0$$
$$(y-5)^2\left[y(y-5) - 14\right] = 0$$
$$(y-5)^2\left(y^2 - 5y - 14\right) = 0$$
$$(y-5)^2(y-7)(y+2) = 0$$
$$\boxed{y = -2, 5, 7}$$

81.

$$x^{9/4} - 2x^{5/4} - 3x^{1/4} = 0$$
$$x^{1/4}\left[x^2 - 2x - 3\right] = 0$$
$$x^{1/4}(x-3)(x+1) = 0$$
$$\boxed{x = 0, 3, \cancel{-1}}$$

83.

$$t^{5/3} - 25t^{-1/3} = 0$$
$$t^{-1/3}\left[t^2 - 25\right] = 0$$
$$t^{-1/3}(t-5)(t+5) = 0$$
$$\boxed{t = \pm 5}$$

(Note: $t^{-1/3} = 0$ has no solution.)

85.

$$y^{3/2} - 5y^{1/2} + 6y^{-1/2} = 0$$
$$y^{-1/2}\left[y^2 - 5y + 6\right] = 0$$
$$y^{-1/2}(y-3)(y-2) = 0$$
$$\boxed{y = 2, 3}$$

(Note: $y^{-1/2} = 0$ has no solution.)

87. Solve $d(t) = 3$. (Note: The right-side is 3, and not 3,000,000, because $d(t)$ is measured in millions.)

$$3\sqrt{t+1} - 0.75t = 3$$
$$3\sqrt{t+1} = 3 + 0.75t$$
$$\left(3\sqrt{t+1}\right)^2 = \left(3 + 0.75t\right)^2$$
$$9t + 9 = 9 + 4.5t + 0.5625t^2$$
$$0.5625t^2 - 4.5t = 0$$
$$t(0.5625t - 4.5) = 0$$
$$t = 0,\ \tfrac{4.5}{0.5625} = 8$$

So, this occurs in January and September.

89. Solve $\sqrt{\frac{wh}{3,600}} = BSA$ for h, when $w = 72$ and $BSA = 1.8$.

$$\sqrt{\frac{72h}{3,600}} = 1.8$$
$$\frac{\sqrt{72h}}{60} = 1.8$$
$$\sqrt{72h} = (1.8)(60)$$
$$72h = 108^2$$
$$h = \frac{11,664}{72} = 162$$

So, the height of such a female is 162 cm.

91.

$$C = \sqrt{10+a}$$
$$C = 9$$
$$9 = \sqrt{10+a}$$
$$81 = 10 + a$$
$$\boxed{a = 71 \text{ years old}}$$

93.

$$P = 5\sqrt{t^2+1} + 50$$
$$P = 85$$
$$85 = 5\sqrt{t^2+1} + 50$$
$$35 = 5\sqrt{t^2+1}$$
$$7 = \sqrt{t^2+1}$$
$$49 = t^2 + 1$$
$$t^2 = 48$$
$$t = \sqrt{48}$$
$$t = 4\sqrt{3} \ (t \text{ must be} \geq 0)$$
$$t \cong 7 \text{ months}$$
$$\boxed{\text{Oct 2004}}$$

95.

$$T = \frac{\sqrt{d}}{4} + \frac{d}{1100}, \ T = 3$$
$$3 = \frac{\sqrt{d}}{4} + \frac{d}{1100}$$
$$\text{LCD} = 1100$$
$$3300 = 275\sqrt{d} + d$$
$$d + 275\sqrt{d} - 3300 = 0$$
$$\text{Let } u = \sqrt{d}$$
$$u^2 + 275u - 3300 = 0$$
$$u = \frac{-275 \pm \sqrt{275^2 + 4 \cdot 1 \cdot 3300}}{2(1)}$$
$$u = -286.5, 11.5$$
$$\sqrt{d} = 11.5$$
$$\boxed{d = 132 \text{ ft}}$$

97.

$$1 = 2\pi\sqrt{\frac{L}{9.8}}$$

$$\left(\frac{1}{2\pi}\right)^2 = \frac{L}{9.8}$$

$$0.24824\,\text{m} \approx \frac{9.8}{4\pi^2} = L$$

Convert to centimeters:

$$\frac{0.24824\ \cancel{m} \mid 100\ \text{cm}}{\mid 1\ \cancel{m}} \approx \boxed{25\,\text{cm}}$$

99.

$$18 = 30\sqrt{1-\frac{v^2}{c^2}}$$

$$\frac{3}{5} = \frac{18}{30} = \sqrt{1-\frac{v^2}{c^2}}$$

$$\left(\frac{3}{5}\right)^2 = 1-\frac{v^2}{c^2}$$

$$\frac{16}{25} = \frac{v^2}{c^2}$$

$$v^2 = \frac{16}{25}c^2$$

$$v = \frac{4}{5}c$$

So, $\boxed{\text{80\% of the speed of light}}$.

101. $t = 5$ is extraneous; there is no solution.

103. Forgot about the substitution $u = x^{1/3}$.

$$x^{1/3} = -4, 5$$

$$\boxed{x = -64, 125}$$

105. True

Let $u = (2x-1)^3$

$u^2 + 4u + 3 = 0$ (quadratic)

107. False

109. Solve $\sqrt{x^2} = x$.

If $x \geq 0$, then $\sqrt{x^2} = x$, while if $x < 0$, then $\sqrt{x^2} = -x$. So, the solution set is $\boxed{[0, \infty)}$.

111.

Let $u = 3x^2 + 2x$

$$u = \sqrt{u}$$

$$u = 0, 1$$

$$3x^2 + 2x = 0$$

$$x(3x+2) = 0$$

$$\boxed{x = 0, -2/3}$$

$$3x^2 + 2x = 1$$

$$3x^2 + 2x - 1 = 0$$

$$(3x-1)(x+1) = 0$$

$$\boxed{x = -1, 1/3}$$

113.

$$\sqrt{x+6} + \sqrt{11+x} = 5\sqrt{3+x}$$

$$(x+6) + 2\sqrt{x+6}\sqrt{11+x} + (11+x) = 25(3+x)$$

$$2x + 17 + 2\sqrt{x+6}\sqrt{11+x} = 75 + 25x$$

$$2\sqrt{x+6}\sqrt{11+x} = 58 + 23x$$

$$4(x+6)(11+x) = 529x^2 + 2668x + 3364$$

$$4(x^2 + 17x + 66) = 529x^2 + 2668x + 3364$$

$$4x^2 + 68x + 264 = 529x^2 + 2668x + 3364$$

$$525x^2 + 2600x + 3100 = 0$$

$$21x^2 + 104x + 124 = 0$$

$$(21x + 62)(x + 2) = 0$$

$$\cancel{x = -\frac{62}{21}},\ \boxed{x = -2}$$

115.

$\sqrt{x-3} = 4 - \sqrt{x+2}$

$x - 3 = 16 - 8\sqrt{x+2} + x + 2$

$-21 = -8\sqrt{x+2}$

$441 = 64(x+2) = 64x + 128$

$313 = 64x$

$\boxed{x = \frac{313}{64} \cong 4.891}$

117.

$-4 = \sqrt{x+3}$

$16 = x + 3$

$x = 13$ (Extraneous)

$\boxed{\text{no solution}}$

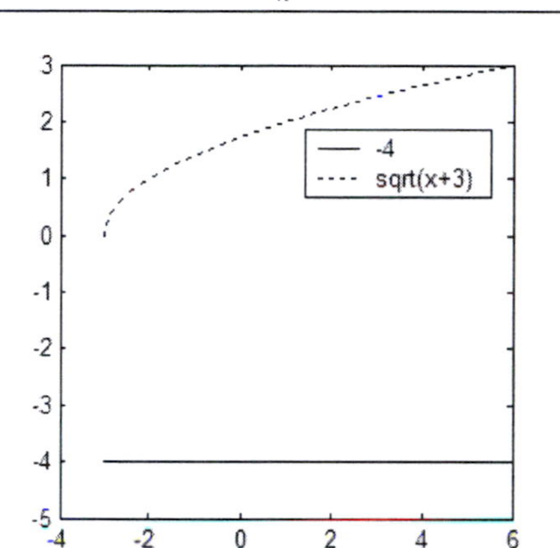

119.

$x^{1/2} = -4x^{1/4} + 21$

$x^{1/2} + 4x^{1/4} - 21 = 0$

Let $u = x^{1/4}$ to obtain

$u^2 + 4u - 21 = 0$

$(u+7)(u-3) = 0$

$u = -7, 3$

$x^{1/4} = -7$ $\quad$ $x^{1/4} = 3$

$\boxed{\text{no solution}}$ $\quad$ $\boxed{x = 81}$

Graphically, let:

$y1 = x^{1/2}, \ y2 = -4x^{1/4} + 21$.

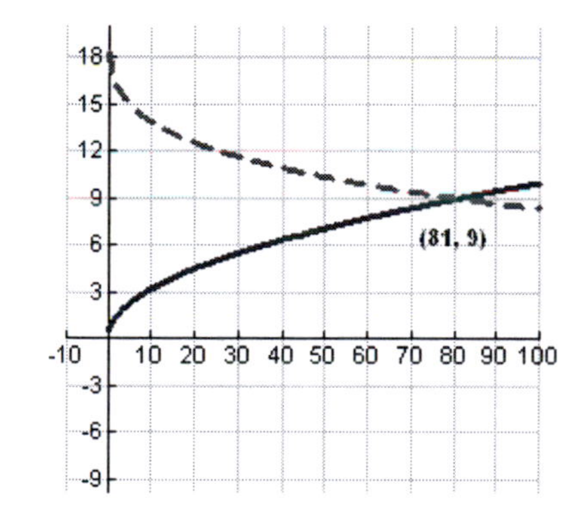

Yes, the two solutions agree.

121.

$$x^{-2} = 3x^{-1} - 10$$

$$x^{-2} - 3x^{-1} + 10 = 0$$

Let $u = x^{-1}$ to obtain

$$u^2 - 3u + 10 = 0$$

$$u = \frac{3 \pm \sqrt{9 - 4(10)(1)}}{2} = \frac{3 \pm i\sqrt{31}}{2}$$

So, there are no real solutions. As such, we expect the graphs to not intersect.

Graphically, let:
$y1 = x^{-2}$, $y2 = 3x^{-1} - 10$.

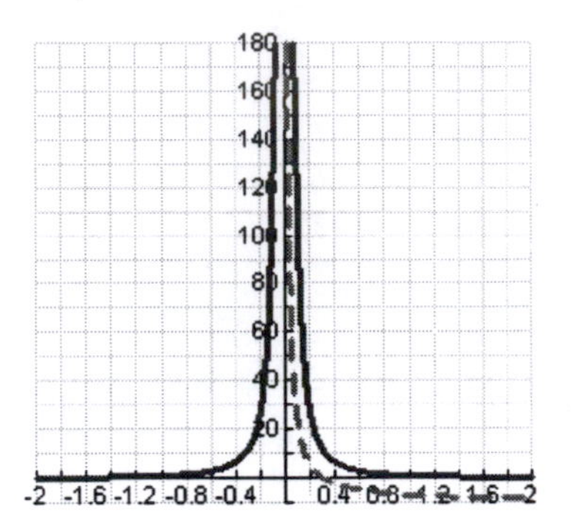

Yes, the two solutions agree.

Section 1.5 Solutions ---

1. $[3,\infty)$ … 0 1 2 3 4 5 6…	**3.** $(-\infty,-5]$ … −7 −6 −5 −4 −3 −2 −1 0…
5. $[-2,3)$ … −3 −2 −1 0 1 2 3 4 …	**7.** $(-3,5]$ 
9. $[0,0]$ … −3 −2 −1 0 1 2 3 …	**11.** $[4,6]$ …3 4 5 6 7 …
13. $[-8,-6]$ 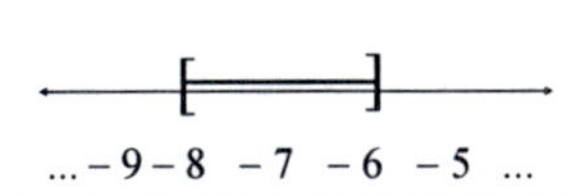	**15.** $\varnothing$ … −3 −2 −1 0 1 2 3 …
17. $\{x : 0 \le x < 2\}$	**19.** $\{x : -7 < x < -2\}$
21. $\{x : x \le 6\}$	**23.** $\{x : -\infty < x < \infty\}$
25. $-3 < x \le 7 \quad (-3,7]$	**27.** $3 \le x < 5 \quad [3,5)$
29. $-2 \le x \quad [-2,\infty)$	**31.** $-\infty < x < 8 \quad (-\infty,8)$
33. $(-5,3)$ … −5 −4 −3 −2 −1 0 1 2 3 …	**35.** $[-6,5)$ 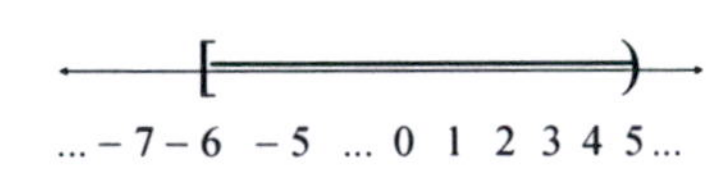

37. $[-1,1]$

... −2 −1 0 1 2 3 ...

39. $[1,4)$

... 0 1 2 3 4 ...

41. $[-1,2)$

... −2 −1 0 1 2 3...

43. $(-\infty,4)\cup(4,\infty)$

...2 3 4 5 6 7 ...

45. $(-\infty,-3]\cup[3,\infty)$

... −3 −2 −1 0 1 2 3 ...

47. $(-3,2]$

... −4 −3 −2 −1 0 1 2 3 4 ...

49. $\varnothing$

... −3 −2 −1 0 1 2 3 ...

51. $(-\infty,2)\cup[3,5)$

53. $(-\infty,-4]\cup(2,5]$

55. $[-4,-2)\cup(3,7]$

57. $(-6,-3]\cup[0,4)$

59.

$$x-3<7$$
$$x<10$$
$$\boxed{(-\infty,10)}$$

61.

$$3x-2\le 4$$
$$3x\le 6$$
$$x\le 2$$
$$\boxed{(-\infty,2]}$$

63.

$$-5p\ge 10$$

Divide by -5 and flip sign

$$p\le -2$$
$$\boxed{(-\infty,-2]}$$

65.

$$3-2x\le 7$$
$$-2x\le 4$$
$$x\ge -2$$
$$\boxed{[-2,\infty)}$$

67.

$$-1.8x+2.5>3.4$$
$$-1.8x>0.9$$
$$x<\frac{0.9}{-1.8}=-0.5$$
$$\boxed{(-\infty,-0.5)}$$

69.

$$3(t+1)>2t$$
$$3t+3>2t$$
$$t+3>0$$
$$t>-3$$
$$\boxed{(-3,\infty)}$$

71.

$$7-2(1-x)>5+3(x-2)$$
$$7-2+2x>5+3x-6$$
$$5+2x>3x-1$$
$$5>x-1$$
$$x<6$$
$$\boxed{(-\infty,6)}$$

73.

$$\frac{x+2}{3}-2\geq\frac{x}{2}$$

LCD = 6

$$2(x+2)-2(6)\geq x(3)$$
$$2x+4-12\geq 3x$$
$$-8\geq x \text{ or } x\leq -8$$
$$\boxed{(-\infty,-8]}$$

75.

$$\frac{t-5}{3}\leq -4$$

LCD = 3

$$t-5\leq -4(3)$$
$$t-5\leq -12$$
$$t\leq -7$$
$$\boxed{(-\infty,-7]}$$

77.

Multiply by LCD = 6

$$4y-3(5-y)<10y-6(2+y)$$
$$4y-15+3y<10y-12-6y$$
$$7y-15<4y-12$$
$$3y-15<-12$$
$$3y<3$$
$$y<1$$
$$\boxed{(-\infty,1)}$$

79.

$$-2<x+3<5$$
$$-5<x<2$$
$$\boxed{(-5,2)}$$

81.

$$-8\leq 4+2x<8$$
$$-12\leq 2x<4$$

Divide by 2

$$-6\leq x<2$$
$$\boxed{[-6,2)}$$

83.

$$-3<1-x\leq 9$$
$$-4<-x\leq 8$$

Divide by -1

Flip the signs

$$-8\leq x<4$$
$$\boxed{[-8,4)}$$

85.

$$0<2-\frac{1}{3}y<4$$
$$-2<-\frac{1}{3}y<2$$

Multiply by -3

Flip the signs

$$-6<y<6$$
$$\boxed{(-6,6)}$$

87.

$$\frac{1}{2}\leq\frac{1+y}{3}\leq\frac{3}{4}$$

Multiply by 3

$$\frac{3}{2}\leq 1+y\leq\frac{9}{4}$$
$$\frac{1}{2}\leq y\leq\frac{5}{4}$$
$$\boxed{\left[\frac{1}{2},\frac{5}{4}\right]}$$

89.

$$-0.7\leq 0.4x+1.1\leq 1.3$$
$$-1.8\leq 0.4x\leq 0.2$$
$$-\frac{1.8}{0.4}\leq x\leq\frac{0.2}{0.4}$$
$$-4.5\leq x\leq 0.5$$
$$\boxed{[-4.5,\ 0.5]}$$

91. Low weight:

$$\underbrace{110}_{1^{st}\ 5\text{ feet}} + \underbrace{2}_{2\text{ lbs}} \underbrace{(9)}_{9\text{ inches}} = 128$$

High weight:

$$\underbrace{110}_{1^{st}\ 5\text{ feet}} + \underbrace{6}_{6\text{ lbs}} \underbrace{(9)}_{9\text{ inches}} = 164$$

$$\boxed{128 \le w \le 164}$$

93.

$\text{Revenue} = 100x\ (x = \#\text{ dresses})$

$\text{Cost} = 4000 + 20x$

$\text{Profit} = \text{Revenue} - \text{Cost}$

$= 100x - (4000 + 20x) > 0$

$100x - 4000 - 20x > 0$

$80x > 4000$

$x > 50$

$\boxed{\text{More than 50 dresses}}$

95. Solve: $5{,}000 + 1.75x \ge 10{,}000$

(Note: We changed from 10 to 10,000 on the right-side of the inequality because $R(x)$ is measured in thousands of dollars.)

$$1.75x \ge 5{,}000$$

$$x \ge 2{,}857.14$$

So, must sell at least 285,700 units.

97. Use the formula

$THR = (HR_{\max} - HR_{rest}) \times I + HR_{rest}$

with $HR_{rest} = 65$, $HR_{\max} = 170$.

Solve for I first when $THR = 100$ and then when $THR = 140$:

$$100 = (170 - 65)I + 65$$
$$35 = 105I$$
$$I \approx 0.33$$

So, about 33%.

$$140 = (170 - 65)I + 65$$
$$75 = 105I$$
$$I \approx 0.71$$

So, about 71%.

So, can consider workouts between 33% and 71% intensity.

99. Cell Phone Charge: $50 + 0.22x$

(x = minutes over 800 used)

$67.16 \le 50 + 0.22x \le 96.86$

$17.16 \le 0.22x \le 46.86$

$78 \le x \le 213$

Least minutes: $800 + 78 = \boxed{878}$

Most minutes: $800 + 213 = \boxed{1013}$

101. Let x = grade on the 4th exam.

$$\frac{67+77+84+x}{4} \geq 80$$

$$67+77+84+x \geq 320$$

$$228+x \geq 320$$

$$x \geq \boxed{92}$$

103. Let x = invoice price.

$$\frac{27{,}999}{1.30} < x < \frac{27{,}999}{1.15}$$

$$\boxed{\$21{,}537.69 < x < \$24{,}346.96}$$

105. $0.9\, r_T \leq r_R \leq 1.1\, r_T$

107. $0.85L \leq B \leq 0.95L$

109. Let x = number of times play.
We want the smallest value of x for which

$$160 + 10x \leq 55x.$$

Solving yields:

$$160 \leq 45x$$

$$3.56 \approx \frac{160}{45} \leq x$$

So, they would need to play $\boxed{\text{4 times}}$ in order to make the membership a better deal.

111. Let T = amount of tax paid.
Least amount of tax = \$4,386.25
Greatest amount of tax = \$15,698.75

So, the range of taxes is:

$$\boxed{4{,}386.25 \leq T \leq 15{,}698.75}$$

113.
Mixed up parenthesis and brackets $[-1,4)$

115. Forgot to flip the sign when dividing by -3. Answer should be $[2,\infty)$.

117. True. In fact, the two inequalities are equivalent.

119. **a, b**

121. **a, b**

123. **c**

125. Mentally, realize that $x \leq -x$ holds only when the left-side is negative or zero. Hence, the solution set is $(-\infty, 0]$.

127. Observe that

$$ax + b < ax - c$$

$$b < -c$$

This is false because we are assuming that $0 < b < c$, so that $-c < b$. Hence, the inequality has $\boxed{\text{no solution}}$.

129.

a)

$$2.7x + 3.1 < 9.4x - 2.5$$

$$2.7x + 5.6 < 9.4x$$

$$5.6 < 6.7x$$

$$x > 0.83582 \text{ (rounded)}$$

b)

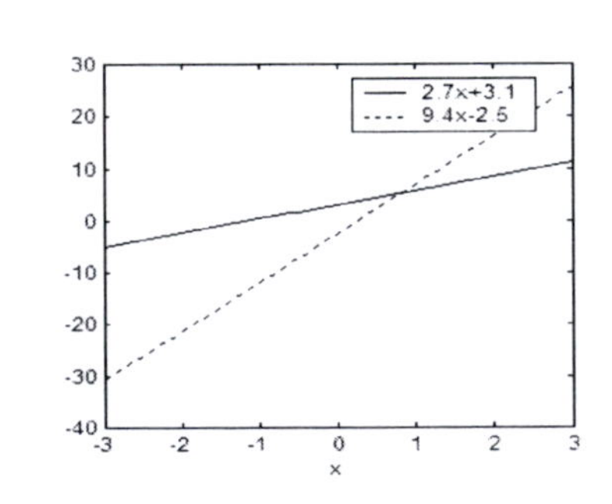

c) Agree

131.

a)

$$x-3<2x-1<x+4$$
$$-3<x-1<4$$
$$-2<x<5$$
$$(-2,5)$$

c) Agree

b)

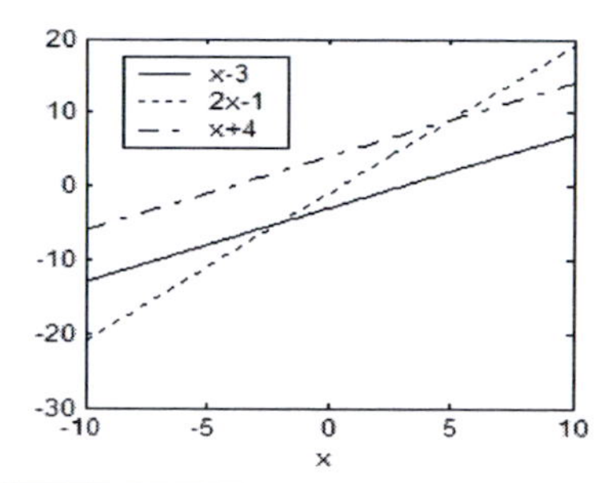

133.

a)

$$x+3<x+5$$
$$3<5$$

true for any $x \in (-\infty,\infty)$

c) Agree

b)

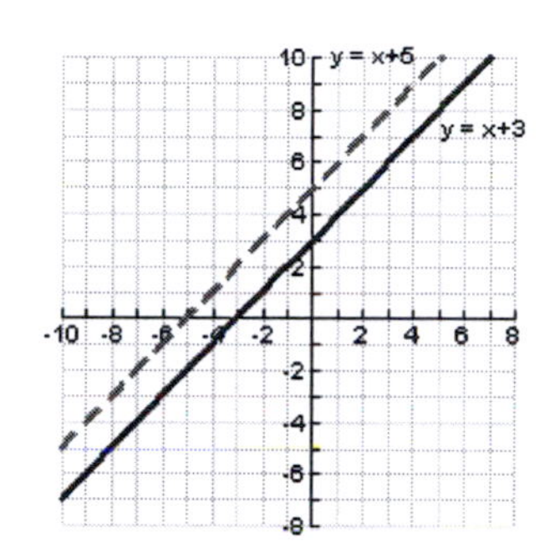

Section 1.6 Solutions

1. $(x-5)(x+2) \geq 0$

CP's: $x=-2,\ 5$

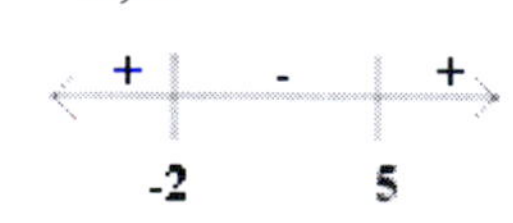

$\boxed{(-\infty,-2]\cup[5,\infty)}$

3. $u^2-5u-6 \leq 0$

$(u-6)(u+1) \leq 0$

CP's: $u=6,-1$

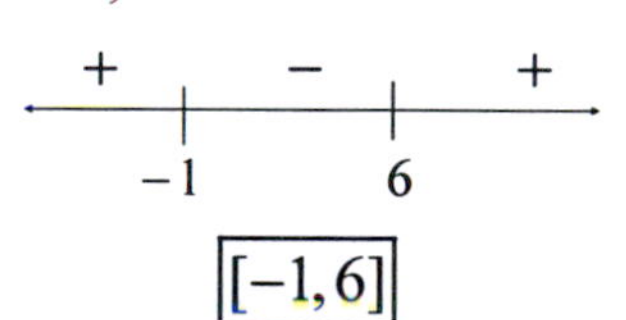

$\boxed{[-1,6]}$

5. $p^2+4p+3<0$

$(p+3)(p+1)<0$

CP's: $p=-3,-1$

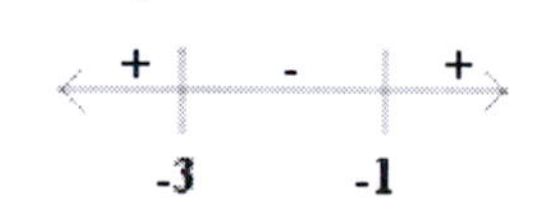

$\boxed{(-3,-1)}$

7. $2t^2-t-3 \leq 0$

$(2t-3)(t+1) \leq 0$

CP's: $t=-1,3/2$

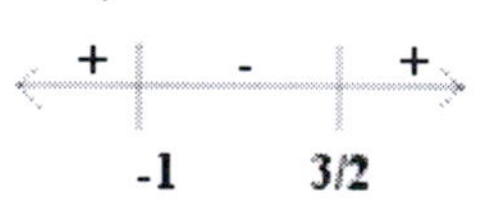

$\boxed{[-1,3/2]}$

9. $6v^2 - 5v + 1 < 0$

$(3v-1)(2v-1) < 0$

CP's: $v = 1/3, 1/2$

+ | - | +

1/3 1/2

$\boxed{(1/3, 1/2)}$

11. $2s^2 - 5s - 3 \geq 0$

$(2s+1)(s-3) \geq 0$

CP's: $s = -1/2, 3$

+ | - | +

-1/2 3

$\boxed{(-\infty, -1/2] \cup [3, \infty)}$

13. $y^2 + 2y - 4 \geq 0$ Note: Can't factor

To find CP's solve $y^2 + 2y - 4 = 0$

$$y = \frac{-2 \pm \sqrt{2^2 - 4(1)(-4)}}{2(1)}$$

$$y = \frac{-2 \pm \sqrt{20}}{2}$$

$$y = \frac{-2 \pm 2\sqrt{5}}{2} = -1 \pm \sqrt{5}$$

+ | - | +

-1-√5 -1+√5

$\boxed{\left(-\infty, -1-\sqrt{5}\right] \cup \left[-1+\sqrt{5}, \infty\right)}$

15. $x^2 - 4x < 6$

$x^2 - 4x - 6 < 0$

CPs: Use quadratic formula:

$$x = \frac{4 \pm \sqrt{16 - 4(1)(-6)}}{2} = \frac{4 \pm 2\sqrt{10}}{2}$$

$$= 2 \pm \sqrt{10}$$

+ | − | +

$2-\sqrt{10}$ $2+\sqrt{10}$

$\boxed{\left(2-\sqrt{10}, 2+\sqrt{10}\right)}$

17. $u^2 - 3u \geq 0$

$u(u-3) \geq 0$

CP's: $u = 0, 3$

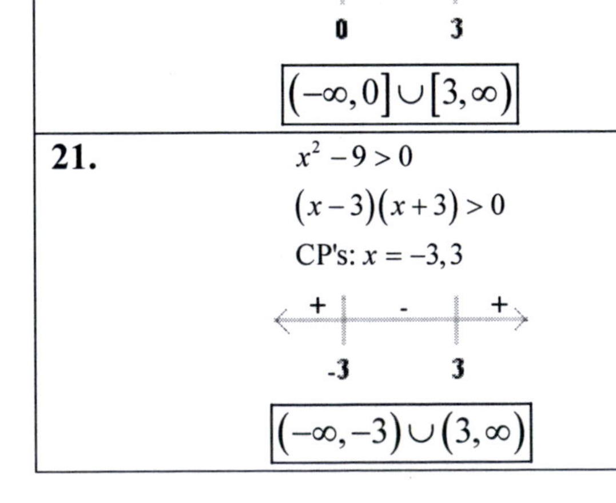

+ | - | +

0 3

$\boxed{(-\infty, 0] \cup [3, \infty)}$

19. $x^2 - 2x \leq 0$

$x(x-2) \leq 0$

CP's: $x = 0, 2$

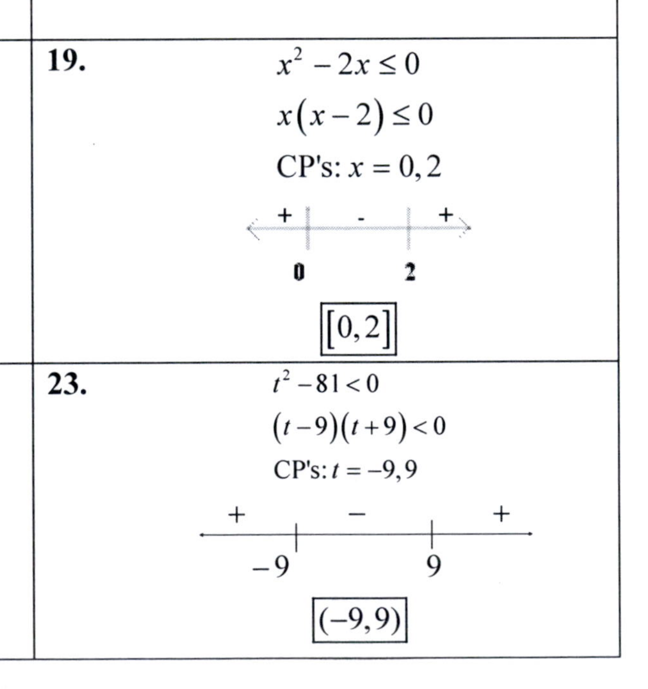

+ | - | +

0 2

$\boxed{[0, 2]}$

21. $x^2 - 9 > 0$

$(x-3)(x+3) > 0$

CP's: $x = -3, 3$

+ | - | +

-3 3

$\boxed{(-\infty, -3) \cup (3, \infty)}$

23. $t^2 - 81 < 0$

$(t-9)(t+9) < 0$

CP's: $t = -9, 9$

+ | − | +

−9 9

$\boxed{(-9, 9)}$

25. $z^2+16>0$ No critical points $z^2+16>0$ for all z $\boxed{\mathbb{R}}$ (consistent)	**27.** $y^2<-4$ $\boxed{\text{no real solution}}$ $\left(\begin{array}{c}\text{A real number squared}\\ \text{is always non-negative.}\end{array}\right)$
29. $\frac{-3}{x}\le 0 \quad x=0$ is CP + - 0 $\boxed{(0,\infty)}$	**31.** $\frac{y}{y+3}>0 \quad$ CP's: $y=-3,\,0$ + - + -3 0 $\boxed{(-\infty,-3)\cup(0,\infty)}$
33. $\frac{t+3}{t-4}\ge 0$ CPs: $-3,4$ + − + -3 4 $\boxed{(-\infty,-3]\cup(4,\infty)}$	**35.** $\frac{s+1}{(2-s)(2+s)}\ge 0$ CP's: $s=-2,\,-1,\,2$ + - + - -2 -1 2 $\boxed{(-\infty,-2)\cup[-1,2)}$
37. $\frac{x-3}{x^2-25}\ge 0$ $\frac{x-3}{(x-5)(x+5)}\ge 0$ CPs: $3,\,\pm 5$ − + − + -5 3 5 $\boxed{(-5,3]\cup(5,\infty)}$	**39.** $2u^2+u<3$ $2u^2+u-3<0$ $(2u+3)(u-1)<0$ CP's: $u=-3/2,\,1$ + - + -3/2 1 $\boxed{(-3/2,1)}$

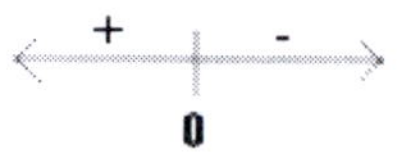

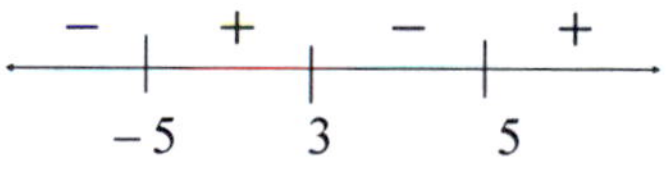

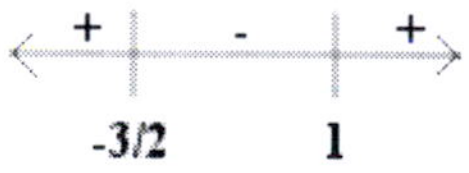

41.

$$\frac{3t^2}{t+2}-5t\geq 0$$

$$\frac{3t^2-5t(t+2)}{t+2}\geq 0$$

$$\frac{3t^2-5t^2-10t}{t+2}\geq 0$$

$$\frac{-2t^2-10t}{t+2}\geq 0$$

$$\frac{-2t(t+5)}{t+2}\geq 0 \quad \text{CP's: } t=-5,-2,0$$

$$\boxed{(-\infty,-5]\cup(-2,0]}$$

43.

$$\frac{3p-2p^2}{4-p^2}-\frac{(3+p)}{(2-p)}<0$$

$$\frac{p(3-2p)}{(2-p)(2+p)}-\frac{(3+p)}{(2-p)}<0$$

$$\frac{p(3-2p)-(3+p)(2+p)}{(2-p)(2+p)}<0$$

$$\frac{3p-2p^2-6-5p-p^2}{(2-p)(2+p)}>0$$

$$\frac{-3p^2-2p-6}{(2-p)(2+p)}<0$$

$$\frac{3p^2+2p+6}{(2-p)(2+p)}>0$$

CP's: $p=-2,\ 2$

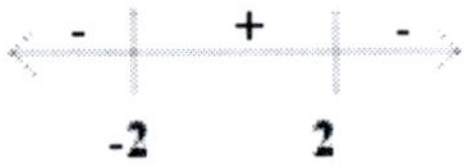

$$\boxed{(-2,2)}$$

45.

$$\frac{x^2}{5+x^2}<0$$

$\boxed{\text{No solution}}$

47.

$$\frac{x^2+10}{x^2+16}>0$$

$\boxed{\mathbb{R}}$ (consistent)

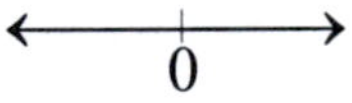

49.

$$\frac{(v-3)(v+3)}{(v-3)}\geq 0 \quad \boxed{v\neq 3}$$

$$v+3\geq 0$$

$$v\geq -3$$

$$\boxed{[-3,3)\cup(3,\infty)}$$

51.

$$\frac{2}{t-3}+\frac{1}{t+3}\geq 0$$

$$\frac{2(t+3)+(t-3)}{(t-3)(t+3)}\geq 0$$

$$\frac{3t+3}{(t-3)(t+3)}\geq 0$$

$$\frac{3(t+1)}{(t-3)(t+3)}\geq 0$$

CPs: $-1,\ \pm 3$

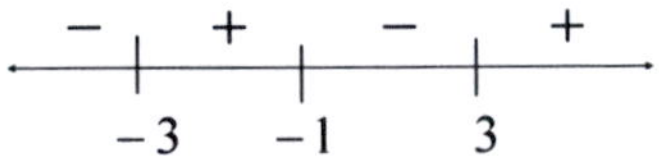

$$\boxed{(-3,-1]\cup(3,\infty)}$$

53.

$$\frac{3}{x+4}-\frac{1}{x-2}\le 0$$

$$\frac{3(x-2)-(x+4)}{(x+4)(x-2)}\le 0$$

$$\frac{2x-10}{(x+4)(x-2)}\le 0$$

$$\frac{2(x-5)}{(x+4)(x-2)}\le 0$$

CPs: $-4, 2, 5$

Sign chart: $-$ | $+$ | $-$ | $+$ at -4, 2, 5

$\boxed{(-\infty,-4)\cup(2,5]}$

55.

$$\frac{1}{p+4}+\frac{1}{p-4}-\frac{p^2-48}{p^2-16}>0$$

$$\frac{(p-4)+(p+4)-\left(p^2-48\right)}{(p+4)(p-4)}>0$$

$$\frac{-\left(p^2-2p-48\right)}{(p+4)(p-4)}>0$$

$$\frac{-(p-8)(p+6)}{(p+4)(p-4)}>0$$

CPs: $-6, \pm 4, 8$

Sign chart: $-$ | $+$ | $-$ | $+$ | $-$ at -6, -4, 4, 8

$\boxed{(-6,-4)\cup(4,8)}$

57.

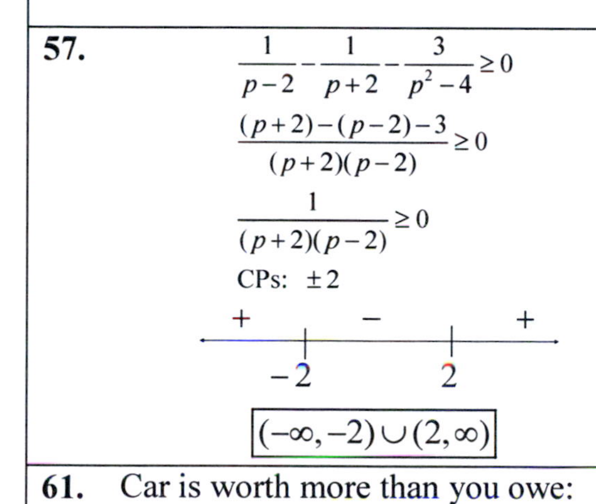

$$\frac{1}{p-2}-\frac{1}{p+2}-\frac{3}{p^2-4}\ge 0$$

$$\frac{(p+2)-(p-2)-3}{(p+2)(p-2)}\ge 0$$

$$\frac{1}{(p+2)(p-2)}\ge 0$$

CPs: ± 2

Sign chart: $+$ | $-$ | $+$ at -2, 2

$\boxed{(-\infty,-2)\cup(2,\infty)}$

59.

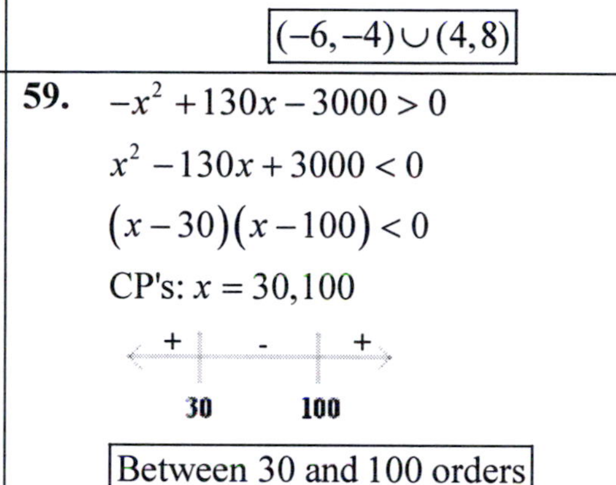

$-x^2+130x-3000>0$

$x^2-130x+3000<0$

$(x-30)(x-100)<0$

CP's: $x=30, 100$

Sign chart: + | - | + at 30, 100

Between 30 and 100 orders

61. Car is worth more than you owe:

$\frac{t}{t-3}>0$ CP's: $t=0,3$

Sign chart: + | - | + at 0, 3

$\boxed{(3,\infty)}$ Greater than 3 years

You owe more than it's worth:

$\frac{t}{t-3}<0$ CP's: $t=0,3$

Sign chart: + | - | + at 0, 3

$\boxed{(0,3)}$ First 3 years

63. $h=-16t^2+1200t$

bullet is in the air if $h>0$

$-16t^2+1200t>0$

$-16t(t-75)>0$

CP's: $t=0,75$

$(0,75)$

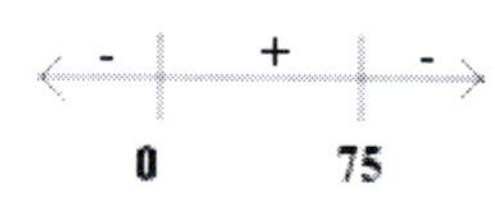

Bullet is in the air for 75 sec

65. Area $= l \cdot w$

$P = 2l + 2w = 100$

$l = \dfrac{100 - 2w}{2}$

$A = l \cdot w = \left(\dfrac{100-2w}{2}\right)(w)$

$50w - w^2 \geq 600$

$w^2 - 50w + 600 \leq 0$

$(w-20)(w-30) \leq 0$

CP's: $w = 20, 30$

$[20, 30]$

$20 \leq \text{width} \leq 30$

$20 \leq \text{length} \leq 30$

$\boxed{\text{Between 20 and 30 feet}}$

67.

$$-5(x+3)(x-24) < 460$$
$$-5x^2 + 105x + 360 < 460$$
$$-5x^2 + 105x - 100 < 0$$
$$x^2 - 21x + 20 > 0$$
$$(x-20)(x-1) > 0$$

The solution set is $(-\infty, 1) \cup (20, \infty)$.

So, a price increase less than \$1 or greater than \$20 per bottle.

69.

$400 \pm 7 = 393, 407$

$\dfrac{1{,}360{,}000}{407} \leq \text{price per acre} \leq \dfrac{1{,}360{,}000}{393}$

$\$3{,}341.52 \leq \text{price per acre} \leq \$3460.56$

$\boxed{\$3{,}342 \text{ to } \$3{,}461 \text{ per acre}}$

71.

Cannot divide by x.

$x^2 - 3x > 0$

$x(x-3) > 0$

$(-\infty, 0) \cup (3, \infty)$

73. $\dfrac{(x-2)(x+2)}{(x+2)} > 0$ $\boxed{x \neq 2}$

$x - 2 > 0$

$x > 2$

Should have considered $x = -2$ a CP

75. False $(-a, a)$

77. Assume that $ax^2 + bx + c < 0$. If $b^2 - 4ac < 0$, then either there are infinitely many solutions or no real solution.

79. $x^2 + a^2 \geq 0$

True for all real values of x

$\boxed{\mathbb{R}}$

81.

$$\frac{x^2+a^2}{x^2+b^2} \geq 0$$

$\boxed{\mathbb{R}}$

83.

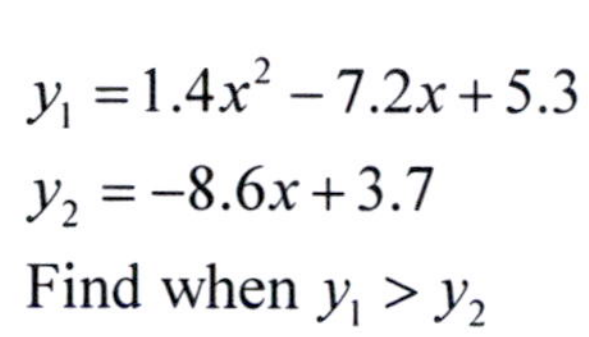

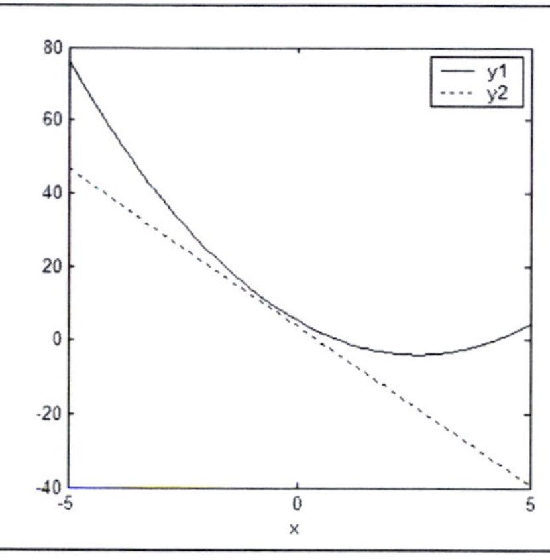

85.

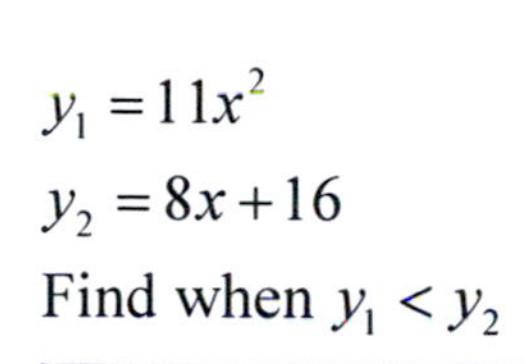

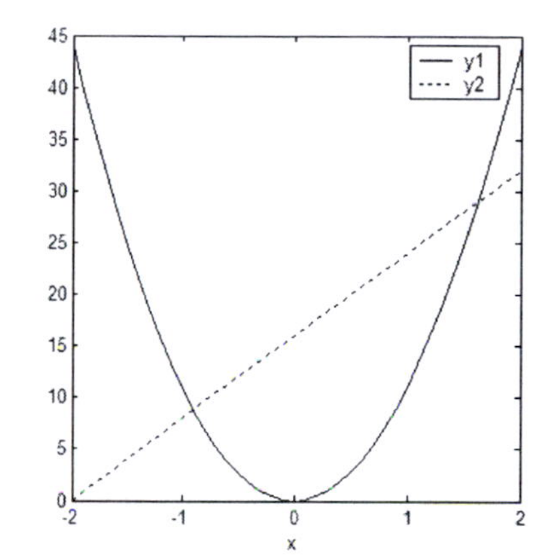

87.

$y_1 = x$

$y_2 = x^2 - 3x$

$y_3 = 6 - 2x$

Find when $y_1 < y_2 < y_3$.

$\boxed{(-2, 0)}$

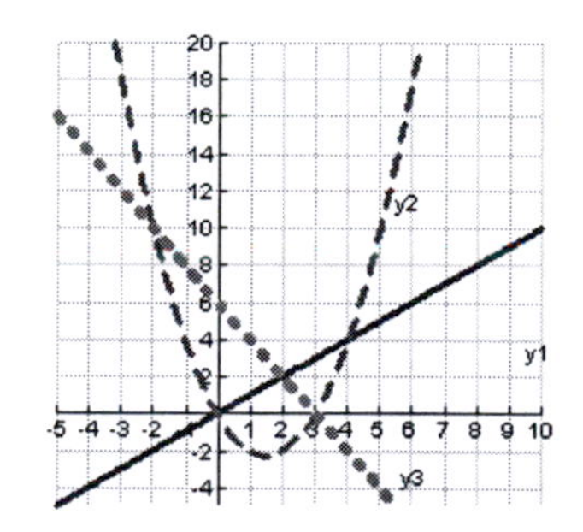

89.

$$y_1 = \frac{2p}{5-p}$$

$$y_2 = 1$$

Find when $y_1 > y_2$.

$\boxed{\left(\frac{5}{3}, 5\right)}$

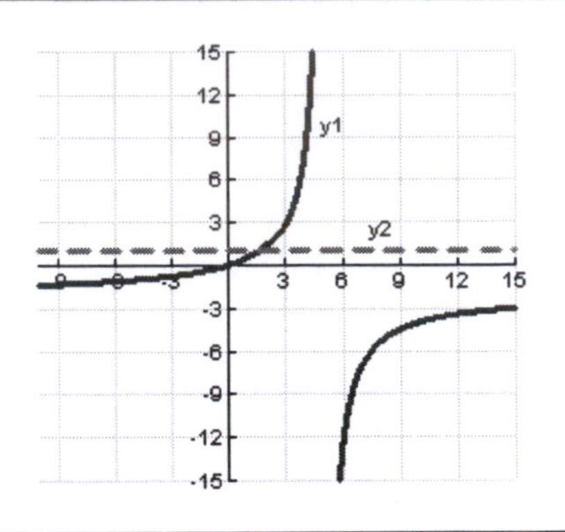

Section 1.7 Solutions --

1.	$x=-3$ or $x=3$		**3.**	No solution (absolute value is always non-negative)	
5.	$t+3=-2$ $\boxed{t=-5}$	$t+3=2$ $\boxed{t=-1}$	**7.**	$p-7=3$ $\boxed{p=10}$	$p-7=-3$ $\boxed{p=4}$
9.	$4-y=-1$ $\boxed{y=5}$	$4-y=1$ $\boxed{y=3}$	**11.**	$3x=-9$ $\boxed{x=-3}$	$3x=9$ $\boxed{x=3}$
13.	$2x+7=-9$ $2x=-16$ $\boxed{x=-8}$	$2x+7=9$ $2x=2$ $\boxed{x=1}$	**15.**	$3t-9=3$ $3t=12$ $\boxed{t=4}$	$3t-9=-3$ $3t=6$ $\boxed{t=2}$
17.	$7-2x=-9$ $2x=16$ $\boxed{x=8}$	$7-2x=9$ $2x=-2$ $\boxed{x=-1}$	**19.**	$1-3y=1$ $-3y=0$ $\boxed{y=0}$	$1-3y=-1$ $-3y=-2$ $\boxed{y=2/3}$
21.	$4.7-2.1x=-3.3$ $2.1x=8$ $\boxed{x=\frac{80}{21}}$	$4.7-2.1x=3.3$ $2.1x=1.4$ $\boxed{x=\frac{2}{3}}$	**23.**	$\frac{2}{3}x-\frac{4}{7}=-\frac{5}{3}$ LCD = 21 $14x-12=-35$ $14x=-23$ $\boxed{x=-23/14}$	$\frac{2}{3}x-\frac{4}{7}=\frac{5}{3}$ LCD = 21 $14x-12=35$ $14x=47$ $\boxed{x=47/14}$

25. $\|x-5\|=8$ $x-5=8$ $\quad$ $x-5=-8$ $\boxed{x=13}$ $\quad$ $\boxed{x=-3}$	**27.** $3\|x-2\|+1=19$ $3\|x-2\|=18$ $\|x-2\|=6$ $x-2=6$ *or* $x-2=-6$ $x=-4,8$
29. $5=7-\|2-x\|$ $-2=-\|2-x\|$ $2=\|2-x\|$ $2-x=2$ *or* $2-x=-2$ $x=0,4$	**31.** $2\|p+3\|=20$ $\|p+3\|=10$ $p+3=10$ $\quad$ $p+3=-10$ $\boxed{p=7}$ $\quad$ $\boxed{p=-13}$
33. $5\|y-2\|-10=4\|y-2\|-3$ $\|y-2\|=7$ $y-2=7$ $\quad$ $y-2=-7$ $\boxed{y=9}$ $\quad$ $\boxed{y=-5}$	**35.** $4-x^2=-1$ $\quad$ $4-x^2=1$ $x^2=5$ $\quad$ $x^2=3$ $\boxed{x=\pm\sqrt{5}}$ $\quad$ $\boxed{x=\pm\sqrt{3}}$
37. $x^2+1=-5$ $\quad$ $x^2+1=5$ $x^2=-6$ $\quad$ $x^2=4$ no solution $\quad$ $\boxed{x=\pm 2}$	**39.** $-7<x<7$ $(-7,7)$
41. $y\le -5$ or $y\ge 5$ $(-\infty,-5]\cup[5,\infty)$	**43.** $-7<x+3<7$ $-10<x<4$ $(-10,4)$
45. $x-4<-2$ or $x-4>2$ $x<2$ $\quad$ $x>6$ $(-\infty,2)\cup(6,\infty)$	**47.** $-1\le 4-x\le 1$ $-5\le -x\le -3$ $3\le x\le 5$ $[3,5]$
49. $\mathbb{R}$	

51. $\|2t+3\|<5$ $-5<2t+3<5$ $-8<2t<2$ $-4<t<1$ $\boxed{(-4,1)}$	**53.** $\|7-2y\|\geq 3$ $7-2y\geq 3$ or $7-2y\leq -3$ $-2y\geq -4$ or $-2y\leq -10$ $y\leq 2$ or $y\geq 5$ $\boxed{(-\infty,2]\cup[5,\infty)}$
55. $\mathbb{R}$	
57. $2\|4x\|-9\geq 3$ $2\|4x\|\geq 12$ $\|4x\|\geq 6$ $4x\geq 6$ or $4x\leq -6$ $x\geq 3/2$ or $x\leq -3/2$ $\boxed{(-\infty,-3/2]\cup[3/2,\infty)}$	**59.** $2\|x+1\|-3\leq 7$ $2\|x+1\|\leq 10$ $\|x+1\|\leq 5$ $-5\leq x+1\leq 5$ $-6\leq x\leq 4$ $\boxed{[-6,4]}$
61. $3-2\|x+4\|<5$ $-2\|x+4\|<2$ $\|x+4\|>-1$ $\boxed{(-\infty,\infty)}$	**63.** $9-\|2x\|<3$ $-\|2x\|<-6$ $\|2x\|>6$ $2x>6$ or $2x<-6$ $x>3$ or $x<-3$ $\boxed{(-\infty,-3)\cup(3,\infty)}$
65. $-\frac{1}{2}<1-2x<\frac{1}{2}$ $-\frac{3}{2}<-2x<-\frac{1}{2}$ $\frac{3}{4}>x>\frac{1}{4}$ $\boxed{(1/4,3/4)}$	**67.** $-1.8<2.6x+5.4<1.8$ $-7.2<2.6x<-3.6$ $-2.769<x<-1.385$ $\boxed{(-2.769,-1.385)}$

69.

$$x^2 - 1 \le 8$$
$$x^2 - 9 \le 0$$
$$(x-3)(x+3) \le 0$$

CP's: $x = -3, 3$

$$-3 \le x \le 3$$
$$\boxed{[-3,3]}$$

71.

$$|x-2| < 7$$

73.

$$|x - 3/2| \ge 1/2$$

75.

$$|x-a| \le 2$$

77. $|T - 83| \le 15$

79. In order to win the hole, $d < 4$.
In order to have a tie, $d = 4$.

81.

$$|(200+5x)-(210+4.8x)| < 5$$
$$|-10+0.2x| < 5$$
$$-5 < -10 + 0.2x < 5$$
$$5 < 0.2x < 15$$
$$25 < x < 75$$

So, where the number of units sold is between 25 and 75.

83. $x - 3 = -7$ also yields a solution $x = -4$

85. Didn't switch signs when dividing by -2. The answer is [2,3].

87. True

89. False

91.

$$-b < x - a < b$$
$$a - b < x < a + b$$
$$\boxed{(a-b, a+b)}$$

93. $\mathbb{R}$

95. $x - a = -b$ | $x - a = b$

$\boxed{x = a - b}$ | $\boxed{x = a + b}$

97. No solution

99.

$y_1 = |x-7|$

$y_2 = x-7$

$\boxed{x \geq 7}$

Agree

101.

$y_1 = |3x^2 - 7x + 2|$

$y_2 = 8$

$(-\infty, -\frac{2}{3}) \cup (3, \infty)$

Agree

103.

$y_1 = \left|\dfrac{x}{x+1}\right|$

$y_2 = 1$

Find when $y_1 < y_2$.

$\boxed{(-\frac{1}{2}, \infty)}$

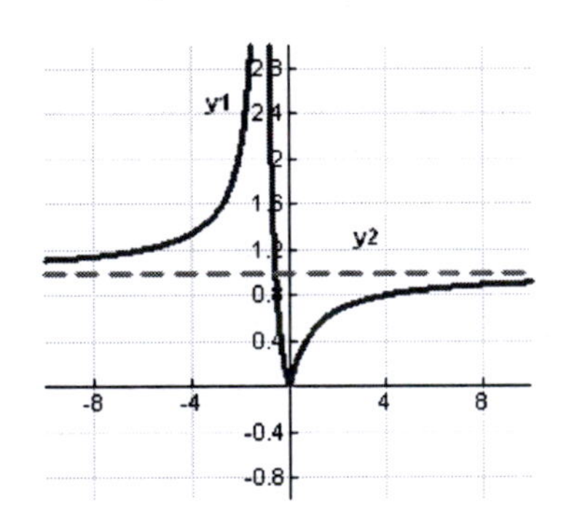

Chapter 1 Review Solutions ---

1. $7x - 4 = 12$ $7x = 16$ $\boxed{x = 16/7}$	**3.** $20p + 14 = 6 - 5p$ $25p = -8$ $\boxed{p = -8/25}$

5. $3x+21-2=4x-8$ $\boxed{x=27}$	**7.** $14-[-3y+12+9]=8y+12-6+4$ $14+3y-21=8y+10$ $-17=5y$ $\boxed{y=-17/5}$
9. $b\neq 0$ $12-3b=6+4b$ $6=7b$ $\boxed{b=6/7}$	**11.** $\text{LCD}=28$ $4(13x)-28x=7x-2(3)$ $52x-28x=7x-6$ $17x=-6$ $\boxed{x=-6/17}$
13. $x\neq 0$ $\text{LCD}=x$ $1-4x=3-5x$ $-2=-x$ $2=x$ $\boxed{x=2}$	**15.** $t\neq -4,0$ $\text{LCD}=t(t+4)$ $2t-7(t+4)=6$ $2t-7t-28=6$ $-5t=34$ $\boxed{t=-34/5}$
17. $x\neq 0$ $\text{LCD}=2x$ $3-12=18x$ $-9=18x$ $\boxed{x=-\frac{1}{2}}$	**19.** $7x-2+4x=3[5-2x]+12$ $11x-2=15-6x+12$ $17x=29$ $\boxed{x=29/17}$
21. $3x-2[3y+12-7]=y-2x+6x-18$ $3x-6y-10=y-2x+6x-18$ $-x+8=7y$ $\boxed{x=8-7y}$	**23.** Let x = total distance Drives: 16 miles Bus: $\frac{3}{4}x$ Taxi: $\frac{1}{12}x$ $16+\frac{3}{4}x+\frac{1}{12}x=x$ $\text{LCD}=12$ $192+9x+x=12x$ $2x=192$ $\boxed{x=96\text{ miles}}$

25. $x = \text{number}$ $12 + \frac{1}{4}x = \frac{1}{3}x$ $\text{LCD} = 12$ $144 + 3x = 4x$ $\boxed{x = 144}$	**27.** $P = 2l + 2w$ $l = 1 + 2w$ $P = 2(1 + 2w) + 2w$ $P = 20$ $20 = 2 + 4w + 2w$ $6w = 18$ $\boxed{w = 3 \text{ inches}}$ $\boxed{l = 7 \text{ inches}}$
29. $x = \text{amount invested @ 20\%}$ $25000 - x = \text{amount invested @ 8\%}$ $\text{Earned interest} = 27600 - 25000 = 2600$ $0.2x + 0.08(25000 - x) = 2600$ $0.2x + 2000 - 0.08x = 2600$ $0.12x = 600$ $x = 5000$ $\boxed{\$5,000 @ 20\%}$ $\boxed{\$20,000 @ 8\%}$	**31.** $x = \text{ml of 5\%}$ $150 - x = \text{ml of 10\%}$ $0.05x + 0.10(150 - x) = 0.08(150)$ $0.05x + 15 - 0.10x = 12$ $-0.05x = -3$ $x = 60$ $\boxed{60 \text{ ml of 5\%}}$ $\boxed{90 \text{ ml of 10\%}}$
33. $x = \text{final exam grade}$ $\frac{3x + 95 + 82 + 90}{6} \geq 90$ $3x + 267 \geq 540$ $3x \geq 273$ $\boxed{\text{At least 91}}$	**35.** $b^2 - 4b - 21 = 0$ $(b - 7)(b + 3) = 0$ $\boxed{b = -3, 7}$
37. $x^2 - 8x = 0$ $x(x - 8) = 0$ $\boxed{x = 0, 8}$	**39.** $q^2 = 169$ $q = \pm\sqrt{169}$ $\boxed{q = \pm 13}$
41. $2x - 4 = \pm\sqrt{-64}$ $2x - 4 = \pm 8i$ $2x = 4 \pm 8i$ $\boxed{x = 2 \pm 4i}$	**43.** $x^2 - 4x = 12$ $x^2 - 4x + 4 = 12 + 4$ $(x - 2)^2 = 16$ $x - 2 = \pm 4$ $x = 2 \pm 4$ $\boxed{x = -2, 6}$

45.

$$x^2 - x = 8$$
$$x^2 - x + \frac{1}{4} = 8 + \frac{1}{4}$$
$$\left(x - \frac{1}{2}\right)^2 = \frac{33}{4}$$
$$x - \frac{1}{2} = \pm\sqrt{\frac{33}{4}}$$
$$\boxed{x = \frac{1 \pm \sqrt{33}}{2}}$$

47.

$$3t^2 - 4t - 7 = 0$$
$$a = 3, b = -4, c = -7$$
$$t = \frac{-(-4) \pm \sqrt{(-4)^2 - 4(3)(-7)}}{2(3)}$$
$$t = \frac{4 \pm \sqrt{100}}{6} = \frac{4 \pm 10}{6}$$
$$\boxed{t = -1, \frac{7}{3}}$$

49.

$$8f^2 - \frac{1}{3}f - \frac{7}{6} = 0$$
$$\text{LCD} = 6$$
$$48f^2 - 2f - 7 = 0$$
$$a = 48, b = -2, c = -7$$
$$f = \frac{-(-2) \pm \sqrt{(-2)^2 - 4(48)(-7)}}{2(48)}$$
$$f = \frac{2 \pm \sqrt{1348}}{96}$$
$$f = \frac{2 \pm 2\sqrt{337}}{96}$$
$$\boxed{f = \frac{1 \pm \sqrt{337}}{48}}$$

51.

$$a = 5, b = -3, c = -3$$
$$q = \frac{-(-3) \pm \sqrt{(-3)^2 - 4(5)(-3)}}{2(5)}$$
$$\boxed{q = \frac{3 \pm \sqrt{69}}{10}}$$

53.

$$(2x - 5)(x + 1) = 0$$
$$\boxed{x = -1, \frac{5}{2}}$$

55.

$$7x^2 + 19x - 6 = 0$$
$$(7x - 2)(x + 3) = 0$$
$$\boxed{x = -3, 2/7}$$

57.

$$r^2 = \frac{S}{\pi h}$$
$$r = \pm\sqrt{\frac{S}{\pi h}}$$
$$\boxed{r = \sqrt{\frac{S}{\pi h}}} \quad \left(\text{negative radius is non-physical}\right)$$

59.

$$vt = h + 16t^2$$
$$\boxed{v = \frac{h + 16t^2}{t} = \frac{h}{t} + 16t}$$

61.

$$A=\frac{1}{2}bh$$
$$b=h+3 \quad A=2$$
$$2=\frac{1}{2}(h+3)h$$
$$4=h^2+3h$$
$$h^2+3h-4=0$$
$$(h+4)(h-1)=0$$
$$h=-4,1 \text{ (height must be positive)}$$
$$\boxed{h=1\text{ ft}, b=4\text{ ft}}$$

63.

$$2x-4=2^3=8$$
$$2x=12$$
$$\boxed{x=6}$$

65.

$$2x-7=3^5$$
$$2x=7+243=250$$
$$\boxed{x=125}$$

67.

$$(x-4)^2=x^2+5x+6$$
$$x^2-8x+16=x^2+5x+6$$
$$13x=10$$
$$\cancel{x=\frac{10}{13}} \quad \left(\text{This answer would make the first } \sqrt{\ } \text{ equal to a negative number}\right)$$
$$\boxed{\text{no solution}}$$

69.

$$x+3=4-4\sqrt{3x+2}+3x+2$$
$$-2x-3=-4\sqrt{3x+2}$$
$$2x+3=4\sqrt{3x+2}$$
$$(2x+3)^2=16(3x+2)$$
$$4x^2+12x+9=48x+32$$
$$4x^2-36x-23=0$$
$$x=\frac{36\pm\sqrt{36^2-4(4)(-23)}}{2(4)}$$
$$x=\frac{36\pm\sqrt{1664}}{8}\cong-0.6, 9.6$$
$$\boxed{x\cong-0.6} \text{ (9.6 doesn't check)}$$

71.

$$x^2-4x+4=49-x^2$$
$$2x^2-4x-45=0$$
$$x=\frac{-(-4)\pm\sqrt{(-4)^2-4(2)(-45)}}{2(2)}$$
$$x=\frac{4\pm\sqrt{376}}{4}$$
$$x\cong\cancel{-3.85}, 5.85$$
$$\boxed{x\cong5.85}$$

73.

$$x^2 = 3 - x$$
$$x^2 + x - 3 = 0$$
$$x = \frac{-1 \pm \sqrt{1 - 4(1)(-3)}}{2(1)}$$
$$x = \frac{-1 \pm \sqrt{13}}{2} \cong -2.303, \text{~~1.3~~}$$
$$\boxed{x \cong -2.303}$$

75.

$$(3x-2)^2 - 11(3x-2) + 28 = 0$$
Let $u = 3x - 2$
$$u^2 - 11u + 28 = 0$$
$$(u-4)(u-7) = 0$$
$$u = 4, 7$$
$3x - 2 = 4$ $3x - 2 = 7$

$3x = 6 \Rightarrow \boxed{x = 2}$ $3x = 9 \Rightarrow \boxed{x = 3}$

77.

$$u = \frac{x}{1-x} \quad \boxed{x \neq 1}$$
$$u^2 + 2u - 15 = 0$$
$$(u+5)(u-3) = 0$$
$$u = -5, 3$$

$-5 = \frac{x}{1-x}$ $3 = \frac{x}{1-x}$

$-5 + 5x = x$ $3 - 3x = x$

$4x = 5$ $4x = 3$

$\boxed{x = \frac{5}{4}}$ $\boxed{x = \frac{3}{4}}$

79.

$$y^{-2} - 5y^{-1} + 4 = 0$$
Let $u = y^{-1}$
$$u^2 - 5u + 4 = 0$$
$$(u-4)(u-1) = 0$$
$$u = 4, 1$$
So, we have:

$y^{-1} = 4 \Rightarrow \boxed{y = \frac{1}{4}}$

$y^{-1} = 1 \Rightarrow \boxed{y = 1}$

81.

$$2x^{2/3} + 3x^{1/3} - 5 = 0$$
Let $u = x^{1/3}$
$$2u^2 + 3u - 5 = 0$$
$$(2u+5)(u-1) = 0$$
$$u = -\frac{5}{2}, 1$$

$x^{1/3} = -\frac{5}{2}$ $x^{1/3} = 1$

$x = \left(-\frac{5}{2}\right)^3$ $\boxed{x = 1}$

$\boxed{x = -\frac{125}{8}}$

83.

$$x^{-2/3} + 3x^{-1/3} + 2 = 0$$
Let $u = x^{-1/3}$.
$$u^2 + 3u + 2 = 0$$
$$(u+2)(u+1) = 0$$
$$u = -2, -1$$
So, we have:

$x^{-1/3} = -2 \Rightarrow x = (-2)^{-3} = \boxed{-\frac{1}{8}}$

$x^{-1/3} = -1 \Rightarrow x = (-1)^{-3} = \boxed{-1}$

85.

Let $u = x^2$

$$u^2 + 5u - 36 = 0$$
$$(u+9)(u-4) = 0$$
$$u = -9, 4$$

$-9 = x^2$ $\quad$ $4 = x^2$

$\boxed{x = \pm 3i}$ $\quad$ $\boxed{x = \pm 2}$

87.

$$x^3 + 4x^2 - 32x = 0$$
$$x\left(x^2 + 4x - 32\right) = 0$$
$$x(x+8)(x-4) = 0$$
$$\boxed{x = 0, -8, 4}$$

89.

$$p^3 - 3p^2 - 4p + 12 = 0$$
$$\left(p^3 - 3p^2\right) - 4(p-3) = 0$$
$$p^2(p-3) - 4(p-3) = 0$$
$$\left(p^2 - 4\right)(p-3) = 0$$
$$(p-2)(p+2)(p-3) = 0$$
$$\boxed{p = \pm 2, 3}$$

91.

$$p(2p-5)^2 - 3(2p-5) = 0$$
$$(2p-5)\left[p(2p-5) - 3\right] = 0$$
$$(2p-5)\left(2p^2 - 5p - 3\right) = 0$$
$$(2p-5)(2p+1)(p-3) = 0$$
$$\boxed{p = -\tfrac{1}{2}, \tfrac{5}{2}, 3}$$

93.

$$y - 81y^{-1} = 0$$
$$y - \frac{81}{y} = 0$$
$$\frac{y^2 - 81}{y} = 0$$
$$\frac{(y-9)(y+9)}{y} = 0$$
$$\boxed{y = \pm 9}$$

95. $(-\infty, -4]$

97. $[2, 6]$

99. $x > -6$

101. $-3 \le x \le 7$

103. $x \ge -4$

$[-4, \infty)$

105. $(4, \infty)$

... 0 1 2 3 4 5 6 ...

107. $[8, 12]$

... 7 8 9 10 11 12 ...

109.

$$3x < 5$$
$$x < 5/3$$
$$\boxed{(-\infty, 5/3)}$$

... 0 $\frac{1}{3}$ $\frac{2}{3}$ 1 $\frac{4}{3}$ $\frac{5}{3}$ ...

111.

$$4x-4>2x-7$$
$$2x>-3$$
$$x>-3/2$$
$$\boxed{(-3/2,\infty)}$$

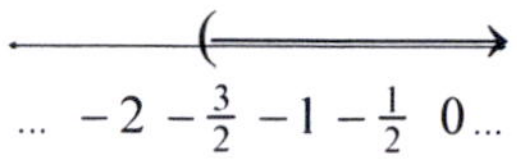

$$\ldots\ -2\ -\tfrac{3}{2}\ -1\ -\tfrac{1}{2}\ 0\ldots$$

113.

$$6<2+x\le 11$$
$$4<x\le 9$$
$$\boxed{(4,9]}$$

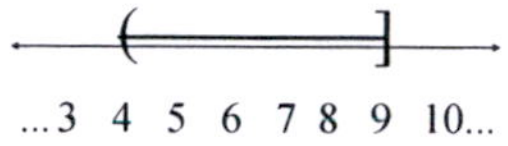

$$\ldots 3\ 4\ 5\ 6\ 7\ 8\ 9\ 10\ldots$$

115.

$$\text{LCD}=12$$
$$8\le 2(1+x)\le 9$$
$$8\le 2+2x\le 9$$
$$6\le 2x\le 7$$
$$3\le x\le 7/2$$
$$\boxed{[3,7/2]}$$

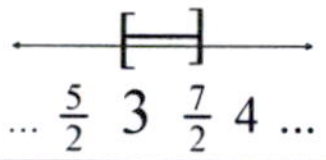

$$\ldots\ \tfrac{5}{2}\ 3\ \tfrac{7}{2}\ 4\ \ldots$$

117.

$$\frac{72+65+69+70+x}{5}\ge 70$$
$$x+276\ge 350$$
$$\boxed{x\ge 74}$$

So, the lowest score is 74.

119.

$$x^2-36\le 0$$
$$(x-6)(x+6)\le 0$$
$$\text{CP's: } x=-6,6$$

+ | - | +

-6 6

$$\boxed{[-6,6]}$$

121.

$$x^2-4x\ge 0$$
$$x(x-4)\ge 0$$
$$\text{CP's: } x=0, x=4$$

+ | - | +

0 4

$$\boxed{(-\infty,0]\cup[4,\infty)}$$

123.

$$x^2-7x>0$$
$$x(x-7)>0$$
$$\text{CP's: } x=0,7$$

+ | - | +

0 7

$$\boxed{(-\infty,0)\cup(7,\infty)}$$

125.

$$4x^2-12>13x$$
$$4x^2-13x-12>0$$
$$(4x+3)(x-4)>0$$
$$\text{CPs: } x=-\tfrac{3}{4},\ 4$$

+ − +

$-\frac{3}{4}$ 4

$$\boxed{(-\infty,-\tfrac{3}{4})\cup(4,\infty)}$$

127.

$$\frac{x}{x-3} < 0 \quad \boxed{x \neq 3}$$

CP's: $x = 0, 3$

Sign chart: $+$ | 0 | $-$ | 3 | $+$

$\boxed{(0,3)}$

129.

$$\frac{x^2 - 3x}{3} - \frac{18(3)}{3} \geq 0$$

$$\frac{x^2 - 3x - 54}{3} \geq 0$$

$$\frac{(x-9)(x+6)}{3} \geq 0$$

CP's: $x = -6, 9$

Sign chart: $+$ | -6 | $-$ | 9 | $+$

$\boxed{(-\infty, -6] \cup [9, \infty)}$

131.

$$\frac{3}{x-2} - \frac{1}{x-4} \leq 0$$

$$\frac{3(x-4)-(x-2)}{(x-2)(x-4)} \leq 0$$

$$\frac{2x-10}{(x-2)(x-4)} \leq 0$$

$$\frac{2(x-5)}{(x-2)(x-4)} \leq 0$$

CPs: $x = 2, 4, 5$

Sign chart: $-$ | 2 | $+$ | 4 | $-$ | 5 | $+$

$\boxed{(-\infty, 2) \cup (4, 5]}$

133.

$$\frac{x^2+9}{x-3} \geq 0$$

CP: 3 (since $x^2 + 9 > 0$, for all x)

Sign chart: $-$ | 3 | $+$

$(3, \infty)$

135.

$$|x-3| = -4$$

$\boxed{\text{no solution}}$

137.

$3x - 4 = -1.1$	$3x - 4 = 1.1$
$3x = 2.9$	$3x = 5.1$
$\boxed{x = 0.9667}$	$\boxed{x = 1.7}$

139. $-4 < x < 4$

$\boxed{(-4,4)}$

141. $x+4 < -7$ $\qquad x+4 > 7$

$x < -11$ $\qquad x > 3$

$\boxed{(-\infty,-11)\cup(3,\infty)}$

143.

$$|2x| > 6$$

$2x < -6$ $\qquad 2x > 6$

$x < -3$ $\qquad x > 3$

$\boxed{(-\infty,-3)\cup(3,\infty)}$

145. $\mathbb{R}$

147. $|T-85| \le 10$ or $75 \le T \le 95$

149.

$y1 = 0.031x + 0.017(4000-x)$

$y2 = 103.14$

$\boxed{x = 2,510}$

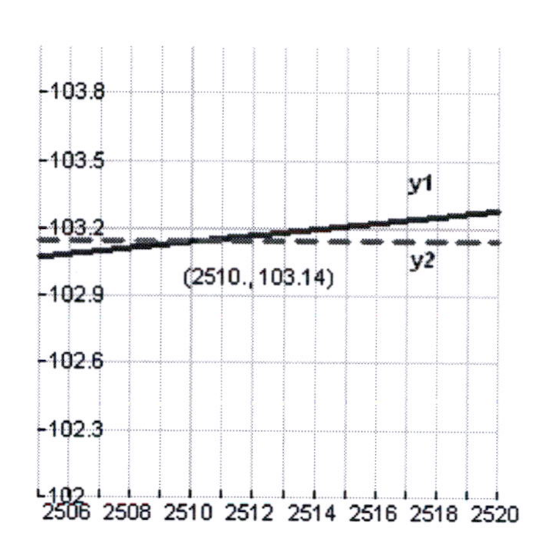

151. (a) Consider $x^2+4x-b=0$. **(1)** For $b=5$, **(1)** factors as $(x-1)(x+5)=0$, so that $x=-5,1$.

Graphically, we let $y1 = x^2+4x$, $y2 = 5$ and look for the intersection points of the graphs:

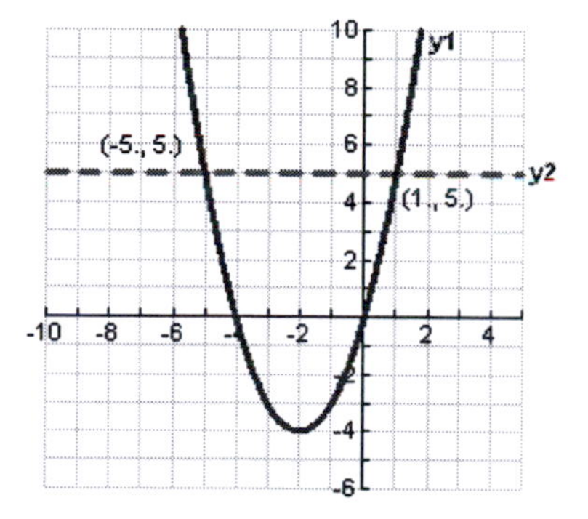

Note that they intersect at precisely the x-values obtained algebraically. So, yes, these values agree with the points of intersections.

(b) We do the same thing now for different values of b.

$\underline{b=-5}$:

$$x^2+4x+5=0$$

$$x = \frac{-4 \pm \sqrt{16-4(5)}}{2} = \frac{-4 \pm 2i}{2} = -2 \pm i$$

$\underline{b=0}$:

$$x^2+4x=0$$

$$x(x+4)=0$$

$$x = 0,-4$$

So, we don't expect the graphs to intersect. Indeed, we have:

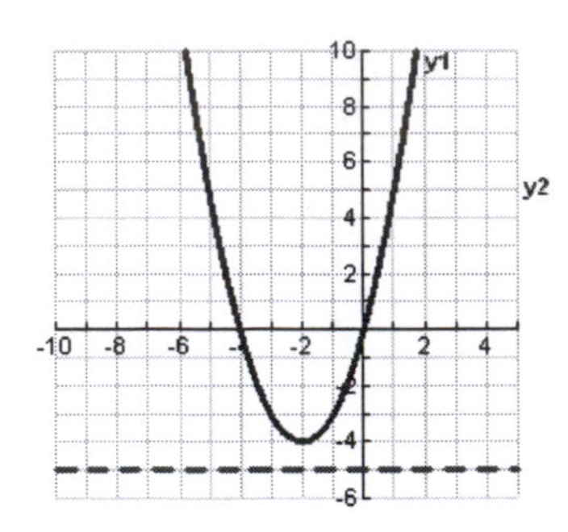

$\underline{b=7}$:

$$x^2+4x-7=0$$

$$x=\frac{-4\pm\sqrt{16+4(7)}}{2}=\frac{-4\pm2\sqrt{11}}{2}=-2\pm\sqrt{11}$$

So, we expect the graphs to intersect twice as in part **(a)**. Indeed, we have:

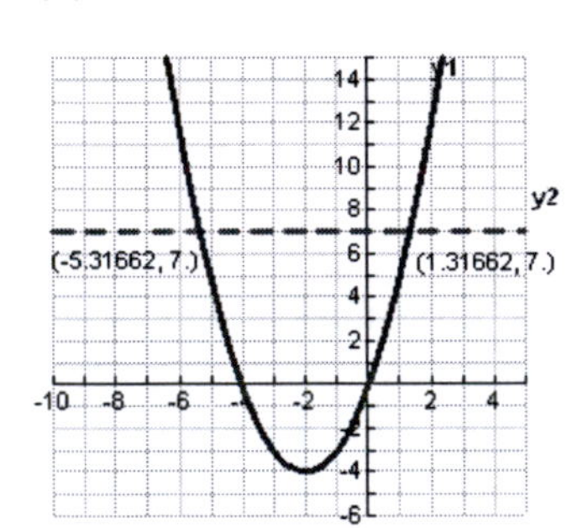

So, we expect the graphs to intersect twice as in part **(a)**. Indeed, we have:

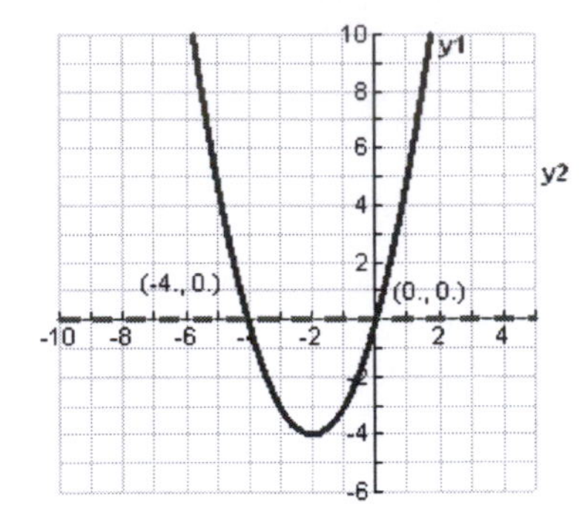

$\underline{b=12}$:

$$x^2+4x-12=0$$

$$(x+6)(x-2)=0$$

$$x=-6,2$$

So, we expect the graphs to intersect twice as in part **(a)**. Indeed, we have:

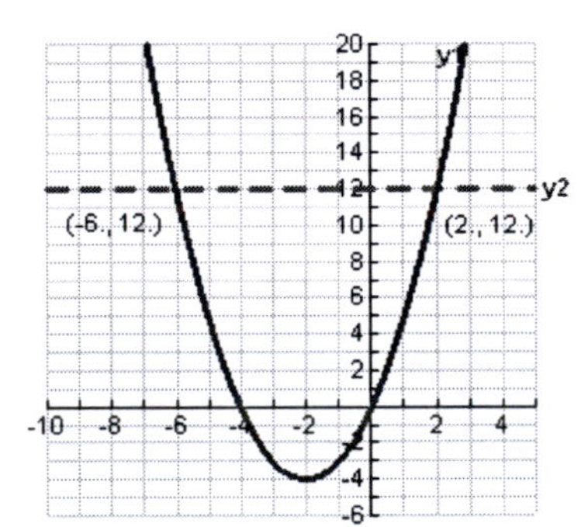

153.

$$2x^{1/4}=-x^{1/2}+6$$

$$x^{1/2}+2x^{1/4}-6=0$$

Let $u=x^{1/4}$ to obtain

$$u^2+2u-6=0$$

$$u=\frac{-2\pm\sqrt{4+4(6)}}{2}=-1\pm\sqrt{7}$$

$x^{1/4}=-1-\sqrt{7}$ $\boxed{\text{no solution}}$

$x^{1/4}=-1+\sqrt{7}$ $\boxed{x=\left(-1+\sqrt{7}\right)^4\approx 7.34}$

Graphically, let

$y1=2x^{1/4}$, $y2=-x^{1/2}+6$

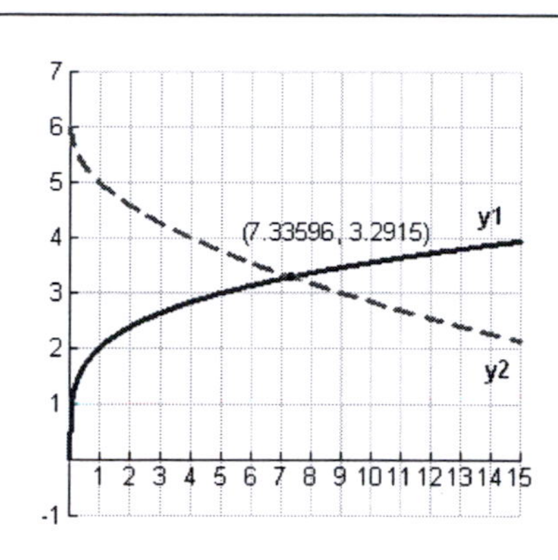

155. a)

$$-0.61x + 7.62 > 0.24x - 5.47$$
$$13.09 > 0.85x$$
$$15.4 > x$$
$$\boxed{(-\infty, 15.4)}$$

b) Graphically, let
$y1 = -0.61x + 7.62,\ y2 = 0.24x - 5.47$

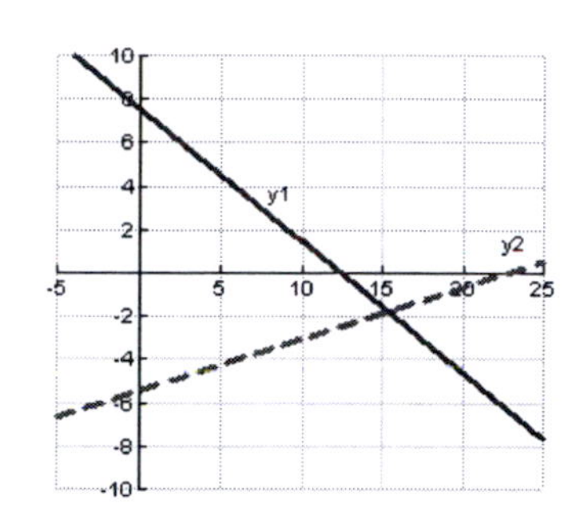

c) Agree

157.

$y1 = 0.2x^2 - 2$

$y2 = 0.05x + 3.25$

Find when $y1 > y2$

$\boxed{(-\infty, -5) \cup (5.25, \infty)}$

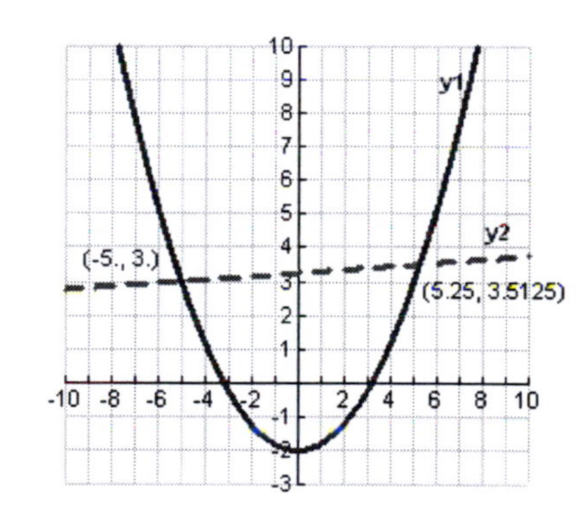

<table>
<tr><td>

159.

$y1 = \frac{3p}{7-2p}$

$y2 = 1$

Find when $y1 > y2$

$\boxed{\left(\frac{7}{5}, \frac{7}{2}\right)}$

</td><td>

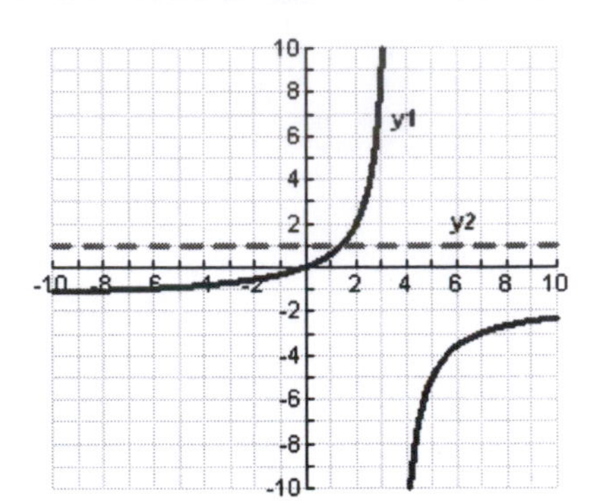

</td></tr>
<tr><td>

161.

$y1 = \left|1.6x^2 - 4.5\right|$

$y2 = 3.2$

Find when $y1 < y2$

$\boxed{(-2.19, -0.9) \cup (0.9, 2.19)}$

</td><td>

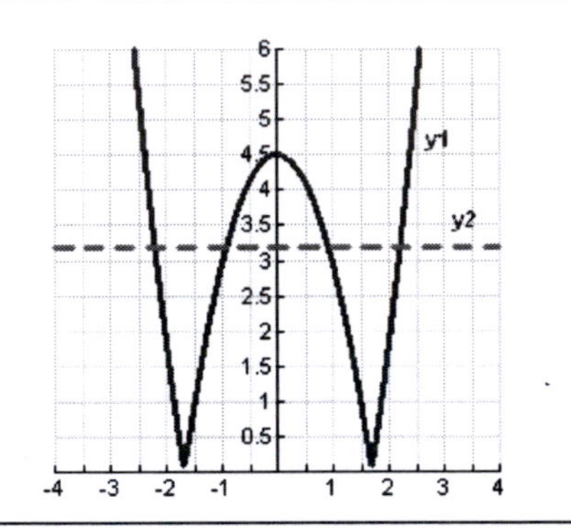

</td></tr>
</table>

Chapter 1 Practice Test Solutions ---

<table>
<tr><td>

1.

$4p - 7 = 6p - 1$

$-6 = 2p$

$\boxed{-3 = p}$

</td><td>

3.

$3t = t^2 - 28$

$t^2 - 3t - 28 = 0$

$(t-7)(t+4) = 0$

$\boxed{t = -4, 7}$

</td></tr>
<tr><td>

5.

$6x^2 - 13x - 8 = 0$

$(3x-8)(2x+1) = 0$

$\boxed{x = -\frac{1}{2}, \frac{8}{3}}$

</td><td>

7.

$\frac{5}{y-3} + 1 - \frac{30}{y^2-9} = 0$

$\frac{5(y+3) + (y^2-9) - 30}{(y-3)(y+3)} = 0$

$\frac{y^2 + 5y - 24}{(y-3)(y+3)} = 0$

$\frac{(y+8)\cancel{(y-3)}}{\cancel{(y-3)}(y+3)} = 0$

$\boxed{y = -8}$

</td></tr>
</table>

9. $\sqrt{2x+1}+x=7$ $\sqrt{2x+1}=7-x$ $2x+1=(7-x)^2$ $2x+1=49-14x+x^2$ $x^2-16x+48=0$ $(x-12)(x-4)=0$ $\boxed{x=4}$, ~~12~~	**11.** $3y-2=9-6\sqrt{3y+1}+3y+1$ $-12=-6\sqrt{3y+1}$ $\sqrt{3y+1}=2$ $3y+1=4$ $3y=3$ $\boxed{y=1}$
13. $x^{7/3}-8x^{4/3}+12x^{1/3}=0$ $x^{1/3}\left(x^2-8x+12\right)=0$ $x^{1/3}(x-6)(x-2)=0$ $\boxed{x=0,2,6}$	**15.** $P=2L+2W$ $P-2W=2L$ $\boxed{L=\frac{P-2W}{2}}$
17. $3x+19\geq 5x-15$ $34\geq 2x$ $17\geq x$ $\boxed{(-\infty,17]}$	**19.** $\frac{2}{5}<\frac{x+8}{4}\leq\frac{1}{2}$ $8<5(x+8)\leq 10$ $-32<5x\leq -30$ $-\frac{32}{5}<x\leq -6$ $\boxed{\left(-\frac{32}{5},-6\right]}$
21. $3p^2-p-4\geq 0$ $(3p-4)(p+1)\geq 0$ CP's: $p=\frac{4}{3},-1$ $\boxed{(-\infty,-1]\cup\left[\frac{4}{3},\infty\right)}$	**23.** $\frac{x-3}{2x+1}\leq 0$ CPs: $x=-\frac{1}{2},3$ 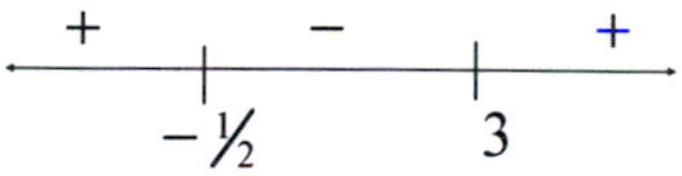 $\boxed{\left(-\frac{1}{2},3\right]}$

25. Let x = height of piling

Sand: $\frac{1}{4}x$

Water: 150

Air: $\frac{3}{5}x$

$$\frac{1}{4}x + 150 + \frac{3}{5}x = x$$

LCD = 20

$$5x + 3000 + 12x = 20x$$

$$3x = 3000$$

$$\boxed{x = 1000 \text{ ft}}$$

27. Let x = number of minutes in excess of 600

Charges $= 49 + 0.17x$

$$53.59 \le 49 + 0.17x \le 69.74$$

$$4.59 \le 0.17x \le 20.74$$

$$27 \le x \le 122$$

$+600$ base

$$\boxed{627 \le x \le 722}$$

29.

$$y1 = \frac{1}{0.75x} - \frac{0.45}{x}$$

$$y2 = \frac{1}{9}$$

$$\boxed{x = 7.95}$$

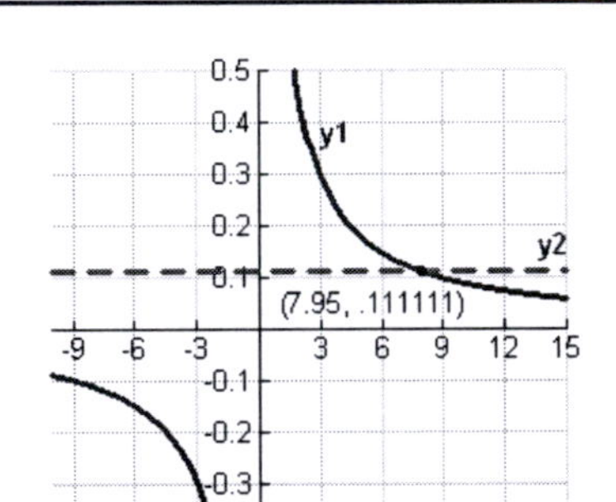

Chapter 1 Cumulative Review--

1.

$$5\cdot(7-3\cdot4+2) = 5\cdot(7-12+2) = 5\cdot(-5+2) = 5\cdot(-3) = \boxed{-15}$$

3.

$$\frac{(x^2y^{-2})^3}{(x^2y)^{-3}} = \frac{x^6y^{-6}}{x^{-6}y^{-3}} = \boxed{\frac{x^{12}}{y^3}}$$

5.

$$x^2(x+5)(x-3) = x^2(x^2+2x-15) = \boxed{x^4+2x^3-15x^2}$$

7.

$$2a^3 + 2000 = 2\left(a^3 + \underbrace{1000}_{=10^3}\right) = \boxed{2(a+10)(a^2-10a+100)}$$

9.

$$\frac{6x}{x-2}-\frac{5x}{x+2}=\frac{6x(x+2)-5x(x-2)}{x^2-4}$$
$$=\frac{6x^2+12x-5x^2+10x}{x^2-4}$$
$$=\boxed{\frac{x^2+22x}{x^2-4}}$$

where $x \neq -2, 2$

11.

$$\tfrac{2}{7}x=\tfrac{1}{8}x+9$$
$$16x=7x+504$$
$$9x=504$$
$$\boxed{x=56}$$

13.

$$\frac{6x}{5}-\frac{8x}{3}=4-\frac{7x}{15}$$
$$18x-40x=60-7x$$
$$-22x=60-7x$$
$$-15x=60$$
$$\boxed{x=-4}$$

15.

Tim rate: 1/9 job in one hour

Chelsea and Tim combined rate:

1/5 job in one hour

Let x = number of hours it takes Chelsea to complete job by herself

Solve:

$$\tfrac{1}{9}+\tfrac{1}{x}=\tfrac{1}{5}$$
$$5x+45=9x$$
$$45=4x \Rightarrow 11.25=x$$

It takes Chelsea $\boxed{\text{11.25 hours}}$ by herself.

17.

$$x^2+12x+40=0$$
$$\left(x^2+12x+36\right)+40-36=0$$
$$(x+6)^2+4=0$$
$$(x+6)^2=-4$$
$$x+6=\pm\sqrt{-4}=\pm 2i$$
$$\boxed{x=-6\pm 2i}$$

19.

$$\sqrt{4-x}=x-4$$
$$4-x=(x-4)^2$$
$$4-x=x^2-8x+16$$
$$x^2-7x+12=0$$
$$(x-4)(x-3)=0$$
$$x=\cancel{3},\boxed{4}$$

21.

$$0<4-x\le 7$$
$$-4<-x\le 3$$
$$4>x\ge -3$$
$$\boxed{[-3,4)}$$

23.

$$\frac{x+2}{9-x^2}\ge 0$$
$$\frac{x+2}{(3-x)(3+x)}\ge 0$$

CPs: $x=-2,\pm 3$

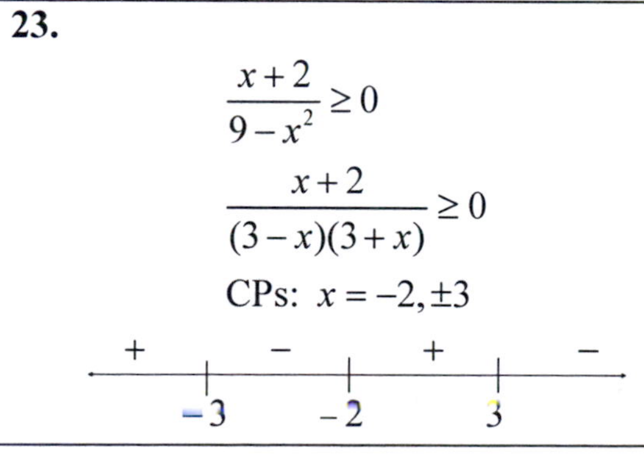

	$\boxed{(-\infty,-3)\cup[-2,3)}$
25. $\left\|\frac{1}{5}x+\frac{2}{3}\right\|=\frac{7}{15}$ $\frac{\|3x+10\|}{15}=\frac{7}{15}$ $\|3x+10\|=7$ $3x+10=7$ or $3x+10=-7$ $3x=-3$ $\quad 3x=-17$ $\boxed{x=-1}$ $\quad \boxed{x=-\frac{17}{3}}$	
27. $y1=\left\|\frac{3x}{x-2}\right\|$ $y2=1$ Find when $y1<y2$ $\boxed{(-1,\frac{1}{2})}$	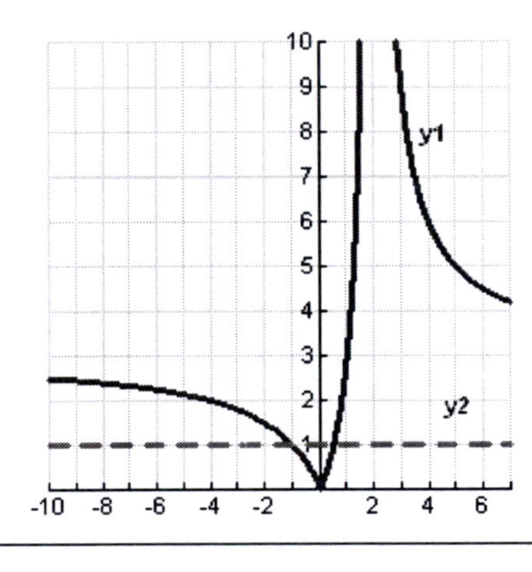

CHAPTER 2

Section 2.1 Solutions

1. $(4,2)$	**3.** $(-3,0)$	**5.** $(0,-3)$

7.

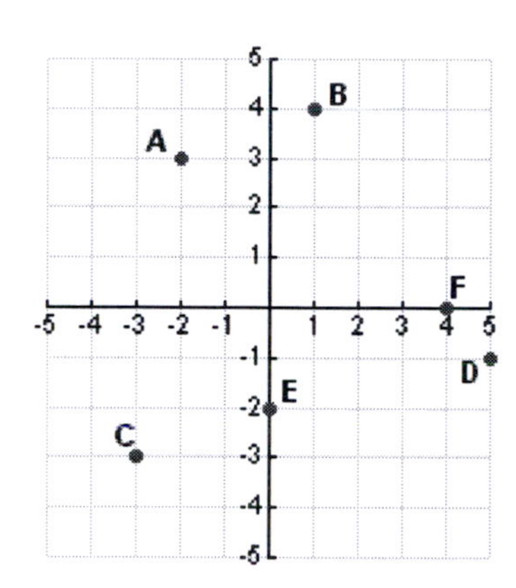

A is in Quadrant II.
B is in Quadrant I.
C is in Quadrant III.
D is in Quadrant IV.
E is on the y-axis.
F is on the x-axis.

9.

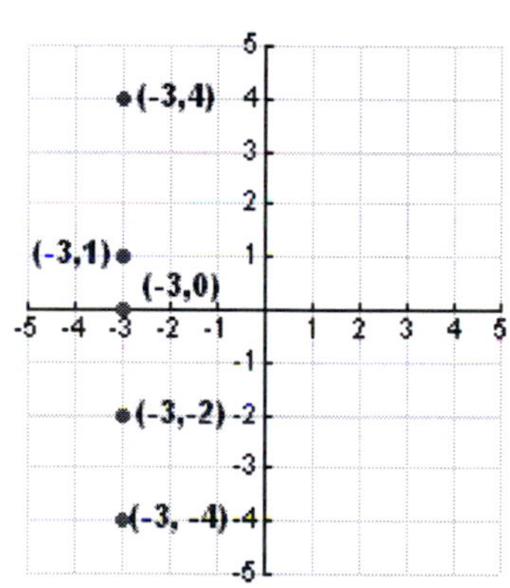

The line being described is x = - 3.

11.

$$d=\sqrt{(1-5)^2+(3-3)^2}=\sqrt{16}=\boxed{4}$$

$$M=\left(\frac{1+5}{2},\frac{3+3}{2}\right)=\boxed{(3,3)}$$

13.

$$d=\sqrt{(-1-3)^2+(4-0)^2}=\sqrt{32}=\boxed{4\sqrt{2}}$$

$$M=\left(\frac{-1+3}{2},\frac{4+0}{2}\right)=\boxed{(1,2)}$$

15.

$$d=\sqrt{(-10-(-7))^2+(8-(-1))^2}$$

$$=\sqrt{90}=\boxed{3\sqrt{10}}$$

$$M=\left(\frac{-10+(-7)}{2},\frac{8+(-1)}{2}\right)=\boxed{\left(\frac{-17}{2},\frac{7}{2}\right)}$$

17.

$$d=\sqrt{(-3-(-7))^2+(-1-2)^2}=\sqrt{25}=\boxed{5}$$

$$M=\left(\frac{-3+(-7)}{2},\frac{-1+2}{2}\right)=\boxed{\left(-5,\frac{1}{2}\right)}$$

19.

$$d=\sqrt{(-6-(-2))^2+(-4-(-8))^2}=\sqrt{32}=\boxed{4\sqrt{2}}$$

$$M=\left(\frac{-6+(-2)}{2},\frac{-4+(-8)}{2}\right)=\boxed{(-4,-6)}$$

21.

$$d=\sqrt{\left(-\frac{1}{2}-\frac{7}{2}\right)^2+\left(\frac{1}{3}-\frac{10}{3}\right)^2}=\sqrt{25}=\boxed{5}$$

$$M=\left(\frac{-\frac{1}{2}+\frac{7}{2}}{2},\frac{\frac{1}{3}+\frac{10}{3}}{2}\right)=\boxed{\left(\frac{3}{2},\frac{11}{6}\right)}$$

23.

$$d=\sqrt{\left(-\frac{2}{3}-\frac{1}{4}\right)^2+\left(-\frac{1}{5}-\frac{1}{3}\right)^2}=\sqrt{\left(-\frac{11}{12}\right)^2+\left(-\frac{8}{15}\right)^2}=\sqrt{\frac{11^2\cdot 15^2+8^2\cdot 12^2}{12^2\cdot 15^2}}=\boxed{\frac{\sqrt{4049}}{60}}$$

$$M=\left(\frac{-\frac{2}{3}+\frac{1}{4}}{2},\frac{-\frac{1}{5}+\frac{1}{3}}{2}\right)=\boxed{\left(-\frac{5}{24},\frac{1}{15}\right)}$$

25.

$$d=\sqrt{(-1.5-2.1)^2+(3.2-4.7)^2}$$

$$=\sqrt{(-3.6)^2+(-1.5)^2}=\boxed{3.9}$$

$$M=\left(\frac{-1.5+2.1}{2},\frac{3.2+4.7}{2}\right)=\boxed{(0.3,\ 3.95)}$$

27.

$$d=\sqrt{(-14.2-16.3)^2+(15.1-(-17.5))^2}$$

$$=\sqrt{(-30.5)^2+(32.6)^2}=\sqrt{1993.01}\cong\boxed{44.64}$$

$$M=\left(\frac{-14.2+16.3}{2},\frac{15.1+(-17.5)}{2}\right)$$

$$=\boxed{(1.05,\ -1.2)}$$

29.

$$d=\sqrt{\left(\sqrt{3}-\sqrt{3}\right)^2+\left(5\sqrt{2}-\sqrt{2}\right)^2}$$

$$=\sqrt{(0)^2+\left(4\sqrt{2}\right)^2}=\boxed{4\sqrt{2}}$$

$$M=\left(\frac{\sqrt{3}+\sqrt{3}}{2},\frac{5\sqrt{2}+\sqrt{2}}{2}\right)$$

$$=\left(\frac{2\sqrt{3}}{2},\frac{6\sqrt{2}}{2}\right)=\boxed{\left(\sqrt{3},3\sqrt{2}\right)}$$

31.

$$d=\sqrt{\left(1-\left(-\sqrt{2}\right)\right)^2+\left(\sqrt{3}-(-2)\right)^2}$$

$$=\sqrt{\left(1+\sqrt{2}\right)^2+\left(\sqrt{3}+2\right)^2}=\boxed{\sqrt{10+2\sqrt{2}+4\sqrt{3}}}$$

$$M=\left(\frac{1+\left(-\sqrt{2}\right)}{2},\frac{\sqrt{3}+(-2)}{2}\right)=\boxed{\left(\frac{1-\sqrt{2}}{2},\frac{-2+\sqrt{3}}{2}\right)}$$

33.

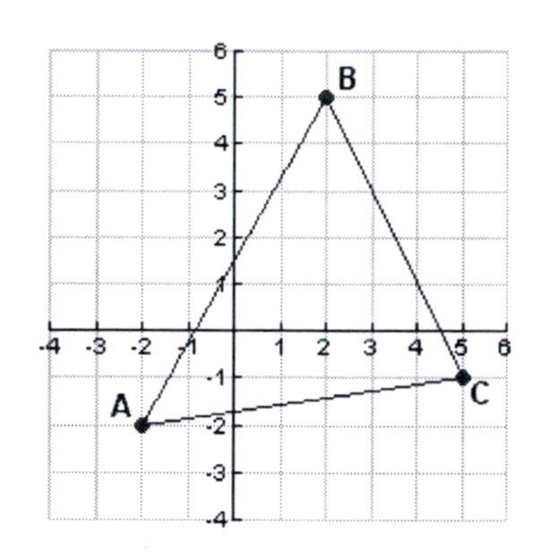

$$d(A,B)=\sqrt{(-2-2)^2+(-2-5)^2}=\sqrt{65}$$
$$d(B,C)=\sqrt{(2-5)^2+(5-(-1))^2}=\sqrt{45}$$
$$d(A,C)=\sqrt{(-2-5)^2+(-2-(-1))^2}=\sqrt{50}$$

The perimeter of the triangle rounded to two decimal places is $\boxed{21.84}$.

35.

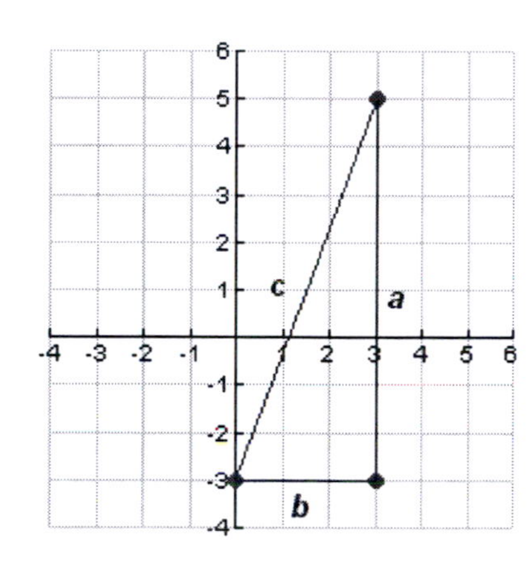

$$a=\sqrt{(3-3)^2+(5-(-3))^2}=8$$
$$b=\sqrt{(0-3)^2+(-3-(-3))^2}=3$$
$$c=\sqrt{(0-3)^2+(-3-5)^2}=\sqrt{8^2+3^2}$$

Observe that $c=\sqrt{a^2+b^2}$ above, so it is a $\boxed{\text{right triangle}}$. It is not isosceles since all sides have different length.

37.

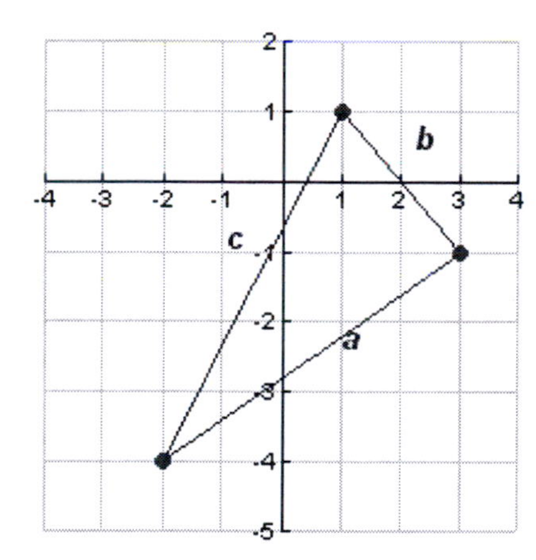

$$a=\sqrt{(3-(-2))^2+(-1-(-4))^2}=\sqrt{34}$$
$$b=\sqrt{(1-3)^2+(1-(-1))^2}=\sqrt{8}$$
$$c=\sqrt{(1-(-2))^2+(1-(-4))^2}=\sqrt{34}$$

Note that $a^2+b^2\neq c^2$, $b^2+c^2\neq a^2$, and $a^2+c^2\neq b^2$, so this is not a right triangle. It is $\boxed{\text{isosceles}}$ since at least two sides have the same length.

39.

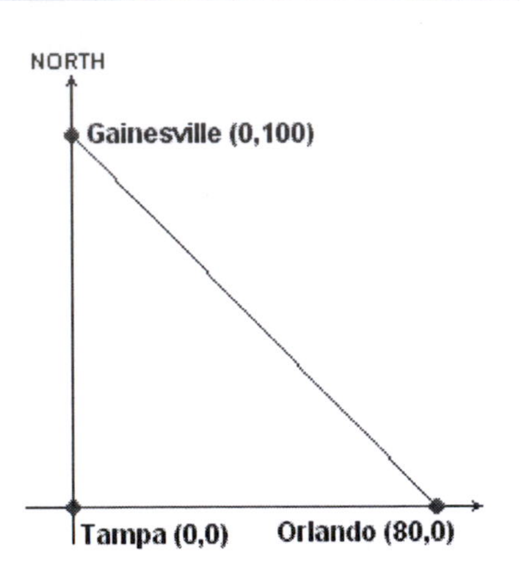

The distance from Gainesville to Orlando is:

$$\sqrt{(0-80)^2+(100-0)^2}=\sqrt{16{,}400}$$
$$=20\sqrt{41}\cong \boxed{128.06 \text{ miles}}$$

41. Assume that Columbia is located at (0,0). Then, Atlanta is located at $(-215,0)$ and Savannah is located at $(0,-160)$. So, $d(\text{Atlanta, Savannah})$ is:

$$\sqrt{(-215-0)^2+(0-(-160))^2}\cong \boxed{268 \text{ miles}}$$

43. $M=\left(\dfrac{2002+2004}{2},\dfrac{400+260}{2}\right)$

$=(2003,330)$

So, the estimated revenue in 2003 is $\boxed{\$330 \text{ million}}$.

45.

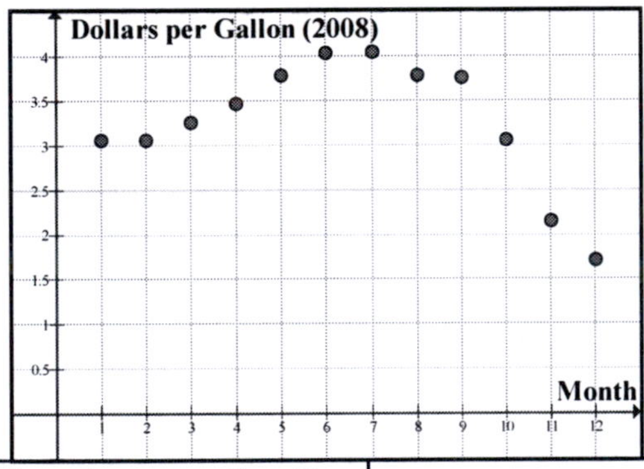

47. Substituted the values incorrectly. It should have been $\sqrt{(9-2)^2+(10-7)^2}$. So, $d=\sqrt{58}$.

49. Substituted the values incorrectly. It should have been $\left(\dfrac{-3+7}{2},\dfrac{4+9}{2}\right)=\left(2,\frac{13}{2}\right)$.

51. True. Follows directly from the distance formula.

53. True. If (x_1,y_1) and (x_2,y_2) are in Quadrant I, then x_1,y_1,x_2,y_2 are all

55. $d=\sqrt{(a-b)^2+(b-a)^2}=\sqrt{2(a-b)^2}$
$=\boxed{\sqrt{2}\,|a-b|}$

$\boxed{M=\left(\frac{a+b}{2},\frac{b+a}{2}\right)}$

positive. So, $\frac{x_1+x_2}{2}$ and $\frac{y_1+y_2}{2}$ are both positive, thereby placing the midpoint $\left(\frac{x_1+x_2}{2},\frac{y_1+y_2}{2}\right)$ in Quadrant I.

57. Since the midpoint is given by $M=\left(\frac{x_1+x_2}{2},\frac{y_1+y_2}{2}\right)$, the distance from (x_1,y_1) to M is as follows:

$$d=\sqrt{\left(x_1-\frac{x_1+x_2}{2}\right)^2+\left(y_1-\frac{y_1+y_2}{2}\right)^2}$$
$$=\sqrt{\left(\frac{2x_1-x_1-x_2}{2}\right)^2+\left(\frac{2y_1-y_1-y_2}{2}\right)^2}$$
$$=\sqrt{\left(\frac{x_1-x_2}{2}\right)^2+\left(\frac{y_1-y_2}{2}\right)^2}$$
$$=\frac{1}{2}\sqrt{(x_1-x_2)^2+(y_1-y_2)^2}$$

Similarly, the distance from (x_2,y_2) to M is as follows:

$$d=\sqrt{\left(x_2-\frac{x_1+x_2}{2}\right)^2+\left(y_2-\frac{y_1+y_2}{2}\right)^2}$$
$$=\sqrt{\left(\frac{2x_2-x_1-x_2}{2}\right)^2+\left(\frac{2y_2-y_1-y_2}{2}\right)^2}$$
$$=\sqrt{\left(\frac{x_2-x_1}{2}\right)^2+\left(\frac{y_2-y_1}{2}\right)^2}$$
$$=\frac{1}{2}\sqrt{(x_1-x_2)^2+(y_1-y_2)^2}$$

59. Let $P_1=(a,b)$, $P_2=(c,d)$, $d_1=$ distance from P_1 to P_2, and $d_2=$ distance from P_2 to P_1. Observe that

$$d_1=\sqrt{(a-c)^2+(b-d)^2} \quad d_2=\sqrt{(c-a)^2+(d-b)^2}\,.$$

Since

$$(c-a)^2=\left(-(a-c)\right)^2=(a-c)^2$$
$$(d-b)^2=\left(-(b-d)\right)^2=(b-d)^2$$

we see that $d_1=d_2$, so that it does not matter what point is labeled "first" in the distance formula.

61.

$$d=\sqrt{(3.7-(-2.3))^2+(6.2-4.1)^2}\cong\boxed{6.357}$$

$$M=\left(\frac{3.7-2.3}{2},\frac{6.2+4.1}{2}\right)=\boxed{(0.7,5.15)}$$

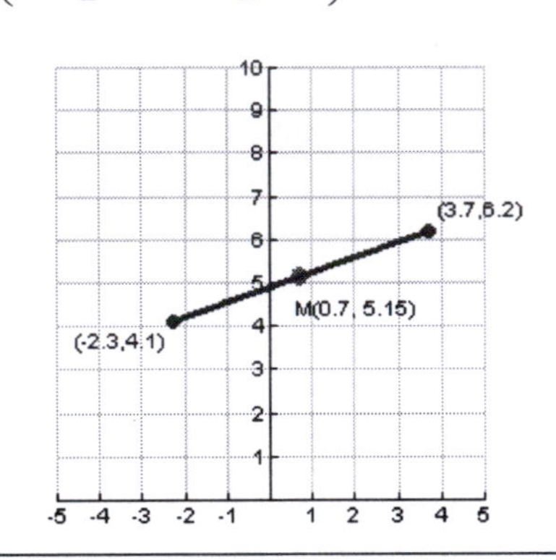

63.

$$d=\sqrt{(3.3-1.1)^2+(4.4-2.2)^2}=\sqrt{2(2.2)^2}\cong\boxed{3.111}$$

$$M=\left(\frac{3.3+1.1}{2},\frac{4.4+2.2}{2}\right)=\boxed{(2.2,3.3)}$$

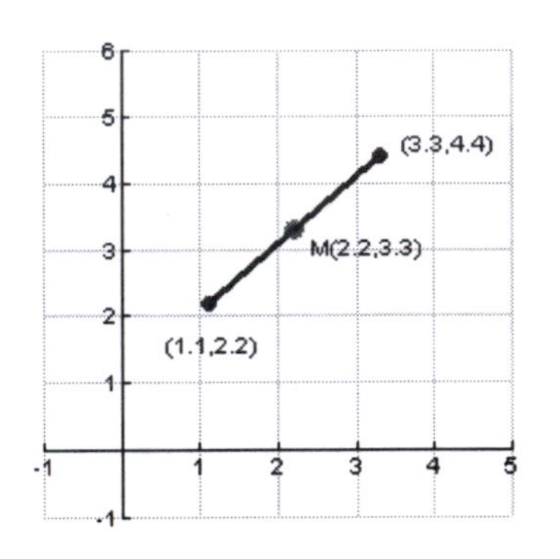

Section 2.2 Solutions

1. a. $3(1)-5\neq 2$ No

b. $3(-2)-5=-11$ Yes

3. a. $\frac{2}{5}(5)-4=-2$ Yes

b. $\frac{2}{5}(-5)-4\neq 6$ No

5. a. $(-1)^2-2(-1)+1=4$ Yes

b. $(0)^2-2(0)+1=1\neq -1$ No

7. a. $\sqrt{7+2}=3$ Yes

b. $\sqrt{-6+2}=2i\neq 4$ No

9.

x	$y=2+x$	(x,y)
-2	0	$(-2,0)$
0	2	$(0,2)$
1	3	$(1,3)$

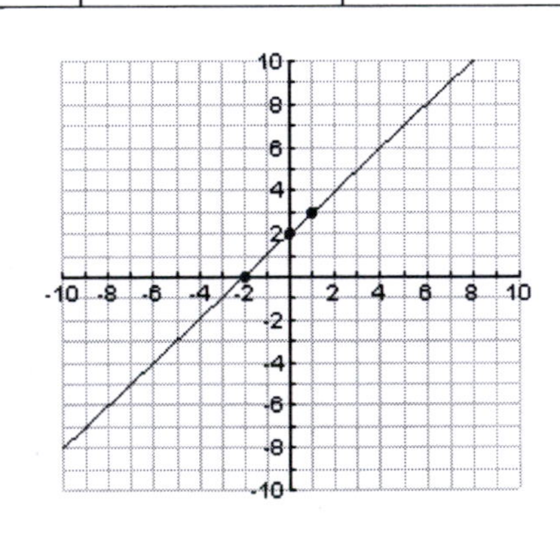

11.

x	$y=x^2-x$	(x,y)
-1	2	$(-1,2)$
0	0	$(0,0)$
$\frac{1}{2}$	$-\frac{1}{4}$	$(\frac{1}{2},-\frac{1}{4})$
1	0	$(1,0)$
2	2	$(2,2)$

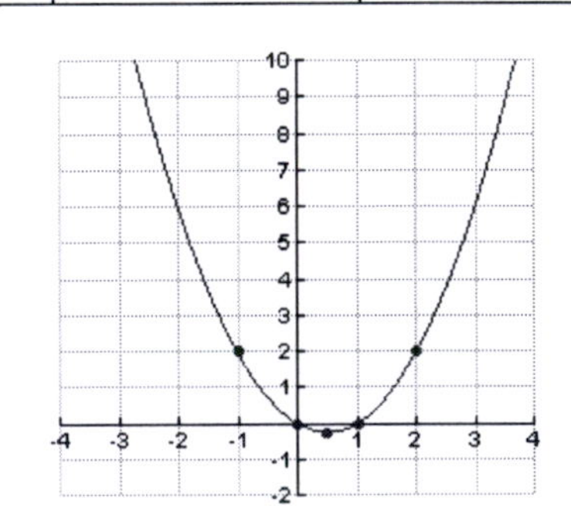

13.

x	$y=\sqrt{x-1}$	(x,y)
1	0	$(1,0)$
2	1	$(2,1)$
5	2	$(5,2)$
10	3	$(10,3)$

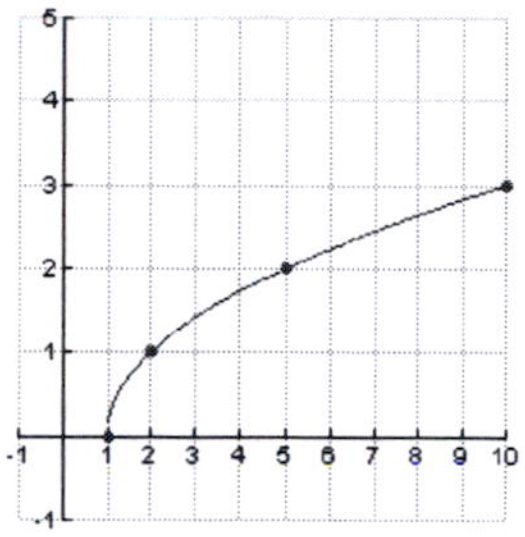

15.

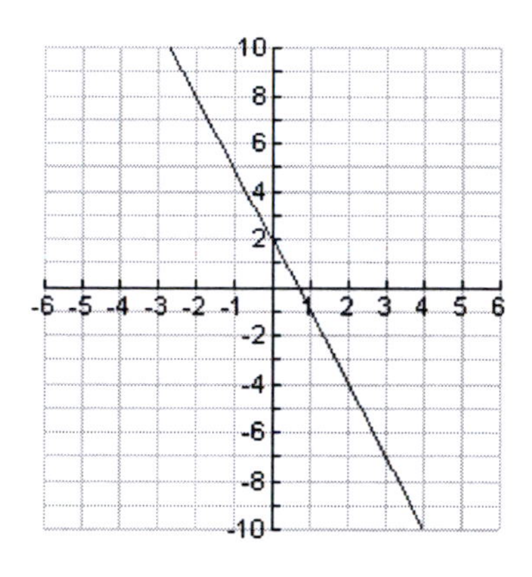

17.

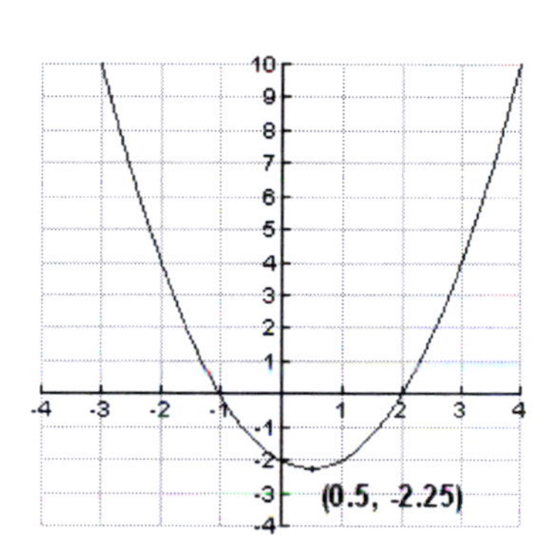

19.

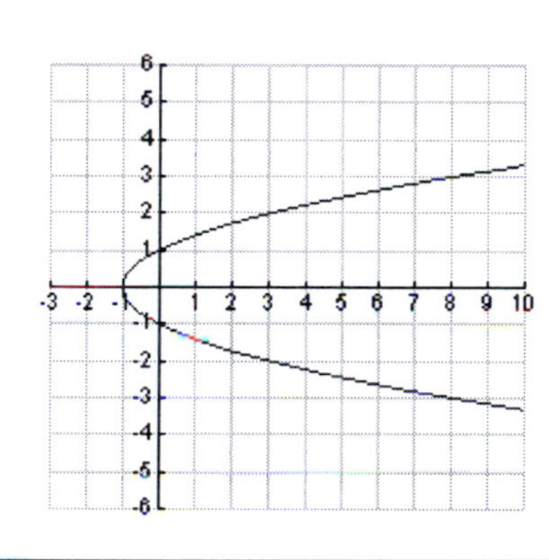

21.

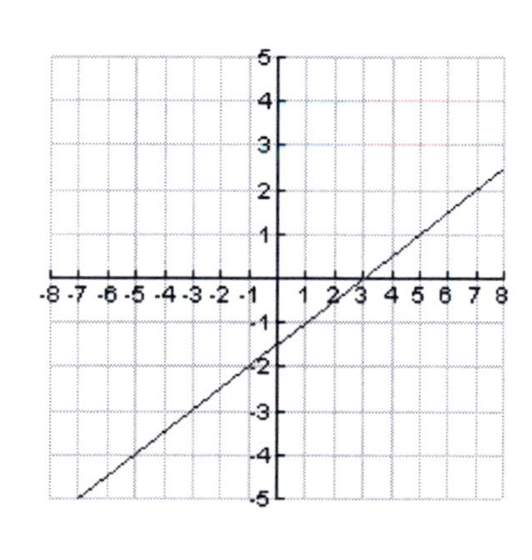

23.

<u>x-intercept</u>:

$$2x-0=6 \Rightarrow x=3 \quad \text{So, } (3,0).$$

<u>y-intercept</u>:

$$2(0)-y=6 \Rightarrow y=-6 \quad \text{So, } (0,-6).$$

25.

<u>x-intercepts</u>:

$$x^2-9=0 \Rightarrow (x-3)(x+3)=0 \Rightarrow x=\pm 3.$$

So, $(\pm 3,0)$.

<u>y-intercept</u>:

$$(0)^2-9=y \Rightarrow y=-9 \quad \text{So, } (0,-9).$$

27.
<u>x-intercept</u>:

$$\sqrt{x-4}=0 \Rightarrow x=4 \quad \text{So, } (4,0).$$

<u>y-intercept</u>:

$$\underbrace{\sqrt{0-4}}_{\text{undefined}}=y. \quad \text{So, no } y-\text{intercept.}$$

29.
<u>x-intercept</u>:

$$\frac{1}{x^2+4}=0 \text{ has no solution. So, no } x-\text{intercept.}$$

<u>y-intercept</u>:

$$\frac{1}{0^2+4}=y \Rightarrow y=\tfrac{1}{4} \quad \text{So, } (0,\tfrac{1}{4}).$$

31.
<u>x-intercepts</u>:

$$4x^2+0=16 \Rightarrow x^2-4=0 \Rightarrow (x-2)(x+2)=0$$
$$\Rightarrow x=\pm 2. \quad \text{So, } (\pm 2,0).$$

<u>y-intercept</u>:

$$4(0)^2+y^2=16 \Rightarrow y^2-16=0 \Rightarrow (y-4)(y+4)=0$$
$$\Rightarrow y=\pm 4. \quad \text{So, } (0,\pm 4).$$

33. d

35. a

37. b

39. $(-1,-3)$

41. $(-7,10)$

43. $(3,2)$, $(-3,2)$, $(-3,-2)$

45. <u>x-axis symmetry</u> (Replace y by $-y$):

$$x=(-y)^2+4$$
$$x=y^2+4$$

Yes, since equivalent to the original.

<u>y-axis symmetry</u> (Replace x by $-x$):

$$-x=y^2+4$$
$$x=-\left(y^2+4\right)$$

No, since not equivalent to the original.

<u>symmetry about origin</u> (Replace y by $-y$ and x by $-x$):

$$-x=(-y)^2+4=y^2+4$$
$$x=-\left(y^2+4\right)$$

No, since not equivalent to the original.

47. <u>x-axis symmetry</u> (Replace y by $-y$):

$$-y=x^3+x$$
$$y=-\left(x^3+x\right)$$

No, since not equivalent to the original.

<u>y-axis symmetry</u> (Replace x by $-x$):

$$y=(-x)^3+(-x)$$
$$y=-\left(x^3+x\right)$$

No, since not equivalent to the original.

<u>symmetry about origin</u> (Replace y by $-y$ and x by $-x$):

$$-y=(-x)^3+(-x)=-\left(x^3+x\right)$$
$$y=x^3+x$$

Yes, since equivalent to the original.

49. x-axis symmetry (Replace y by $-y$):

$$x = |-y| = |-1||y|$$
$$x = |y|$$

Yes, since equivalent to the original.

y-axis symmetry (Replace x by $-x$):

$$-x = |y|$$
$$x = -|y|$$

No, since not equivalent to the original.

symmetry about origin (Replace y by $-y$ and x by $-x$):

$$-x = |-y| = |-1||y| = |y|$$
$$x = -|y|$$

No, since not equivalent to the original.

51. x-axis symmetry (Replace y by $-y$):

$$x^2 - (-y)^2 = 100$$
$$x^2 - y^2 = 100$$

Yes, since equivalent to the original.

y-axis symmetry (Replace x by $-x$):

$$(-x)^2 - y^2 = 100$$
$$x^2 - y^2 = 100$$

Yes, since equivalent to the original.

symmetry about origin (Replace y by $-y$ and x by $-x$):

$$(-x)^2 - (-y)^2 = 100$$
$$x^2 - y^2 = 100$$

Yes, since equivalent to the original.

53. x-axis symmetry (Replace y by $-y$):

$$-y = x^{2/5}$$
$$y = -x^{2/5}$$

No, since not equivalent to the original.

y-axis symmetry (Replace x by $-x$):

$$y = (-x)^{2/5}$$
$$y = x^{2/5}$$

Yes, since equivalent to the original.

symmetry about origin (Replace y by $-y$ and x by $-x$):

$$-y = (-x)^{2/5} = x^{2/5}$$
$$y = -x^{2/5}$$

No, since not equivalent to the original.

55. x-axis symmetry (Replace y by $-y$):

$$x^2 + (-y)^3 = 1$$
$$x^2 - y^3 = 1$$

No, since not equivalent to the original.

y-axis symmetry (Replace x by $-x$):

$$(-x)^2 + y^3 = 1$$
$$x^2 + y^3 = 1$$

Yes, since equivalent to the original.

symmetry about origin (Replace y by $-y$ and x by $-x$):

$$(-x)^2 + (-y)^3 = 1$$
$$x^2 - y^3 = 1$$

No, since not equivalent to the original.

57. <u>x-axis symmetry</u> (Replace y by $-y$):

$$-y = \frac{2}{x}$$
$$y = -\frac{2}{x}$$

No, since not equivalent to the original.

<u>y-axis symmetry</u> (Replace x by $-x$):

$$y = \frac{2}{-x}$$
$$y = -\frac{2}{x}$$

No, since not equivalent to the original.

<u>symmetry about origin</u> (Replace y by $-y$ and x by $-x$):

$$-y = \frac{2}{-x} = -\frac{2}{x}$$
$$y = \frac{2}{x}$$

Yes, since equivalent to the original.

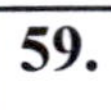

59.

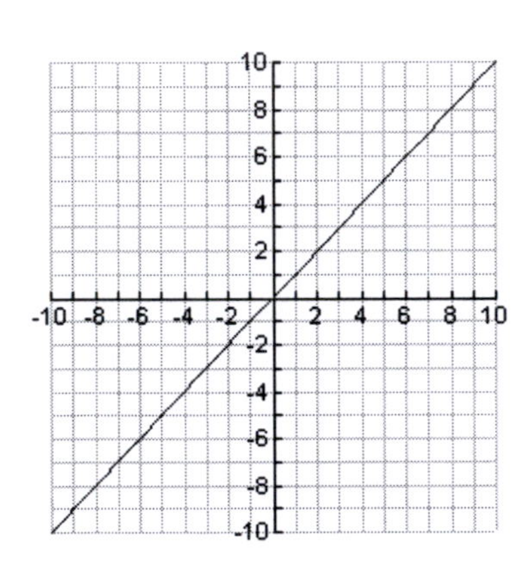

61.

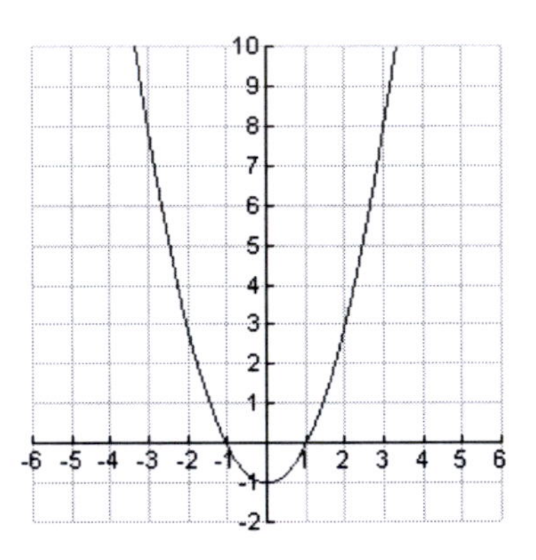

63.

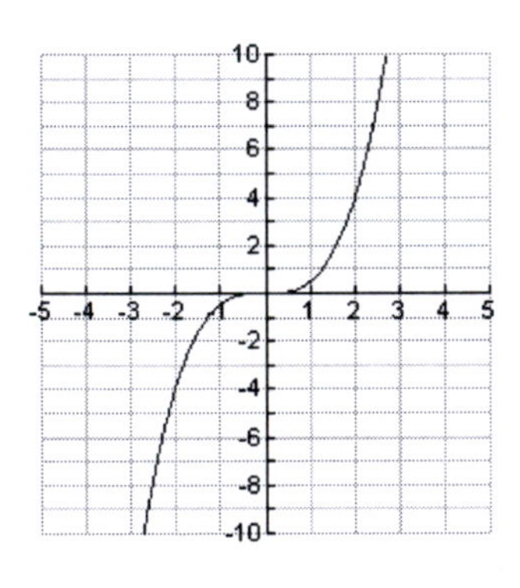

65.

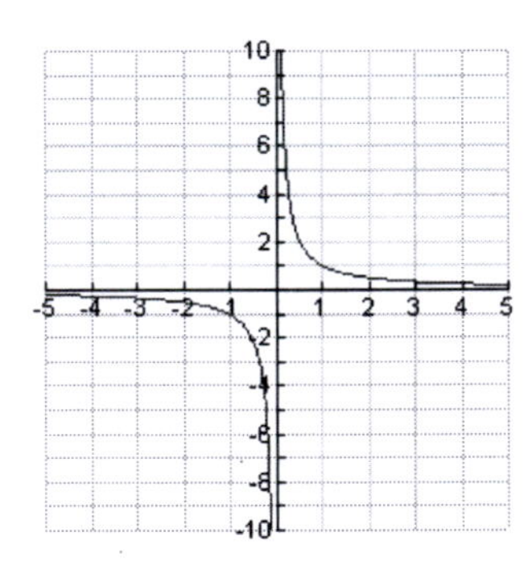

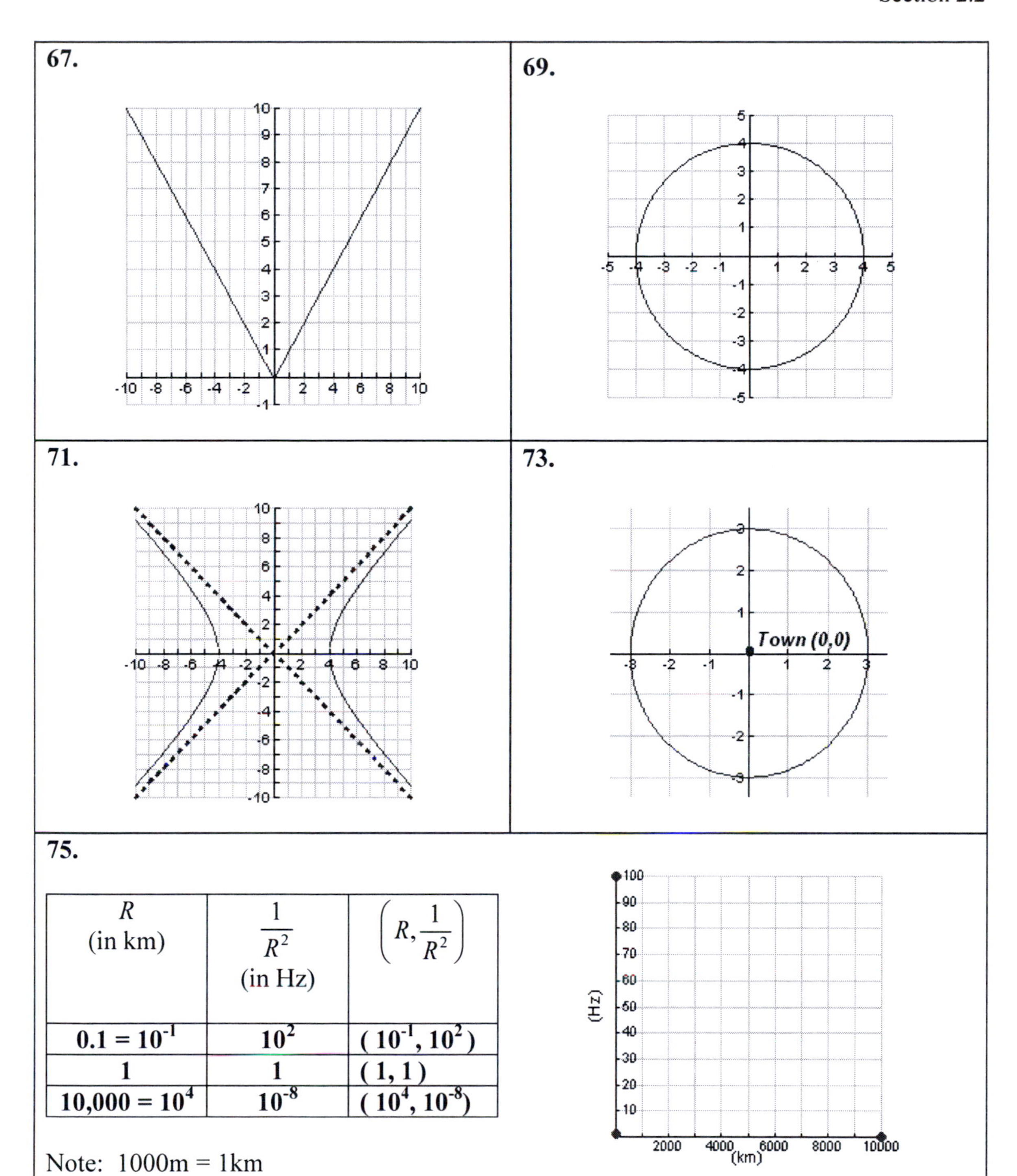

R (in km)	$\frac{1}{R^2}$ (in Hz)	$\left(R, \frac{1}{R^2}\right)$
$0.1 = 10^{-1}$	**10^{2}**	**$(10^{-1}, 10^{2})$**
1	**1**	**$(1, 1)$**
$10{,}000 = 10^{4}$	**10^{-8}**	**$(10^{4}, 10^{-8})$**

Note: 1000m = 1km

77. Consider the following graph:

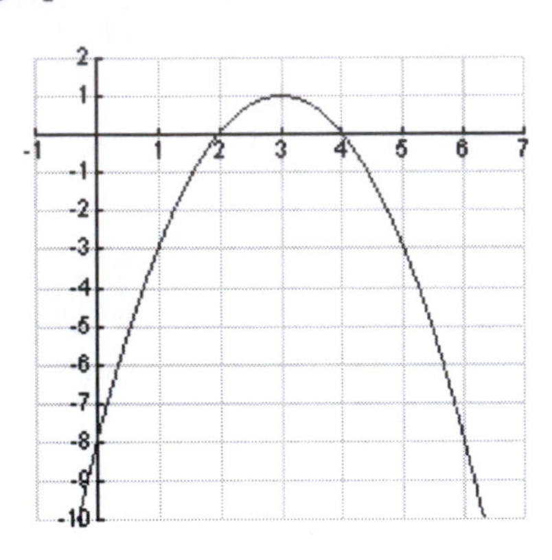

Either 2000 or 4000 units need to be sold to break even. The range of units that correspond to making a profit correspond to those x-values for which $y > 0$. This occurs for $\boxed{2000 < x < 4000}$.

79. a. The domain is the set of x such that $0.01x - 0.01 \geq 0$. Solving this inequality yields

$$0.01x \geq 0.01$$

$$x \geq 1$$

This means that the demand model is defined when at least 1,000 units per day are demanded.

b.

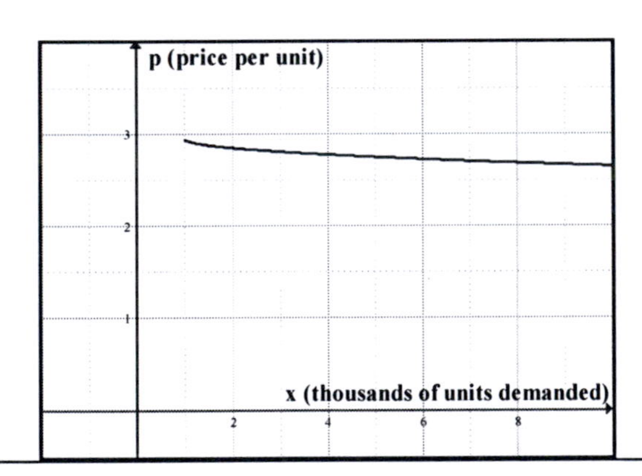

81. The equation is not linear – you need more than two points to plot the graph.

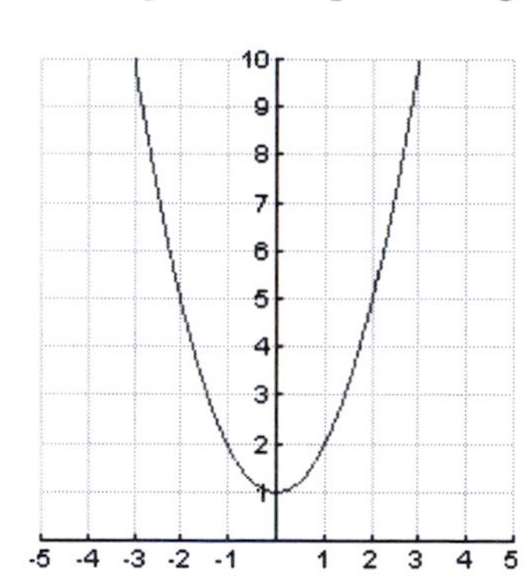

83.
To test for symmetry about the y-axis, one should replace x by $-x$, NOT y by $-y$. Doing so here yields the equation $-x = |y|$, which is not equivalent to the original equation $x = |y|$.

85. False The correct conclusion would be that the point $(a, -b)$ must also be on the graph. For instance, $y^2 = x$ is symmetric about the x-axis and $(1,1)$ is on the graph, but $(-1,1)$ is not.

87. True

89. <u>x-axis symmetry</u> (Replace y by $-y$):

$$-y = \frac{ax^2 + b}{cx^3}$$

$$y = -\left(\frac{ax^2 + b}{cx^3}\right)$$

No, since not equivalent to the original.

<u>y-axis symmetry</u> (Replace x by $-x$):

$$y = \frac{a(-x)^2 + b}{c(-x)^3} = \frac{ax^2 + b}{-cx^3}$$

$$y = -\left(\frac{ax^2 + b}{cx^3}\right)$$

No, since not equivalent to the original.

<u>symmetry about origin</u> (Replace y by $-y$ and x by $-x$):

$$-y = \frac{a(-x)^2 + b}{c(-x)^3} = \frac{ax^2 + b}{-cx^3} = -\left(\frac{ax^2 + b}{cx^3}\right) \text{ so that } y = \frac{ax^2 + b}{cx^3}$$

Yes, since equivalent to the original.

91.

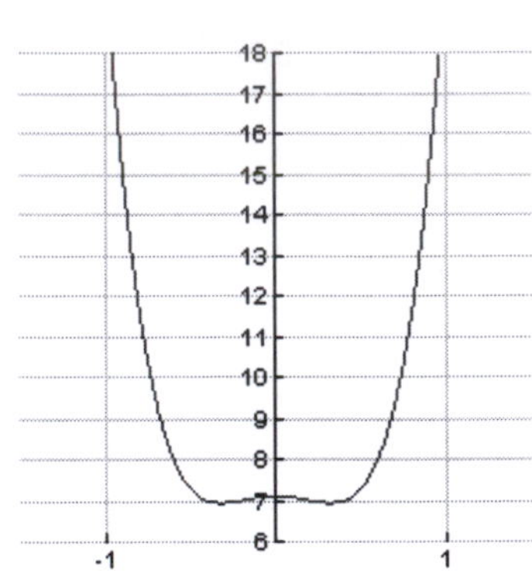

Symmetric with respect to the y-axis.

93.

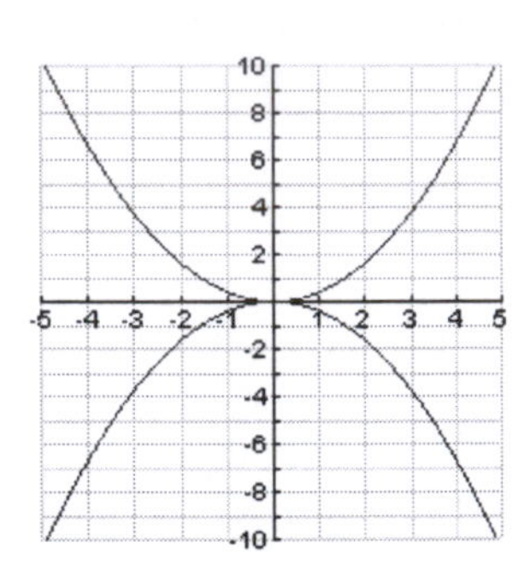

Symmetric with respect to the x-axis, y-axis, and the origin.

95.

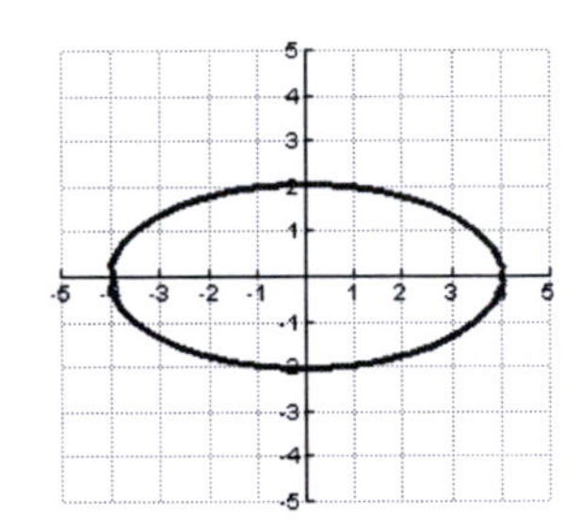

Symmetric with respect to the x-axis, y-axis, and the origin.

Section 2.3 Solutions ---

1. $m = \dfrac{3-6}{1-2} = \boxed{3}$

3. $m = \dfrac{5-(-3)}{-2-2} = \boxed{-2}$

5. $m = \dfrac{9-(-10)}{-7-3} = \boxed{-\dfrac{19}{10}}$

7. $m = \dfrac{-1.7-5.2}{0.2-3.1} = \dfrac{6.9}{2.9} \cong \boxed{2.379}$

9. $m = \dfrac{-\frac{1}{4}-\left(-\frac{3}{4}\right)}{\frac{2}{3}-\frac{5}{6}} = -\dfrac{\frac{1}{2}}{\frac{1}{6}} = \boxed{-3}$

11. x-intercept: $(0.5, 0)$

y-intercept: $(0, -1)$

slope: $m = \dfrac{-3-3}{-1-2} = 2$

rising

13. x-intercept: $(1, 0)$

y-intercept: $(0, 1)$

slope: $m = \dfrac{3-(-1)}{-2-2} = -1$

falling

15. x-intercept: None

y-intercept: $(0, 1)$

slope: $m = 0$

horizontal

17.

x-intercept:

$0 = 2x - 3$
$\frac{3}{2} = x$ So, $\left(\frac{3}{2}, 0\right)$.

y-intercept: $y = 2(0) - 3 = -3$. So, $(0, -3)$.

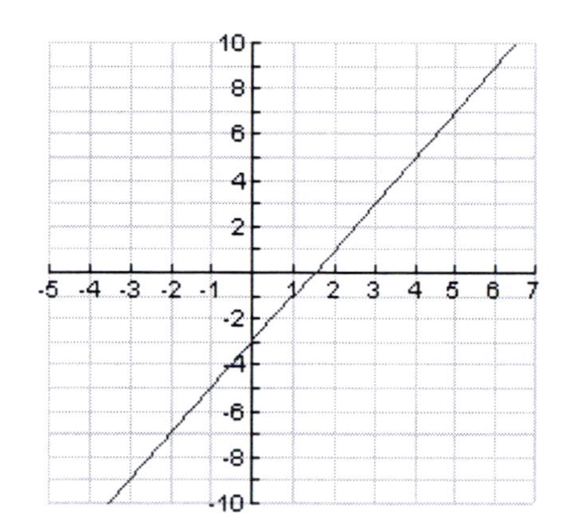

19.

x-intercept:

$0 = -\frac{1}{2}x + 2$
$4 = x$ So, $(4, 0)$.

y-intercept: $y = -\frac{1}{2}(0) + 2 = 2$. So, $(0, 2)$.

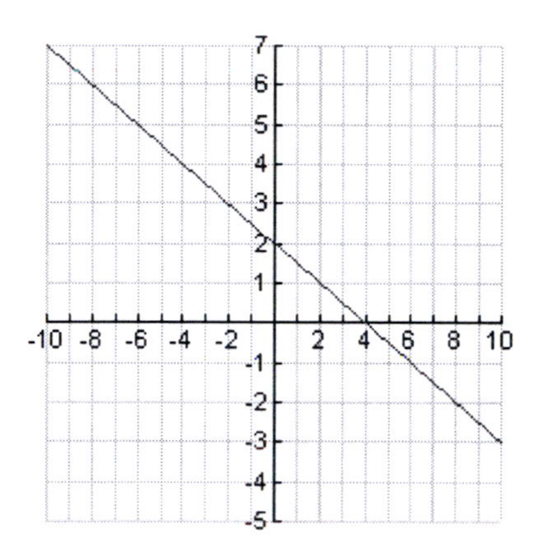

21.

x-intercept:

$4 = 2x - 3(0)$
$2 = x$ So, $(2, 0)$.

y-intercept:

$4 = 2(0) - 3y$
$-\frac{4}{3} = y$. So, $\left(0, -\frac{4}{3}\right)$.

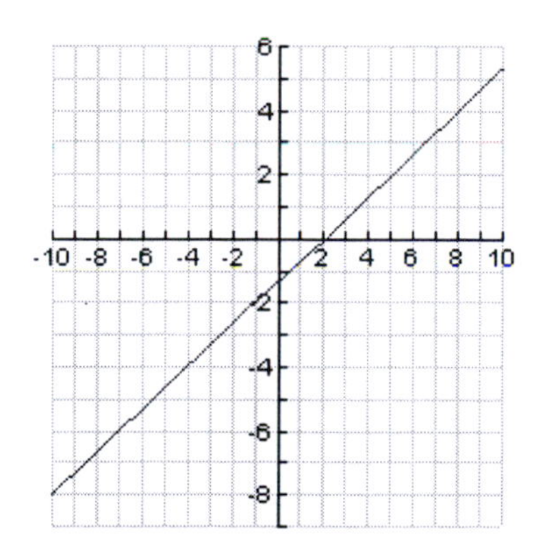

23.

x-intercept:

$-1 = \frac{1}{2}x + \frac{1}{2}(0)$
$-2 = x$ So, $(-2, 0)$.

y-intercept:

$-1 = \frac{1}{2}(0) + \frac{1}{2}y$
$-2 = y$. So, $(0, -2)$.

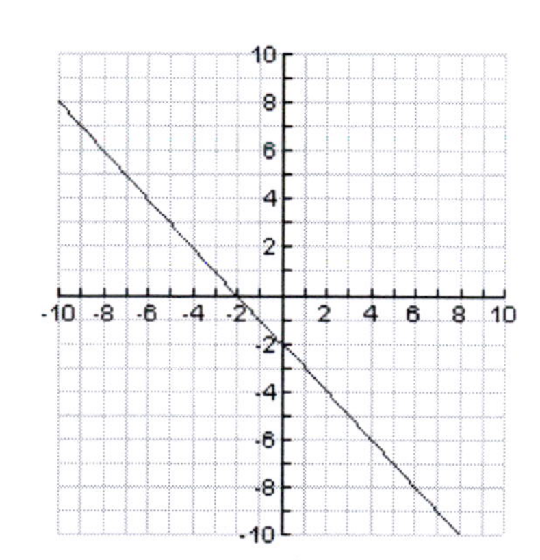

25.
x-intercept: $(-1,0)$
y-intercept: None

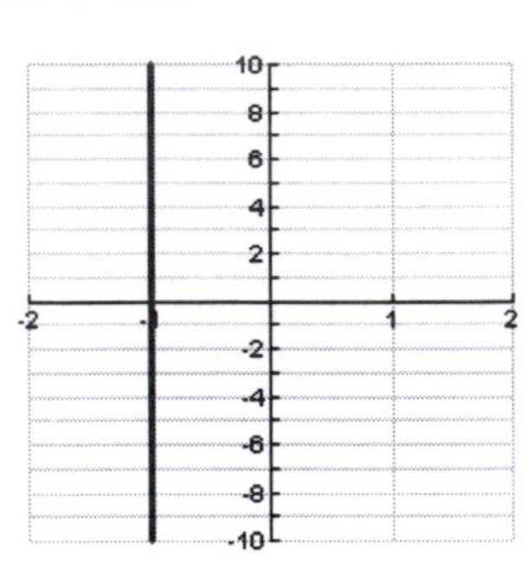

27.
x-intercept: None
y-intercept: $(0,1.5)$

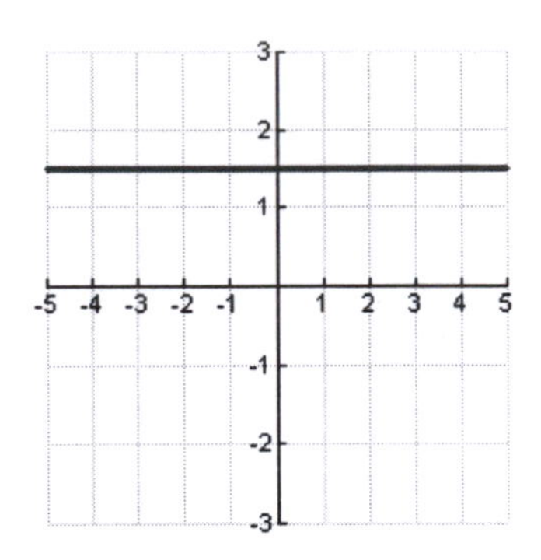

29.
x-intercept: $\left(-\frac{7}{2},0\right)$
y-intercept: None

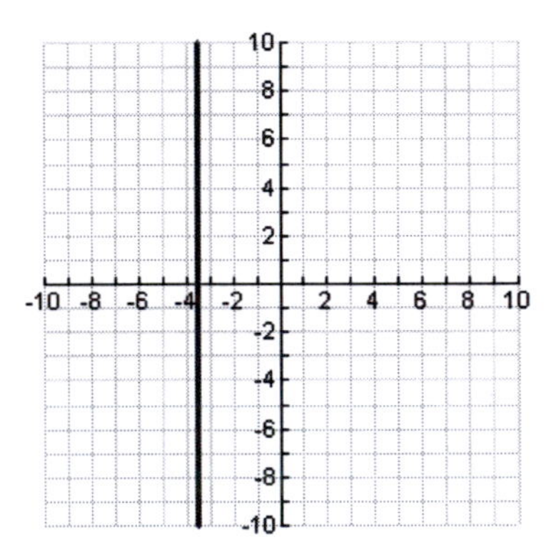

31.
$y=\frac{2}{5}x-2 \quad m=\frac{2}{5} \quad y-$intercept: $(0,-2)$

33.
$y=-\frac{1}{3}x+2 \quad m=-\frac{1}{3} \quad y-$intercept: $(0,2)$

35.
$y=4x-3 \quad m=4 \quad y-$intercept: $(0,-3)$

37.
$y=-2x+4 \quad m=-2 \quad y-$intercept: $(0,4)$

39.
$y=\frac{2}{3}x-2 \quad m=\frac{2}{3} \quad y-$intercept: $(0,-2)$

41.
$y=-\frac{3}{4}x+6 \quad m=-\frac{3}{4} \quad y-$intercept: $(0,6)$

43. $y=2x+3$

45. $y=-\frac{1}{3}x+0=-\frac{1}{3}x$

47. $y=2$

49. $x=\frac{3}{2}$

51. Find b:
$$-3=5(-1)+b$$
$$2=b$$
So, the equation is $\boxed{y=5x+2}$.

53. Find b:
$$2=-3(-2)+b$$
$$-4=b$$
So, the equation is $\boxed{y=-3x-4}$.

55. Find b:
$$-1=\frac{3}{4}(1)+b$$
$$-\frac{7}{4}=b$$
So, the equation is $\boxed{y=\frac{3}{4}x-\frac{7}{4}}$.

57. Since $m=0$, the line is horizontal.
So, the equation is $\boxed{y=4}$.

59. Since m is undefined, the line is vertical. So, the equation is $\boxed{x=-1}$.

61. slope:

$$m=\frac{-1-2}{-2-3}=\frac{3}{5}$$

y-intercept:

Use the point $(-2,-1)$ to find b:

$$-1=\tfrac{3}{5}(-2)+b$$
$$\tfrac{1}{5}=b$$

So, the equation is $\boxed{y=\tfrac{3}{5}x+\tfrac{1}{5}}$.

63. slope:

$$m=\frac{-1-(-6)}{-3-(-2)}=-5$$

y-intercept:

Use the point $(-3,-1)$ to find b:

$$-1=-5(-3)+b$$
$$-16=b$$

So, the equation is $\boxed{y=-5x-16}$.

65. slope:

$$m=\frac{-37-(-42)}{20-(-10)}=\frac{1}{6}$$

y-intercept:

Use the point $(20,-37)$ to find b:

$$-37=\tfrac{1}{6}(20)+b$$
$$-\tfrac{121}{3}=b$$

So, the equation is $\boxed{y=\tfrac{1}{6}x-\tfrac{121}{3}}$.

67. slope: $m=\dfrac{4-(-5)}{-1-2}=-3$

y-intercept:

Use the point $(-1,4)$ to find b:

$$4=-3(-1)+b$$
$$1=b$$

So, the equation is $\boxed{y=-3x+1}$.

69.

slope: $m=\dfrac{\frac{3}{4}-\frac{9}{4}}{\frac{1}{2}-\frac{3}{2}}=\dfrac{3}{2}$

y-intercept:

Use the point $\left(\tfrac{1}{2},\tfrac{3}{4}\right)$ to find b:

$$\tfrac{3}{4}=\tfrac{3}{2}\left(\tfrac{1}{2}\right)+b$$
$$0=b$$

So, the equation is $\boxed{y=\tfrac{3}{2}x}$.

71. Since m is undefined, the line is vertical. So, the equation is $\boxed{x=3}$.

73. Since $m=\dfrac{7-7}{3-9}=0$, the line is horizontal. So, the equation is $\boxed{y=7}$.

75. The slope is $m=\dfrac{6-0}{0-(-5)}=\dfrac{6}{5}$.

Since $(0,6)$ is the y-intercept, $b=6$.

So, the equation is $\boxed{y=\tfrac{6}{5}x+6}$.

77. Since m is undefined, the line is vertical. So, the equation is $\boxed{x=-6}$.

79. The slope is undefined. So, the equation of the line passing through the points $\left(\tfrac{2}{5},-\tfrac{3}{4}\right)$ and $\left(\tfrac{2}{5},\tfrac{1}{2}\right)$ is $\boxed{x=\tfrac{2}{5}}$.

81. We identify the slope and y-intercept, and then express the equation in slope-intercept form as $\boxed{y=x-1}$.

83. We identify the slope and y-intercept, and then express the equation in slope-intercept form as $\boxed{y=-2x+3}$.

85. We identify the slope and y-intercept, and then express the equation in slope-intercept form as $\boxed{y=-\frac{1}{2}x+1}$.

87. slope:
Since parallel to $y=2x-1$, $m=2$.
y-intercept:
Use the point $(-3,1)$ to find b:

$$1=2(-3)+b$$
$$7=b$$

So, the equation is $\boxed{y=2x+7}$.

89.
slope:
Since perpendicular to $2x+3y=12$ (whose slope is $-\frac{2}{3}$), $m=\frac{3}{2}$.
y-intercept:
Since $(0,0)$ is assumed to be on the line and is the y-intercept, $b=0$.
So, the equation is $\boxed{y=\frac{3}{2}x}$.

91. Since parallel to x-axis, the line is horizontal. So, its equation is of the form $y=b$. Since $(3,5)$ is assumed to be on the line, the equation must be $\boxed{y=5}$.

93. Since perpendicular to y-axis, the line is horizontal. So, its equation is of the form $y=b$. Since $(-1,2)$ is assumed to be on the line, the equation must be $\boxed{y=2}$.

95. slope:
Since parallel to $\frac{1}{2}x-\frac{1}{3}y=5$ (whose slope is $\frac{3}{2}$), $m=\frac{3}{2}$.
y-intercept:
Use the point $(-2,-7)$ to find b:

$$-7=\tfrac{3}{2}(-2)+b$$
$$-4=b$$

So, the equation is $\boxed{y=\frac{3}{2}x-4}$.

97. Note that $8x+10y=-45$ is equivalent to $y=-\frac{4}{5}x-\frac{9}{2}$. The slope is $-\frac{4}{5}$. So, the slope of a line perpendicular to it is $\frac{5}{4}$. Using this as the slope, together with the point $\left(-\frac{2}{3},\frac{2}{3}\right)$, we have the equation:

$$y-\tfrac{2}{3}=\tfrac{5}{4}\left(x+\tfrac{2}{3}\right)$$
$$y-\tfrac{2}{3}=\tfrac{5}{4}x+\tfrac{5}{6}$$
$$y=\tfrac{5}{4}x+\tfrac{3}{2}$$

99. Note that $-15x+35y=7$ is equivalent to $y=\frac{3}{7}x+\frac{1}{5}$. The slope is $\frac{3}{7}$. So, the slope of a line parallel to it is $\frac{3}{7}$. Using this as the slope, together with the point $\left(\frac{7}{2},4\right)$, we have the equation:

$$y-4=\tfrac{3}{7}\left(x-\tfrac{7}{2}\right)$$
$$y-4=\tfrac{3}{7}x-\tfrac{3}{2}$$
$$y=\tfrac{3}{7}x+\tfrac{5}{2}$$

101. Let h = number of hours a job lasts. The cost (in dollars) is given by $C(h)=1200+25h$. So, for $h=32$, $C=2000$. This means that a $\boxed{\text{32-hour job will cost \$2,000}}$.

103. Let L = monthly loan payment and x = # times filled up gas tank.
Then, $500 = L + 25x$. So, if $x = 5$, then $500 = L + 125$, so that $L = 375$.
The monthly payment is $\boxed{\$375}$.

105. Let x = number of units sold.

Operating Cost: C(x) = 1,300 + 3.50x
Revenue: R(x) = 7.25x

Find the value of x such that C(x) = R(x).

$$1{,}300 + 3.50x = 7.25x$$
$$1{,}300 = 3.75x$$
$$x = \frac{1{,}300}{3.75} \approx 347 \text{ units}$$

107. Assume that F and C are related by $F = mC + b$.
We are given that $(25, 77)$ and $(20, 68)$ (where the points are listed in the form (C, F)) satisfy the equation. We compute m and b as follows:

slope: $m = \dfrac{77 - 68}{25 - 20} = \dfrac{9}{5}$ y-intercept: Use $(25, 77)$ to find b:

$$77 = \tfrac{9}{5}(25) + b$$
$$32 = B$$

So, the equation relating F and C is given by $\boxed{F = \tfrac{9}{5}C + 32}$.

The temperature for which $F = C$ can be found by substituting $F = C$ into the equation:

$$C = \tfrac{9}{5}C + 32$$
$$-\tfrac{4}{5}C = 32$$
$$-40 = C$$

Thus, $\boxed{-40 \text{ degrees Celsius} = -40 \text{ degrees Fahrenheit}}$.

109. Since $(1900, 67)$ and $(2000, 69)$ are assumed to be on the line, the slope of the line is $m = \dfrac{69 - 67}{2000 - 1900} = \dfrac{2}{100} = \dfrac{1}{50}$. So, the rate of change (in inches per year) is $\boxed{\dfrac{1}{50}}$.

111. Note: 6 pounds 4 ounces = 100 ounces 6 pounds 10 ounces = 106 ounces
Since $(1900,100)$ and $(2000,106)$ are assumed to be on the line, the slope of the line is

$m=\dfrac{106-100}{2000-1900}=0.06$. So, the rate of change (in ounces per year) is $\boxed{0.06}$.

To determine the expected weight in the year 2040, we need the equation of the line. So far, we know the form of the equation is $y=0.06x+b$. Use $(1900,100)$ to find b:

$$100=0.06(1900)+b$$
$$-14=b$$

So, the equation is $y=0.06x-14$. Therefore, in 2040, $y=0.06(2040)-14=108.4$ oz.
Since 108.4 oz. = 6.775 pounds, and 0.775 pounds = 12.4 oz., we expect a baby to weigh $\boxed{\text{6 pounds 12.4 oz}}$ in 2040.

113. The y-intercept represents the flat monthly fee of $\boxed{\$35}$ since $x = 0$ implies that no long distance minutes were used.

115. The rate of change is $\dfrac{3.8-5.2}{4-0}=-\dfrac{1.4}{4}=-0.35$ in./yr.

Using the point (0, 5.2), and assuming that $x = 0$ corresponds to 2003, the equation that governs rainfall per year is given by

$$y-5.2=-0.35(x-0) \Rightarrow y=-0.35x+5.2.$$

Since 2010 corresponds to $x = 7$, we see that $y=-0.35(7)+5.2=2.75$. So, the expected average rainfall for July in 2010 is $\boxed{\text{2.75 inches}}$.

117. The rate of change is $\dfrac{392-380}{5-0}=2.4$ plastic bags per year (in billions).

Using the point (0, 380), and assuming that $x = 0$ corresponds to 2000, the equation that governs this trend is given by

$$y-380=2.4(x-0) \Rightarrow y=2.4x+380.$$

Hence, in 2010, you can expect the average number of plastic bags used to be $380+2.4(10)=\boxed{\text{404 billion}}$.

119. a. (1, 16(1) + 15.93) = (1, 31.93), (2, 16(2) + 19.18) = (2, 51.18), (5, 111.83)

b. $m=\dfrac{31.93-0}{1-0}=31.93$ This means that when you buy a single bottle, the cost per bottle of Hoison sauce is \$31.93.

c. $m=\dfrac{51.18-0}{2-0}=25.59$ This means that when you buy 2 bottles of Hoison sauce, the cost per bottle of Hoison sauce is \$25.59.

d. $m=\dfrac{111.83-0}{5-0}=22.37$ This means that when you buy 5 bottles of Hoison sauce, the cost per bottle of Hoison sauce is \$22.37.

121. The computations used to calculate the x- and y-intercepts should be reversed. So, the x-intercept is $(3,0)$ and the y-intercept is $(0,-2)$.
123. The denominator and numerator in the slope computation should be switched, resulting in the slope being undefined.
125. True. Since the equation $mx+b=0$ has at most one solution.
127. False. The lines are perpendicular.
129. Any vertical line is perpendicular to a line with slope 0.
131. Since $B \neq 0$, $Ax+By=C$ can be written as $y=-\frac{A}{B}x+\frac{C}{B}$. Since the desired line is parallel to this line, its slope $m=-\frac{A}{B}$. Use the point $(-B, A+1)$ on the line to find the y-intercept: $$A+1=-\tfrac{A}{B}(-B)+b$$ $$A+1=A+b$$ $$b=1$$ So, the equation of the line in this case is $y=-\frac{A}{B}x+1$.
133. We know that the point $(-A, B-1)$ is on the line. Since it is perpendicular to $Ax+By=C$, and the slope of this line is $-\frac{A}{B}$ (assuming $B \neq 0$), the slope of the desired line is $\frac{B}{A}$ (assuming $A \neq 0$). So, the equation of the desired line is $$y-(B-1)=\tfrac{B}{A}(x+A)$$ $$y=(B-1)+\tfrac{B}{A}x+B$$ $$\boxed{y=\tfrac{B}{A}x+(2B-1)}$$
135. Let $y_1=mx+b_1$ and $y_2=mx+b_2$, assuming that $b_1 \neq b_2$. At a point of intersection of these two lines, $y_1=y_2$. This is equivalent to $mx+b_1=mx+b_2$, which implies $b_1=b_2$, which contradicts our assumption. Hence, there are no points of intersection.

137. Slope of $y_1 = 17x + 22$ is 17.
Slope of $y_2 = -\frac{1}{17}x - 13$ is $-\frac{1}{17}$.
Since $17\left(-\frac{1}{17}\right) = -1$, $\boxed{\text{perpendicular}}$.

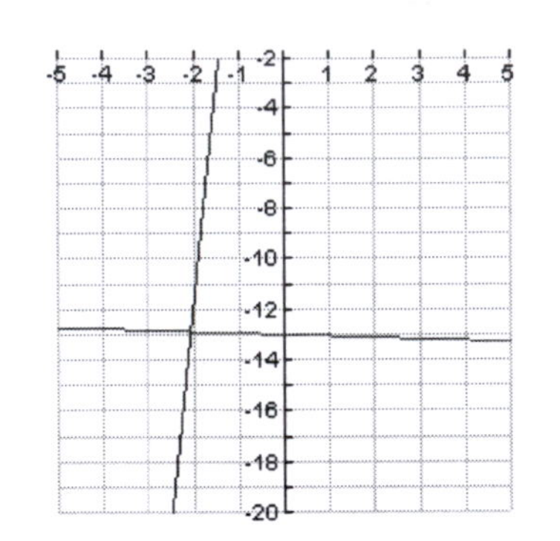

139. Slope of $y_1 = 0.25x + 3.3$ is 0.25.
Slope of $y_2 = -4x + 2$ is -4.
Since $0.25(-4) = -1$, they are $\boxed{\text{perpendicular}}$.

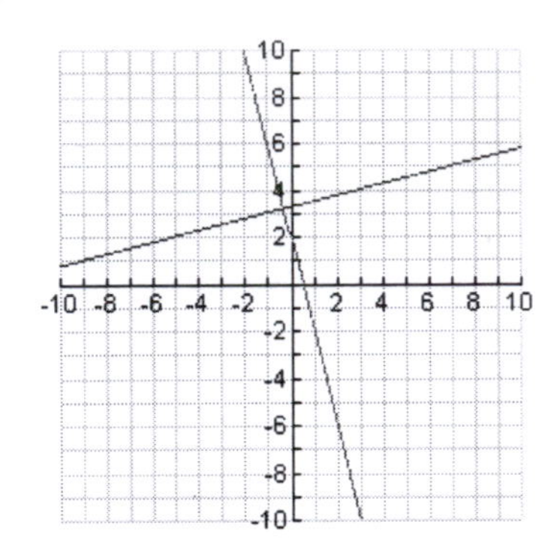

141. Slope of $y_1 = 0.16x + 2.7$ is 0.16. Slope of $y_2 = 6.25x - 1.4$ is 6.25.
Since $(0.16)(6.25) \neq -1$ and $0.16 \neq 6.25$, they are $\boxed{\text{neither}}$ parallel nor perpendicular.

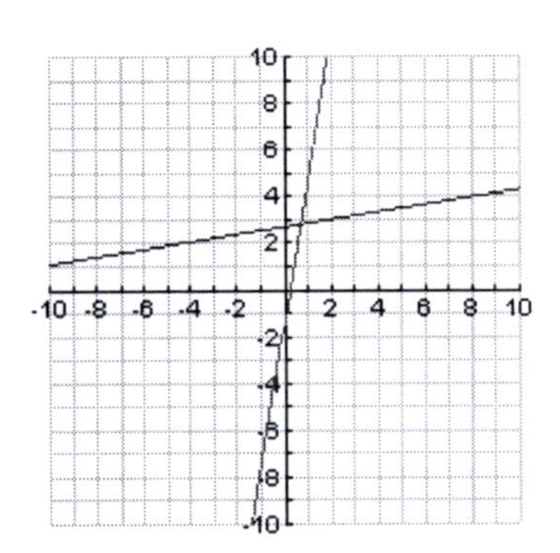

Section 2.4 Solutions --

1. $(x-1)^2 + (y-2)^2 = 9$	**3.** $(x-(-3))^2 + (y-(-4))^2 = 10^2$ $(x+3)^2 + (y+4)^2 = 100$
5. $(x-5)^2 + (y-7)^2 = 81$	**7.** $(x-(-11))^2 + (y-12)^2 = 13^2$ $(x+11)^2 + (y-12)^2 = 169$

9. $(x-0)^2+(y-0)^2=2^2$

$x^2+y^2=4$

11. $(x-0)^2+(y-2)^2=3^2$

$x^2+(y-2)^2=9$

13. $(x-0)^2+(y-0)^2=\left(\sqrt{2}\right)^2$

$x^2+y^2=2$

15. $(x-5)^2+(y-(-3))^2=\left(2\sqrt{3}\right)^2$

$(x-5)^2+(y+3)^2=12$

17. $\left(x-\frac{2}{3}\right)^2+\left(y-\left(-\frac{3}{5}\right)\right)^2=\left(\frac{1}{4}\right)^2$

$\left(x-\frac{2}{3}\right)^2+\left(y+\frac{3}{5}\right)^2=\frac{1}{16}$

19. $(x-1.3)^2+(y-2.7)^2=(3.2)^2$

$(x-1.3)^2+(y-2.7)^2=10.24$

21. The center is $(1,3)$ and the radius is $\sqrt{25}=5$.

23. The equation can be written as

$$(x-2)^2+(y-(-5))^2=(7)^2.$$

So, center is $(2,-5)$ and radius is 7.

25. The center is $(4,9)$ and the radius is $\sqrt{20}=2\sqrt{5}$.

27. The center is $\left(\frac{2}{5},\frac{1}{7}\right)$ and the radius is $\sqrt{\frac{4}{9}}=\frac{2}{3}$.

29. The equation can be written as

$$(x-1.5)^2+(y-(-2.7))^2=\left(\sqrt{1.69}\right)^2.$$

So, center is $(1.5,-2.7)$ and radius is $\sqrt{1.69}=1.3$.

31. The equation can be written as

$$(x-0)^2+(y-0)^2=\left(\sqrt{50}\right)^2.$$

So, center is $(0,0)$ and radius is $\sqrt{50}=5\sqrt{2}$.

33. Completing the square gives us:

$$x^2+y^2+4x+6y-3=0$$

$$x^2+4x \quad +y^2+6y=3$$

$$\left(x^2+4x+4\right)+\left(y^2+6y+9\right)=3+4+9$$

$$(x+2)^2+(y+3)^2=16$$

$$(x-(-2))^2+(y-(-3))^2=16$$

So, center is $(-2,-3)$ and radius is $\sqrt{16}=4$.

35. Completing the square gives us:

$$x^2+y^2+6x+8y-75=0$$
$$x^2+6x\quad+y^2+8y=75$$
$$\left(x^2+6x+9\right)+\left(y^2+8y+16\right)=75+9+16$$
$$(x+3)^2+(y+4)^2=100$$
$$(x-(-3))^2+(y-(-4))^2=100$$

So, center is $(-3,-4)$ and radius is $\sqrt{100}=10$.

37. Completing the square gives us:

$$x^2+y^2-10x-14y-7=0$$
$$x^2-10x\quad+y^2-14y=7$$
$$\left(x^2-10x+25\right)+\left(y^2-14y+49\right)=7+25+49$$
$$(x-5)^2+(y-7)^2=81$$

So, center is $(5,7)$ and radius is 9.

39. Completing the square gives us:

$$x^2+y^2-2y-15=0$$
$$x^2+\left(y^2-2y+1\right)-15-1=0$$
$$x^2+(y-1)^2=16$$

So, the center is (0,1) and radius is 4.

41. Completing the square gives us:

$$x^2+y^2-2x-6y+1=0$$
$$x^2-2x\quad+y^2-6y=-1$$
$$\left(x^2-2x+1\right)+\left(y^2-6y+9\right)=-1+1+9$$
$$(x-1)^2+(y-3)^2=9$$

So, center is $(1,3)$ and radius is 3.

43. Completing the square gives us:

$$x^2+y^2-10x+6y+22=0$$
$$x^2-10x\quad+y^2+6y=-22$$
$$\left(x^2-10x+25\right)+\left(y^2+6y+9\right)=-22+25+9$$
$$(x-5)^2+(y+3)^2=12$$
$$(x-5)^2+(y-(-3))^2=12$$

So, center is $(5,-3)$ and radius is $\sqrt{12}=2\sqrt{3}$.

45. Completing the square gives us:

$$x^2+y^2-6x-4y+1=0$$
$$x^2-6x\quad+y^2-4y=-1$$
$$\left(x^2-6x+9\right)+\left(y^2-4y+4\right)=-1+9+4$$
$$(x-3)^2+(y-2)^2=12$$

So, center is $(3,2)$ and radius is $\sqrt{12}=2\sqrt{3}$.

47. Completing the square gives us:

$$x^2+y^2-x+y+\tfrac{1}{4}=0$$
$$x^2-x\quad+y^2+y=-\tfrac{1}{4}$$
$$\left(x^2-x+\tfrac{1}{4}\right)+\left(y^2+y+\tfrac{1}{4}\right)=-\tfrac{1}{4}+\tfrac{1}{4}+\tfrac{1}{4}$$
$$\left(x-\tfrac{1}{2}\right)^2+\left(y+\tfrac{1}{2}\right)^2=\tfrac{1}{4}$$
$$\left(x-\tfrac{1}{2}\right)^2+\left(y-\left(-\tfrac{1}{2}\right)\right)^2=\tfrac{1}{4}$$

So, center is $\left(\tfrac{1}{2},-\tfrac{1}{2}\right)$ and radius is $\sqrt{\tfrac{1}{4}}=\tfrac{1}{2}$.

49. Completing the square gives us:

$$x^2 + y^2 - 2.6x - 5.4y - 1.26 = 0$$
$$x^2 - 2.6x \quad + y^2 - 5.4y = 1.26$$
$$\left(x^2 - 2.6x + 1.69\right) + \left(y^2 - 5.4y + 7.29\right) = 1.26 + 1.69 + 7.29$$
$$(x-1.3)^2 + (y-2.7)^2 = 10.24$$

So, center is $(1.3, 2.7)$ and radius is $\sqrt{10.24} = 3.2$.

51. The center is $(-1,-2)$.
The radius is the distance from $(-1,-2)$ to $(1,0)$, which is given by

$$\sqrt{(-1-1)^2 + (-2-0)^2} = \sqrt{8}.$$

So, the equation is

$$(x-(-1))^2 + (y-(-2))^2 = \left(\sqrt{8}\right)^2$$
$$(x+1)^2 + (y+2)^2 = 8.$$

53. The center is $(-2,3)$.
The radius is the distance from $(-2,3)$ to $(3,7)$, which is given by

$$\sqrt{(-2-3)^2 + (3-7)^2} = \sqrt{41}.$$

So, the equation is

$$(x-(-2))^2 + (y-3)^2 = 41$$
$$(x+2)^2 + (y-3)^2 = 41.$$

55. The center is $(-2,-5)$.
The radius is the distance from $(-2,-5)$ to $(1,-9)$, which is given by

$$\sqrt{(-2-1)^2 + (-5-(-9))^2} = 5.$$

So, the equation is

$$(x-(-2))^2 + (y-(-5))^2 = 25$$
$$(x+2)^2 + (y+5)^2 = 25.$$

57. Assume that the phone tower is located at $(0,0)$. Then, the coordinates of your home are $(33,95)$. The distance from the phone tower is then $\sqrt{95^2 + 33^2} \cong 100.568$ miles, which exceeds the reception area. So, you cannot use the cell phone at home.

59. Assume that the stake is located at $(0,0)$. In order to stay within the boundary of the yard, the leash must be no longer than 50 ft. The outer perimeter for the dog would then be described by $(x-0)^2 + (y-0)^2 = 50^2$, that is $x^2 + y^2 = 2500$.

61. The center of the inner circle is $(0,0)$. Since the diameter is 3000 ft., the radius is 1500 ft. So, the equation of the inner circle is $(x-0)^2 + (y-0)^2 = 1500^2$, that is $x^2 + y^2 = 2,250,000$.

63. Location of tower = center of reception area = (0,0). We also know the radius is 200 miles. Hence, the equation of the circle is $\boxed{x^2 + y^2 = 40,000}$.

65. a.

$$\text{Tower 1}: \ (x-2.5)^2+(y-2.5)^2=3.5^2$$
$$\text{Tower 2}: \ (x-2.5)^2+(y-7.5)^2=3.5^2$$
$$\text{Tower 3}: \ (x-7.5)^2+(y-2.5)^2=3.5^2$$
$$\text{Tower 4}: \ (x-7.5)^2+(y-7.5)^2=3.5^2$$

b.
This placement of cell phone towers will provide cell phone coverage for the entire 10 mile by 10 mile square.

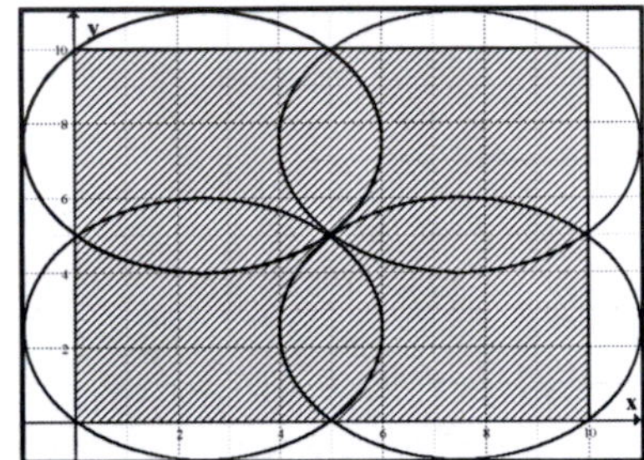

67. The center should be $(4,-3)$.

69. The standard form of the equation of a circle requires that the right-side be non-negative. Since the radius would have to be $\sqrt{-16}$, which is not a real number, the result cannot be a circle.

71. True. Since the graph is comprised of infinitely many points.

73. True. Since the left-side is always non-negative.

75. Completing the square gives us:

$$x^2+y^2+10x-6y+34=0$$
$$x^2+10x \quad +y^2-6y=-34$$
$$\left(x^2+10x+25\right)+\left(y^2-6y+9\right)=-34+25+9$$
$$(x+5)^2+(y-3)^2=0$$

Thus, the graph consists of the single point $(-5,3)$.

77. The diameter is equal to the distance between $(5,2)$ and $(1,-6)$, which is given by $\sqrt{(5-1)^2+(2-(-6))^2}=\sqrt{80}=4\sqrt{5}$. So, the radius is $\frac{1}{2}\left(4\sqrt{5}\right)=2\sqrt{5}$.
The center is the midpoint of the segment joining the points $(5,2)$ and $(1,-6)$, which is given by $\left(\frac{5+1}{2},\frac{2+(-6)}{2}\right)=(3,-2)$. Therefore, the equation is then given by:

$$(x-3)^2+(y-(-2))^2=\left(2\sqrt{5}\right)^2$$
$$(x-3)^2+(y+2)^2=20$$

79. Completing the square gives us:

$$x^2+y^2+ax+by+c=0$$
$$x^2+ax\quad+y^2+by=-c$$
$$\left(x^2+ax+\tfrac{a2}{4}\right)+\left(y^2+by+\tfrac{b2}{4}\right)=-c+\tfrac{a2}{4}+\tfrac{b2}{4}$$
$$\left(x+\tfrac{a}{2}\right)^2+\left(y+\tfrac{b}{2}\right)^2=-c+\tfrac{a2}{4}+\tfrac{b2}{4}$$

If $-c+\frac{a2}{4}+\frac{b2}{4}=0$, which is equivalent to $4c=a^2+b^2$, then the graph will consist of the single point $\left(-\frac{a}{2},-\frac{b}{2}\right)$.

81. Completing the square gives us:

$$x^2+y^2-2ax=100-a^2$$
$$\left(x^2-2ax+a^2\right)+y^2-a^2=100-a^2$$
$$(x-a)^2+y^2=100$$

So, the center is $(a, 0)$ and radius is 10.

83. The equation is given by

$$(x-2)^2+(y+5)^2=-20,$$

so there is no graph.

85. Completing the square gives us:

$$x^2+y^2+10x-6y+34=0$$
$$x^2+10x\quad+y^2-6y=-34$$
$$\left(x^2+10x+25\right)+\left(y^2-6y+9\right)=-34+25+9$$
$$(x+5)^2+(y-3)^2=0$$

Thus, the graph consists of the single point $(-5,3)$.

87. a. Completing the square yields

$$\left(x^2-11x\quad\right)+\left(y^2+3y\quad\right)=7.19$$
$$\left(x^2-11x+30.25\right)+\left(y^2+3y+2.25\right)=7.19+30.25+2.25=39.69$$
$$(x-5.5)^2+(y+1.5)^2=39.69\quad\textbf{(1)}$$

b. Solving **(1)** for y yields:

$$(y+1.5)^2 = 39.69-(x-5.5)^2$$

$$y+1.5 = \pm\sqrt{39.69-(x-5.5)^2}$$

$$y = -1.5 \pm\sqrt{39.69-(x-5.5)^2}$$

c. The graph in both parts **a** and **b** are the same, and are as follows:

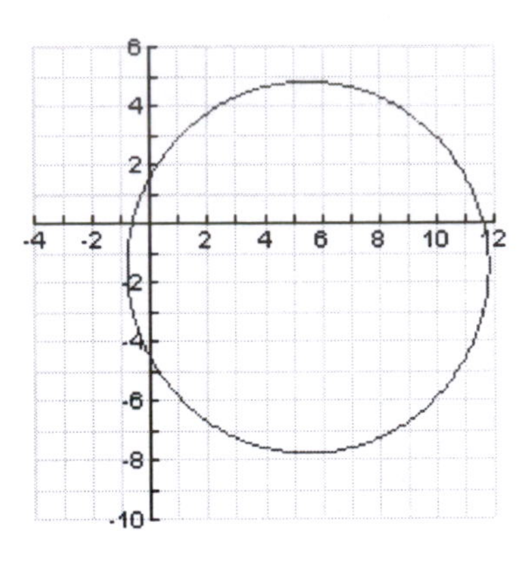

Section 2.5 --

1. Negative linear association because the data closely cluster around what is reasonably described as a line with negative slope.

3. Although the data seem to be comprised of two *linear* segments, the overall data set cannot be described as having a positive or negative direction of association. Moreover, the pattern of the data is not linear, per se; rather, it is nonlinear and conforms to an identifiable curve (an upside down V called the *absolute value* function).

5. B because the association is positive, thereby eliminating choices A and C. And, since the data are closely clustered around a linear curve, the bigger of the two correlation coefficients, 0.80, is more appropriate.

7. C because the association is negative, thereby eliminating choices B and D. And, the data are more loosely clustered around a linear curve than are those pictured in #6. So, the correlation coefficient is the negative choice closer to 0.

9. **a.)**

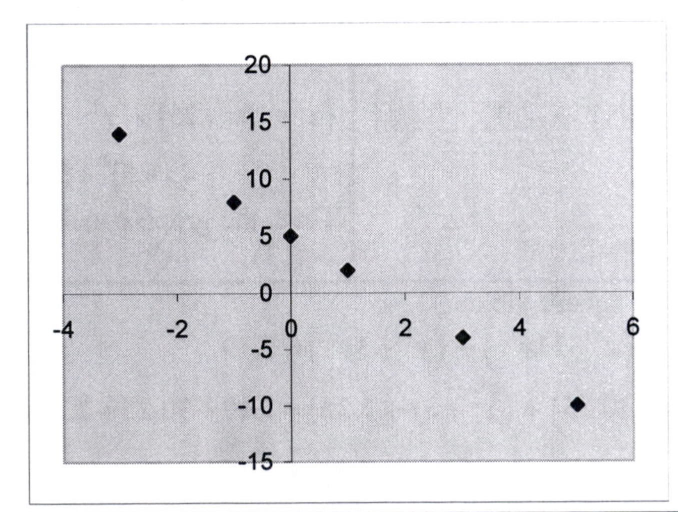

	b.)	The data seem to be nearly perfectly aligned to a line with negative slope. So, it is reasonable to guess that the correlation coefficient is very close to –1.
	c.)	The equation of the best fit line is $y = -3x + 5$ with a correlation coefficient of $r = -1$.
	d.)	There is a perfect negative linear association between x and y.
11.	**a.)**	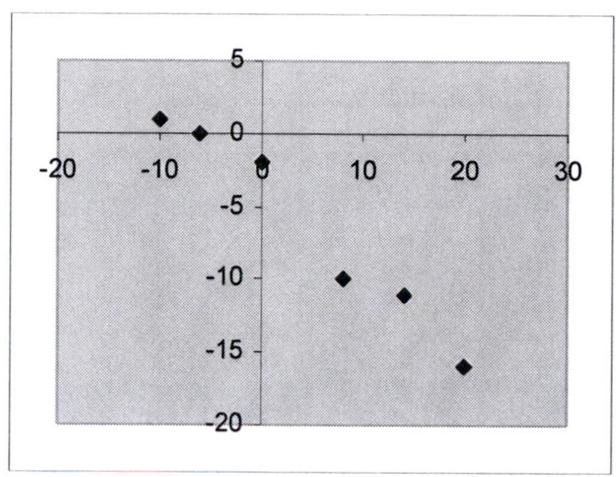
	b.)	The data tend to fall from left to right, so that the correlation coefficient should be negative. Also, the data do not seem to stray too far from a linear curve, so the r value should be reasonably close to –1, but not equal to it. A reasonable guess would be around – 0.90.
	c.)	The equation of the best fit line is approximately $y = -0.5844x - 3.801$ with a correlation coefficient of about $r = 0.9833$.
	d.)	There is a strong (but not perfect) negative linear relationship between x and y.
13.	**a.)**	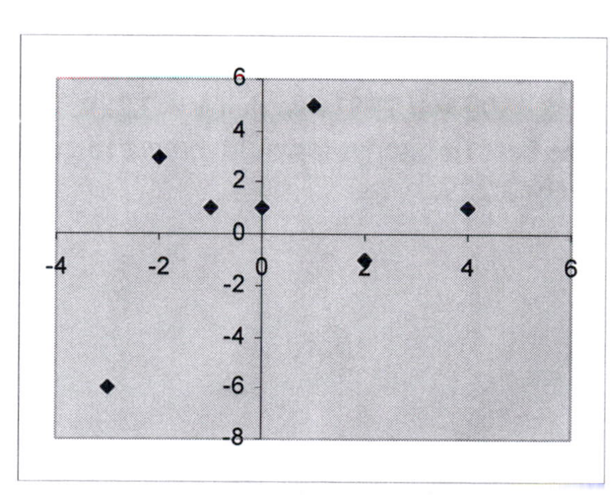

	b.)	The data seem to rise from left to right, but it is difficult to be certain about this relationship since the data stray considerably away from an identifiable line. As such, it is reasonable to guess that r is a rather small value close to 0, say around 0.30.
	c.)	The equation of the best fit line is approximately $y = 0.5x + 0.5$ with a correlation coefficient of about 0.349.
	d.)	There is a very loose (bordering on unidentifiable) positive linear relationship between x and y.

15. **a.)**

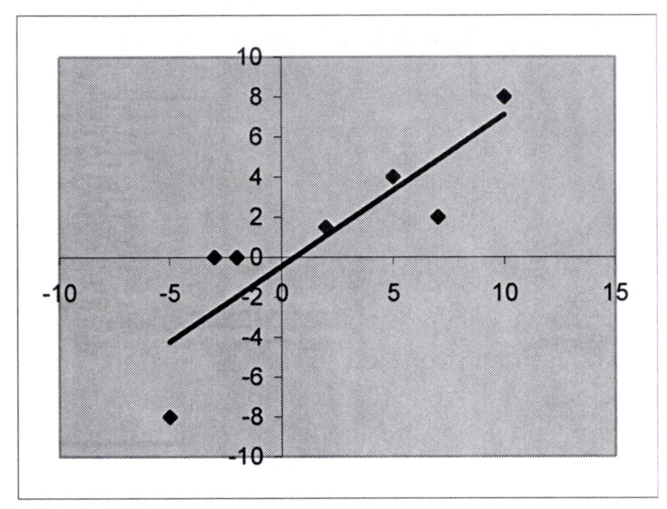

The equation of the best fit line is about $y = 0.7553x - 0.4392$ with a correlation coefficient of about $r = 0.868$.

b.) The values $x = 0$ and $x = -6$ are within the range of the data set, so that using the best fit line for predictive purposes is reasonable. This is not the case for the values $x = 12$ and $x = -15$. The predicted value of y when $x = 0$ is approximately -0.4392, and the predicted value of y when $x = -6$ is -4.971.

c.) Solve the equation $2 = 0.7553x - 0.4392$ for x to obtain:

$$2.4392 = 0.7553x \text{ so that } x = 3.229.$$

So, using the best fit line, you would expect to get a y-value of 2 when x is approximately 3.229.

17. **a.)**

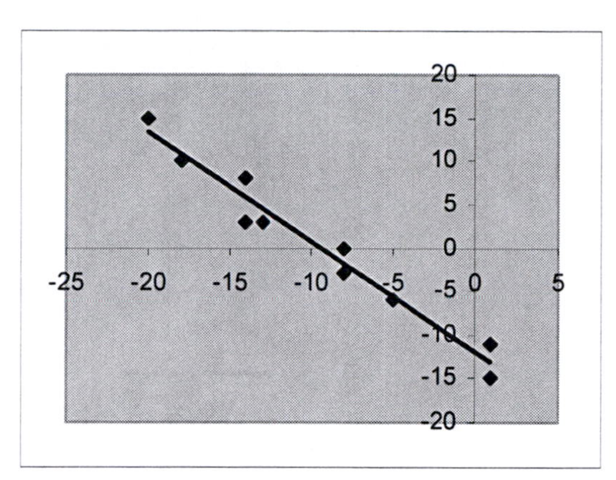

The equation of the best fit line is about $y = -1.2631x - 11.979$ with a correlation coefficient of about $r = -0.980$.

b.) The values $x = -15$, -6, and 0 are within the range of the data set, so that using the best fit line for predictive purposes is reasonable. This is not the case for the value $x = 12$. The predicted value of y when $x = -15$ is approximately 6.9675, the predicted value of y when $x = -6$ is about -4.4004, and the predicted value of y when $x = 0$ is -11.979.

c.) Solve the equation $2 = -1.2631x - 11.979$ for x to obtain:

$$13.979 = -1.2631x \text{ so that } x = -11.067.$$

So, using the best fit line, you would expect to get a y-value of 2 when x is approximately -11.067

19. **a.)** The scatterplot for the entire data set is:

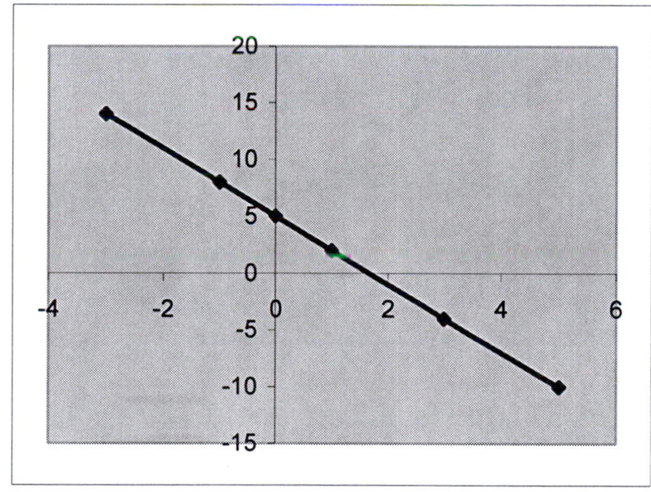

The equation of the best fit line is $y = -3x + 5$ with a correlation coefficient of $r = -1$.

b.) The scatterplot for the data set obtained by removing the starred data point (5, -10) is:

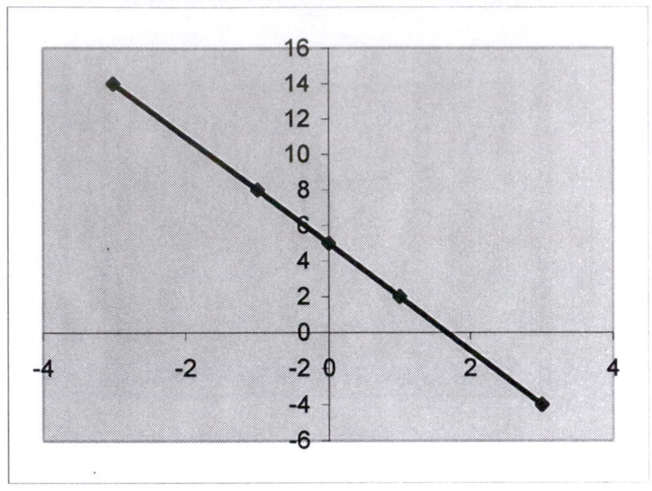

The equation of the best fit line of this modified data set is $y = -3x + 5$ with a correlation coefficient of $r = -1$.

c.) Removing the data point did not result in the slightest change in either the equation of the best fit line or the correlation coefficient. This is reasonable since the relationship between x and y in the original data set is perfectly linear, so that all of the points lie ON the same line. As such, removing one of them has no effect on the line itself.

21. a.) The scatterplot for the entire data set is:

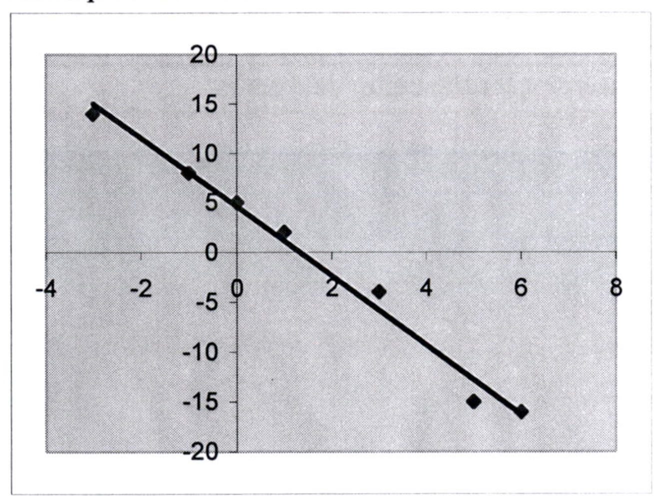

The equation of the best fit line is $y = -3.4776x + 4.6076$ with a correlation coefficient of about $r = -0.993$.

b.) The scatterplot for the data set obtained by removing the starred data points (3, -4) and (6, -16) is:

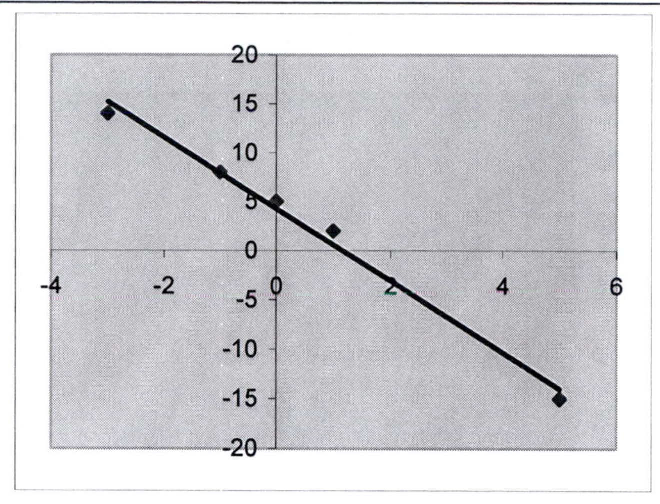

The equation of the best fit line of this modified data set is $y = -3.6534x + 4.2614$ with a correlation coefficient of $r = -0.995$.

c.) Removing the data point did change both the best fit line and the correlation coefficient, but only very slightly.

23. **a.)**

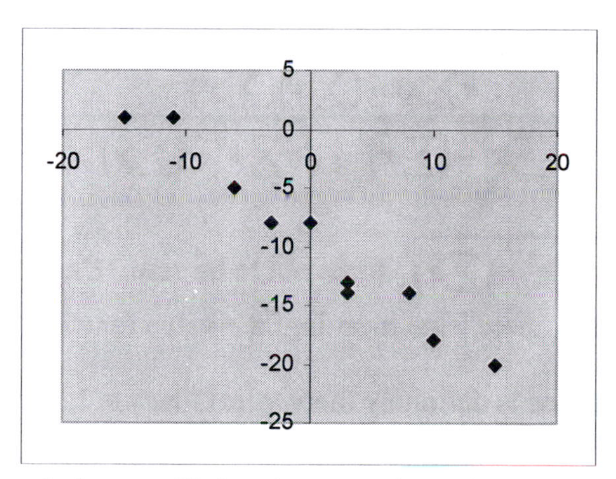

b.) The correlation coefficient is approximately $r = -0.980$. This is identical to the r-value from Problem 17. This makes sense because simply interchanging the x and y-values does not change how the points cluster together in the xy-plane.

c.) The equation of the best fit line for the paired data (y, x) is $x = -0.7607y - 9.4957$.

d.) It is not reasonable to use the best fit line in (c) to find the predicted value of x when $y = 23$ because this value falls outside the range of the given data. However, it is okay to use the best fit line to find the predicted values of x when $y = 2$ or $y = -16$. Indeed, the predicted value of x when $y = 2$ is about -11.0171, and the predicted value of x when $y = -16$ is about 2.6755.

25. First, note that the scatterplot is given by

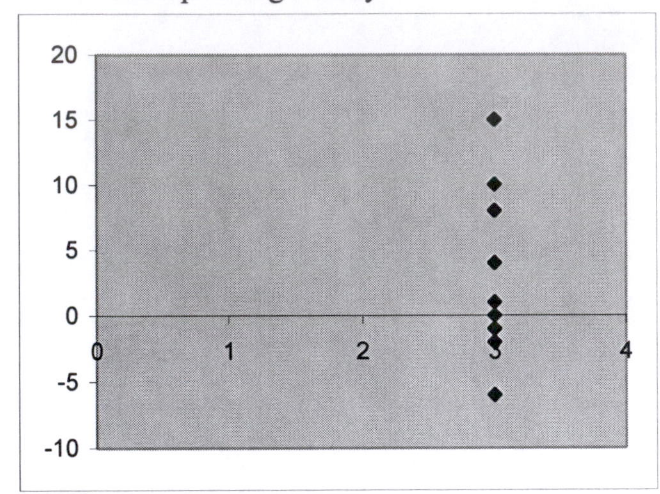

The paired data all lie identically on the vertical line $x = 3$. As such, you might think that the square of the correlation coefficient would be 1 and the best fit line is, in fact, $x = 3$. However, since there is absolutely no *variation* in the x-values for this data set, it turns out that in the formula for the correlation coefficient

$$r = \frac{n\sum xy - \left(\sum x\right)\left(\sum y\right)}{\sqrt{n\sum x^2 - \left(\sum x\right)^2} \cdot \sqrt{n\sum y^2 - \left(\sum y\right)^2}}$$

the quantity $\sqrt{n\sum x^2 - \left(\sum x\right)^2}$ turns out to be zero. (Check this on Excel for this data set!) As such, there is no meaningful r-value for this data set.

Also, the best fit line is definitely the vertical line $x = 3$, but the technology cannot provide it because its slope is undefined.

27. The y-intercept 1.257 is mistakenly interpreted as the slope. The correct interpretation is that for every unit increase in x, the y-value increases by about 5.175.

29. **a)** Here is a table listing all of the correlation coefficients between each of the events and the total points:

Event	r
100m	-0.714
Long Jump	0.768
Shot Put	0.621
High Jump	0.627
400m	-0.704
1500m	-0.289
110m hurdle	-0.653
Discus	0.505
Pole Vault	0.283
Javeline	0.421

Long jump has the strongest relationship to the total points.

b) The correlation coefficient between *long jump* and *total events* is $r =$ 0.768.

c) The equation of the best fit line between the two events in (b) is $y = 838.70x + 1957.77$.

d) Evaluate the equation in (c) at $x = 40$ to get the total points are about 35,506.

31. **a)** The following is a scatterplot illustrating the relationship between ***left thumb length*** and ***total both scores***.

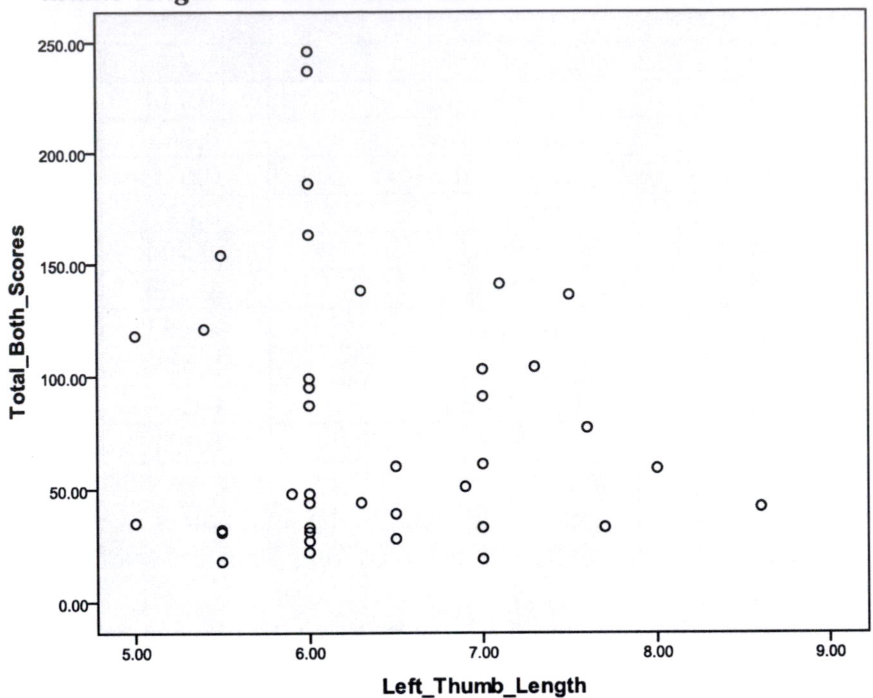

b) r = - 0.105

c) The correlation coefficient (r = -0.105) indicates a weak relationship between left thumb length and total both scores.

d) The equation of the best fit line for these two events is y = -7.62x + 127.73.

e) No, given the weak correlation coefficient between total both scores and left thumb length, one could not use the best fit line to produce accurate predictions.

33. **a.)** A scatterplot illustrating the relationship between ***% residents immunized*** and ***% residents with influenza*** is shown below.

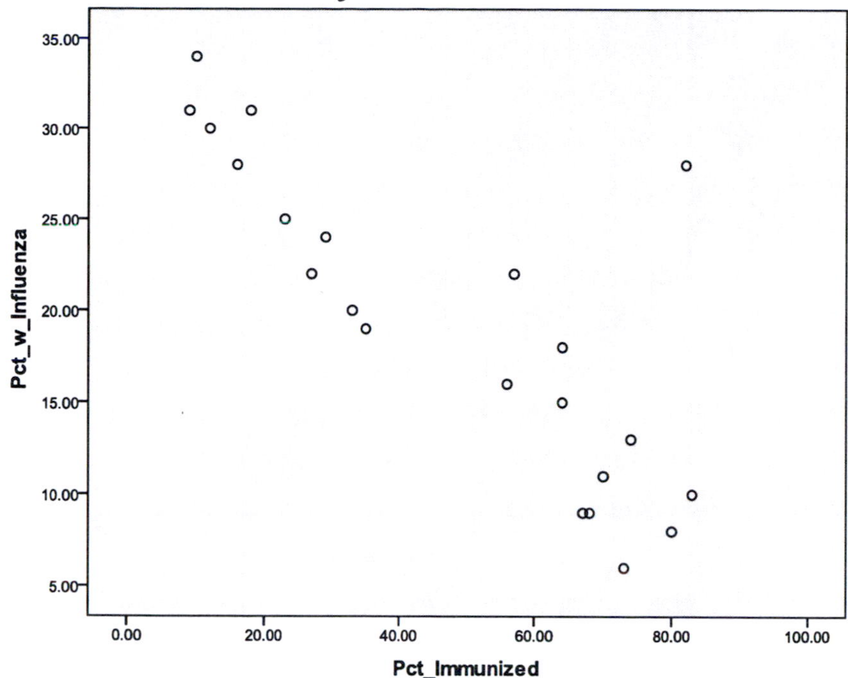

b) The correlation coefficient between ***% residents immunized*** and ***% residents with influenza*** is r = -0.812

c) Based on the correlation coefficient (r = -0.812), we would believe that there is a strong relationship between % residents immunized and % residents with influenza.

d) The equation of the best fit line that describes the relationship between ***% residents immunized*** and ***% residents with influenza*** is y = -0.27x + 32.20.

e) Since r = -0.812 indicates a strong relationship between % residents immunized and % residents with influenza, we can make a reasonably accurate prediction. However it will not be completely accurate.

35. a) A scatterplot illustrating the relationship between ***average wait times*** and ***average rating of enjoyment*** is shown below.

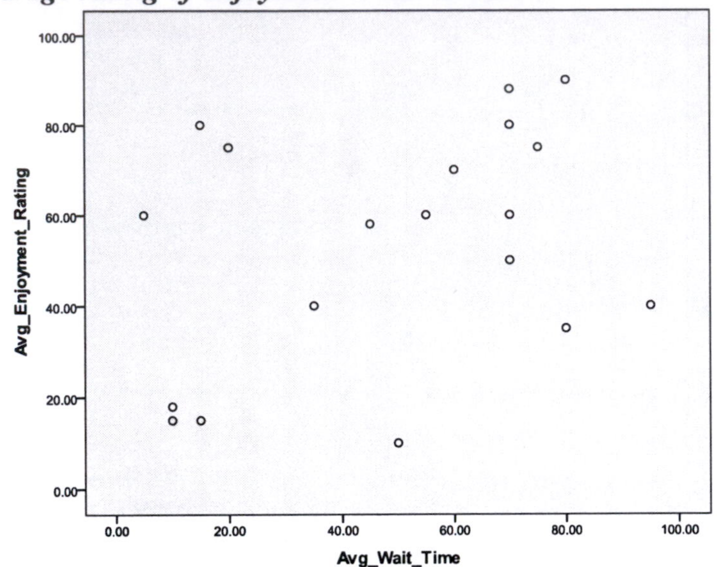

b) The correlation coefficient between ***average wait times*** and ***average rating of enjoyment*** is r = 0.348.

c) The correlation coefficient (r = 0.348) indicates a somewhat weak relationship between average wait times and average rating of enjoyment.

d) The equation of the best fit line that describes the relationship between ***average wait times*** and ***average rating of enjoyment*** is y = 0.31x + 37.83.

e) No, given the somewhat weak correlation coefficient between average wait times and average rating of enjoyment, one could not use the best fit line to produce accurate predictions.

37. **a)** A scatterplot illustrating the relationship between ***average wait times*** and ***average rating of enjoyment*** for ***Park 2*** is shown below.

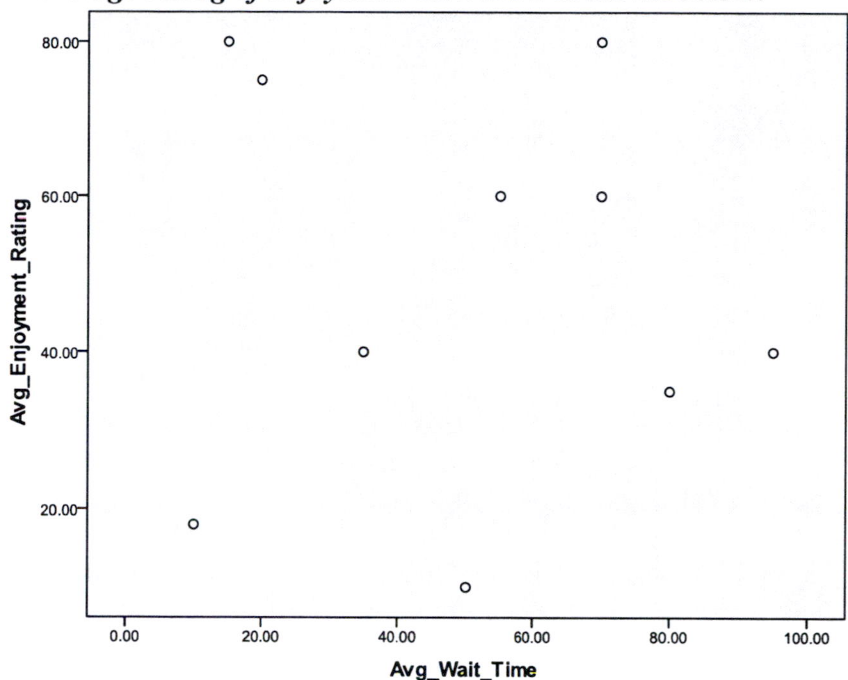

b) The correlation coefficient between ***average wait times*** and ***average rating of enjoyment*** is r = -0.064.

c) The correlation coefficient (r = -0.064) indicates a weak relationship between average wait times and average rating of enjoyment for Park 2.

d) The equation of the best fit line that describes the relationship between ***average wait times*** and ***average rating of enjoyment*** is y = -0.05x + 52.53.

e) No, given the weak correlation coefficient between average wait times and average rating of enjoyment for Park 2, one could not use the best fit line to produce accurate predictions.

39. **a.)** The scatterplot for this data set is given by

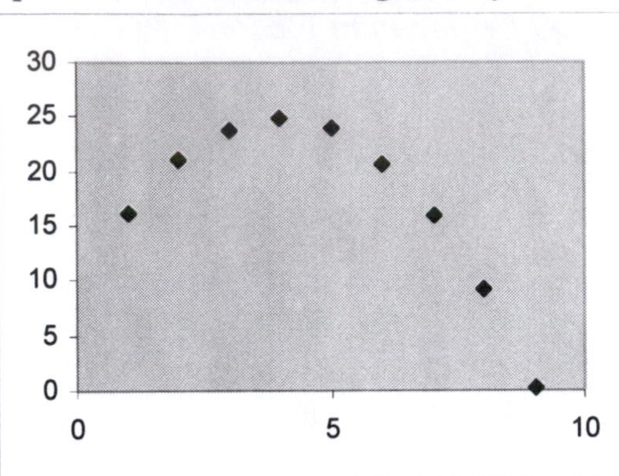

b.) The equation of the best fit *line* is $y = -1.9867x + 27.211$ with a correlation coefficient of $r = -0.671$. This line does not seem to accurately describe the data because some of the points rise as you move left to right, while others fall as you move left to right; a line cannot capture both types of behavior simultaneously. Also, r being negative has no meaning here.

c.) The best fit is provided by QuadReg. The associated equation of the best fit *quadratic* curve, the correlation coefficient, and associated scatterplot are:

```
QuadReg
 y=ax²+bx+c
 a=-.9683982684
 b=7.697316017
 c=9.457142857
 R²=.999878327
```

41. **a.)** The scatterplot for this data set is given by

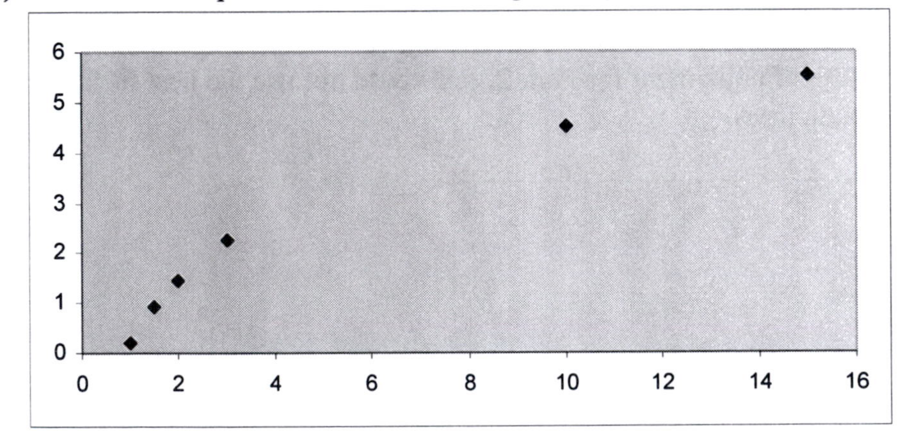

b.) The equation of the best fit *line* is $y = 0.3537x + 0.5593$ with a correlation coefficient of $r = 0.971$. This line seems to provide a very good fit for this data, although not perfect.

c.) The best fit is provided by LnReg. The associated equation of the best fit *logarithmic* curve, the correlation coefficient, and associated scatterplot are:

```
LnReg
 y=a+blnx
 a=.1458167857
 b=1.938869447
 r²=.9988475957
 r=.9994236318
```

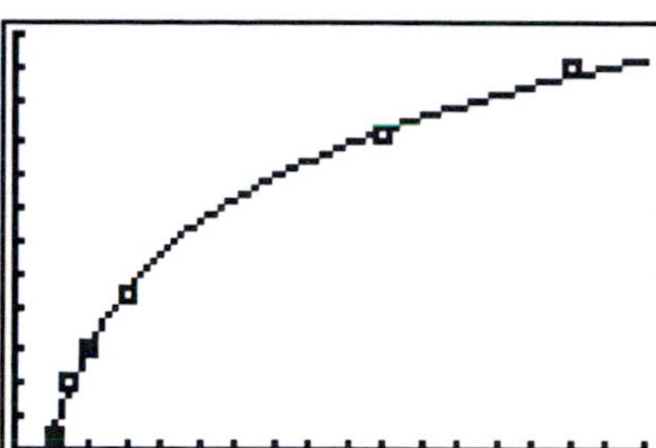

Chapter 2 Review Solutions --

1 & 3.

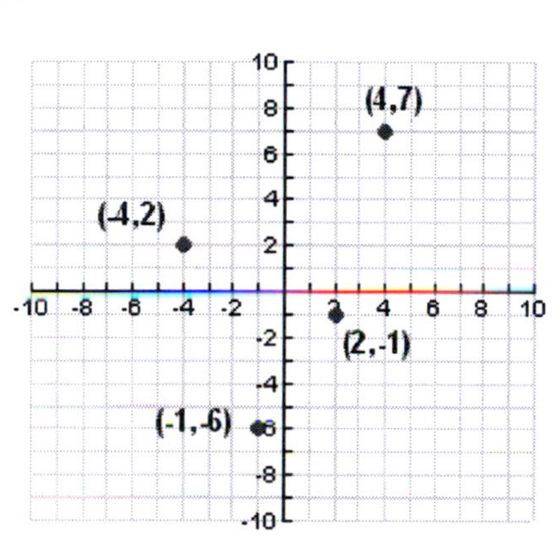

$(-4, 2)$ is in Quadrant II.
$(-1, -6)$ is in Quadrant III.

5. $\sqrt{(-2-4)^2+(0-3)^2} = \sqrt{45} = \boxed{3\sqrt{5}}$

7. $\sqrt{(-4-2)^2+(-6-7)^2} = \boxed{\sqrt{205}}$

9. $\left(\frac{2+3}{2}, \frac{4+8}{2}\right) = \boxed{\left(\frac{5}{2}, 6\right)}$

11. $\left(\frac{2.3+5.4}{2}, \frac{3.4+7.2}{2}\right) = \boxed{(3.85, 5.3)}$

13. The distance equals $\sqrt{(-5-10)^2+(-20-30)^2} = \sqrt{2725} \cong \boxed{52.20}$.

15.

x-intercepts: $x^2+4(0)^2 = 4 \Rightarrow x = \pm 2$

So, $(\pm 2, 0)$.

y-intercepts: $0^2+4y^2 = 4 \Rightarrow y = \pm 1$

So, $(0, \pm 1)$.

17.

x-intercepts: $\sqrt{x^2-9} = 0 \Rightarrow x = \pm 3$

So, $(\pm 3, 0)$.

y-intercepts: $3i = \sqrt{0^2-9} = y$. Since this is not real, there is no y-intercept.

19. <u>x-axis symmetry</u> (Replace y by $-y$):

$$x^2 + (-y)^3 = 4$$
$$x^2 - y^3 = 4$$

No, since not equivalent to the original.

<u>y-axis symmetry</u> (Replace x by $-x$):

$$(-x)^2 + y^3 = 4$$
$$x^2 + y^3 = 4$$

Yes, since equivalent to the original.

<u>symmetry about origin</u> (Replace y by $-y$ and x by $-x$):

$$(-x)^2 + (-y)^3 = 4$$
$$x^2 - y^3 = 4$$

No, since not equivalent to the original.

21. <u>x-axis symmetry</u> (Replace y by $-y$):

$$x(-y) = 4$$
$$-xy = 4$$

No, since not equivalent to the original.

<u>y-axis symmetry</u> (Replace x by $-x$):

$$(-x)y = 4$$
$$-xy = 4$$

No, since not equivalent to the original.

<u>symmetry about origin</u> (Replace y by $-y$ and x by $-x$):

$$(-x)(-y) = 4$$
$$xy = 4$$

Yes, since equivalent to the original.

23.

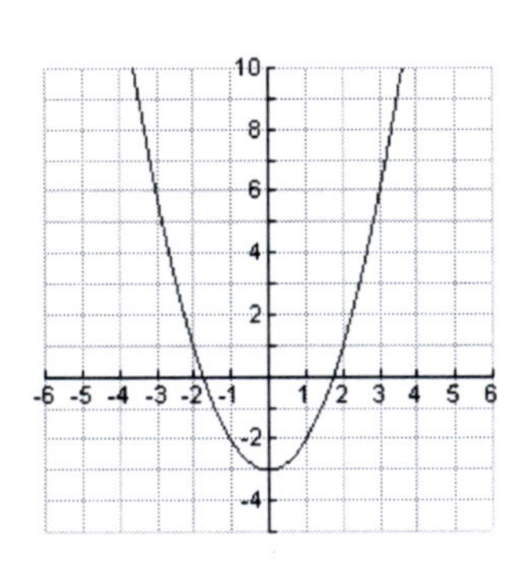

25.

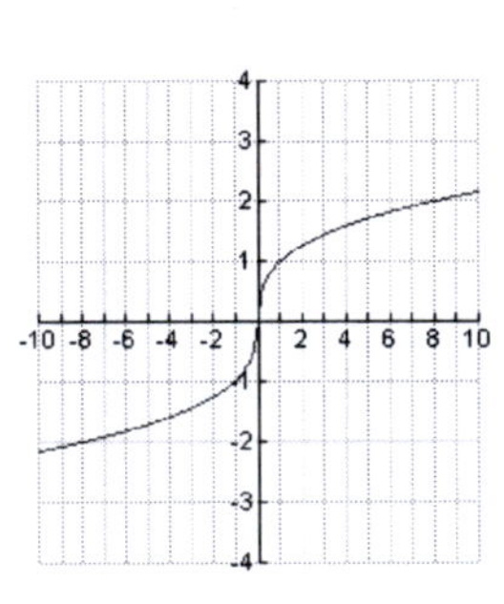

27.

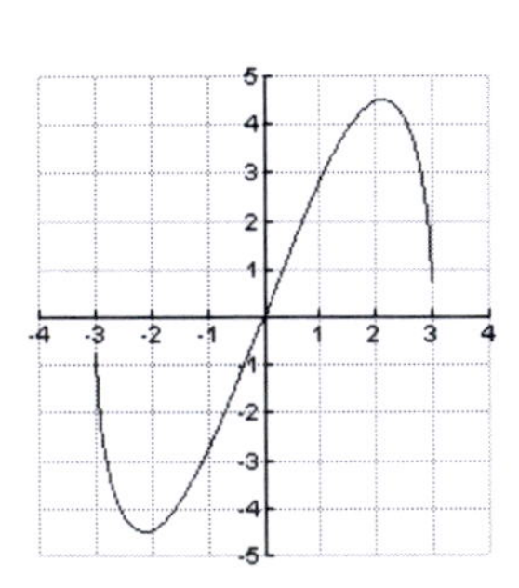

29.

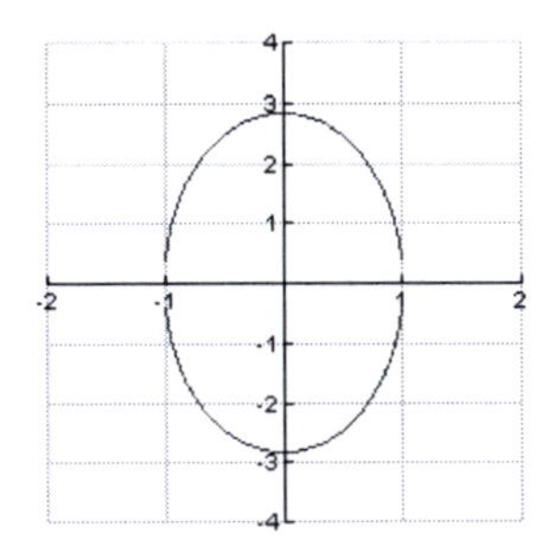

31. $y = -3x + 6 \quad m = -3 \quad y-\text{intercept: } (0,6)$

33. $y = -\frac{3}{2}x - \frac{1}{2} \quad m = -\frac{3}{2} \quad y-\text{intercept: } (0,-\frac{1}{2})$

35.

x-intercept:

$$0 = 4x - 5 \qquad \text{So, } \left(\tfrac{5}{4}, 0\right).$$
$$\tfrac{5}{4} = x$$

y-intercept: $y = 4(0) - 5 = -5$. So, $(0,-5)$.

slope: $m = 4$

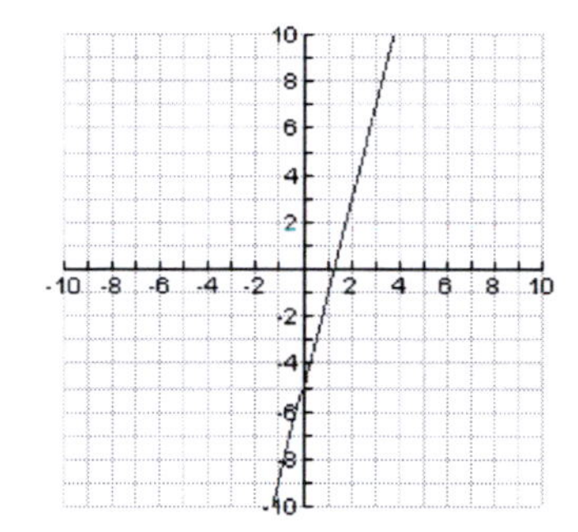

37.

x-intercept:

$x + 0 = 4$ So, $(4,0)$.

y-intercept: $0 + y = 4$. So, $(0,4)$.

slope:

$$x + y = 4 \qquad \text{So, } m = -1.$$
$$y = -x + 4$$

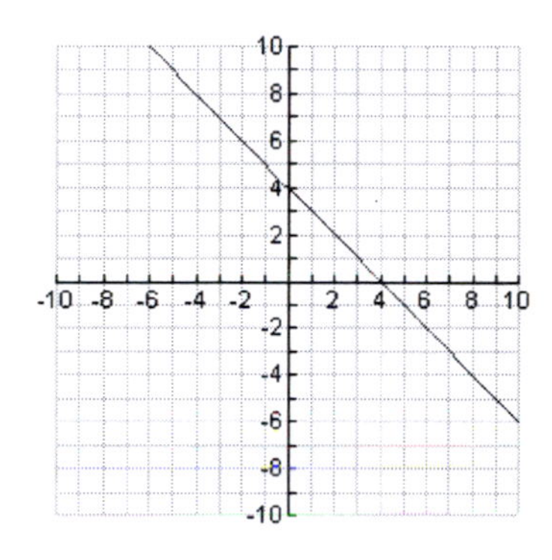

39.

x-intercept: None

y-intercept: $(0,2)$

slope: $m = 0$

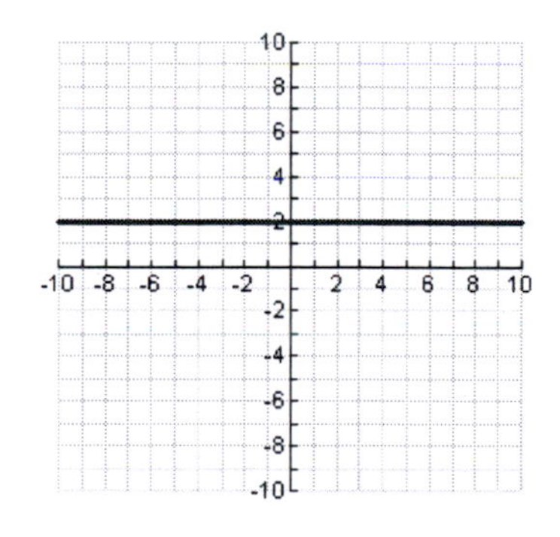

41. $y = 4x - 3$

43. $x = -3$

45. Find b:

$$4 = -2(-3) + b$$
$$-2 = b$$

So, the equation is $\boxed{y = -2x - 2}$.

47. Since $m = 0$, the line is horizontal. So, the equation is $\boxed{y = 6}$.

49.

slope: $m = \frac{-2-3}{-4-2} = \frac{5}{6}$

y-intercept:

Use the point $(-4,-2)$ to find b:

$$-2 = \tfrac{5}{6}(-4) + b$$
$$\tfrac{4}{3} = b$$

So, the equation is $\boxed{y = \tfrac{5}{6}x + \tfrac{4}{3}}$.

51.

slope: $m = \dfrac{\frac{1}{2} - \frac{5}{2}}{-\frac{3}{4} - \left(-\frac{7}{4}\right)} = -2$

y-intercept:

Use the point $\left(-\tfrac{3}{4}, \tfrac{1}{2}\right)$ to find b:

$$\tfrac{1}{2} = -2\left(-\tfrac{3}{4}\right) + b$$
$$-1 = b$$

So, the equation is $\boxed{y = -2x - 1}$.

53.

slope:

Since parallel to $2x - 3y = 6$ (whose slope is $\frac{2}{3}$), $m = \frac{2}{3}$.

y-intercept:

Use the point $(-2,-1)$ to find b:

$$-1 = \tfrac{2}{3}(-2) + b$$
$$\tfrac{1}{3} = b$$

So, the equation is $\boxed{y = \tfrac{2}{3}x + \tfrac{1}{3}}$.

55.

slope:

Since perpendicular to $\frac{2}{3}x - \frac{1}{2}y = 12$ (whose slope is $\frac{4}{3}$), $m = -\frac{3}{4}$.

y-intercept:

Use the point $\left(-\tfrac{3}{4}, \tfrac{5}{2}\right)$ to find b:

$$\tfrac{5}{2} = -\tfrac{3}{4}\left(-\tfrac{3}{4}\right) + b$$
$$\tfrac{31}{16} = b$$

So, the equation is $\boxed{y = -\tfrac{3}{4}x + \tfrac{31}{16}}$.

57. Since $(1020, 1324)$ and $(950, 1240)$ are assumed to be on the line, the slope of the line is $m = \frac{1324-1240}{1020-950} = 1.2$. Use the point $(1020, 1324)$ to find y-intercept b:

$$1324 = 1.2(1020) + b$$
$$100 = b$$

Thus, the equation is $\boxed{y = 1.2x + 100}$, where x is the pretest score and y is the posttest score.

59.

$$(x-(-2))^2 + (y-3)^2 = 6^2$$
$$(x+2)^2 + (y-3)^2 = 36$$

61. $\left(x - \tfrac{3}{4}\right)^2 + \left(y - \tfrac{5}{2}\right)^2 = \tfrac{4}{25}$

63. The center is $(-2,-3)$ and the radius is $\sqrt{81} = 9$.

65. The center is $\left(-\tfrac{3}{4}, \tfrac{1}{2}\right)$ and the radius is $\sqrt{\tfrac{16}{36}} = \tfrac{4}{6} = \tfrac{2}{3}$.

67. Completing the square gives us:

$$x^2+y^2-4x+2y+11=0$$
$$x^2-4x \quad +y^2+2y=-11$$
$$\left(x^2-4x+4\right)+\left(y^2+2y+1\right)=-11+4+1$$
$$\left(x-2\right)^2+\left(y+1\right)^2=-6$$

Since the left-side is always non-negative, there is no graph. Hence, this is not a circle and so there is no center or radius in this case.

69. Completing the square gives us:

$$9x^2+9y^2-6x+12y-76=0$$
$$x^2-\tfrac{2}{3}x \quad +y^2+\tfrac{4}{3}y=\tfrac{76}{9}$$
$$\left(x^2-\tfrac{2}{3}x+\tfrac{1}{9}\right)+\left(y^2+\tfrac{4}{3}y+\tfrac{4}{9}\right)=\tfrac{76}{9}+\tfrac{1}{9}+\tfrac{4}{9}$$
$$\left(x-\tfrac{1}{3}\right)^2+\left(y+\tfrac{2}{3}\right)^2=9$$
$$\left(x-\tfrac{1}{3}\right)^2+\left(y-(-\tfrac{2}{3})\right)^2=9$$

So, center is $\left(\tfrac{1}{3},-\tfrac{2}{3}\right)$ and radius is 3.

71. The center is $(2,7)$. The radius is the distance from $(2,7)$ to $(3,6)$, which is given by $\sqrt{(2-3)^2+(7-6)^2}=\sqrt{2}$. So, the equation is $\boxed{(x-2)^2+(y-7)^2=2}$.

73. The triangle is depicted as follows:

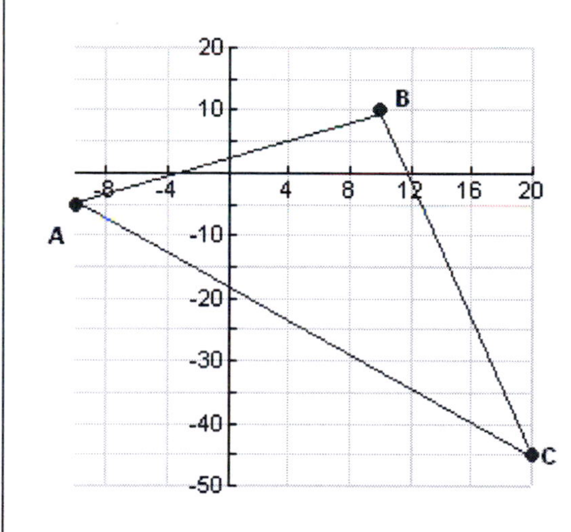

Observe that

$$m_{AB}=\frac{10-(-5)}{10-(-10)}=\frac{3}{4}$$
$$m_{BC}=\frac{10-(-45)}{10-20}=-\frac{11}{2}$$
$$m_{AC}=\frac{-5-(-45)}{-10-20}=-\frac{4}{3}$$

Since $m_{AB}\cdot m_{AC}=-1$, we know that AB is perpendicular to AC. Thus, the triangle is $\boxed{\text{right}}$.

Next, observe that

$$d(A,B)=\sqrt{(10-(-10))^2+(10-(-5))^2}=25$$
$$d(A,C)=\sqrt{(20-(-10))^2+(-45-(-5))^2}=50$$
$$d(B,C)=\sqrt{(10-(-45))^2+(10-20)^2}\approx 55.9$$

So, it is not isosceles.

75.

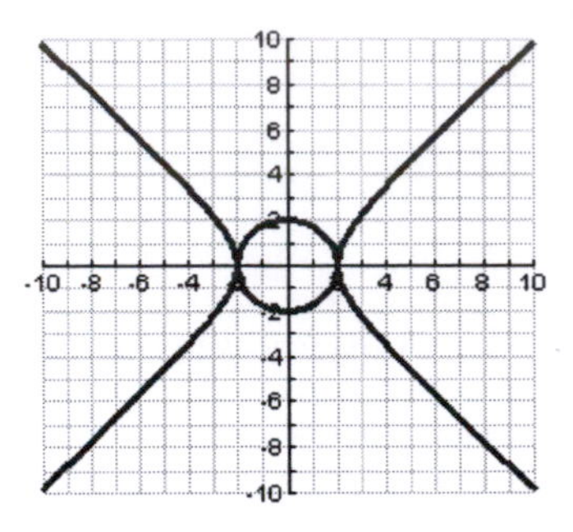

Symmetric with respect to x-axis, y-axis, and origin

77. Slope of $y_1 = 0.875x + 1.5$ is 0.875.

Slope of $y_2 = -\frac{8}{7}x - \frac{9}{14}$ is $-\frac{8}{7}$.

Since $(0.875)(-\frac{8}{7}) = -1$, the lines are **perpendicular**.

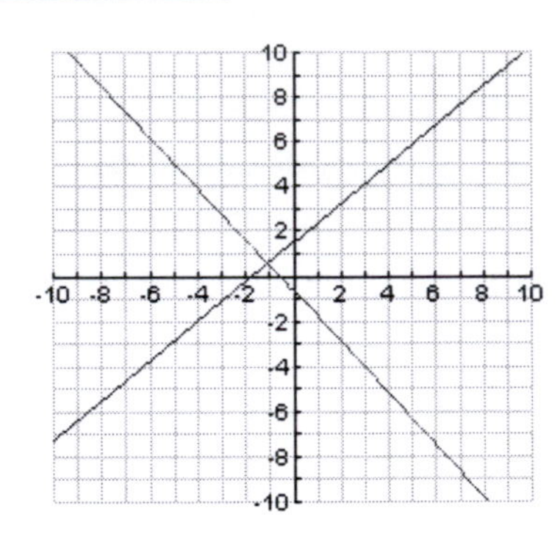

79. Completing the square yields

$$\left(9x^2-6x\quad\right)+\left(9y^2+12y\quad\right)=76$$
$$9\left(x^2-\tfrac{2}{3}x\quad\right)+9\left(y^2+\tfrac{4}{3}y\quad\right)=76$$
$$9\left(x^2-\tfrac{2}{3}x+\tfrac{1}{9}\right)+\left(y^2+\tfrac{4}{3}y+\tfrac{4}{9}\right)=76+1+4=81$$
$$\left(x-\tfrac{1}{3}\right)^2+\left(y+\tfrac{2}{3}\right)^2=9 \quad \textbf{(1)}$$

Solving **(1)** for y yields:

$$\left(x-\tfrac{1}{3}\right)^2+\left(y+\tfrac{2}{3}\right)^2=9$$
$$\left(y+\tfrac{2}{3}\right)^2=9-\left(x-\tfrac{1}{3}\right)^2$$
$$y+\tfrac{2}{3}=\pm\sqrt{9-\left(x-\tfrac{1}{3}\right)^2}$$
$$y=-\tfrac{2}{3}\pm\sqrt{9-\left(x-\tfrac{1}{3}\right)^2}$$

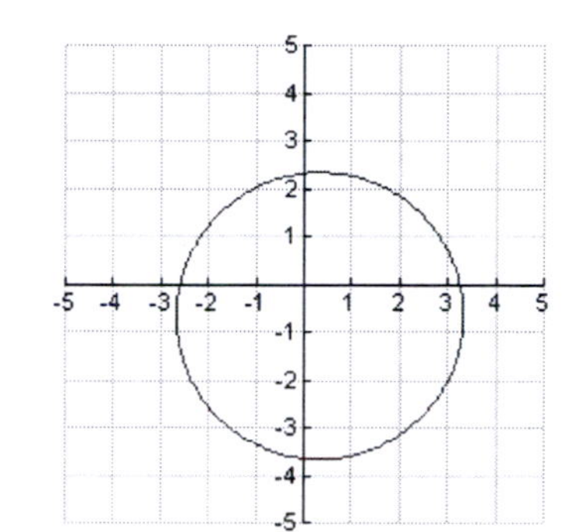

The graph agrees with the graph in Exercise 69.

Chapter 2 Practice Test Solutions --

1. $d=\sqrt{(-7-2)^2+(-3-(-2))^2}=\boxed{\sqrt{82}}$

3. $d=\sqrt{(-2-3)^2+(4-6)^2}=\boxed{\sqrt{29}}$

$$M=\left(\frac{-2+3}{2},\frac{4+6}{2}\right)=\boxed{\left(\tfrac{1}{2},5\right)}$$

5. Solve for y:

$$5=\sqrt{(3-6)^2+(y-5)^2}$$
$$25=9+(y-5)^2$$
$$25=9+y^2-10y+25$$
$$0=y^2-10y+9$$
$$0=(y-9)(y-1)$$
$$\boxed{1,9=y}$$

7. <u>x-axis symmetry</u> (Replace y by $-y$):

$$x-(-y)^2=5$$
$$x-y^2=5$$

Yes, since equivalent to the original.

<u>y-axis symmetry</u> (Replace x by $-x$):

$$-x-y^2=5$$

No, since not equivalent to the original.

symmetry about origin (Replace y by $-y$ and x by $-x$):

$$-x-(-y)^2=5$$

$$-x-y^2=5$$

No, since not equivalent to the original.

9.

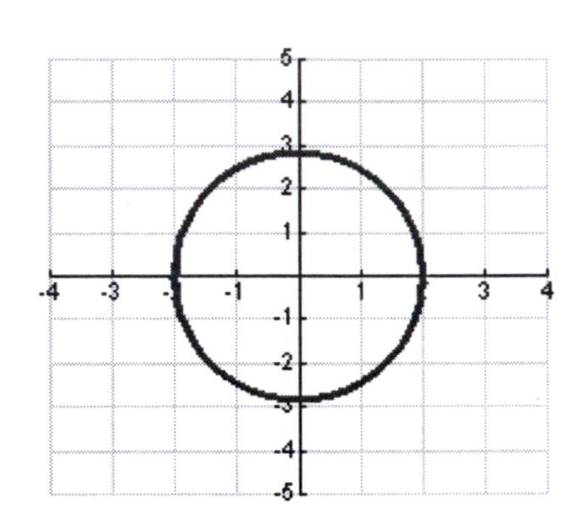

11.

x-intercept: $x-3(0)=6 \Rightarrow x=6$. So, $(6,0)$.

y-intercept: $0-3y=6 \Rightarrow y=-2$.

So, $(0,-2)$.

13.

$\frac{2}{3}x-\frac{1}{4}y=2 \Rightarrow -\frac{1}{4}y=2-\frac{2}{3}x \Rightarrow \boxed{y=\frac{8}{3}x-8}$

15.

slope: $m=\dfrac{9-2}{4-(-3)}=1$

y-intercept: Use the point $(-3,2)$ to find b:

$$2=1(-3)+b$$

$$5=b$$

So, the equation is $\boxed{y=x+5}$.

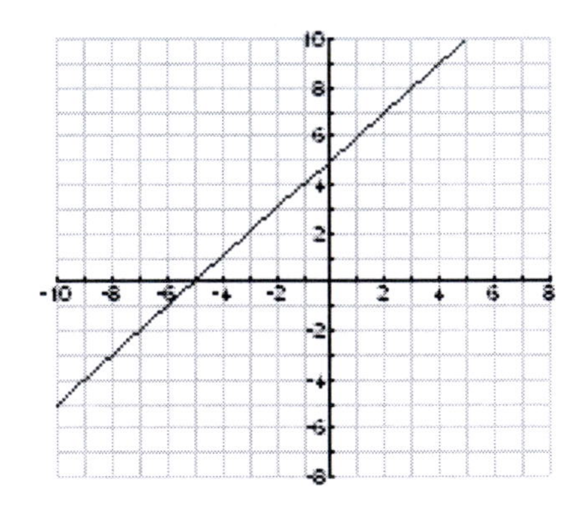

17.

slope: Since perpendicular to $2x-4y=5$ (whose slope is $\frac{1}{2}$), $m=-2$.

y-intercept:

Use the point $(1,1)$ to find b:

$$1=-2(1)+b$$

$$3=b$$

So, the equation is $\boxed{y=-2x+3}$.

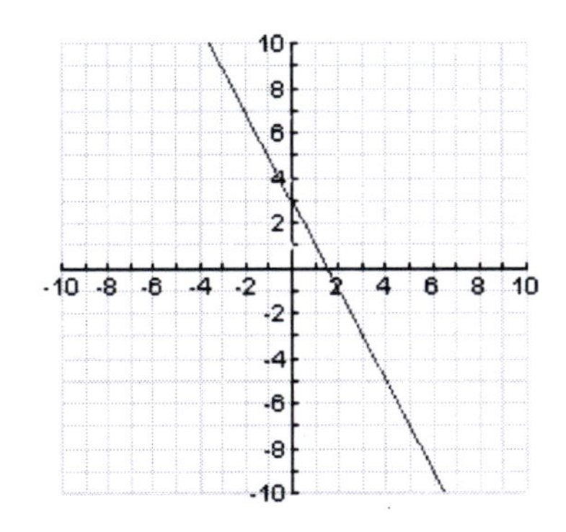

19.

slope: $m = \dfrac{4-(-2)}{1-(-2)} = 2$

y-intercept: Use the point $(1,4)$ to find b:

$$4 = 2(1) + b$$
$$2 = b$$

So, the equation is $\boxed{y = 2x + 2}$.

21.

$$(x-6)^2 + (y-(-7))^2 = 8^2$$
$$(x-6)^2 + (y+7)^2 = 64$$

23. The center is $(4,9)$. The radius is the distance from $(4,9)$ to $(2,5)$, which is given by $\sqrt{(4-2)^2 + (9-5)^2} = \sqrt{20}$. So, the equation is $\boxed{(x-4)^2 + (y-9)^2 = 20}$.

25. The triangle is depicted as follows:

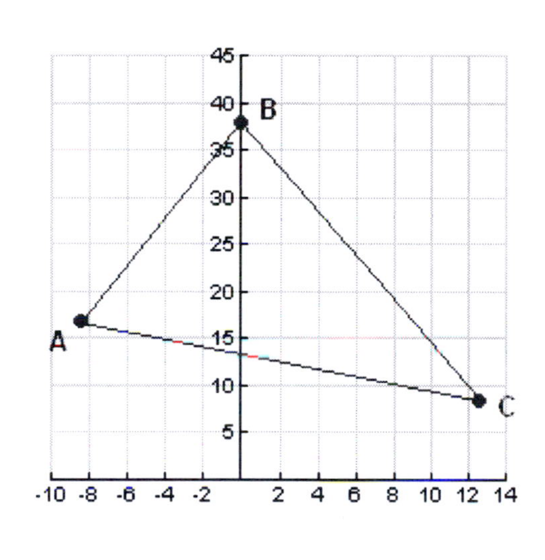

Observe that

$$m_{AB} = \frac{37.8-16.8}{0-(-8.4)} = 2.5$$
$$m_{BC} = \frac{37.8-8.4}{0-12.6} = -2.\overline{3}$$
$$m_{AC} = \frac{16.8-8.4}{-8.4-12.6} = -0.4$$

Since $m_{AB} \cdot m_{AC} = -1$, AB is perpendicular to AC. So, the triangle is $\boxed{\text{right}}$.

Next, observe that

$$d(A,B) = \sqrt{(37.8-16.8)^2 + (0+8.4)^2} = \sqrt{511.56}$$
$$d(B,C) = \sqrt{(37.8-8.4)^2 + (0-12.6)^2} = \sqrt{1023.12}$$
$$d(A,C) = \sqrt{(16.8-8.4)^2 + (-8.4-12.6)^2} = \sqrt{511.56}$$

So, it is also $\boxed{\text{isosceles}}$.

Chapter 2 Cumulative Review

1. $\dfrac{7-2}{7+3} = \dfrac{5}{10} = \boxed{\dfrac{1}{2}}$

3.

$$(x-4)^2(x+4)^2 = [(x-4)(x+4)]^2$$
$$= [x^2 - 16]^2$$
$$= \boxed{x^4 - 32x^2 + 256}$$

5. $\dfrac{\frac{1}{x} - \frac{1}{5}}{\frac{1}{x} + \frac{1}{5}} = \dfrac{\frac{5-x}{5x}}{\frac{5+x}{5x}} = \boxed{\dfrac{5-x}{5+x}}$

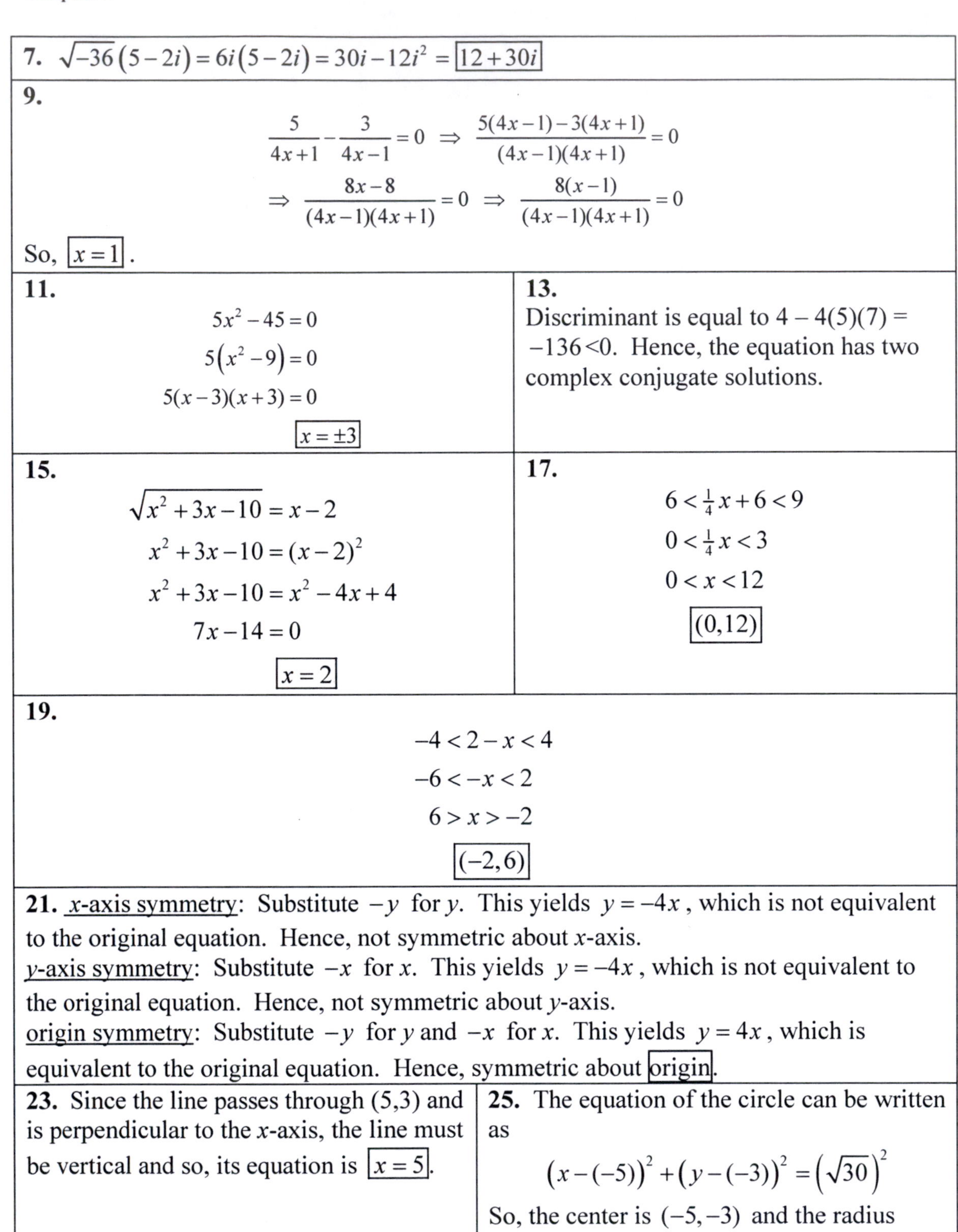

7. $\sqrt{-36}\,(5-2i)=6i(5-2i)=30i-12i^2=\boxed{12+30i}$

9.

$$\frac{5}{4x+1}-\frac{3}{4x-1}=0 \;\Rightarrow\; \frac{5(4x-1)-3(4x+1)}{(4x-1)(4x+1)}=0$$

$$\Rightarrow\; \frac{8x-8}{(4x-1)(4x+1)}=0 \;\Rightarrow\; \frac{8(x-1)}{(4x-1)(4x+1)}=0$$

So, $\boxed{x=1}$.

11.

$$5x^2-45=0$$
$$5\left(x^2-9\right)=0$$
$$5(x-3)(x+3)=0$$
$$\boxed{x=\pm 3}$$

13. Discriminant is equal to 4 – 4(5)(7) = $-136<0$. Hence, the equation has two complex conjugate solutions.

15.

$$\sqrt{x^2+3x-10}=x-2$$
$$x^2+3x-10=(x-2)^2$$
$$x^2+3x-10=x^2-4x+4$$
$$7x-14=0$$
$$\boxed{x=2}$$

17.

$$6<\tfrac{1}{4}x+6<9$$
$$0<\tfrac{1}{4}x<3$$
$$0<x<12$$
$$\boxed{(0,12)}$$

19.

$$-4<2-x<4$$
$$-6<-x<2$$
$$6>x>-2$$
$$\boxed{(-2,6)}$$

21. <u>*x*-axis symmetry</u>: Substitute $-y$ for *y*. This yields $y=-4x$, which is not equivalent to the original equation. Hence, not symmetric about *x*-axis.
<u>*y*-axis symmetry</u>: Substitute $-x$ for *x*. This yields $y=-4x$, which is not equivalent to the original equation. Hence, not symmetric about *y*-axis.
<u>origin symmetry</u>: Substitute $-y$ for *y* and $-x$ for *x*. This yields $y=4x$, which is equivalent to the original equation. Hence, symmetric about $\boxed{\text{origin}}$.

23. Since the line passes through (5,3) and is perpendicular to the *x*-axis, the line must be vertical and so, its equation is $\boxed{x=5}$.

25. The equation of the circle can be written as

$$\left(x-(-5)\right)^2+\left(y-(-3)\right)^2=\left(\sqrt{30}\right)^2$$

So, the center is $(-5,-3)$ and the radius is $\sqrt{30}$.

27.

Slope of $y_1 = 0.32x + 1.5$ is 0.32.

Slope of $y_2 = -\frac{5}{16}x + \frac{1}{14}$ is $-\frac{5}{16}$.

Since $(0.32)(-\frac{5}{16}) = -0.1$, the lines are $\boxed{\text{neither}}$ parallel nor perpendicular.

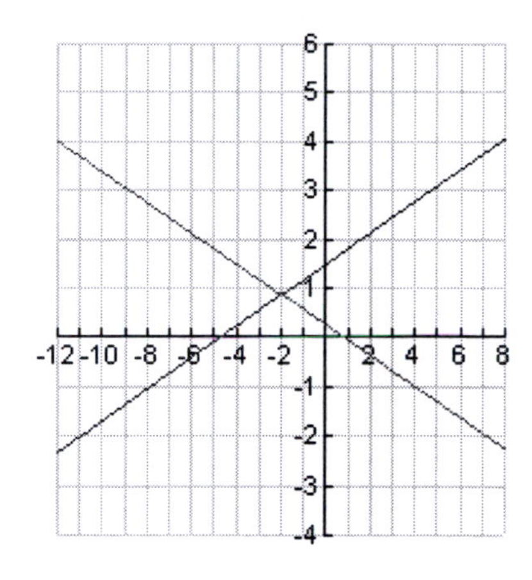

Chapter 2

CHAPTER 3

Section 3.1 Solutions --

1. Function
3. Not a function – 1pm and 4pm both map to two elements in the range.
5. Function
7. Not a function – 0 maps to both -3 and 3.
9. Not a function – 4 maps to both -2 and 2, and 9 maps to both -3 and 3.
11. Function
13. Not a function – Since $\left(1,-2\sqrt{2}\right)$ and $\left(1,2\sqrt{2}\right)$ are both on the graph, it does not pass vertical line test.
15. Not a function – Since $(1,-1)$ and $(1,1)$ are both on the graph, it does not pass the vertical line test.
17. Function
19. Not a function – Since $(0,5)$ and $(0,-5)$ are both on the graph, it does not pass the vertical line test.
21. Function
23. Not a function – Since $(0,-1)$ and $(0,-3)$ are both on the graph, it does not pass the vertical line test.

25. **a)** 5 **b)** 1 **c)** -3	**27.** **a)** 3 **b)** 2 **c)** 5
29. **a)** -5 **b)** -5 **c)** -5	**31.** **a)** 2 **b)** -8 **c)** -5
33. 1	**35.** 1 and -3
37. For all x in the interval $[-4,4]$	**39.** 6
41. $f(-2)=2(-2)-3=\boxed{-7}$	**43.** $g(1)=5+1=\boxed{6}$
45. Using #41 and #43, we see that $f(-2)+g(1)=-7+6=\boxed{-1}$.	**47.** Using #41 and #43, we see that $3f(-2)-2g(1)=3(-7)-2(6)=\boxed{-33}$.
49. Using #41 and #43, we see that $\dfrac{f(-2)}{g(1)}=\boxed{-\dfrac{7}{6}}$.	**51.** $\dfrac{f(0)-f(-2)}{g(1)}=\dfrac{(2(0)-3)-(-7)}{6}=\dfrac{-3+7}{6}=\boxed{\dfrac{2}{3}}$

53.

$$\begin{aligned} f(x+1)-f(x-1) &= [2(x+1)-3]-[2(x-1)-3] \\ &= [2x+2-3]-[2x-2-3] \\ &= [2x-1]-[2x-5] \\ &= 2x-1-2x+5 \\ &= \boxed{4} \end{aligned}$$

55.

$$\begin{aligned} g(x+a)-f(x+a) &= [5+(x+a)]-[2(x+a)-3] \\ &= [5+x+a]-[2x+2a-3] \\ &= 5+x+a-2x-2a+3 \\ &= \boxed{8-x-a} \end{aligned}$$

57.

$$\begin{aligned} \frac{f(x+h)-f(x)}{h} &= \frac{[2(x+h)-3]-[2x-3]}{h} \\ &= \frac{2x+2h-3-2x+3}{h} \\ &= \frac{2h}{h} = \boxed{2} \end{aligned}$$

59.

$$\begin{aligned} \frac{g(t+h)-g(t)}{h} &= \frac{[5+(t+h)]-[5+t]}{h} \\ &= \frac{5+t+h-5-t}{h} = \frac{h}{h} = \boxed{1} \end{aligned}$$

61. It follows directly from the computation in #57 with $x=-2$ that this equals $\boxed{2}$.

63. It follows directly from the computation in #59 with $t=1$ that this equals $\boxed{1}$.

65. The domain is $\mathbb{R}$. This is written using interval notation as $\boxed{(-\infty,\infty)}$.

67. The domain is $\mathbb{R}$. This is written using interval notation as $\boxed{(-\infty,\infty)}$.

69. The domain is the set of all real numbers *x* such that $x-5 \neq 0$, that is $x \neq 5$. This is written using interval notation as $\boxed{(-\infty,5)\cup(5,\infty)}$.

71. The domain is the set of all real numbers *x* such that

$$x^2-4=(x-2)(x+2)\neq 0,$$

that is $x \neq -2,2$. This is written using interval notation as $\boxed{(-\infty,-2)\cup(-2,2)\cup(2,\infty)}$.

73. Since $x^2 + 1 \neq 0$, for every real number x, the domain is $\mathbb{R}$. This is written using interval notation as $\boxed{(-\infty, \infty)}$.
75. The domain is the set of all real numbers x such that $7 - x \geq 0$, that is $7 \geq x$. This is written using interval notation as $\boxed{(-\infty, 7]}$.
77. The domain is the set of all real numbers x such that $2x + 5 \geq 0$, that is $x \geq -\frac{5}{2}$. This is written using interval notation as $\boxed{\left[-\frac{5}{2}, \infty\right)}$.
79. The domain is the set of all real numbers t such that $t^2 - 4 \geq 0$, which is equivalent to $(t-2)(t+2) \geq 0$. CPs are -2, 2 + − + (sign chart: −2, 2) This is written using interval notation as $\boxed{(-\infty, -2] \cup [2, \infty)}$.
81. The domain is the set of all real numbers x such that $x - 3 > 0$, that is $x > 3$. This is written using interval notation as $\boxed{(3, \infty)}$.
83. Since $1 - 2x$ can be any real number, there is no restriction on x, so that the domain is $\boxed{(-\infty, \infty)}$.
85. The only restriction is that $x + 4 \neq 0$, so that $x \neq -4$. So, the domain is $\boxed{(-\infty, -4) \cup (-4, \infty)}$.
87. The domain is the set of all real numbers x such that $3 - 2x > 0$, that is $\frac{3}{2} > x$. This is written using interval notation as $\boxed{\left(-\infty, \frac{3}{2}\right)}$.
89. The domain is the set of all real numbers t such that $t^2 - t - 6 > 0$, which is equivalent to $(t-3)(t+2) > 0$. CPs are -2, 3 + − + (sign chart: −2, 3) This is written using interval notation as $\boxed{(-\infty, -2) \cup (3, \infty)}$.
91. The domain is the set of all real numbers x such that $x^2 - 16 \geq 0$, which is equivalent to $(x-4)(x+4) \geq 0$. CPs are -4, 4 + − + (sign chart: −4, 4) This is written using interval notation as $\boxed{(-\infty, -4] \cup [4, \infty)}$.

93. The function can be written as $r(x) = \frac{x^2}{\sqrt{3-2x}}$. So, the domain is the set of real numbers x such that $3-2x>0$, that is $\frac{3}{2}>x$. This is written using interval notation as $\boxed{\left(-\infty,\frac{3}{2}\right)}$.

95. The domain of any linear function is $\boxed{(-\infty,\infty)}$.

97. Solve $x^2-2x-5=3$.

$$x^2-2x-8=0$$
$$(x-4)(x+2)=0$$
$$\boxed{x=-2,4}$$

99.

$$2x(x-5)^3-12(x-5)^2=0$$
$$2(x-5)^2\left[x(x-5)-6\right]=0$$
$$2(x-5)^2(x^2-5x-6)=0$$
$$2(x-5)^2(x-6)(x+1)=0$$
$$\boxed{x=-1,5,6}$$

101. Let $x =$ number of people and $y =$ cost. Then, $y=45x,\ x>75$. The domain is $(75,\infty)$.

103. Assume: 6am corresponds to $x=6$
noon corresponds to $x=12$

Then, the temperature at 6am is:

$$T(6)=-0.7(6)^2+16.8(6)-10.8$$
$$=64.8^\circ F$$

The temperature at noon is:

$$T(12)=-0.7(12)^2+16.8(12)-10.8$$
$$=90^\circ F$$

105.

$$P(10)=10+\sqrt{400{,}000-100(10)}$$
$$\cong \$641.66$$
$$P(100)=10+\sqrt{400{,}000-100(100)}$$
$$\cong \$634.50$$

107. Start with a square piece of cardboard with dimensions 10 in. × 10 in.. Then, cut out 4 square corners with dimensions x in. × x in., as shown in the diagram:

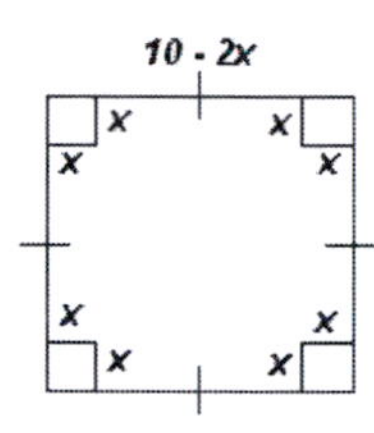

Upon bending all four corners up, a box of height x is formed. Notice that all four sides of the base of the resulting box have length $10-2x$. The volume of the box, $V(x)$, is given by:

$$V(x) = (\text{Length})\cdot(\text{Width})\cdot(\text{Height}) = (10-2x)(10-2x)(x) = x(10-2x)^2$$

The domain is $(0,5)$. (For any other values of x, one cannot form a box.)

109. $E(4)\approx 84$ Yen, $E(7)\approx 84$ Yen, $E(8)\approx 83$ Yen

111. $P(14) = -\frac{1}{4}(14^2)+7(14)+180 = 229$ people

113. a. Let x = Length of the window (in inches)
Then, width = $x-\frac{27}{8}$ (inches)
So, area of the window is $A(x) = x\left(x-\frac{27}{8}\right) = x^2-\frac{27}{8}x$.
b. $A(4.5) = 4.5\left(4.5-\frac{27}{8}\right) = 5.0625$. This means that the area of the window is approximately 5 square inches.
c. $A(8.5) = 8.5\left(8.5-\frac{27}{8}\right) = 43.5625$. This would be the area of the window in the envelope. However, this is not possible for an envelope with dimensions 4 inches by 8 inches because the window would be larger than the entire envelope.

115. Yes. For because every input (year), there corresponds to exactly one output (federal funds rate).

117. (1989, 4000), (1993, 6000), (1997, 6000), (2001, 8000), (2005,11000)

119. a) $F(50)$ = number of tons of carbon emitted by natural gas in 1950 = 0
b) $g(50)$ = number of tons of coal emitted by natural gas in 1950 = 1000
c) $H(50) = 2000$

121. Should apply the vertical line test to determine if the relationship describes a function. The given relationship IS a function in this case.

123. $f(x+1)\neq f(x)+f(1)$, in general. You cannot distribute the function f through the input at which you are evaluating it.

125. $G(-1+h)\neq G(-1)+G(h)$, in general.

127. False. Consider the function $f(x)=\sqrt{9-x^2}$ on its domain $[-3,3]$. The vertical line test $x=4$ doesn't intersect the graph, but it still defines a function.

129. False. This simply means that a particular horizontal line intersects the graph twice. Consider $f(x)=x^2$ on $\mathbb{R}$. In this case, $f(a)=f(-a)$, for all real numbers a.

131.

$$\begin{aligned} f(1) &= A(1)^2-3(1)=-1 \\ A-3 &= -1 \\ A &= 2 \end{aligned}$$

133. $F(-2)=\dfrac{C-(-2)}{D-(-2)}=\dfrac{C+2}{D+2}$ is undefined only if $\boxed{D=-2}$. So,

$F(-1)=\dfrac{C-(-1)}{D-(-1)}=\dfrac{C+1}{D+1}=\dfrac{C+1}{-2+1}=-(C+1)=4$ implies that $\boxed{C=-5}$.

135. The domain is the set of all real numbers x such that $x^2-a^2=(x-a)(x+a)\neq 0$, which is equivalent to $x\neq \pm a$. So, the domain is $\boxed{(-\infty,-a)\cup(-a,a)\cup(a,\infty)}$.

137.

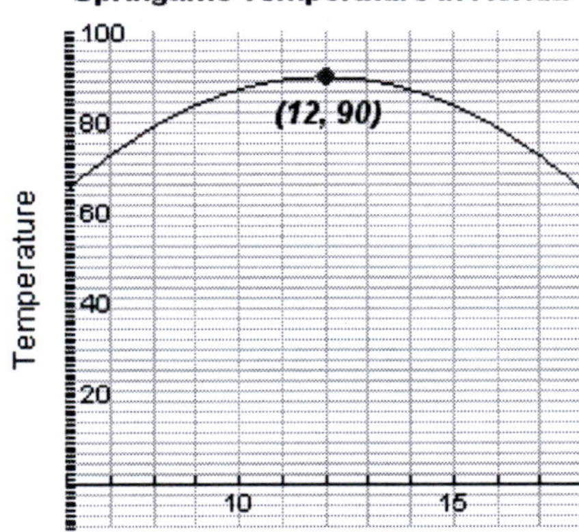

The time of day when it is warmest is Noon ($x=12$) and the temperature is approximately 90 degrees.

This model is only valid on the interval $[6,18]$ since the values of T outside the interval $[6,18]$ are too small to be considered temperatures in Florida.

139.

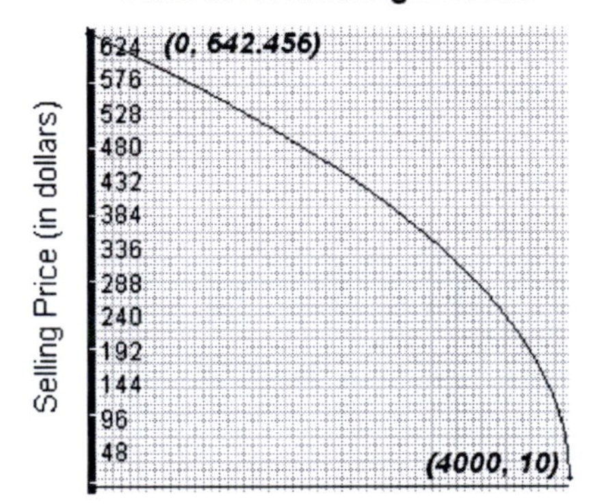

The lowest price is \$10 and the highest price is \$642.38. These agree with the values of Exercise 105.

141.

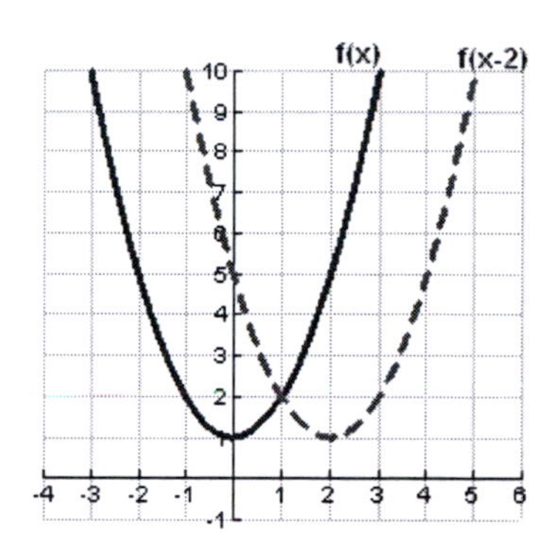

The graph of y2 can be obtained by shifting the graph of y1 two units to the right.

Section 3.2 Solutions

1. $G(-x) = -x+4 \neq x+4 = G(x)$ So, not even.

$-G(-x) = -(-x+4) = x-4 \neq G(x)$ So, not odd. Thus, $\boxed{\text{neither}}$.

3. $f(-x) = 3(-x)^2+1 = 3x^2+1 = f(x)$ So, $\boxed{\text{even}}$. Thus, f cannot be odd.

5. $g(-t) = 5(-t)^3 - 3(-t) = -5t^3+3t$

$= -\left(5t^3-3t\right) \neq g(t)$

So, not even.

$-g(-t) = -\left(-\left(5t^3-3t\right)\right) = g(t)$

So, $\boxed{\text{odd}}$.

7. $h(-x) = (-x)^2 + 2(-x) = x^2 - 2x \neq h(x)$

So, not even.

$-h(-x) = -\left(x^2-2x\right) = -x^2+2x \neq h(x)$

So, not odd. Thus, $\boxed{\text{neither}}$.

9. $h(-x) = (-x)^{1/3} - (-x)$

$= -\left(x^{1/3} - x\right) \neq h(x)$

So, not even.

$-h(-x) = -\left(-\left(x^{1/3}-x\right)\right) = x^{1/3} - x = h(x)$

So, $\boxed{\text{odd}}$.

11. $f(-x) = |-x|+5 = |-1||x|+5$

$= |x|+5 = f(x)$

So, $\boxed{\text{even}}$. Thus, f cannot be odd.

13. $f(-x) = |-x| = |-1||x| = f(x)$ So, $\boxed{\text{even}}$. Thus, f cannot be odd.

15. $G(-t) = |(-t)-3| = |-(t+3)|$

$= |t+3| \neq G(t)$

So, not even.

$-G(-t) = -|t+3| \neq G(t)$

So, not odd. Thus, neither.

17. $G(-t) = \sqrt{-t-3} = \sqrt{-(t+3)} \neq G(t)$

So, not even.

$-G(-t) = \underbrace{-\sqrt{-(t+3)}}_{\text{Note: Cannot distribute } -1 \text{ here}} \neq G(t)$

So, not odd. Thus, neither.

19. $g(-x) = \sqrt{(-x)^2 + (-x)}$

$= \sqrt{x^2 - x} \neq g(x)$

So, not even.

$-g(-x) = -\sqrt{x^2 - x} \neq g(x)$

So, not odd. Thus, neither.

21. $h(-x) = \frac{1}{-x} + 3 \neq h(x)$

So, not even.

$-h(-x) = -\left(\frac{1}{-x} + 3\right) = \frac{1}{x} - 3 \neq h(x)$

So, not odd. Thus, neither.

23. Call the function h.

h is not even since $h(1) = 4 \neq 0 = h(-1)$.

h is not odd since $h(1) = 4 \neq 0 = -h(-1)$.

Thus, neither.

25.

Domain	$(-\infty, \infty)$
Range	$[-1, \infty)$
Increasing	$(-1, \infty)$
Decreasing	$(-3, -2)$
Constant	$(-\infty, -3) \cup (-2, -1)$

d) 0
e) -1
f) 2

27.

Domain	$[-7, 2]$
Range	$[-5, 4]$
Increasing	$(-4, 0)$
Decreasing	$(-7, -4) \cup (0, 2)$
Constant	nowhere

d) 4
e) 1
f) -5

29.

Domain	$(-\infty,\infty)$
Range	$(-\infty,\infty)$
Increasing	$(-\infty,-3)\cup(4,\infty)$
Decreasing	nowhere
Constant	$(-3,4)$

d) 2
e) 2
f) 2

31.

Domain	$(-\infty,\infty)$
Range	$[-4,\infty)$
Increasing	$(0,\infty)$
Decreasing	$(-\infty,0)$
Constant	nowhere

d) -4
e) 0
f) 0

33.

Domain	$(-\infty,0)\cup(0,\infty)$
Range	$(-\infty,0)\cup(0,\infty)$
Increasing	$(-\infty,0)\cup(0,\infty)$
Decreasing	nowhere
Constant	nowhere

d) undefined
e) 3
f) -3

35.

Domain	$(-\infty,0)\cup(0,\infty)$
Range	$(-\infty,5)\cup[7]$
Increasing	$(-\infty,0)$
Decreasing	$(5,\infty)$
Constant	$(0,5)$

d) undefined
e) 3
f) 7

37.

$$\frac{\left[(x+h)^2-(x+h)\right]-\left[x^2-x\right]}{h}=$$

$$\frac{x^2+2hx+h^2-x-h-x^2+x}{h}=$$

$$\frac{\cancel{h}(2x+h-1)}{\cancel{h}}=\boxed{2x+h-1}$$

39.

$$\frac{\left[(x+h)^2+3(x+h)\right]-\left[x^2+3x\right]}{h}=$$

$$\frac{x^2+2hx+h^2+3x+3h-x^2-3x}{h}=$$

$$\frac{\cancel{h}(2x+h+3)}{\cancel{h}}=\boxed{2x+h+3}$$

41.

$$\frac{\left[(x+h)^2-3(x+h)+2\right]-\left[x^2-3x+2\right]}{h}=$$

$$\frac{x^2+2hx+h^2-3x-3h+2-x^2+3x-2}{h}=$$

$$\frac{\cancel{h}(2x+h-3)}{\cancel{h}}=\boxed{2x+h-3}$$

43.

$$\frac{\left[-3(x+h)^2+5(x+h)-4\right]-\left[-3x^2+5x-4\right]}{h}=$$

$$\frac{-3x^2-6hx-3h^2+5x+5h-4+3x^2-5x+4}{h}=$$

$$\frac{\cancel{h}(-6x-3h+5)}{\cancel{h}}=\boxed{-6x-3h+5}$$

45. $\dfrac{3^3-1^3}{3-1}=\dfrac{27-1}{2}=\boxed{13}$

47. $\dfrac{|3|-|1|}{3-1}=\boxed{1}$

49. $\dfrac{(1-2(3))-(1-2(1))}{3-1}=\dfrac{-5-(-1)}{2}=\boxed{-2}$

51. $\dfrac{|5-2(3)|-|5-2(1)|}{3-1}=\dfrac{|-1|-3}{2}=\boxed{-1}$

53.

Domain	$(-\infty,\infty)$
Range	$(-\infty,2]$
Increasing	$(-\infty,2)$
Decreasing	nowhere
Constant	$(2,\infty)$

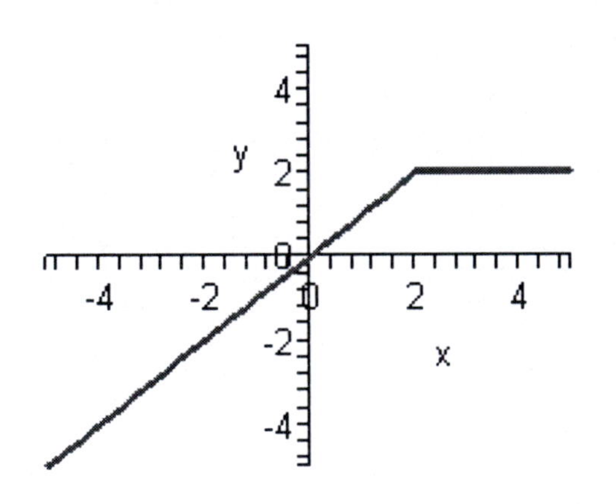

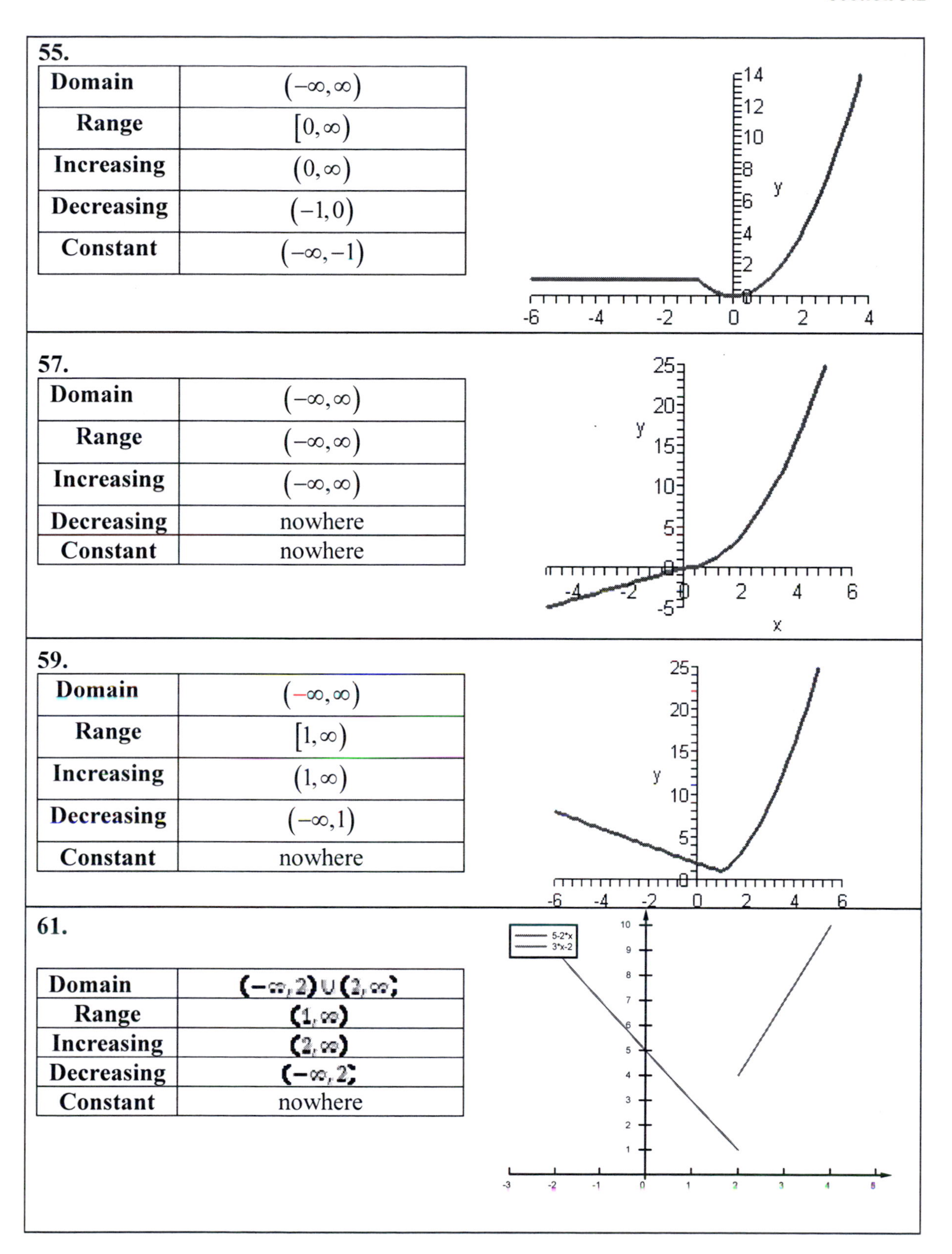

55.

Domain	$(-\infty,\infty)$
Range	$[0,\infty)$
Increasing	$(0,\infty)$
Decreasing	$(-1,0)$
Constant	$(-\infty,-1)$

57.

Domain	$(-\infty,\infty)$
Range	$(-\infty,\infty)$
Increasing	$(-\infty,\infty)$
Decreasing	nowhere
Constant	nowhere

59.

Domain	$(-\infty,\infty)$
Range	$[1,\infty)$
Increasing	$(1,\infty)$
Decreasing	$(-\infty,1)$
Constant	nowhere

61.

Domain	$(-\infty,2)\cup(2,\infty)$
Range	$(1,\infty)$
Increasing	$(2,\infty)$
Decreasing	$(-\infty,2)$
Constant	nowhere

63.

Domain	$(-\infty,\infty)$
Range	$[-1,3]$
Increasing	$(-1,3)$
Decreasing	nowhere
Constant	$(-\infty,-1)\cup(3,\infty)$

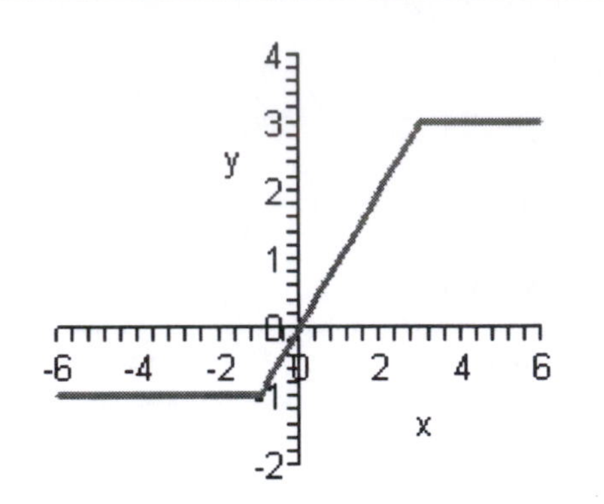

65.

Domain	$(-\infty,\infty)$
Range	$[1,4]$
Increasing	$(1,2)$
Decreasing	nowhere
Constant	$(-\infty,1)\cup(2,\infty)$

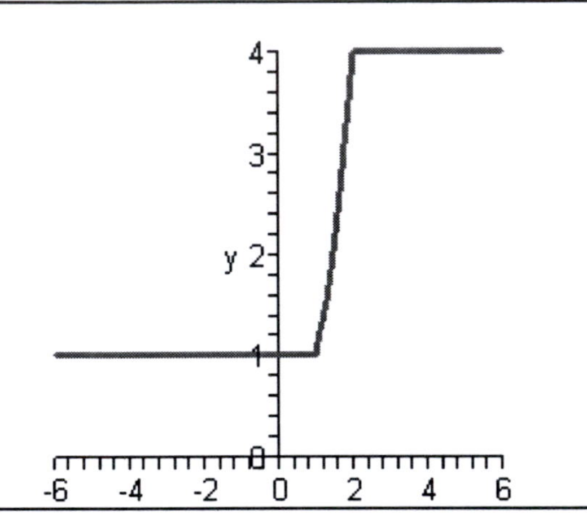

67.

Domain	$(-\infty,-2)\cup(-2,\infty)$
Range	$(-\infty,\infty)$
Increasing	$(-2,1)$
Decreasing	$(-\infty,-2)\cup(1,\infty)$
Constant	nowhere

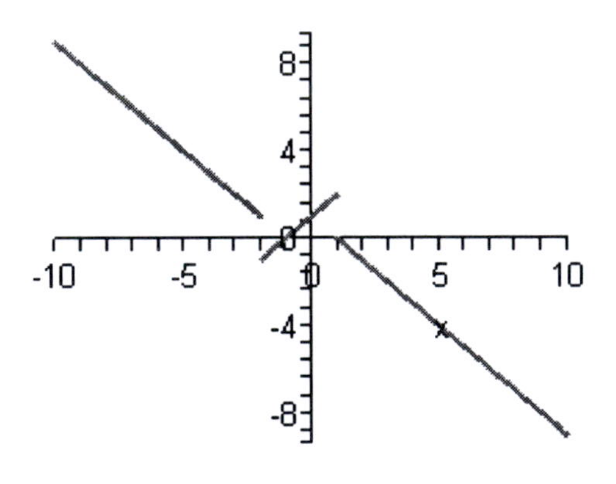

Notes on Graph: There should be open holes at $(-2,1)$, $(-2,-1)$, and $(1,2)$, and a closed hole at $(1,0)$.

69.

Domain	$(-\infty,\infty)$
Range	$[0,\infty)$
Increasing	$(0,\infty)$
Decreasing	nowhere
Constant	$(-\infty,0)$

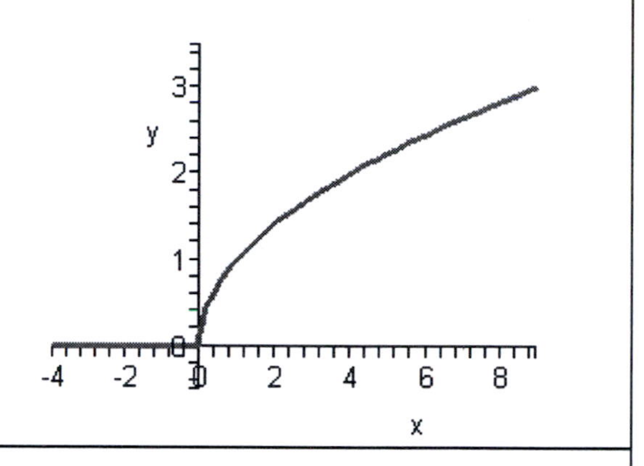

71.

Domain	$(-\infty,\infty)$
Range	$(-\infty,\infty)$
Increasing	nowhere
Decreasing	$(-\infty,0)\cup(0,\infty)$
Constant	nowhere

Notes on Graph: There should be a closed hole at (0,0).

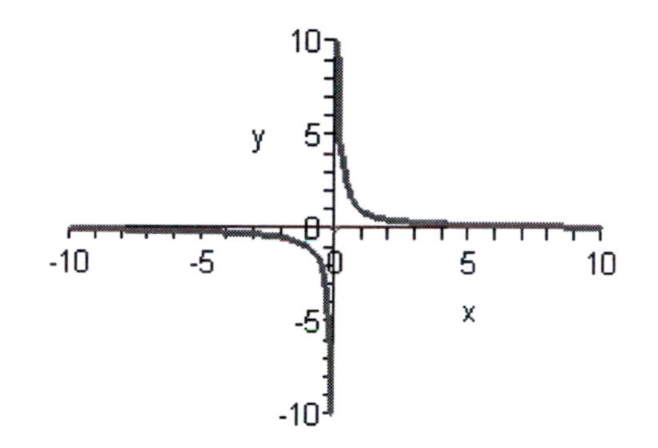

73.

Domain	$(-\infty,1)\cup(1,\infty)$
Range	$(-\infty,-1)\cup(-1,\infty)$
Increasing	$(-1,1)$
Decreasing	$(-\infty,-1)\cup(1,\infty)$
Constant	nowhere

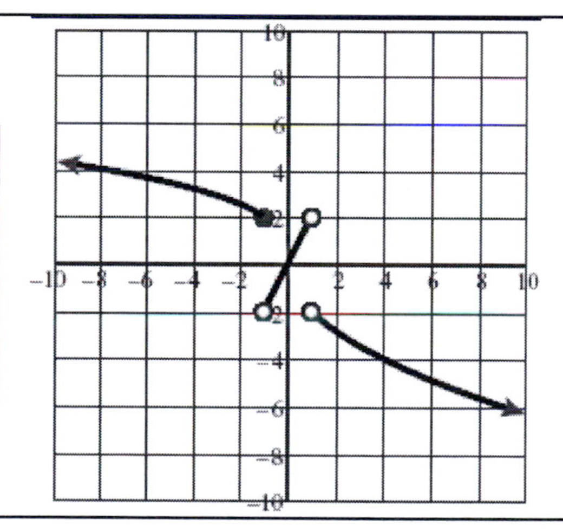

75.

Domain	$(-\infty,\infty)$
Range	$(-\infty,2)\cup[4,\infty)$
Increasing	$(-\infty,-2)\cup(0,2)\cup(2,\infty)$
Decreasing	$(-2,0)$
Constant	nowhere

Notes on Graph: There should be open holes at $(-2,2)$, $(2,2)$ and closed holes at $(-2,1)$, $(2,4)$.

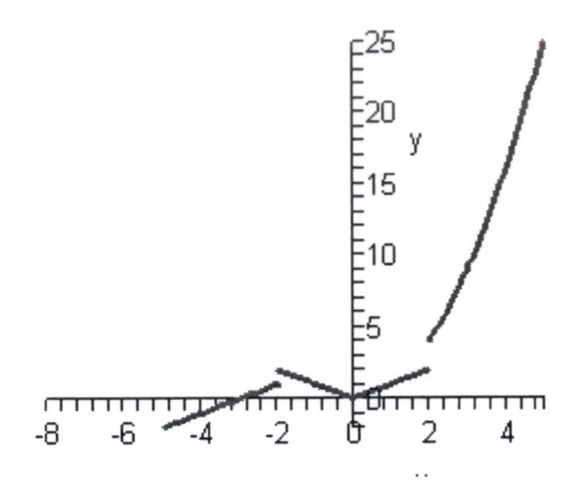

77.

Domain	$(-\infty,1)\cup(1,\infty)$
Range	$(-\infty,1)\cup(1,\infty)$
Increasing	$(-\infty,1)\cup(1,\infty)$
Decreasing	nowhere
Constant	nowhere

Notes on Graph: There should be an open hole at $(1,1)$.

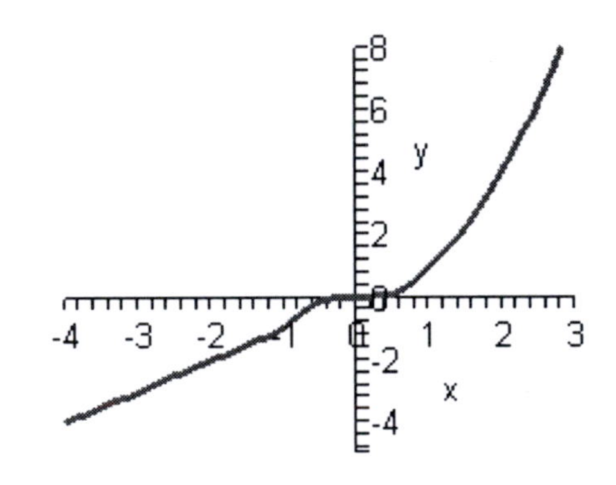

79. Profit is increasing from $t = 10$ to $t = 12$, which corresponds to Oct. to Dec. Profit is decreasing from $t = 1$ to $t = 10$, which corresponds to Jan to Oct. Profit never remains constant.

81. Let x = number of T-shirts ordered. The cost function is given by

$$C(x)=\begin{cases}10x, & 0\le x\le 50\\ 9x, & 50<x\le 100\\ 8x, & x>100\end{cases}$$

83. Let x = number of boats entered. The cost function is given by

$$C(x)=\begin{cases}250x, & 0\le x\le 10\\ \underbrace{2500}_{\text{Cost for first 10 boats}}+175\cdot\underbrace{(x-10)}_{\text{\# of boats beyond first 10}}, & x>10\end{cases}=\begin{cases}250x, & 0\le x\le 10\\ 175x+750, & x>10\end{cases}$$

85. Let x = number of people attending the reception. The cost function is given by

$$C(x)=\begin{cases}\underbrace{1000}_{\text{Fee for reserving dining room}}+35x, & 0\le x\le 100\\ 1000+\underbrace{3500}_{\text{Cost for first 100 guests}}+25\cdot\underbrace{(x-100)}_{\text{\# of guests beyond first 100}}, & x>100\end{cases}$$

Simplifying the terms above yields

$$C(x)=\begin{cases}1000+35x, & 0\le x\le 100\\ 2000+25x, & x>100\end{cases}$$

87. Let x = number of books sold.
Since a single book sells for \$20, the amount of money earned for x books is $20x$.
Then, the amount of royalties due to the author (as a function of x) is given by:

$$R(x)=\begin{cases}\underbrace{50,000}_{\text{Amount upfront}}+\underbrace{0.15(20x)}_{\text{Amount from 15\% royalties}}, & 0\le x\le 100,000\\ 50,000+\underbrace{0.15(2,000,000)}_{\text{Royalties from first 100,000 books}}+\underbrace{0.20(20)(x-100,000)}_{\text{20\% royalties on books beyond initial 100,000}}, & 100,000<x\end{cases}$$

Simplifying the terms above yields

$$R(x)=\begin{cases}50,000+3x, & 0\le x\le 100,000\\ -50,000+4x, & x>100,000\end{cases}$$

89. Let x = number of stained glass units sold.
Total monthly cost is given by: $C(x)=\underbrace{100}_{\text{Business Costs}}+\underbrace{700}_{\text{Studio Rent}}+\underbrace{35x}_{\text{Cost of materials for } x \text{ units}}=800+35x$.

Revenue for x units sold is given by: $R(x)=100x$

So, the total profit is given by: $P(x)=R(x)-C(x)=100x-(800+35x)=65x-800$

91. Observe that

$$f(x)=\begin{cases}0.80, & 0\le x<1\\ 0.80+0.17, & 1\le x<2\\ 0.80+0.17(2), & 2\le x<3\\ \vdots & \\ 0.80+0.17(n), & n\le x<n+1\end{cases}$$

Using the greatest integer function, we have $\boxed{f(x)=0.80+0.17[\![x]\!],\ x\ge 0}$.

93. $f(t)=3(-1)^{[\![t]\!]},\ t\ge 0$

95. **a)** $\dfrac{1500-500}{1950-1900}=\boxed{20 \text{ per year}}$

b) $\dfrac{7000-1500}{2000-1950}=\boxed{110 \text{ per year}}$

97. $\dfrac{h(2)-h(1)}{2-1}=\dfrac{\left(-16(2)^2+48(2)\right)-\left(-16(1)^2+48(1)\right)}{2-1}\boxed{=0 \text{ ft/sec}}$

99. The first quarter starts at $t = 1$ and ends at $t = 90$. So, the average rate of change in $d(t)$ during the first quarter is

$$\frac{d(90)-d(1)}{90-1} = \frac{\left(3\sqrt{90^2+1}-2.75(90)\right)-\left(3\sqrt{1^2+1}-2.75(1)\right)}{89} \approx 0.236$$

So, demand is increasing at an approximate rate of 236 units over the first quarter.

101. Should exclude the origin since $x = 0$ is not in the domain. The range should be $(0, \infty)$.

103. The portion of $C(x)$ for $x > 30$ should be: $15 + \underbrace{x-30}_{\text{Number miles beyond first 30}}$

105. True. This corresponds to $y = mx + b$ with $m = 1, b = 0$.

107. False. For instance, $f(x) = x^3$ is always increasing.

109. The individual pieces used to form *f*, namely ax, bx^2, are continuous on $\mathbb{R}$. So, the only *x*-value with which we need to be concerned regarding the continuity of *f* is $x = 2$. For *f* to be continuous at 2, we need $a(2) = b(2)^2$, which is the same as $\boxed{a = 2b}$.

111. The graph is $\boxed{\text{odd}}$.

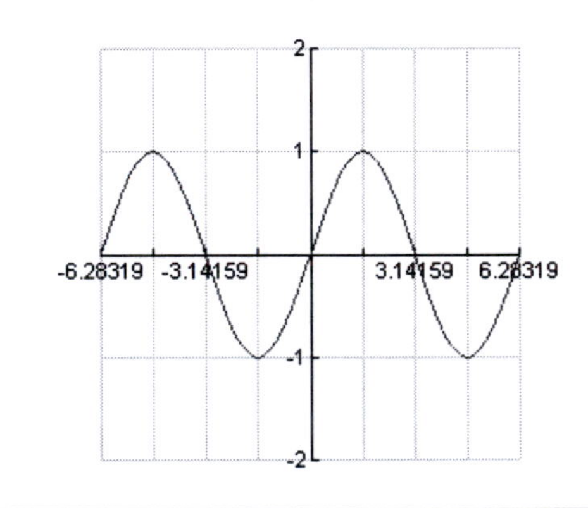

113. The graph is $\boxed{\text{odd}}$.

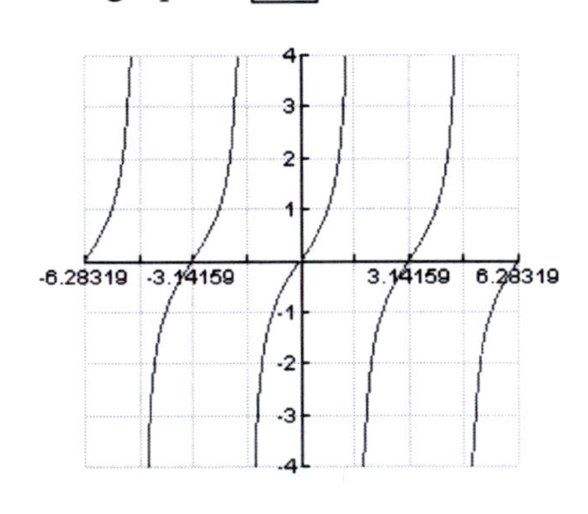

115.

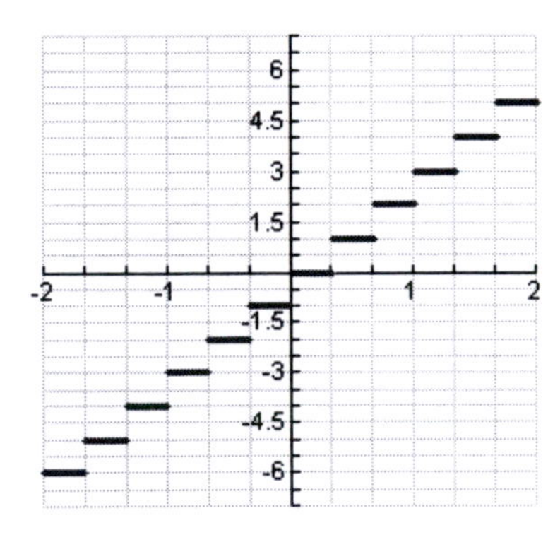

Domain: $\mathbb{R}$ Range: The set of integers

Section 3.3 Solutions ---

1. l Shift the graph of x^2 up 1 unit.	**3. a** Shift the graph of x^2 right 1 unit, then reflect over x-axis.
5. b Shift the graph of x^2 left 1 unit, then reflect over x-axis.	**7. i** Shift the graph of $\sqrt{x}$ right 1 unit, then shift up 1 unit.
9. c Shift the graph of $\sqrt{x}$ right 1 unit, then reflect over y-axis, and then shift down 1 unit.	**11. g** Reflect the graph of $\sqrt{x}$ over y-axis, then reflect over x-axis, and then shift up 1 unit.
13. $y=\lvert x\rvert+3$	**15.** $y=\lvert -x\rvert=\lvert x\rvert$ (since $\lvert -x\rvert=\lvert -1\rvert\lvert x\rvert=\lvert x\rvert$)
17. $y=3\lvert x\rvert$	**19.** $y=x^3-4$
21. $y=(x+1)^3+3$	**23.** $y=-x^3$

25.

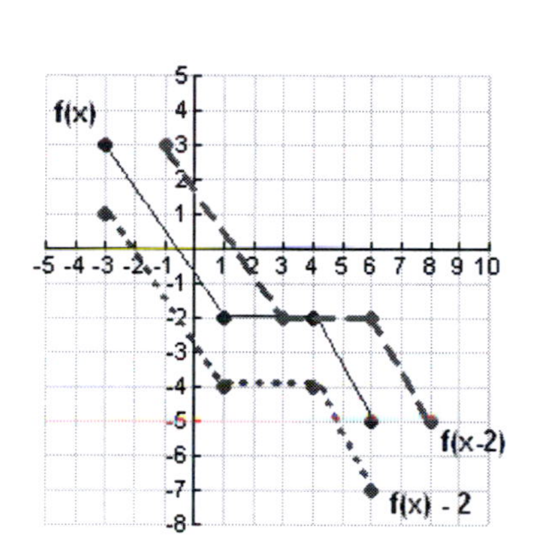

27.

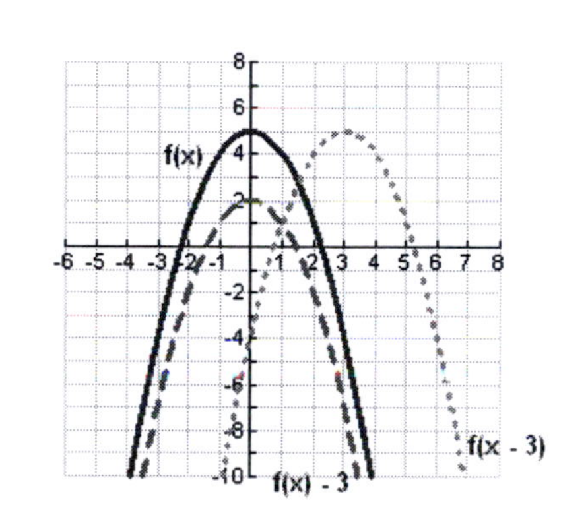

29.

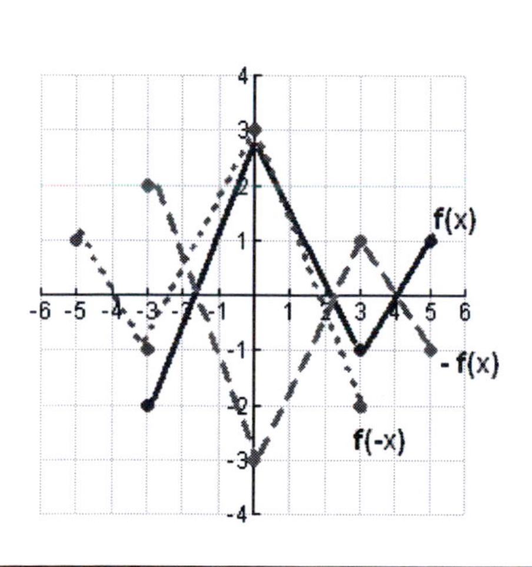

31.

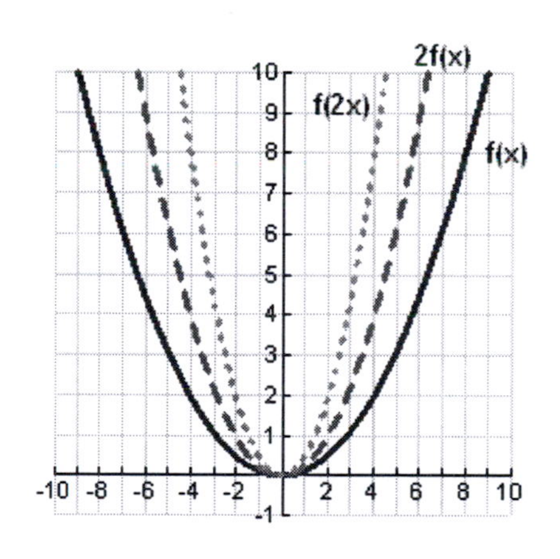

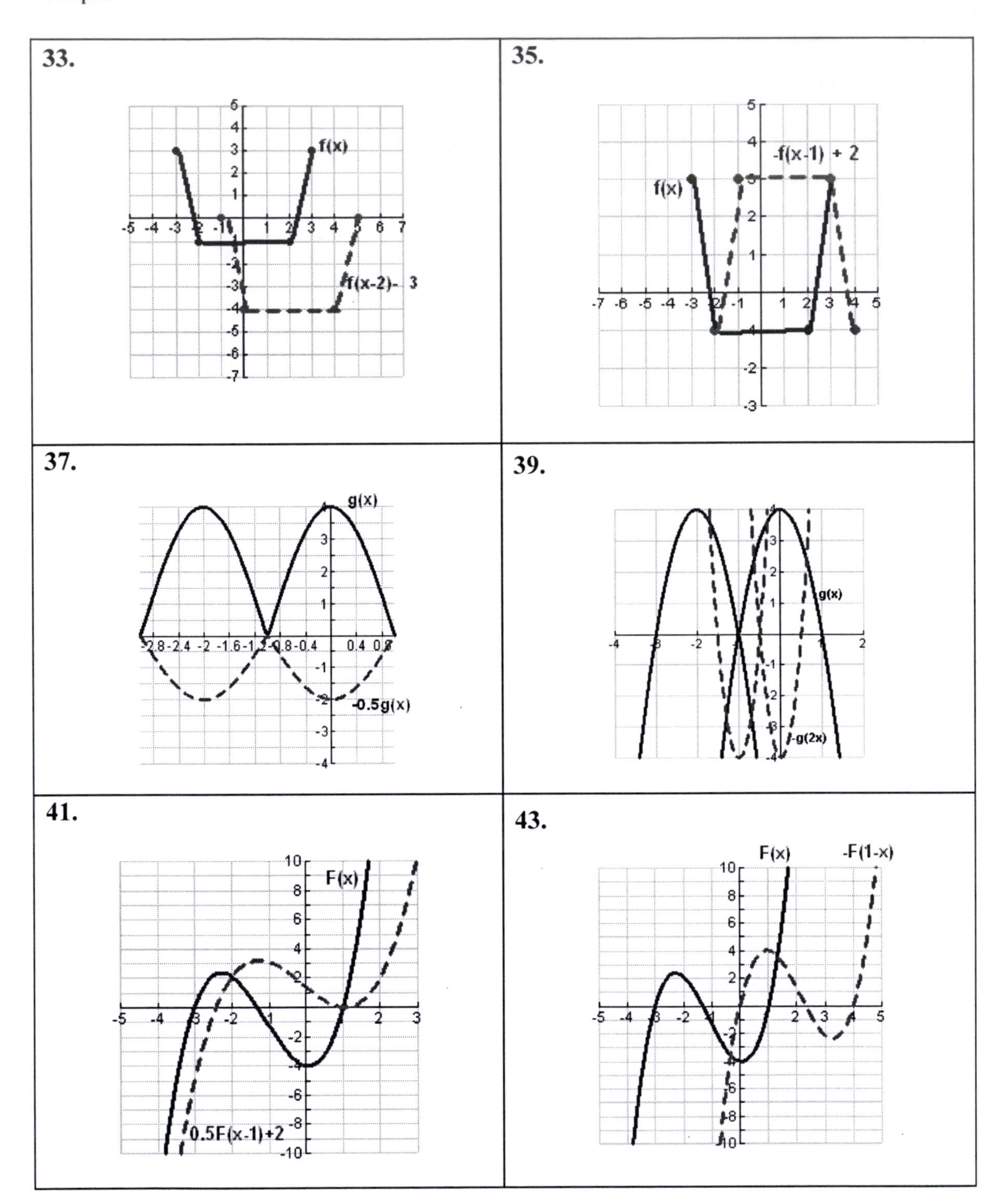
33.
f(x)
f(x-2)- 3
35.
f(x)
-f(x-1) + 2
37.
g(x)
-0.5g(x)
39.
g(x)
-g(2x)
41.
F(x)
0.5F(x-1)+2
43.
F(x)
-F(1-x)

45.

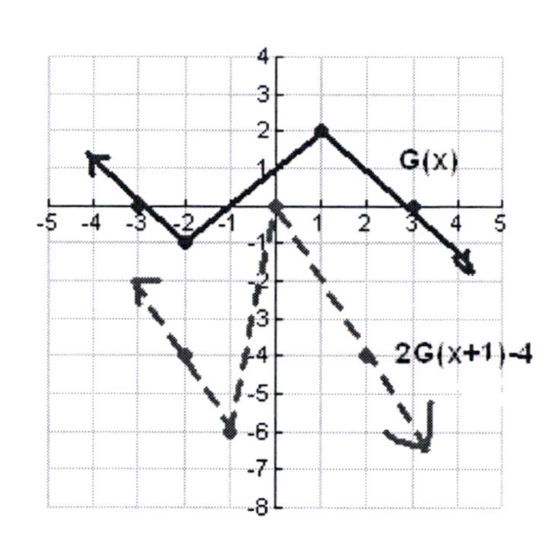

47.

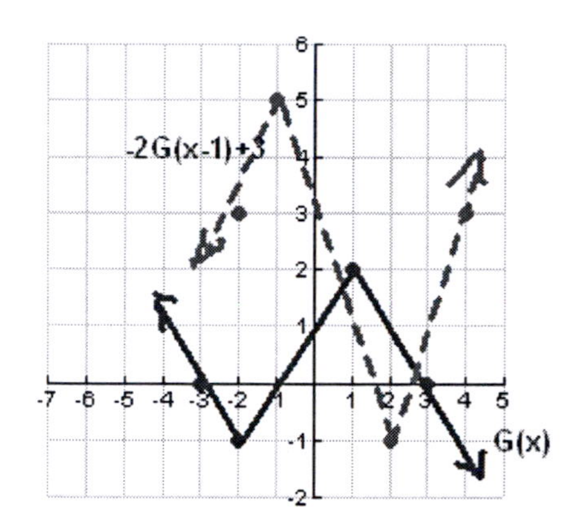

49. Shift the graph of x^2 down 2 units.

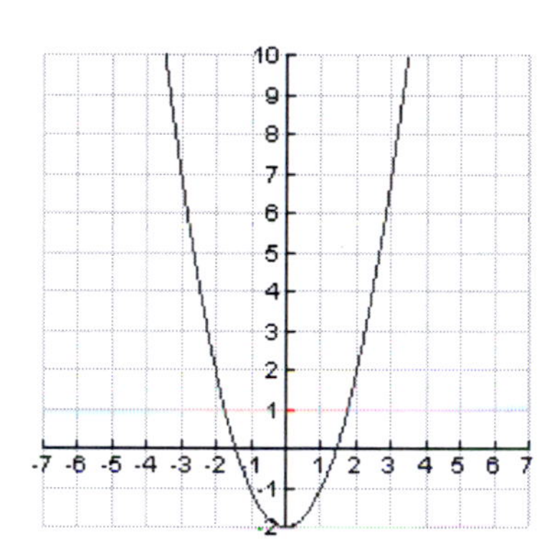

51. Shift the graph of x^2 left 1 unit.

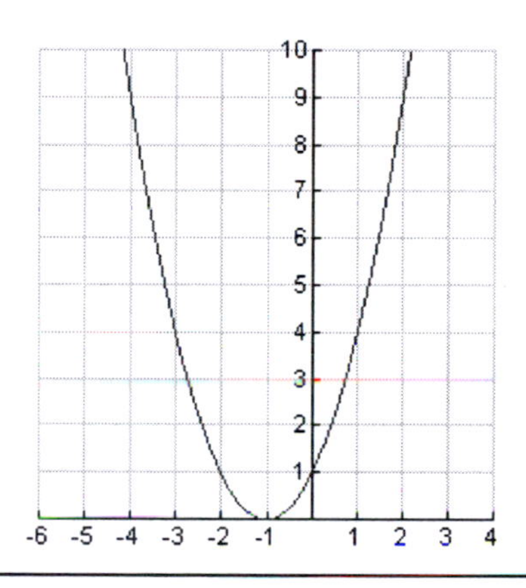

53. Shift the graph of x^2 right 3 units, and up 2 units.

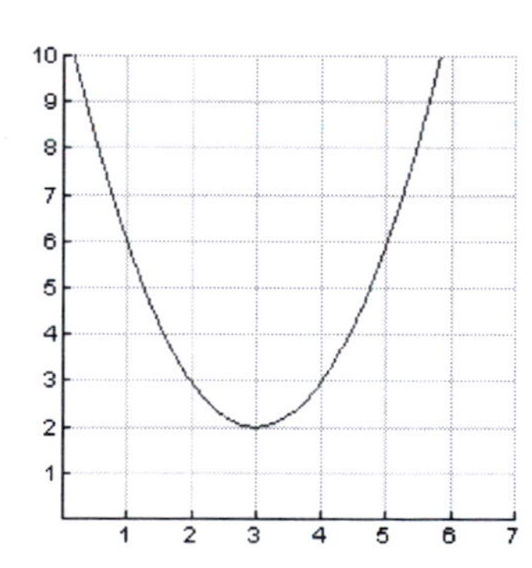

55. Shift the graph of x^2 right 1 unit, and then reflect over x-axis.

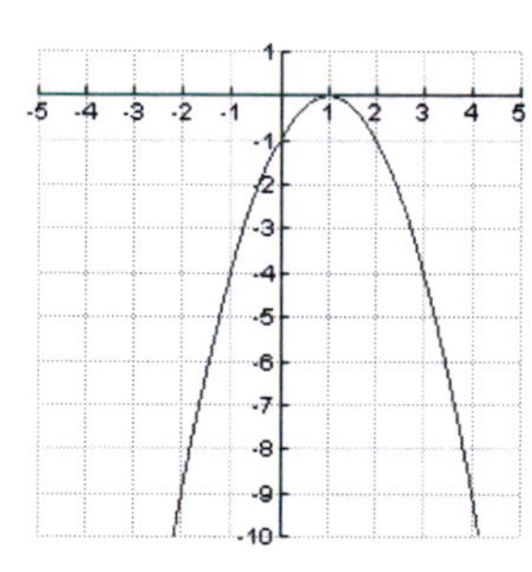

57. Reflect the graph of $|x|$ over y-axis. (This yields the same graph as $|x|$ since $|-x| = |-1||x| = |x|$.)

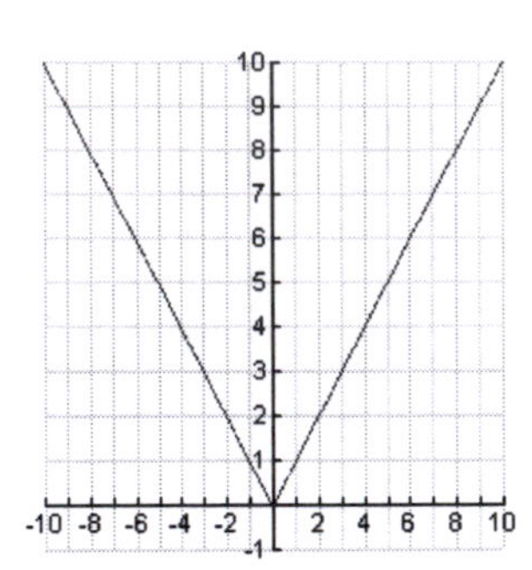

59. Reflect the graph of $|x|$ over x-axis, then shift left 2 units and down 1 unit.

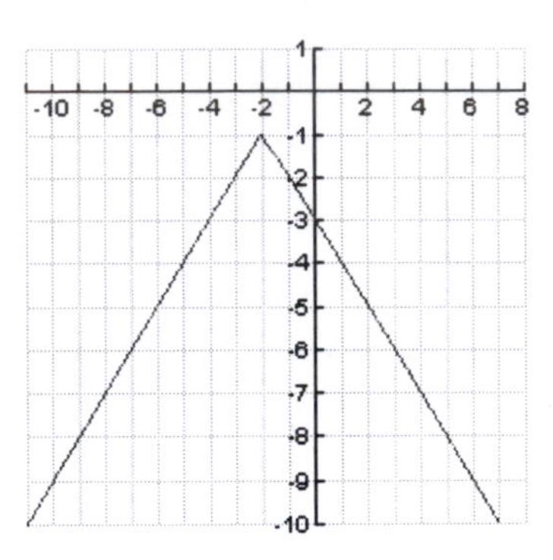

61. Vertically stretch the graph of x^2 by a factor of 2, then shift up 1unit.

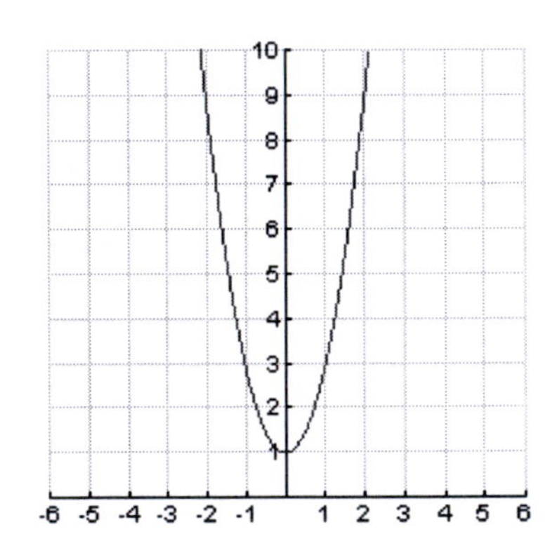

63. Shift the graph of $\sqrt{x}$ right 2 units, then reflect over x-axis.

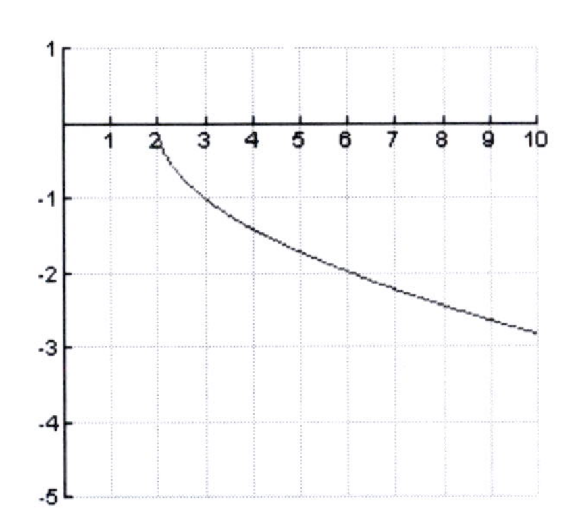

65. Reflect the graph of $\sqrt{x}$ over x-axis, then shift left 2 units and down 1 unit.

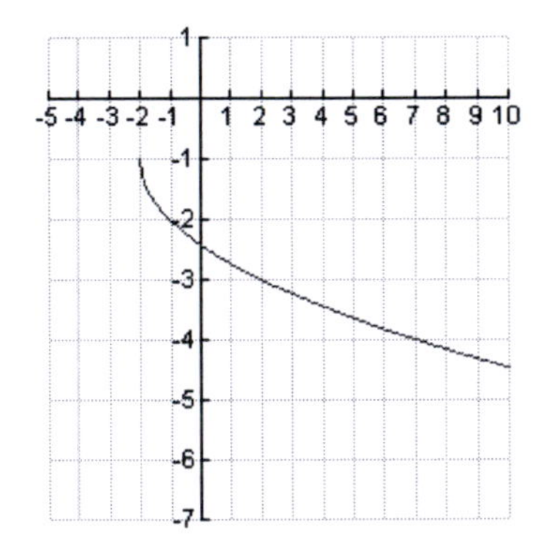

67. Shift the graph of $\sqrt[3]{x}$ right 1 unit, then up 2 units.

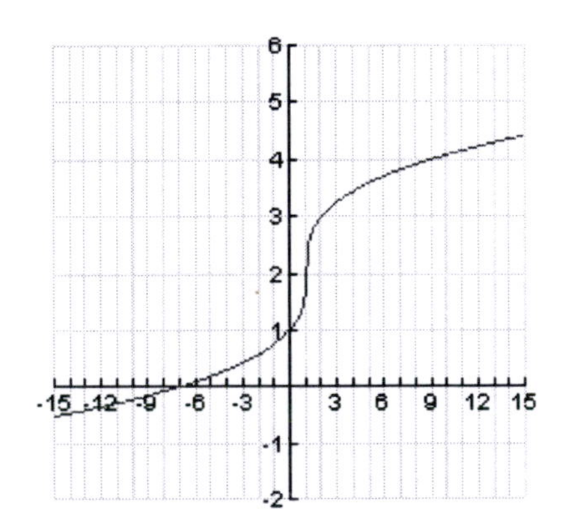

69. Shift the graph of $\frac{1}{x}$ left 3 units, then up 2 units.

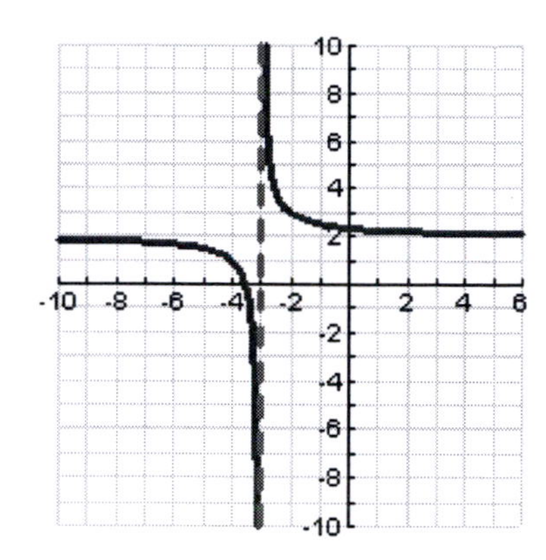

71. Shift the graph of $\frac{1}{x}$ left 2 units, then reflect over x-axis, and then shift up 2 units.

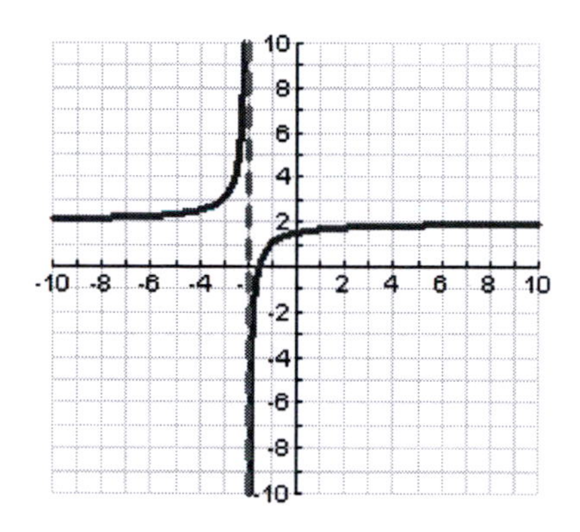

73. Reflect the graph of $\sqrt{x}$ over y-axis, then expand vertically by a factor of 5.

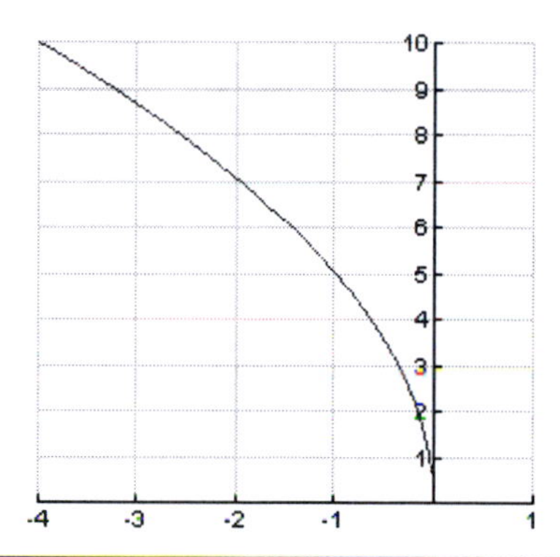

75.

Completing the square yields

$$\begin{aligned} f(x) &= x^2 - 6x + 11 \\ &= (x^2 - 6x + 9) + 11 - 9 \\ &= (x-3)^2 + 2 \end{aligned}$$

So, shift the graph of x^2 right 3 units, then up 2 units.

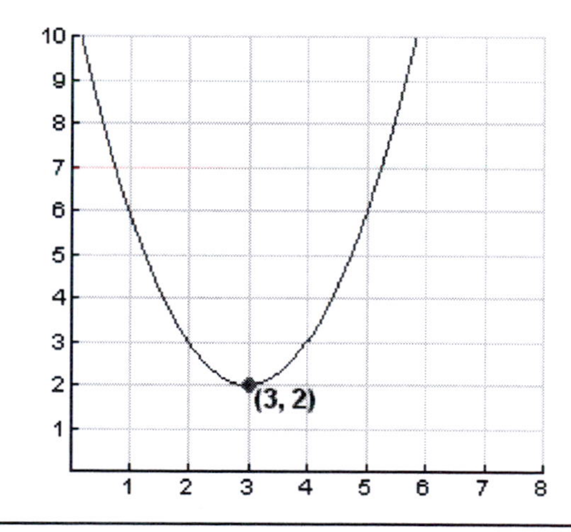

77.

Completing the square yields

$$
\begin{aligned}
f(x) &= -\left(x^2+2x\right)\\
&= -\left(x^2+2x+1\right)+1\\
&= -\left(x+1\right)^2+1
\end{aligned}
$$

So, reflect the graph of x^2 over x-axis, then shift left 1 unit, then up 1 unit.

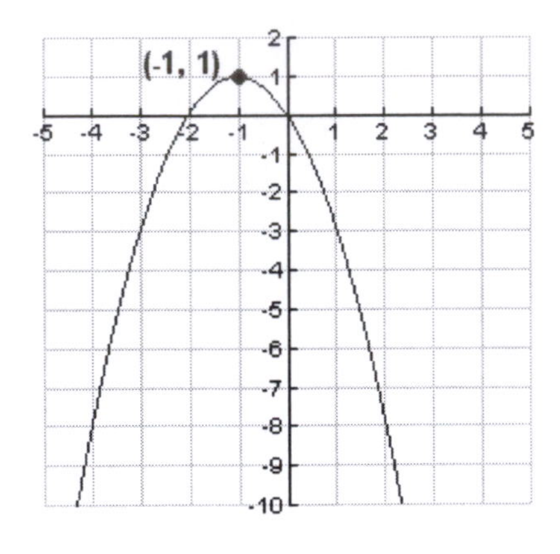

79.

Completing the square yields

$$
\begin{aligned}
f(x) &= 2x^2-8x+3\\
&= 2\left(x^2-4x\right)+3\\
&= 2\left(x^2-4x+4\right)+3-8\\
&= 2\left(x-2\right)^2-5
\end{aligned}
$$

So, vertically stretch the graph of x^2 by a factor of 2, then shift right 2 units, then down 5 units.

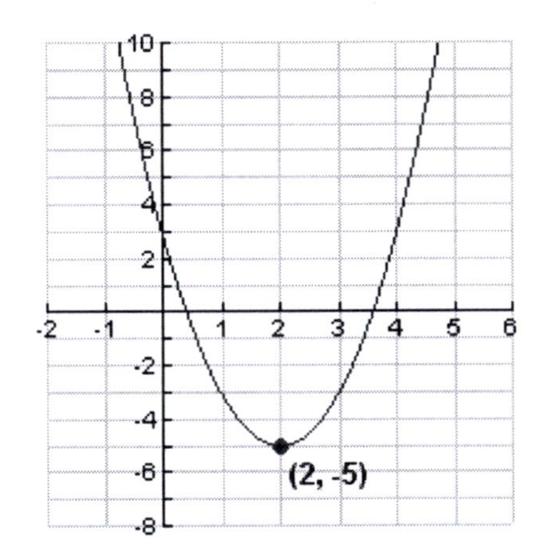

81. Let x = number of hours worked per week. Then, the salary is given by $\boxed{S(x)=10x}$ (in dollars). After 1 year, taking into account the raise, the new salary is $\boxed{\overline{S}(x)=10x+50}$.

83. The 2006 taxes would be: $T(x)=0.33(x-6500)$

85. **a.** Use h = 162 to get $BSA(w)=\sqrt{\dfrac{162w}{3,600}}=\sqrt{\dfrac{9w}{200}}$.

b. If she loses 3 kg, the new function is $BSA(w-3)=\sqrt{\dfrac{9(w-3)}{200}}$.

87. (b) is wrong – shift right 3 units.

89. (b) should be deleted since $|3-x|=|x-3|$. The correct sequence of steps would be: $(a)\to(c)^*\to(d)$, where (c)* : Shift to the right 3

91. True. Since $|-x| = |-1||x| = |x|$.

93. True.

95. The graph of $y = f(x-3)+2$ is the graph of $y = f(x)$ shifted right 3 units, then up 2 units. So, if the point (a,b) is on the graph of $y = f(x)$, then the point $(a+3, b+2)$ is on the graph of the translation $y = f(x-3)+2$.

97. a.

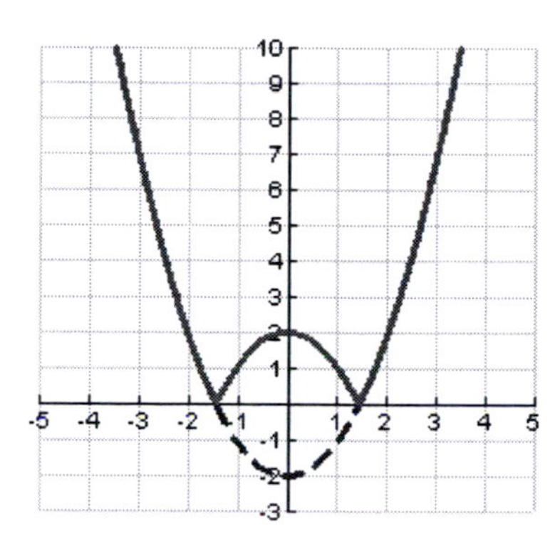

b.

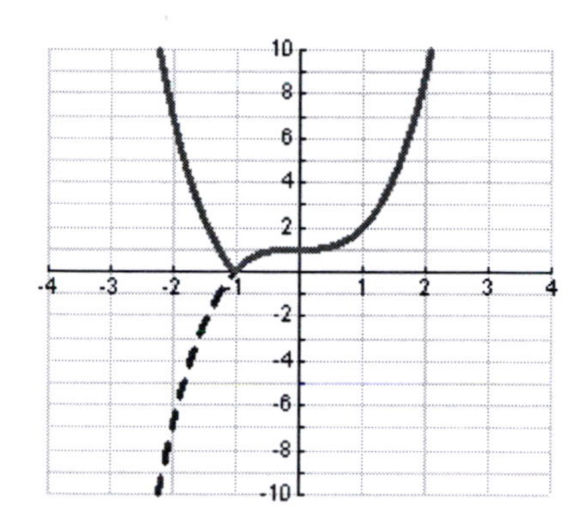

Any part of the graph of $y = f(x)$ that is below the x-axis is reflected above it for the graph of $y = |f(x)|$.

99. a.

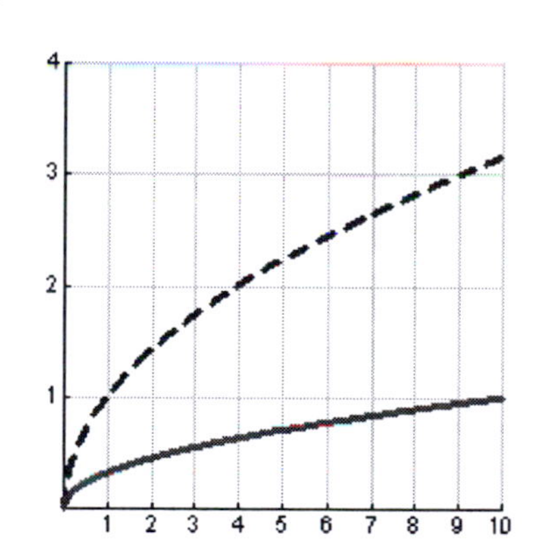

b.

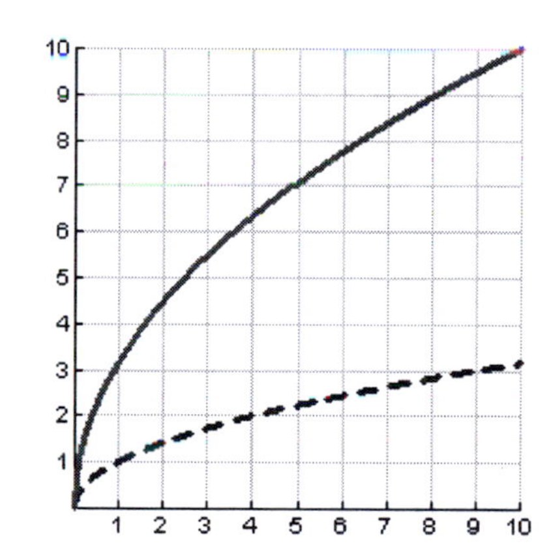

If $a > 1$, then the graph is a horizontal compression.
If $0 < a < 1$, then the graph is a horizontal expansion.

101. The graph of f is as follows:

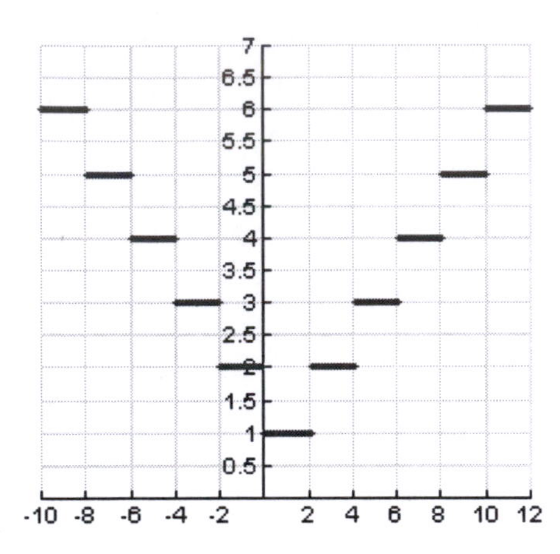

Each horizontal line in the graph of $y = [\![x]\!]$ is stretched by a factor of 2. Any portion of the graph that is below the x-axis is reflected above it. Also, there is a vertical shift up of one unit.

Section 3.4 Solutions --

1.

$$\begin{aligned} f(x)+g(x) &= (2x+1)+(1-x) \\ &= x+2 \\ f(x)-g(x) &= (2x+1)-(1-x) \\ &= 2x+1-1+x \\ &= 3x \\ f(x)\cdot g(x) &= (2x+1)(1-x) \\ &= 2x+1-2x^2-x \\ &= -2x^2+x+1 \\ \frac{f(x)}{g(x)} &= \frac{2x+1}{1-x} \end{aligned}$$

Domains:

$$\left.\begin{array}{r} dom(f+g) \\ dom(f-g) \\ dom(fg) \end{array}\right\} = (-\infty,\infty)$$

$$dom\left(\frac{f}{g}\right) = (-\infty,1)\cup(1,\infty)$$

3.

$$\begin{aligned} f(x)+g(x) &= (2x^2-x)+(x^2-4) \\ &= 3x^2-x-4 \\ f(x)-g(x) &= (2x^2-x)-(x^2-4) \\ &= 2x^2-x-x^2+4 \\ &= x^2-x+4 \\ f(x)\cdot g(x) &= (2x^2-x)\cdot(x^2-4) \\ &= 2x^4-x^3-8x^2+4x \\ \frac{f(x)}{g(x)} &= \frac{2x^2-x}{x^2-4} \end{aligned}$$

Domains:

$dom(f+g),\ dom(f-g),\ dom(fg)\} = (-\infty,\infty)$

$$dom\left(\frac{f}{g}\right) = (-\infty,-2)\cup(-2,2)\cup(2,\infty)$$

5.

$$f(x)+g(x)=\tfrac{1}{x}+x=\frac{1+x^2}{x}$$

$$f(x)-g(x)=\tfrac{1}{x}-x=\frac{1-x^2}{x}$$

$$f(x)\cdot g(x)=\tfrac{1}{x}\cdot x=1$$

$$\frac{f(x)}{g(x)}=\frac{\frac{1}{x}}{x}=\frac{1}{x^2}$$

Domains:

$$\left.\begin{array}{r} dom(f+g) \\ dom(f-g) \\ dom(fg) \\ dom\left(\dfrac{f}{g}\right) \end{array}\right\}=(-\infty,0)\cup(0,\infty)$$

7.

$$f(x)+g(x)=\sqrt{x}+2\sqrt{x}=3\sqrt{x}$$

$$f(x)-g(x)=\sqrt{x}-2\sqrt{x}=-\sqrt{x}$$

$$f(x)\cdot g(x)=\sqrt{x}\cdot 2\sqrt{x}=2x$$

$$\frac{f(x)}{g(x)}=\frac{\sqrt{x}}{2\sqrt{x}}=\frac{1}{2}$$

Domains:

$$\left.\begin{array}{r} dom(f+g) \\ dom(f-g) \\ dom(fg) \end{array}\right\}=[0,\infty)$$

$$dom\left(\frac{f}{g}\right)=(0,\infty)$$

9.

$$f(x)+g(x)=\sqrt{4-x}+\sqrt{x+3}$$

$$f(x)-g(x)=\sqrt{4-x}-\sqrt{x+3}$$

$$f(x)\cdot g(x)=\sqrt{4-x}\cdot\sqrt{x+3}$$

$$\frac{f(x)}{g(x)}=\frac{\sqrt{4-x}}{\sqrt{x+3}}=\frac{\sqrt{4-x}\sqrt{x+3}}{x+3}$$

Domains:
Must have both $4-x\geq 0$ and $x+3\geq 0$. So,

$$\left.\begin{array}{r} dom(f+g) \\ dom(f-g) \\ dom(fg) \end{array}\right\}=[-3,4].$$

For the quotient, must have both $4-x\geq 0$ and $x+3>0$. So, $dom\left(\frac{f}{g}\right)=(-3,4]$.

11.

$$(f\circ g)(x)=2\left(x^2-3\right)+1=2x^2-6+1=2x^2-5$$

$$(g\circ f)(x)=(2x+1)^2-3=4x^2+4x+1-3=4x^2+4x-2$$

Domains: $dom(f\circ g)=(-\infty,\infty)=dom(g\circ f)$

13.

$$(f\circ g)(x)=\frac{1}{(x+2)-1}=\frac{1}{x+1}$$

$$(g\circ f)(x)=\frac{1}{x-1}+2=\frac{1+2(x-1)}{x-1}=\frac{1+2x-2}{x-1}=\frac{2x-1}{x-1}$$

Domains: $dom(f\circ g)=(-\infty,-1)\cup(-1,\infty)$, $dom(g\circ f)=(-\infty,1)\cup(1,\infty)$

15.

$$(f\circ g)(x)=\left|\frac{1}{x-1}\right|=\frac{1}{|x-1|}\qquad (g\circ f)(x)=\frac{1}{|x|-1}$$

<u>Domains</u>: $dom(f\circ g)=(-\infty,1)\cup(1,\infty)\quad dom(g\circ f)=(-\infty,-1)\cup(-1,1)\cup(1,\infty)$

17.

$$(f\circ g)(x)=\sqrt{(x+5)-1}=\sqrt{x+4}$$
$$(g\circ f)(x)=\sqrt{x-1}+5$$

<u>Domains</u>:

$dom(f\circ g)$: Must have $x+4\geq 0$. So, $dom(f\circ g)=[-4,\infty)$.

$dom(g\circ f)$: Must have $x-1\geq 0$. So, $dom(g\circ f)=[1,\infty)$.

19.

$$(f\circ g)(x)=\left[(x-4)^{1/3}\right]^3+4=x-4+4=x$$
$$(g\circ f)(x)=\left[\left(x^3+4\right)-4\right]^{1/3}=\left[x^3\right]^{1/3}=x$$

<u>Domains</u>: $dom(f\circ g)=(-\infty,\infty)=dom(g\circ f)$

21.

$$\begin{aligned}(f+g)(2)&=f(2)+g(2)\\&=\left[2^2+10\right]+\sqrt{2-1}\\&=14+1=\boxed{15}\end{aligned}$$

23.

$$\begin{aligned}(f-g)(2)&=f(2)-g(2)\\&=\left[2^2+10\right]-\sqrt{2-1}\\&=14-1=\boxed{13}\end{aligned}$$

25.

$$\begin{aligned}(f\cdot g)(4)&=f(4)\cdot g(4)\\&=\left[4^2+10\right]\cdot\sqrt{4-1}=\boxed{26\sqrt{3}}\end{aligned}$$

27.

$$\left(\frac{f}{g}\right)(10)=\frac{f(10)}{g(10)}=\frac{10^2+10}{\sqrt{10-1}}=\boxed{\tfrac{110}{3}}$$

29.

$$f(g(2))=f\Big(\underbrace{\sqrt{2-1}}_{=1}\Big)=1^2+10=\boxed{11}$$

31.

$$g(f(-3))=g\Big(\underbrace{(-3)^2+10}_{=19}\Big)=\sqrt{19-1}=\boxed{3\sqrt{2}}$$

33. 0 is not in the domain of g, so that $g(0)$ is not defined. Hence, $f(g(0))$ is $\boxed{\text{undefined}}$.

35. $f(g(-3))$ is not defined since $g(-3)$ in not defined.

37.

$$(f\circ g)(4)=f(g(4))=f\left(\sqrt{4-1}\right)$$
$$=f\left(\sqrt{3}\right)=\left(\sqrt{3}\right)^2+10=\boxed{13}$$

39.

$$f(g(1))=f\Big(\underbrace{2(1)+1}_{=3}\Big)=\boxed{\tfrac{1}{3}}$$
$$g(f(2))=g\left(\tfrac{1}{2}\right)=2\left(\tfrac{1}{2}\right)+1=\boxed{2}$$

41. $f(g(1)) = f\big(\underbrace{1^2+2}_{=3}\big)$ Since 3 is not in the domain of f, this is $\boxed{\text{undefined}}$. Likewise, $g(f(2))$ is $\boxed{\text{undefined}}$ since 2 is not in the domain of f.

43.

$$f(g(1)) = f\big(\underbrace{1+3}_{=4}\big) = \frac{1}{|4-1|} = \boxed{\tfrac{1}{3}}$$

$$g(f(2)) = g\left(\frac{1}{\underbrace{|2-1|}_{=1}}\right) = 1+3 = \boxed{4}$$

45.

$$f(g(1)) = f\big(\underbrace{1^2+5}_{=6}\big) = \sqrt{6-1} = \boxed{\sqrt{5}}$$

$$g(f(2)) = g\big(\underbrace{\sqrt{2-1}}_{=1}\big) = 1^2+5 = \boxed{6}$$

47. $f(g(1))$ is $\boxed{\text{undefined}}$ since $g(1)$ is not defined.

$g(f(2)) = g\left(\dfrac{1}{2^2-3}\right) = g(1)$, which is not defined. So, this is also $\boxed{\text{undefined}}$.

49.

$$f(g(1)) = f\left(1^2+2(1)+1\right) = f(4)$$
$$= (4-1)^{1/3} = \boxed{\sqrt[3]{3}}$$
$$g(f(2)) = g\left((2-1)^{1/3}\right) = g(1)$$
$$= 1^2+2(1)+1 = \boxed{4}$$

51.

$$f(g(x)) = \cancel{2}\left(\frac{x-1}{\cancel{2}}\right)+1 = x-1+1 = x$$

$$g(f(x)) = \frac{(2x+1)-1}{2} = \frac{2x}{2} = x$$

53.

$$f(g(x)) = \sqrt{(x^2+1)-1} = \sqrt{x^2} = \underbrace{|x| = x}_{\text{Since } x \geq 1}$$

$$g(f(x)) = \left(\sqrt{x-1}\right)^2+1 = (x-1)+1 = x$$

55.

$$f(g(x)) = \frac{1}{\frac{1}{x}} = x \qquad g(f(x)) = \frac{1}{\frac{1}{x}} = x$$

57.

$$f(g(x)) = 4\left(\frac{\sqrt{x+9}}{2}\right)^2 - 9 = 4\left(\frac{x+9}{4}\right) - 9 = x$$

$$g(f(x)) = \frac{\sqrt{(4x^2-9)+9}}{2} = \frac{\sqrt{4x^2}}{2} = \frac{2x}{2} = x$$

59.

$$f(g(x)) = \frac{1}{\frac{x+1}{x} - 1} = \frac{1}{\frac{x+1-x}{x}} = \frac{1}{\frac{1}{x}} = x$$

$$g(f(x)) = \frac{\frac{1}{x-1} + 1}{\frac{1}{x-1}} = \frac{\frac{1+x-1}{x-1}}{\frac{1}{x-1}} = \frac{\frac{x}{\cancel{x-1}}}{\frac{1}{\cancel{x-1}}} = x$$

61. $f(x) = 2x^2 + 5x \quad g(x) = 3x - 1$

63. $f(x) = \frac{2}{|x|} \quad g(x) = x - 3$

65. $f(x) = \dfrac{3}{\sqrt{x} - 2} \quad g(x) = x + 1$

67. $F(C(K)) = \frac{9}{5}\left(K - 273.15\right) + 32$

69. Let x = number of linear feet of fence purchased.

a. Let l = length of each side of the square pen $\left(= \frac{x}{4}\right)$

A = area of the square pen $= l^2$.

So, $(A \circ l)(x) = A\left(\frac{x}{4}\right) = \left(\frac{x}{4}\right)^2$.

b. $A(100) = \left(\frac{100}{4}\right)^2 = 625 \text{ ft}^2$

c. $A(200) = \left(\frac{200}{4}\right)^2 = 2500 \text{ ft}^2$

71. First, solve $p = 3000 - \frac{1}{2}x$ for x: $\quad x = 2(3000 - p) = 6000 - 2p$

a. $C(x(p)) = C\left(6000 - 2p\right) = 2000 + 10(6000 - 2p) = 62{,}000 - 20p$

b. $R(x(p)) = 100(6000 - 2p) = 600{,}000 - 200p$

c. Profit $P = R - C$. So,

$$P(x(p)) = R(x(p)) - C(x(p)) = \left(600{,}000 - 200p\right) - \left(62{,}000 - 20p\right)$$
$$= 538{,}000 - 180p$$

73. a. $\left(C \circ n\right)(t) = C(n(t)) = 10\left(50t - t^2\right) + 1375 = -10t^2 + 500t + 1375$

b. $\left(C \circ n\right)(16) = C(n(16)) = -10(16)^2 + 500(16) + 1375 = 6815$

This is the cost of production on a day when the assembly line was running for 16 hours; this cost is \$6,815,000.

75. a. $A(r(t)) = \pi\left(10t - 0.2t^2\right)^2$

b. $A(r(7)) = \pi\left(10(7) - 0.2(7)^2\right)^2 = 11{,}385$ square miles

77. $A(t) = \pi\left[150\sqrt{t}\right]^2 = 22{,}500\pi t \text{ ft}^2$

79. Let h = height of the fireworks above ground.

Then, the distance between the family and the fireworks is given by

$$d(h)=\sqrt{(h-0)^2+(0-2)^2}=\sqrt{h^2+4}.$$

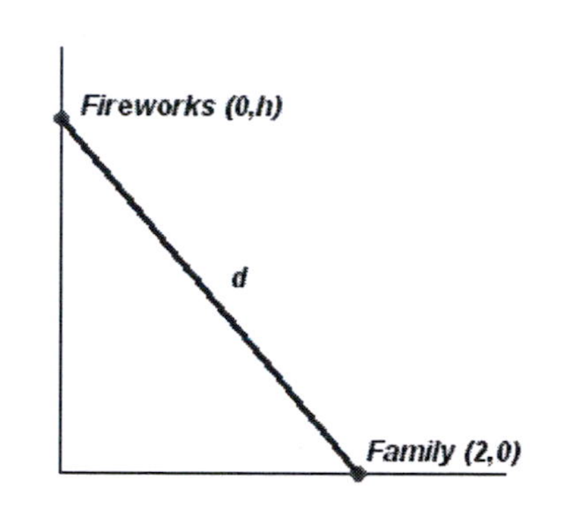

81. Must exclude -2 from the domain.

83. $(f\circ g)(x)=f(g(x))$, not $f(x)\cdot g(x)$

85. The mistake made was that $(f+g)$ was multiplied by 2 when it ought to have been evaluated at 2.

87. False. The domain of the sum, difference, or product of two functions is the intersection of their domains; the domain of the quotient is the set obtained by intersecting the two domains and then excluding all values where the denominator equals 0.

89. True

91.

$(g\circ f)(x)=\dfrac{1}{(x+a)-a}=\dfrac{1}{x}$ Domain: $x\neq 0,a$

93. $(g\circ f)(x)=\left(\sqrt{x+a}\right)^2-a=x+a-a=x$

Domain: Must have $x+a\geq 0$, so that $x\geq -a$. So, domain is $[-a,\infty)$.

95.

Notes on the graph:
The dotted curve is the graph of y_1, while the thick, solid curve is the graph of y_1+y_2.

Domain of y_2 is [-7, 9].

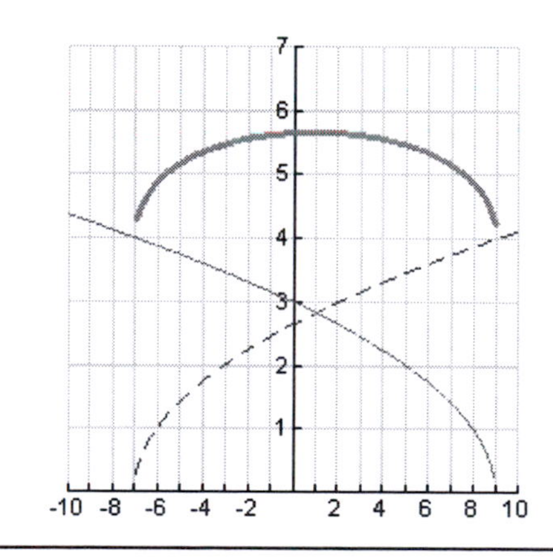

97. $y1 = \sqrt{x^2 - 3x - 4}$ $y2 = \dfrac{1}{x^2 - 14}$ $y3 = \dfrac{1}{(y1)^2 - 14}$ Domain of $y3 = \boxed{(-\infty,-3)\cup(-3,-1]\cup[4,6)\cup(6,\infty)}$ The graph of y3 only is shown to the right.	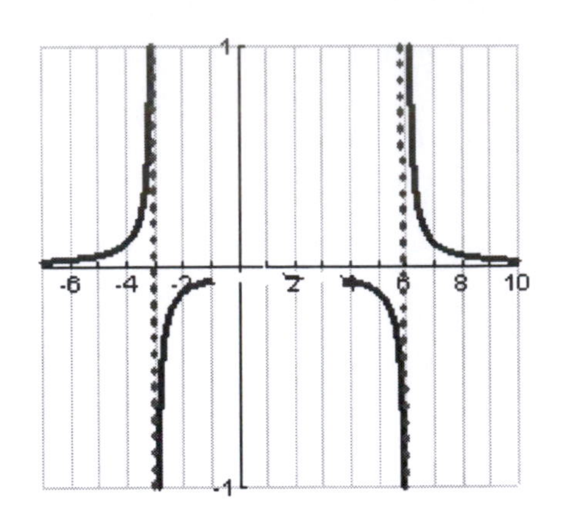

Section 3.5 Solutions ---

1. Is a function. Not one-to-one April and October both map to $78^\circ F$.	**3.** Is a function. One-to-one	**5.** Is a function. One-to-one
7. Not a function since 4 maps to both 2 and -2.	**9.** Is a function. Not one-to-one 0, 2, -2 all map to 1 in the range, for instance.	**11.** Is a function. Not one-to-one Doesn't pass the horizontal line test. Both $(-1,1)$, $(0,1)$ are on the graph.

13. Is a function. One-to-one

15. Is a function. Not one-to-one since it doesn't pass the horizontal line test.

17. Not one-to-one. Both $(0,3)$, $(6,3)$ lie on the graph.

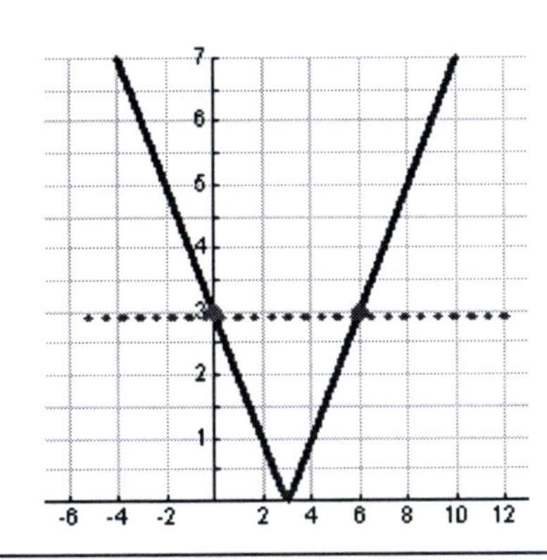

19.

$$f(x_1) = f(x_2) \Rightarrow \frac{1}{x_1 - 1} = \frac{1}{x_2 - 1}$$
$$\Rightarrow x_2 - 1 = x_1 - 1$$
$$\Rightarrow x_2 = x_1$$

One-to-one

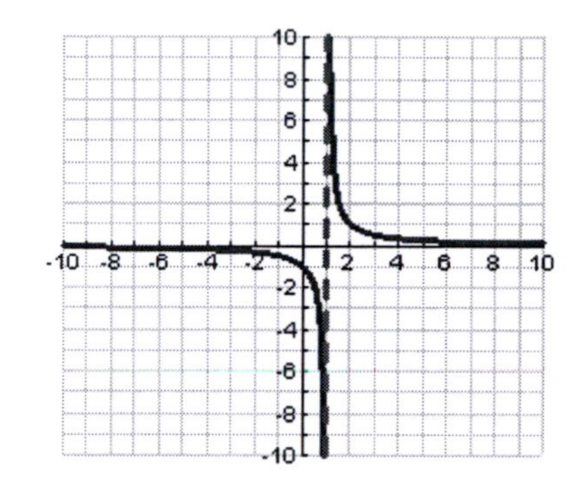

21. f is not one-to-one since, for example, $f(-1) = f(1) = -3$.

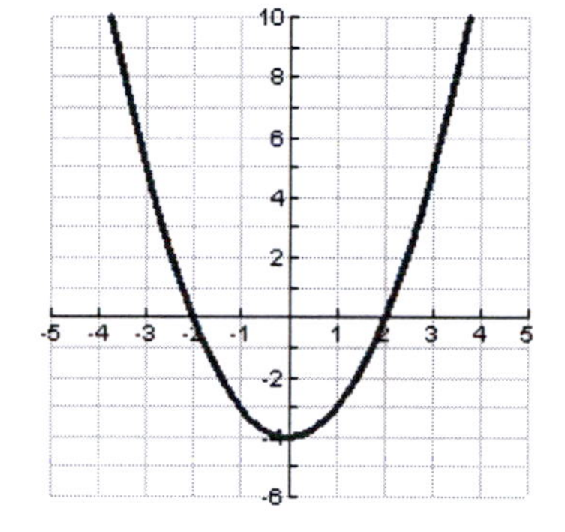

23.

$$f(x_1) = f(x_2) \Rightarrow x_1^3 - 1 = x_2^3 - 1$$
$$\Rightarrow x_1^3 = x_2^3$$
$$\Rightarrow x_1 = x_2$$

One-to-one

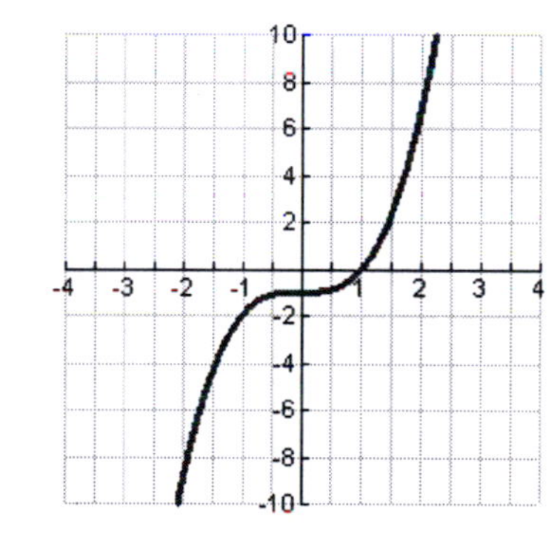

25.

Given: $f(x) = 2x + 1,\ f^{-1}(x) = \dfrac{x-1}{2}$

$f\left(f^{-1}(x)\right) = \not{2}\left(\dfrac{x-1}{\not{2}}\right) + 1 = x - 1 + 1 = x$

$f^{-1}\left(f(x)\right) = \dfrac{(2x+1)-1}{2} = \dfrac{2x}{2} = x$

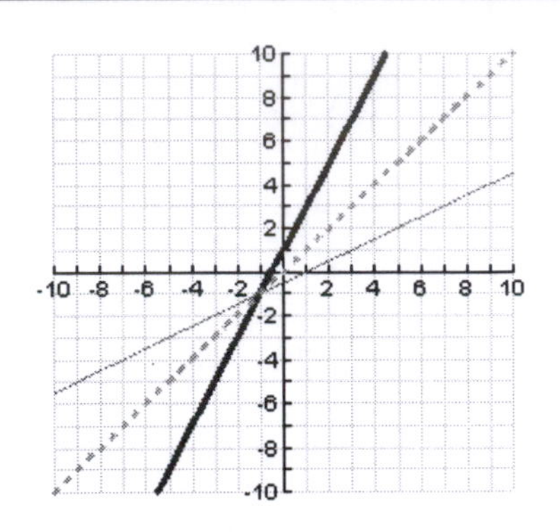

Notes on the Graphs:
Thick, solid curve is the graph of *f*.
Thin, solid curve is the graph of f^{-1}.
Thick, dotted curve is the graph of $y = x$.

27.

Given: $f(x) = \sqrt{x-1},\ x \geq 1$

$f^{-1}(x) = x^2 + 1,\ x \geq 0$

$f\left(f^{-1}(x)\right) = \sqrt{\left(x^2+1\right)-1} = \sqrt{x^2} = \underbrace{|x| = x}_{\text{Since } x \geq 0}$

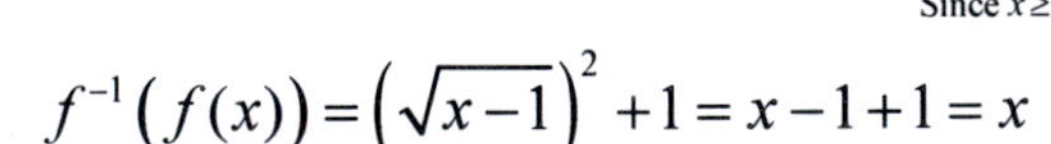

$f^{-1}\left(f(x)\right) = \left(\sqrt{x-1}\right)^2 + 1 = x - 1 + 1 = x$

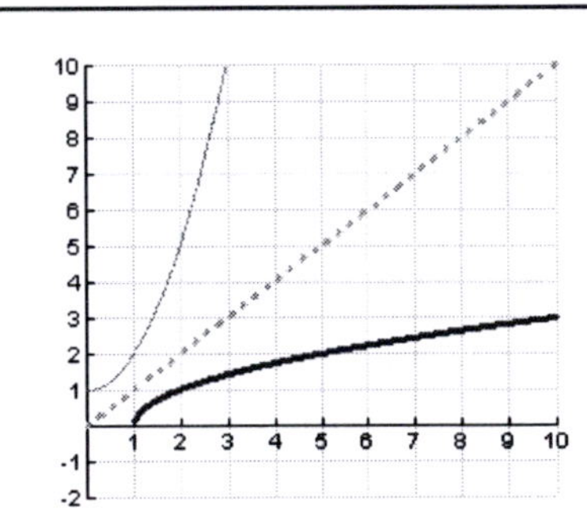

Notes on the Graphs:
Thick, solid curve is the graph of *f*.
Thin, solid curve is the graph of f^{-1}.
Thick, dotted curve is the graph of $y = x$.

29.

Given: $f(x)=\frac{1}{x},\ f^{-1}(x)=\frac{1}{x}$

$$f\left(f^{-1}(x)\right)=\frac{1}{\frac{1}{x}}=x$$

$$f^{-1}\left(f(x)\right)=\frac{1}{\frac{1}{x}}=x$$

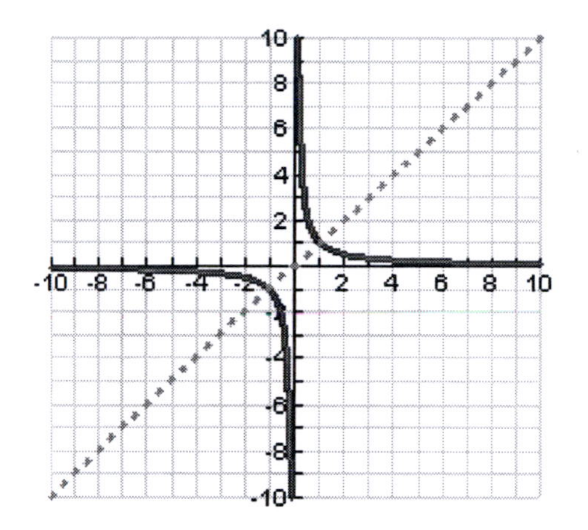

Notes on the Graphs:
Thick, solid curve is the graph of *f*.
Thin, solid curve is the graph of f^{-1}.
Thick, dotted curve is the graph of $y=x$.

31.

Given: $f(x)=\frac{1}{2x+6},\ f^{-1}(x)=\frac{1}{2x}-3$

$$f\left(f^{-1}(x)\right)=\frac{1}{2\left(\frac{1}{2x}-3\right)+6}=\frac{1}{\frac{1}{x}-6+6}$$

$$=\frac{1}{\frac{1}{x}}=x$$

$$f^{-1}\left(f(x)\right)=\frac{1}{2\left[\frac{1}{2x+6}\right]}-3=\frac{1}{\frac{1}{x+3}}-3$$

$$=x+3-3=x$$

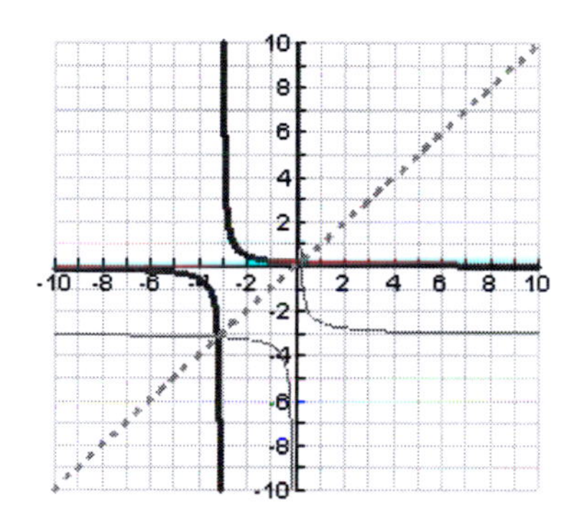

Notes on the Graphs:
Thick, solid curve is the graph of *f*.
Thin, solid curve is the graph of f^{-1}.
Thick, dotted curve is the graph of $y=x$.

33.

Given: $f(x)=\dfrac{x+3}{x+4},\ f^{-1}(x)=\dfrac{3-4x}{x-1}$

$$f\left(f^{-1}(x)\right)=\frac{\frac{3-4x}{x-1}+3}{\frac{3-4x}{x-1}+4}=\frac{\frac{3-4x+3(x-1)}{x-1}}{\frac{3-4x+4(x-1)}{x-1}}$$

$$=\frac{\frac{\cancel{3}-4x+3x-\cancel{3}}{x-1}}{\frac{3-\cancel{4x}+\cancel{4x}-4}{x-1}}=\frac{\frac{-x}{x-1}}{\frac{-1}{x-1}}$$

$$=\frac{-x}{\cancel{x-1}}\cdot\frac{\cancel{x-1}}{-1}=x$$

$$f^{-1}\left(f(x)\right)=\frac{3-4\left(\frac{x+3}{x+4}\right)}{\left(\frac{x+3}{x+4}\right)-1}=\frac{3-\frac{4x+12}{x+4}}{\frac{x+3}{x+4}-1}$$

$$=\frac{\frac{3x+\cancel{12}-4x-\cancel{12}}{x+4}}{\frac{\cancel{x}+3-\cancel{x}-4}{x+4}}=\frac{\frac{-x}{x+4}}{\frac{-1}{x+4}}$$

$$=\frac{-x}{\cancel{x+4}}\cdot\frac{\cancel{x+4}}{-1}=x$$

Notes on the Graphs:
Thick, solid curve is the graph of f.
Thin, solid curve is the graph of f^{-1}.
Thick, dotted curve is the graph of $y=x$.

35.

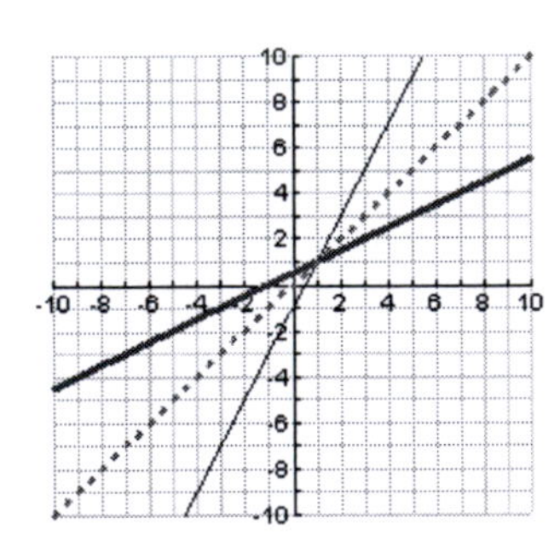

Notes on the Graphs:
Thin, solid curve is the graph of f.
Thick, dotted curve is the graph of $y=x$.
Thick, solid curve is the graph of f^{-1}.

37.

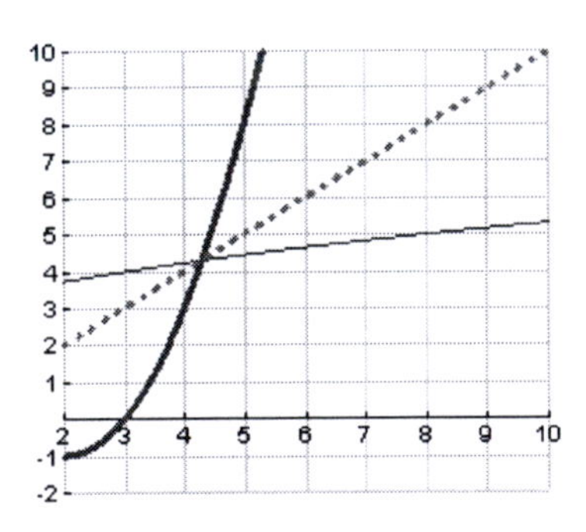

Notes on the Graphs:
Thin, solid curve is the graph of f.
Thick, dotted curve is the graph of $y=x$.
Thick, solid curve is the graph of f^{-1}.

39.

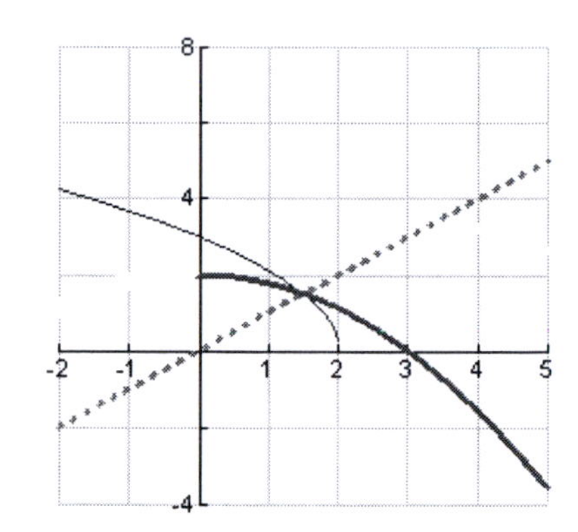

Notes on the Graphs:
Thin, solid curve is the graph of *f*.
Thick, dotted curve is the graph of $y = x$.
Thick, solid curve is the graph of f^{-1}.

41.

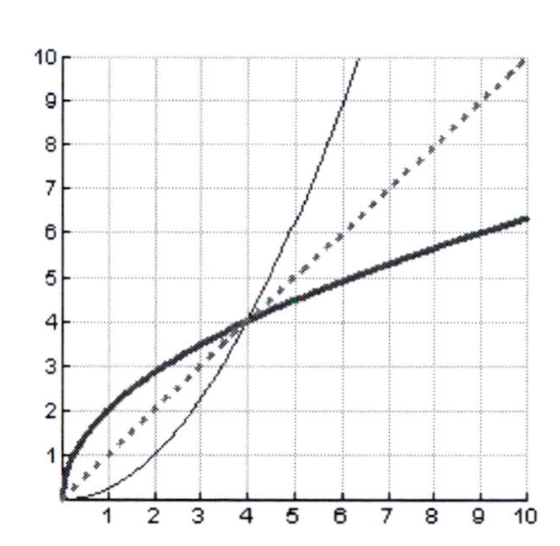

Notes on the Graphs:
Thin, solid curve is the graph of *f*.
Thick, dotted curve is the graph of $y = x$.
Thick, solid curve is the graph of f^{-1}.

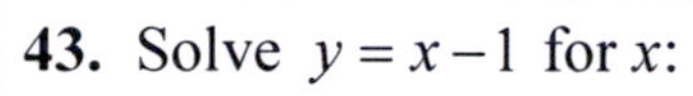

43. Solve $y = x - 1$ for *x*:

$$x = y + 1$$

Thus, $f^{-1}(x) = x + 1$.

Domains:
$dom(f) = rng(f^{-1}) = (-\infty, \infty)$
$rng(f) = dom(f^{-1}) = (-\infty, \infty)$

45. Solve $y = -3x + 2$ for *x*:

$$x = -\tfrac{1}{3}(y - 2)$$

Thus, $f^{-1}(x) = -\tfrac{1}{3}(x - 2) = -\tfrac{1}{3}x + \tfrac{2}{3}$.

Domains:
$dom(f) = rng(f^{-1}) = (-\infty, \infty)$
$rng(f) = dom(f^{-1}) = (-\infty, \infty)$

47. Solve $y = x^3 + 1$ for *x*:

$$x = \sqrt[3]{y - 1}$$

Thus, $f^{-1}(x) = \sqrt[3]{x - 1}$.

Domains:
$dom(f) = rng(f^{-1}) = (-\infty, \infty)$
$rng(f) = dom(f^{-1}) = (-\infty, \infty)$

49. Solve $y = \sqrt{x - 3}$ for *x*:

$$x = y^2 + 3$$

Thus, $f^{-1}(x) = x^2 + 3$.

Domains:
$dom(f) = rng(f^{-1}) = [3, \infty)$
$rng(f) = dom(f^{-1}) = [0, \infty)$

51. Solve $y = x^2 - 1$ for *x*:

$$x = \sqrt{y + 1}$$

Thus, $f^{-1}(x) = \sqrt{x + 1}$.

Domains:
$dom(f) = rng(f^{-1}) = [0, \infty)$
$rng(f) = dom(f^{-1}) = [-1, \infty)$

53. Solve $y=(x+2)^2-3$ for x:

$$y+3=(x+2)^2$$
$$\sqrt{y+3}=x+2 \text{ (since } x\geq -2)$$
$$-2+\sqrt{y+3}=x$$

Thus, $f^{-1}(x)=-2+\sqrt{x+3}$.

Domains:
$dom(f)=rng(f^{-1})=[-2,\infty)$
$rng(f)=dom(f^{-1})=[-3,\infty)$

55. Solve $y=\frac{2}{x}$ for x:

$$xy=2$$
$$x=\frac{2}{y}$$

Thus, $f^{-1}(x)=\frac{2}{x}$.

Domains:
$dom(f)=rng(f^{-1})=(-\infty,0)\cup(0,\infty)$
$rng(f)=dom(f^{-1})=(-\infty,0)\cup(0,\infty)$

57. Solve $y=\frac{2}{3-x}$ for x:

$$(3-x)y=2$$
$$3y-xy=2$$
$$xy=3y-2$$
$$x=\frac{3y-2}{y}$$

Thus, $f^{-1}(x)=\frac{3x-2}{x}=3-\frac{2}{x}$.

Domains:
$dom(f)=rng(f^{-1})=(-\infty,3)\cup(3,\infty)$
$rng(f)=dom(f^{-1})=(-\infty,0)\cup(0,\infty)$

59. Solve $y=\frac{7x+1}{5-x}$ for x:

$$y(5-x)=7x+1$$
$$5y-xy=7x+1$$
$$-7x-xy=1-5y$$
$$-x(7+y)=1-5y$$
$$x=\frac{5y-1}{7+y}$$

Thus, $f^{-1}(x)=\frac{5x-1}{x+7}$.

Domains:
$dom(f)=rng(f^{-1})=(-\infty,5)\cup(5,\infty)$
$rng(f)=dom(f^{-1})=(-\infty,-7)\cup(-7,\infty)$

61. Not one-to-one

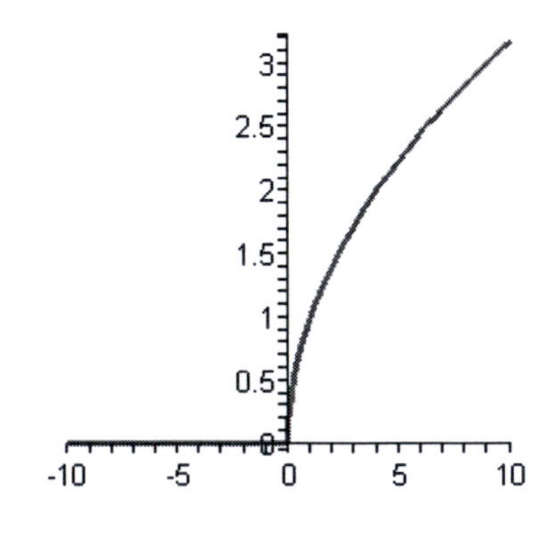

63. One-to-one
The portion of the graph for non-negative x values is as follows. The graph for negative x-values is merely a reflection of this graph over the origin.

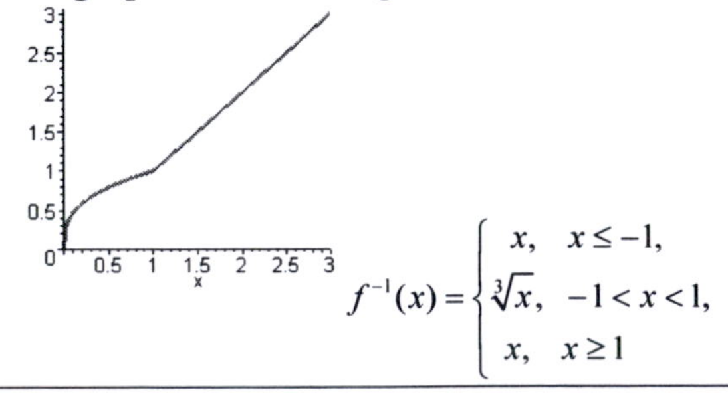

$$f^{-1}(x)=\begin{cases} x, & x\leq -1, \\ \sqrt[3]{x}, & -1<x<1, \\ x, & x\geq 1 \end{cases}$$

65. Solve $y = \frac{9}{5}x + 32$ for x:

$$y - 32 = \frac{9}{5}x$$
$$\frac{5}{9}(y - 32) = x$$

So, $f^{-1}(x) = \frac{5}{9}(x - 32)$. The inverse function represents the conversion from degrees Fahrenheit to degrees Celsius.

67. Let x = number of boats entered. The cost function is

$$C(x) = \begin{cases} 250x, & 0 \le x \le 10 \\ \underbrace{2500 + 175(x - 10)}_{= 175x + 750}, & x > 10 \end{cases}$$

To calculate $C^{-1}(x)$, we calculate the inverse of each piece separately:

For $0 \le x \le 10$: Solve $y = 250x$ for x: $x = \frac{y}{250}$. So, $C^{-1}(x) = \frac{x}{250}$, for $0 \le x \le 2500$.

For $x > 10$: Solve $y = 175x + 750$ for x: $x = \frac{y - 750}{175}$. So, $C^{-1}(x) = \frac{x - 750}{175}$, for $x > 2500$.

Thus, the inverse function is given by:

$$C^{-1}(x) = \begin{cases} \frac{x}{250}, & 0 \le x \le 2500 \\ \frac{x - 750}{175}, & x > 2500 \end{cases}$$

69. Let x = number of hours worked. Then, the take home pay is given by

$$E(x) = \underbrace{10x}_{\text{\$10 per hour, for } x \text{ hours}} - \underbrace{0.25(10x)}_{\text{Amount withheld for taxes}} = 7.5x.$$

To calculate E^{-1}, solve $y = 7.5x$ for x: $x = \frac{y}{7.5}$. So, $E^{-1}(x) = \frac{x}{7.5}$, $x \ge 0$.

The inverse function tells you how many hours you need to work to attain a certain take home pay.

71. Since we are only looking at 1 24-hour period, the domain is [0, 24].
For the range, note that $T(t)$ is always increasing (being a translate of t^3) and so, its minimum is $T(0)$ and its maximum is $T(24)$. So, the range is the approximate interval [97.5528, 101.70].

73. The domain of $T^{-1}(t)$ is the range of $T(t)$ and is [97.5528, 101.70].
The range of $T^{-1}(t)$ is the domain of $T(t)$ and is [0, 24].

75. Not a function since the graph does not pass the vertical line test.

77. Must restrict the domain to a portion on which f is one-to-one, say $x \ge 0$. Then, the calculation will be valid.

79. False. In fact, no even function can be one-to-one since the condition $f(x) = f(-x)$ implies that the horizontal line test is violated.

81. False. Consider $f(x) = x$. Then, $f^{-1}(x) = x$ also.

83. $(b, 0)$ since the x and y coordinates of all points on the graph of f are switched to get the corresponding points on the graph of f^{-1}.

85. The equation of the unit circle is $x^2 + y^2 = 1$. The portion in Quadrant I is given by

$$y = \sqrt{1-x^2},\ 0 \le x \le 1,\ 0 \le y \le 1.$$

To calculate the inverse of this function, solve for *x*:

$$y^2 = 1 - x^2, \text{ which gives us } x = \sqrt{1-y^2}$$

So, $f^{-1}(x) = \sqrt{1-x^2}$, $0 \le x \le 1$. The domain and range of both are $[0,1]$.

87. As long as $m \ne 0$ (that is, while the graph of *f* is not a horizontal line), it is one-to-one.

89. Not one-to-one

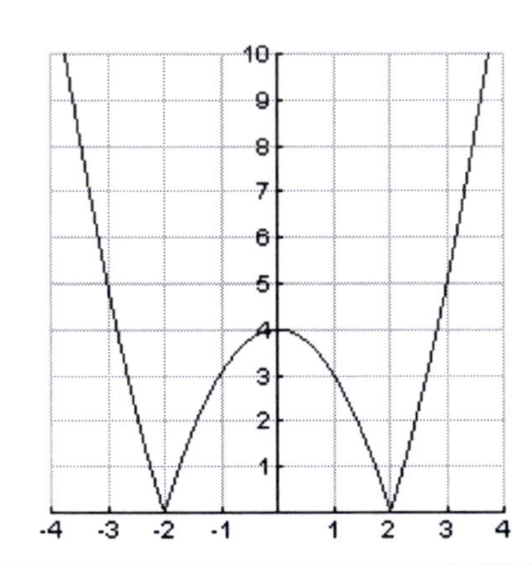

91. Not one-to-one

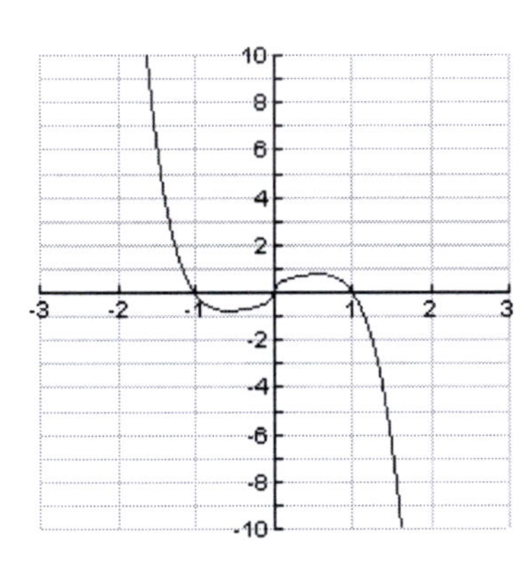

93. Not inverses. Had we restricted the domain of the parabola to $[0,\infty)$, then they would have been.

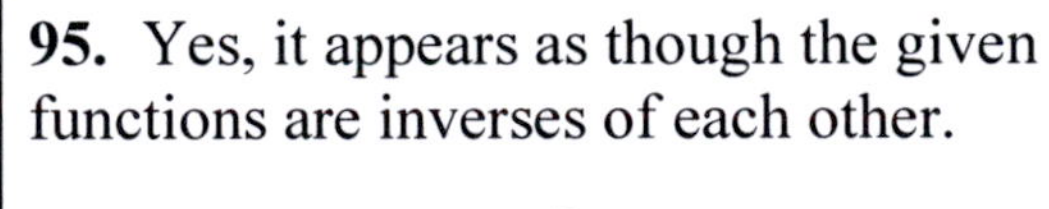

95. Yes, it appears as though the given functions are inverses of each other.

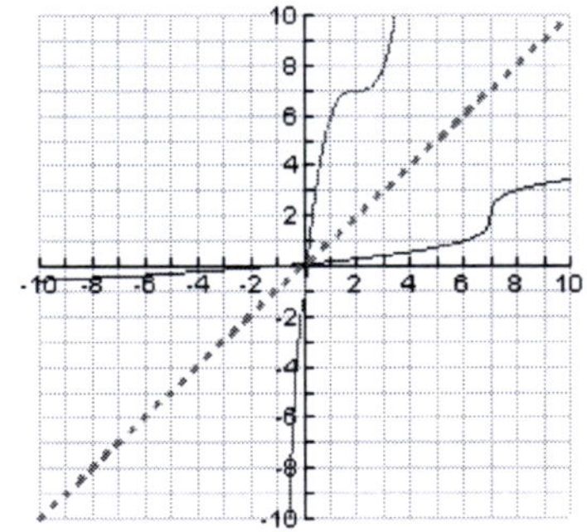

Section 3.6 Solutions --

1. $y = kx$	**3.** $V = kx^3$
5. $z = km$	**7.** $f = \dfrac{k}{\lambda}$

9. $F = \frac{kw}{L}$

11. $v = kgt$

13. $R = \frac{k}{PT}$

15. $y = k\sqrt{x}$

17. The general equation is $d = kt$. Using the fact that $d = k(1) = r$, we see that $k = r$. So, $\boxed{d = rt}$.

19. The general equation is $V = klw$. Using the fact that $V = k(2)(3) = 6h$ we see that $k = h$. So, $\boxed{V = lwh}$.

21. The general equation is $d = kr^2$. Using the fact that $A = k(3)^2 = 9\pi$, we see that $k = \pi$. So, $\boxed{A = \pi r^2}$.

23. The general equation is $V = khr^2$. Using the fact that $V = k\left(\frac{4}{\pi}\right)(2)^2 = 1$, we see that $k = \frac{\pi}{16}$. So, $\boxed{V = \frac{\pi}{16} hr^2}$.

25. The general equation is $V = \frac{k}{P}$. Using the fact that $V = \frac{k}{400} = 1000$, we see that $k = 400{,}000$. Hence, $\boxed{V = \frac{400{,}000}{P}}$.

27. The general equation is $F = \frac{k}{\lambda L}$. Using the fact that $F = \frac{k}{(10^{-6}m)(10^5 m)} = 20\pi$, we see that $k = 2\pi$. So, $\boxed{F = \frac{2\pi}{\lambda L}}$.

29. The general equation is $t = \frac{k}{s}$. Using the fact that $t = \frac{k}{8} = 2.4$, we see that $k = 2.4(8) = 19.2$. Hence, $\boxed{t = \frac{19.2}{s}}$.

31. The general equation is $R = \frac{k}{I^2}$. Using the fact that $R = \frac{k}{(3.5)^2} = 0.4$, we see that $k = (3.5)^2(0.4) = 4.9$. Hence, $\boxed{R = \frac{4.9}{I^2}}$.

33. The general equation is $R=\frac{kL}{A}$. Using the fact that $R=\frac{k(20)}{(0.4)}=0.5$, we see that $50k=0.5$, so that $k=0.01$. Hence, $\boxed{R=\frac{0.01L}{A}}$.

35. The general equation is $F=\frac{km_1m_2}{d^2}$. Using the fact that $F=\frac{k(8)(16)}{(0.4)^2}=20$, we see that $k=\frac{20(0.4)^2}{(8)(16)}=0.025$. Hence, $\boxed{F=\frac{0.025m_1m_2}{d^2}}$.

37. Assume that $W=kH$. We need to determine k. Using Jason's data, we see that $172.50=23k$, so that $k=7.5$. So, $\boxed{W=7.5H}$. (Note that Valerie's data also satisfies this equation.)

39. Let S = speed of the object, and M = Mach number. We are given that $S=kM$. We also know that when $S=760$ mph (at sea level), $M=1$. As such, $k=760$. Hence, for U.S. Navy Blue Angels, $S=760(1.7)=\boxed{1,292\text{ mph}}$.

41. We are given that $F=kH$. Using the fact that $F=11$ when $H=6.8$, we see that $11=6.8k$, so that $k=1.618$. Hence, $\boxed{F=1.618H}$.

43. Assume Hooke's law holds: $F=kx$. Using the fact that $F=30\text{N}$ when $x=10$ cm, we see that $k=3\ \text{N}/\text{cm}$. So, $F=3x$. As such, $72\text{N}=\left(3\ \text{N}/\text{cm}\right)x$, so that $\boxed{x=24\text{ cm}}$.

45. Here, $c=kx$. We must first find k:

$$17.70=k(236),\text{ so that } k=\frac{17.70}{236}.$$

The cost for 500 minutes is $c=\frac{17.70}{236}(500)=\$37.50$

47. Let D = demand for Levi's jeans and P = price for Levi's jeans.

We are told that $D=\frac{k}{P}$. Using the given information for Flare 519 jeans (namely that $P=20$ when $D=300{,}000$) yields $300{,}000=\frac{k}{20}$ so that $k=6{,}000{,}000$.

Thus, for Vintage Flare jeans, the demand D is given by: $D=\frac{6{,}000{,}000}{300}=\boxed{20{,}000}$.

49. Use the formula $I = \dfrac{k}{D^2}$. Using the data for Earth, we obtain:

$$1400 \, {}^{w}\!/_{m^2} = \frac{k}{(150{,}000 \text{ km})^2} \text{ so that}$$

$$k = \left(1400 \, {}^{w}\!/_{m^2}\right)(150{,}000 \text{ km})^2 = \left(1400 \, {}^{w}\!/_{m^2}\right)(150{,}000{,}000 \, m)^2 = 3.15\times10^{19} \, w$$

Hence, the intensity for Mars is given by:

$$I = \frac{3.15\times10^{19}\, w}{(228{,}000 \, km)^2} = \frac{3.15\times10^{19}\, w}{(228{,}000{,}000 \, m)^2} \approx \boxed{600 \, {}^{w}\!/_{m^2}}$$

51. Use the formula $I = kPt$.

Bank of America:

$750 = k(25{,}000)(2)$, so that $k = 0.015$, which corresponds to 1.5%.

Navy Federal Credit Union:

$1500 = k(25{,}000)(2)$, so that $k = 0.03$, which corresponds to 3%.

53. Use the formula $P = \dfrac{kT}{V}$ with $T = 300\text{K}$, $P = 1$ atm., and $V = 4$ ml to obtain

$$k = \frac{PV}{T} = \frac{(1 \text{ atm})(4)}{300} = \frac{4}{300}. \text{ Thus, } P = \frac{4}{300}\left(\frac{275}{4}\right) = \boxed{\frac{11}{12} \text{ or } 0.92 \text{ atm}}.$$

55. Should be y is inversely proportional to x.

57. True. $A = \frac{1}{2}bh$, so area is directly proportional to both base and height.

59. b

61. Use the equation $\sigma_{p_1}^2 = \alpha C_n^2 k^{7/6} L^{11/6}$ with the following information (all converted to meters):

$C_n^2 = 1.0\times10^{-13}$, $L = 2000\text{m}$, $\lambda = 1.55\times10^{-6}\text{m}$ (so that $k = \dfrac{2\pi}{1.55\times10^{-6}\text{m}}$), and $\sigma_{p_1}^2 = 7.1$.

Substituting this information into the equation yields α:

$$7.1 = \alpha\left(1.0\times10^{-13}\right)\left(\frac{2\pi}{1.55\times10^{-6}}\right)^{7/6} 2000^{11/6}$$

so that

$$\alpha = \frac{7.1}{\left(1.0\times10^{-13}\right)\left(\frac{2\pi}{1.55\times10^{-6}}\right)^{7/6} 2000^{11/6}} \approx 1.23.$$

Thus, the equation is given by $\boxed{\sigma_{p_1}^2 = 1.23 C_n^2 k^{7/6} L^{11/6}}$.

63. (a) The least squares regression line is $y = 2.93x + 201.72$, and is plotted as seen below:

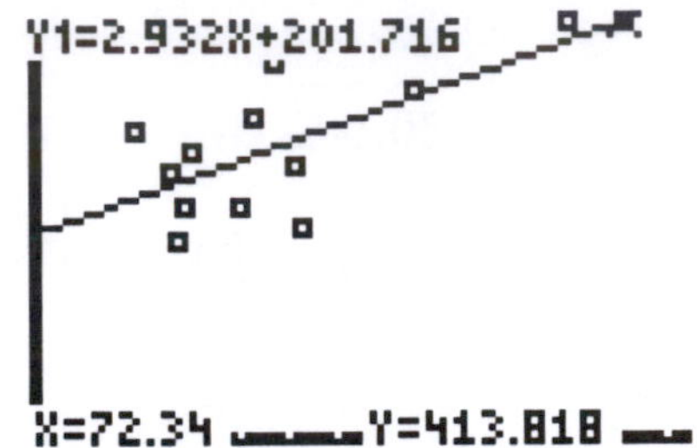

(b) The variation constant is 120.07 and the equation of the direct variation is $y = 120.074x^{0.259}$.

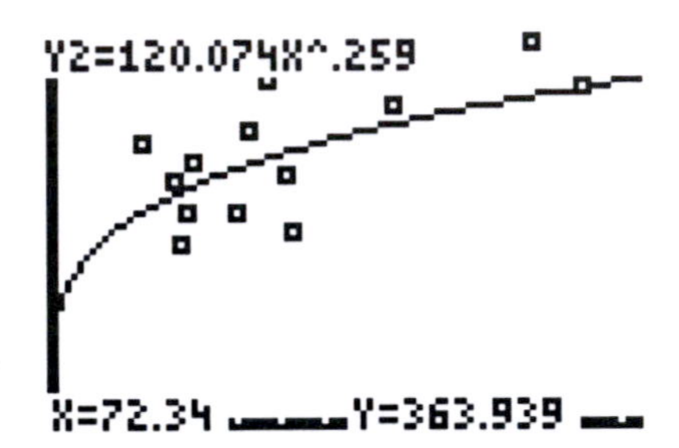

(c) When the oil price is $72.70 per barrel in September 2006, the predicted stock index obtained from the least squares regression line is 415, and the value from the equation of direct variation is 364. In this case, the least squares regression line provides a closer approximation to the actual value, 417.

65. (a) The least squares regression line is $y = -141.73x + 2{,}419.35$. The following is the sequence of commands, and screen captures, to use on the TI-8*.

```
LinReg(ax+b) L4,      LinReg
L3,Y1                   y=ax+b
                        a=-141.732
                        b=2419.351

Plot1 Plot2 Plot3       WINDOW
On Off                   Xmin=0
Type:                    Xmax=10
                         Xscl=1
Xlist:L4                 Ymin=0
Ylist:L3                 Ymax=2500
Mark: □ + ·              Yscl=1
                         Xres=1
```

Then, the graph of the least squares regression line, with the scatterplot, is given by:

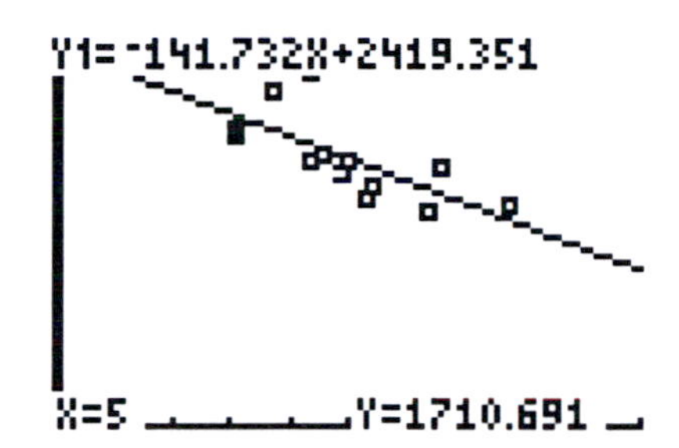

(b) The variation constant is 3,217.69 and the equation of the inverse variation is $y = \dfrac{3217.69}{x^{0.41}}$. The following is the sequence of commands, and screen captures, to use on the TI-8*.

```
PwrReg L4,L3,Y2     PwrReg
                     y=a*x^b
                     a=3217.691
                     b=-.410
```

The graph of the curve of inverse variation,with the scatter plot, is given by:

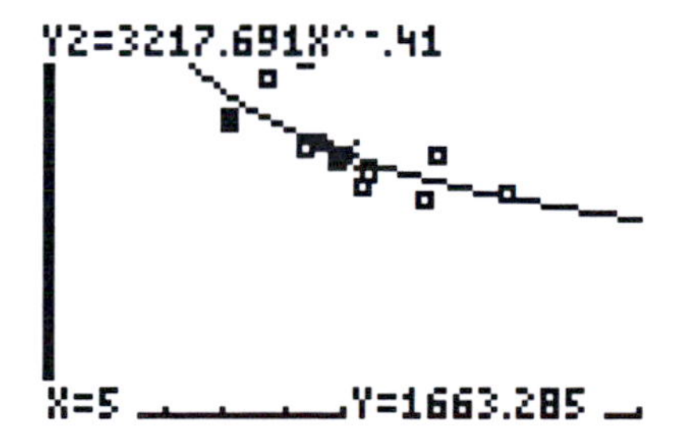

Continued onto the next page…

(c) When the 5-year maturity rate is 5.02% in September 2006, the predicted number of housing units obtained from the least squares regression line is 1708, and the equation of inverse variation is 1661. The equation of the least squares regression line provides a closer approximation to the actual value, 1861. The picture of the least squares line,with the scatter plot, as well as the computations using the TI-8* are as follows:

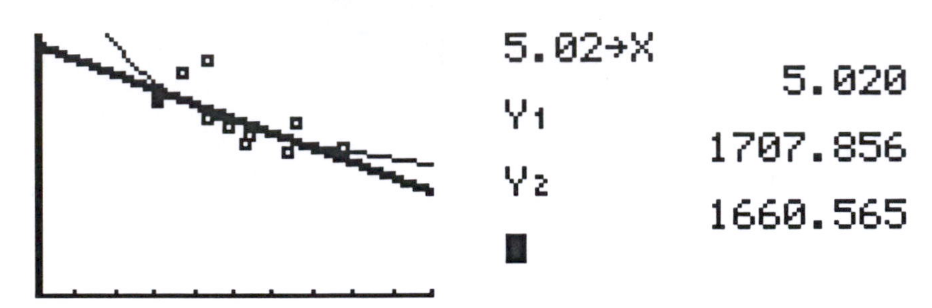

67. (a) The equation is approximately $y = 0.218x + 0.898$

(b) About $2.427 per gallon. Yes, it is very close to the actual price at $2.425 per gallon.

(c) $3.083

Chapter 3 Review Solutions --

1. Yes	**3.** Yes
5. No, since both $(0,6)$ and $(0,-6)$ satisfy the equation, so that the graph fails the vertical line test.	
7. Yes	**9.** No, since the graph fails the vertical line test.
11. (a) 2 **(b)** 4 **(c)** when $x = -3,\ 4$	**13. (a)** 0 **(b)** -2 **(c)** when $x \approx -5, 2$
15. $f(3) = 4(3) - 7 = \boxed{5}$	**17.** $f(-7)\cdot g(3) = \left(4(-7)-7\right)\cdot\left\|3^2+2(3)+4\right\| = -35\left\|19\right\| = \boxed{-665}$

19. $\dfrac{f(2)-F(2)}{g(0)} = \dfrac{\left(4(2)-7\right)-\left(2^2+4(2)-3\right)}{4} = \dfrac{1-9}{4} = \boxed{-2}$

21.

$$\frac{f(3+h)-f(3)}{h} = \frac{\left(4(3+h)-7\right)-\left(4(3)-7\right)}{h} = \frac{5+4h-5}{h} = \boxed{4}$$

23. $(-\infty,\infty)$

25. $(-\infty,-4)\cup(-4,\infty)$

27. We need $x-4\geq 0$, so the domain is $[4,\infty)$.

29. Solve $2=f(5)=\dfrac{D}{5^2-16}$ for D: $2=\frac{D}{9}$, so that $\boxed{D=18}$.

31.

$f(-x)=2(-x)-7=-(2x+7)\neq f(x)$

So, not even.

$-f(-x)=-(-(2x+7))=2x+7\neq f(x)$

So, not odd.

Thus, $\boxed{\text{neither}}$.

33.

$h(-x)=(-x)^3-7(-x)=-\left(x^3-7x\right)\neq h(x)$

So, not even.

$-h(-x)=-\left(-\left(x^3-7x\right)\right)=x^3-7x=h(x)$

So, $\boxed{\text{odd}}$.

35.

$f(-x)=(-x)^{1/4}+(-x)=(-x)^{1/4}-x\neq f(x)$

So, not even.

$-f(-x)=-\left(x^{1/4}-x\right)=-x^{1/4}+x\neq f(x)$

So, not odd.

Thus, $\boxed{\text{neither}}$.

37.

$f(-x)=\dfrac{1}{(-x)^3}+3(-x)=-\left(\dfrac{1}{x^3}+3x\right)\neq f(x)$

So, not even.

$-f(-x)=-\left(-\left(\dfrac{1}{x^3}+3x\right)\right)=\dfrac{1}{x^3}+3x=f(x)$

So, $\boxed{\text{odd}}$.

39.

Domain	$[-4,7]$
Range	$[-2,4]$
Increasing	$(3,7)$
Decreasing	$(0,3)$
Constant	$(-4,0)$

41. $\dfrac{\left(4-2^2\right)-\left(4-0^2\right)}{2}=\boxed{-2}$

43. Domain: $(-\infty, \infty)$ Range: $(0, \infty)$

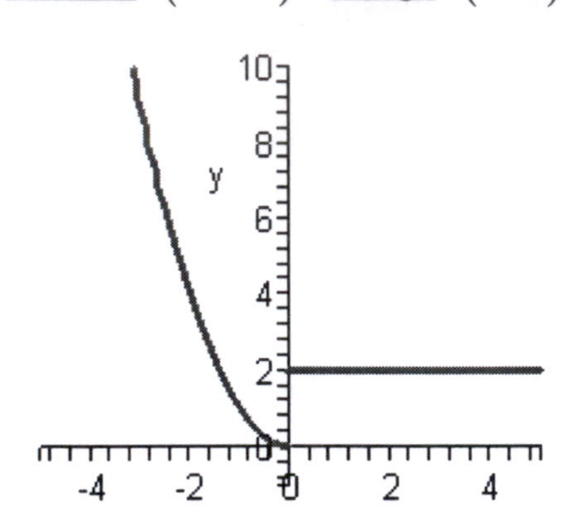

Notes on the graph: There is an open hole at (0,0), and a closed hole at (0,2).

45. Domain: $(-\infty, \infty)$ Range: $[-1, \infty)$

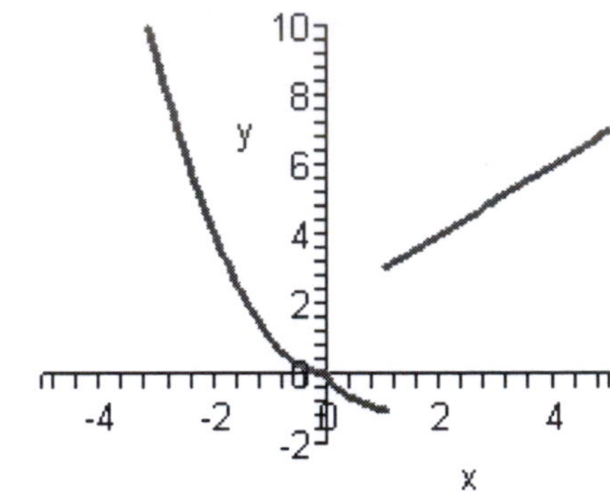

Notes on the graph: There is an open hole at (1,3), and a closed hole at $(1,-1)$.

47. Let x = number of 30-minute periods. Then,

$$C(x) = \begin{cases} 25, & x \le 2, \\ 25 + 10.50(x-2), & x > 2 \end{cases}.$$

49. Reflect the graph of x^2 over x-axis, then shift right 2 units and then up 4 units.

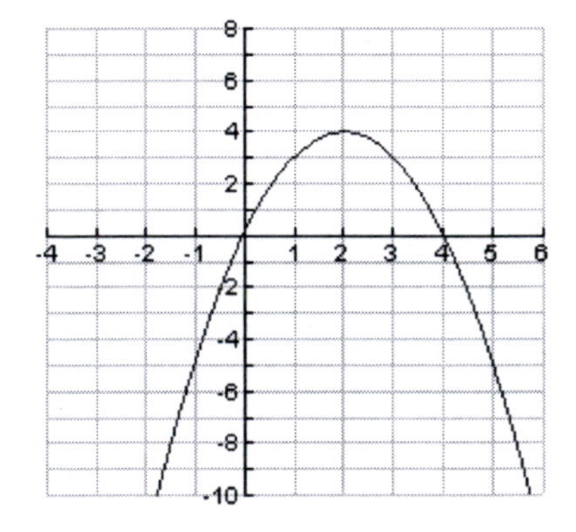

51. Shift the graph of $\sqrt[3]{x}$ right 3 units, and then up 2 units.

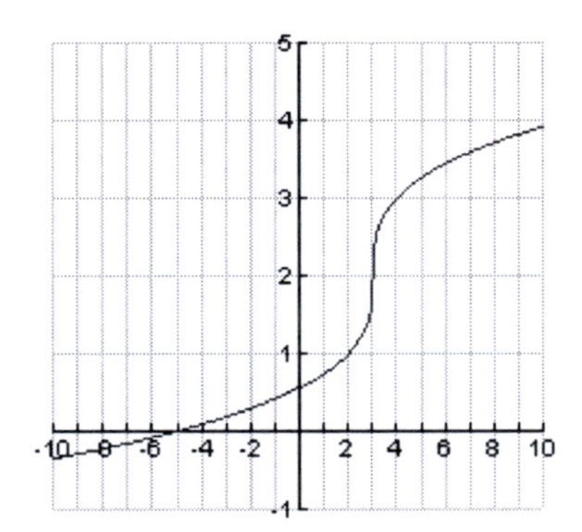

53. Reflect the graph of x^3 over x-axis, and then contract vertically by a factor of 2.

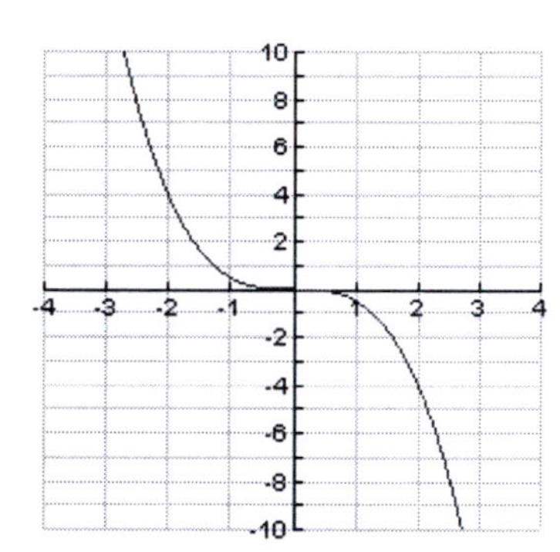

55.

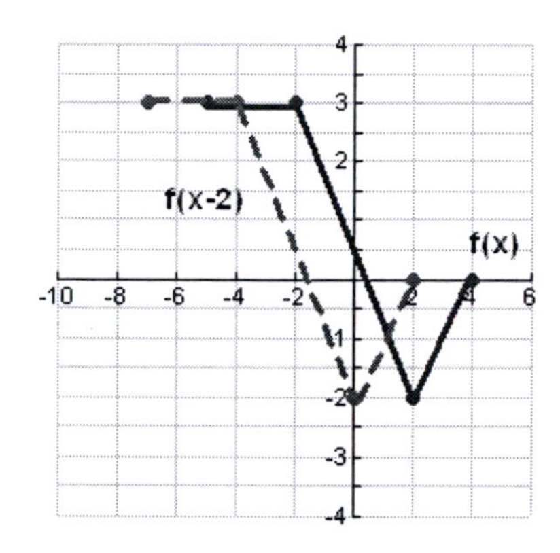

57.

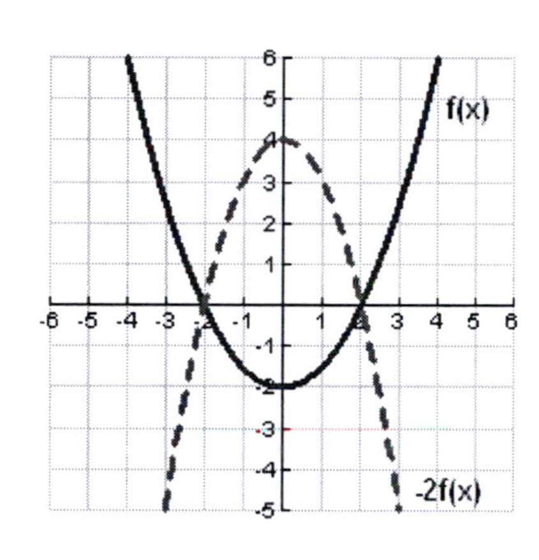

59. $y=\sqrt{x+3}$ Domain: $[-3,\infty)$

61. $y=\sqrt{x-2}+3$ Domain: $[2,\infty)$

63. $y=5\sqrt{x}-6$ Domain: $[0,\infty)$

65. $y=(x^2+4x+4)-8-4=(x+2)^2-12$

Domain: $\mathbb{R}$ or $(-\infty,\infty)$

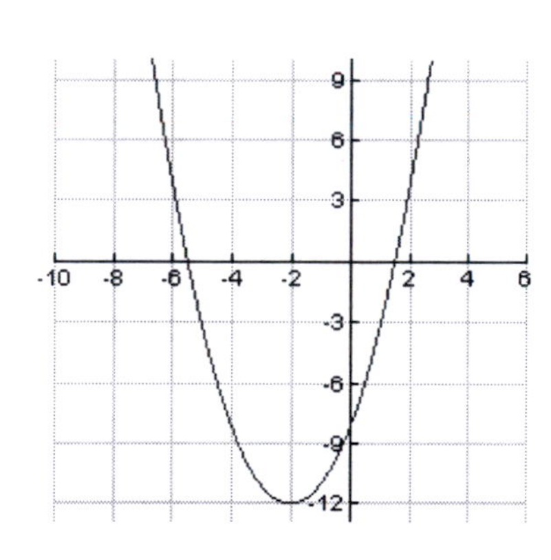

67.

$g(x)+h(x)=(-3x-4)+(x-3)=-2x-7$

$g(x)-h(x)=(-3x-4)-(x-3)=-4x-1$

$g(x)\cdot h(x)=(-3x-4)\cdot(x-3)=-3x^2+5x+12$

$\dfrac{g(x)}{h(x)}=\dfrac{-3x-4}{x-3}$

Domains:

$$\left.\begin{array}{r} dom(g+h) \\ dom(g-h) \\ dom(gh) \end{array}\right\}=(-\infty,\infty)$$

$$dom\left(\frac{g}{h}\right)=(-\infty,3)\cup(3,\infty)$$

69.

$$g(x)+h(x)=\tfrac{1}{x^2}+\sqrt{x}$$
$$g(x)-h(x)=\tfrac{1}{x^2}-\sqrt{x}$$
$$g(x)\cdot h(x)=\tfrac{1}{x^2}\cdot\sqrt{x}=\tfrac{1}{x^{3/2}}$$
$$\frac{g(x)}{h(x)}=\frac{\frac{1}{x^2}}{\sqrt{x}}=\frac{1}{x^{5/2}}$$

Domains:

$$\left.\begin{array}{r} dom(g+h) \\ dom(g-h) \\ dom(gh) \\ dom\left(\dfrac{g}{h}\right) \end{array}\right\}=(0,\infty)$$

71.

$$g(x)+h(x)=\sqrt{x-4}+\sqrt{2x+1}$$
$$g(x)-h(x)=\sqrt{x-4}-\sqrt{2x+1}$$
$$g(x)\cdot h(x)=\sqrt{x-4}\cdot\sqrt{2x+1}$$
$$\frac{g(x)}{h(x)}=\frac{\sqrt{x-4}}{\sqrt{2x+1}}$$

Domains:
Must have both $x-4\geq 0$ and $2x+1\geq 0$. So,

$$\left.\begin{array}{r} dom(f+g) \\ dom(f-g) \\ dom(fg) \end{array}\right\}=[4,\infty).$$

For the quotient, must have both $x-4\geq 0$ and $2x+1>0$. So,

$$dom\left(\frac{f}{g}\right)=[4,\infty).$$

73.

$$(f\circ g)(x)=3(2x+1)-4=6x-1$$
$$(g\circ f)(x)=2(3x-4)+1=6x-7$$

Domains:
$dom(f\circ g)=(-\infty,\infty)=dom(g\circ f)$

75.

$$(f\circ g)(x)=\frac{2}{\frac{1}{4-x}+3}=\frac{2}{\frac{1+3(4-x)}{4-x}}=\frac{2(4-x)}{13-3x}=\frac{8-2x}{13-3x}$$
$$(g\circ f)(x)=\frac{1}{4-\frac{2}{x+3}}=\frac{1}{\frac{4(x+3)-2}{x+3}}=\frac{x+3}{4x+10}$$

Domains:
$dom(f\circ g)=(-\infty,4)\cup\left(4,\tfrac{13}{3}\right)\cup\left(\tfrac{13}{3},\infty\right)$
$dom(g\circ f)=(-\infty,-3)\cup\left(-3,-\tfrac{5}{2}\right)\cup\left(-\tfrac{5}{2},\infty\right)$

77.

$$(f\circ g)(x)=\sqrt{x^2-4-5}=\sqrt{(x-3)(x+3)}$$
$$(g\circ f)(x)=\left(\sqrt{x-5}\right)^2-4=x-9$$

Domains:
$\underline{dom(f\circ g)}$: Need $(x-3)(x+3)\geq 0$.

CPs: ± 3

+	−3	−	3	+

So, $dom(g\circ f)=(-\infty,-3]\cup[3,\infty)$.

$\underline{dom(g\circ f)}$: Need $x-5\geq 0$. Thus, $dom(g\circ f)=[5,\infty)$.

79.

$g(3) = 6(3) - 3 = 15$

$f(g(3)) = f(15) = 4(15)^2 - 3(15) + 2 = 857$

$f(-1) = 4(-1)^2 - 3(-1) + 2 = 9$

$g(f(-1)) = g(9) = 6(9) - 3 = 51$

81.

$g(3) = |5(3) + 2| = 17$

$f(g(3)) = f(17) = \dfrac{17}{|2(17) - 3|} = \dfrac{17}{31}$

$f(-1) = \dfrac{-1}{|2(-1) - 3|} = -\dfrac{1}{5}$

$g(f(-1)) = g\left(-\dfrac{1}{5}\right) = \left|5\left(-\dfrac{1}{5}\right) + 2\right| = 1$

83.

$f(g(3)) = \left(\sqrt[3]{3-4}\right)^2 - \left(\sqrt[3]{3-4}\right) + 10$

$= (-1)^2 - (-1) + 10 = \boxed{12}$

$g(f(-1)) = g\left((-1)^2 + 1 + 10\right)$

$= \sqrt[3]{12 - 4} = \boxed{2}$

85. Let $f(x) = 3x^2 + 4x + 7,\quad g(x) = x - 2$.
Then, $h(x) = f(g(x))$.

87. Let $f(x) = \dfrac{1}{\sqrt{x}},\quad g(x) = x^2 + 7$.
Then, $h(x) = f(g(x))$.

89. The area of a circle with radius $r(t)$ is given by:

$$A(t) = \pi\left(r(t)\right)^2 = \pi\left(25\sqrt{t+2}\right)^2 = 625\pi(t+2)\ \text{in}^2$$

91. Yes

93. No, since both (2,3) and (3,3) lie on the graph of the function.

95. Yes

97. Yes

99. One-to-one

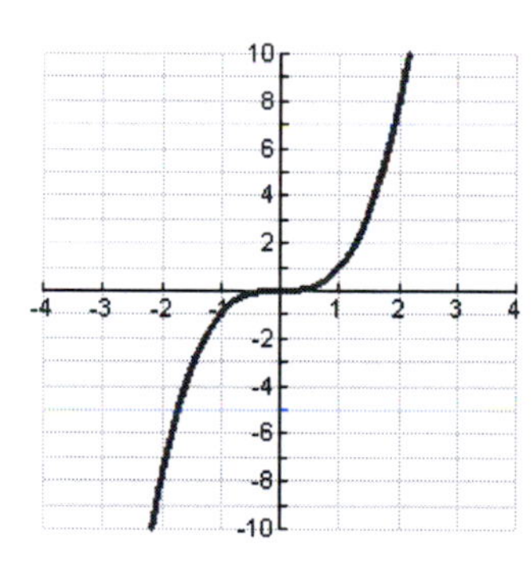

101.

$$f\left(f^{-1}(x)\right) = \not{3}\left(\frac{x-4}{\not{3}}\right) + 4 = x - 4 + 4 = x$$

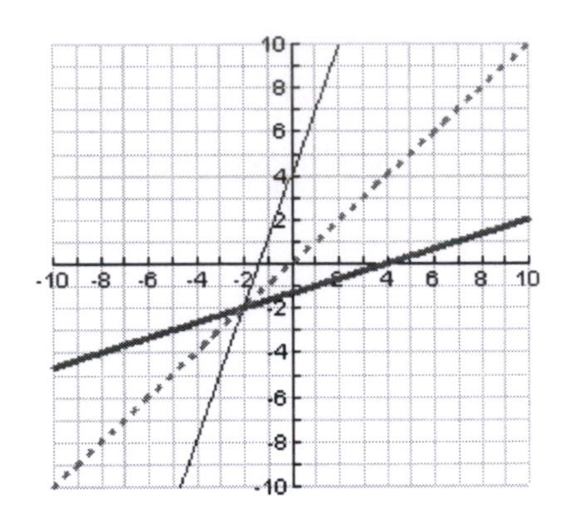

103.

$$f\left(f^{-1}(x)\right) = \sqrt{\left(x^2 - 4\right) + 4} = \sqrt{x^2} = x,$$
since $x \geq 0$.

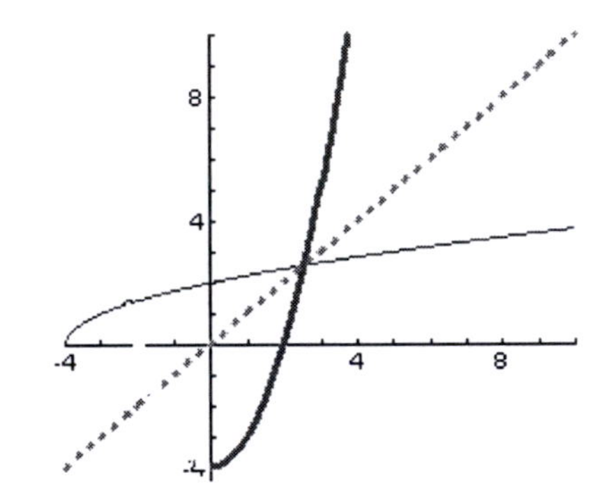

105. Solve $y = 2x + 1$ for x:

$$x = \tfrac{1}{2}(y-1)$$

Thus, $f^{-1}(x) = \tfrac{1}{2}(x-1) = \dfrac{x-1}{2}$.

Domains:
$dom(f) = rng\left(f^{-1}\right) = (-\infty, \infty)$
$rng(f) = dom\left(f^{-1}\right) = (-\infty, \infty)$

107. Solve $y = \sqrt{x+4}$ for x:

$$x = y^2 - 4$$

Thus, $f^{-1}(x) = x^2 - 4$.

Domains:
$dom(f) = rng\left(f^{-1}\right) = [-4, \infty)$
$rng(f) = dom\left(f^{-1}\right) = [0, \infty)$

109. Solve $y = \frac{x+6}{x+3}$ for x:

$$(x+3)y = x+6$$
$$xy + 3y = x+6$$
$$xy - x = 6 - 3y$$
$$x(y-1) = 6 - 3y$$
$$x = \frac{6-3y}{y-1}$$

Thus, $f^{-1}(x) = \frac{6-3x}{x-1}$.

Domains:

$dom(f) = rng(f^{-1}) = (-\infty,-3)\cup(-3,\infty)$

$rng(f) = dom(f^{-1}) = (-\infty,1)\cup(1,\infty)$

111. Let x = total dollars worth of products sold. Then, $S(x) = 22{,}000 + 0.08x$. Solving $y = 22{,}000 + 0.08x$ for x yields: $x = \frac{1}{0.08}(y - 22{,}000)$

Thus, $S^{-1}(x) = \frac{x - 22{,}000}{0.08}$. This inverse function tells you the sales required to earn a desired income.

113. The general equation is $C = kr$. Using the fact that $C = k(1) = 2\pi$, we see that $k = 2\pi$. So, $\boxed{C = 2\pi r}$.

115. The general equation is $A = kr^2$. Using the fact that $A = k5^2 = 25\pi$, we see that $k = \pi$. So, $\boxed{A = \pi r^2}$.

117. Assume that $W = kH$. We need to determine k. Using Cole's data, we see that $229.50 = 27k$, so that $k = 8.5$. So, $\boxed{W = 8.5H}$. (Note that Dickson's data also satisfies this equation.)

119. The graph of f is as follows:

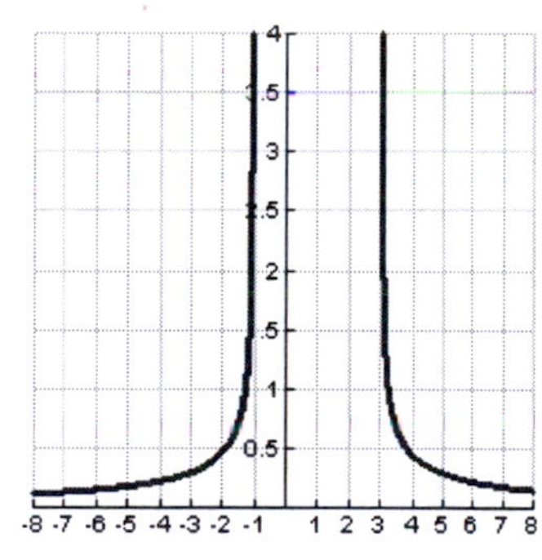

Domain of f: $\boxed{(-\infty,-1)\cup(3,\infty)}$

121.

Domain	$(-\infty,2)\cup(2,\infty)$
Range	$\{-1,0,1\}\cup(2,\infty)$
Increasing	$(2,\infty)$
Decreasing	$(-\infty,-1)$
Constant	$(-1,0)\cup(0,1)\cup(1,2)$

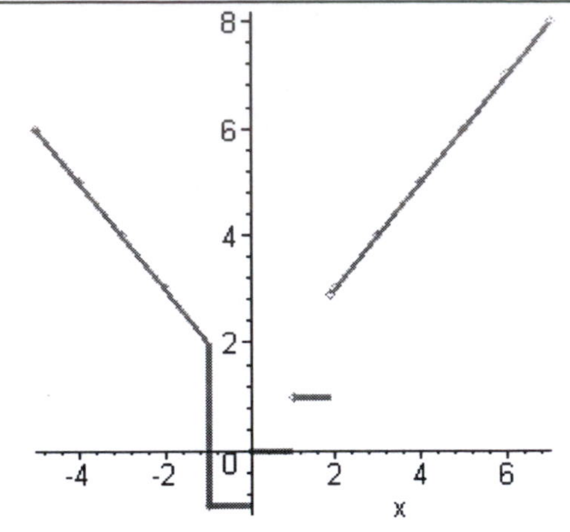

Note on the graph: The vertical line at $x=-1$ should NOT be a part of the graph.

123. The graphs of f and g are as follows:

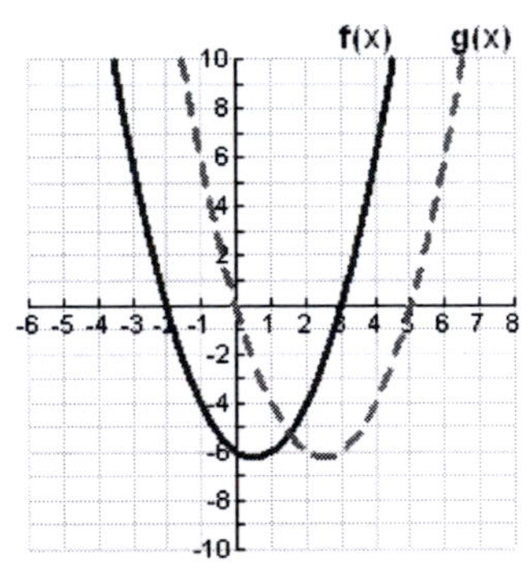

The graph of f can be obtained by shifting the graph of g two units to the left. That is, $f(x)=g(x+2)$.

125.

$y1=\sqrt{2x+3}$

$y2=\sqrt{4-x}$

$y3=\dfrac{y1}{y2}$

Domain of $y3=\boxed{[-1.5,4)}$

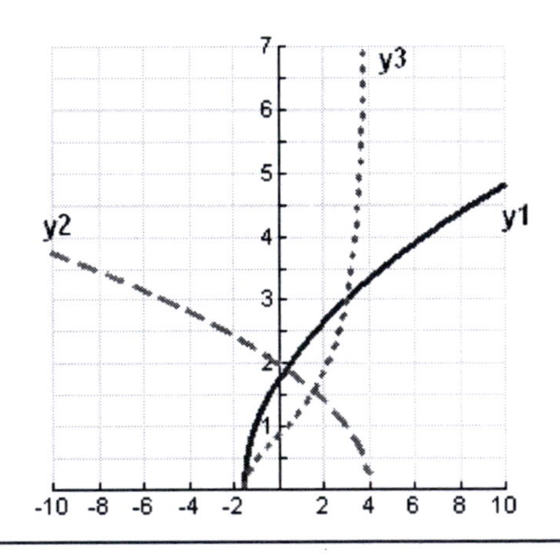

127. Yes, the function is one-to-one.

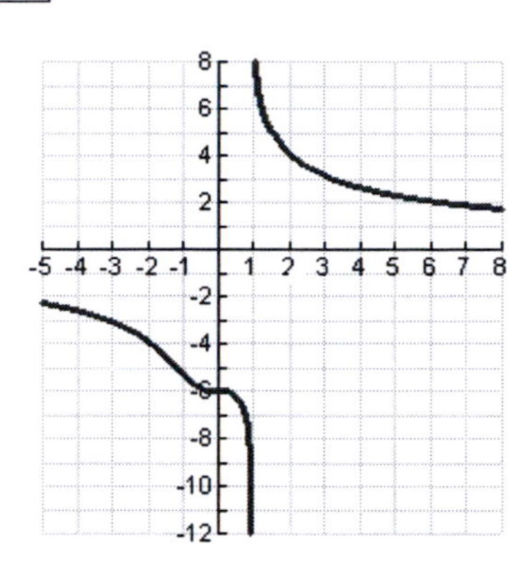

129. (a) The line is approximately given by $y = 64.03x + 127.06$.
(b) No, it is not close to the actual price of $517.20.
(c) $767.36

Chapter 3 Practice Test Solutions

1. b (Not one-to-one since both (0,3) and (−3,3) lie on the graph.)

3. c

5. $\left(\frac{f}{g}\right)(x) = \frac{\sqrt{x-2}}{x^2+11}$ Domain: $[2,\infty)$

7. $g(f(x)) = \left(\sqrt{x-2}\right)^2 + 11 = x - 2 + 11 = x + 9$

Domain: $[2,\infty)$

9. $f\left(g\left(\sqrt{7}\right)\right) = f\left(\left(\sqrt{7}\right)^2 + 11\right) = f(18) = \sqrt{18-2} = \boxed{4}$

11.

$$f(-x) = 9(-x)^3 + 5(-x) - 3 = -\left[9x^3 + 5x + 3\right] \neq f(x)$$

So, not even.

$$-f(-x) = -\left(-\left[9x^3 + 5x + 3\right]\right) = 9x^3 + 5x + 3 \neq f(x)$$

So, not odd. Thus, neither.

13. $f(x) = -\sqrt{x-3} + 2$

Reflect the graph of $\sqrt{x}$ over the x-axis, then shift right 3 units, and then up 2 units.

Domain: $[3,\infty)$ Range: $(-\infty, 2]$

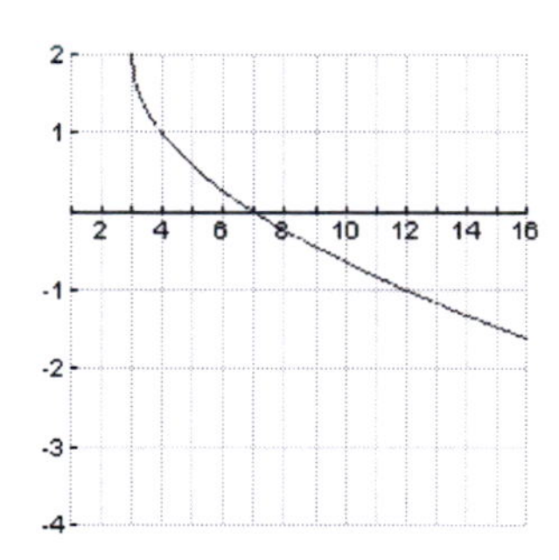

15. 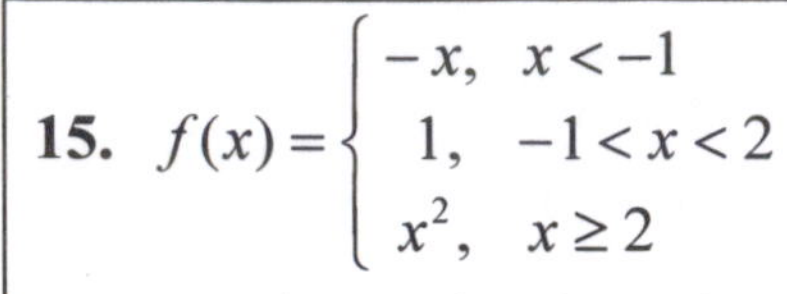

$$f(x)=\begin{cases} -x, & x<-1 \\ 1, & -1<x<2 \\ x^2, & x\ge 2 \end{cases}$$

Domain: $(-\infty,-1)\cup(-1,\infty)$

Range: $[1,\infty)$

Open holes at (-1,1) and (2,1); closed hole at (2,4)

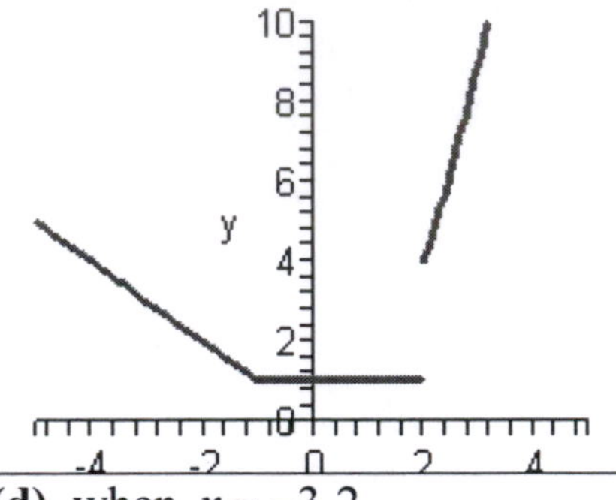

17. **(a)** -2 **(b)** 4 **(c)** -3 **(d)** when $x=-3,2$

19.

$$\frac{\left(3(x+h)^2-4(x+h)+1\right)-\left(3x^2-4x+1\right)}{h}=\frac{3x^2+6xh+3h^2-4x-4h+1-3x^2+4x-1}{h}$$

$$=\frac{h\left(6x+3h-4\right)}{h}=\boxed{6x+3h-4}$$

21.

$$\frac{\left(64-16(2)^2\right)-\left(64-16(0)^2\right)}{2}=\frac{0-64}{2}=\boxed{-32}$$

23. Solve $y=\sqrt{x-5}$ for x:

$$y^2=x-5$$

$$y^2+5=x$$

Thus, $f^{-1}(x)=x^2+5$.

Domains:

$dom(f)=rng(f^{-1})=[5,\infty)$

$rng(f)=dom(f^{-1})=[0,\infty)$

25. Solve $y=\dfrac{2x+1}{5-x}$ for x:

$$(5-x)y=2x+1$$

$$5y-xy=2x+1$$

$$5y-1=x(y+2)$$

$$x=\frac{5y-1}{y+2}$$

Thus, $f^{-1}(x)=\dfrac{5x-1}{x+2}$.

Domains:

$dom(f)=rng(f^{-1})=(-\infty,5)\cup(5,\infty)$

$rng(f)=dom(f^{-1})=(-\infty,-2)\cup(-2,\infty)$

27. Can restrict to $[0,\infty)$ so that f will have an inverse. Also, one could restrict to any interval of the form $[a,\infty)$ or $(-\infty,-a]$, where a is a positive real number, to ensure f is one-to-one.

29. Let x = original price of a suit. Then, the sale price is $x-0.40x=0.60x$. Hence, the checkout price is $S(x)=0.60x-0.30(0.60x)=0.42x$.

31. Consider $f(x)=-\sqrt{1-x^2}$, $-1\le x\le 0$. (The graph of *f* is the quarter unit circle in the third quadrant.) To find its inverse, solve $y=-\sqrt{1-x^2}$ for *x*:

$$y=-\sqrt{1-x^2}$$
$$(-y)^2=1-x^2$$
$$x^2=1-y^2$$
$$x=-\sqrt{1-y^2} \text{ since } -1\le x\le 0$$

So, $f^{-1}(x)=-\sqrt{1-x^2}$ (The graph looks identical to that of *f*.)

33. Let *x* = number of minutes.
Then,

$$C(x)=\begin{cases} 15, & 0\le x\le 30 \\ 15+\underbrace{1(x-30)}_{\text{Amount for minutes beyond the initial 30.}}, & x>30 \end{cases}$$

$$=\begin{cases} 15, & 0\le x\le 30 \\ x-15, & x>30 \end{cases}$$

35. The general equation is $F=\frac{km}{P}$.

Using the fact that $F=\frac{k(2)}{3}=20$, we see that $k=20\left(\frac{3}{2}\right)=30$. So, $F=\frac{30m}{P}$.

37. Yes, the function is one-to-one.

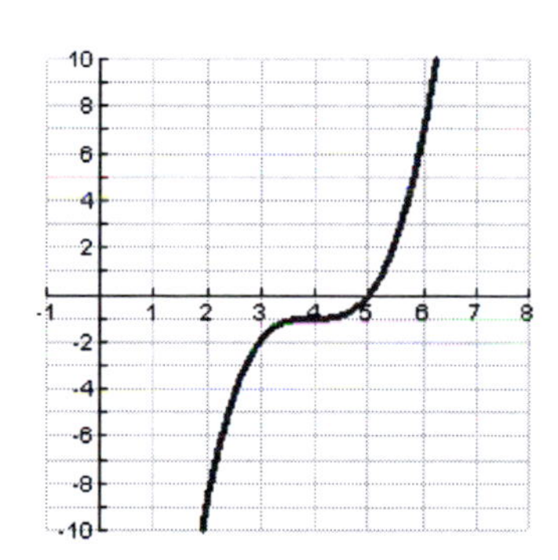

Chapter 3 Cumulative Review

1.

$$\frac{2}{3-\sqrt{5}}=\frac{2}{3-\sqrt{5}}\cdot\frac{3+\sqrt{5}}{3+\sqrt{5}}=\frac{6+2\sqrt{5}}{4}=\boxed{\frac{3+\sqrt{5}}{2}}$$

3.

$$\frac{-x^3-4x}{x+2}=\frac{-x(x^2+4)}{x+2}$$

Domain: $(-\infty,-2)\cup(-2,\infty)$

5. $(8-9i)(8+9i)=64+81=\boxed{145+0i}$

7. Since $\frac{35.70}{59.50}=0.60$, the percent markdown is $\boxed{40\%}$.

9.

$$\frac{x^2}{2} - x = \frac{1}{5}$$
$$5x^2 - 10x = 2$$
$$5\left(x^2 - 2x \quad\right) - 2 = 0$$
$$5\left(x^2 - 2x + 1\right) - 2 - 5 = 0$$
$$5(x-1)^2 = 7$$
$$(x-1)^2 = \tfrac{7}{5}$$
$$x = 1 \pm \sqrt{\tfrac{7}{5}} = \boxed{1 \pm \tfrac{\sqrt{35}}{5}}$$

11.

$$x^4 - x^2 - 12 = 0$$

Let $u = x^2$.

$$u^2 - u - 12 = 0$$
$$(u-4)(u+3) = 0$$
$$u = -3, 4$$

So, we have:

$$x^2 = -3 \Rightarrow x = \boxed{\pm i\sqrt{3}}$$
$$x^2 = 4 \Rightarrow x = \boxed{\pm 2}$$

13.

$$\frac{x}{x-5} < 0$$

CPs: $x = 0, 5$

$$\begin{array}{ccccc} + & | & - & | & + \\ & 0 & & 5 & \end{array}$$

So, the solution set is $\boxed{(0,5)}$.

15.

$$d = \sqrt{(5.2-(-2.7))^2 + (6.3-(-1.4))^2}$$
$$= \sqrt{(7.9)^2 + (7.7)^2} \approx \boxed{11.03}$$
$$M = \left(\frac{-2.7+5.2}{2}, \frac{-1.4+6.3}{2}\right) = \boxed{(1.25, 2.45)}$$

17. Since $m = 0$, the line is horizontal. Hence, its equation is $\boxed{y = -3}$.

19. Since the center is $(-2,-1)$, the general equation is $(x-(-2))^2 + (y-(-1))^2 = r^2$. Now, use the fact that $(-4,3)$ lies on the circle to find r^2. Indeed, observe that

$$(-4-(-2))^2 + (3-(-1))^2 = 20 = r^2.$$

So, the equation is $\boxed{(x+2)^2 + (y+1)^2 = 20}$.

21. $(-\infty, 1) \cup (1, \infty)$

23. $g(f(-1)) = g(|6-(-1)|) = 7^2 - 3 = \boxed{46}$

25. The general equation is $r = \dfrac{k}{t}$. Using t $r = \dfrac{k}{3} = 45$, we see that $k = 135$. So, $\boxed{r = \dfrac{135}{t}}$.

27. Let $f(x) = x^2 - 3x, \quad g(x) = x^2 + x - 2$.
Observe that

$$g(h(x)) = (h(x))^2 + (h(x)) - 2 = x^2 - 3x$$

implies that

$$\boxed{h(x) = x - 2}.$$

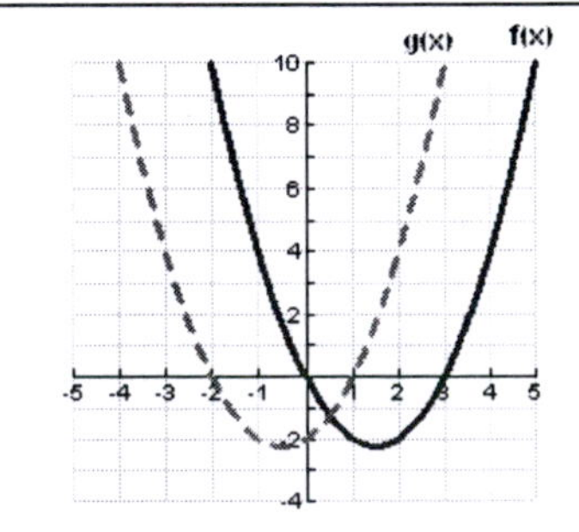

CHAPTER 4

Section 4.1 Solutions --

1. b Vertex $(-2,-5)$ and opens up	**3. a** Vertex $(-3,2)$ and opens down

5. Since the coefficient of x^2 is positive, the parabola opens up, so it must be *b* or *d*. Since the coefficient of x is positive, the x-coordinate of the vertex is negative. So, the graph is **b.**

7. Since the coefficient of x^2 is negative, the parabola opens down, so it must be *a* or *c*. In comparison to #8, this one will grow more slowly in the negative y-direction. So, the graph is **c**.

9.

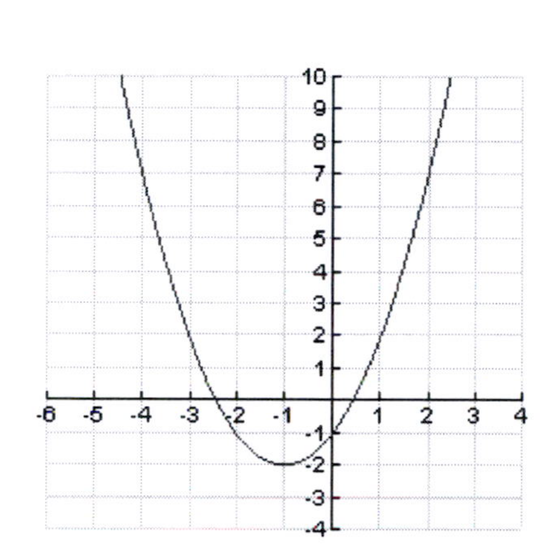

11.

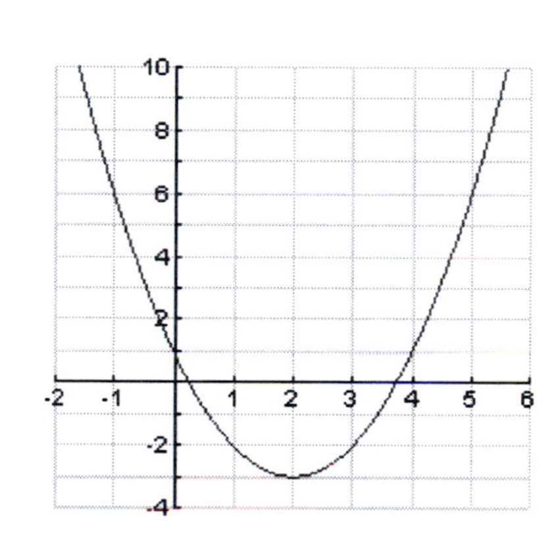

13.

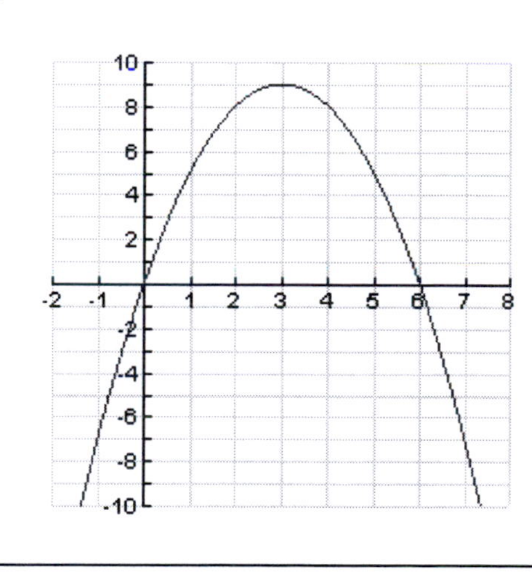

15.

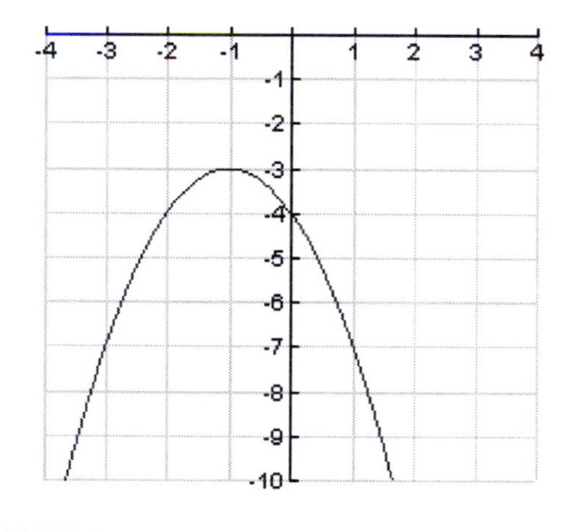

17.

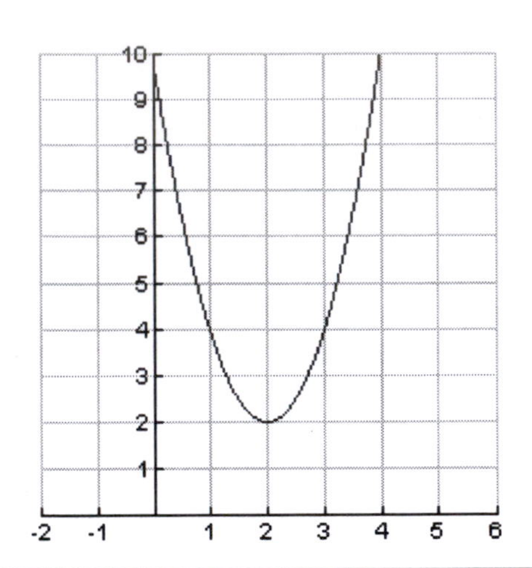

19.

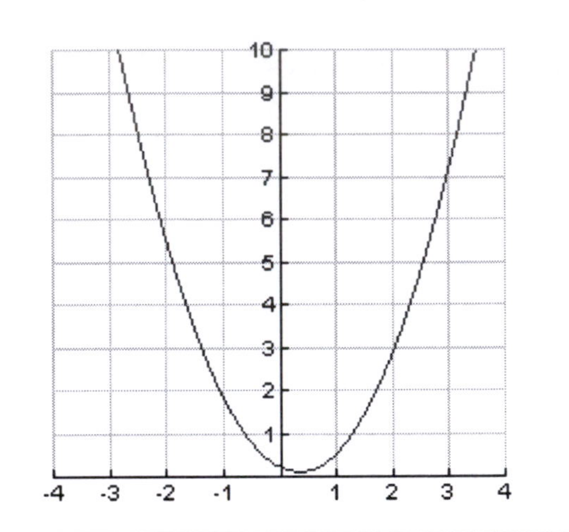

21.

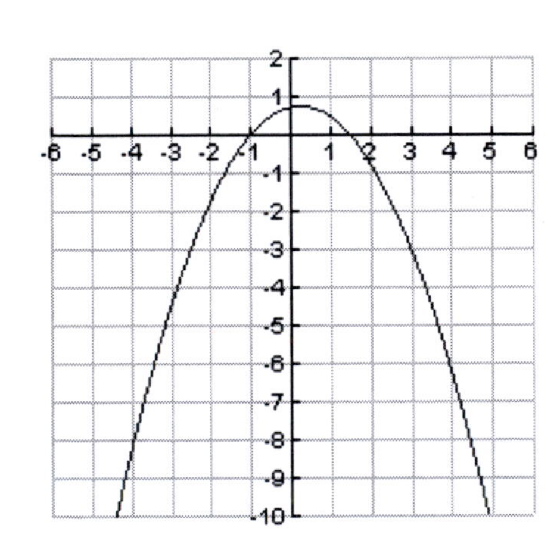

23.

$$f(x) = \left(x^2 + 6x + 9\right) - 3 - 9$$
$$= \left(x + 3\right)^2 - 12$$

25.

$$f(x) = -\left(x^2 + 10x \quad\right) + 3$$
$$= -\left(x^2 + 10x + 25\right) + 3 + 25$$
$$= -\left(x + 5\right)^2 + 28$$

27.

$$f(x) = 2\left(x^2 + 4x \quad\right) - 2$$
$$= 2\left(x^2 + 4x + 4\right) - 2 - 8$$
$$= 2\left(x + 2\right)^2 - 10$$

29.

$$f(x) = -4\left(x^2 - 4x \quad\right) - 7$$
$$= -4\left(x^2 - 4x + 4\right) - 7 + 16$$
$$= -4\left(x - 2\right)^2 + 9$$

31.

$$f(x) = \tfrac{1}{2}\left(x^2 - 8x \quad\right) + 3$$
$$= \tfrac{1}{2}\left(x^2 - 8x + 16\right) + 3 - 8$$
$$= \tfrac{1}{2}\left(x - 4\right)^2 - 5$$

33.

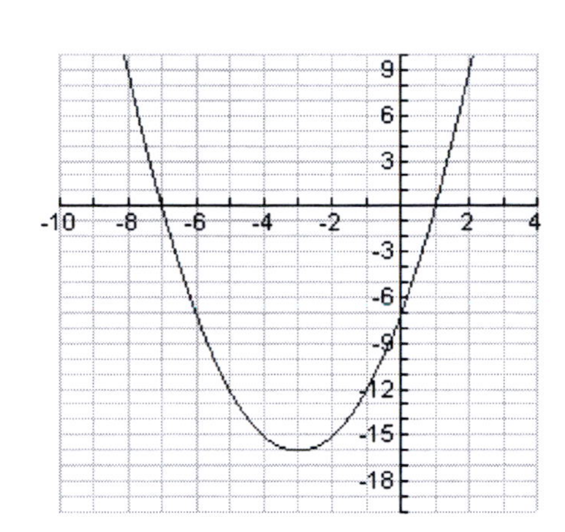

35.

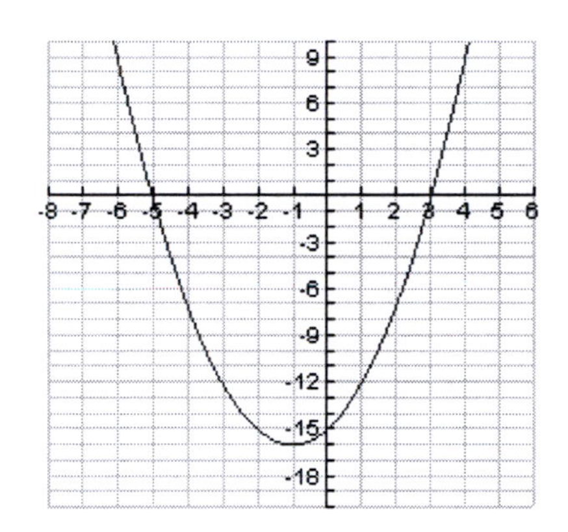

37.

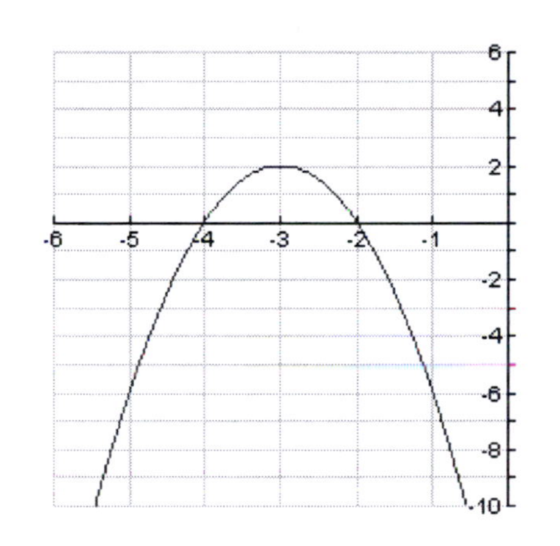

39.

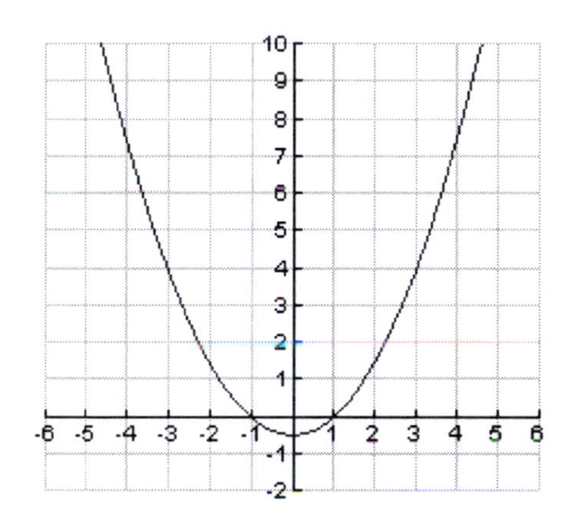

41.

$$\left(-\frac{b}{2a}, c-\frac{b^2}{4a}\right)=\left(-\frac{-2}{2(33)}, 15-\frac{(-2)^2}{4(33)}\right)$$
$$=\left(\frac{1}{33}, \frac{494}{33}\right)$$

43.

$$\left(-\frac{b}{2a}, c-\frac{b^2}{4a}\right)=\left(-\frac{-7}{2(\frac{1}{2})}, 5-\frac{(-7)^2}{4(\frac{1}{2})}\right)$$
$$=\left(7, -\frac{39}{2}\right)$$

45.

$$\left(-\frac{b}{2a}, c-\frac{b^2}{4a}\right)=\left(-\frac{-0.3}{2(-0.002)}, 1.7-\frac{(-0.3)^2}{4(-0.002)}\right)$$
$$=\left(-75, 12.95\right)$$

47.

$$\left(-\frac{b}{2a}, c-\frac{b^2}{4a}\right)=\left(-\frac{\frac{3}{7}}{2(-\frac{2}{5})}, 2-\frac{(\frac{3}{7})^2}{4(-\frac{2}{5})}\right)$$
$$=\left(\frac{15}{28}, \frac{829}{392}\right)$$

49. Since the vertex is $(-1,4)$, the function has the form $y=a(x+1)^2+4$. To find a, use the fact that the point (0,2) is on the graph:

$$2=a(0+1)^2+4$$
$$-2=a$$

So, the function is $y=-2(x+1)^2+4$.

51. Since the vertex is $(2,5)$, the function has the form $y=a(x-2)^2+5$. To find a, use the fact that the point (3,0) is on the graph:

$$0=a(3-2)^2+5$$
$$-5=a$$

So, the function is $y=-5(x-2)^2+5$.

53. Since the vertex is $(-1,-3)$, the function has the form $y=a(x+1)^2-3$. To find a, use the fact that the point $(-4,2)$ is on the graph:

$$2=a(-4+1)^2-3$$
$$2=9a-3$$
$$\tfrac{5}{9}=a$$

So, the function is $y=\frac{5}{9}(x+1)^2-3$.

55. Since the vertex is $(\frac{1}{2},\frac{-3}{4})$, the function has the form $y=a(x-\frac{1}{2})^2-\frac{3}{4}$. To find a, use the fact that the point $(\frac{3}{4},0)$ is on the graph:

$$0=a(\tfrac{3}{4}-\tfrac{1}{2})^2-\tfrac{3}{4}$$
$$0=\tfrac{1}{16}a-\tfrac{3}{4}$$
$$12=a$$

So, the function is $y=12(x-\frac{1}{2})^2-\frac{3}{4}$.

57. Since the vertex is $(2.5,-3.5)$, the function has the form $y=a(x-2.5)^2-3.5$. To find a, use the fact that the point $(4.5,1.5)$ is on the graph:

$$1.5=a(4.5-2.5)^2-3.5$$
$$1.5=4a-3.5$$
$$\tfrac{5}{4}=a$$

So, the function is $y=\frac{5}{4}(x-2.5)^2-3.5$.

59. Completing the square will enable you to identify the vertex of the parabola, which is precisely where the maximum occurs.

$$\begin{aligned}P(x)&=-0.0001\left(x^2-700{,}000x\right)+12{,}500\\&=-0.0001\left(x^2-700{,}000x+350{,}000^2\right)+12{,}500+12{,}250{,}000\\&=-0.0001\left(x-350{,}000\right)^2+12{,}262{,}500\end{aligned}$$

a. Maximum profit occurs when 350,000 units are sold.
b. The maximum profit is $P(350{,}000)=\$12{,}262{,}500$.

61. Complete the square to identify the vertex. Since the coefficient of t^2 is negative, $W(t)$ will be increasing to the left of the vertex and decreasing to the right of it.

$$\begin{aligned} W(t) &= -\tfrac{2}{3}\left(t^2 - \tfrac{39}{10}t\right) + \tfrac{433}{5} \\ &= -\tfrac{2}{3}\left(t^2 - \tfrac{39}{10}t + \left(\tfrac{39}{20}\right)^2\right) + \tfrac{433}{5} + \tfrac{2}{3}\left(\tfrac{39}{20}\right)^2 \\ &= -\tfrac{2}{3}\left(t - \tfrac{39}{20}\right)^2 + \tfrac{17,827}{200} \end{aligned}$$

So, gaining weight through most of January 2010 and then losing weight during the second to eighteenth months, namely Feb 2010 to June 2011.

63. a. The maximum occurs at the vertex, which is $(-5, 40)$. So, the maximum height is $\boxed{120 \text{ feet}}$.

b. If the height of the ball is assumed to be zero when the ball is kicked, and is zero when it lands, then we need to simply compute the x-intercepts of h and determine the distance between them. To this end, solve

$$0 = -\tfrac{8}{125}(x+5)^2 + 40 \;\Rightarrow\; \underbrace{40\left(\tfrac{125}{8}\right)}_{=625} = (x+5)^2 \;\Rightarrow\; \pm 25 = x + 5 \;\Rightarrow\; x = -30, 20$$

So, the distance the ball covers is $\boxed{50 \text{ yards}}$.

65. Let x = length and y = width.

The total amount of fence is given by: $4x + 3y = 10,000$ so that $y = \dfrac{10,000 - 4x}{3}$ **(1)**.

The combined area of the two identical pens is $2xy$. Substituting **(1)** in for y, we see that the area is described by the function:

$$A(x) = 2x\left(\frac{10,000 - 4x}{3}\right) = -\tfrac{8}{3}x^2 + \tfrac{20,000}{3}x$$

Since this parabola opens downward (since the coefficient of x^2 is negative), the maximum occurs at the x-coordinate of the vertex, namely

$$x = -\tfrac{b}{2a} = \frac{-\frac{20,000}{3}}{2\left(-\frac{8}{3}\right)} = 1250.$$

The corresponding width of the pen (from **(1)**) is $y = \dfrac{10,000 - 4(1250)}{3} \approx 1666.67$

So, each of the two pens would have area $\boxed{\cong 2,083,333 \text{ sq. ft.}}$

67. a. Completing the square on h yields

$$h(t) = -16\left(t^2 - 2t\right) + 100$$
$$= -16\left(t^2 - 2t + 1\right) + 100 + 16$$
$$= -16(t-1)^2 + 116$$

So, it takes 1 second to reach maximum height of 116 ft.

b. Solve $h(t) = 0$.

$$-16(t-1)^2 + 116 = 0$$
$$(t-1)^2 = \tfrac{116}{16}$$
$$t - 1 = \pm\sqrt{\tfrac{116}{16}}$$
$$t = 1 \pm \sqrt{\tfrac{116}{16}}$$

Since time must be positive, we conclude that the rock hits the water after about $t = 1 + \sqrt{\tfrac{116}{16}} \cong 3.69$ seconds (assuming the time started at $t = 0$).

69.

$$A(x) = -0.0003\left(x^2 - 31{,}000x\right) - 46{,}075$$
$$= -0.0003\left(x - 15{,}500\right)^2 - 46{,}075 + 72{,}075$$
$$= -0.0003(x - 15{,}500)^2 + 26{,}000$$

Also, we need the x-intercepts to determine the horizontal distance. Observe

$$-0.0003(x - 15{,}500)^2 + 26{,}000 = 0$$
$$(x - 15{,}500)^2 = \frac{26{,}000}{0.0003} \approx 86{,}666{,}667$$
$$x - 15{,}500 \approx \pm 9309.49$$
$$x \approx 15{,}500 \pm 9309.49$$
$$= 6{,}309.49 \text{ and } 24{,}809.49$$

So, the maximum altitude is 26,000 ft. over a horizontal distance of 8,944 ft.

71. a. Maximum profit occurs at the x-coordinate of the vertex of $P(x)$. Completing the square yields

$$P(x) = -0.4\left(x^2 + 200x \quad\right) + 20{,}000$$
$$= -0.4\left(x^2 + 200x + 10{,}000\right) + 20{,}000 + 4{,}000 = -0.4(x - 100)^2 + 24{,}000$$

So, the profit is maximized when 100 boards are sold.

b. The maximum profit is the y-coordinate of the vertex, namely $\boxed{\$24{,}000}$.

73. First, completing the square yields

$$P(x) = (100 - x)x - 1000 - 20x$$
$$= -x^2 + 80x - 1000 = -\left(x^2 - 80x\right) - 1000 = -\left(x^2 - 80x + 1600\right) - 1000 + 1600$$
$$= -(x-40)^2 + 600$$

Now, solve $-(x-40)^2 + 600 = 0$:

$$-(x-40)^2 + 600 = 0$$
$$(x-40)^2 = 600$$
$$x - 40 = \pm\sqrt{600} \text{ so that } x = 40 \pm \sqrt{600} \approx 15.5,\ 64.5$$

So, $\boxed{\text{15 to 16 units to break even, or 64 or 65 units to break even.}}$

75. We are given that the vertex is $(h,k) = (0,\ 16)$ and that $(9,\ 100)$ is on the graph.

a. We need to find a such that the equation governing the situation is

$$y = a(x-0)^2 + 16. \quad \textbf{(1)}$$

To do this, we use the fact that (9, 100) satisfies **(1)**: $100 = a(9-0)^2 + 16 \Rightarrow a = \frac{28}{27}$

Thus, the equation is $\boxed{y = \frac{28}{27}x^2 + 16}$, where y is measured in millions.

b. Note that the year 2010 corresponds to $t = 14$. Observe that

$$y(14) = \tfrac{28}{27}(14)^2 + 16 \approx \boxed{219 \text{ million}}.$$

77. a. We are given that the vertex is $(h,k) = (225,\ 400)$, that $(50,\ 93.75)$ is on the graph, and that the graph opens down $(a < 0)$ since the peak occurs at the vertex. We need to find a such that the equation governing the situation is

$$y = a(t-225)^2 + 400. \quad \textbf{(1)}$$

To do this, we use the fact that (50, 93.75) satisfies **(1)**:

$$93.75 = a(50-225)^2 + 400 \Rightarrow a = -0.01$$

Thus, the equation is $\boxed{y = -0.01(t-225)^2 + 400}$.

b. We must find the value(s) of t for which $0 = -0.01(t-225)^2 + 400$. To this end,

$$0 = -0.01(t-225)^2 + 400 \Rightarrow (t-225)^2 = 40{,}000$$
$$\Rightarrow t - 225 = \pm\sqrt{40{,}000} = \pm 200$$
$$\Rightarrow t = 225 \pm 200 = 25, 425$$

So, it takes $\boxed{\text{425 minutes}}$ for the drug to be eliminated from the bloodstream.

79. Step 2 is wrong: Vertex is $(-3,-1)$

Step 4 is wrong: The x-intercepts are $(-2,0)$, $(-4,0)$. So, should graph $y = (x+3)^2 - 1$.

81. Step 2 is wrong: $\left(-x^2 + 2x\right) = -\left(x^2 - 2x\right)$

83. True. $f(x) = a(x-h)^2 + k$, so that $f(0) = ah^2 + k$.

85. False. The graph would not pass the vertical line test in such case, and hence wouldn't define a function.

87. Completing the square yields

$$\begin{aligned} f(x) &= ax^2 + bx + c \\ &= a\left(x^2 + \tfrac{b}{a}x\right) + c \\ &= a\left(x^2 + \tfrac{b}{a}x + \left(\tfrac{b}{2a}\right)^2\right) + c - a\left(\tfrac{b}{2a}\right)^2 \\ &= a\left(x + \tfrac{b}{2a}\right)^2 + c - \tfrac{b^2}{4a} = a\left(x + \tfrac{b}{2a}\right)^2 + \tfrac{4ac - b^2}{4a} \end{aligned}$$

So, the vertex is $\left(-\frac{b}{2a}, c - \frac{b^2}{4a}\right)$. Observe that $f\left(-\frac{b}{2a}\right) = a\underbrace{\left(-\frac{b}{2a} + \frac{b}{2a}\right)^2}_{=0} + c - \frac{b^2}{4a} = c - \frac{b^2}{4a}$.

89. a. Let x = width of rectangular pasture, y = length of rectangular pasture.
Then, the total amount of fence is described by $2x + 2y = 1000$ (so that $y = 500 - x$ **(1)**).
The area of the pasture is xy. Substituting in **(1)** yields $x(500 - x) = -x^2 + 500x$.
The maximum area occurs at the y-coordinate of the vertex (since the coefficient of x^2 is negative); in this case this value is $c - \frac{b^2}{4a} = 0 - \frac{500^2}{4(-1)} = 62,500$.
So the maximum area is 62,500 sq. ft.

b. Let x = radius of circular pasture.
Then, the total amount of fence is described by $2\pi x = 1000$ (so that $x = \frac{1000}{2\pi} = \frac{500}{\pi}$ **(2)**).
The area of the pasture is πx^2, which in this case must be (by **(2)**) $\pi\left(\frac{500}{\pi}\right)^2 = \frac{500^2}{\pi}$.
So, the area of the pasture must be approximately 79,577 sq. ft.

91. a. Vertex:

$$\left(-\frac{5.7}{2(-0.002)}, -23-\frac{(5.7)^2}{4(-0.002)}\right) = (1425, 4038.25)$$

b. y-intercept: $f(0) = -23$, so $(0, -23)$

c. x-intercepts: Solve

$$-0.002x^2 + 5.7x - 23 = 0.$$

Using part **a.**, the standard form is $-0.002(x-1425)^2 + 4038.25 = 0$. So, solving this equation yields:

$$(x-1425)^2 = 2{,}019{,}125$$

$$x = 1425 \pm \sqrt{2{,}019{,}125}$$

$$\cong 4.04,\ 2845.96$$

So, the x-intercepts are

$$(4.04, 0),\ (2845.96, 0).$$

d. Axis of symmetry: $x = 1425$

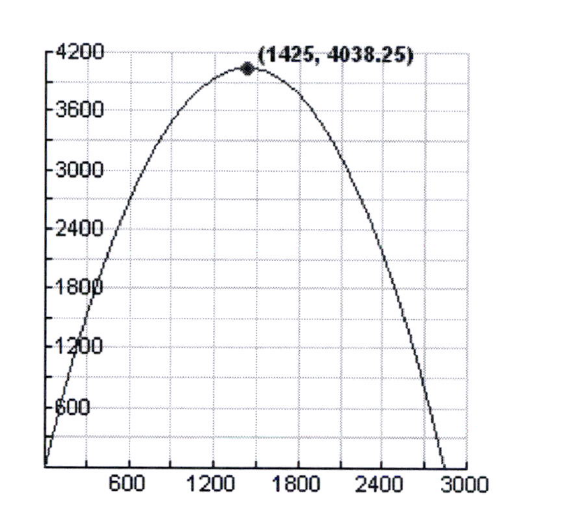

93. a. The equation obtained using this particular calculator feature is

$$f(x) = -2x^2 + 12.8x + 4.32.$$

b. Completing the square yields

$$\begin{aligned} f(x) &= -2x^2 + 12.8x + 4.32 \\ &= -2\left(x^2 - 6.4x\right) + 4.32 \\ &= -2\left(x^2 - 6.4x + 10.24\right) + 4.32 + 20.48 \\ &= -2(x-3.2)^2 + 24.8 \end{aligned}$$

So, the vertex is $(3.2, 24.8)$.

c. The graph is to the right. Yes, they agree with the given values.

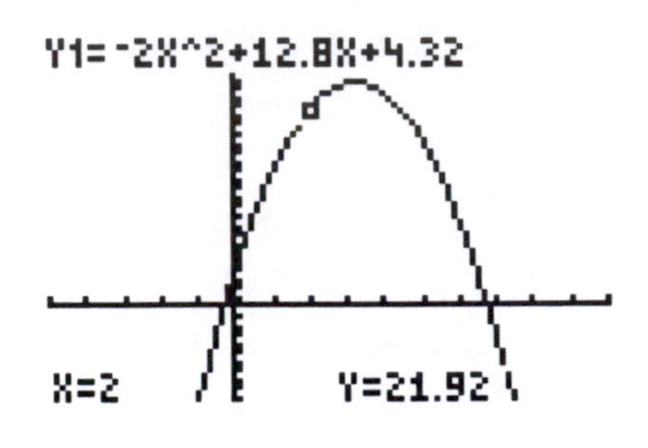

95. (a)

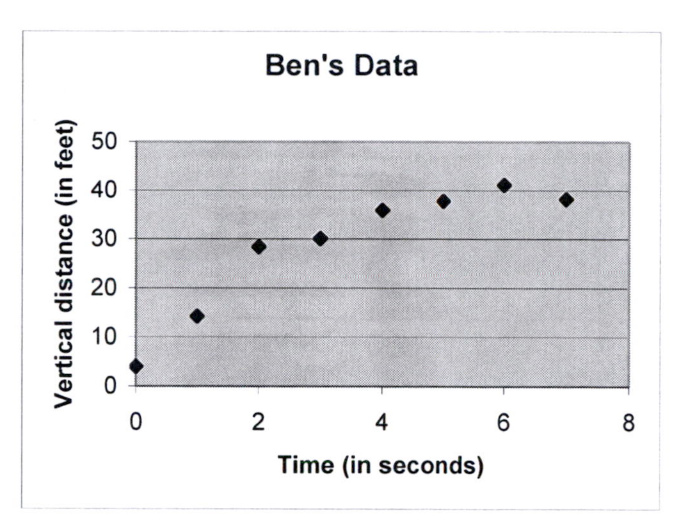

(b) The equation of the best fit parabola is $y = -1.0589x^2 + 12.268x + 4.3042$ with $r^2 = 0.9814$. This is shown below on the scatterplot:

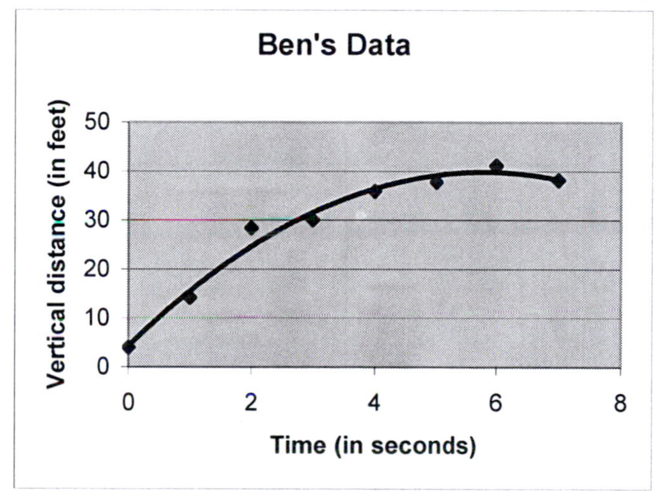

(c) **(i)** Using the equation from part (b), the initial height from which the pumpkin was thrown is about 4.3 feet, and the initial velocity is about 12.3 feet per second.

(ii) The maximum height occurs at the vertex and is approximately 39.8 feet; this occurs about 5.85 seconds into the flight.

(iii) The pumpkin lands approximately 12 seconds after it is launched.

Section 4.2 Solutions ---

1. Polynomial with degree 2	**3.** Polynomial with degree 5
5. Not a polynomial (due to the term $x^{1/2}$)	**7.** Not a polynomial (due to the term $x^{1/3}$)
9. Not a polynomial (due to the terms $\frac{1}{x}, \frac{1}{x^2}$)	
11. h linear function	**13. b** Parabola that opens up
15. e $x^3 - x^2 = x^2(x-1)$ So, there are two x-intercepts: the graph is tangent at 0 and crosses at 1.	**17. c** $-x^4 + 5x^3 = -x^3(x-5)$ So, there are two x-intercepts (0, 5) and the graph crosses at each of them.
19. 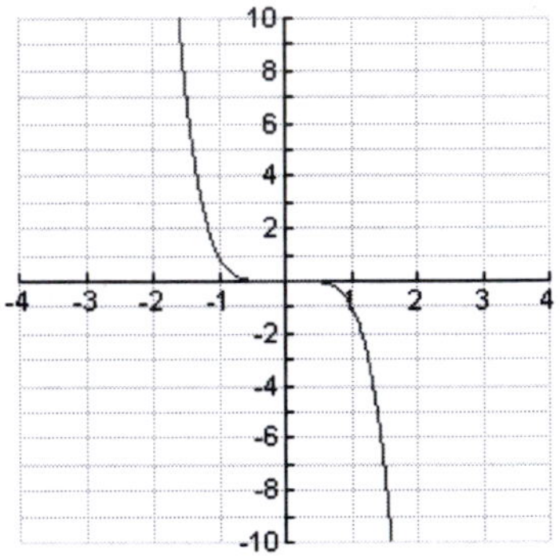	**21.** 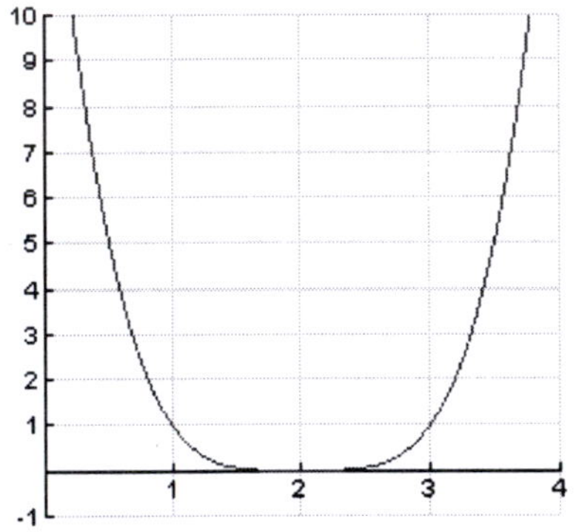
23. 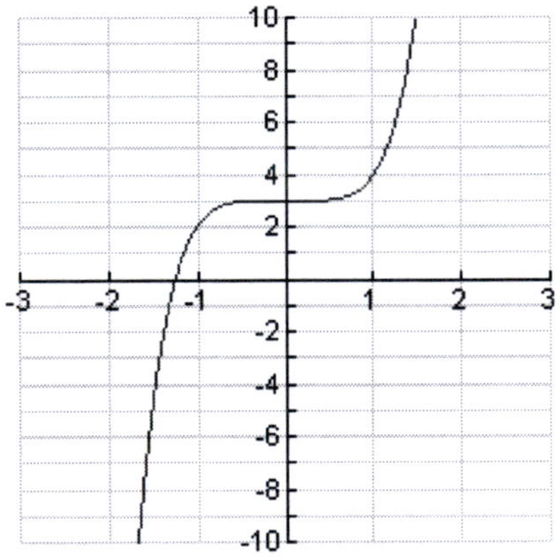	**25.** 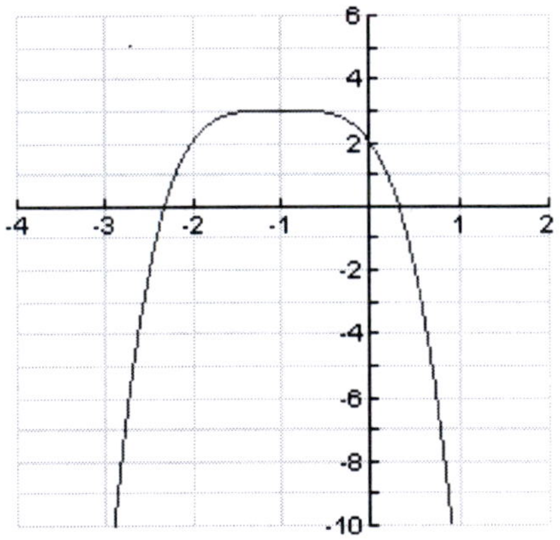
27. 3 (multiplicity 1) −4 (multiplicity 3)	**29.** 0 (multiplicity 2) 7 (multiplicity 2) −4 (multiplicity 1)

31. 0 (multiplicity 2)
1 (multiplicity 2)
Note: $x^2 + 4 = 0$ has no real solutions.

33.

$$8x^3 + 6x^2 - 27x = x\left(8x^2 + 6x - 27\right)$$
$$= x(2x-3)(4x+9)$$

So, the zeros are:
0 (multiplicity 1)
$\frac{3}{2}$ (multiplicity 1)
$-\frac{9}{4}$ (multiplicity 1)

35.

$$-2.7x^3 - 8.1x^2 = -2.7x^2(x+3)$$

So, the zeros are:
0 (multiplicity 2)
-3 (multiplicity 1)

37.

$$\tfrac{1}{3}x^6 + \tfrac{2}{5}x^4 = \tfrac{1}{3}x^4 \underbrace{\left(x^2 + \tfrac{6}{5}\right)}_{\text{Always positive}}$$

So, the only zero is:
0 (multiplicity 4)

39. $P(x) = x(x+3)(x-1)(x-2)$

41. $P(x) = x(x+5)(x+3)(x-2)(x-6)$

43.

$$P(x) = (2x+1)(3x-2)(4x-3)$$

45.

$$P(x) = \left(x-(1-\sqrt{2})\right)\left(x-(1+\sqrt{2})\right)$$
$$= x^2 - 2x - 1$$

47. $P(x) = x^2(x+2)^3$

49. $P(x) = (x+3)^2(x-7)^5$

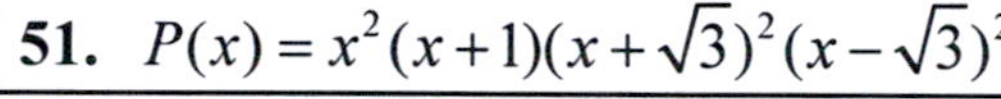

51. $P(x) = x^2(x+1)(x+\sqrt{3})^2(x-\sqrt{3})^2$

53. $f(x) = -\left(x^2 + 6x + 9\right) = -(x+3)^2$

a. Zeros: -3 (multiplicity 2)
b. Touches at -3
c. y-intercept: $f(0) = -9$, so $(0,-9)$
d. End behavior: Behaves like $y = -x^2$.
Even degree and leading coefficient negative, so graph falls without bound to the left and right.

e.

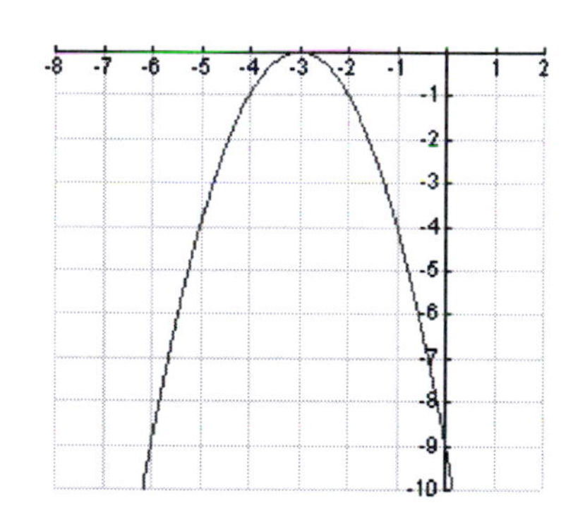

e.

55. $f(x) = (x-2)^3$

a. Zeros: 2 (multiplicity 3)

b. Crosses at 2

c. y-intercept: $f(0) = -8$, so $(0, -8)$

d. End behavior: Behaves like $y = x^3$.
Odd degree and leading coefficient positive, so graph falls without bound to the left and rises to the right.

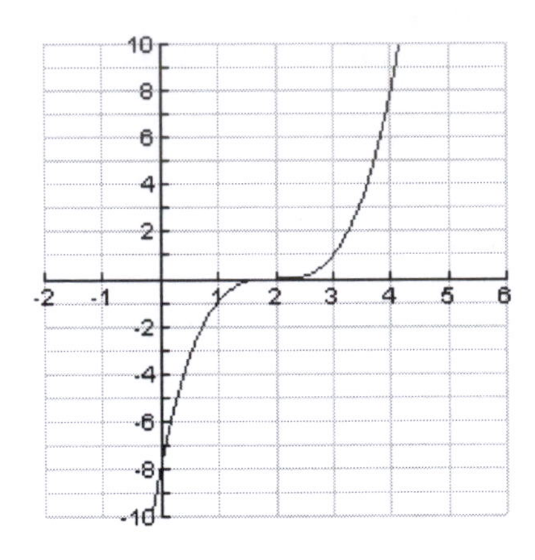

57. $f(x) = x^3 - 9x = x(x-3)(x+3)$

a. Zeros: 0, 3, -3 (multiplicity 1)

b. Crosses at each zero

c. y-intercept: $f(0) = 0$, so $(0, 0)$

d. End behavior: Behaves like $y = x^3$.
Odd degree and leading coefficient positive, so graph falls without bound to the left and rises to the right.

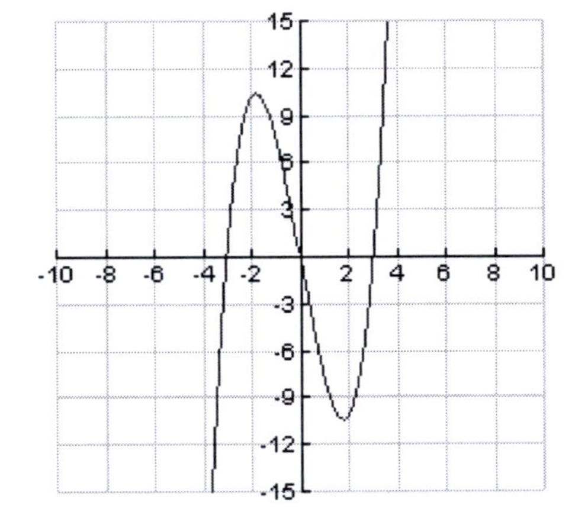

59. $f(x) = -x^3 + x^2 + 2x = -x(x-2)(x+1)$

a. Zeros: 0, 2, -1 (multiplicity 1)

b. Crosses at each zero

c. y-intercept: $f(0) = 0$, so $(0, 0)$

d. End behavior: Behaves like $y = -x^3$.
Odd degree and leading coefficient negative, so graph falls without bound to the right and rises to the left.

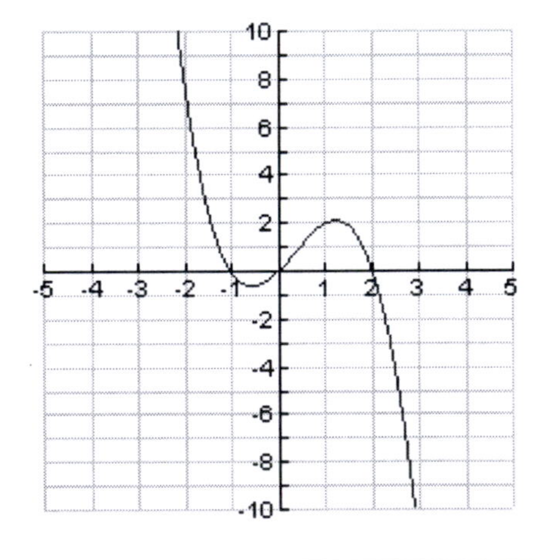

61. $f(x) = -x^4 - 3x^3 = -x^3(x+3)$

a. <u>Zeros</u>: 0 (multiplicity 3)
-3 (multiplicity 1)

b. Crosses at both 0 and -3

c. <u>y-intercept</u>: $f(0) = 0$, so $(0, 0)$

d. <u>End behavior</u>: Behaves like $y = -x^4$.
Even degree and leading coefficient negative, so graph falls without bound to left and right.

e.

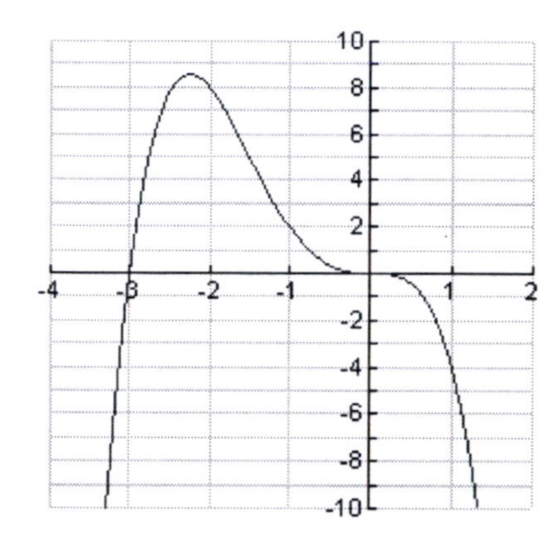

63.

$$f(x) = 12x^6 - 36x^5 - 48x^4$$
$$= 12x^4\left(x^2 - 3x - 4\right)$$
$$= 12x^4(x-4)(x+1)$$

a. <u>Zeros</u>: 0 (multiplicity 4),
4 (multiplicity 1), -1 (multiplicity 1)

b. Touches at 0 and crosses at 4 and -1.

c. <u>y-intercept</u>: $f(0) = 0$, so $(0, 0)$

d. <u>End behavior</u>: Behaves like $y = x^6$.
Even degree and leading coefficient positive, so graph rises without bound to the left and right.

e.

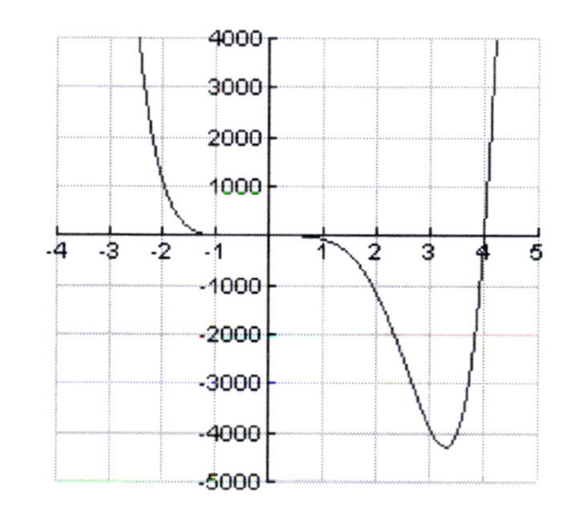

65.

$$f(x) = 2x^5 - 6x^4 - 8x^3$$
$$= 2x^3(x-4)(x+1)$$

a. <u>Zeros</u>: 0 (multiplicity 3),
4(multiplicity 1), -1 (multiplicity 1)

b. Crosses at each zero

c. <u>y-intercept</u>: $f(0) = 0$, so $(0, 0)$

d. <u>End behavior</u>: Behaves like $y = x^5$. Odd degree and leading coefficient positive, so graph falls without bound to the left and rises to the right.

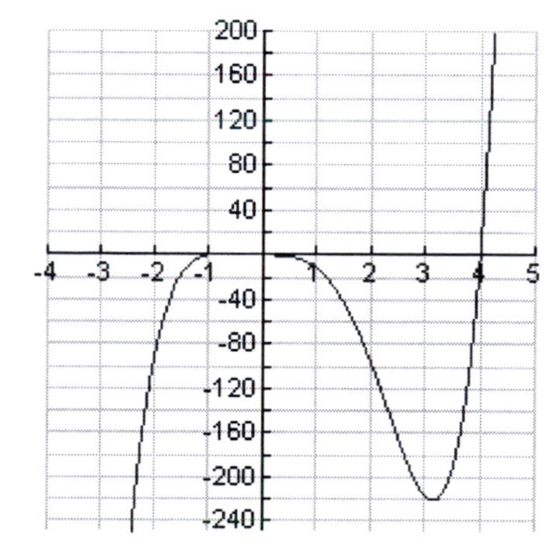

67.

$$f(x) = x^3 - x^2 - 4x + 4$$
$$= (x^3 - x^2) - 4(x-1)$$
$$= x^2(x-1) - 4(x-1)$$
$$= (x-2)(x+2)(x-1)$$

a. Zeros: 1, 2, -2 (multiplicity 1)

b. Crosses at each zero

c. *y*-intercept: $f(0) = 4$, so $(0, 4)$

d. End behavior: Behaves like $y = x^3$.

Odd degree and leading coefficient positive, so graph falls without bound to the left and rises to the right.

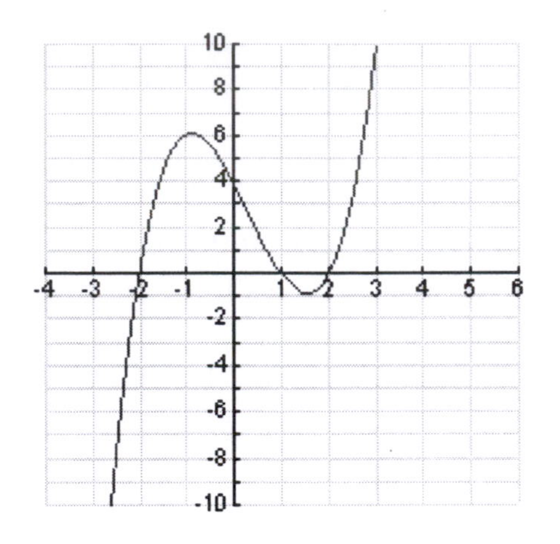

69. $f(x) = -(x+2)^2(x-1)^2$

a. Zeros: -2 (multiplicity 2)
1 (multiplicity 2)

b. Touches at both -2 and 1

c. *y*-intercept: $f(0) = -4$, so $(0, -4)$

d. End behavior: Behaves like $y = -x^4$.

Even degree and leading coefficient negative, so graph falls without bound to left and right.

e.

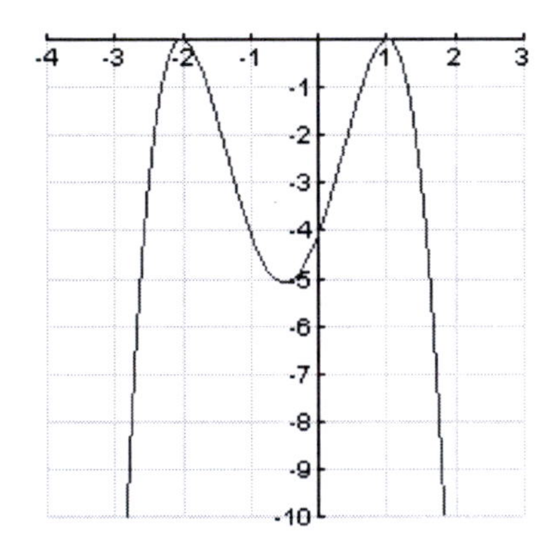

71. $f(x) = x^2(x-2)^3(x+3)^2$

a. Zeros: 0 (multiplicity 2)
2 (multiplicity 3)
-3 (multiplicity 2)

b. Touches at both 0 and -3, and crosses at 2.

c. *y*-intercept: $f(0) = 0$, so $(0, 0)$

d. End behavior: Behaves like $y = x^7$.

Odd degree and leading coefficient positive, so graph falls without bound to the left and rises to the right.

e.

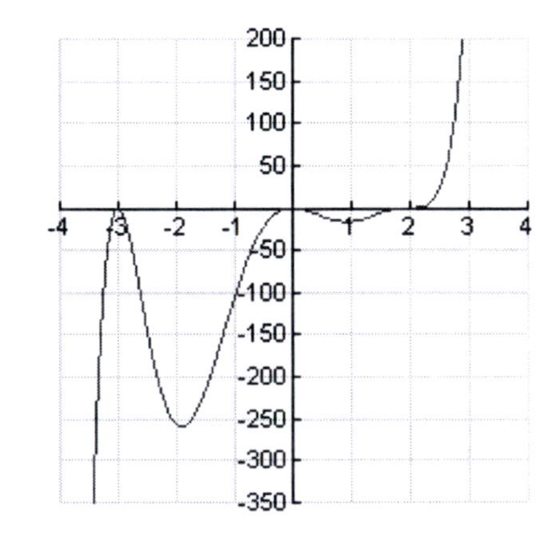

73. a. zeros: -3 (multiplicity 1), -1 (multiplicity 2), 2 (multiplicity 1)
b. degree of polynomial: even
c. sign of leading coefficient: negative
d. y-intercept: (0,6)
e. $f(x) = -(x+1)^2(x-2)(x+3)$.

75. a. zeros: 0 (multiplicity 2), -2 (multiplicity 2), $\frac{3}{2}$ (multiplicity 1)
b. degree of polynomial: odd
c. sign of leading coefficient: positive
d. y-intercept: (0,0)
e. $f(x) = x^2(2x-3)(x+2)^2$.

77. a. Revenue for the company is increasing when advertising costs are less than $400,000. Revenue for the company is decreasing when advertising costs are between $400,000 and $600,000.

b. The zeros of the revenue function occur when $0 and $600,000 are spent on advertising. When either $0 or $600,000 is spent on advertising the company's revenue is $0.

79. The velocity of air in the trachea is increasing when the radius of the trachea is between 0 and 0.45 cm and decreasing when the radius of the trachea is between 0 and 0.65 cm.

81. From the data, the turning points occur between months 3 & 4, 4 & 5, 5 & 6, 6 & 7, and 7 & 8. So, since there are at least 5 turning points, the degree of the polynomial must be at least 6.

Note: There could be more turning points if there was more oscillation during these month-long periods, but the data doesn't reveal such behavior.

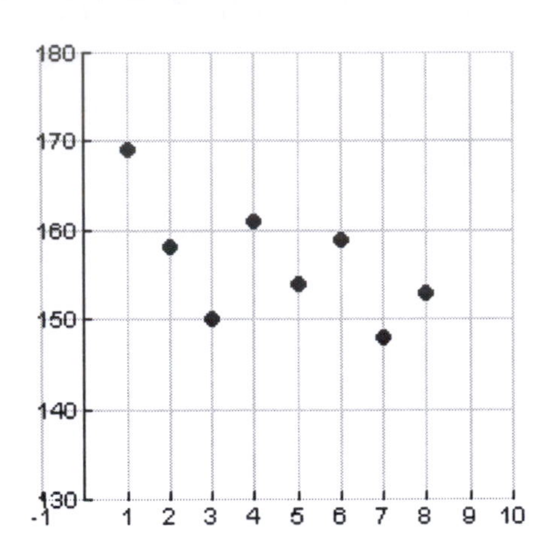

83. From the data, there is a turning point at 2. Since it is a third degree polynomial, one would expect the price to go down.

85. The given graph has three turning points, and is assumed to be defined only on the interval [2000,2008]. Since it seems to touch the axis at 2004, the lowest degree polynomial that would work is 4.

87. If h is a zero of a polynomial, then $(x-h)$ is a factor of it. So, in this case the function would be:
$f(x)=(x+2)(x+1)(x-3)(x-4)$

89. The zeros are correct. But, it is a fifth degree polynomial. The graph should touch at 1 (since even multiplicity) and cross at -2. The graph should look like:

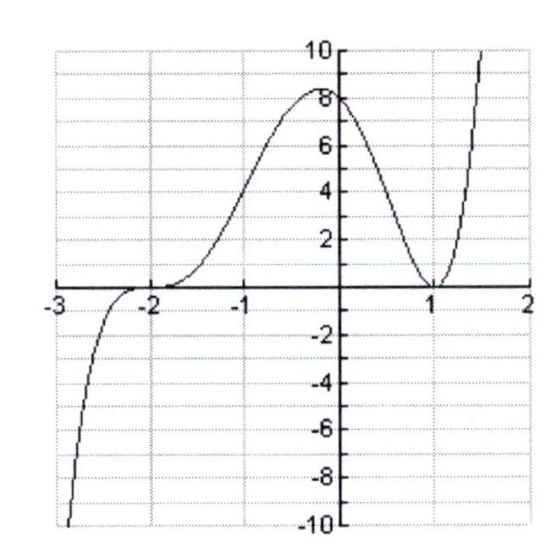

91. False. A polynomial has general form
$f(x)=a_n x^n + a_{n-1}x^{n-1} + \ldots + a_1 x + a_0$.
So, $f(0)=a_0$.

93. True.

95. A polynomial of degree n can have at most n zeros.

97. It touches at -1, so the multiplicity of this zero must be 2, 4, or 6.
It crosses at 3, so the multiplicity of this zero must be 1, 3, or 5.
Thus, the following polynomials would work:

$$f(x)=(x+1)^2(x-3)^5,\quad g(x)=(x+1)^4(x-3)^3,\quad h(x)=(x+1)^6(x-3)$$

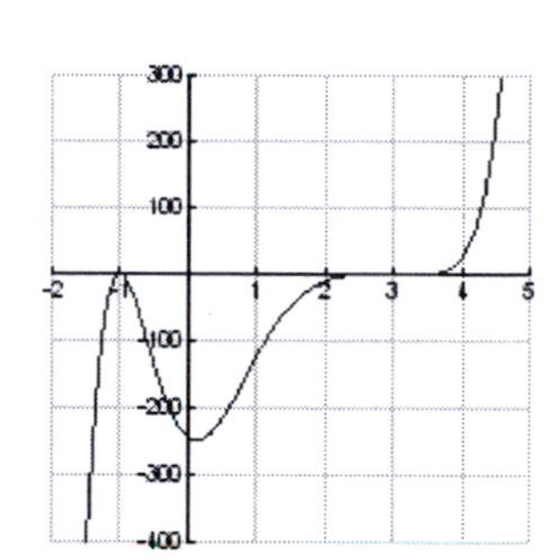

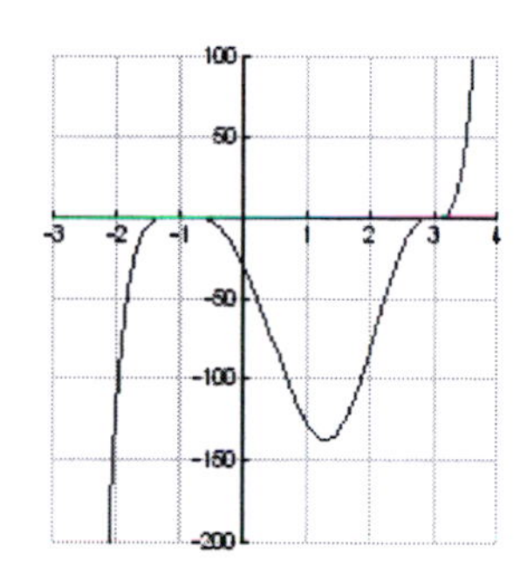

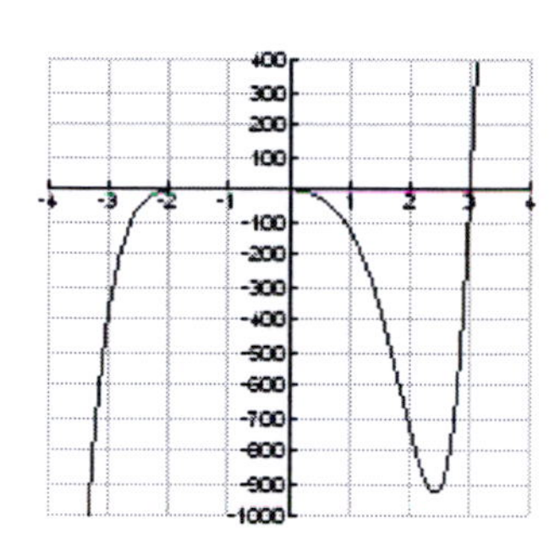

99. Observe that $x^3+(b-a)x^2-abx=x\left(x^2+(b-a)x-ab\right)=x(x-a)(x+b)$. So, the zeros are 0, a, and $-b$.

101. Note that $x^4+2x^2+1=\underbrace{\left(x^2+1\right)^2}_{\text{Always positive}}$.

So, there are no real zeros, and hence no x-intercepts. The graph is as follows:

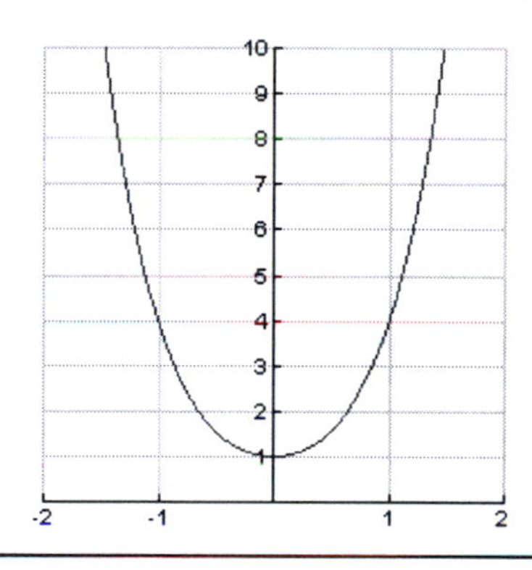

103. From the following graph of both $f(x)=-2x^5-5x^4-3x^3$ (solid curve) and $y=-2x^5$ (dotted curve), we do conclude that they have the same end behavior.

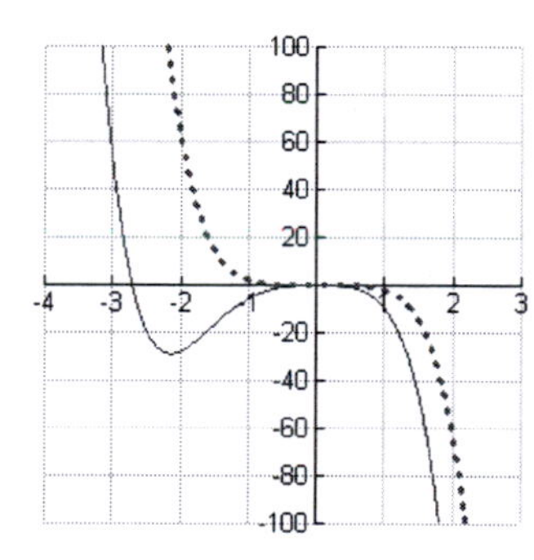

105.
$f(x) = x^4 - 15.9x^3 + 1.31x^2 + 292.905x + 445.7025$

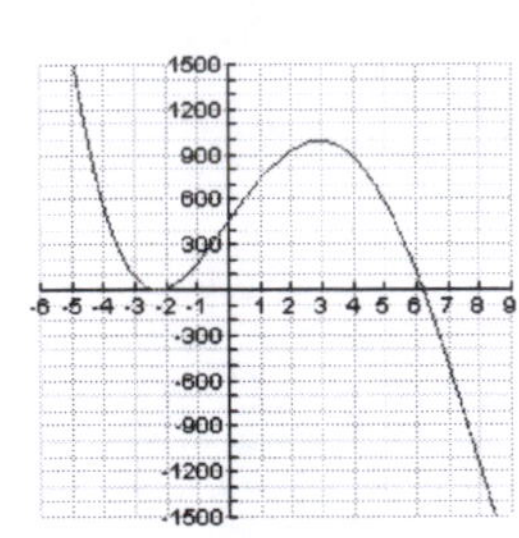

The following are estimates:
x-intercepts: $(-2.25, 0)$, $(6.2, 0)$, $(14.2, 0)$
zeros: -2.25 (multiplicity 2),
6.2 (multiplicity 1), 14.2 (multiplicity 1)

107. The graph is:

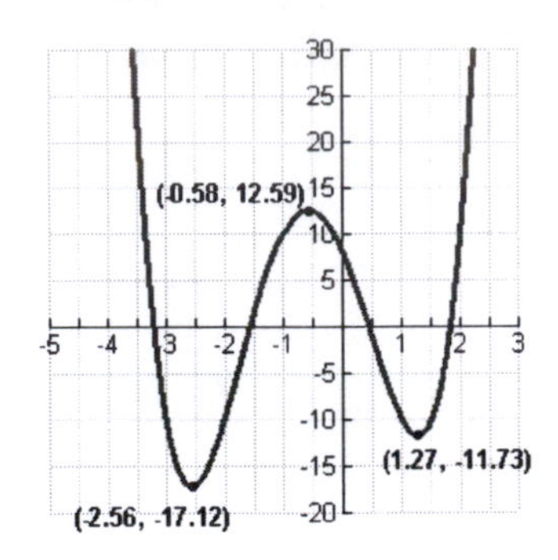

The coordinates of the relative extrema are (-2.56,-17.12), (-0.58,12.59), (1.27,-11.73).

Section 4.3 Solutions ---

1.

$$\begin{array}{r|l} & 2x-1 \\ \hline x+3 & 2x^2+5x-3 \\ & -(2x^2+6x) \\ \hline & -x-3 \\ & -(-x-3) \\ \hline & 0 \end{array}$$

So, $\boxed{Q(x) = 2x-1, \quad r(x) = 0}$.

3.

$$\begin{array}{r|l} & x-3 \\ \hline x-2 & x^2-5x+6 \\ & -(x^2-2x) \\ \hline & -3x+6 \\ & -(-3x+6) \\ \hline & 0 \end{array}$$

So, $\boxed{Q(x) = x-3, \quad r(x) = 0}$.

5.

$$\begin{array}{r|l} & 3x-3 \\ \hline x-2 & 3x^2-9x-5 \\ & -(3x^2-6x) \\ \hline & -3x-5 \\ & -(-3x+6) \\ \hline & -11 \end{array}$$

So, $\boxed{Q(x) = 3x-3, \quad r(x) = -11}$.

7.

$$\begin{array}{r|l} & 3x-28 \\ \hline x+5 & 3x^2-13x-10 \\ & -(3x^2+15x) \\ \hline & -28x-10 \\ & -(-28x-140) \\ \hline & 130 \end{array}$$

So, $\boxed{Q(x) = 3x-28, \quad r(x) = 130}$.

9.

$$\begin{array}{r} x-4 \\ x+4\overline{)\,x^2+0x-4} \\ \underline{-(x^2+4x)} \\ -4x-4 \\ \underline{-(-4x-16)} \\ 12 \end{array}$$

So, $\boxed{Q(x)=x-4,\quad r(x)=12}$.

11.

$$\begin{array}{r} 3x+5 \\ 3x-5\overline{)\,9x^2+0x-25} \\ \underline{-(9x^2-15x)} \\ 15x-25 \\ \underline{-(15x-25)} \\ 0 \end{array}$$

So, $\boxed{Q(x)=3x+5,\quad r(x)=0}$.

13.

$$\begin{array}{r} 2x-3 \\ 2x+3\overline{)\,4x^2+0x-9} \\ \underline{-(4x^2+6x)} \\ -6x-9 \\ \underline{-(-6x-9)} \\ 0 \end{array}$$

So, $\boxed{Q(x)=2x-3,\quad r(x)=0}$.

15.

$$\begin{array}{r} 4x^2+4x+1 \\ 3x+2\overline{)\,12x^3+20x^2+11x+2} \\ \underline{-(12x^3+8x^2)} \\ 12x^2+11x \\ \underline{-(12x^2+8x)} \\ 3x+2 \\ \underline{-(3x+2)} \\ 0 \end{array}$$

So, $\boxed{Q(x)=4x^2+4x+1,\quad r(x)=0}$.

17.

$$\begin{array}{r} 2x^2-x-\frac{1}{2} \\ 2x+1\overline{)\,4x^3+0x^2-2x+7} \\ \underline{-(4x^3+2x^2)} \\ -2x^2-2x \\ \underline{-(-2x^2-x)} \\ -x+7 \\ \underline{-(-x-\frac{1}{2})} \\ \frac{15}{2} \end{array}$$

So, $\boxed{Q(x)=2x^2-x-\frac{1}{2},\quad r(x)=\frac{15}{2}}$.

19.

$$\begin{array}{r} 4x^2-10x-6 \\ x-\frac{1}{2}\overline{)\,4x^3-12x^2-x+3} \\ \underline{-(4x^3-2x^2)} \\ -10x^2-x \\ \underline{-(-10x^2+5x)} \\ -6x+3 \\ \underline{-(-6x+3)} \\ 0 \end{array}$$

So, $\boxed{Q(x)=4x^2-10x-6,\quad r(x)=0}$.

21.

$$\begin{array}{r} -2x^2-3x-9 \\ x^3-3x^2+0x+1 \overline{\big)\,-2x^5+3x^4+0x^3-2x^2+0x+0} \\ \underline{-(-2x^5+6x^4+0x^3-2x^2)} \\ -3x^4+0x^3+0x^2+0x \\ \underline{-(-3x^4+9x^3+0x^2-3x)} \\ -9x^3+0x^2+3x+0 \\ \underline{-(-9x^3+27x^2+0x-9)} \\ -27x^2+3x+9 \end{array}$$

So, $\boxed{Q(x)=-2x^2-3x-9,\quad r(x)=-27x^2+3x+9}$.

23.

$$\begin{array}{r} x^2+0x+1 \\ x^2+0x-1 \overline{\big)\,x^4+0x^3+0x^2+0x-1} \\ \underline{-(x^4+0x^3-x^2)} \\ x^2+0x-1 \\ \underline{-(x^2+0x-1)} \\ 0 \end{array}$$

So, $\boxed{Q(x)=x^2+1,\quad r(x)=0}$.

25.

$$\begin{array}{r} x^2+x+\frac{1}{6} \\ 6x^2+x-2 \overline{\big)\,6x^4+7x^3+0x^2-22x+40} \\ \underline{-(6x^4+x^3-2x^2)} \\ 6x^3+2x^2-22x \\ \underline{-(6x^3+x^2-2x)} \\ x^2-20x+40 \\ \underline{-(x^2+\frac{1}{6}x-\frac{1}{3})} \\ -\frac{121}{6}x+\frac{121}{3} \end{array}$$

So, $\boxed{Q(x)=x^2+x+\frac{1}{6},\quad r(x)=-\frac{121}{6}x+\frac{121}{3}}$.

27.

$$\begin{array}{r} -3x^3+5.2x^2+3.12x-0.128 \\ x-0.6\overline{\big)\,-3x^4+7x^3+0x^2-2x+1} \\ \underline{-(-3x^4+1.8x^3)} \\ 5.2x^3+0x^2 \\ \underline{-(5.2x^3-3.12x^2)} \\ 3.12x^2-2x \\ \underline{-(3.12x^2-1.872x)} \\ -0.128x+1 \\ \underline{-(-0.128x+0.0768)} \\ 0.9232 \end{array}$$

So, $\boxed{Q(x)=-3x^3+5.2x^2+3.12x-0.128,\quad r(x)=0.9232}$.

29.

$$\begin{array}{r} x^2-0.6x+0.09 \\ x^2+1.4x+0.49\overline{\big)\,x^4+0.8x^3-0.26x^2-0.168x+0.0441} \\ \underline{-(x^4+1.4x^3+0.49x^2)} \\ -0.6x^3-0.75x^2-0.168x \\ \underline{-(-0.6x^3-0.84x^2-0.294x)} \\ 0.09x^2+0.126x+0.0441 \\ \underline{-(0.09x^2+0.126x+0.0441)} \\ 0 \end{array}$$

So, $\boxed{Q(x)=x^2-0.6x+0.09,\quad r(x)=0}$.

31.

$$\begin{array}{r|rrr} -2 & 3 & 7 & 2 \\ & & -6 & -2 \\ \hline & 3 & 1 & 0 \end{array}$$

So, $\boxed{Q(x)=3x+1,\quad r(x)=0}$.

33.

$$\begin{array}{r|rrr} -1 & 7 & -3 & 5 \\ & & -7 & 10 \\ \hline & 7 & -10 & 15 \end{array}$$

So, $\boxed{Q(x)=7x-10,\quad r(x)=15}$.

35.

$$\begin{array}{r|rrrrr} -2 & -1 & -2 & 3 & 4 & -4 \\ & & 2 & 0 & -6 & 4 \\ \hline & -1 & 0 & 3 & -2 & 0 \end{array}$$

So, $\boxed{Q(x)=-x^3+3x-2,\quad r(x)=0}$.

37.

$$\begin{array}{r|rrrrr} -1 & 1 & 0 & 0 & 0 & 1 \\ & & -1 & 1 & -1 & 1 \\ \hline & 1 & -1 & 1 & -1 & 2 \end{array}$$

So, $\boxed{Q(x)=x^3-x^2+x-1,\quad r(x)=2}$.

39.

$$\begin{array}{r|rrrrr} -2 & 1 & 0 & 0 & 0 & -16 \\ & & -2 & 4 & -8 & 16 \\ \hline & 1 & -2 & 4 & -8 & 0 \end{array}$$

So, $\boxed{Q(x)=x^3-2x^2+4x-8,\quad r(x)=0}$.

41.

$$\begin{array}{r|rrrr} -\frac{1}{2} & 2 & -5 & -1 & 1 \\ & & -1 & 3 & -1 \\ \hline & 2 & -6 & 2 & 0 \end{array}$$

So, $\boxed{Q(x)=2x^2-6x+2,\quad r(x)=0}$.

43.

$$\begin{array}{r|rrrrr} \frac{2}{3} & 2 & -3 & 7 & 0 & -4 \\ & & \frac{4}{3} & -\frac{10}{9} & \frac{106}{27} & \frac{212}{81} \\ \hline & 2 & -\frac{5}{3} & \frac{53}{9} & \frac{106}{27} & -\frac{112}{81} \end{array}$$

So,

$\boxed{Q(x)=2x^3-\frac{5}{3}x^2+\frac{53}{9}x+\frac{106}{27},\quad r(x)=-\frac{112}{81}}$.

45.

$$\begin{array}{r|rrrrr} -1.5 & 2 & 9 & -9 & -81 & -81 \\ & & -3 & -9 & 27 & 81 \\ \hline & 2 & 6 & -18 & -54 & 0 \end{array}$$

So, $\boxed{Q(x)=2x^3+6x^2-18x-54,\quad r(x)=0}$.

47.

$$\begin{array}{r|rrrrrrrr} 1 & 1 & 0 & 0 & -8 & 0 & 3 & 0 & 1 \\ & & 1 & 1 & 1 & -7 & -7 & -4 & -4 \\ \hline & 1 & 1 & 1 & -7 & -7 & -4 & -4 & -3 \end{array}$$

So, $\boxed{Q(x)=x^6+x^5+x^4-7x^3-7x^2-4x-4,\quad r(x)=-3}$.

49.

$$\begin{array}{r|rrrrrrr} \sqrt{5} & 1 & 0 & -49 & 0 & -25 & 0 & 1225 \\ & & \sqrt{5} & 5 & -44\sqrt{5} & -220 & -245\sqrt{5} & -1225 \\ \hline & 1 & \sqrt{5} & -44 & -44\sqrt{5} & -245 & -245\sqrt{5} & 0 \end{array}$$

So, $\boxed{Q(x)=x^5+\sqrt{5}x^4-44x^3-44\sqrt{5}x^2-245x-245\sqrt{5},\quad r(x)=0}$.

51.

$$\begin{array}{r} 2x-7 \\ 3x-1\overline{\big)\,6x^2-23x+7} \\ \underline{-(6x^2-2x)} \\ -21x+7 \\ \underline{-(-21x+7)} \\ 0 \end{array}$$

So, $\boxed{Q(x)=2x-7,\quad r(x)=0}$.

53.

$$\begin{array}{r|rrrr} 1 & 1 & -1 & -9 & 9 \\ & & 1 & 0 & -9 \\ \hline & 1 & 0 & -9 & 0 \end{array}$$

So, $\boxed{Q(x)=x^2-9,\quad r(x)=0}$.

55.

$$\begin{array}{r|rrrrrr} 2 & 1 & 0 & 4 & 2 & 0 & -1 \\ & & 2 & 4 & 16 & 36 & 72 \\ \hline & 1 & 2 & 8 & 18 & 36 & 71 \end{array}$$

So,

$$\boxed{Q(x) = x^4 + 2x^3 + 8x^2 + 18x + 36, \; r(x) = 71}.$$

57.

$$\begin{array}{r} x^2 + 0x + 1 \\ x^2 + 0x - 1 \overline{\big) x^4 + 0x^3 + 0x^2 + 0x - 25} \\ \underline{-(x^4 + 0x^3 - x^2)} \\ x^2 + 0x - 25 \\ \underline{-(x^2 + 0x - 1)} \\ -24 \end{array}$$

So, $\boxed{Q(x) = x^2 + 1, \quad r(x) = -24}$.

59.

$$\begin{array}{r|rrrrrrrr} 1 & 1 & 0 & 0 & 0 & 0 & 0 & 0 & -1 \\ & & 1 & 1 & 1 & 1 & 1 & 1 & 1 \\ \hline & 1 & 1 & 1 & 1 & 1 & 1 & 1 & 0 \end{array}$$

So,

$$\boxed{Q(x) = x^6 + x^5 + x^4 + x^3 + x^2 + x + 1, \quad r(x) = 0}.$$

61. Area = length × width. So, solving for width, we see that width = Area ÷ length. So, we have:

$$\begin{array}{r} 3x^2 + 2x + 1 \\ 2x^2 + 0x - 1 \overline{\big) 6x^4 + 4x^3 - x^2 - 2x - 1} \\ \underline{-(6x^4 + 0x^3 - 3x^2)} \\ 4x^3 + 2x^2 - 2x \\ \underline{-(4x^3 + 0x^2 - 2x)} \\ 2x^2 + 0x - 1 \\ \underline{-(2x^2 + 0x - 1)} \\ 0 \end{array}$$

Thus, the width (in terms of x) is $\boxed{3x^2 + 2x + 1 \text{ feet}}$.

63. Distance = Rate × Time. So, solving for Time, we have: Time = Distance ÷ Rate. So, we calculate $\left(x^3 + 60x^2 + x + 60\right) \div (x + 60)$ using synthetic division:

$$\begin{array}{r|rrrr} -60 & 1 & 60 & 1 & 60 \\ & & -60 & 0 & -60 \\ \hline & 1 & 0 & 1 & 0 \end{array}$$

So, the time is $\boxed{x^2 + 1 \text{ hours}}$.

65. Should have subtracted each term in the long division rather than adding them.

67. Forgot the "0" placeholder.

69. True.

71. False. Only use when the divisor has degree 1.

73.

$$\begin{array}{r|cccc} -b & 1 & 2b-a & b^2-2ab & -ab^2 \\ & & -b & -b^2+ab & ab^2 \\ \hline & 1 & b-a & -ab & 0 \end{array}$$

Since the remainder is 0 upon using synthetic division, YES, $(x+b)$ is a factor of $x^3+(2b-a)x^2+\left(b^2-2ab\right)x-ab^2$.

75.

$$\begin{array}{r} x^{2n}+2x^n+1 \\ x^n-1\overline{\big)x^{3n}+x^{2n}-x^n-1} \\ \underline{-(x^{3n}-x^{2n})} \\ 2x^{2n}-x^n \\ \underline{-(2x^{2n}-2x^n)} \\ x^n-1 \\ \underline{-(x^n-1)} \\ 0 \end{array}$$

So, $\boxed{Q(x)=x^{2n}+2x^n+1,\quad r(x)=0}$

77. Long division gives us

$$\begin{array}{r} 2x-1 \\ x^2+0x+5\overline{\big)\,2x^3-x^2+10x-5} \\ \underline{-(2x^3+0x^2+10x)} \\ -x^2+0x-5 \\ \underline{-(-x^2+0x-5)} \\ 0 \end{array}$$

So, the graph is a line, as shown:

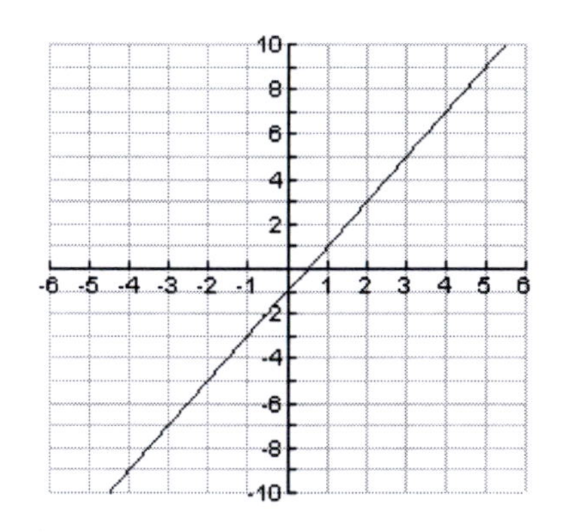

79. Using synthetic division gives us:

$$\begin{array}{r|ccccc} -2 & 1 & 2 & 0 & -1 & -2 \\ & & -2 & 0 & 0 & 2 \\ \hline & 1 & 0 & 0 & -1 & 0 \end{array}$$

So, the quotient is the cubic $y=x^3-1$ with a hole at -2, as shown to the right:

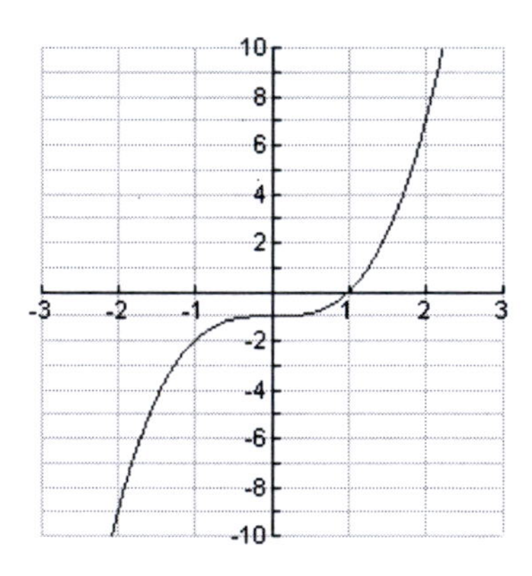

81. Long division gives us

$$\begin{array}{r} -3x^2+8x-5 \\ 2x+3\overline{\big)\,-6x^3+7x^2+14x-15} \\ \underline{-(-6x^3-9x^2)} \\ 16x^2+14x \\ \underline{-(16x^2+24x)} \\ -10x-15 \\ \underline{-(-10x-15)} \\ 0 \end{array}$$

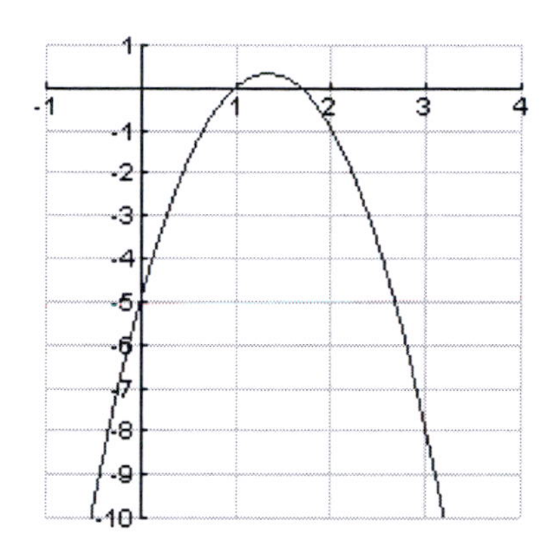

So, it is a quadratic function with a hole at -3/2.

Section 4.4 Solutions

1.

$$\begin{array}{r|rrrrr} 1 & 3 & -2 & 7 & 0 & -8 \\ & & 3 & 1 & 8 & 8 \\ \hline & 3 & 1 & 8 & 8 & 0 \end{array}$$

So, $f(1)=0$.

3.

$$\begin{array}{r|rrrr} 1 & 2 & 1 & 0 & 1 \\ & & 2 & 3 & 3 \\ \hline & 2 & 3 & 3 & 4 \end{array}$$

So, $g(1)=4$.

5.

$$\begin{array}{r|rrrrr} -2 & 3 & -2 & 7 & 0 & -8 \\ & & -6 & 16 & -46 & 92 \\ \hline & 3 & -8 & 23 & -46 & 84 \end{array}$$

So, $f(-2)=84$.

7.

$$\begin{array}{r|rrrr} -7 & 1 & 2 & -29 & 42 \\ & & -7 & 35 & -42 \\ \hline & 1 & -5 & 6 & 0 \end{array}$$

Yes, -7 is a zero of $P(x)$.

9.

$$\begin{array}{r|rrrr} -3 & 1 & -1 & -8 & 12 \\ & & -3 & 12 & -12 \\ \hline & 1 & -4 & 4 & 0 \end{array}$$

Yes, -3 is a zero of $P(x)$.

11.

$$\begin{array}{r|rrrr} 1 & 1 & 0 & -13 & 12 \\ & & 1 & 1 & -12 \\ \hline & 1 & 1 & -12 & 0 \end{array}$$

So,

$$\begin{aligned} P(x) &= (x-1)(x^2+x-12) \\ &= (x-1)(x+4)(x-3) \end{aligned}.$$

The zeros are 1, 3 and -4.

13.

$$\begin{array}{r|rrrr} \frac{1}{2} & 2 & 1 & -13 & 6 \\ & & 1 & 1 & -6 \\ \hline & 2 & 2 & -12 & 0 \end{array}$$

So,

$$\begin{aligned} P(x) &= (x-\tfrac{1}{2})(2x^2+2x-12) \\ &= 2(x-\tfrac{1}{2})(x^2+x-6) \\ &= 2(x-\tfrac{1}{2})(x+3)(x-2) \\ &= (2x-1)(x+3)(x-2) \end{aligned}.$$

The zeros are $\frac{1}{2}, -3$ and 2.

15. Since -3 and 5 are both zeros, we know that $(x+3)$ and $(x-5)$ are factors of $P(x)$ and hence, must divide $P(x)$ evenly. (Note: $(x+3)(x-5) = x^2-2x-15$.)

$$\begin{array}{r}
x^2+4 \\
x^2-2x-15\overline{\smash{\big)}\, x^4-2x^3-11x^2-8x-60} \\
\underline{-(x^4-2x^3-15x^2)} \\
4x^2-8x-60 \\
\underline{-(4x^2-8x-60)} \\
0
\end{array}$$

So, $P(x) = (x^2+4)(x^2-2x-15) = \boxed{(x^2+4)(x-5)(x+3)}$. The real zeros are 5 and -3.

17.

Since -3 and 1 are both zeros, we know that $(x+3)$ and $(x-1)$ are factors of $P(x)$ and hence, must divide $P(x)$ evenly. (Note: $(x+3)(x-1) = x^2+2x-3$.)

$$\begin{array}{r}
x^2-2x+2 \\
x^2+2x-3\overline{\smash{\big)}\, x^4+0x^3-5x^2+10x-6} \\
\underline{-(x^4+2x^3-3x^2)} \\
-2x^3-2x^2+10x \\
\underline{-(-2x^3-4x^2+6x)} \\
2x^2+4x-6 \\
\underline{-(2x^2+4x-6)} \\
0
\end{array}$$

So, $P(x) = (x^2+2x-3)(x^2-2x+2) = (x-1)(x+3)(x^2-2x+2)$. The real zeros are 1 and -3.

19.

$$\begin{array}{r|rrrrr} -2 & 1 & 6 & 13 & 12 & 4 \\ & & -2 & -8 & -10 & -4 \\ \hline -2 & 1 & 4 & 5 & 2 & 0 \\ & & -2 & -4 & -2 & \\ \hline & 1 & 2 & 1 & 0 & \end{array}$$

So, $P(x)=(x+2)^2(x^2+2x+1)=(x+2)^2(x+1)^2$ and the zeros are -2 and -1, both with multiplicity 2.

21.

Factors of 4: $\pm 1, \pm 2, \pm 4$

Factors of 1: ± 1

Possible rational zeros: $\pm 1, \pm 2, \pm 4$

23.

Factors of 12: $\pm 1, \pm 2, \pm 3, \pm 4, \pm 6, \pm 12$

Factors of 1: ± 1

Possible rational zeros:
$\pm 1, \pm 2, \pm 3, \pm 4, \pm 6, \pm 12$

25.

Factors of 8: $\pm 1, \pm 2, \pm 4, \pm 8$

Factors of 2: $\pm 1, \pm 2$

Possible rational zeros:
$\pm \frac{1}{2}, \pm 1, \pm 2, \pm 4, \pm 8$

27.

Factors of -20: $\pm 1, \pm 2, \pm 4, \pm 5, \pm 10, \pm 20$

Factors of 5: $\pm 1, \pm 5$

Possible rational zeros:
$\pm 1, \pm 2, \pm 4, \pm 5, \pm 10, \pm 20, \pm \frac{1}{5}, \pm \frac{2}{5}, \pm \frac{4}{5}$

29.

Factors of 8: $\pm 1, \pm 2, \pm 4, \pm 8$

Factors of 1: ± 1

Possible rational zeros: $\pm 1, \pm 2, \pm 4, \pm 8$

Testing: $P(1)=P(-1)=P(2)=P(-4)=0$

31.

Factors of -3: $\pm 1, \pm 3$

Factors of 2: $\pm 1, \pm 2$

Possible rational zeros: $\pm 1, \pm 3, \pm \frac{1}{2}, \pm \frac{3}{2}$

Testing: $P(1)=P(3)=P\left(\frac{1}{2}\right)=0$

33.

Number of sign variations for $P(x)$: 1

$P(-x)=P(x)$, so

Number of sign variations for $P(-x)$: 1

Since $P(x)$ is degree 4, there are 4 zeros, the real ones of which are classified to the right:

Positive Real Zeros	Negative Real Zeros
1	1

35.

Number of sign variations for $P(x)$: 1

$P(-x)=(-x)^5-1=-x^5-1$, so

Number of sign variations for $P(-x)$: 0

Since $P(x)$ is degree 5, there are 5 zeros, the real ones of which are classified to the right:

Positive Real Zeros	Negative Real Zeros
1	0

37.

Number of sign variations for $P(x)$: 2

$P(-x)=-x^5+3x^3+x+2$, so

Number of sign variations for $P(-x)$: 1

Since $P(x)$ is degree 5, there are 5 zeros, the real ones of which are classified to the right:

Positive Real Zeros	Negative Real Zeros
2	1
0	1

39.

Number of sign variations for $P(x)$: 1

$P(-x)=-9x^7-2x^5+x^3+x$, so

Number of sign variations for $P(-x)$: 1

Since $P(x)$ is degree 7, there are 7 zeros. But, 0 is also a zero. So, we classify the remaining real zeros to the right:

Positive Real Zeros	Negative Real Zeros
1	1

41.

Number of sign variations for $P(x)$: 2

$P(-x)=P(x)$, so

Number of sign variations for $P(-x)$: 2

Since $P(x)$ is degree 6, there are 6 zeros, the real ones of which are classified to the right:

Positive Real Zeros	Negative Real Zeros
2	2
0	2
2	0
0	0

43.

Number of sign variations for $P(x)$: 4

$P(-x)=-3x^4-2x^3-4x^2-x-11$, so

Number of sign variations for $P(-x)$: 0

Since $P(x)$ is degree 4, there are 4 zeros, the real ones of which are classified to the right:

Positive Real Zeros	Negative Real Zeros
4	0
2	0
0	0

45.

a. Number of sign variations for $P(x)$: 0

$P(-x) = -x^3 + 6x^2 - 11x + 6$, so

Number of sign variations for $P(-x)$: 3

Since $P(x)$ is degree 3, there are zeros, the real ones of which are classified as:

Positive Real Zeros	Negative Real Zeros
0	3
0	1

b. Factors of 6: $\pm 1, \pm 2, \pm 3, \pm 6$

Factors of 1: ± 1

Possible rational zeros: $\pm 1, \pm 2, \pm 3, \pm 6$

c. Note that $P(-1) = P(-2) = P(-3) = 0$. So, the rational zeros are $-1, -2, -3$. These are the only zeros since P has degree 3.

d. $P(x) = (x+1)(x+2)(x+3)$

47.

a. Number of sign variations for $P(x)$: 2

$P(-x) = -x^3 - 7x^2 + x + 7$, so

Number of sign variations for $P(-x)$: 1

Since $P(x)$ is degree 3, there are 3 zeros, the real ones of which are classified as:

Positive Real Zeros	Negative Real Zeros
2	1
0	1

b. Factors of 7: $\pm 1, \pm 7$

Factors of 1: ± 1

Possible rational zeros: $\pm 1, \pm 7$

c. Note that $P(-1) = P(1) = P(7) = 0$. So, the rational zeros are $-1, 1, 7$. These are the only zeros since P has degree 3.

d. $P(x) = (x+1)(x-1)(x-7)$

49.

a. Number of sign variations for $P(x)$: 1

$P(-x) = x^4 - 6x^3 + 3x^2 + 10x$, so

Number of sign variations for $P(-x)$: 2

Since $P(x)$ is degree 4, there are 4 zeros, one of which is 0. We classify the remaining real zeros below:

Positive Real Zeros	Negative Real Zeros
1	2
1	0

b. $P(x) = x\left(x^3 + 6x^2 + 3x - 10\right)$ We list the possible nonzero rational zeros below:

Factors of -10: $\pm 1, \pm 2, \pm 5, \pm 10$

Factors of 1: ± 1

Possible rational zeros: $\pm 1, \pm 2, \pm 5, \pm 10$

c. Note that

$$P(0) = P(1) = P(-2) = P(-5) = 0.$$

So, the rational zeros are $0, 1, -2, -5$. These are the only zeros since P has degree 4.

d. $P(x) = x(x-1)(x+2)(x+5)$

51.

a. Number of sign variations for $P(x)$: 4

$P(-x) = x^4 + 7x^3 + 27x^2 + 47x + 26$, so

Number of sign variations for $P(-x)$: 0

Since $P(x)$ is degree 4, there are 4 zeros, the real ones of which are classified as:

Positive Real Zeros	Negative Real Zeros
4	0
2	0
0	0

b. Factors of 26: $\pm 1, \pm 2, \pm 13, \pm 26$

Factors of 1: ± 1

Possible rational zeros:

$\pm 1, \pm 2, \pm 13, \pm 26$

c. Note that $P(1) = P(2) = 0$. After testing the others, it is found that the only rational zeros are 1, 2. So, we at least know that $(x-1)(x-2) = x^2 - 3x + 2$ divides $P(x)$ evenly. To find the remaining zeros, we long divide:

$$\begin{array}{r} x^2 - 4x + 13 \\ x^2 - 3x + 2 \overline{\big) x^4 - 7x^3 + 27x^2 - 47x + 26} \\ \underline{-(x^4 - 3x^3 + 2x^2)} \\ -4x^3 + 25x^2 - 47x \\ \underline{-(-4x^3 + 12x^2 - 8x)} \\ 13x^2 - 39x + 26 \\ \underline{-(13x^2 - 39x + 26)} \\ 0 \end{array}$$

Since $x^2 - 4x + 13$ is irreducible, the real zeros are 1 and 2.

d. $P(x) = (x-1)(x-2)\left(x^2 - 4x + 13\right)$

53.

a. Number of sign variations for $P(x)$: 2

$P(-x) = -10x^3 - 7x^2 + 4x + 1$, so

Number of sign variations for $P(-x)$: 1

Since $P(x)$ is degree 3, there are 3 zeros, the real ones of which are classified as:

Positive Real Zeros	Negative Real Zeros
2	1
0	1

b. Factors of 1: ± 1

Factors of 10: $\pm 1, \pm 2, \pm 5, \pm 10$

Possible rational zeros: $\pm 1, \pm \frac{1}{2}, \pm \frac{1}{5}, \pm \frac{1}{10}$

c. Note that $P(1) = P(-\frac{1}{2}) = P(\frac{1}{5}) = 0$. So, the rational zeros are $-1, -\frac{1}{2}, \frac{1}{5}$. These are the only zeros since P has degree 3.

d. $P(x) = (x-1)(2x+1)(5x-1)$

55.

a. Number of sign variations for $P(x)$: 1

$P(-x) = -6x^3 + 17x^2 - x - 10$, so

Number of sign variations for $P(-x)$: 2

Since $P(x)$ is degree 3, there are 3 zeros, the real ones of which are classified as:

Positive Real Zeros	Negative Real Zeros
1	2
1	0

b. Factors of -10: $\pm1, \pm2, \pm5, \pm10$

Factors of 6: $\pm1, \pm2, \pm3, \pm6$

Possible rational zeros:

$\pm1, \pm2, \pm5, \pm10, \pm\frac{1}{2}, \pm\frac{1}{3}, \pm\frac{1}{6}, \pm\frac{2}{3}, \pm\frac{5}{2}, \pm\frac{5}{3}, \pm\frac{5}{6}, \pm\frac{10}{3}$

c. Note that $P(-1) = P(-\frac{5}{2}) = P(\frac{2}{3}) = 0$.

So, the rational zeros are $-1, -\frac{5}{2}, \frac{2}{3}$. These are the only zeros since P has degree 3.

d.
$$P(x) = (x+1)(2x+5)(3x-2) = 6(x+1)\left(x+\tfrac{5}{2}\right)\left(x-\tfrac{2}{3}\right)$$

57.

a. Number of sign variations for $P(x)$: 4

$P(-x) = x^4 + 2x^3 + 5x^2 + 8x + 4$, so

Number of sign variations for $P(-x)$: 0

Since $P(x)$ is degree 4, there are 4 zeros, the real ones of which are classified as:

Positive Real Zeros	Negative Real Zeros
4	0
2	0
0	0

b. Factors of 4: $\pm1, \pm2, \pm4$

Factors of 1: ±1

Possible rational zeros: $\pm1, \pm2, \pm4$

c. Note that $P(1) = 0$. After testing the others, it is found that the only rational zeros is 1. Hence, by **a**, 1 must have multiplicity 2 or 4. So, we know that at least $(x-1)^2 = x^2 - 2x + 1$ divides $P(x)$ evenly. To find the remaining zeros, we long divide:

$$\begin{array}{r} x^2 + 4 \\ x^2 - 2x + 1 \overline{\big) x^4 - 2x^3 + 5x^2 - 8x + 4} \\ \underline{-(x^4 - 2x^3 + x^2)} \\ 4x^2 - 8x + 4 \\ \underline{-(4x^2 - 8x + 4)} \\ 0 \end{array}$$

Since $x^2 + 4$ is irreducible, the only real zero is 1 (multiplicity 2).

d. $P(x) = (x-1)^2(x^2+4)$

59.

a. Number of sign variations for $P(x)$: 1

$P(-x) = P(x)$, so

Number of sign variations for $P(-x)$: 1

Since $P(x)$ is degree 6, there are 6 zeros, the real ones of which are classified as:

Positive Real Zeros	Negative Real Zeros
1	1

b. Factors of -36:

$\pm 1, \pm 2, \pm 3, \pm 4, \pm 6, \pm 9, \pm 12, \pm 18, \pm 36$

Factors of 1: ± 1

Possible rational zeros:

$\pm 1, \pm 2, \pm 3, \pm 4, \pm 6, \pm 9, \pm 12, \pm 18, \pm 36$

c. Note that $P(-1) = P(1) = 0$. From **a**, there can be no other rational zeros for P.

We know that $(x+1)(x-1) = x^2 - 1$ divides $P(x)$ evenly. To find the remaining zeros, we long divide:

$$\begin{array}{r} x^4 + 13x^2 + 36 \\ x^2 - 1 \overline{\big)\, x^6 + 0x^5 + 12x^4 + 0x^3 + 23x^2 + 0x - 36} \\ \underline{-(x^6 + 0x^5 - x^4)} \qquad\qquad\qquad\qquad \\ 13x^4 + 0x^3 + 23x^2 \qquad\quad \\ \underline{-(13x^4 + 0x^3 - 13x^2)} \qquad\quad \\ 36x^2 + 0x - 36 \\ \underline{-(36x^2 + 0x - 36)} \\ 0 \end{array}$$

Observe that

$$x^4 + 13x^2 + 36 = (x^2 + 9)(x^2 + 4),$$

both of which are irreducible. So, the real zeros are: -1 and 1

d. $P(x) = (x+1)(x-1)(x^2+9)(x^2+4)$

61.

a. Number of sign variations for $P(x)$: 4

$P(-x) = 4x^4 + 20x^3 + 37x^2 + 24x + 5$, so

Number of sign variations for $P(-x)$: 0

Since $P(x)$ is degree 4, there are 4 zeros, the real ones of which are classified as:

Positive Real Zeros	Negative Real Zeros
4	0
2	0
0	0

b. Factors of 5: $\pm 1, \pm 5$

Factors of 4: $\pm 1, \pm 2, \pm 4$

Possible rational zeros:

$\pm 1, \pm 5, \pm \frac{1}{2}, \pm \frac{1}{4}, \pm \frac{5}{2}, \pm \frac{5}{4}$

c. Note that $P(\frac{1}{2}) = 0$. After testing, we conclude that there the only rational zero is $\frac{1}{2}$, which has multiplicity 2 or 4. So, we know that at least $(x - \frac{1}{2})^2$ divides $P(x)$ evenly:

$$\begin{array}{r|rrrrr} \frac{1}{2} & 4 & -20 & 37 & -24 & 5 \\ & & 2 & -9 & 14 & -5 \\ \hline \frac{1}{2} & 4 & -18 & 28 & -10 & 0 \\ & & 2 & -8 & 10 & \\ \hline & 4 & -16 & 20 & 0 & \end{array}$$

So,

$$\begin{aligned} P(x) &= (x - \tfrac{1}{2})^2 (4x^2 - 16x + 20) \\ &= 4(x - \tfrac{1}{2})^2 (x^2 - 4x + 5) \\ &= (2x-1)^2 (x^2 - 4x + 5) \end{aligned}$$

63.

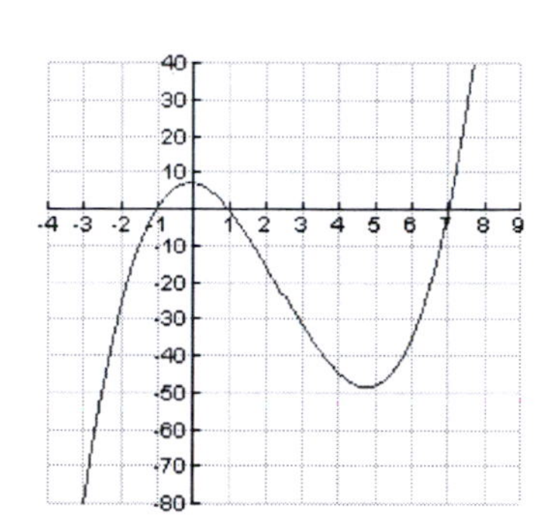

65.

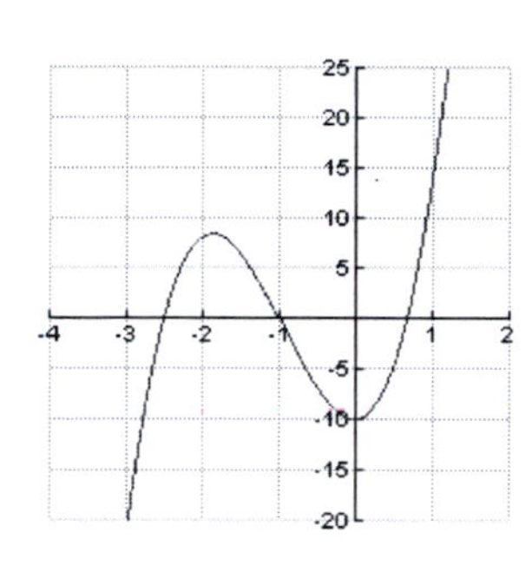

67. Observe that $f(1)=2,\ f(2)=-4$. So, by Intermediate Value Theorem, there must exist a zero in the interval (1,2). Graphically, we approximate this zero to be approximately $\boxed{1.34}$, as seen below:

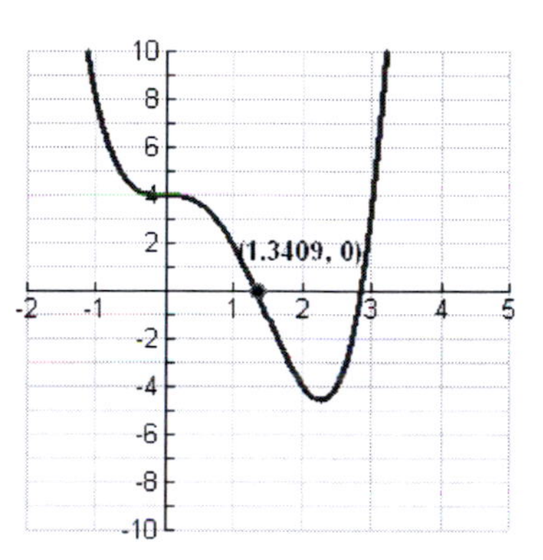

69. Observe that $f(0)=-1,\ f(1)=9$. So, by Intermediate Value Theorem, there must exist a zero in the interval (0,1). Graphically, we approximate this zero to be approximately $\boxed{0.22}$, as seen below:

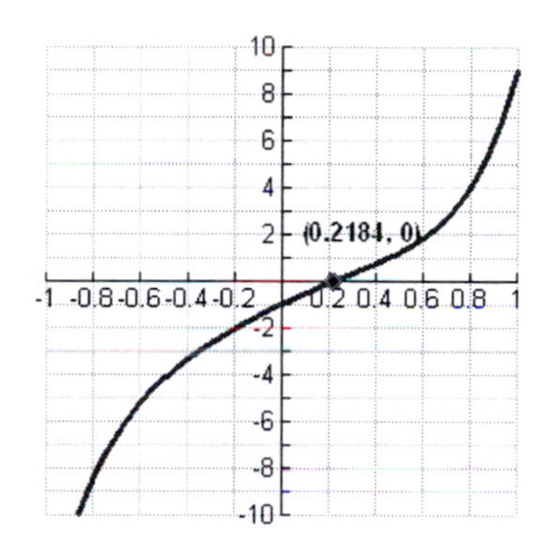

71. Observe that $f(-1)=2,\ f(0)=-3$. So, by Intermediate Value Theorem, there must exist a zero in the interval (-1, 0). Graphically, we approximate this zero to be approximately $\boxed{-0.43}$, as seen below:

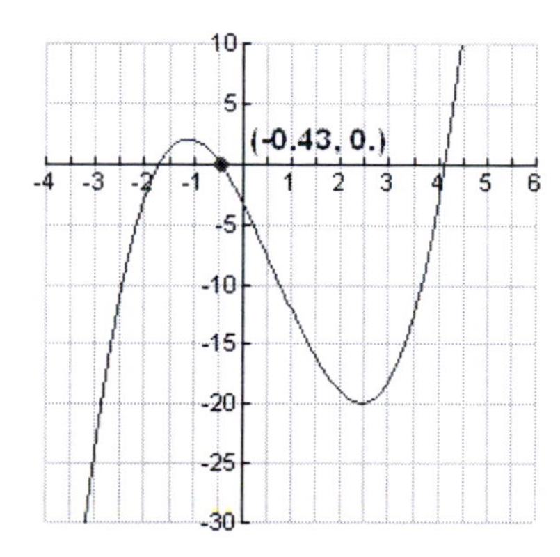

73. a. $P(x)=(46-3x^2)-(20+2x)=-3x^2-2x+26,\quad x\geq 0$.

b. Find all values of x such that: $P(x)=-3x^2-2x+26=0$:

$$x=\frac{2\pm\sqrt{(-2)^2-4(-3)(26)}}{2(-3)}=\frac{2\pm\sqrt{316}}{-6}\cong\frac{2\pm 17.78}{-6}\cong 2.63,\quad \cancel{-3.30}$$

So, approximately 263 subscribers will break even.

75.

$$\begin{aligned}P(x)&=xp(x)-C(x)\\&=x(28-0.0002x)-(20x+1{,}500)\\&=28x-0.0002x^2-20x-1{,}500\\&=-0.0002x^2+8x-1{,}500\end{aligned}$$

By Descartes Reule of Signs, there are either 0 or 2 positive real zeros.

77. Solve $C(t)=0$.

$$\begin{aligned}15.4-0.05t^2&=0\\1{,}540-5t^2&=0\\5(308-t^2)&=0\end{aligned}$$

$$t=\pm\sqrt{308}\approx 17.55 \text{ hours}$$

So, it takes about 18 hours to eliminate the drug from the bloodstream.

79. It is true that one can get 5 negative zeros here, but there may be just 1 or 3.

Positive Real Zeros	Negative Real Zeros
0	5
0	3
0	1

81. True

83. False. For instance, $f(x)=\left(x^2+1\right)\left(x^2+2\right)$ cannot be factored over the reals.

85.

$$\begin{array}{r|rrrr} a & 1 & -(a+b+c) & (ab+ac+bc) & -abc\\ & & a & -ab-ac & abc\\ \hline & 1 & -b-c & bc & 0\end{array}$$

So, $P(x)=(x-a)(x^2-(b+c)x+bc)=(x-a)(x-b)(x-c)$. So, the other zeros are b, c.

87.

Factors of -32:
$\pm1, \pm2, \pm4, \pm8, \pm16, \pm32$
Factors of 1: ±1
Possible rational zeros:
$\pm1, \pm2, \pm4, \pm8, \pm16, \pm32$

From the graph, we conclude that the only rational zero is 2.

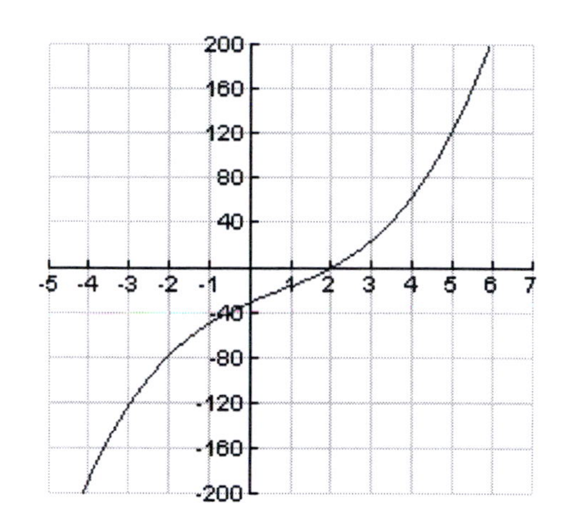

89. Consider the graph to the right of
$P(x) = 12x^4 + 25x^3 + 56x^2 - 7x - 30$.

a. From the graph, there are two real zeros, namely $-\frac{3}{4}$ and $\frac{2}{3}$. As such, we know that both $(3x-2)$ and $(4x+3)$ are factors of $P(x)$ and so, $(3x-2)(4x+3) = 12x^2 + x - 6$ divides it evenly. Indeed, observe that

$$\begin{array}{r} x^2 + 2x + 5 \\ 12x^2 + x - 6 \overline{)12x^4 + 25x^3 + 56x^2 - 7x - 30} \\ \underline{-(12x^4 + x^3 - 6x^2)} \\ 24x^3 + 62x^2 - 7x \\ \underline{-(24x^3 + 2x^2 - 12x)} \\ 60x^2 + 5x - 30 \\ \underline{-(60x^2 + 5x - 30)} \\ 0 \end{array}$$

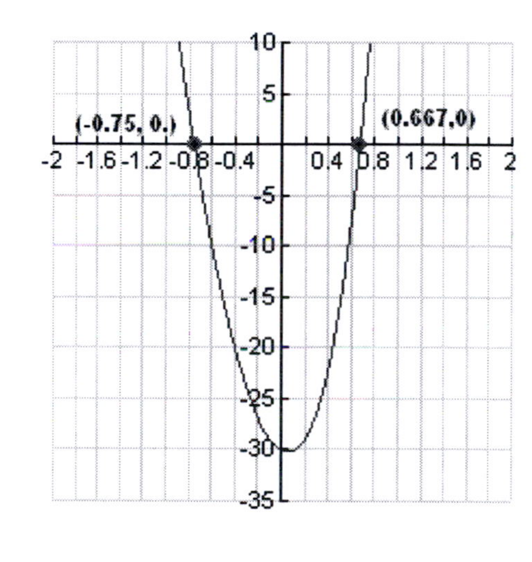

b. Since $x^2 + 2x + 5$ is irreducible, $P(x)$ factors as $(3x-2)(4x+3)\left(x^2 + 2x + 5\right)$.

91. Consider the graph to the right of
$P(x)=12x^4+25x^3+56x^2-7x-30$.
From the graph, there are two real zeros, namely $-\frac{3}{4}$ and $\frac{2}{3}$. As such, we know that both $(3x-2)$ and $(4x+3)$ are factors of $P(x)$ and so, $(3x-2)(4x+3)=12x^2+x-6$ divides it evenly. Indeed, observe that

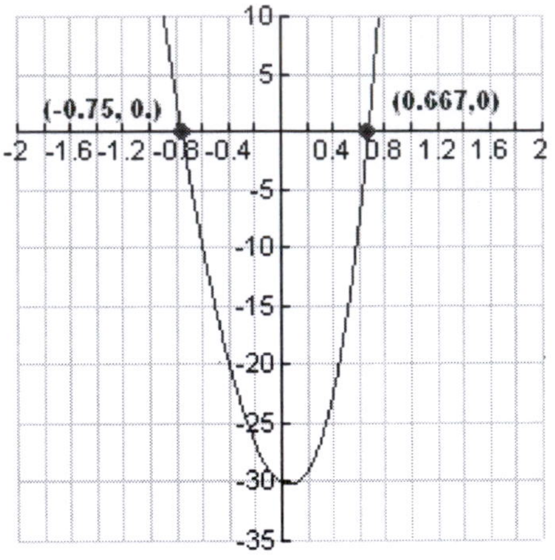

$$\begin{array}{r}
x^2+2x+5 \\
12x^2+x-6\overline{)12x^4+25x^3+56x^2-7x-30} \\
\underline{-(12x^4+x^3-6x^2)} \\
24x^3+62x^2-7x \\
\underline{-(24x^3+2x^2-12x)} \\
60x^2+5x-30 \\
\underline{-(60x^2+5x-30)} \\
0
\end{array}$$

Since x^2+2x+5 is irreducible, $P(x)$ factors as $(3x-2)(4x+3)\left(x^2+2x+5\right)$.

Section 4.5 Solutions ---

1. $P(x)=(x+2i)(x-2i)$. Zeros are $\pm 2i$.

3. Note that the zeros are
$x^2-2x+2=0 \Rightarrow$
$$x=\frac{2\pm\sqrt{4-4(2)}}{2}=\frac{2\pm 2i}{2}=1\pm i$$
So, $P(x)=(x-(1-i))(x-(1+i))$.

5. Observe that
$$P(x)=\left(x^2-4\right)\left(x^2+4\right)$$
$$=(x-2)(x+2)(x-2i)(x+2i)$$
So, the zeros are $\pm 2, \pm 2i$.

7. Observe that
$$P(x)=\left(x^2-5\right)\left(x^2+5\right)$$
$$=\left(x-\sqrt{5}\right)\left(x+\sqrt{5}\right)\left(x-i\sqrt{5}\right)\left(x+i\sqrt{5}\right)$$
So, the zeros are $\pm\sqrt{5}, \pm i\sqrt{5}$.

9. If i is a zero, then so is its conjugate $-i$. Since $P(x)$ has degree 3, this is the only missing zero.

11. Since $2i$ and $3-i$ are zeros, so are their conjugates $-2i$ and $3+i$, respectively. Since $P(x)$ has degree 4, these are the only missing zeros.

13. Since $1-3i$ and $2+5i$ are zeros, so are their conjugates $1+3i$ and $2-5i$, respectively. Since $P(x)$ has degree 6 and 0 is a zero with multiplicity 2, these are the only missing zeros.

15. Since $-i$ and $1-i$ are zeros, so are their conjugates i and $1+i$, respectively. Since $1-i$ has multiplicity 2, so does its conjugate. Since $P(x)$ has degree 6, these are the only missing zeros.

17. Let $P(x)$ be the desired polynomial. Since 0 is a zero of P, x is a factor of P. Also, since $1\pm 2i$ is a conjugate pair, the following must divide into $P(x)$ evenly:

$$(x-(1-2i))(x-(1+2i)) = x^2-2x+5$$

So, $P(x)$ is given by $x\left(x^2-2x+5\right) = x^3-2x^2+5x$.

19. Let $P(x)$ be the desired polynomial. Since 1 is a zero of P, $x-1$ is a factor of P. Also, since $1\pm 5i$ is a conjugate pair, the following must divide into $P(x)$ evenly:

$$(x-(1-5i))(x-(1+5i)) = x^2-2x+26$$

So, $P(x)$ is given by $(x-1)\left(x^2-2x+26\right) = x^3-3x^2+28x-26$.

21. Let $P(x)$ be the desired polynomial.
Since $1\pm i$ is a conjugate pair, the following must divide into $P(x)$ evenly:

$$(x-(1-i))(x-(1+i)) = x^2-2x+2$$

Also, since $\pm 3i$ is a conjugate pair, the following must divide into $P(x)$ evenly:

$$(x-3i)(x+3i) = x^2+9$$

So, $P(x)$ is given by $\left(x^2+9\right)\left(x^2-2x+2\right) = x^4-2x^3+11x^2-18x+18$.

23. Since $-2i$ is a zero of $P(x)$, so is its conjugate $2i$. As such, $(x-2i)(x+2i) = x^2+4$ divides $P(x)$ evenly. Indeed, observe that

$$\begin{array}{r} x^2-2x-15 \\ x^2+0x+4 \,\big)\, x^4-2x^3-11x^2-8x-60 \\ \underline{-(x^4+0x^3+4x^2)} \\ -2x^3-15x^2-8x \\ \underline{-(-2x^3+0x^2-8x)} \\ -15x^2-60 \\ \underline{-(-15x^2-60)} \\ 0 \end{array}$$

So,

$$\begin{aligned} P(x) &= (x-2i)(x+2i)\left(x^2-2x-15\right) \\ &= (x-2i)(x+2i)(x-5)(x+3) \end{aligned}$$

So, the zeros are $\pm 2i$, -3, 5.

25. Since i is a zero of $P(x)$, so is its conjugate $-i$. As such, $(x-i)(x+i) = x^2+1$ divides $P(x)$ evenly. Indeed,

$$\begin{array}{r} x^2-4x+3 \\ x^2+0x+1 \,\big)\, x^4-4x^3+4x^2-4x+3 \\ \underline{-(x^4+0x^3+x^2)} \\ -4x^3+3x^2-4x \\ \underline{-(-4x^3+0x^2-4x)} \\ 3x^2+3 \\ \underline{-(3x^2+3)} \\ 0 \end{array}$$

So,

$$\begin{aligned} P(x) &= (x-i)(x+i)\left(x^2-4x+3\right) \\ &= (x-i)(x+i)(x-3)(x-1) \end{aligned}$$

So, the zeros are $\pm i$, 1, 3.

27. Since $-3i$ is a zero of $P(x)$, so is its conjugate $3i$. As such, $(x-3i)(x+3i)=$ x^2+9 divides $P(x)$ evenly. Indeed, observe that

$$x^2+0x+9\overline{\smash{\big)}x^4-2x^3+10x^2-18x+9}\quad \text{quotient: } x^2-2x+1$$
$$\underline{-(x^4+0x^3+9x^2)}$$
$$-2x^3+x^2-18x$$
$$\underline{-(-2x^3+0x^2-18x)}$$
$$x^2+9$$
$$\underline{-(x^2+9)}$$
$$0$$

So,

$$P(x)=(x-3i)(x+3i)\left(x^2-2x+1\right)$$
$$=(x-3i)(x+3i)(x-1)^2$$

So, the zeros are $\pm 3i$ and 1 (multiplicity 2).

29. Since $1+i$ is a zero of $P(x)$, so is its conjugate $1-i$. As such, $(x-(1+i))(x-(1-i))=$ x^2-2x+2 divides $P(x)$ evenly. Indeed, observe that

$$x^2-2x+2\overline{\smash{\big)}x^4+0x^3-9x^2+18x-14}\quad \text{quotient: } x^2+2x-7$$
$$\underline{-(x^4-2x^3+2x^2)}$$
$$2x^3-11x^2+18x$$
$$\underline{-(2x^3-4x^2+4x)}$$
$$-7x^2+14x-14$$
$$\underline{-(-7x^2+14x-14)}$$
$$0$$

Now, we find the roots of x^2+2x-7:

$$x=\frac{-2\pm\sqrt{4-4(-7)}}{2}=-1\pm 2\sqrt{2}$$

So, $P(x)=(x-(1+i))(x-(1-i))\cdot$
$(x-(-1-2\sqrt{2}))(x-(-1+2\sqrt{2}))$

So, the zeros are $1\pm i,\ -1\pm 2\sqrt{2}$.

31. Since $3-i$ is a zero of $P(x)$, so is its conjugate $3+i$. As such, $(x-(3+i))(x-(3-i))=$ $x^2-6x+10$ divides $P(x)$ evenly. Indeed, observe that

$$x^2-6x+10\overline{\smash{\big)}x^4-6x^3+6x^2+24x-40}\quad \text{quotient: } x^2-4$$
$$\underline{-(x^4-6x^3+10x^2)}$$
$$-4x^2+24x-40$$
$$\underline{-(-4x^2+24x-40)}$$
$$0$$

So,

$$P(x)=(x-(3+i))(x-(3-i))(x-2)(x+2)$$

So, the zeros are $3\pm i,\ \pm 2$.

33. Since $2-i$ is a zero of $P(x)$, so is its conjugate $2+i$. As such, $(x-(2+i))(x-(2-i))=$ x^2-4x+5 divides $P(x)$ evenly. Indeed, observe that

$$x^2-4x+5\overline{\smash{\big)}x^4-9x^3+29x^2-41x+20}\quad \text{quotient: } x^2-5x+4$$
$$\underline{-(x^4-4x^3+5x^2)}$$
$$-5x^3+24x^2-41x$$
$$\underline{-(-5x^3+20x^2-25x)}$$
$$4x^2-16x+20$$
$$\underline{-(4x^2-16x+20)}$$
$$0$$

So,

$$P(x)=(x-(2+i))(x-(2-i))(x-1)(x-4)$$

So, the zeros are $2\pm i,\ 1,4$.

35.

$$\begin{aligned} x^3 - x^2 + 9x - 9 &= (x^3 - x^2) + 9(x-1) \\ &= x^2(x-1) + 9(x-1) \\ &= (x^2+9)(x-1) \\ &= (x+3i)(x-3i)(x-1) \end{aligned}$$

37.

$$\begin{aligned} x^3 - 5x^2 + x - 5 &= (x^3 - 5x^2) + (x-5) \\ &= x^2(x-5) + (x-5) \\ &= (x^2+1)(x-5) \\ &= (x+i)(x-i)(x-5) \end{aligned}$$

39.

$$\begin{aligned} x^3 + x^2 + 4x + 4 &= (x^3 + x^2) + 4(x+1) = x^2(x+1) + 4(x+1) \\ &= (x^2+4)(x+1) = (x+2i)(x-2i)(x+1) \end{aligned}$$

41. Consider $P(x) = x^3 - x^2 - 18$.
By the Rational Zero Theorem, the only possible rational roots are
$\pm 1, \pm 2, \pm 3, \pm 6, \pm 9, \pm 18$. Note that

3	1	−1	0	−18
		3	6	18
	1	2	6	0

So, $P(x) = (x-3)(x^2+2x+6)$. Now, we find the roots of x^2+2x+6:

$$x = \frac{-2 \pm \sqrt{4-4(6)}}{2} = -1 \pm i\sqrt{5}$$

So, $P(x) = (x-3)(x-(-1+i\sqrt{5}))(x-(-1-i\sqrt{5}))$.

43. Consider $P(x) = x^4 - 2x^3 - 11x^2 - 8x - 60$.
By the Rational Zero Theorem, the only possible rational roots are

$\pm 1, \pm 2, \pm 3, \pm 4, \pm 5, \pm 6,$

$\pm 10, \pm 12, \pm 15, \pm 20, \pm 30, \pm 60.$

Note that

−3	1	−2	−11	−8	−60
		−3	15	−12	60
	1	−5	4	−20	0

So,

$$\begin{aligned} P(x) &= (x+3)(x^3 - 5x^2 + 4x - 20) \\ &= (x+3)[x^2(x-5) + 4(x-5)] \\ &= (x+3)(x^2+4)(x-5) \\ &= (x+3)(x-5)(x+2i)(x-2i) \end{aligned}$$

45. Consider $P(x) = x^4 - 4x^3 - x^2 - 16x - 20$.
By the Rational Zero Theorem, the only possible rational roots are

$\pm 1, \pm 2, \pm 4, \pm 5, \pm 10, \pm 20$

Note that

−1	1	−4	−1	−16	−20
		−1	5	−4	20
	1	−5	4	−20	0

So,

$$\begin{aligned} P(x) &= (x+1)(x^3 - 5x^2 + 4x - 20) \\ &= (x+1)[x^2(x-5) + 4(x-5)] \\ &= (x+1)(x^2+4)(x-5) \\ &= (x+1)(x-5)(x+2i)(x-2i) \end{aligned}$$

47. Consider $P(x)=x^4-7x^3+27x^2-47x+26$.
By the Rational Zero Theorem, the only possible rational roots are

$$\pm 1,\ \pm 2, \pm 13,\ \pm 26$$

Note that

$$\begin{array}{r|rrrrr} 1 & 1 & -7 & 27 & -47 & 26 \\ & & 1 & -6 & 21 & -26 \\ \hline 2 & 1 & -6 & 21 & -26 & 0 \\ & & 2 & -8 & 26 & \\ \hline & 1 & -4 & 13 & 0 & \end{array}$$

So, $P(x)=(x-1)(x-2)\left(x^2-4x+13\right)$.

Next, we find the roots of $x^2-4x+13$:

$$x=\frac{4\pm\sqrt{16-4(13)}}{2}=2\pm 3i$$

So,

$$P(x)=(x-1)(x-2)(x-(2-3i))(x-(2+3i))$$

49. Consider $P(x)=-x^4-3x^3+x^2+13x+10$.
By the Rational Zero Theorem, the only possible rational roots are

$$\pm 1,\ \pm 2, \pm 5,\ \pm 10$$

Note that

$$\begin{array}{r|rrrrr} -1 & 1 & 3 & -1 & -13 & -10 \\ & & -1 & -2 & 3 & 10 \\ \hline 2 & 1 & 2 & -3 & -10 & 0 \\ & & 2 & 8 & 10 & \\ \hline & 1 & 4 & 5 & 0 & \end{array}$$

So, $P(x)=-(x+1)(x-2)\left(x^2+4x+5\right)$.

Next, we find the roots of x^2+4x+5:

$$x=\frac{-4\pm\sqrt{16-4(5)}}{2}=-2\pm i$$

So,

$$P(x)=-(x+1)(x-2)(x-(-2-i))(x-(-2+i))$$

51. Consider $P(x)=x^4-2x^3+5x^2-8x+4$.
By the Rational Zero Theorem, the only possible rational roots are $\pm 1,\ \pm 2,\ \pm 4$.
Note that

$$\begin{array}{r|rrrrr} 1 & 1 & -2 & 5 & -8 & 4 \\ & & 1 & -1 & 4 & -4 \\ \hline & 1 & -1 & 4 & -4 & 0 \end{array}$$

So,

$$\begin{aligned} P(x) &= (x-1)\left(x^3-x^2+4x-4\right) \\ &= (x-1)\left[x^2(x-1)+4(x-1)\right]. \\ &= (x-1)\left(x^2+4\right)(x-1) \\ &= (x-1)^2(x+2i)(x-2i) \end{aligned}$$

53. Consider $P(x)=x^6+12x^4+23x^2-36$.
By the Rational Zero Theorem, the only possible rational roots are

$$\pm 1,\ \pm 2,\ \pm 4,\ \pm 6,\ \pm 9, \pm 18,\ \pm 36$$

Note that

$$\begin{array}{r|rrrrrrr} 1 & 1 & 0 & 12 & 0 & 23 & 0 & -36 \\ & & 1 & 1 & 13 & 13 & 36 & 36 \\ \hline -1 & 1 & 1 & 13 & 13 & 36 & 36 & 0 \\ & & -1 & 0 & -13 & 0 & -36 & \\ \hline & 1 & 0 & 13 & 0 & 36 & 0 & \end{array}$$

So,

$$\begin{aligned} P(x) &= (x-1)(x+1)\left(x^4+13x^2+36\right) \\ &= (x-1)(x+1)\left(x^2+4\right)\left(x^2+9\right) \\ &= (x-1)(x+1)(x-2i)(x+2i)(x-3i)(x+3i) \end{aligned}$$

55. Consider $P(x)=4x^4-20x^3+37x^2-24x+5$.
By the Rational Zero Theorem, the only possible rational roots are

$$\pm 1, \pm 5, \pm \tfrac{1}{2}, \pm \tfrac{5}{2}, \pm \tfrac{1}{4}, \pm \tfrac{5}{4}$$

Note that

$$\begin{array}{r|rrrrr} \tfrac{1}{2} & 4 & -20 & 37 & -24 & 5 \\ & & 2 & -9 & 14 & -5 \\ \hline \tfrac{1}{2} & 4 & -18 & 28 & -10 & 0 \\ & & 2 & -8 & 10 & \\ \hline & 4 & -16 & 20 & 0 & \end{array}$$

So, $P(x)=4\left(x-\tfrac{1}{2}\right)^2\left(x^2-4x+5\right)$.

Next, we find the roots of x^2-4x+5:

$$x=\frac{4\pm\sqrt{16-4(5)}}{2}=2\pm i$$

So, $P(x)=4\left(x-\tfrac{1}{2}\right)^2(x-(2-i))(x-(2+i))$
$=(2x-1)^2(x-(2-i))(x-(2+i))$

57. Consider

$$P(x)=3x^5-2x^4+9x^3-6x^2-12x+8.$$

By the Rational Zero Theorem, the only possible rational roots are

$$\pm 1, \pm 2, \pm 4, \pm 8, \pm \tfrac{1}{3}, \pm \tfrac{2}{3}, \pm \tfrac{4}{3}, \pm \tfrac{8}{3}$$

Note that

$$\begin{array}{r|rrrrrr} -1 & 3 & -2 & 9 & -6 & -12 & 8 \\ & & -3 & 5 & -14 & 20 & -8 \\ \hline 1 & 3 & -5 & 14 & -20 & 8 & 0 \\ & & 3 & -2 & 12 & -8 & \\ \hline \tfrac{2}{3} & 3 & -2 & 12 & -8 & 0 & \\ & & 2 & 0 & 8 & & \\ \hline & 3 & 0 & 12 & 0 & & \end{array}$$

So,

$$P(x)=(x-1)(x+1)(x-\tfrac{2}{3})(3x^2+12)$$
$$=(x-1)(x+1)(3x-2)(x-2i)(x+2i)$$

59. Yes. In such case, $P(x)$ never touches the x-axis (since crossing it would require $P(x)$ to have a real root) and is always above it since the leading coefficient is positive, indicating that the end behavior should resemble that of $y=x^{2n}$, for some positive integer n. So, profit is always positive and increasing.

61. No. In such case, it crosses the x-axis and looks like $y=-x^3$. So, profit is decreasing.

63. Since the profit function is a third-degree polynomial we know that the function has three zeros and at most two turning points. Looking at the graph we can see there is one real zero where $t \le 0$. There are no real zeros when $t > 0$, therefore the other two zeros must be complex conjugates. Therefore, the company always has a profit greater than approximately 5.1 million dollars and, in fact, the profit will increase towards infinity as t increases.

65. Since the concentration function is a third degree polynomial, we know the function has three zeros and at most two turning points. Looking at the graph we can see there will be one real zero at some time $t \ge 8$. The remaining zeros are a pair of complex conjugates. Therefore, the concentration of the drug in the blood stream will decrease to zero as the hours go by. Note that the concentration will not approach negative infinity since concentration is a non-negative quantity.

67. Step 2 is an error. In general, the additive inverse of a real root need not be a root. This is being confused with the fact that complex roots occur in conjugate pairs.

69. False. For example, consider $P(x)=(x-1)(x+3)$. Note that 1 is a zero of P, but -1 is not.

71. True. It has n complex zeros.

73. No. Complex zeros occur in conjugate pairs. So, the collection of complex solutions contributes an even number of zeros, thereby requiring there to be at least one real zero.

75. Since bi is a zero of multiplicity 3, its conjugate $-bi$ is also a zero of multiplicity 3. Hence,

$$P(x) = (x-bi)^3(x+bi)^3 = \left[(x-bi)(x+bi)\right]^3$$
$$= \left(x^2+b^2\right)^3 = x^6 + 3b^2x^4 + 3b^4x^2 + b^6$$

77. The possible combinations of zeros of $P(x)$ are:

Real Zeros	Complex Zeros
0	4
2	2
4	0

From the graph of P to the right, we note that all roots are complex.

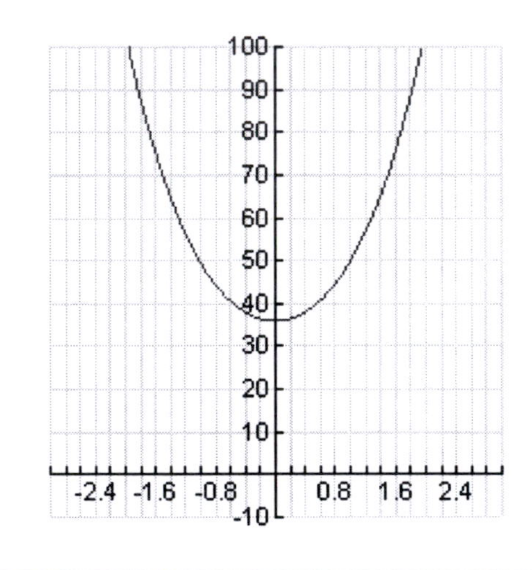

79. Consider the graph of $P(x)$ to the right. Note that the only real zero is $\frac{3}{5}$, so that there must be four complex zeros. Observe that

$$\begin{array}{r|rrrrrr} 0.6 & -5 & 3 & -25 & 15 & -20 & 12 \\ & & -3 & 0 & -15 & 0 & -12 \\ \hline & -5 & 0 & -25 & 0 & -20 & 0 \end{array}$$

So,

$$P(x) = (x-0.6)\left(-5x^4 - 25x^2 - 20\right)$$
$$= -5(x-0.6)\left(x^4 + 5x^2 + 4\right)$$
$$= -5(x-0.6)\left(x^2+4\right)\left(x^2+1\right)$$
$$= -5(x-0.6)(x-2i)(x+2i)(x-i)(x+i)$$

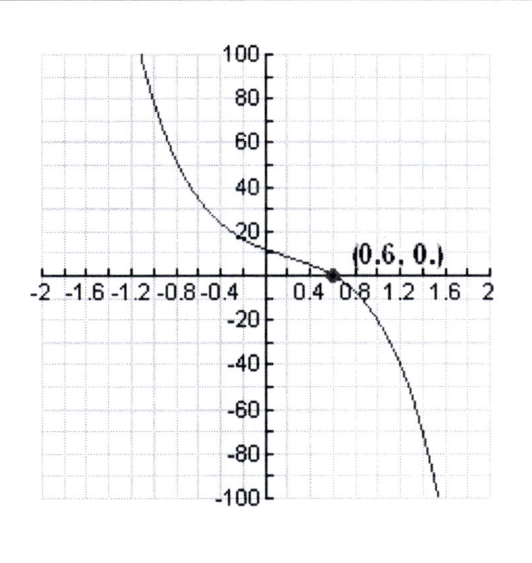

Section 4.6 Solutions --

1. Domain: $(-\infty,-3)\cup(-3,\infty)$	**3.** Domain: $(-\infty,-\frac{1}{3})\cup(-\frac{1}{3},\frac{1}{2})\cup(\frac{1}{2},\infty)$

5. Note that $x^2+x-12=(x+4)(x-3)$.
<u>Domain</u>: $(-\infty,-4)\cup(-4,3)\cup(3,\infty)$

7. Note that x^2+16 is never zero.
<u>Domain</u>: $(-\infty,\infty)$

9. Note that $2\left(x^2-x-6\right)=2(x-3)(x+2)$. <u>Domain</u>: $(-\infty,-2)\cup(-2,3)\cup(3,\infty)$

11. <u>Vertical Asymptote</u>:
$x+2=0$, so $x=-2$ is the VA.
<u>Horizontal Asymptote</u>:
Since the degree of the numerator is less than degree of the denominator, $y=0$ is the HA.

13. <u>Vertical Asymptote</u>:
$x+5=0$, so $x=-5$ is the VA.
<u>Horizontal Asymptote</u>:
Since the degree of the numerator is greater than degree of the denominator, there is no HA.

15. <u>Vertical Asymptote</u>:
$6x^2+5x-4=(2x-1)(3x+4)=0$, so
$x=\frac{1}{2}$, $x=-\frac{4}{3}$ are the VAs.
<u>Horizontal Asymptote</u>:
Since the degree of the numerator is greater than degree of the denominator, there is no HA.

17. <u>Vertical Asymptote</u>:
$x^2+\frac{1}{9}$ is never 0, so there is no VA.
<u>Horizontal Asymptote</u>:
Since the degree of the numerator equals the degree of the denominator, $y=\frac{1}{3}$ is the HA.

19. <u>Vertical Asymptote</u>:
$(x-0.5)(0.2x+0.3)=0$, so
$x=0.5$, $x=-1.5$ are the VAs.
<u>Horizontal Asymptote</u>:
Since the degree of the numerator equals the degree of the denominator, $y=\frac{12}{7}\approx 1.71$ is the HA.

21. To find the slant asymptote, we use synthetic division:

$$\begin{array}{r|rrr} -4 & 1 & 10 & 25 \\ & & -4 & -24 \\ \hline & 1 & 6 & 1 \end{array}$$

So, the slant asymptote is $y=x+6$.

23. To find the slant asymptote, we use synthetic division:

$$\begin{array}{r|rrr} 5 & 2 & 14 & 7 \\ & & 10 & 120 \\ \hline & 2 & 24 & 127 \end{array}$$

So, the slant asymptote is $y=2x+24$.

25. To find the slant asymptote, we use long division:

$$\begin{array}{r} 4x+\frac{11}{2} \\ 2x^3-x^2+3x-1 \overline{)\,8x^4+7x^3+0x^2+2x-5} \\ \underline{-(8x^4-4x^3+12x^2-4x)} \\ 11x^3-12x^2+6x-5 \\ \underline{-(11x^3-\frac{11}{2}x^2+\frac{33}{2}x-\frac{11}{2})} \\ -\frac{13}{2}x^2-\frac{21}{2}x+\frac{1}{2} \end{array}$$

So, the slant asymptote is $y = 4x + \frac{11}{2}$.

27. b Vertical Asymptote: $x = 4$ Horizontal Asymptote: $y = 0$

29. a Vertical Asymptotes: $x = 2,\ x = -2$ Horizontal Asymptote: $y = 3$

31. e Vertical Asymptotes: $x = 2,\ x = -2$ Horizontal Asymptote: $y = -3$

33. Vertical Asymptote: $x = -1$
Horizontal Asymptote: $y = 0$

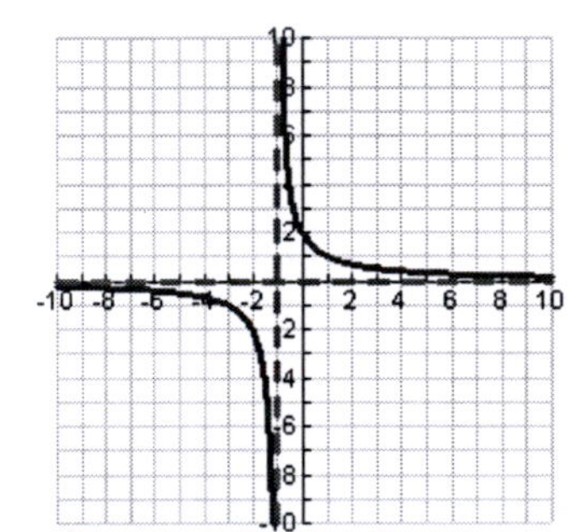

35. Vertical Asymptote: $x = 1$
Horizontal Asymptote: $y = 2$
Intercept: (0,0)

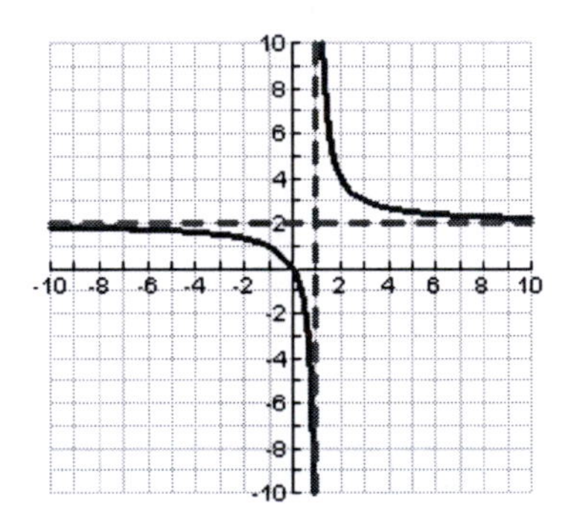

37. Vertical Asymptote: $x = 0$
Horizontal Asymptote: $y = 1$
Intercept: (1,0)

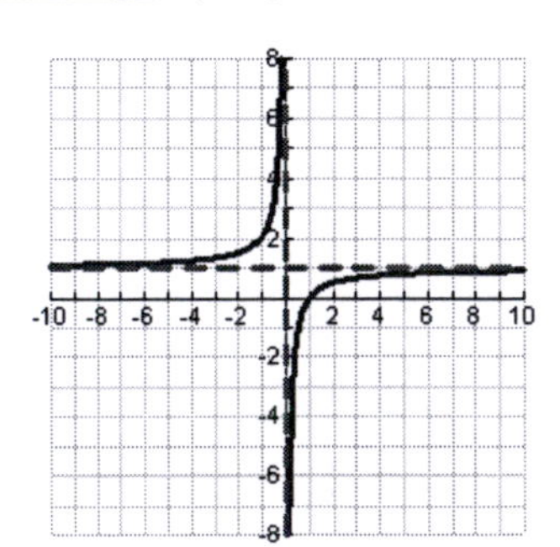

39. Vertical Asymptotes: $x = 0,\ x = -2$
Horizontal Asymptote: $y = 2$
Intercepts: $(3,0),\ (-1,0)$

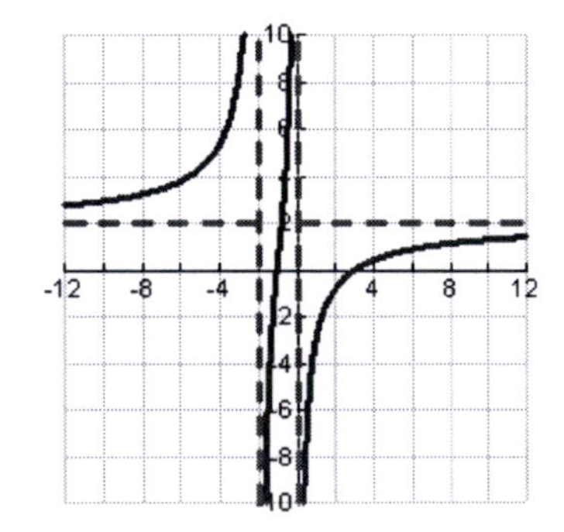

41. Vertical Asymptote: $x = -1$

Intercept: (0,0)

Slant Asymptote: $y = x - 1$

$$
\begin{array}{r}
x-1 \\
x+1 \overline{\smash{\big)}\, x^2 + 0x + 0} \\
\underline{-(x^2 + x)} \\
-x + 0 \\
\underline{-(-x - 1)} \\
1
\end{array}
$$

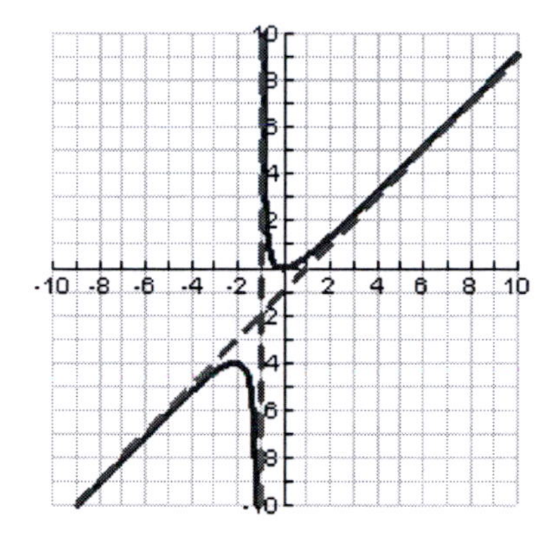

43. Vertical Asymptotes: $x = -2$, $x = 2$

Intercepts: $(0,0)$, $(-\frac{1}{2},0)$, $(1,0)$

Slant Asymptote: $y = 2x - 1$

$$
\begin{array}{r}
2x-1 \\
x^2 + 0x - 4 \overline{\smash{\big)}\, 2x^3 - x^2 - x + 0} \\
\underline{-(2x^3 + 0x^2 - 8x)} \\
-x^2 + 7x + 0 \\
\underline{-(-x^2 + 0x + 4)} \\
7x - 4
\end{array}
$$

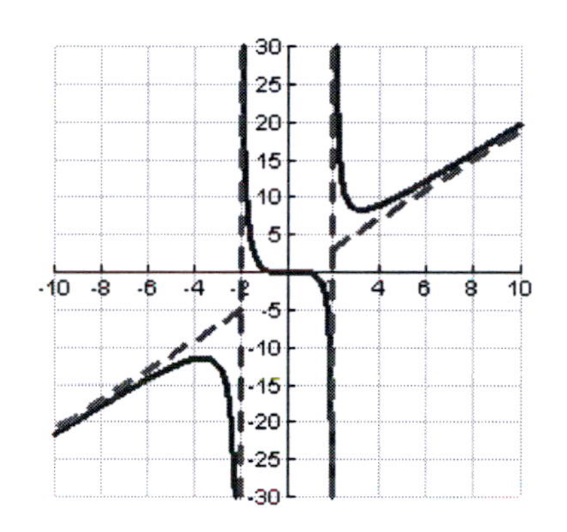

45. Vertical Asymptote: $x = 1$, $x = -1$

Horizontal Asymptote: $y = 1$

Intercept: $(0, -1)$

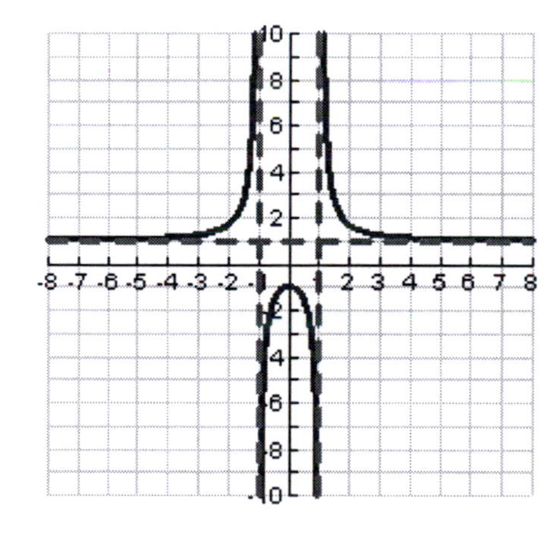

47. Vertical Asymptote: $x = -\frac{1}{2}$
Horizontal Asymptote: $y = \frac{7}{4}$
Intercept: (0,0)

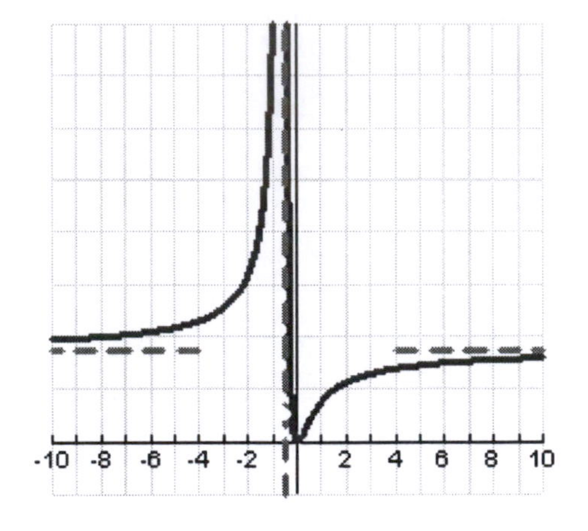

49. Vertical Asymptotes: $x = -\frac{1}{2}$, $x = \frac{1}{2}$
Horizontal Asymptote: $y = 0$
Intercepts: (0,1), $(\frac{1}{3},0)$, $(-\frac{1}{3},0)$

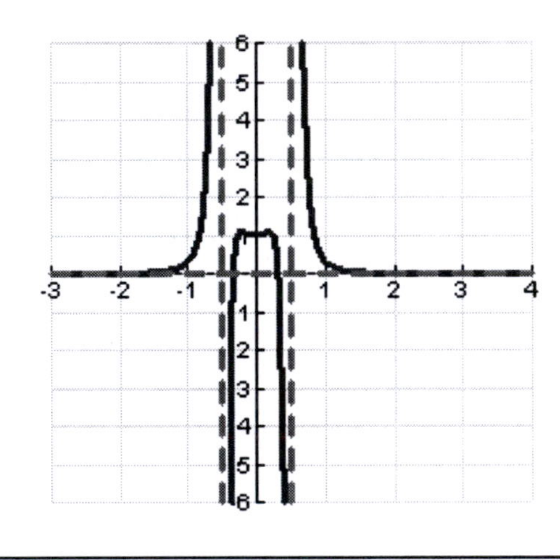

51. Vertical Asymptotes: $x = 0$
Slant Asymptote: $y = 3x$

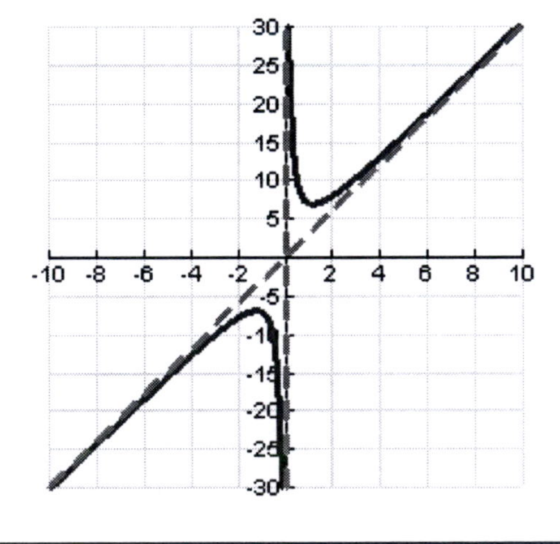

53. Observe that

$$f(x) = \frac{(x-1)^2}{(x-1)(x+1)} = \frac{x-1}{x+1}.$$

Open hole: (1,0)
Vertical Asymptote: $x = -1$
Horizontal Asymptote: $y = 1$
Intercepts: $(0,-1)$ and $(1,0)$

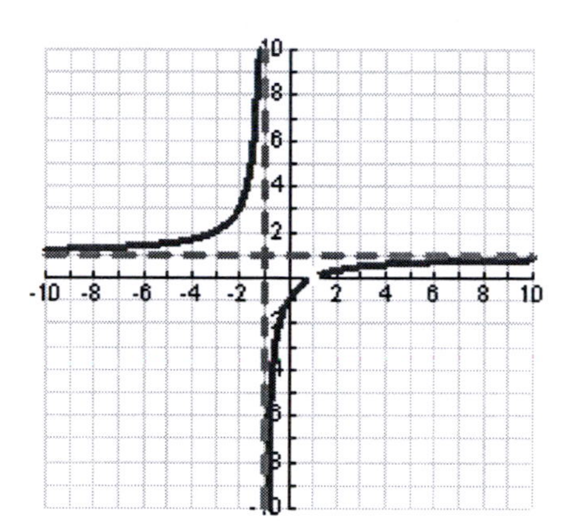

55. Observe that

$$f(x) = \frac{(x-1)(x-2)(x+2)}{(x-2)(x^2+1)} = \frac{(x-1)(x+2)}{x^2+1}.$$

Open hole: (2,0)
Vertical Asymptote: None
Horizontal Asymptote: $y = 1$
Intercepts: $(0,-2)$, $(1,0)$, and $(-2,0)$

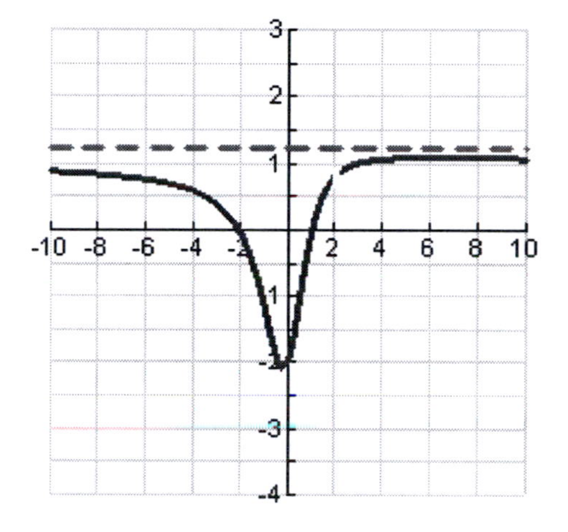

57. Observe that

$$f(x) = \frac{3x(x-1)}{x(x-2)(x+2)} = \frac{3x-3}{(x-2)(x+2)}.$$

Open hole: (0,0)
Vertical Asymptote: $x = 2$, $x = -2$
Horizontal Asymptote: $y = 0$
Intercepts: $(1,0)$

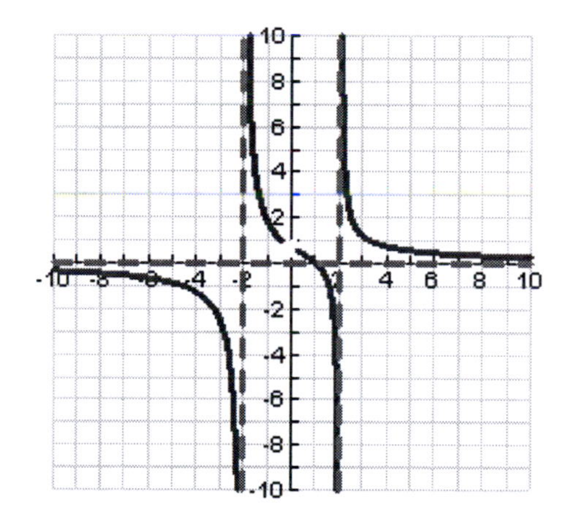

59. a. x-intercept: (2,0)
y-intercept: (0,0.5)
b. horizontal asymptote: $y = 0$
vertical asymptotes: $x = -1$, $x = 4$
c. $f(x) = \dfrac{x-2}{(x+1)(x-4)}$

61. a. x-intercept: (0,0)
y-intercept: (0,0)
b. horizontal asymptote: $y = -3$
vertical asymptotes: $x = -4$, $x = 4$
c. $f(x) = \dfrac{-3x^2}{(x+4)(x-4)}$

63. a. $C(1) = \frac{2(1)}{1^2+100} = \frac{2}{101} \cong 0.0198$
b. Since 1 hour = 60 minutes, we have
$C(60) = \frac{2(60)}{60^2+100} \cong 0.0324$.
c. Since 5 hours = 300 minutes, we have
$C(300) = \frac{2(300)}{300^2+100} \cong 0$.
d. The horizontal asymptote is $y = 0$. So, after several days, C is approximately 0.

65. a. $N(0) = 52$ wpm
b. $N(12) \cong 107$ wpm
c. Since 3 years = 36 months,
$N(36) \cong 120$wpm
d. The horizontal asymptote is $y = 130$.
So, expect to type approximately 130 wpm as time goes on.

67. The horizontal asymptote is $y = 10$. So most adult cats eventually eat about 10 ounces of food.

69. Let w = width of the pen, l = length of the pen.
Then, the area is given by $wl = 500$, which is equivalent to $l = \frac{500}{w}$ **(1)**.
The amount of fence needed is given by the perimeter of the pen, namely $2w + 2l$.
Substituting in **(1)**, we obtain the equivalent expression $2w + 2\left(\frac{500}{w}\right) = \frac{2w^2+1000}{w}$.

71. Solve for x:

$$\frac{-x^3 + 10x^2}{x} = 16$$

$$-x^2 + 10x - 16 = 0$$

$$(x-8)(x-2) = 0 \Rightarrow x = 2,8$$

So, must sell either 2,000 or 8,000 units to get this average profict. This yields an average value of $16 per unit.

73. $C(15) = \dfrac{22(14)}{15^2+1} + 24 \approx 25.4 \text{ mcg/mL}$.

The times t for which this concentration is achieved are found by solving $C(t) = 25.4$, as follows:

$$\frac{22(t-1)}{t^2+1} + 24 = 25.4$$

$$22(t-1) - 1.4\left(t^2+1\right) = 0$$

$$1.4t^2 - 22t + 23.4 = 0$$

$$t = \frac{22 \pm \sqrt{22^2 - 4(1.4)(23.4)}}{2(1.4)} \approx 1,\ 15$$

So, there are two times, 1 hours and 15 hours, after taking the medication at which the concentration of the drug in the bloodstream is approximately 25.4 mcg/mL. The first time, approximately 1 hour, occurs as the concentration of the drug is increasing to a level high enough that the body will be able to maintain a concentration of approximately 25 mcg/mL throughout the day. The second time, approximately 15 hours, occurs many hours later in the day as the concentration of the medication in the bloodstream drops.

75. $f(x) = \dfrac{x-1}{x^2-1} = \dfrac{\cancel{x-1}}{\cancel{(x-1)}(x+1)} = \dfrac{1}{x+1}$

with a hole at $x = 1$. So, $x = 1$ is not a vertical asymptote.

77. In Step 2, the ratio of the leading coefficients should be $\frac{-1}{1}$. So, the horizontal asymptote is $y = -1$.

79. True. The only way to have a slant asymptote is for the degree of the numerator to be greater than the degree of the denominator (by 1). In such case, there is no horizontal asymptote.

81. False. This would require the denominator to equal 0, causing the function to be undefined.

83. Vertical Asymptotes: $x = c,\ x = -d$ Horizontal Asymptote: $y = 1$

85. Two such possibilities are:

$$y = \frac{4x^2}{(x+3)(x-1)} \text{ and } y = \frac{4x^5}{(x+3)^3(x-1)^2}$$

87. $f(x)=\dfrac{x-4}{(x-4)(x+2)}=\dfrac{1}{x+2}$ has a hole at $x=4$ and a vertical asymptote at $x=-2$.

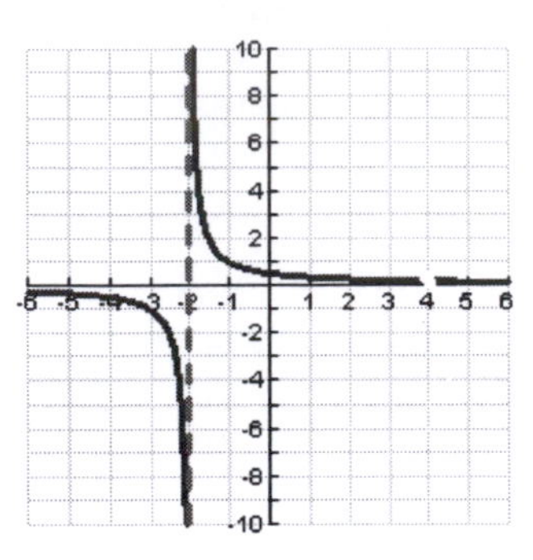

89. $f(x)=\dfrac{x-2(3x+1)}{x(3x+1)}=\dfrac{-5x-2}{x(3x+1)}$

Vertical asymptotes: $x=0,\ x=-\frac{1}{3}$

Horizontal asymptote: $y=0$

Intercepts: $\left(-\frac{2}{5},0\right)$

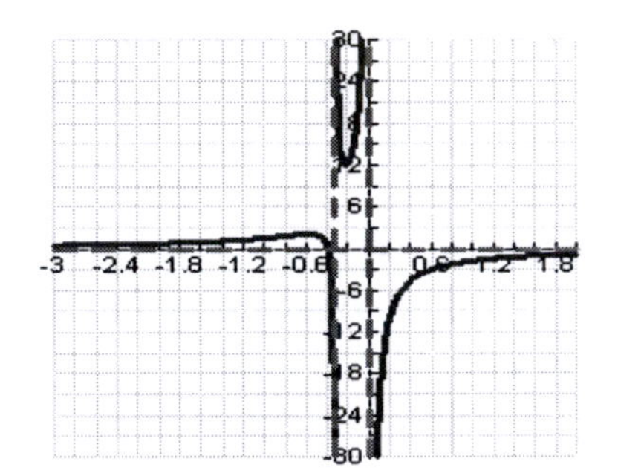

91.

a. Asymptotes for *f*:
vertical asymptote: $x=3$
horizontal asymptote: $y=0$
Asymptotes for *g*:
vertical asymptote: $x=3$
horizontal asymptote: $y=2$
Asymptotes for h:
vertical asymptote: $x=3$
horizontal asymptote: $y=-3$

b. The graphs of *f* and *g* are below. As $x\to\pm\infty$, $f(x)\to 0$ and $g(x)\to 2$.

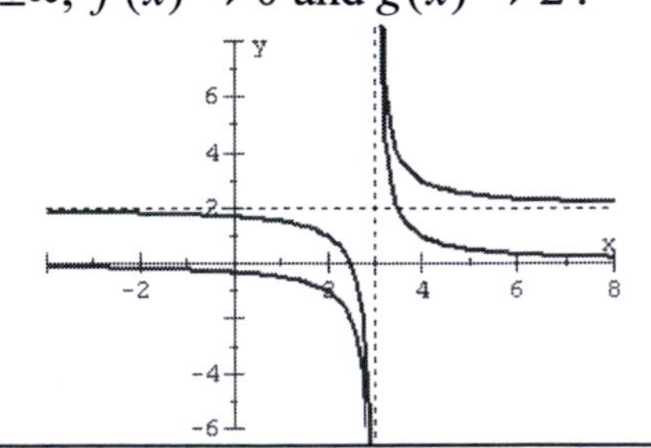

c. The graphs of *g* and *h* are below. As

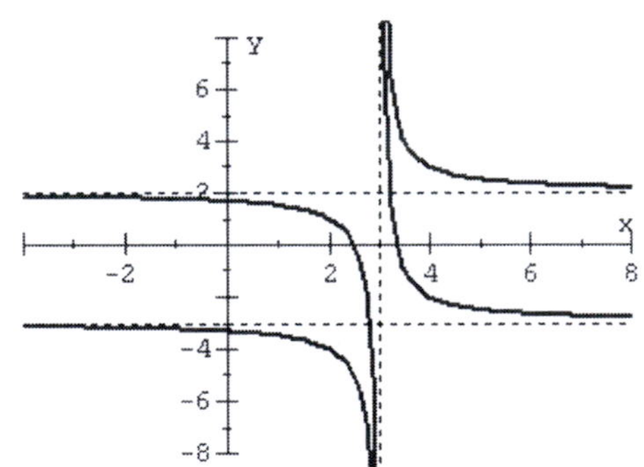

$x\to\pm\infty$, $g(x)\to 2$ and $h(x)\to -3$

d. $g(x)=\dfrac{2x-5}{x-3}$, $h(x)=\dfrac{-3x+10}{x-3}$.

Yes, if the degree of the numerator is the same as the degree of the denominator, then the horizontal asymptote is the ratio of the leading coefficients for both *g* and *h*.

Chapter 4 Review Solutions --

1. b Opens down, vertex is $(-6,3)$

3. a Opens up, vertex is $\left(-\frac{1}{2},-\frac{25}{4}\right)$.

$$x^2+x-6=(x^2+x+\tfrac{1}{4})-6-\tfrac{1}{4}$$
$$=(x+\tfrac{1}{2})^2-\tfrac{25}{4}$$

5.

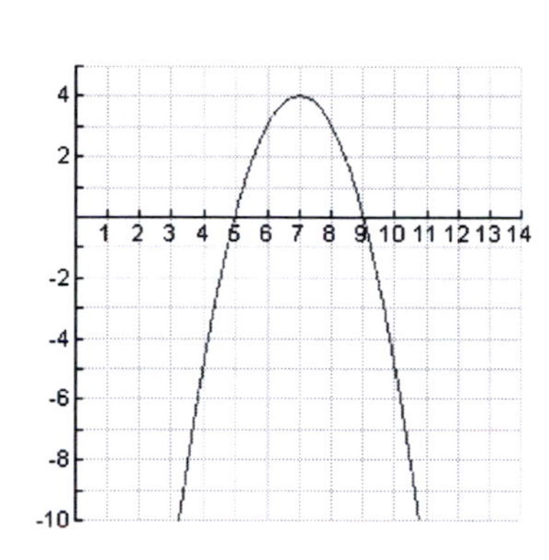

7.

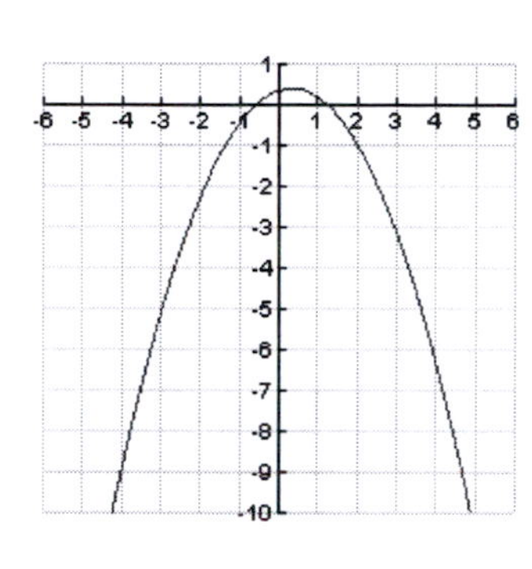

9.

$$x^2-3x-10=(x^2-3x+\tfrac{9}{4})-10-\tfrac{9}{4}$$
$$=(x-\tfrac{3}{2})^2-\tfrac{49}{4}$$

11.

$$4x^2+8x-7=4(x^2+2x\ \ )-7$$
$$=4(x^2+2x+1)-7-4$$
$$=4(x+1)^2-11$$

13.

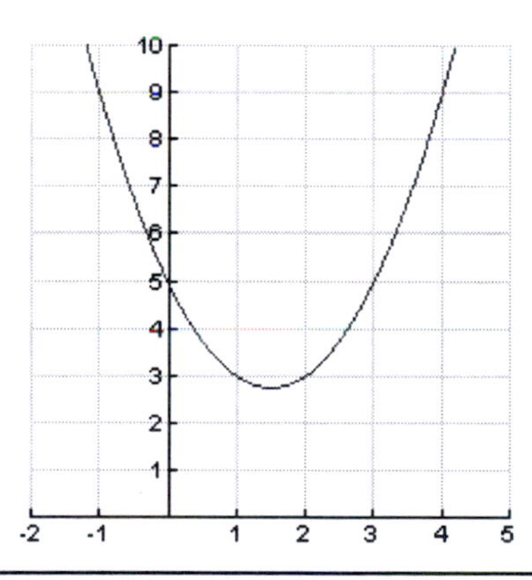

15.

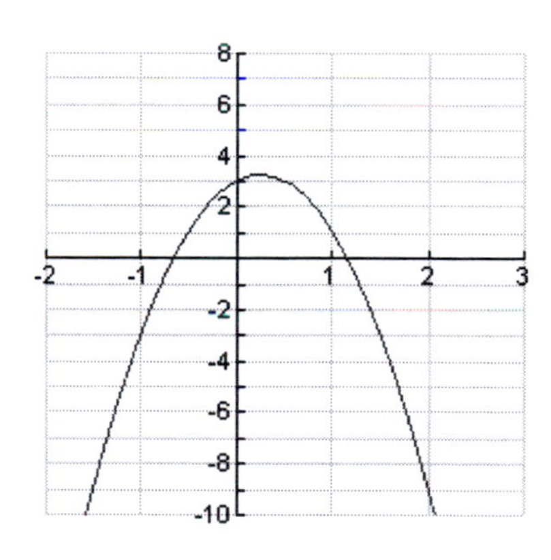

17. Vertex is

$$\left(-\frac{b}{2a},c-\frac{b^2}{4a}\right)=\left(-\frac{-5}{2(13)},12-\frac{(-5)^2}{4(13)}\right)$$
$$=\left(\frac{5}{26},\frac{599}{52}\right)$$

19. Vertex is

$$\left(-\frac{b}{2a},c-\frac{b^2}{4a}\right)=\left(-\frac{-0.12}{2(-0.45)},3.6-\frac{(-0.12)^2}{4(-0.45)}\right)$$
$$=\left(-\frac{2}{15},\frac{451}{125}\right)$$

21. Since the vertex is $(-2,3)$, the function has the form $y = a(x+2)^2 + 3$. To find *a*, use the fact that the point (1,4) is on the graph:

$$\begin{aligned} 4 &= a(1+2)^2 + 3 \\ 4 &= 9a + 3 \\ \tfrac{1}{9} &= a \end{aligned}$$

So, the function is $y = \frac{1}{9}(x+2)^2 + 3$.

23. Since the vertex is $(2.7, 3.4)$, the function has the form $y = a(x-2.7)^2 + 3.4$. To find *a*, use the fact that the point (3.2, 4.8) is on the graph:

$$\begin{aligned} 4.8 &= a(3.2-2.7)^2 + 3.4 \\ 4.8 &= 0.25a + 3.4 \\ 5.6 &= a \end{aligned}$$

So, the function is $y = 5.6(x-2.7)^2 + 3.4$.

25.

a.

$$\begin{aligned} P(x) &= R(x) - C(x) \\ &= \left(-2x^2 + 12x - 12\right) - \left(\tfrac{1}{3}x + 2\right) \\ &= -2x^2 + \tfrac{35}{3}x - 14 \end{aligned}$$

b. Solve $P(x) = 0$.

$$\begin{aligned} -2x^2 + \tfrac{35}{3}x - 14 &= 0 \\ -2\left(x^2 - \tfrac{35}{6}x \right) - 14 &= 0 \\ -2\left(x^2 - \tfrac{35}{6}x + \tfrac{1225}{144}\right) - 14 + \tfrac{1225}{72} &= 0 \\ -2(x - \tfrac{35}{12})^2 + \tfrac{217}{72} &= 0 \\ (x - \tfrac{35}{12})^2 &= \tfrac{217}{144} \\ x &= \tfrac{35}{12} \pm \sqrt{\tfrac{217}{144}} \\ &= \tfrac{35 \pm \sqrt{217}}{12} \\ &\cong 4.1442433,\ 1.68909 \end{aligned}$$

c.

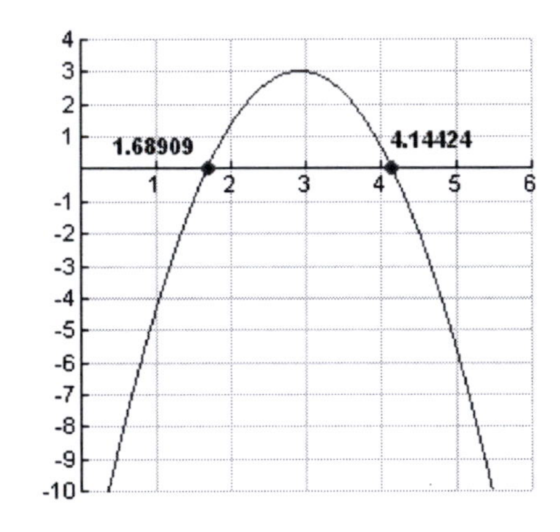

d. The range is approximately $(1.6891, 4.144)$, which corresponds to where the graph is above the *x*-axis.

27. Area is

$$\begin{aligned} A(x) = \tfrac{1}{2}(x+2)(4-x) &= -\tfrac{1}{2}(x^2 - 2x \quad) + 4 \\ &= -\tfrac{1}{2}(x^2 - 2x + 1) + 4 + \tfrac{1}{2} \\ &= -\tfrac{1}{2}(x-1)^2 + \tfrac{9}{2} \end{aligned}$$

Note that $A(x)$ has a maximum at $x = 1$ (since its graph is a parabola that opens down). The corresponding dimensions are both base and height are 3 units.

29. Polynomial with degree 6

31. Not a polynomial (due to the term $x^{1/4}$)

33. d linear

35. a 4th degree polynomial, looks like $y = x^4$ for *x* very large.

37. Reflect the graph of x^7 over the x-axis

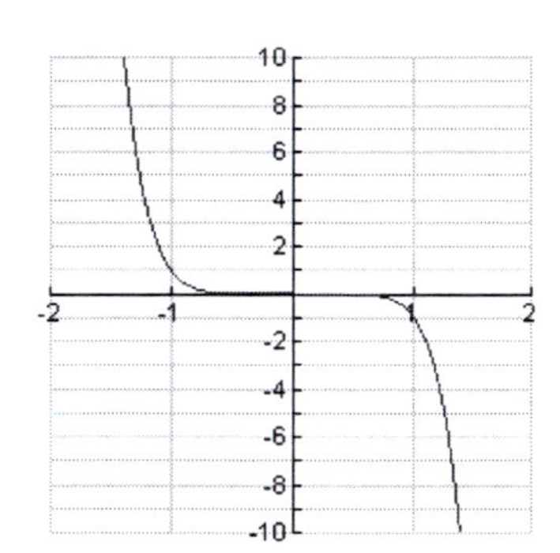

39. Shift the graph of x^4 down 2 units.

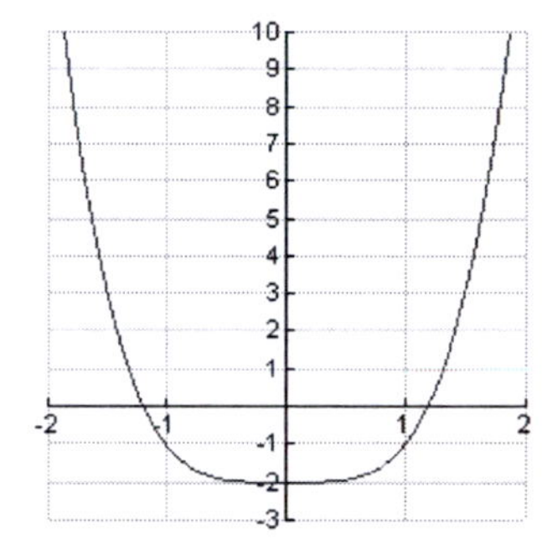

41. 6 (multiplicity 5) -4 (multiplicity 2)

43. $x^5 - 13x^3 + 36x = x(x^2 - 9)(x^2 - 4) = x(x+3)(x-3)(x+2)(x-2)$

So, the zeros are $0, -2, 2, 3, -3$, all with multiplicity 1.

45. $x(x+3)(x-4)$

47. $x(x+\frac{2}{5})(x-\frac{3}{4}) = x(5x+2)(4x-3)$

49. $(x+2)^4(x-3)^2 = x^4 - 2x^3 - 11x^2 + 12x + 36$

51. $f(x) = x^2 - 5x - 14 = (x-7)(x+2)$

a. Zeros: $-2, 7$ (both multiplicity 1)

b. Crosses at both $-2, 7$

c. y-intercept: $f(0) = -14$, so $(0, -14)$

d. Long-term behavior: Behaves like $y = x^2$.

Even degree and leading coefficient positive, so graph rises without bound to the left and right.

e.

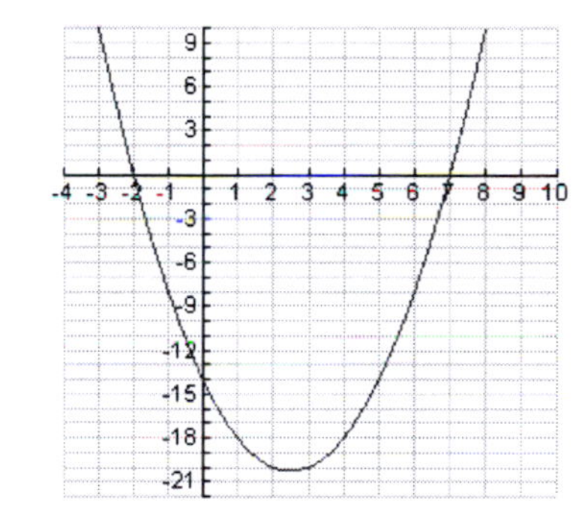

53. $f(x)=6x^7+3x^5-x^2+x-4$

a. Zeros: We first try to apply the Rational Root Test:

Factors of -4: $\pm1, \pm2, \pm4$

Factors of 6: $\pm1, \pm2, \pm3, \pm6$

Possible rational zeros:

$$\pm1, \pm2, \pm4, \pm\tfrac{1}{2}, \pm\tfrac{1}{3}, \pm\tfrac{2}{3}, \pm\tfrac{4}{3}, \pm\tfrac{1}{6}$$

Unfortunately, it can be shown that none of these are zeros of f. To get a feel for the possible number of irrational and complex root, we apply Descartes' Rule of Signs:

Number of sign variations for $f(x)$: 3

$f(-x)=-6x^7-3x^5-x^2-x-4$, so

Number of sign variations for $f(-x)$: 0

Since $f(x)$ is degree 7, there are 7 zeros, classified as:

Positive Real Zeros	Negative Real Zeros	Imaginary Zeros
3	0	4
1	0	6

Need to actually graph the polynomial to determine the approximate zeros. From **e.**, we see there is a zero at approximately (0.8748,0) with multiplicity 1.

b. From the analysis in part **a**, we know the graph crosses at its only real zero.

c. y-intercept: $f(0)=-4$, so $(0,-4)$

d. Long-term behavior: Behaves like $y=x^7$.

Odd degree and leading coefficient positive, so graph falls without bound to the left and rises to the right.

e.

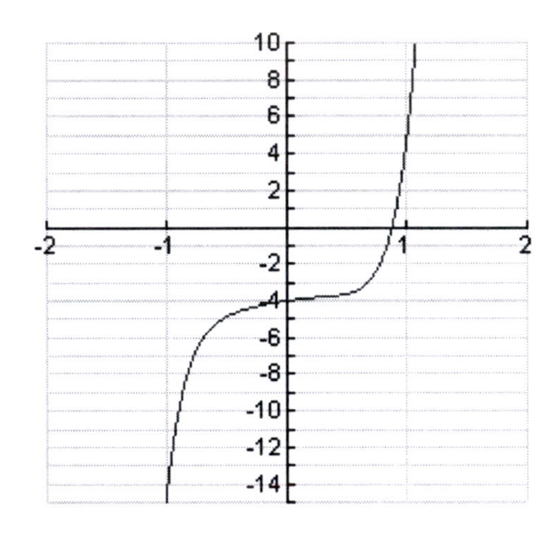

55. $f(x)=(x-1)(x-3)(x-7)$

a.

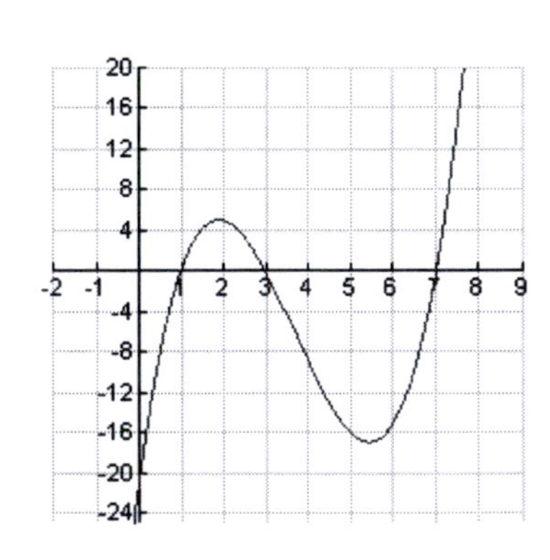

b. Zeros: 1, 3, 7 (all with multiplicity 1)

c. Want intervals where the graph of f is above the x-axis. From **a.**, this occurs on the intervals $(1,3)$ and $(7,\infty)$.

So, between 1 and 3 hours, and more than 7 hours would be financially beneficial.

57.

$$\begin{array}{r} x+4 \\ x-2 \overline{\smash{\big)}\, x^2+2x-6} \\ \underline{-(x^2-2x)} \\ 4x-6 \\ \underline{-(4x-8)} \\ 2 \end{array}$$

So, $\boxed{Q(x)=x+4,\quad r(x)=2}$.

59.

$$\begin{array}{r} 2x^3-4x^2-2x-\tfrac{7}{2} \\ 2x-4 \overline{\smash{\big)}\, 4x^4-16x^3+12x^2+x-9} \\ \underline{-(4x^4-8x^3)} \\ -8x^3+12x^2 \\ \underline{-(-8x^3+16x^2)} \\ -4x^2+x \\ \underline{-(-4x^2+8x)} \\ -7x-9 \\ \underline{-(-7x+14)} \\ -23 \end{array}$$

So,

$\boxed{Q(x)=2x^3-4x^2-2x-\tfrac{7}{2},\quad r(x)=-23}$.

61.

$$\begin{array}{r|rrrrr} \underline{-2} & 1 & 4 & 5 & -2 & -8 \\ & & -2 & -4 & -2 & 8 \\ \hline & 1 & 2 & 1 & -4 & 0 \end{array}$$

So, $Q(x)=x^3+2x^2+x-4,\quad r(x)=0$.

63.

$$\begin{array}{r|rrrrrrr} \underline{-8} & 1 & 0 & 0 & 0 & 0 & 0 & -64 \\ & & -8 & 64 & -512 & 4096 & -32,768 & 262,144 \\ \hline & 1 & -8 & 64 & -512 & 4096 & -32,768 & 262,080 \end{array}$$

So, $Q(x)=x^5-8x^4+64x^3-512x^2+4096x-32,768,\quad r(x)=262,080$.

65.

$$\begin{array}{r} x+3 \\ 5x^2-7x+3 \overline{\smash{\big)}\, 5x^3+8x^2-22x+1} \\ \underline{-(5x^3-7x^2+3x)} \\ 15x^2-25x+1 \\ \underline{-(15x^2-21x+9)} \\ -4x-8 \end{array}$$

So, $Q(x)=x+3,\quad r(x)=-4x-8$.

67.

$$\begin{array}{r|rrrr} \underline{-1} & 1 & -4 & 2 & -8 \\ & & -1 & 5 & -7 \\ \hline & 1 & -5 & 7 & -15 \end{array}$$

So, $Q(x)=x^2-5x+7,\quad r(x)=-15$.

69. Area = length × width. So, solving for length, we see that length = Area ÷ width. So, in this case,

$$\text{length} = \frac{6x^4 - 8x^3 - 10x^2 + 12x - 16}{2x - 4} = \frac{3x^4 - 4x^3 - 5x^2 + 6x - 8}{x - 2}.$$

We compute this quotient using synthetic division:

$$\begin{array}{r|rrrrr} 2 & 3 & -4 & -5 & 6 & -8 \\ & & 6 & 4 & -2 & 8 \\ \hline & 3 & 2 & -1 & 4 & 0 \end{array}$$

Thus, the length (in terms of x) is $\boxed{3x^3 + 2x^2 - x + 4 \text{ feet}}$.

71.

$$\begin{array}{r|rrrrrr} -2 & 6 & 1 & 0 & -7 & 1 & -1 \\ & & -12 & 22 & -44 & 102 & -206 \\ \hline & 6 & -11 & 22 & -51 & 103 & -207 \end{array}$$

So, $f(-2) = -207$.

73.

$$\begin{array}{r|rrrr} 1 & 1 & 2 & 0 & -3 \\ & & 1 & 3 & 3 \\ \hline & 1 & 3 & 3 & 0 \end{array}$$

So, $g(1) = 0$.

75.

$P(-3) = (-3)^3 - 5(-3)^2 + 4(-3) + 2 = -82$.

So, it is not a zero.

77. $P(1) = 2(1)^4 - 2(1) = 0$

Yes, it is a zero.

79. $P(x) = x\left(x^3 - 6x^2 + 32\right)$ Observe that since -2 is a zero, synthetic division yields:

$$\begin{array}{r|rrrr} -2 & 1 & -6 & 0 & 32 \\ & & -2 & 16 & -32 \\ \hline & 1 & -8 & 16 & 0 \end{array}$$

So,

$$P(x) = x(x+2)(x^2 - 8x + 16) = x(x+2)(x-4)^2.$$

81. $P(x) = x^2\left(x^3 - x^2 - 8x + 12\right)$

We need to factor $x^3 - x^2 - 8x + 12$. To do so, we begin by applying the Rational Root Test:

Factors of 12: $\pm 1, \pm 2, \pm 3, \pm 4, \pm 6, \pm 12$

Factors of 1: ± 1

Possible rational zeros:

$$\pm 1, \pm 2, \pm 3, \pm 4, \pm 6, \pm 12$$

Observe that both -3 (multiplicity 1) and 2 (multiplicity 2) are zeros. So,

$$P(x) = x^2(x+3)(x-2)^2.$$

83.

Number of sign variations for $P(x)$: 1

$P(-x) = x^4 - 3x^3 - 16$, so

Number of sign variations for $P(-x)$: 1

Positive Real Zeros	Negative Real Zeros
1	1

85. Number of sign variations for $P(x)$: 5

$P(-x) = -x^9 + 2x^7 + x^4 + 3x^3 - 2x - 1$, so

Number of sign variations for $P(-x)$: 2

Positive Real Zeros	Negative Real Zeros
5	2
5	0
3	2
3	0
1	2
1	0

87. Factors of 6: $\pm1, \pm2, \pm3, \pm6$

Factors of 1: ±1

Possible rational zeros: $\pm1, \pm2, \pm3, \pm6$

89. Factors of 64:

$\pm1, \pm2, \pm4, \pm8, \pm16, \pm32, \pm64$

Factors of 2: $\pm1, \pm2$

Possible rational zeros:

$\pm1, \pm2, \pm4, \pm8, \pm16, \pm32, \pm64, \pm\frac{1}{2}$

91. Factors of 1: ±1

Factors of 2: $\pm1, \pm2$

Possible rational zeros: $\pm1, \pm\frac{1}{2}$

The only rational zero is $\frac{1}{2}$.

93. Factors of -16: $\pm1, \pm2, \pm4, \pm8, \pm16$

Factors of 1: ±1

Possible rational zeros:

$\pm1, \pm2, \pm4, \pm8, \pm16$

The rational zeros are 1, 2, 4, and -2.

95.

a. Number of sign variations for $P(x)$: 1

$P(-x) = -x^3 - 3x - 5$, so

Number of sign variations for $P(-x)$: 0

Since $P(x)$ is degree 3, there are 3 zeros, the real ones of which are classified as:

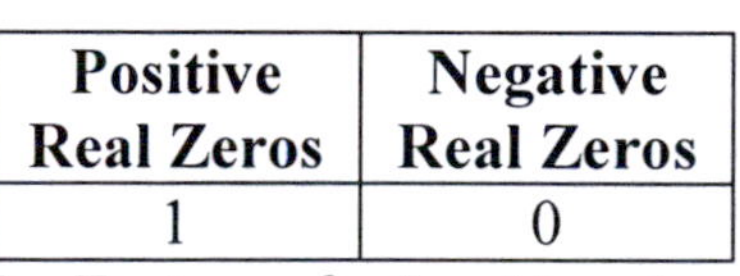

Positive Real Zeros	Negative Real Zeros
1	0

b. Factors of -5: $\pm1, \pm5$

Factors of 1: ±1

Possible rational zeros: $\pm1, \pm5$

c. -1 is a lower bound, 5 is an upper bound

d. There are no rational zeros.

e. Not possible to accurately factor since we do not have the zeros.

f.

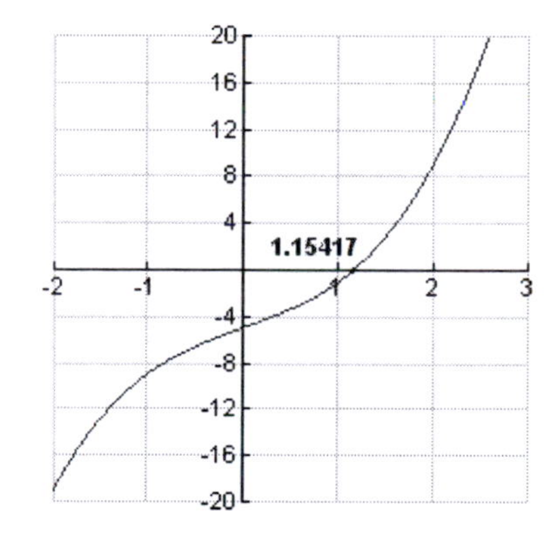

97.

a. Number of sign variations for $P(x)$: 3

$P(-x) = -x^3 - 9x^2 - 20x - 12$, so

Number of sign variations for $P(-x)$: 0

Positive Real Zeros	Negative Real Zeros
3	0
1	0

b. Factors of -12: $\pm 1, \pm 2, \pm 3, \pm 4, \pm 6, \pm 12$

Factors of 1: ± 1

Possible rational zeros:

$\pm 1, \pm 2, \pm 3, \pm 4, \pm 6, \pm 12$

c. -4 is a lower bound, 12 is upper bound (check by synthetic division)

d. Observe that

$$\begin{array}{r|rrrr} 1 & 1 & -9 & 20 & -12 \\ & & 1 & -8 & 12 \\ \hline & 1 & -8 & 12 & 0 \end{array}$$

$P(x) = (x-1)(x^2 - 8x + 12) = (x-1)(x-6)(x-2)$

Hence, the zeros are 1, 2, and 6.

e. $P(x) = (x-1)(x-6)(x-2)$

f.

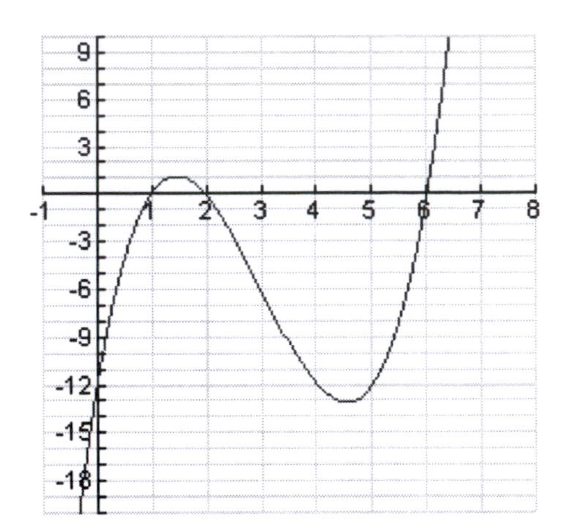

99.

a. Number of sign variations for $P(x)$: 2

$P(-x) = x^4 + 5x^3 - 10x^2 - 20x + 24$, so

Number of sign variations for $P(-x)$: 2

Positive Real Zeros	Negative Real Zeros
0	0
0	2
2	2
2	0

b. Factors of 24:

$\pm 1, \pm 2, \pm 3, \pm 4, \pm 6, \pm 8, \pm 12, \pm 24$

Factors of 1: ± 1

Possible rational zeros:

$\pm 1, \pm 2, \pm 3, \pm 4, \pm 6, \pm 8, \pm 12, \pm 24$

c. -4 is a lower bound, 8 is an upper bound (check by synthetic division)

d. Observe that

$$\begin{array}{r|rrrrr} 2 & 1 & -5 & -10 & 20 & 24 \\ & & 2 & -6 & -32 & -24 \\ \hline -1 & 1 & -3 & -16 & -12 & 0 \\ & & -1 & 4 & 12 & \\ \hline & 1 & -4 & -12 & 0 & \end{array}$$

So,

$$P(x) = (x-2)(x+1)(x^2 - 4x + 12)$$
$$= (x-2)(x+1)(x+2)(x-6)$$

Hence, the zeros are -2, -1, 1, and 6.

e. $P(x) = (x-2)(x+1)(x+2)(x-6)$

f.

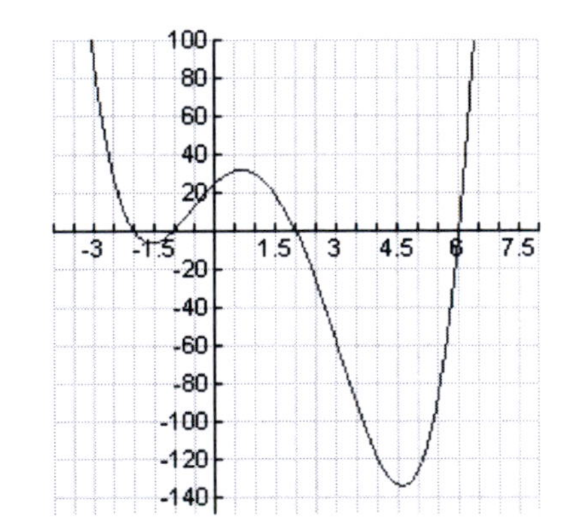

101. $x^2+25=(x-5i)(x+5i)$

103. Note that the zeros are

$$x^2-2x+5=0 \Rightarrow x=\frac{2\pm\sqrt{4-4(5)}}{2}=\frac{2\pm 4i}{2}=1\pm 2i$$

So, $P(x)=(x-(1-2i))(x-(1+2i))$.

105. Since $-2i$ and $3+i$ are zeros, so are their conjugates $2i$ and $3-i$, respectively. Since $P(x)$ has degree 4, these are the only missing zeros.

107. Since i is a zero, then so is its conjugate $-i$. Also, since $2-i$ is a zero of multiplicity 2, then its conjugate $2+i$ is also a zero of multiplicity 2. These are all of the zeros.

109. Since i is a zero of $P(x)$, so is its conjugate $-i$. As such, $(x-i)(x+i)=x^2+1$ divides $P(x)$ evenly. Indeed, observe that

$$\begin{array}{r}
x^2-3x-4 \\
x^2+0x+1\overline{)\,x^4-3x^3-3x^2-3x-4} \\
\underline{-(x^4+0x^3+x^2)} \\
-3x^3-4x^2-3x \\
\underline{-(-3x^3+0x^2-3x)} \\
-4x^2+0x-4 \\
\underline{-(-4x^2+0x-4)} \\
0
\end{array}$$

So,

$$\begin{aligned}P(x)&=(x-i)(x+i)\left(x^2-3x-4\right)\\&=(x-i)(x+i)(x-4)(x+1)\end{aligned}$$

111. Since $-3i$ is a zero of $P(x)$, so is its conjugate $3i$. As such, $(x-3i)(x+3i)=x^2+9$ divides $P(x)$ evenly. Indeed, observe that

$$\begin{array}{r}
x^2-2x+2 \\
x^2+0x+9\overline{)\,x^4-2x^3+11x^2-18x+18} \\
\underline{-(x^4+0x^3+9x^2)} \\
-2x^3+2x^2-18x \\
\underline{-(-2x^3+0x^2-18x)} \\
2x^2+0x+18 \\
\underline{-(2x^2+0x+18)} \\
0
\end{array}$$

Next, we find the roots of x^2-2x+2:

$$x=\frac{2\pm\sqrt{4-4(2)}}{2}=1\pm i$$

So,

$$P(x)=(x-3i)(x+3i)(x-(1+i))(x-(1-i)).$$

113. $P(x)=x^4-81=\left(x^2-9\right)\left(x^2+9\right)=(x-3)(x+3)(x-3i)(x+3i)$

115. $x^3-x^2+4x-4=x^2(x-1)+4(x-1)=\left(x^2+4\right)(x-1)=(x-2i)(x+2i)(x-1)$

117. <u>Vertical Asymptote</u>: $x = -2$
<u>Horizontal Asymptote</u>:
Since the degree of the numerator equals the degree of the denominator, $y = -1$ is the HA.

119. <u>Vertical Asymptote</u>: $x = -1$
<u>Slant Asymptote</u>: $y = 4x - 4$
To find the slant asymptote, we use long division:

$$\begin{array}{r} 4x - 4 \\ x+1 \overline{\big) \; 4x^2 + 0x + 0} \\ \underline{-(4x^2 + 4x)} \\ -4x + 0 \\ \underline{-(-4x - 4)} \\ 4 \end{array}$$

121. No vertical asymptotes, Horizontal asymptote: $y = 2$

123.

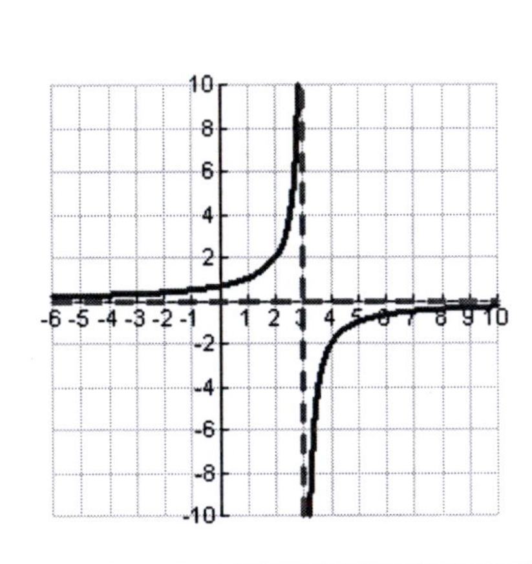

125.

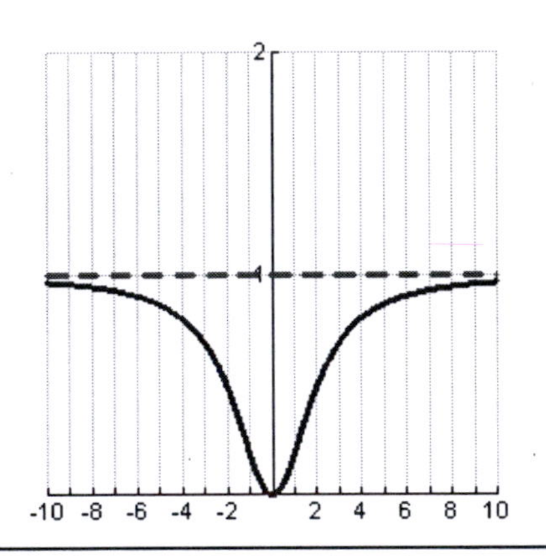

127. Note the hole at $x = -7$.

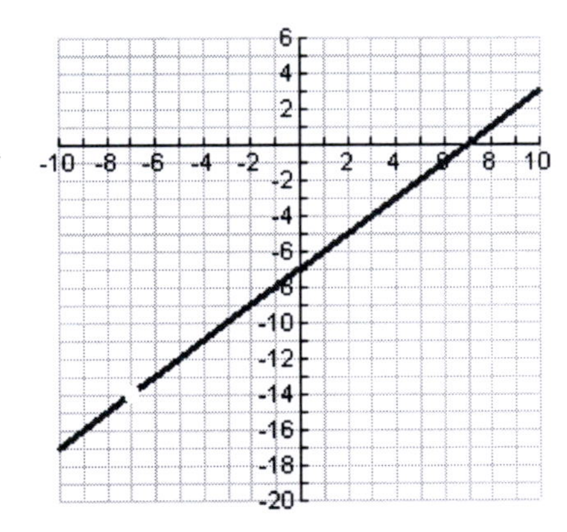

129. The graph of f is to the right. The following are identified graphically:

a. Vertex is (480, -1211)
b. y-intercept: (0, -59)
c. x-intercepts: (-12.14,0), (972.14,0)
d. axis of symmetry is the line $x = 480$.

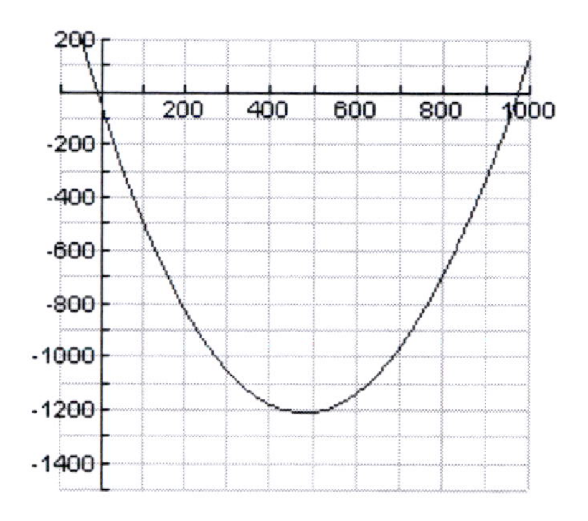

131. The graph of f is as follows:

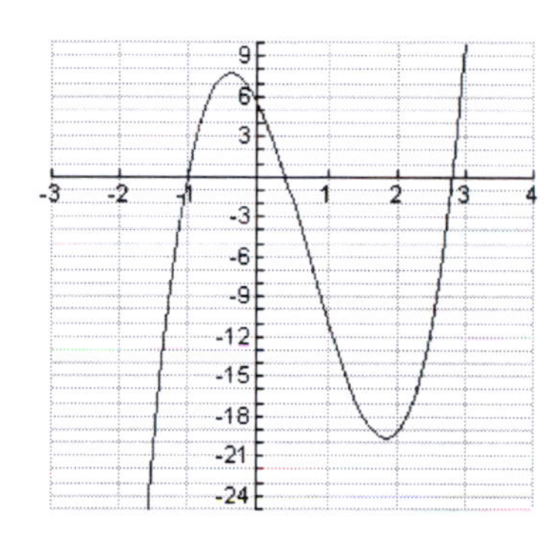

x-intercepts: (-1,0), (0.4, 0), (2.8,0)   zeros: -1, 0.4, 2.8, each with multiplicity 1

133. The graph is to the right.
It is a linear function, as the following long division verifies:

$$\begin{array}{r} 5x-4 \\ 3x^2-7x+2\overline{)15x^3-47x^2+38x-8} \\ \underline{-(15x^3-35x^3+10x^2)} \\ -12x^2+28x-8 \\ \underline{-(-12x^2+28x-8)} \\ 0 \end{array}$$

Note: This graph has holes at $x = \frac{1}{3}$ and $x = 2$.

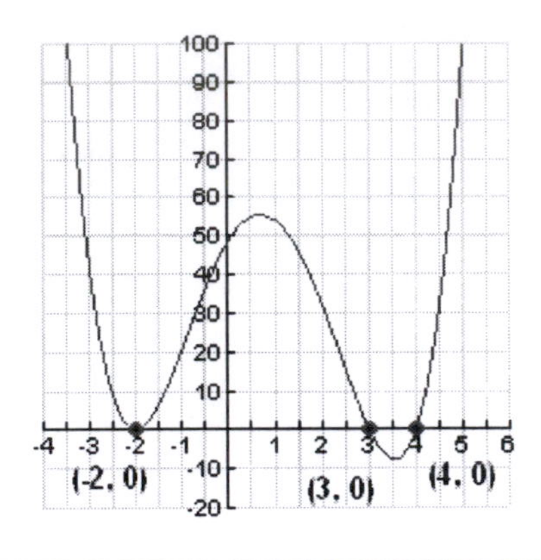

135. The graph is to the right.
a. The zeros are -2 (multiplicity 2), 3, and 4.
b. $P(x) = (x+2)^2(x-3)(x-4)$

137. The graph is to the right.
Note that the only real zero is $\frac{7}{2}$, and so we have:

$$\begin{array}{r|rrrr} \frac{7}{2} & 2 & 1 & -2 & -91 \\ & & 7 & 28 & 91 \\ \hline & 2 & 8 & 26 & 0 \end{array}$$

So,

$$P(x) = \left(x - \tfrac{7}{2}\right)(2x^2 + 8x + 26)$$
$$= (2x-7)(x^2 + 4x + 13)$$

Need to find the roots of $x^2 + 4x + 13$:

$$x = \frac{-4 \pm \sqrt{16 - 4(13)}}{2} = -2 \pm 3i$$

Thus, $P(x) = (2x-7)(x+2-3i)(x+2+3i)$.

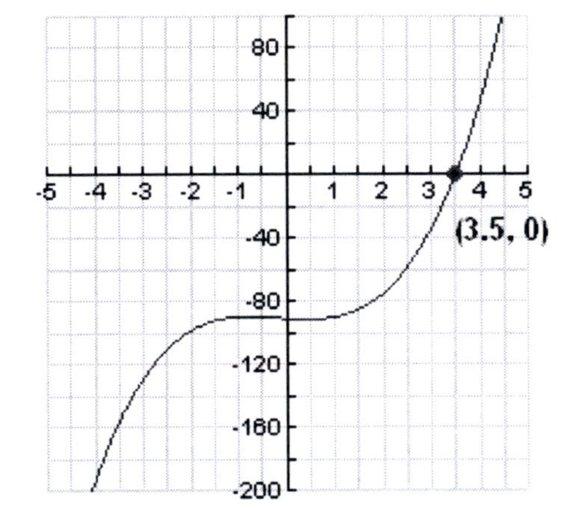

139. From the graph, to the right, we see that indeed f is one-to-one and hence has an inverse (which is also graphed on the same set of axes). The inverse is as follows:

$$y = \frac{2x-3}{x+1} \xrightarrow{\text{Switch } x \text{ and } y} x = \frac{2y-3}{y+1}$$

Solve for y: $xy + x = 2y - 3 \Rightarrow$

$$xy - 2y = -3 - x \Rightarrow y = \frac{-3-x}{x-2}$$

Thus, $f^{-1}(x) = \frac{-3-x}{x-2} = \frac{3+x}{2-x}$.

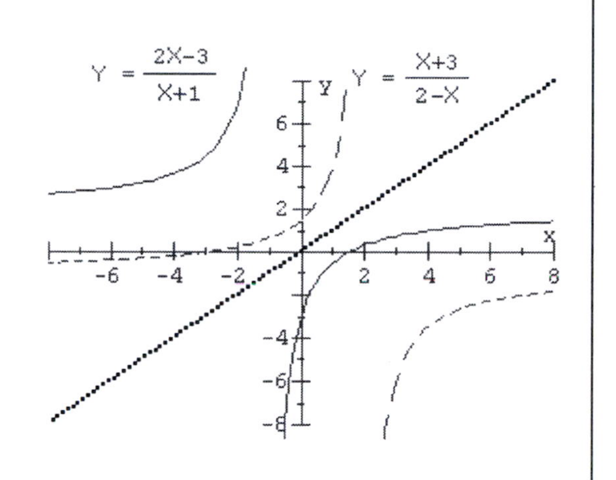

Chapter 4 Practice Test Solutions

1.

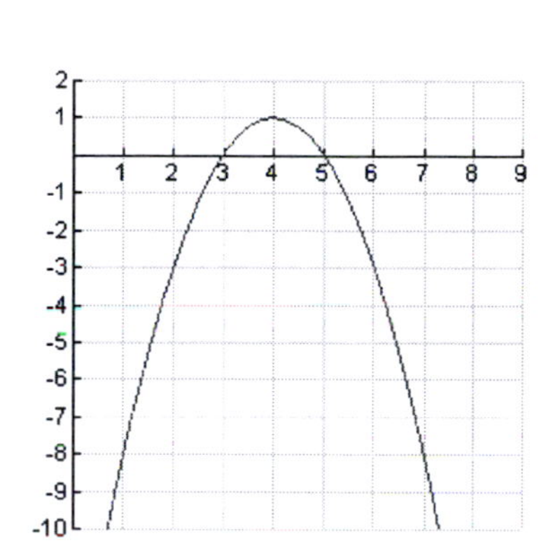

3.

$$y = -\tfrac{1}{2}\left(x^2 - 6x\right) - 4$$
$$= -\tfrac{1}{2}(x^2 - 6x + 9) - 4 + \tfrac{9}{2}$$
$$= -\tfrac{1}{2}(x-3)^2 + \tfrac{1}{2}$$

So, the vertex is $(3, \tfrac{1}{2})$.

5. $f(x) = x(x-2)^3(x-1)^2$

7.

$$\begin{array}{r}
-2x^2 - 2x - \tfrac{11}{2} \\
2x^2 - 3x + 1 \overline{\big) \; -4x^4 + 2x^3 - 7x^2 + 5x - 2} \\
\underline{-(-4x^4 + 6x^3 - 2x^2)} \\
-4x^3 - 5x^2 + 5x \\
\underline{-(-4x^3 + 6x^2 - 2x)} \\
-11x^2 + 7x - 2 \\
\underline{-(-11x^2 + \tfrac{33}{2}x - \tfrac{11}{2})} \\
-\tfrac{19}{2}x + \tfrac{7}{2}
\end{array}$$

So, $\boxed{Q(x) = -2x^2 - 2x - \tfrac{11}{2}, \quad r(x) = -\tfrac{19}{2}x + \tfrac{7}{2}}$.

9.

$$\begin{array}{r|rrrrr} 3 & 1 & 1 & -13 & -1 & 12 \\ & & 3 & 12 & -3 & -12 \\ \hline & 1 & 4 & -1 & -4 & 0 \end{array}$$

Yes, $x-3$ is a factor of $x^4+x^3-13x^2-x+12$.

11.

$$\begin{array}{r|rrrr} 7 & 1 & -6 & -9 & 14 \\ & & 7 & 7 & -14 \\ \hline & 1 & 1 & -2 & 0 \end{array}$$

So,
$P(x)=(x-7)(x^2+x-2)=(x-7)(x+2)(x-1)$

13. Yes, a complex zero cannot be an x-intercept.

15.
Factors of 12: $\pm 1, \pm 2, \pm 3, \pm 4, \pm 6, \pm 12$ Factors of 3: $\pm 1, \pm 3$
Possible rational zeros: $\pm 1, \pm 2, \pm 3, \pm 4, \pm 6, \pm 12, \pm\frac{1}{3}, \pm\frac{2}{3}, \pm\frac{4}{3}$

17.

$$\begin{aligned} P(x) &= 2x^3-3x^2+8x-12 \\ &= x^2(2x-3)+4(2x-3) \\ &= (x^2+4)(2x-3) \\ &= (x+2i)(x-2i)(2x-3) \end{aligned}$$

The only real zero is $3/2$, the other two are complex, namely $\pm 2i$.

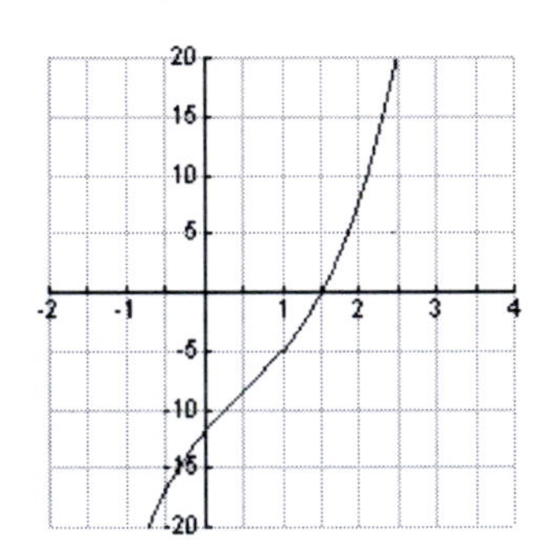

19. The polynomial can have degree 3 since there are 2 turning points. See the graph below:

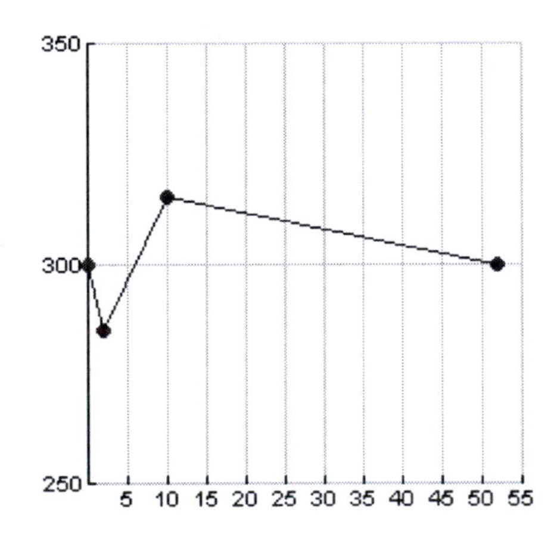

21. Since there are 2 turning points (seen on the graph to the right), the polynomial must be at least degree 3.

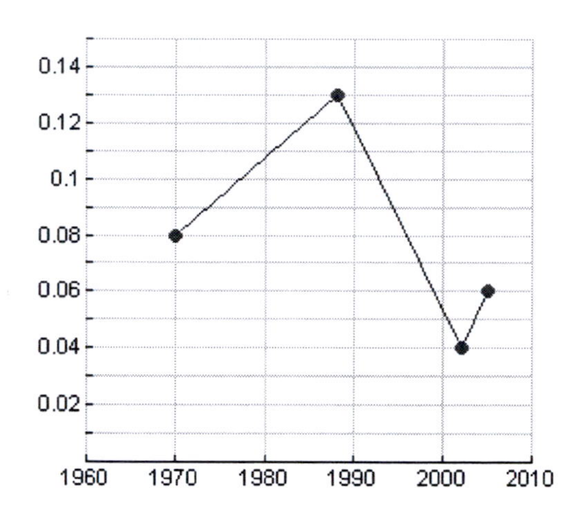

23. Observe that

$$g(x) = \frac{x}{x^2 - 4} = \frac{x}{(x-2)(x+2)}.$$

x-intercept: (0,0)
y-intercept: (0,0)
Vertical Asymptote: $x = \pm 2$
Horizontal Asymptote: $y = 0$
Slant Asymptote: None

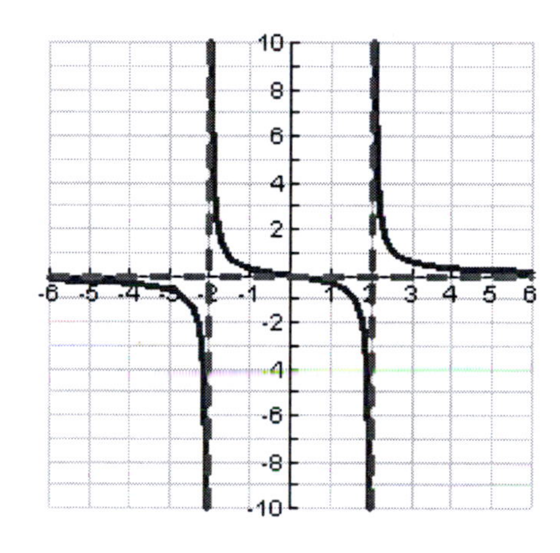

25. Observe that

$$F(x) = \frac{x-3}{x^2 - 2x - 8} = \frac{x-3}{(x-4)(x+2)}.$$

x-intercept: (3,0)
y-intercept: $(0, \frac{3}{8})$
Vertical Asymptote: $x = -2,\ x = 4$
Horizontal Asymptote: $y = 0$
Slant Asymptote: None

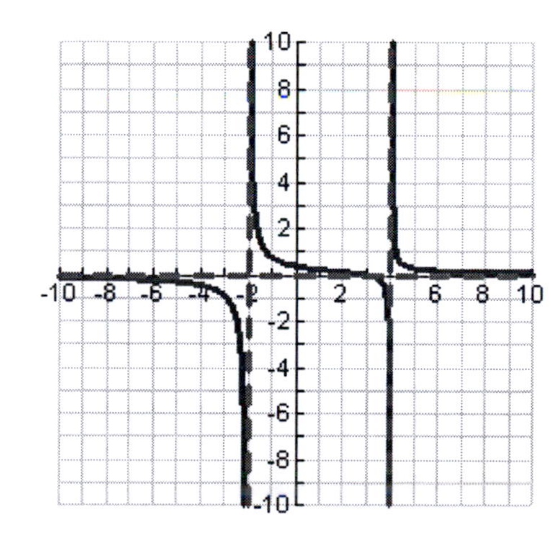

27. a & b. The equation we obtain using the calculator is $y = x^2 - 3x - 7.99$. We complete the square to put this into standard form: $y = (x-1.5)^2 - 10.24$

c. x-intercepts: (-1.7,0) and (4.7,0)

d. The graph is to the right. Yes, they agree.

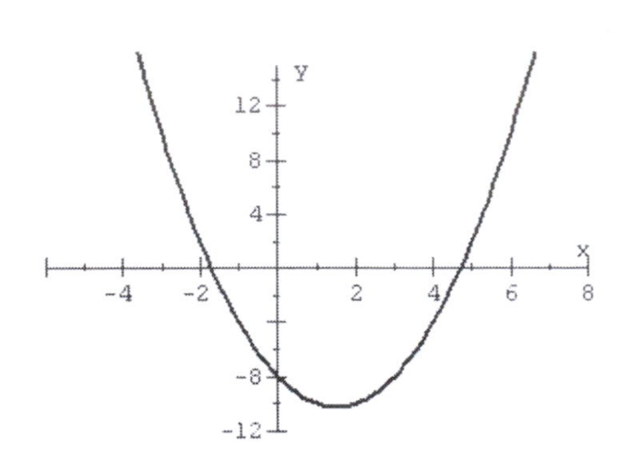

Chapter 4 Cumulative Review

1. $\dfrac{\left(5x^{-1}y^{-2}\right)^3}{\left(-10x^2y^2\right)^2} = \dfrac{125x^{-3}y^{-6}}{100x^4y^4} = \boxed{\dfrac{5}{4x^7y^{10}}}$

3.

$$\frac{4x^2-36}{x^2} \cdot \frac{x^3+6x^2}{x^2+9x+18} = \frac{4(x-3)(x+3)}{x^2} \cdot \frac{x^2(x+6)}{(x+6)(x+3)} = 4(x-3) = \boxed{4x-12}$$

5. Let x = number of minutes to do the job together. Then

$$\frac{1}{x} = \frac{1}{75} + \frac{1}{60} \Rightarrow 300 = 4x + 5x$$

$$x = \tfrac{300}{9} = \boxed{33.\overline{3}\text{ min.}}$$

7.

$$\sqrt{16+x^2} = x+2$$
$$16 + x^2 = (x+2)^2$$
$$16 + x^2 = x^2 + 4x + 4$$
$$12 = 4x \quad \Rightarrow \quad \boxed{3 = x}$$

9. Observe that $x - 3y = 8 \Rightarrow y = \frac{1}{3}x - \frac{8}{3}$

So, slope is $m = \frac{1}{3}$. So, the equation of the line so far is $y = \frac{1}{3}x + b$. Now, use (4,1) to find b:

$$1 = \tfrac{1}{3}(4) + b \Rightarrow b = -\tfrac{1}{3}$$

So, the equation is $y = \frac{1}{3}x - \frac{1}{3}$.

11. Since the center is (0,6), the equation of the circle thus far is

$$x^2 + (y-6)^2 = r^2.$$

Use the fact that (1,5) is on the circle to now find the r^2:

$$1^2 + (5-6)^2 = r^2 \Rightarrow r^2 = 2.$$

So, the equation is

$$x^2 + (y-6)^2 = 2$$

13. Observe that

$$g(-x) = \sqrt{-x+10} \neq g(x) \qquad -g(-x) = -\sqrt{-x+10} \neq g(x)$$

So, neither.

15. Translate the graph of $y=\sqrt{x}$ to the right 1 unit and then up 3 units.

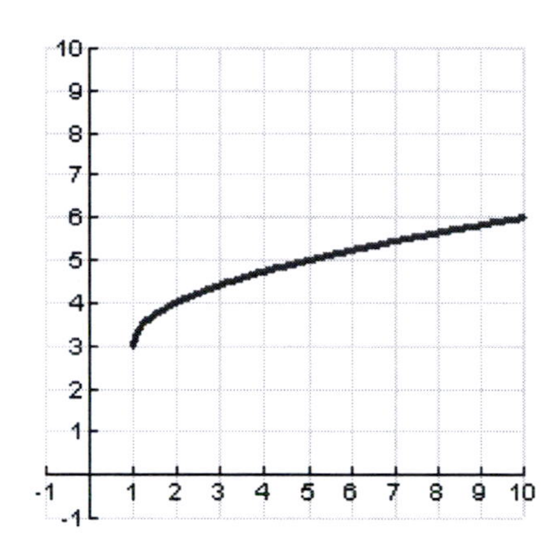

17.
$f(-1)=7-2(-1)^2=5$
$g(f(-1))=2(5)-10=0$

19. Since the vertex is (-2,3), the equation so far is $f(x)=a(x+2)^2+3$. Use (-1,4) to find a:

$$4=a(-1+2)^2+3=a+3 \Rightarrow a=1$$

So, $f(x)=(x+2)^2+3$.

21. Observe that

$$\begin{array}{r|l} & 4x^2+4x+1 \\ -5x+3 & -20x^3-8x^2+7x-5 \\ & \underline{-(-20x^3+12x^2)} \\ & \quad -20x^2+7x \\ & \quad \underline{-(-20x^2+12x)} \\ & \qquad -5x-5 \\ & \qquad \underline{-(-5x+3)} \\ & \qquad\quad -8 \end{array}$$

So, $Q(x)=4x^2+4x+1,\ r(x)=-8$

23. $P(x)=12x^3+29x^2+7x-6$
Factors of -6: $\pm1, \pm2, \pm3, \pm6$
Factors of 12: $\pm1, \pm2, \pm3, \pm4, \pm6, \pm12$
Possible rational zeros:

$$\pm1, \pm2, \pm3, \pm4, \pm6, \pm\tfrac{1}{2}, \pm\tfrac{1}{3},$$
$$\pm\tfrac{1}{4}, \pm\tfrac{1}{6}, \pm\tfrac{1}{12}, \pm\tfrac{2}{3}, \pm\tfrac{3}{2}, \pm\tfrac{3}{4}$$

Note that

$$\begin{array}{r|rrrr} -2 & 12 & 29 & 7 & -6 \\ & & -24 & -10 & 6 \\ \hline & 12 & 5 & -3 & 0 \end{array}$$

So, $P(x)=(x+2)(12x^2+5x-3)=(x+2)(4x+3)(3x-1)$
So, the zeros are $-2, -\frac{3}{4}, \frac{1}{3}$.

25. vertical asymptotes: $x=\pm2$ Horizontal asymptote: $y=0$

27. Observe that

$$f(x) = \frac{3(x+1)}{x(2x-3)}$$

x-intercepts: (-1,0)
y-intercept: none
Vertical Asymptote: $x = 0, x = \frac{3}{2}$
Horizontal Asymptote: $y = 0$

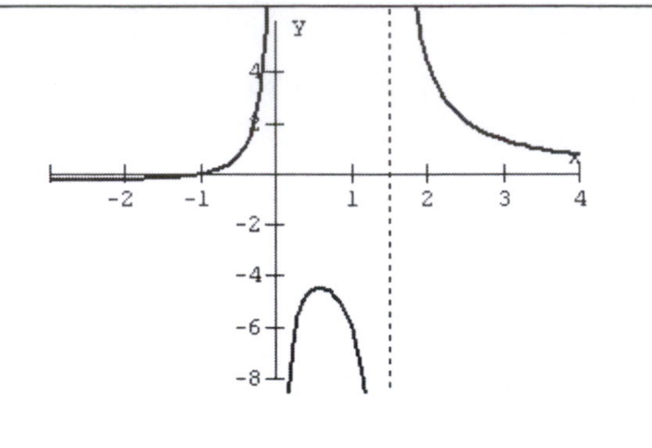

The graph of f is to the right. Yes, they agree.

CHAPTER 5

Section 5.1 Solutions --

1. 16	**3.** $\frac{1}{5^2}=\frac{1}{25}$
5. $8^{2/3}=\left(\sqrt[3]{8}\right)^2=2^2=4$	**7.** $\left(\frac{1}{9}\right)^{-3/2}=9^{3/2}=\left(\sqrt{9}\right)^3=3^3=27$
9. $-5^0=-1$	**11.** 5.2780
13. 9.7385	**15.** 7.3891
17. 0.0432	**19.** $f(3)=3^3=27$
21. $g(-1)=\left(\frac{1}{16}\right)^{-1}=16$	**23.** $g\left(-\frac{1}{2}\right)=\left(\frac{1}{16}\right)^{-1/2}=16^{1/2}=4$
25. $f(e)=3^e\approx 19.81$	**27. f** $y=\frac{1}{5}\left(5^x\right)$ Rising a factor of 5 slower than 5^x
29. e $y=-\left(5^x\right)$ The reflection of 5^x over the x-axis	**31. b** $y=-\left(5^{-x}\right)+1$ The reflection of 5^{-x} over the x-axis and then shifted up 1 unit.

33. <u>y-intercept</u>: $f(0)=6^0=1$, so (0,1).
<u>HA</u>: $y=0$
<u>Domain</u>: $(-\infty,\infty)$ <u>Range</u>: $(0,\infty)$
<u>Other points</u>: (-1,1/6), (1,6)

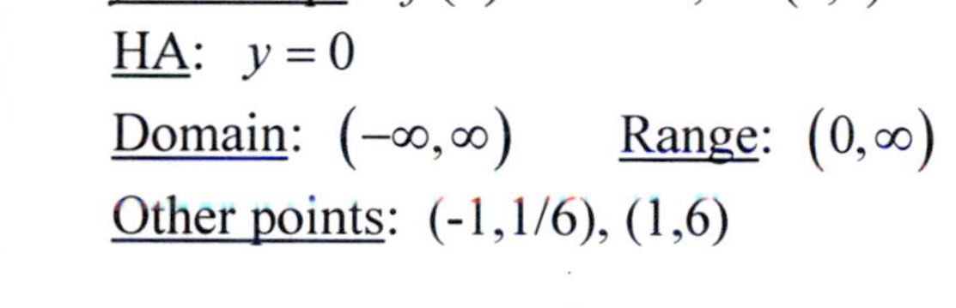

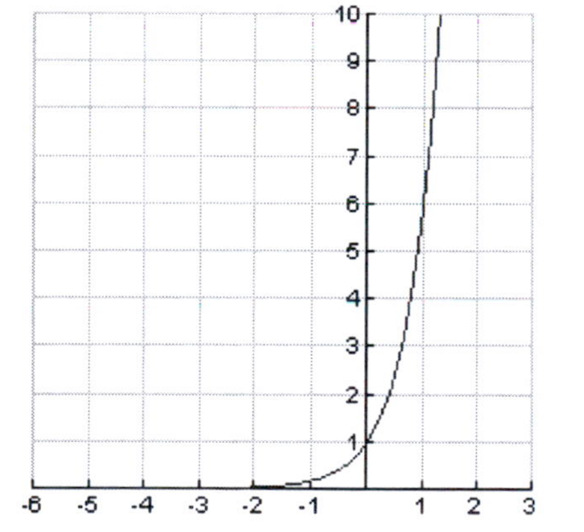

35. <u>y-intercept</u>: $f(0)=10^0=1$, so (0,1).
<u>HA</u>: $y=0$
Reflect graph of $y=10^x$ over y-axis.
<u>Domain</u>: $(-\infty,\infty)$ <u>Range</u>: $(0,\infty)$
<u>Other points</u>: (1, 0.1), (-1,10)

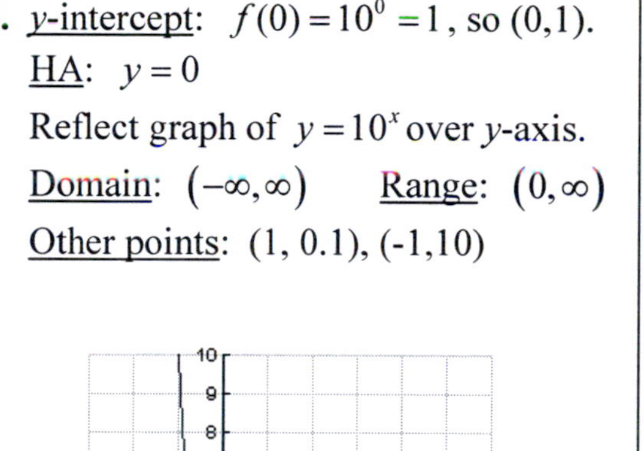

37. y-intercept: $f(0)=e^{0}=1$, so (0,1).
HA: $y=0$
Reflect graph of $y=e^{x}$ over y-axis.
Domain: $(-\infty,\infty)$ Range: $(0,\infty)$
Other points: (1,1/*e*), (-1, *e*)

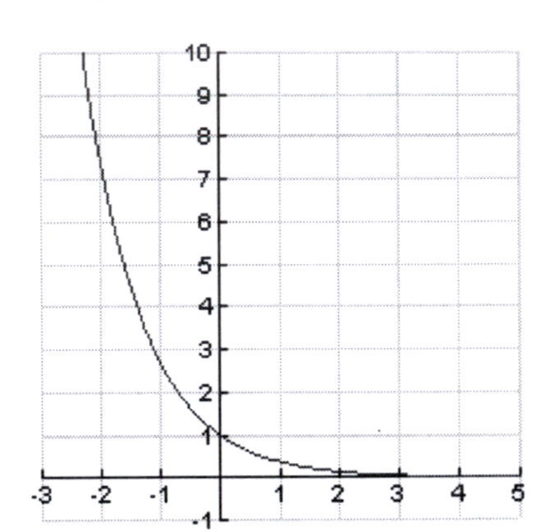

39. y-intercept: $f(0)=2^{0}-1=0$, so (0,0).
HA: $y=-1$
Shift graph of $y=2^{x}$ down 1 unit
Domain: $(-\infty,\infty)$ Range: $(-1,\infty)$
Other points: (2,3), (1,1)

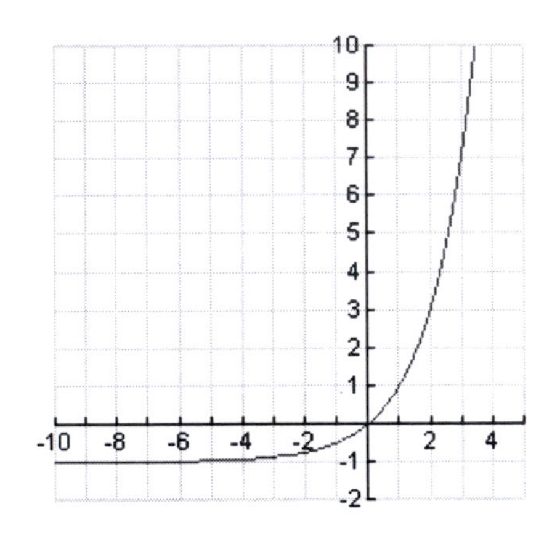

41. y-intercept: $f(0)=2-e^{0}=1$, so (0,1).
HA: $y=2$
Reflect graph of e^{x} over the x-axis, then shift up 2 units.
Domain: $(-\infty,\infty)$ Range: $(-\infty,2)$
Other points: (1, 2-*e*), (-1, 2 – 1/*e*)

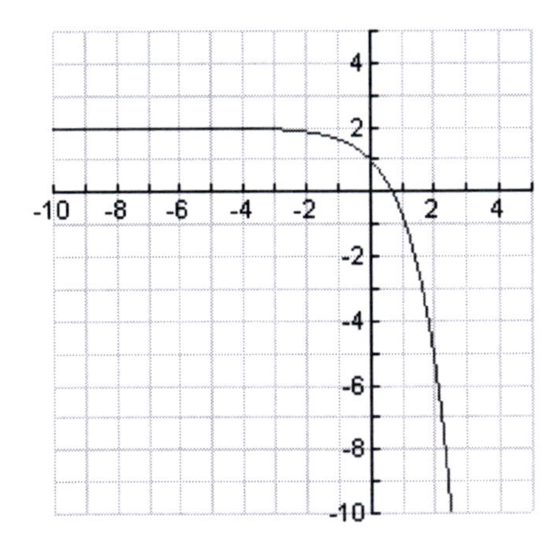

43. y-intercept: $f(0)=e^{0+1}-4=e-4$, so $(0,e-4)$.
HA: $y=-4$
Shift the graph of e^{x} left 1 unit, then down 4 units.
Domain: $(-\infty,\infty)$ Range: $(-4,\infty)$
Other points: (-1,-3), (1, $e^{2}-4$)

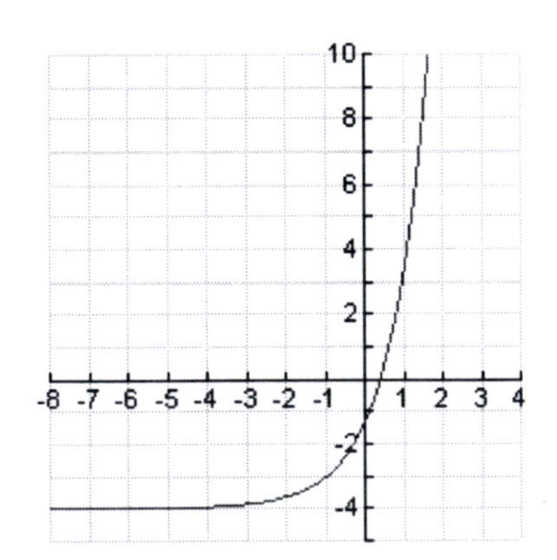

45. <u>y-intercept</u>: $f(0) = 3 \cdot e^0 = 3$, so $(0,3)$.
<u>HA</u>: $y = 0$
Expand the graph of $y = e^x$ horizontally by a factor of 2, then expand vertically by a factor of 3.
<u>Domain</u>: $(-\infty, \infty)$ <u>Range</u>: $(0, \infty)$
<u>Other points</u>: $(2, 3e)$, $(1, 3\sqrt{e})$

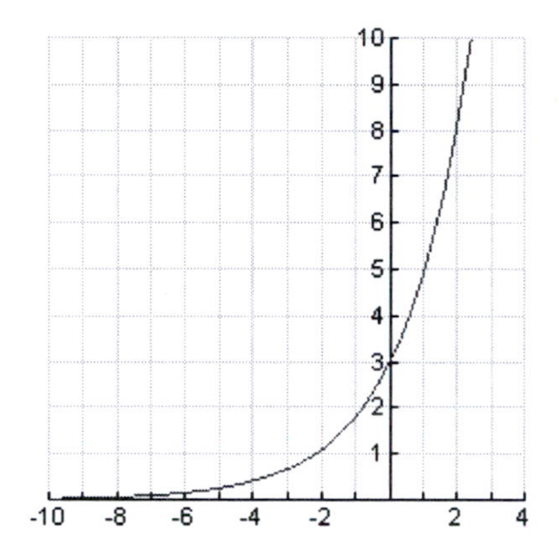

47. <u>y-intercept</u>: $f(0) = 1 + \left(\frac{1}{2}\right)^{0-2} = 1 + 2^2 = 5$, so $(0,5)$.
<u>HA</u>: $y = 1$
Shift the graph of $y = \left(\frac{1}{2}\right)^x$ right 2 units, then up 1 unit.
<u>Domain</u>: $(-\infty, \infty)$ <u>Range</u>: $(1, \infty)$
<u>Other points</u>: $(0,5)$, $(2,2)$

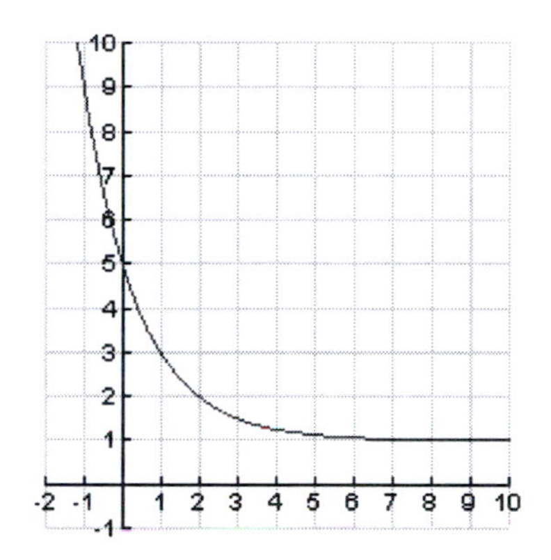

49. Use $P(t) = P_0\left(2^{t/d}\right)$.
Here,
$P_0 = 7.1$, $d = 88$, $t = 48$ $(2090 - 2002)$.
Thus, $P(48) = 7.1\left(2^{48/88}\right) \cong 10.4$.
So, the expected population in 2050 is approximately 10.4 million.

51. Use $P(t) = P_0\left(2^{t/d}\right)$.
Here,
$P_0 = 1500$, $d = 5$ (doubling time), $t = 30$.
Thus, $P(30) = 1500\left(2^{30/5}\right) \cong 96{,}000$. So, the cost per acre 30 years after initial investment is \$96,000.

53. Use $A(t) = A_0\left(\frac{1}{2}\right)^{t/h}$
Here,
$h = 119.77$ days, $A_0 = 200\ mg$, $t = 30$ days
Thus, $A(30) = 200\left(\frac{1}{2}\right)^{30/119.77} \cong 168$.
So, 168 mg remain after 30 days.

55. Use $A(t) = A_0\left(\frac{1}{2}\right)^{t/h}$
Here,
$h = 10$ years, $A_0 = 8000$, $t = 14$ years
Thus, $A(14) = 8000\left(\frac{1}{2}\right)^{14/10} \cong 3031$.
So, the value after 14 years is \$3031.

57. Use $A(t) = P\left(1+\frac{r}{n}\right)^{nt}$. Here,

$P = 3200,\ r = 0.025,\ n = 4,\ t = 3$ years.

Thus, $A(4) = 3200\left(1+\frac{0.025}{4}\right)^{4(3)} \cong 3448.42$.

So, the amount in the account after 4 years is $3,448.42.

59. Use $A(t) = P\left(1+\frac{r}{n}\right)^{nt}$. Here,

$A(18) = 32,000,\ r = 0.05,\ n = 365,\ t = 18$

Thus, solving the above formula for P, we have $P = \dfrac{32,000}{\left(1+\frac{0.05}{365}\right)^{365(18)}} \cong 13,011.03$.

So, the initial investment should be $13,011.03.

61. Use $A(t) = Pe^{rt}$.

Here,

$P = 3200,\ r = 0.02,\ t = 15$.

Thus, $A(15) = 3200e^{(0.02)(15)} \cong 4319.55$.

So, the amount in the account after 15 years is $4319.55.

63. Use $A(t) = Pe^{rt}$. Here,

$A(20) = 38,000,\ r = 0.05,\ t = 20$.

Thus, solving the above formula for P, we have $P = \dfrac{38,000}{e^{(0.05)(20)}} \cong 13,979.42$.

So, the initial investment should be $13,979.42.

65. Use $C(t) = C_0 e^{-rt}$ with $r = 0.020,\ C_0 = 5$ mg/L, $t = 20$ hours.

$C(20) = 5e^{-0.020(20)} \approx 3.4$ mg/L.

67.

p (price per unit)	$D(p)$—approximate demand for product in units
1.00	1,955,000
5.00	1,020,500
10.00	452,810
20.00	89,147
40.00	3,455
60.00	134
80.00	5
90.00	1

69. The mistake is that $4^{-1/2} \neq 4^2$. Rather, $4^{-1/2} = \dfrac{1}{4^{1/2}} = \dfrac{1}{2}$.

71. $r = 0.025$ rather than 2.5

73. False. $(0,-1)$ is the y-intercept.

75. True. $3^{-x} = (3^{-1})^x = \left(\frac{1}{3}\right)^x$

77.

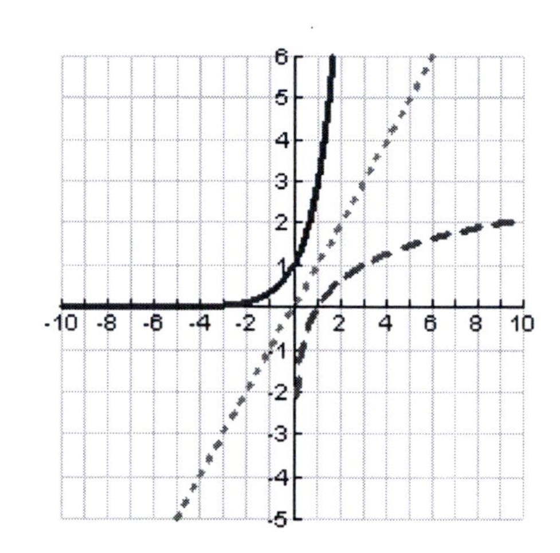

Note on Graphs: Solid curve is $y = 3^x$ and the dashed curve is $y = \log_3 x$.

79.

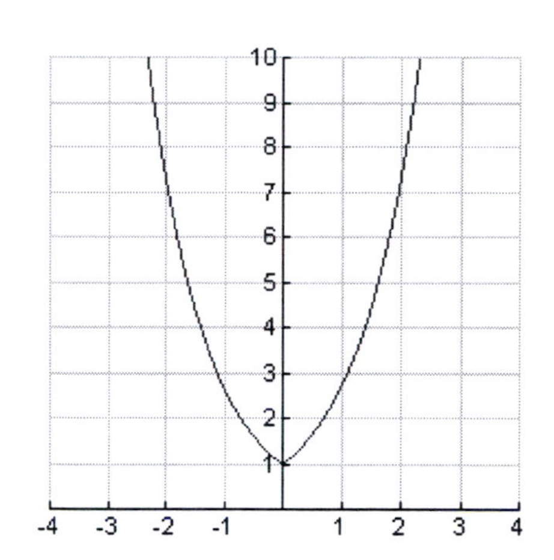

81. y-intercept: $f(0) = be^{-0+1} - a = be - a$ So, $(0, be - a)$.

Horizontal asymptote: For x very large, $be^{-x+1} \approx 0$, so $y = -a$ is the horizontal asymptote.

83. The domain and range for $f(x) = b^{|x|}$, where $b > 1$, are:

Domain: $(-\infty, \infty)$ Range: $[1, \infty)$.

Indeed, note that $f(x)$ is defined piecewise, as follows:

$$b^{|x|} = \begin{cases} b^x, & x \geq 0 \\ b^{-x}, & x < 0 \end{cases}$$

Recall that the graph of b^{-x} is the reflection of the graph of b^x over the y-axis.

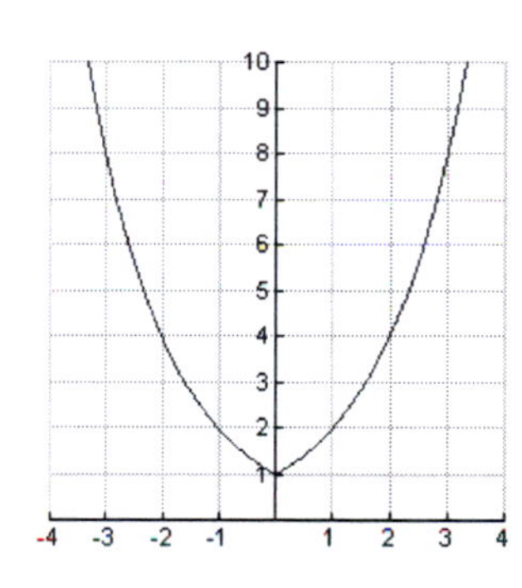

85.

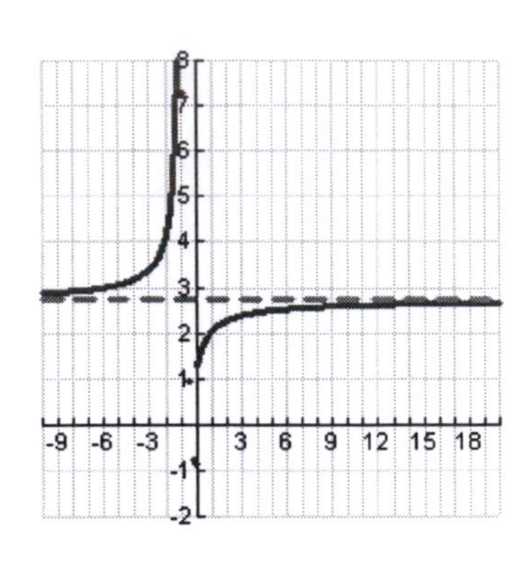

Note on Graphs: Solid curve is $y=(1+\frac{1}{x})^x$ and dashed curve (the horizontal asymptote) is $y=e$.

87. The graphs are close on the interval $(-3,3)$.

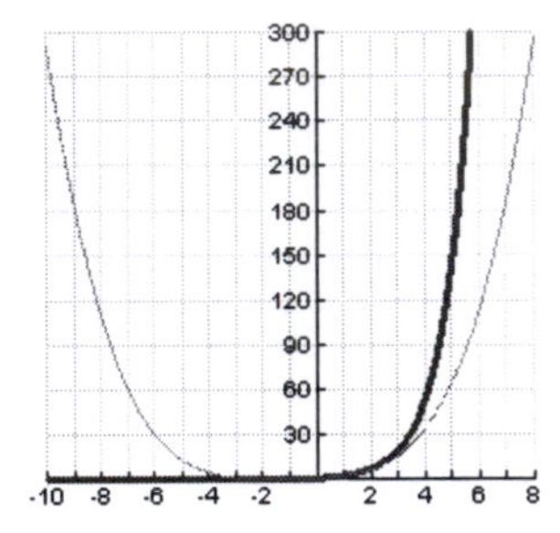

Note on Graphs: Solid curve is $y=e^x$ and thin curve is $y=1+x+\frac{x^2}{2}+\frac{x^3}{6}+\frac{x^4}{24}$.

89. The graphs of *f*, *g*, and *h* are as follows:

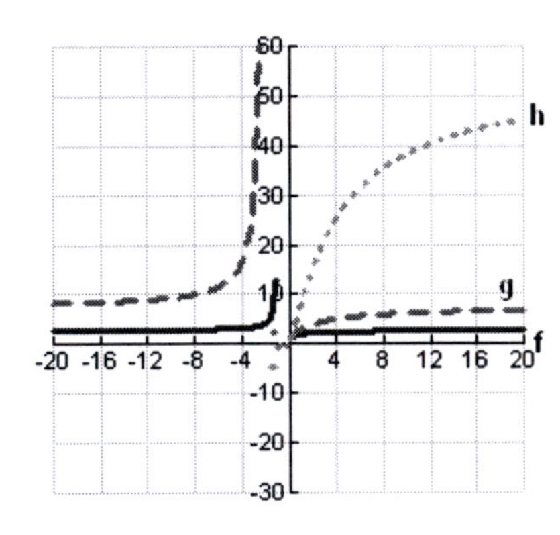

The horizontal asymptotes for *f*, *g*, and *h* are $y=e$, $y=e^2, y=e^4$. As *x* increases, $f(x)\to e$, $g(x)\to e^2, h(x)\to e^4$.

91. (a)

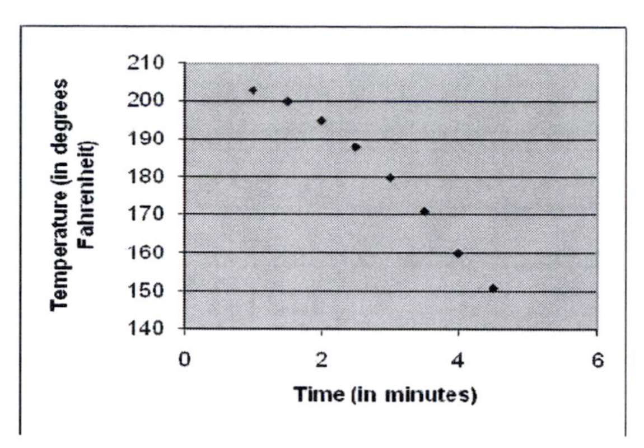

(b) The best fit exponential curve is $y = 228.34(0.9173)^x$ with $r^2 = 0.9628$. This best fit curve is shown below on the scatterplot. The fit is very good, as evidenced by the fact that the square of the correlation coefficient is very close to 1.

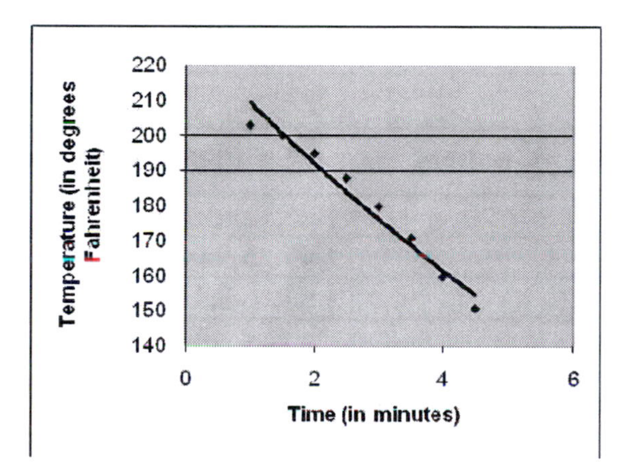

(c) **(i)** Compute the y-value when $x = 6$ to obtain about 136 degrees Fahrenheit.

(ii) The temperature of the soup the moment it was taken out of the microwave is the y-value at $x = 0$, namely about 228 degrees Fahrenheit.

(d) The shortcoming of this model for large values of x is that the curve approaches the x-axis, not 72 degrees. As such, it is no longer useful for describing the temperature beyond the x-value at which the temperature is 72 degrees.

Section 5.2 Solutions ---

1. $5^3 = 125$	**3.** $81^{\frac{1}{4}} = 3$
5. $2^{-5} = \frac{1}{32}$	**7.** $10^{-2} = 0.01$
9. $10^4 = 10{,}000$	**11.** $\left(\frac{1}{4}\right)^{-3} = 4^3 = 64$
13. $e^{-1} = \frac{1}{e}$	**15.** $e^0 = 1$
17. $e^x = 5$	**19.** $x^z = y$
21. $\log_8(512) = 3$	**23.** $\log(0.00001) = -5$
25. $\log_{225}(15) = \frac{1}{2}$	**27.** $\log_{2/5}\left(\frac{8}{125}\right) = 3$
29. $\log_{1/27}(3) = -\frac{1}{3}$	**31.** $\ln 6 = x$
33. $\log_y x = z$	**35.** $\log_2(1) = 0$
37. $\log_5(3125) = x \quad \Rightarrow \quad 5^x = 3125 \quad \Rightarrow \quad x = 5$	
39. $\log_{10}(10^7) = 7$	
41. $\log_{1/4}(4096) = x \;\Rightarrow\; \left(\frac{1}{4}\right)^x = 4096 \;\Rightarrow\; 4^{-x} = 4096 \;\Rightarrow\; x = -6$	
43. undefined	**45.** undefined
47. 1.46	**49.** 5.94
51. undefined	**53.** -8.11
55. Must have $x+5>0$, so that the domain is $(-5,\infty)$.	
57. Must have $5-2x>0$, so that the domain is $\left(-\infty,\frac{5}{2}\right)$.	
59. Must have $7-2x>0$, so that the domain is $\left(-\infty,\frac{7}{2}\right)$.	
61. Must have $\lvert x\rvert>0$, so that the domain is $(-\infty,0)\cup(0,\infty)$.	
63. Must have $x^2+1>0$ (which always occurs), so that the domain is $\mathbb{R}$.	
65. b	
67. c Reflect the graph of $y=\log_5 x$ over the x-axis and then the y-axis.	
69. d Since $\log_5(1-x)-2 = \log_5(-(x-1))-2$, Reflect the graph of $y=\log_5 x$ over the y-axis, then shift right 1 unit, and then shift down 5 units.	

71. Shift the graph of $y = \log x$ right 1 unit. Domain: $(1, \infty)$ Range: $(-\infty, \infty)$

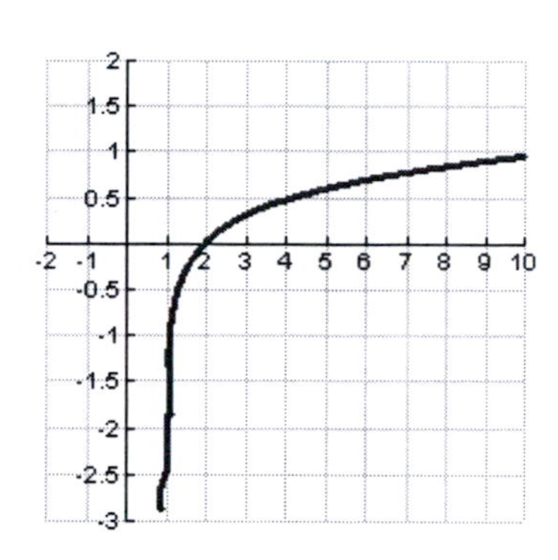

73. Shift the graph of $y = \ln x$ up 2 units. Domain: $(0, \infty)$ Range: $(-\infty, \infty)$

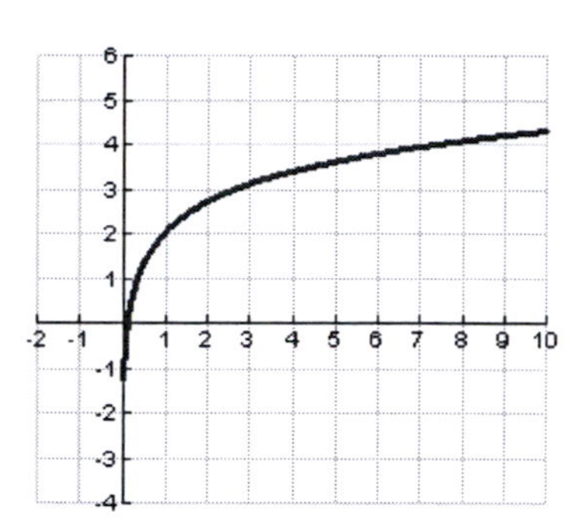

75. Shift the graph of $y = \log_3 x$ left 2 units, then down 1 unit.
Domain: $(-2, \infty)$ Range: $(-\infty, \infty)$

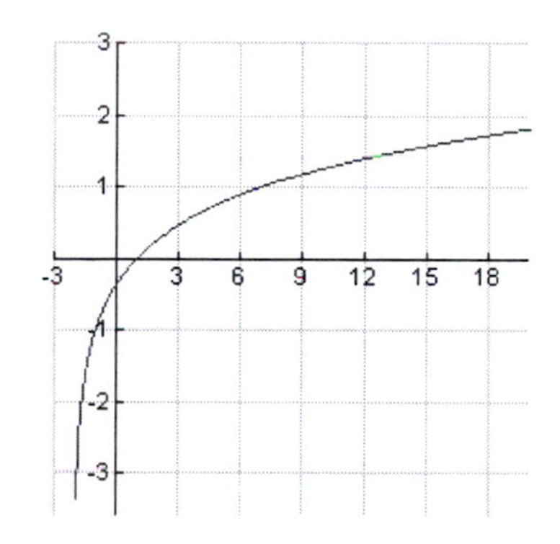

77. Reflect the graph of $y = \log x$ over the x-axis, then shift up 1 unit.
Domain: $(0, \infty)$ Range: $(-\infty, \infty)$

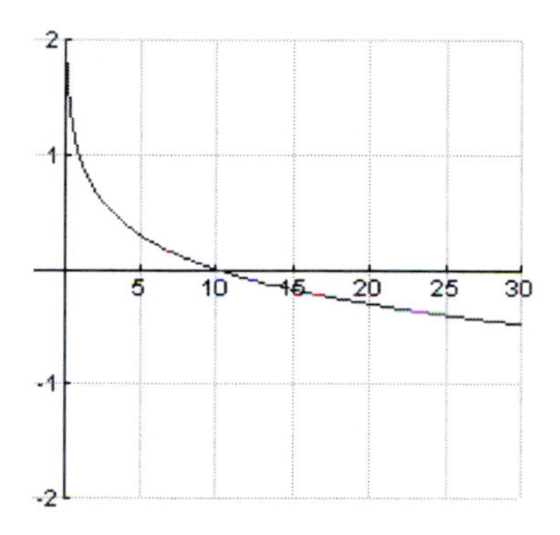

79. Shift the graph of $y = \ln x$ left 4 units.
Domain: $(-4, \infty)$ Range: $(-\infty, \infty)$

81. Compress the graph of $y = \log x$ horizontally by a factor of 2.
Domain: $(0, \infty)$ Range: $(-\infty, \infty)$

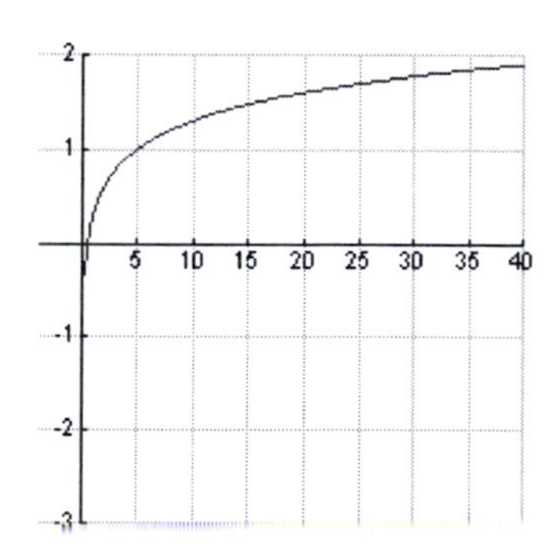

83. Use $D = 10\log\left(\dfrac{I}{1\times10^{-12}}\right)$.

Here,

$$D = 10\log\left(\frac{1\times10^{-6}}{1\times10^{-12}}\right) = 10\log(10^{6})$$
$$= 60\log(10) = 60\,dB$$

85. Use $D = 10\log\left(\dfrac{I}{1\times10^{-12}}\right)$.

Here,

$$D = 10\log\left(\frac{1\times10^{-0.3}}{1\times10^{-12}}\right) = 10\log(10^{11.7})$$
$$= 117\underbrace{\log(10)}_{=1} = 117\,dB$$

87. Use $M = \frac{2}{3}\log\left(\frac{E}{10^{4.4}}\right)$.

Here,

$$M = \tfrac{2}{3}\log\left(\tfrac{1.41\times10^{17}}{10^{4.4}}\right) = \tfrac{2}{3}\log\left(1.41\times10^{12.6}\right)$$
$$\approx 8.5$$

89. Use $M = \frac{2}{3}\log\left(\frac{E}{10^{4.4}}\right)$.

Here,

$$M = \tfrac{2}{3}\log\left(\tfrac{2\times10^{14}}{10^{4.4}}\right) = \tfrac{2}{3}\log\left(2\times10^{9.6}\right)$$
$$\approx 6.6$$

91. Use $pH = -\log_{10}\left[H^{+}\right]$.

Here,

$$pH = -\log_{10}(5.01\times10^{-4})$$
$$\approx 3.3$$

93. Use $pH = -\log_{10}\left[H^{+}\right]$.

Normal Rainwater:

$$pH = -\log_{10}(10^{-5.6}) = 5.6$$

Acid rain/tomato juice:

$$pH = -\log_{10}(10^{-4}) = 4$$

95. Use $pH = -\log_{10}\left[H^{+}\right]$.

Here,

$$pH = -\log_{10}(10^{-3.6})$$
$$= -(-3.6)\log_{10}10 = 3.6$$

97. Use $t = -\dfrac{\ln\left(\frac{C}{500}\right)}{0.0001216}$.

Here,

$$t = -\frac{\ln\left(\frac{100}{500}\right)}{0.0001216} = -\frac{\ln\left(\frac{1}{5}\right)}{0.0001216} \cong 13,236 \text{ yr}$$

99. $dB = 10\log\left(\dfrac{3\times10^{-3}}{1}\right) \approx -25\,dB$.

So, the result is approximately a 25 dB loss.

101. a.

Usage	Wavelength	Frequency
Super Low Frequency—Communication with Submarines	10,000,000 m	30 Hz
Ultra Low Frequency—Communication within Mines	1,000,000 m	300 Hz
Very Low Frequency—Avalanche Beacons	100,000 m	3000 Hz
Low Frequency—Navigation, AM Longwave Broadcasting	10,000 m	30,000 Hz
Medium Frequency—AM Bradcasts, Amatuer Radio	1,000 m	300,000 Hz
High Frequency—Shortwave broadcasts, Citizens Band Radio	100 m	3,000,000 Hz
Very High Frequency—FM Radio, Television	10 m	30,000,000 Hz
Ultra High Frequency—Television, Mobile Phones	0.050 m	6,000,000,000 Hz

b.

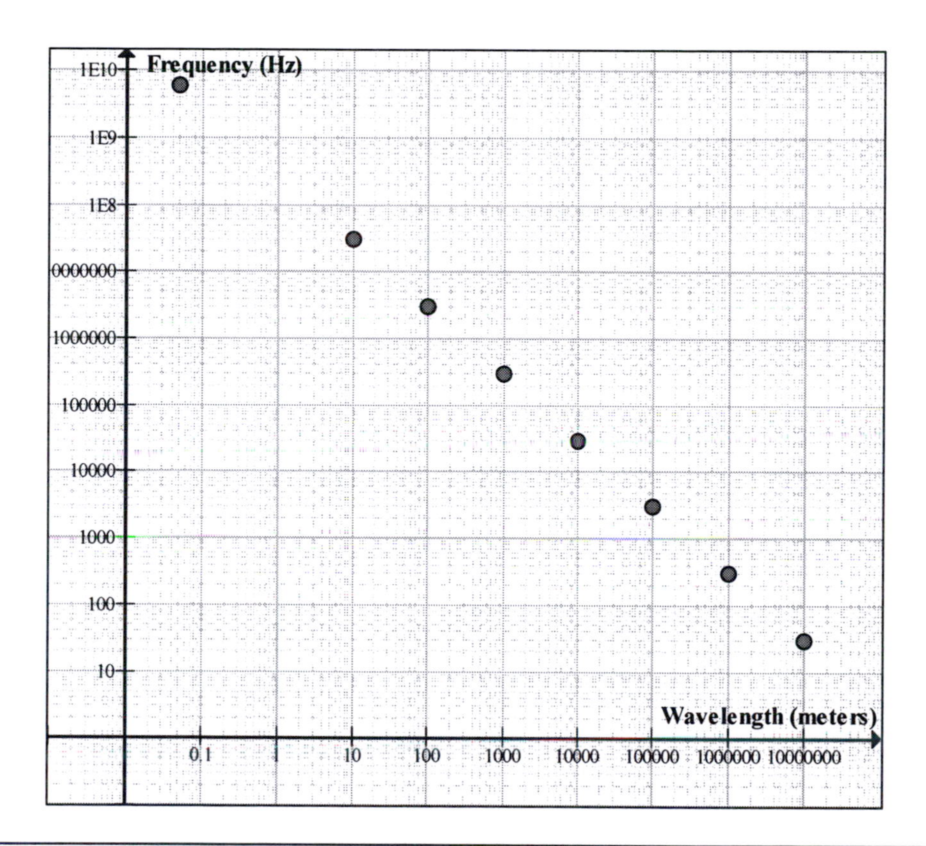

103. $\log_2 4 = x$ is equivalent to $2^x = 4$ (not $x = 2^4$).

105. The domain is the set of all real numbers such that $x+5>0$, which is written as $(-5,\infty)$.

107. False. The domain is all positive real numbers.

109. True.

111. Consider $f(x) = -\ln(x-a)+b$, where a, b are real numbers.

Domain: Must have $x-a>0$, so that the domain is (a,∞).

Range: The graph of $f(x)$ is the graph of $\ln x$ shifted a units to the right, then reflected over the x-axis, and then shifted up b units. Through all of this movement, the range of y-values remains the same as that of $\ln x$, namely $(-\infty,\infty)$.

x-intercept: Solve $-\ln(x-a)+b=0$:

$$-\ln(x-a)+b=0$$
$$\ln(x-a)=b$$
$$x-a=e^b$$
$$x=a+e^b$$

So, the x-intercept is $(a+e^b,0)$.

113.

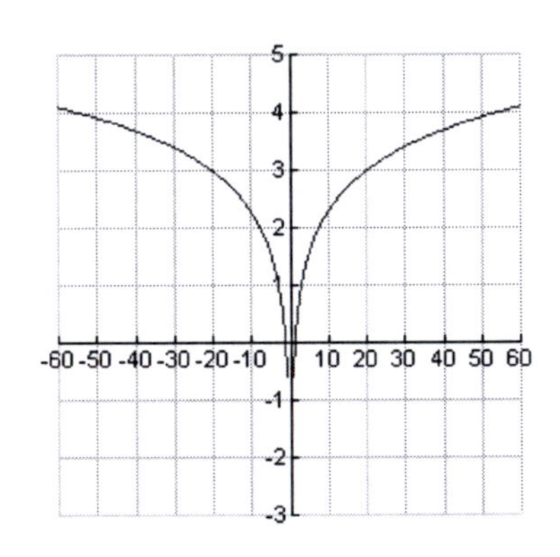

115. The graphs are symmetric about the line $y=x$.

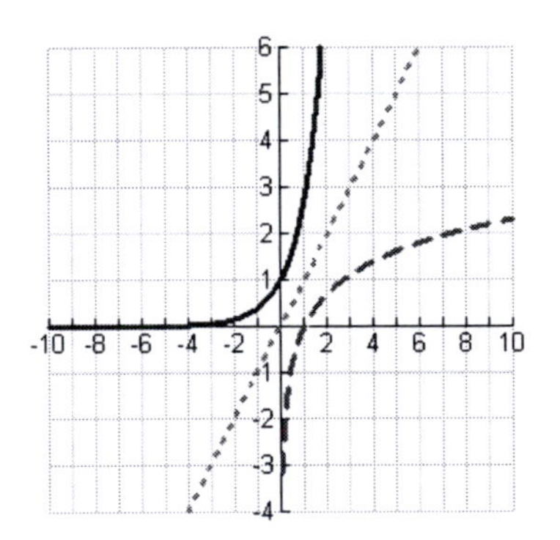

Note on Graphs: Solid curve is $y=e^x$ and dashed curve is $y=\ln x$.

117. The common characteristics are:

- x-intercept for both is (1,0).
- y-axis is the vertical asymptote for both.
- Range is $(-\infty,\infty)$ for both.
- Domain is $(0,\infty)$ for both.

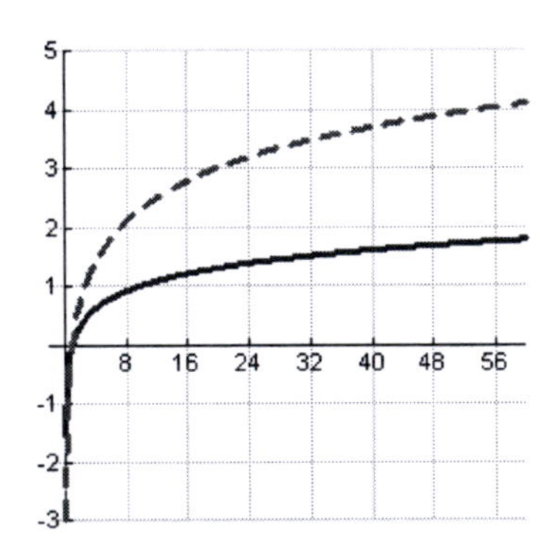

Note on Graphs: Solid curve is $y=\log x$ and dashed curve is $y=\ln x$

119. The graphs of f, g, and h are below. We note that f and g have the same graph with domain $(0, \infty)$.

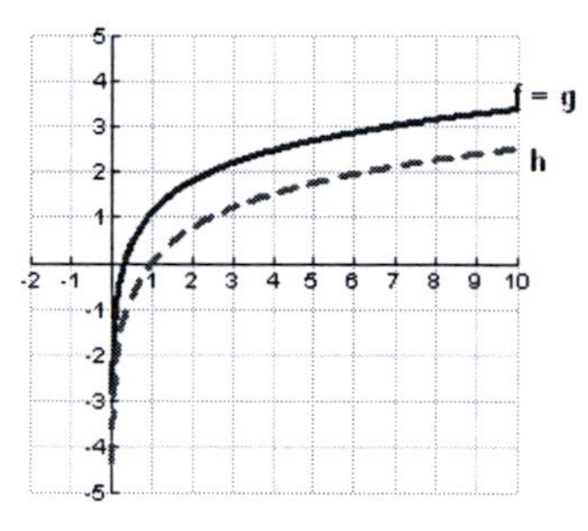

121. a.) 

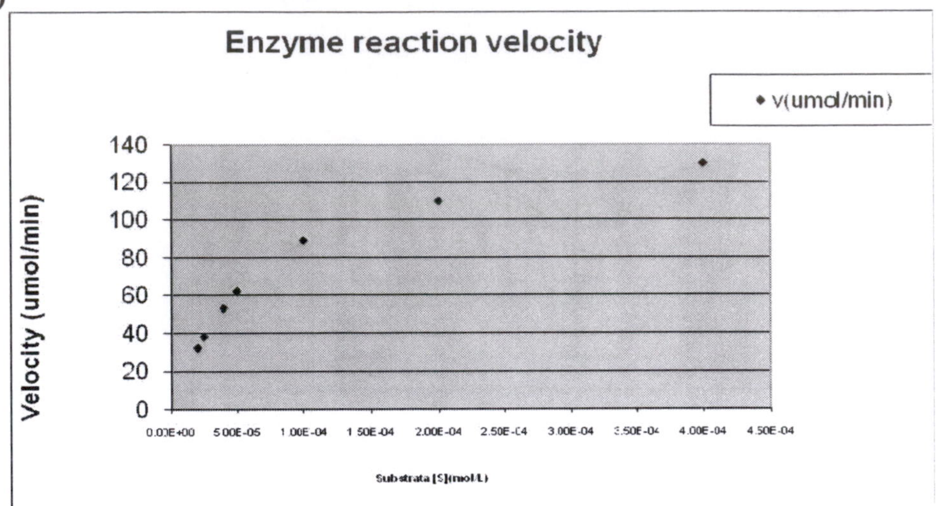

b.) A reasonable estimate for V_{max} is about 156 μmol/min.

c.) K_m is the value of [S] that results in the velocity being half of its maximum value, which by (b) is about 156. So, we need the value of [S] that corresponds to $v = 78$. From the graph, this is very difficult to ascertain because of the very small units. We can simply say that it occurs between 0.0001 and 0.0002. A more accurate estimate can be obtained if a best fit curve is known.

d.) **i.)** The equation of the best fit logarithmic curve is $v = 33.70\ln([S]) + 395.80$ with $r^2 = 0.9984$. It is shown on the scatterplot below.

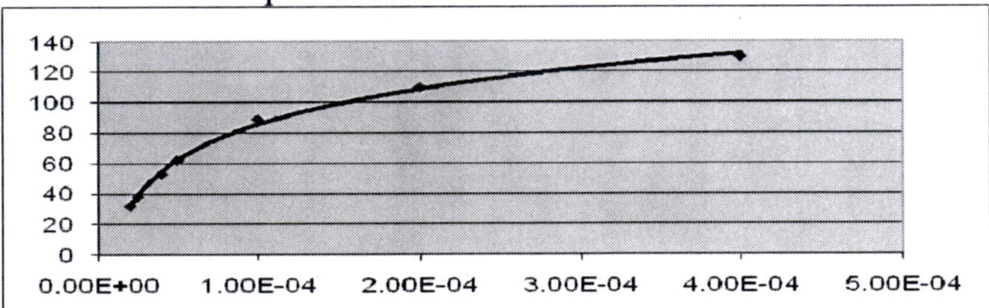

ii.) Using the equation, we must solve the following equation for [S]:

$$100 = 33.70\ln([S]) + 395.80$$

$$-295.80 = 33.70\ln[S]$$

$$-8.77745 = \ln[S]$$

$$0.000154171 \approx [S]$$

Section 5.3 Solutions ---

1. $\log_9 1 = x$ $9^x = 1$ $x = 0$	**3.** $\log_{\frac{1}{2}}\left(\frac{1}{2}\right) = x$ $\left(\frac{1}{2}\right)^x = \frac{1}{2}$ $x = 1$	**5.** $\log_{10} 10^8 = x$ $10^x = 10^8$ $x = 8$
7. $\log_{10} 0.001 = x$ $10^x = 0.001 = 10^{-3}$ $x = -3$	**9.** $\log_2 \sqrt{8} = x$ $2^x = \sqrt{8} = 2^{\frac{3}{2}}$ $x = \frac{3}{2}$	**11.** $8^{\log_8 5} = 5$
13. $e^{\ln(x+5)} = x + 5$	**15.** $5^{3\log_5 2} = 5^{\log_5 2^3} = 2^3 = 8$	**17.** $7^{-2\log_7 3} = 7^{\log_7 3^{-2}} = 3^{-2} = \frac{1}{9}$

19.

$$\log_b\left(x^3 y^5\right) = \log_b\left(x^3\right) + \log_b\left(y^5\right) = 3\log_b(x) + 5\log_b(y)$$

21.

$$\log_b\left(x^{\frac{1}{2}} y^{\frac{1}{3}}\right) = \log_b\left(x^{\frac{1}{2}}\right) + \log_b\left(y^{\frac{1}{3}}\right) = \tfrac{1}{2}\log_b(x) + \tfrac{1}{3}\log_b(y)$$

23.

$$\log_b\left(\frac{r^{\frac{1}{3}}}{s^{\frac{1}{2}}}\right) = \log_b\left(r^{\frac{1}{3}}\right) - \log_b\left(s^{\frac{1}{2}}\right) = \tfrac{1}{3}\log_b(r) - \tfrac{1}{2}\log_b(s)$$

25.

$$\log_b\left(\frac{x}{yz}\right) = \log_b(x) - \log_b(yz) = \log_b(x) - \left[\log_b(y) + \log_b(z)\right] = \log_b(x) - \log_b(y) - \log_b(z)$$

27.

$$\log\left(x^2\sqrt{x+5}\right) = \log\left(x^2\right) + \log\left(\sqrt{x+5}\right) = \log\left(x^2\right) + \log(x+5)^{\frac{1}{2}} = 2\log x + \tfrac{1}{2}\log(x+5)$$

29.

$$\ln\left(\frac{x^3(x-2)^2}{\sqrt{x^2+5}}\right) = \ln\left(x^3(x-2)^2\right) - \ln\left(\sqrt{x^2+5}\right) = \ln\left(x^3\right) + \ln\left((x-2)^2\right) - \ln\left(x^2+5\right)^{\frac{1}{2}} = 3\ln(x) + 2\ln(x-2) - \tfrac{1}{2}\ln\left(x^2+5\right)$$

31.

$$\begin{aligned}\log\left(\frac{x^2-2x+1}{x^2-9}\right) &= \log\left(\frac{(x-1)^2}{(x-3)(x+3)}\right)\\ &= \log\left((x-1)^2\right)-\log\left((x-3)(x+3)\right)\\ &= 2\log(x-1)-\left[\log(x-3)+\log(x+3)\right]\\ &= 2\log(x-1)-\log(x-3)-\log(x+3)\end{aligned}$$

33.

$$\begin{aligned}3\log_b x+5\log_b y &= \log_b\left(x^3\right)+\log_b y^5\\ &= \log_b\left(x^3y^5\right)\end{aligned}$$

35.

$$\begin{aligned}5\log_b u-2\log_b v &= \log_b u^5-\log_b v^2\\ &= \log_b\left(\frac{u^5}{v^2}\right)\end{aligned}$$

37. $\frac{1}{2}\log_b x+\frac{2}{3}\log_b y=\log_b x^{\frac{1}{2}}+\log_b y^{\frac{2}{3}}=\log_b\left(x^{\frac{1}{2}}y^{\frac{2}{3}}\right)$

39.

$$\begin{aligned}2\log u-3\log v-2\log z &= \log u^2-\left[\log v^3+\log z^2\right]\\ &= \log u^2-\log\left(v^3z^2\right)\\ &= \log\left(\frac{u^2}{v^3z^2}\right)\end{aligned}$$

41.

$$\begin{aligned}\ln(x+1)+\ln(x-1)-2\ln(x^2+3) &= \ln(x+1)+\ln(x-1)-\ln(x^2+3)^2\\ &= \ln\left[(x+1)(x-1)\right]-\ln(x^2+3)^2\\ &= \ln\left(\frac{x^2-1}{(x^2+3)^2}\right)\end{aligned}$$

43.

$$\begin{aligned}\tfrac{1}{2}\ln(x+3)-\tfrac{1}{3}\ln(x+2)-\ln x &= \ln(x+3)^{\frac{1}{2}}-\left[\ln(x+2)^{\frac{1}{3}}+\ln x\right]\\ &= \ln(x+3)^{\frac{1}{2}}-\left[\ln\left(x(x+2)^{\frac{1}{3}}\right)\right]\\ &= \ln\left(\frac{(x+3)^{\frac{1}{2}}}{x(x+2)^{\frac{1}{3}}}\right)\end{aligned}$$

45. $\log_5 7=\dfrac{\log 7}{\log 5}\cong 1.2091$

47. $\log_{\frac{1}{2}} 5=\dfrac{\log 5}{\log \frac{1}{2}}\cong -2.3219$

49. $\log_{2.7} 5.2=\dfrac{\log 5.2}{\log 2.7}\cong 1.6599$

51. $\log_{\pi} 10=\dfrac{\log 10}{\log \pi}\cong 2.0115$

53. $\log_{\sqrt{3}} 8 = \dfrac{\log 8}{\log\sqrt{3}} \cong 3.7856$

55. Use $D = 10\log\left(\dfrac{I}{1\times10^{-12}}\right)$.

In this case, $I = \underbrace{(1\times10^{-1})\,W/m^2}_{\text{From music}} + \underbrace{(1\times10^{-6})\,W/m^2}_{\text{From conversation}} = \left(1.00001\times10^{-1}\right)W/m^2$.

So, $D = 10\log\left(\dfrac{1.00001\times10^{-1}}{1\times10^{-12}}\right) \cong 110\,dB$ that you are exposed to.

57. Use $M = \frac{2}{3}\log\left(\dfrac{E}{10^{4.4}}\right)$.

Here, the combined energy is $(4.5\times10^{12}) + (7.8\times10^{8})$ joules. The corresponding magnitude on the Richter scale is $M = \frac{2}{3}\log\left(\dfrac{(4.5\times10^{12}) + (7.8\times10^{8})}{10^{4.4}}\right) \cong 5.5$.

59. Cannot apply the quotient property directly. Observe that

$$3\log 5 - \log 5^2 = 3\log 5 - 2\log 5 = \log 5.$$

61. Cannot apply the product and quotient properties to logarithms with different bases. So, you cannot reduce the given expression further without using the change of base formula.

63. True. $\log e = \dfrac{\ln e}{\ln 10} = \dfrac{1}{\ln 10}$

65. False. $\ln(xy)^3 = 3\ln(xy) = 3(\ln x + \ln y)$, which does not equal $(\ln x + \ln y)^3$, in general.

67. Claim: $\log_b\left(\frac{M}{N}\right) = \log_b M - \log_b N$

Proof: Let $u = \log_b M$, $v = \log_b N$. Then, $b^u = M$, $b^v = N$.

Observe that $\log_b\left(\frac{M}{N}\right) = \log_b\left(\frac{b^u}{b^v}\right) = \log_b\left(b^{u-v}\right) = u - v = \log_b M - \log_b N$. ▪

69.

$$\begin{aligned}\log_b\left(\sqrt{\frac{x^2}{y^3z^{-5}}}\right)^6 &= \log_b\left(\frac{x^2}{y^3z^{-5}}\right)^3 = \log_b\left(\frac{x^6}{y^9z^{-15}}\right)\\ &= \log_b\left(x^6\right) - \log_b\left(y^9z^{-15}\right)\\ &= \log_b\left(x^6\right) - \left[\log_b\left(y^9\right) + \log_b\left(z^{-15}\right)\right]\\ &= 6\log_b x - 9\log_b y + 15\log_b z\end{aligned}$$

71. Yes, they are the same graph.

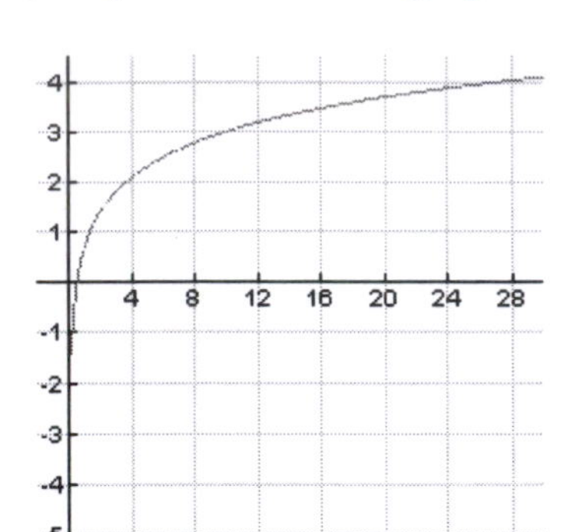

73. No, they are not the same graph.

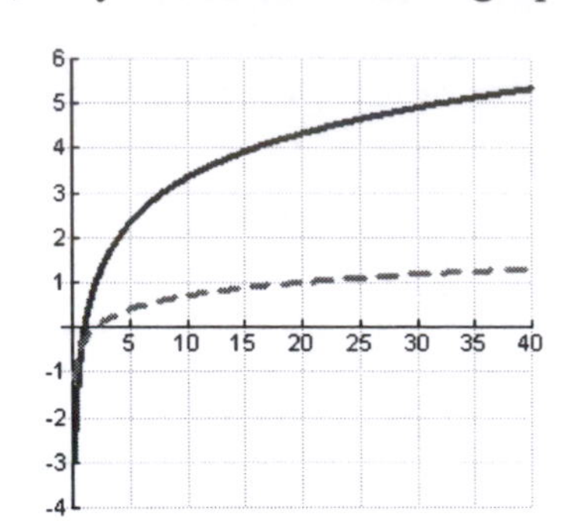

Note on Graphs: Solid curve is $y = \frac{\log x}{\log 2}$ and dashed curve is $y = \log x - \log 2$.

75. No, they are not the same graph, even though the property is true.

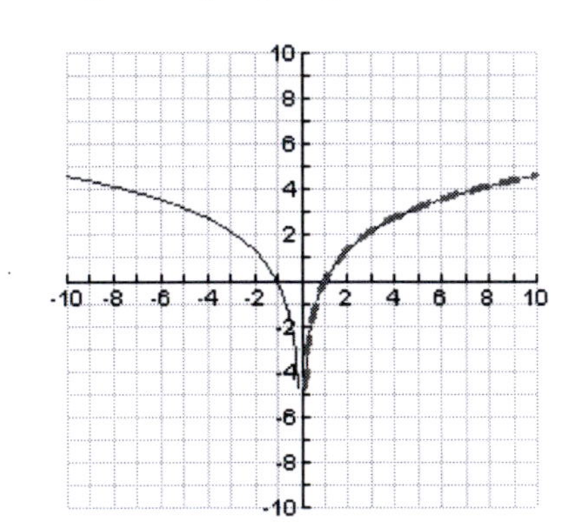

Note on Graphs: The thin curve plus the dashed curve is $y = \ln(x^2)$ (domain all real numbers except 0) and the dashed curve only is $y = 2\ln x$ (domain is $(0,\infty)$).

77. The graphs of $y = \ln x$ and $y = \dfrac{\log x}{\log e}$ do coincide, and are as follows:

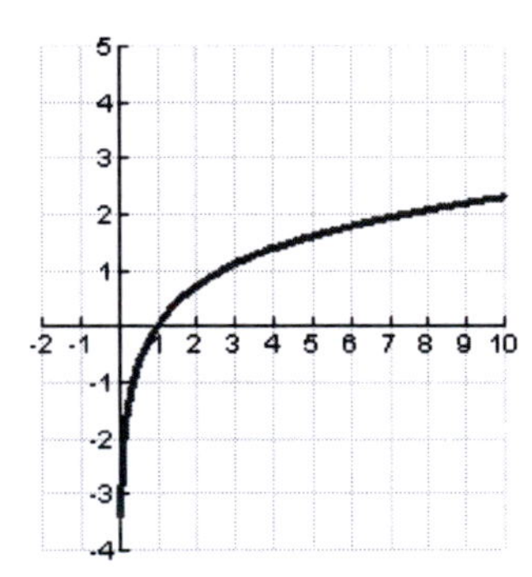

Section 5.4 Solutions ---

1.	**3.**	**5.**
$3^x = 81$ $3^x = 3^4$ $x = 4$	$7^x = \dfrac{1}{49}$ $7^x = 7^{-2}$ $x = -2$	$2^{x^2} = 16$ $2^{x^2} = 2^4$ $x^2 = 4$ $x = \pm 2$

7.

$$\left(\tfrac{2}{3}\right)^{x+1}=\tfrac{27}{8}$$
$$\left(\tfrac{2}{3}\right)^{x+1}=\left(\tfrac{3}{2}\right)^{3}=\left(\tfrac{2}{3}\right)^{-3}$$
$$x+1=-3$$
$$x=-4$$

9.

$$e^{2x+3}=1=e^{0}$$
$$2x+3=0$$
$$x=-\tfrac{3}{2}$$

11.

$$7^{2x-5}=7^{3x-4}$$
$$2x-5=3x-4$$
$$\boxed{x=-1}$$

13.

$$2^{x^2+12}=2^{7x}$$
$$x^2+12=7x$$
$$x^2-7x+12=0$$
$$(x-4)(x-3)=0$$
$$\boxed{x=3,4}$$

15.

$$9^{x}=3^{x^2-4x}$$
$$3^{2x}=\left(3^2\right)^{x}=3^{x^2-4x}$$
$$2x=x^2-4x$$
$$x^2-6x=0$$
$$x(x-6)=0$$
$$\boxed{x=0,6}$$

17.

$$e^{5x-1}=e^{x^2+3}$$
$$5x-1=x^2+3$$
$$x^2-5x+4=0$$
$$(x-4)(x-1)=0$$
$$\boxed{x=1,4}$$

19.

$$10^{2x-3}=81$$
$$\log\left(10^{2x-3}\right)=\log(81)$$
$$2x-3=\log(81)$$
$$2x=3+\log(81)$$
$$x=\tfrac{3+\log(81)}{2}\cong 2.454$$

21.

$$3^{x+1}=5$$
$$\log_3(5)=x+1$$
$$x=\log_3(5)-1\approx 0.465$$

23.

$$27=2^{3x-1}$$
$$\log_2(27)=3x-1$$
$$x=\tfrac{1}{3}\left[\log_2(27)+1\right]\approx 1.918$$

25.

$$3e^{x}-8=7$$
$$3e^{x}=15$$
$$e^{x}=5$$
$$x=\ln 5\approx 1.609$$

27.

$$9-2e^{0.1x}=1$$
$$2e^{0.1x}=8$$
$$e^{0.1x}=4$$
$$0.1x=\ln 4$$
$$x=10\ln 4\approx 13.863$$

29.

$$2\left(3^x\right)-11=9$$
$$2\left(3^x\right)=20$$
$$3^x=10$$
$$x=\log_3(10)\approx 2.096$$

31.

$$e^{3x+4}=22$$
$$\ln\left(e^{3x+4}\right)=\ln(22)$$
$$3x+4=\ln(22)$$
$$3x=-4+\ln(22)$$
$$x=\tfrac{-4+\ln(22)}{3}\cong -0.303$$

33.

$$3e^{2x}=18$$
$$e^{2x}=6$$
$$\ln\left(e^{2x}\right)=\ln(6)$$
$$2x=\ln(6)$$
$$x=\tfrac{\ln 6}{2}\cong 0.896$$

35. Note that $e^{2x}+7e^x-3=0$ is equivalent to $\left(e^x\right)^2+7\left(e^x\right)-3=0$. Let $y=e^x$ and solve $y^2+7y-3=0$ using the quadratic formula:

$$y=\frac{-7\pm\sqrt{7^2-4(1)(-3)}}{2}=\frac{-7\pm\sqrt{61}}{2}$$

So, substituting back in for y, the following two equations must be solved for x:

$$e^x=\frac{-7+\sqrt{61}}{2} \text{ and } e^x=\frac{-7-\sqrt{61}}{2}$$

Since $\frac{-7-\sqrt{61}}{2}<0$, the second equation has no real solution. Solving the first one yields $x=\ln\left(\frac{-7+\sqrt{61}}{2}\right)\cong -0.904$.

37.

$$\left(3^x-3^{-x}\right)^2=0$$
$$3^x-3^{-x}=0$$
$$3^x=3^{-x}$$
$$x=-x$$
$$x=0$$

39.

$$\frac{2}{e^x-5}=1$$
$$2=e^x-5$$
$$e^x=7$$
$$x=\ln(7)\approx 1.946$$

41.

$$\frac{20}{6-e^{2x}}=4$$
$$20=24-4e^{2x}$$
$$4e^{2x}=4$$
$$e^{2x}=1$$
$$2x=0$$
$$x=0$$

43.

$$\frac{4}{10^{2x}-7}=2$$
$$4=2\left(10^{2x}\right)-14$$
$$10^{2x}=9$$
$$2x=\log_{10}(9)$$
$$x\approx 0.477$$

45.

$$\log(2x)=2$$
$$10^2=2x$$
$$100=2x$$
$$50=x$$

47.

$$\log_3(2x+1)=4$$
$$2x+1=3^4$$
$$2x=80$$
$$x=40$$

49.

$$\log_2(4x-1)=-3$$
$$2^{-3}=4x-1$$
$$4x=\tfrac{1}{8}+1$$
$$x=\tfrac{9}{32}$$

51.

$$\ln x^2-\ln 9=0$$
$$\ln x^2=\ln 9$$
$$x^2=9$$
$$x=\pm 3$$

53.

$$\log_5(x-4)+\log_5 x=1$$
$$\log_5\left(x(x-4)\right)=1$$
$$x(x-4)=5$$
$$x^2-4x-5=0$$
$$(x-5)(x+1)=0$$
$$x=5,\ \not{-1}$$

55.

$$\log(x-3)+\log(x+2)=\log(4x)$$
$$\log(x-3)(x+2)=\log(4x)$$
$$(x-3)(x+2)=4x$$
$$x^2-x-6=4x$$
$$x^2-5x-6=0$$
$$(x-6)(x+1)=0$$
$$x=6,\ \not{-1}$$

57.

$$\log(4-x)+\log(x+2)=\log(3-2x)$$
$$\log\left((4-x)(x+2)\right)=\log(3-2x)$$
$$(4-x)(x+2)=3-2x$$
$$-x^2+2x+8=3-2x$$
$$x^2-4x-5=0$$
$$(x-5)(x+1)=0$$
$$x=-1,\ \not{5}$$

59.

$$\log_4(4x)-\log_4\left(\tfrac{x}{4}\right)=3$$
$$\log_4\left(\frac{4x}{\frac{x}{4}}\right)=3$$
$$\log_4(16)=3$$

Since the last line is a false statement, this equation has no solution.

61.

$$\log(2x-5)-\log(x-3)=1$$
$$\log\left(\frac{2x-5}{x-3}\right)=1$$
$$\frac{2x-5}{x-3}=10$$
$$2x-5=10x-30$$
$$8x=25$$
$$x=\tfrac{25}{8}$$

63.

$$\ln x^2 = 5$$
$$x^2 = e^5$$
$$x = \pm\sqrt{e^5} \approx \pm 12.182$$

65.

$$\log(2x+5) = 2$$
$$2x+5 = 10^2$$
$$2x = 95$$
$$x = 47.5$$

67.

$$\ln\left(x^2+1\right) = 4$$
$$x^2+1 = e^4$$
$$x^2 = e^4 - 1$$
$$x = \pm\sqrt{e^4-1} \approx \pm 7.321$$

69.

$$\ln(2x+3) = -2$$
$$2x+3 = e^{-2}$$
$$x = \tfrac{1}{2}\left[-3+e^{-2}\right] \approx -1.432$$

71.

$$\log(2-3x) + \log\left(3-2x\right) = 1.5$$
$$\log\left((2-3x)(3-2x)\right) = 1.5$$
$$(2-3x)\left(3-2x\right) = 10^{1.5} \cong 31.622$$
$$6x^2 - 13x + 6 \cong 31.622$$
$$6x^2 - 13x - 25.622 \cong 0$$

Now, use the quadratic formula:

$x = \dfrac{13 \pm \sqrt{783.93}}{12}$, so that $x \cong \cancel{3.42}, -1.25$

73.

$$\ln x + \ln(x-2) = 4$$
$$\ln x(x-2) = 4$$
$$x(x-2) = e^4$$
$$x^2 - 2x - e^4 = 0$$
$$x = \frac{2 \pm \sqrt{4-4(1)(-e^4)}}{2}$$
$$\approx \cancel{-6.456}, 8.456$$

75.

$$\log_7(1-x) - \log_7(x+2) = \log_7(x)$$
$$\log_7\left(\frac{1-x}{x+2}\right) = \log_7(x)$$
$$\frac{1-x}{x+2} = x$$
$$1-x = x(x+2)$$
$$x^2 + 3x - 1 = 0$$

There are no rational solutions since neither 1 nor −1 work. So, graph to find the real roots:

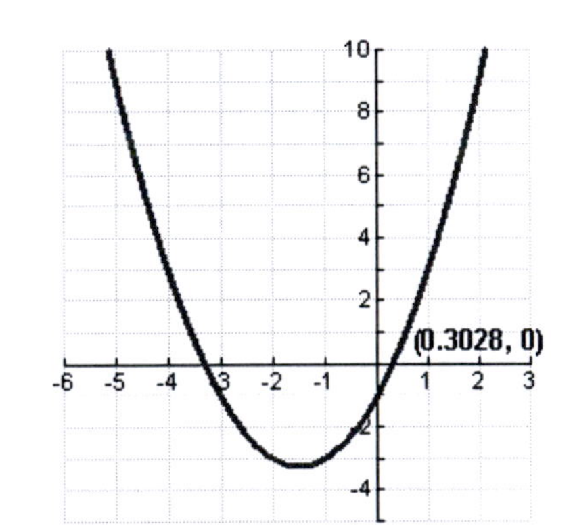

So, the solution is approximately 0.3028.

77.

$$\ln\sqrt{x+4} - \ln\sqrt{x-2} = \ln\sqrt{x+1}$$
$$\tfrac{1}{2}\ln(x+4) - \tfrac{1}{2}\ln(x-2) = \tfrac{1}{2}\ln(x+1)$$
$$\ln\left(\frac{x+4}{x-2}\right) = \ln(x+1)$$
$$\frac{x+4}{x-2} = x+1$$
$$x+4 = x^2 - x - 2$$
$$x^2 - 2x - 6 = 0$$
$$x = \frac{2 \pm \sqrt{4-4(1)(-6)}}{2} = 1 \pm \sqrt{7}$$
$$\approx \cancel{-1.646}, 3.646$$

79. a. $R(0) = 151$ beats per minute.

b. Solve for t:

$$151e^{-0.055t} = 100$$
$$e^{-0.055t} = \tfrac{100}{151}$$
$$-0.055t = \ln\left(\tfrac{100}{151}\right)$$
$$t = -\tfrac{1}{0.055}\ln\left(\tfrac{100}{151}\right) \approx 7 \text{ min.}$$

c. $R(15) = 151e^{-0.055(15)} \approx 66$ beats per minute.

81. Use $A(t) = P\left(1+\frac{r}{n}\right)^{nt}$.

Here,

$$r = 0.035, \quad n = 1.$$

In order to triple, if P is the initial investment, then we seek the time t such that $A(t) = 3P$. So, we solve the following equation:

$$3P = P\left(1+\tfrac{0.035}{1}\right)^{1(t)}$$
$$3 = (1.035)^t$$
$$\log_{1.035} 3 = t$$

So, it takes approximately 31.9 years to triple.

83. Use $A(t) = P\left(1+\frac{r}{n}\right)^{nt}$.

Here,

$A = 20{,}000, \ r = 0.05, \ n = 4, \ P = 7500.$

So, we solve the following equation:

$$20{,}000 = 7500\left(1+\tfrac{0.05}{4}\right)^{4(t)}$$
$$2.667 = (1.0125)^{4t}$$
$$\log_{1.0125} 2.667 = 4t$$
$$\tfrac{1}{4}\log_{1.0125} 2.667 = t$$

So, it takes approximately 19.74 years.

85. Use $M = \frac{2}{3}\log\left(\frac{E}{10^{4.4}}\right)$.

Here, $M = 7.4$. So, substituting in, we can solve for E:

$$7.4 = \tfrac{2}{3}\log\left(\frac{E}{10^{4.4}}\right)$$
$$11.1 = \log\left(\frac{E}{10^{4.4}}\right)$$
$$10^{11.1} = \frac{E}{10^{4.4}}$$
$$\underbrace{10^{11.1} \times 10^{4.4}}_{=10^{15.5}} \approx 3.16 \times 10^{15} = E$$

So, it would generate 3.16×10^{15} joules of energy.

87. Use $D=10\log\left(\frac{I}{10^{-12}}\right)$. Here, $D=120$. So, substituting in, we can solve for *I*:

$$120=10\log\left(\frac{I}{10^{-12}}\right)$$

$$12=\log\left(\frac{I}{10^{-12}}\right)$$

$$10^{12}=\frac{I}{10^{-12}}$$

$$1=10^{12}\times10^{-12}=I$$

So, the intensity is $1\,{}^{W}\!/_{m^2}$.

89. Use $A=A_0e^{-0.5t}$. Here, $A=0.10A_0$, where A_0 is the initial amount of anesthesia. So, substituting into the above equation, we can solve for *t*:

$$0.10A_0=A_0e^{-0.5t}$$

$$0.10=e^{-0.5t}$$

$$\ln(0.10)=-0.5t$$

$$-\tfrac{1}{0.5}\ln(0.10)=t$$

So, it takes about 4.61 hours until 10% of the anesthesia remains in the bloodstream.

91. Use $N=\frac{200}{1+24e^{-0.2t}}$. Here, $N=100$. So, substituting into the above equation, we can solve for *t*:

$$100=\frac{200}{1+24e^{-0.2t}}$$

$$100\left(1+24e^{-0.2t}\right)=200$$

$$1+24e^{-0.2t}=2$$

$$24e^{-0.2t}=1$$

$$e^{-0.2t}=\tfrac{1}{24}$$

$$-0.2t=\ln\left(\tfrac{1}{24}\right)$$

$$t=-\tfrac{1}{0.2}\ln\left(\tfrac{1}{24}\right)\approx15.89$$

So, it takes about 15.9 years.

93. Use $M = \frac{2}{3}\log\left(\frac{E}{10^{4.4}}\right)$.

For P waves: Here, $M = 6.2$. So, substituting in, we can solve for E:

$$6.2 = \tfrac{2}{3}\log\left(\frac{E}{10^{4.4}}\right)$$

$$9.3 = \log\left(\frac{E}{10^{4.4}}\right)$$

$$10^{9.3} = \frac{E}{10^{4.4}}$$

$$\underbrace{10^{9.3} \times 10^{4.4}}_{=10^{13.7}} = E$$

For S waves: Here, $M = 3.3$. So, substituting in, we can solve for E:

$$3.3 = \tfrac{2}{3}\log\left(\frac{E}{10^{4.4}}\right)$$

$$4.95 = \log\left(\frac{E}{10^{4.4}}\right)$$

$$10^{4.95} = \frac{E}{10^{4.4}}$$

$$\underbrace{10^{4.95} \times 10^{4.4}}_{=10^{9.35}} = E$$

So, the combined energy is $\left(10^{13.7} + 10^{9.35}\right)$joules. Hence, the reading on the Richter scale is: $M = \frac{2}{3}\log\left(\frac{10^{9.35} + 10^{13.7}}{10^{4.4}}\right) \approx 6.2$

95. $\ln(4e^x) \neq 4x$. Should first divide both sides by 4, then take the natural log:

$$4e^x = 9$$

$$e^x = \tfrac{9}{4}$$

$$\ln(e^x) = \ln(\tfrac{9}{4})$$

$$x = \ln(\tfrac{9}{4})$$

97. $x = -5$ is not a solution since $\log(-5)$ is not defined.

99. True.

101. False. Since $\log x = \frac{\ln x}{\ln 10}$, $e^{\log x} = e^{\frac{\ln x}{\ln 10}} = \left(e^{\ln x}\right)^{\frac{1}{\ln 10}} = x^{\frac{1}{\ln 10}} \neq x$.

103.

$$\tfrac{1}{3}\log_b(x^3)+\tfrac{1}{2}\log_b(\underbrace{x^2-2x+1}_{=(x-1)^2})=2$$

$$\log_b(x^3)^{\frac{1}{3}}+\log_b(x-1)^{2\cdot\frac{1}{2}}=2$$

$$\log_b x+\log_b(x-1)=2$$

$$\log_b(x(x-1))=2$$

$$b^2=x(x-1)$$

$$x^2-x-b^2=0$$

Now, use the quadratic formula to find the solutions:

$$x=\frac{1\pm\sqrt{1-4(-b^2)}}{2}$$

$$=\frac{1+\sqrt{1+4b^2}}{2},\quad \underbrace{\frac{1-\sqrt{1+4b^2}}{2}}_{\text{This is negative, for any value of } b.\text{ So, it cannot be a solution.}}$$

(The second expression is crossed out.)

105.

$$y=\frac{3000}{1+2e^{-0.2t}}$$

$$y\left(1+2e^{-0.2t}\right)=3000$$

$$y+2ye^{-0.2t}=3000$$

$$2ye^{-0.2t}=3000-y$$

$$e^{-0.2t}=\tfrac{3000-y}{2y}$$

$$-0.2t=\ln\left(\tfrac{3000-y}{2y}\right)$$

$$t=-5\ln\left(\tfrac{3000-y}{2y}\right)$$

107. Consider the function $y=\dfrac{e^x+e^{-x}}{2}$, for $x\geq 0,\ y\geq 1$. (Need this restriction in order for the function to be one-to-one, and hence have an inverse.) Solve $y=\dfrac{e^x+e^{-x}}{2}$ for x.

$$y=\frac{e^x+e^{-x}}{2}$$

$$y=\frac{e^x+\frac{1}{e^x}}{2}=\frac{\frac{(e^x)^2+1}{e^x}}{2}$$

$$2y=\tfrac{(e^x)^2+1}{e^x}$$

$$2ye^x=(e^x)^2+1$$

$$(e^x)^2-2y\left(e^x\right)+1=0$$

Now, solve using the quadratic formula:

$$e^x=\tfrac{2y\pm\sqrt{(-2y)^2-4}}{2}=\tfrac{2y\pm\sqrt{4y^2-4}}{2}$$

$$=\tfrac{2y\pm2\sqrt{y^2-1}}{2}=y\pm\sqrt{y^2-1}$$

Since $dom(f)=(0,\infty)=rng(f^{-1})$, we use only $y+\sqrt{y^2-1}$ here. Now to solve for *x*, take natural log of both sides:

$$e^x=y+\sqrt{y^2-1}$$

$$x=\ln\left(y+\sqrt{y^2-1}\right)$$

Thus, inverse function is given by

$$f^{-1}(x)=\ln\left(x+\sqrt{x^2-1}\right).$$

109. Observe that

$$\ln(3x) = \ln(x^2+1)$$
$$3x = x^2+1$$
$$x^2-3x+1=0$$
$$x = \tfrac{3\pm\sqrt{9-4}}{2} = \tfrac{3\pm\sqrt{5}}{2}$$

These solutions agree with the graphical solution seen to the right.

Note on Graphs: Solid curve is $y=\ln(3x)$ and the thin curve is $y=\ln(x^2+1)$.

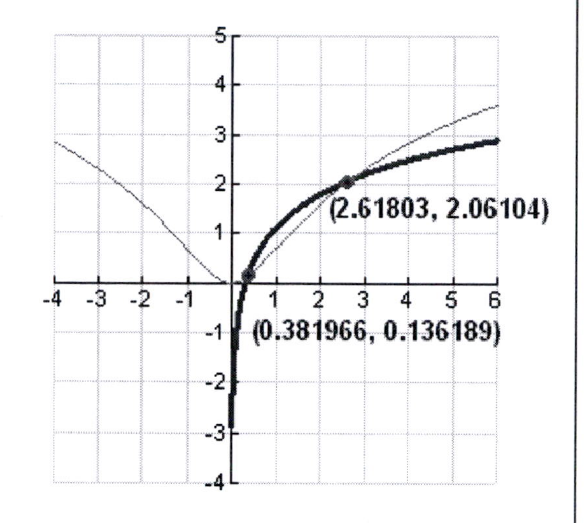

111.

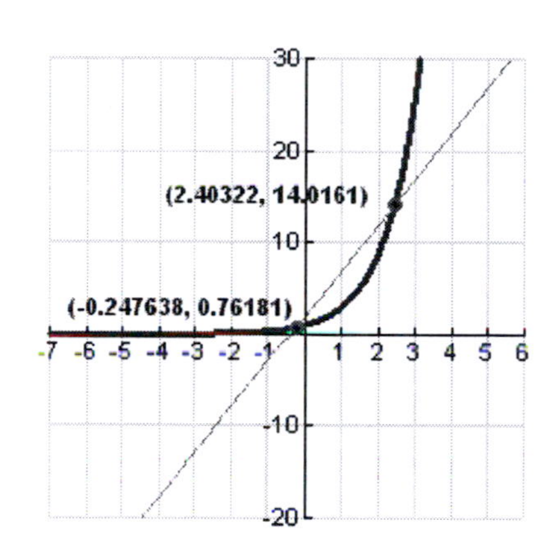

Note on Graphs: Solid curve is $y=3^x$ and the thin curve is $y=5x+2$.

113. The graph of $f(x)=\dfrac{e^x+e^{-x}}{2}$ is given below:

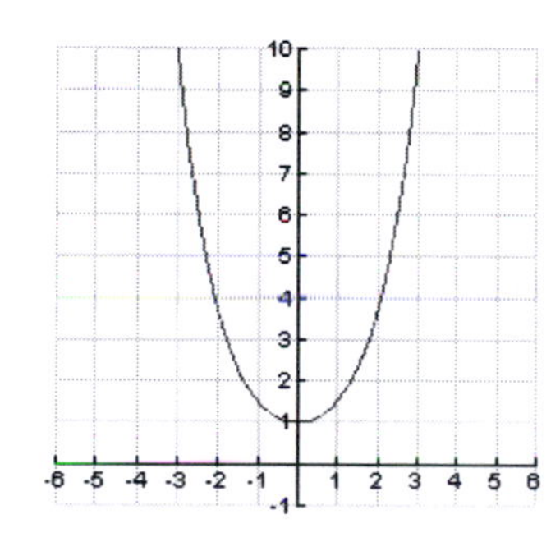

The domain is $(-\infty,\infty)$, and the graph is symmetric about the y-axis. There are no asymptotes.

Section 5.5 Solutions--

1. c (iv)	**3. a (iii)**	**5. f (i)**
7. Use $N=N_0e^{rt}$. Here, $N_0=80$, $r=0.0236$. Determine N when $t=7$ (determined by 2010 – 2003): $N=80e^{0.0236(7)}\approx 94$ million.		
9. Use $N=N_0e^{rt}$. Here, $N_0=103{,}800$, $r=0.12$. Determine t such that $N=200{,}000$: $200{,}000=103{,}800e^{0.12t} \Rightarrow \frac{200{,}000}{103{,}800}=e^{0.12t} \Rightarrow t=\frac{1}{0.12}\ln\left(\frac{200{,}000}{103{,}800}\right)\approx 5.5$ So, the population would hit 200,000 sometime in the year 2008.		

11. Use $N = N_0 e^{rt}$. Here, $N_0 = 487.4$, $r = 0.165$. Determine N when $t = 3$ (corresponds to the number of cell phone subscribers in 2010): $N = 487.4e^{0.165(3)} \approx 799.6$
There are approximately 799.6 subscribers in 2010.

13. Use $N = N_0 e^{rt}$. Here, $N_0 = 185,000$, $r = 0.30$. Determine N when $t = 3$:

$$N = 185,000e^{0.30(3)} \approx 455,027$$

The amount is approximately $455,000.

15. Use $N = N_0 e^{rt}$ (t measured in months). Here, $N_0 = 100$ (million), $r = 0.20$. Determine N when $t = 6$: $N = 100e^{0.20(6)} \approx 332$ million.

17. Use $N = N_0 e^{rt}$ (t measured in years). Here, $N_0 = 1$ (million), $r = 0.025$. Determine N when $t = 12$: $N = 1e^{0.025(12)} \approx 1.45$ million.

19. Use $A = 100e^{-0.5t}$. Observe that $A(4) = 100e^{-0.5(4)} \approx 13.53$ ml

21. a. Solve for k:

$$350 = 750\left(1 - e^{-k(1)}\right)$$

$$\tfrac{350}{750} = 1 - e^{-k}$$

$$-\left(\tfrac{350}{750} - 1\right) = e^{-k}$$

$$\ln\left(1 - \tfrac{350}{750}\right) = -k$$

$$k = -\ln\left(\tfrac{400}{750}\right) = -\ln\left(\tfrac{8}{15}\right) \approx 0.6286$$

b. $S(3) = 750\left(1 - e^{-0.6286(3)}\right) \approx 636$. So, about 636,000 MP3 players.

23. Use $N = N_0 e^{rt}$ **(1)**. We know that $N(5,730) = \frac{1}{2}N_0$ **(2)**. If $N_0 = 5$ (grams), find t such that $N(t) = 2$. To do so, we must first find r. To this end, substitute **(2)** into **(1)** to obtain:

$$\tfrac{1}{2} = e^{r(5,730)} \Rightarrow \tfrac{1}{5,730}\ln\left(\tfrac{1}{2}\right) = r$$

Now, solve for t:

$$2 = 5e^{\frac{1}{5,730}\ln\left(\frac{1}{2}\right)t}$$

$$\ln\left(\tfrac{2}{5}\right) = \tfrac{1}{5,730}\ln\left(\tfrac{1}{2}\right)t$$

$$t = \frac{\ln\left(\frac{2}{5}\right)}{\frac{1}{5,730}\ln\left(\frac{1}{2}\right)} \approx 7575 \text{ years}$$

25. Use $N = N_0 e^{rt}$ **(1)**. We know that $N(4.5\times10^9) = \frac{1}{2}N_0$ **(2)**. Find t such that $N(t) = 0.98N_0$. To do so, we must first find r. To this end, substitute **(2)** into **(1)** to obtain:

$$\tfrac{1}{2} = e^{r(4.5\times10^9)} \Rightarrow \tfrac{1}{4.5\times10^9}\ln\left(\tfrac{1}{2}\right) = r$$

Now, solve for t:

$$0.98\cancel{N_0} = \cancel{N_0}e^{\frac{1}{4.5\times10^9}\ln\left(\frac{1}{2}\right)t}$$

$$t = \frac{\ln(0.98)}{\frac{1}{4.5\times10^9}\ln\left(\frac{1}{2}\right)} \approx 131,158,556 \text{ years old}$$

27. Use $T = T_s + (T_0 - T_s)e^{-kt}$. Here, $T_0 = 325$, $T_s = 72$, and $T(10) = 200$. Find $T(30)$. To do so, we must first find *k*. Observe

$$200 = 72 + (325 - 72)e^{-k(10)}$$
$$128 = 253e^{-10k}$$
$$k = -\tfrac{1}{10}\ln\left(\tfrac{128}{253}\right)$$

Now, $T(30) = 72 + 253e^{-\left(-\frac{1}{10}\ln\left(\frac{128}{253}\right)\right)\cdot 30} \approx 105^\circ\text{F}$.

29. Use $T = T_s + (T_0 - T_s)e^{-kt}$ **(1)**.

Assume $t = 0$ corresponds to 7am. We know that

$$T(0) = 85,\ T(1.5) = 82,\ T_s = 74. \ \textbf{(2)}$$

Find *t* such that $T(t) = 98.6$. We first use **(2)** in **(1)** to find *k* and T_0.

$$85 = 74 + (T_0 - 74)e^0 = T_0 \Rightarrow T_0 = 85$$
$$82 = 74 + (85 - 74)e^{-1.5k} \Rightarrow 8 = 11e^{-1.5k}$$
$$\Rightarrow k = -\tfrac{1}{1.5}\ln\left(\tfrac{8}{11}\right)$$

Now, solve for *t*:

$$98.6 = 74 + (85 - 74)e^{-\left(-\frac{1}{1.5}\ln\left(\frac{8}{11}\right)\right)t}$$
$$24.6 = 11e^{\frac{1}{1.5}\ln\left(\frac{8}{11}\right)t}$$
$$-3.8 \approx \frac{\ln\left(\frac{24.6}{11}\right)}{\frac{1}{1.5}\ln\left(\frac{8}{11}\right)} = t$$

So, the victim died approximately 3.8 hours before 7am. So, by 8:30am, the victim has been dead for about 5.29 hours.

31. Use $N = N_0e^{-rt}$ **(1)**.

We have

$$\left.\begin{aligned} N(0) &= N_0 = 38{,}000 \\ N(1) &= 32{,}000 \end{aligned}\right\} \textbf{(2)}$$

In order to determine $N(4)$, we need to first determine *r*. To this end, substitute **(2)** into **(1)** to obtain:

$$32{,}000 = 38{,}000e^{-r} \quad \Rightarrow \quad r = -\ln\left(\tfrac{32{,}000}{38{,}000}\right)$$

Thus, $N(4) = 38{,}000e^{-\left(-\ln\left(\frac{32{,}000}{38{,}000}\right)\right)\cdot 4} \approx 19{,}100$.

The book value after 4 years is approximately \$19,100.

33. Use $N = \dfrac{100{,}000}{1 + 10e^{-2t}}$.

a. $N(2) = \dfrac{100{,}000}{1 + 10e^{-2(2)}} \approx 84{,}520$

b. $N(30) = \dfrac{100{,}000}{1 + 10e^{-2(30)}} \approx 100{,}000$

c. The highest number of new convertibles that will be sold is 100,000 since the smallest that $1 + 10e^{-2t}$ can be is 1.

35. Use $N = N_0 e^{-rt}$ **(1).**
Assuming that $t = 0$ corresponds to 1997, we have

$$\begin{aligned} N(0) &= N_0 = 2,422 \\ N(6) &= 7,684 \end{aligned} \quad \textbf{(2)}$$

In order to determine $N(16)$, we need to first determine r. To this end, substitute **(2)** into **(1)** to obtain:

$$7,684 = 2,422e^{-r(6)} \Rightarrow r = -\tfrac{1}{6}\ln\left(\tfrac{7,684}{2,422}\right)$$

Thus,

$$N(16) = 2,422e^{-\left(-\frac{1}{6}\ln\left(\frac{7,684}{2,422}\right)\right)\cdot 16} \approx 11,439,406 \text{ cases.}$$

37. Find t such that $\dfrac{10,000}{1+19e^{-1.56t}} = 5000$.

$$10,000 = 5000\left(1+19e^{-1.56t}\right)$$

$$2 = 1+19e^{-1.56t}$$

$$1 = 19e^{-1.56t}$$

$$e^{-1.56t} = \tfrac{1}{19}$$

$$t = -\tfrac{1}{1.56}\ln\left(\tfrac{1}{19}\right) \approx 1.89 \text{ years}$$

39. Consider $I(r) = e^{-r^2}$, whose graph is below. Note that the beam is brightest when $r = 0$.

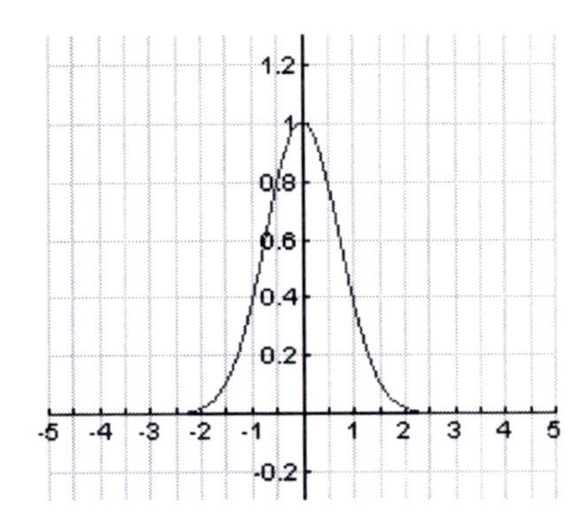

41. Consider $N(x) = 10e^{-\frac{(x-75)^2}{25^2}}$.

a. The graph of N is as follows:

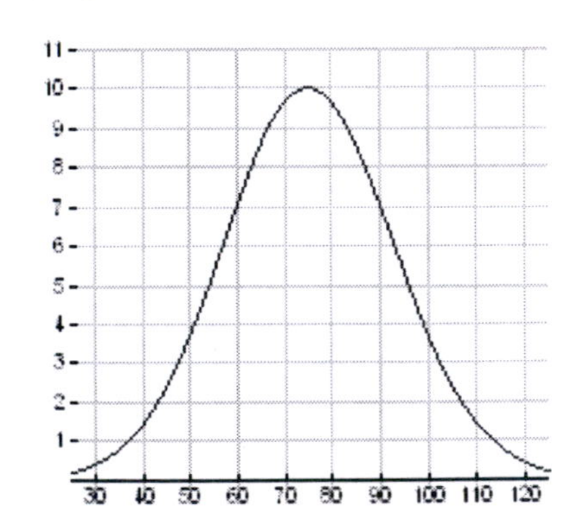

b. Average grade is 75.

c. $N(50) = 10e^{-\frac{(50-75)^2}{25^2}} = 10e^{-1} \approx 4$

d. $N(100) = 10e^{-\frac{(100-75)^2}{25^2}} = 10e^{-1} \approx 4$

43. Use $t=-\dfrac{\ln\left(1-\frac{Pr}{nR}\right)}{n\ln\left(1+\frac{r}{n}\right)}$.

a. Here, $P=80,000,\ r=0.09,\ n=12, R=750$. So, $t=-\dfrac{\ln\left(1-\frac{80,000(0.09)}{12(750)}\right)}{12\ln\left(1+\frac{0.09}{12}\right)}\approx 18$ years.

b. Here, $P=80,000,\ r=0.09,\ n=12, R=1000$. So, $t=-\dfrac{\ln\left(1-\frac{80,000(0.09)}{12(1000)}\right)}{12\ln\left(1+\frac{0.09}{12}\right)}\approx 10$ years.

45. Use $R=P\left(\dfrac{i}{1-(1+i)^{-nt}}\right)$ with

R = 1,467, P= 200,000, $i=\frac{0.08}{12}\approx 0.006667$, and t is measured in years.

a. Solve for t:

$$1,467=200,000\left(\frac{0.006667}{1-(1+0.006667)^{-12t}}\right)$$

$$1,467-1,467(1.006667)^{-12t}=(0.006667)(200,000)$$

$$133.6=1,467(0.92336109)^{t}$$

$$0.0910702=(0.92336109)^{t}$$

$$t=\frac{\ln(0.0910702)}{\ln(0.92336109)}\approx 30 \text{ years}$$

b. Total amount paid = \$1,467(30)(12) = \$528,120. Since the original loan is \$200,000, you've paid \$328,120 in interest.

47. r = 0.07, not 7

49. True

51. False (Since there is a finite number of students at the school to which the lice can spread.)

53. Take a look at a couple of graphs for increasing values of c. For definiteness, let a = k = 1, and take c = 1 and c = 5, respectively. The graphs are:

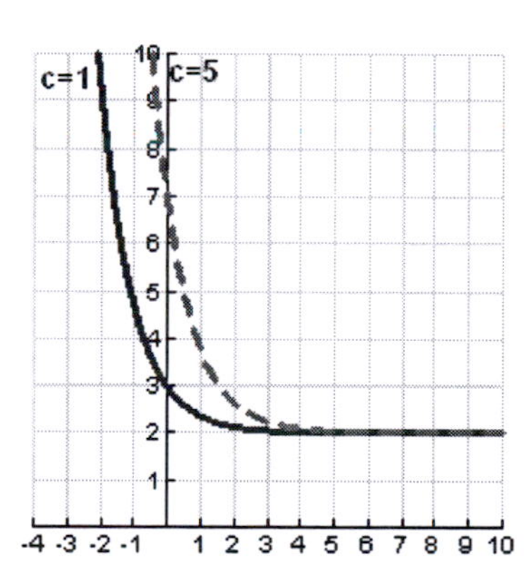

As c increases, the model reaches the carrying capacity in less time.

55. a. The graphs are below:
For the same periodic payment, it will take Wing Shan fewer years to pay off the loan if she can afford to pay biweekly.

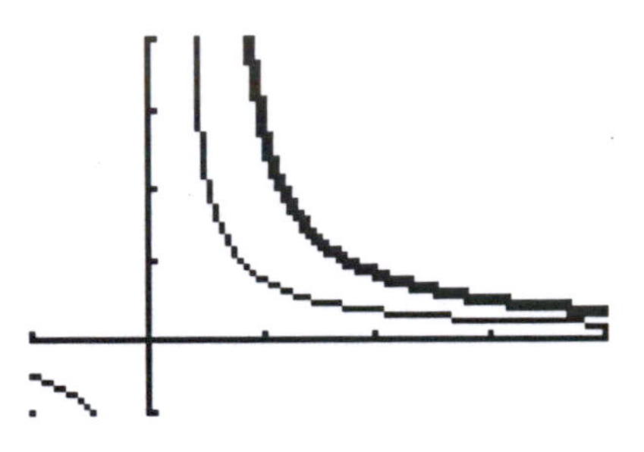

b. 11.58 years
c. 10.33 years
d. 8.54 years, 7.69 years, respectively.

Chapter 5 Review Solutions --

1. 17,559.94	**3.** 5.52
5. 24.53	**7.** 5.89
9. $2^{4-(-2.2)} = 2^{6.2} \cong 73.52$	**11.** $\left(\frac{2}{5}\right)^{1-6\left(\frac{1}{2}\right)} = \left(\frac{2}{5}\right)^{-2} = 6.25$
13. b y-intercept $\left(0,\frac{1}{4}\right)$	**15. c** y-intercept $(0,11)$
17. <u>y-intercept</u>: $(0,-1)$ <u>HA</u>: $y=0$ Reflect the graph of $\left(\frac{1}{6}\right)^x$ over the x-axis. 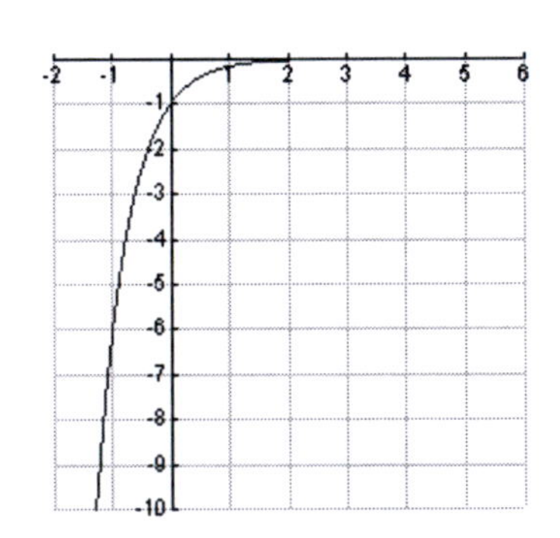	**19.** <u>y-intercept</u>: $(0,2)$. <u>HA</u>: $y=1$ Shift the graph of $\left(\frac{1}{100}\right)^x$ up 1 unit. 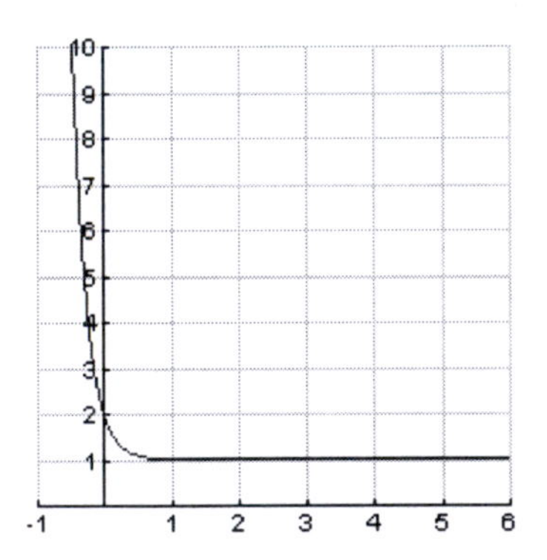

21. y-intercept: $(0,1)$.
HA: $y=0$

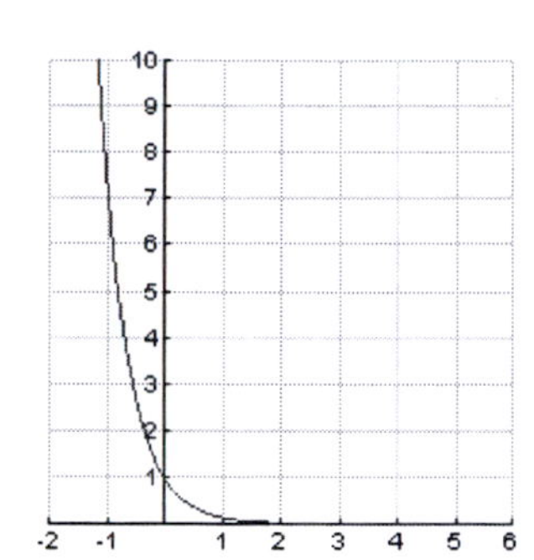

23. y-intercept: $(0,3.2)$.
HA: $y=0$

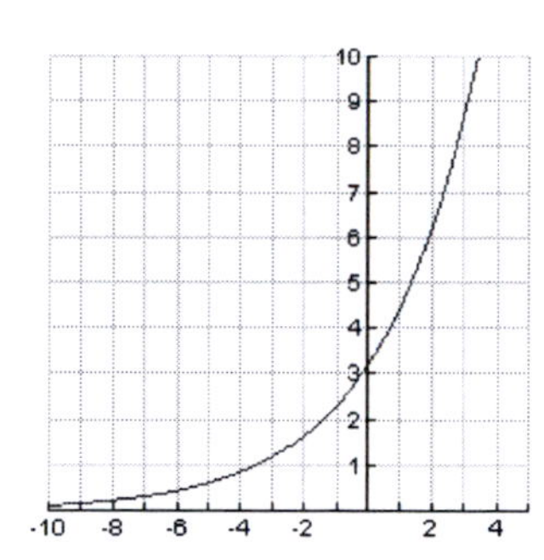

25. Use $A(t)=P\left(1+\frac{r}{n}\right)^{nt}$
Here, $P=4500,\ r=0.045,\ n=2,\ t=7$.
Thus, $A(7)=4500\left(1+\frac{0.045}{2}\right)^{2(7)} \cong 6144.68$.
So, the amount in the account after 7 years is \$6144.68.

27. Use $A(t)=Pe^{rt}$.
Here, $P=13,450,\ r=0.036,\ t=15$.
Thus,
$A(15)=13,450e^{(0.036)(15)} \cong 23,080.29$.
So, the amount in the account after 15 years is \$23,080.29.

29. $4^3=64$

31. $10^{-2}=\frac{1}{100}$

33. $\log_6 216=3$

35. $\log_{\frac{2}{13}}\left(\frac{4}{169}\right)=2$

37.
$$\log_7 1=x$$
$$7^x=1$$
$$x=0$$

39.
$$\log_{\frac{1}{6}} 1296=x$$
$$\left(\tfrac{1}{6}\right)^x=1296=6^4$$
$$x=-4$$

41. 1.51

43. -2.08

45. Must have $x+2>0$, so that the domain is $(-2,\infty)$.

47. Since $x^2+3>0$, for all values of x, the domain is $(-\infty,\infty)$.

49. b

51. d Shift the graph of $\log_7 x$ left 1 unit, then down 3 units. Also, VA is $x=-1$.

53. Shift the graph of $\log_4 x$ right 4 units, then up 2 units.

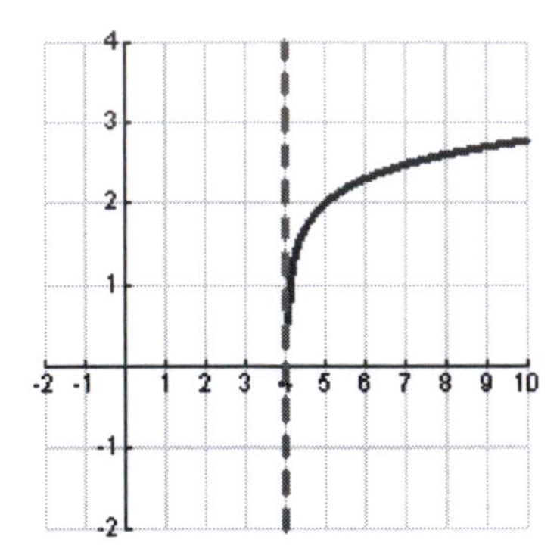

55. Reflect the graph of $\log_4 x$ over the x-axis, then shift down 6 units.

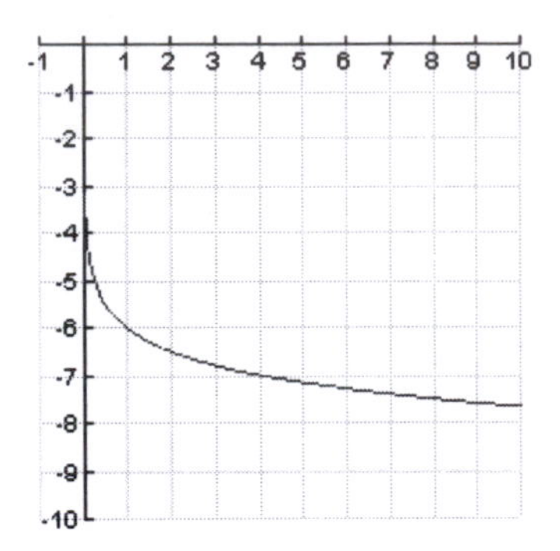

57. Use $pH = -\log_{10}\left[H^+\right]$.

Here,

$$\begin{aligned} pH &= -\log_{10}(3.16\times10^{-7}) \\ &= -\left[\log(3.16)+\log(10^{-7})\right] \\ &= -\left[\log(3.16)-7\right] \\ &\cong 6.5 \end{aligned}$$

59. Use $D = 10\log\left(\dfrac{I}{1\times10^{-12}}\right)$.

Here,

$$\begin{aligned} D &= 10\log\left(\frac{1\times10^{-7}}{1\times10^{-12}}\right) = 10\log(10^5) \\ &= 50\log(10) = 50\,dB \end{aligned}$$

61. 1

63. 6

65.

$$\begin{aligned} \log_c\left(x^a y^b\right) &= \log_c\left(x^a\right)+\log_c\left(y^b\right) \\ &= a\log_c(x)+b\log_c(y) \end{aligned}$$

67.

$$\begin{aligned} \log_j\left(\frac{rs}{t^3}\right) &= \log_j(rs)-\log_j\left(t^3\right) \\ &= \log_j(r)+\log_j(s)-3\log_j(t) \end{aligned}$$

69.

$$\begin{aligned} \log\left(\frac{a^{\frac{1}{2}}}{b^{\frac{3}{2}}c^{\frac{2}{5}}}\right) &= \log\left(a^{\frac{1}{2}}\right)-\log\left(b^{\frac{3}{2}}c^{\frac{2}{5}}\right) \\ &= \log\left(a^{\frac{1}{2}}\right)-\left[\log\left(b^{\frac{3}{2}}\right)+\log\left(c^{\frac{2}{5}}\right)\right] \\ &= \tfrac{1}{2}\log(a)-\tfrac{3}{2}\log(b)-\tfrac{2}{5}\log(c) \end{aligned}$$

71.

$$\begin{aligned} \log_8 3 &= \frac{\log 3}{\log 8} \\ &\cong 0.5283 \end{aligned}$$

73.

$$\begin{aligned} \log_\pi 1.4 &= \frac{\log 1.4}{\log \pi} \\ &\cong 0.2939 \end{aligned}$$

75.

$$\begin{aligned} 4^x &= \frac{1}{256} \\ 4^x &= 4^{-4} \\ x &= -4 \end{aligned}$$

77.

$$\begin{aligned} e^{3x-4} &= 1 = e^0 \\ 3x-4 &= 0 \\ x &= \tfrac{4}{3} \end{aligned}$$

79.

$$\left(\tfrac{1}{3}\right)^{x+2} = 81$$
$$\left(3^{-1}\right)^{x+2} = 3^4$$
$$3^{-x-2} = 3^4$$
$$-x-2 = 4$$
$$x = -6$$

81.

$$e^{2x+3} - 3 = 10$$
$$e^{2x+3} = 13$$
$$2x+3 = \ln 13$$
$$x = \tfrac{-3+\ln 13}{2} \approx -0.218$$

83. Note that $e^{2x} + 6e^x + 5 = 0$ is equivalent to $\left(e^x\right)^2 + 6\left(e^x\right) + 5 = \left(e^x + 5\right)\left(e^x + 1\right) = 0$. Neither $\underbrace{e^x + 5 = 0}_{\text{No solution}}$ or $\underbrace{e^x + 1 = 0}_{\text{No solution}}$ has a real solution. So, the original equation has no solution.

85.

$$\left(2^x - 2^{-x}\right)\left(2^x + 2^{-x}\right) = 0$$
$$\left(2^x\right)^2 - \left(2^{-x}\right)^2 = 0$$
$$2^{2x} - 2^{-2x} = 0$$
$$2^{2x} = 2^{-2x}$$
$$2x = -2x$$
$$x = 0$$

87.

$$\log(3x) = 2$$
$$10^2 = 3x$$
$$100 = 3x$$
$$\tfrac{100}{3} = x$$

89.

$$\log_4(x) + \log_4 2x = 8$$
$$\log_4\left(2x^2\right) = 8$$
$$2x^2 = 4^8$$
$$x^2 = \tfrac{1}{2}\left(4^8\right)$$
$$x = \pm\sqrt{\tfrac{1}{2}}\left(4^4\right)$$
$$x = 128\sqrt{2},\ \cancel{-128\sqrt{2}}$$

91.

$$\ln x^2 = 2.2$$
$$x^2 = e^{2.2}$$
$$x = \pm\sqrt{e^{2.2}} \approx \pm 3.004$$

93.

$$\log_3(2-x) - \log_3(x+3) = \log_3(x)$$
$$\log_3\left(\tfrac{2-x}{x+3}\right) = \log_3(x)$$
$$\tfrac{2-x}{x+3} = x$$
$$2 - x = x(x+3)$$
$$x^2 + 4x - 2 = 0$$
$$x = \tfrac{-4\pm\sqrt{16+8}}{2}$$
$$x = -2+\sqrt{6},\ \cancel{-2-\sqrt{6}}$$
$$\approx 0.449$$

95. Use $A(t) = Pe^{rt}$.

Here,

$$A = 30{,}000,\ r = 0.05,\ t = 1$$

Substituting into the above equation, we can solve for P:

$$30{,}000 = Pe^{0.05(1)}$$
$$28{,}536.88 \cong \tfrac{30{,}000}{e^{0.05}} = P$$

So, the initial investment in 1 year CD should be approximately $28,536.88.

97. Use $A(t) = P\left(1+\frac{r}{n}\right)^{nt}$

Here, we know that $r = 0.042,\ n = 4,\ A = 2P$. We can substitute these in to find t.

$$2P = P\left(1+\tfrac{0.042}{4}\right)^{4(t)}$$
$$2 = (1.0105)^{4t}$$
$$\log_{1.0105} 2 = 4t$$
$$16.6 \cong \tfrac{1}{4}\log_{1.0105} 2 = t$$

So, it takes approximately 16.6 years until the initial investment doubles.

99. Use $N = N_0 e^{rt}$. Here, $N_0 = 2.62,\ r = 0.035$. Determine N when $t = 10$:

$$N = 2.62e^{0.035(10)} \approx 3.72 \text{ million}$$

The population in 2014 is about 3.72 million.

101. Use $N = N_0 e^{rt}$ **(1)**.

We have

$$\begin{aligned} N(0) &= N_0 = 1000 \\ N(3) &= 2500 \end{aligned} \quad \textbf{(2)}$$

In order to determine $N(6)$, we need to first determine r. To this end, substitute **(2)** into **(1)** to obtain $2500 = 1000e^{r(3)} \Rightarrow r = \frac{1}{3}\ln\left(\frac{2500}{1000}\right)$.

Thus, $N(6) = 1000e^{\frac{1}{3}\ln\left(\frac{2500}{1000}\right)\cdot 6} \approx 6250$ bacteria.

103. Use $N = N_0 e^{-rt}$.

We know that $N(28) = \frac{1}{2}N_0$. Assuming that $N_0 = 20$, determine t such that $N = 5$. To do so, we first find r:

$$\tfrac{1}{2}N_0 = N_0 e^{-r(28)} \Rightarrow r = -\tfrac{1}{28}\ln\left(\tfrac{1}{2}\right)$$

Now, solve:

$$5 = 20e^{-\left(-\frac{1}{28}\ln\left(\frac{1}{2}\right)\right)t}$$
$$\tfrac{1}{4} = e^{\frac{1}{28}\ln\left(\frac{1}{2}\right)t}$$
$$t = \frac{\ln\left(\frac{1}{4}\right)}{\frac{1}{28}\ln\left(\frac{1}{2}\right)} \approx 56 \text{ years}$$

105. Use $N = N_0 e^{-rt}$.

Assuming that 2003 occurs at $t = 0$, we know that

$$N_0 = 5600,\ N(1) = 2420.$$

Find $N(7)$. To do so, we first find r:

$$2420 = 5600e^{-r(1)} \Rightarrow r = -\ln\left(\tfrac{2420}{5600}\right)$$

Now, solve:

$$N(7) = 5600e^{-\left(-\ln\left(\frac{2420}{5600}\right)\right)(7)} \approx 16 \text{ fish}$$

107. Use $M = 1000\left(1-e^{-0.035t}\right)$. Since $t = 0$ corresponds to 1998, we have $M(12) = 1000\left(1-e^{-0.035(12)}\right) \approx 343$ mice.

109. The graph is as follows:

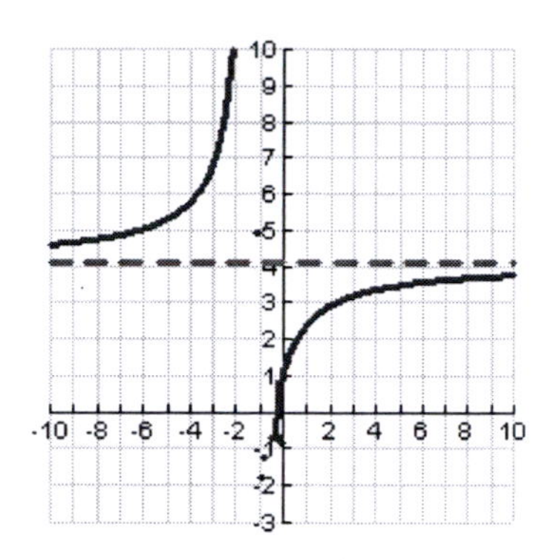

Using the calculator to compute the functional values for large values of x suggests that the HA is about $y = 4.11$. (Note: The exact equation of the HA is $y = e^{\sqrt{2}}$.)

111. Let
$y1 = \log_{2.4}(3x-1),\ y2 = \log_{0.8}(x-1)+3.5$.
The graphs are as follows:

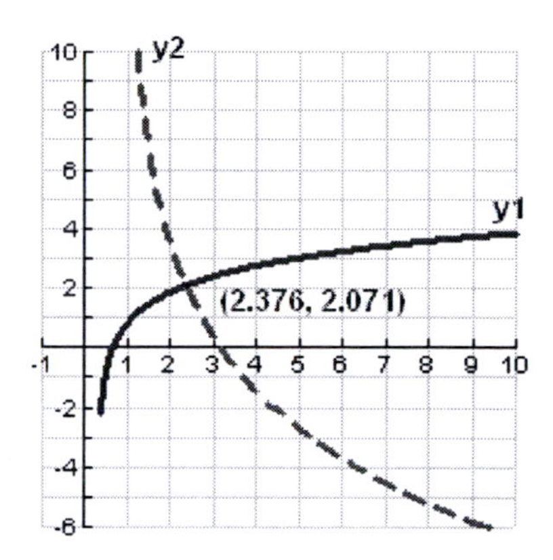

The coordinates of the point of intersection are about (2.376, 2.071).

113. The graphs agree on $(0,\infty)$, as seen below.

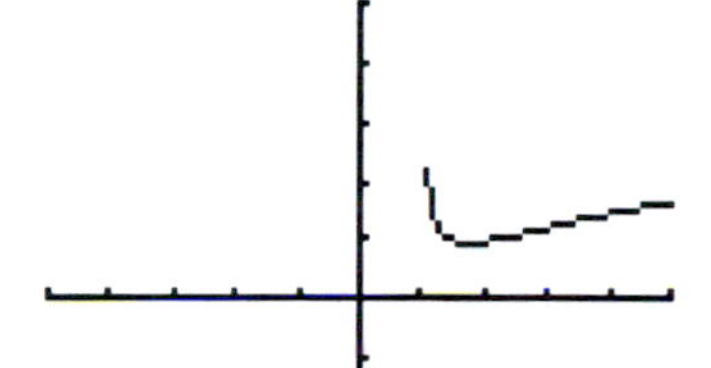

115. The graph is below.
Domain: $(-\infty,\infty)$.
Symmetric about the origin.
Horizontal asymptotes:
$y=-1$ (as $x \to -\infty$), $y = 1$ (as $x \to \infty$)

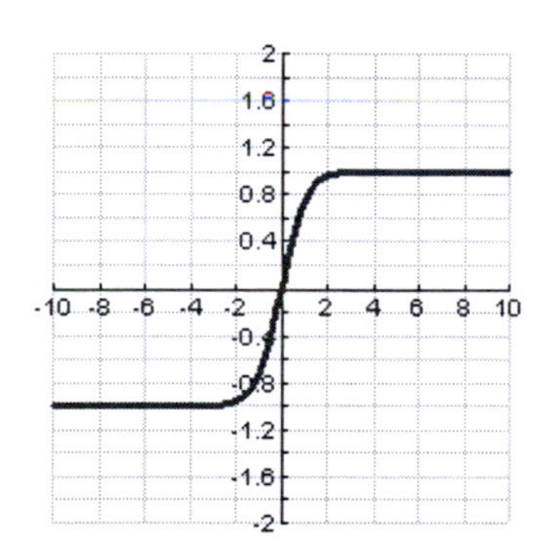

117. a. Using $N = N_0 e^{rt}$ with (0,4) and (18,2), we need to find r:

$$2 = 4e^{18r} \Rightarrow r = \tfrac{1}{18}\ln\left(\tfrac{1}{2}\right) \approx -0.038508$$

So, the equation of dosage is given by

$$N = 4e^{-0.038508t} \approx 4(0.9622)^t.$$

b. $N = 4(0.9622)^t$

c. Yes, they are the same.

Chapter 5 Practice Test

1. $\log 10^{x^3} = x^3 \log 10 = x^3$

3.

$$\log_{1/3} 81 = x$$
$$3^4 = \left(\tfrac{1}{3}\right)^x = 3^{-x}$$
$$-4 = x$$

5.

$$e^{x^2-1} = 42$$
$$x^2 - 1 = \ln 42$$
$$x^2 = 1 + \ln 42$$
$$x = \pm\sqrt{1+\ln 42} \approx \pm 2.177$$

7.

$$27e^{0.2x+1} = 300$$
$$e^{0.2x+1} = \tfrac{300}{27}$$
$$0.2x + 1 = \ln\left(\tfrac{300}{27}\right)$$
$$0.2x = -1 + \ln\left(\tfrac{300}{27}\right)$$
$$x = \frac{-1+\ln\left(\frac{300}{27}\right)}{0.2} \cong 7.04$$

9.

$$3\ln(x-4) = 6$$
$$\ln(x-4) = 2$$
$$x - 4 = e^2$$
$$x = 4 + e^2 \approx 11.389$$

11.

$$\ln(\ln x) = 1$$
$$\ln x = e$$
$$x = e^e \approx 15.154$$

13.

$$\log_6 x + \log_6(x-5) = 2$$
$$\log_6\left(x(x-5)\right) = 2$$
$$x^2 - 5x = 36$$
$$x^2 - 5x - 36 = 0$$
$$(x-9)(x+4) = 0$$
$$x = \cancel{-4}, 9$$

15.

$$\ln x + \ln(x+3) = 1$$
$$\ln\left(x(x+3)\right) = 1$$
$$x^2 + 3x = e$$
$$x^2 + 3x - e = 0$$
$$x = \frac{-3 \pm \sqrt{9 - 4(-e)}}{2}$$
$$\approx 0.729, \cancel{-3.729}$$

17.

$$\frac{12}{1+2e^x} = 6$$
$$12 = 6 + 12e^x$$
$$6 = 12e^x$$
$$x = \ln\left(\tfrac{1}{2}\right) \approx -0.693$$

19. Must have $\frac{x}{x^2-1} > 0$ and $x^2 - 1 \neq 0$

CPs $-1, 0, 1$

$-$ $+$ $-$ $+$

-1 0 1

So, the domain is $(-1,0)\cup(1,\infty)$.

21. y-intercept: $f(0)=3^{-0}+1=2$. So, (0,2).
x-intercept: None
Domain: $(-\infty,\infty)$ Range: $(1,\infty)$
Horizontal Asymptote: $y=1$
The graph is as follows:

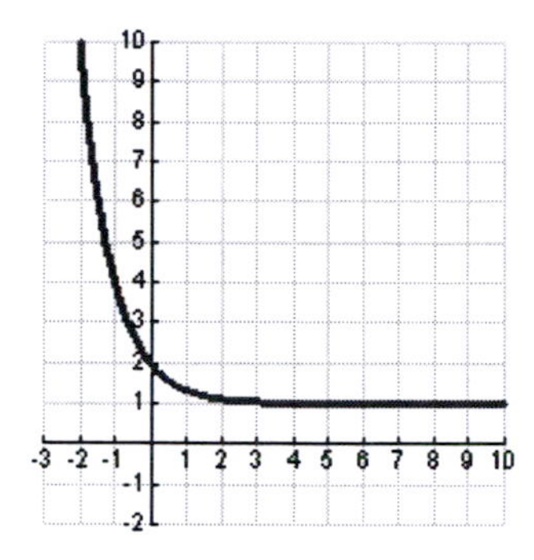

23. y-intercept: None
x-intercept: Must solve $\ln(2x-3)+1=0$.

$$\ln(2x-3)=-1 \Rightarrow 2x-3=\tfrac{1}{e} \Rightarrow x=\frac{3+\frac{1}{e}}{2}$$

So, $\left(\frac{3+\frac{1}{e}}{2},0\right)$.

Domain: $(\tfrac{3}{2},\infty)$ Range: $(-\infty,\infty)$
Vertical Asymptote: $x=\tfrac{3}{2}$
The graph is as follows:

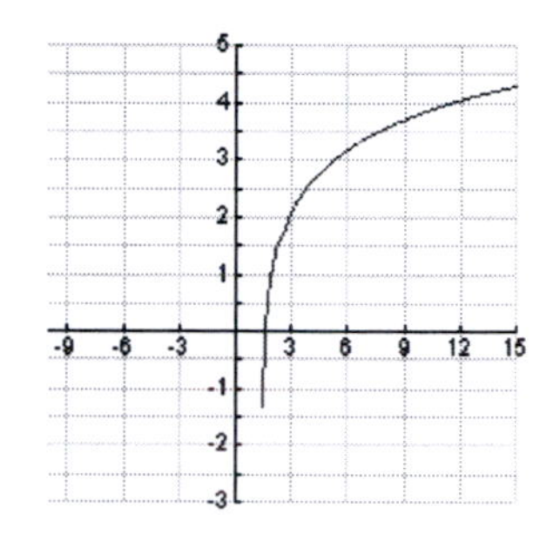

25. Use $A(t)=P\left(1+\frac{r}{n}\right)^{nt}$. Here, $P=5000$, $r=0.06$, $n=4$, $t=8$. Thus, $A(8)=5000\left(1+\frac{0.06}{4}\right)^{4(8)}\cong 8051.62$. So, the amount in the account after 8 years is \$8051.62.

27. Use $D=10\log\left(\frac{I}{1\times10^{-12}}\right)$. Here, $D=10\log\left(\frac{1\times10^{-3}}{1\times10^{-12}}\right)=10\log(10^{9})=90\log(10)=90\,dB$.

29. Use $M=\frac{2}{3}\log\left(\frac{E}{10^{4.4}}\right)$. Here,

For $M=5$:

$$5=\tfrac{2}{3}\log\left(\tfrac{E}{10^{4.4}}\right)$$
$$7.5=\log\left(\tfrac{E}{10^{4.4}}\right)$$
$$10^{7.5}=\tfrac{E}{10^{4.4}}$$
$$\underbrace{10^{7.5}\times10^{4.4}}_{=10^{11.9}}=E$$

So, the energy here is $10^{11.9}\approx7.9\times10^{11}$ joules.

For $M=6$:

$$6=\tfrac{2}{3}\log\left(\tfrac{E}{10^{4.4}}\right)$$
$$9=\log\left(\tfrac{E}{10^{4.4}}\right)$$
$$10^{9}=\tfrac{E}{10^{4.4}}$$
$$\underbrace{10^{9}\times10^{4.4}}_{=10^{13.4}}=E$$

So, the energy here is $10^{13.4}\approx2.5\times10^{13}$ joules.

Thus, the range of energy is $7.9\times10^{11}<E<2.5\times10^{13}$ joules.

31. Use $N = N_0 e^{rt}$ **(1)**.
We have

$$\left.\begin{aligned} N(0) &= N_0 = 200 \\ N(2) &= 500 \end{aligned}\right\} \textbf{(2)}$$

In order to determine $N(8)$, we need to first determine r. To this end, substitute **(2)** into **(1)** to obtain:

$$500 = 200e^{r(2)} \Rightarrow r = \tfrac{1}{2}\ln\left(\tfrac{500}{200}\right)$$

Thus, $N(8) = 200e^{\frac{1}{2}\ln\left(\frac{500}{200}\right)\cdot 8} \approx 7800$ bacteria.

33. Solve $1000 = \dfrac{2000}{1+3e^{-0.4t}}$.

$$1000(1+3e^{-0.4t}) = 2000$$
$$1+3e^{-0.4t} = 2$$
$$3e^{-0.4t} = 1$$
$$e^{-0.4t} = \tfrac{1}{3}$$
$$-0.4t = \ln\left(\tfrac{1}{3}\right)$$
$$t = -\tfrac{1}{0.4}\ln\left(\tfrac{1}{3}\right) \approx 3 \text{ days}$$

35. The graph is below. Domain: $(-\infty,\infty)$.
Symmetric about the origin. No asymptotes

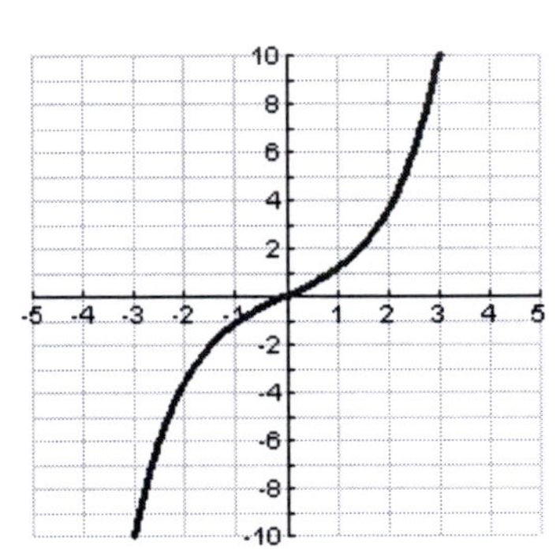

Chapter 5 Cumulative Review ---

1. $\dfrac{x^{1/3}y^{2/5}}{x^{-1/2}y^{-8/5}} = x^{1/3}x^{1/2}y^{2/5}y^{8/5} = x^{5/6}y^2$

3.

$$5x^2 - 4x - 3 = 0$$
$$x = \frac{4 \pm \sqrt{16-4(5)(-3)}}{2(5)}$$
$$= \frac{2 \pm \sqrt{19}}{5}$$

5.

$$\frac{x+4}{5} \le -3$$
$$x+4 \le -15$$
$$x \le -19$$
$$(-\infty,-19]$$

7. The line $4x+3y=6$ is equivalent to $y = -\frac{4}{3}x+2$ and so, has slope $-\frac{4}{3}$. A line perpendicular to it must have slope $\frac{3}{4}$. Since the desired line must pass through (7,6), we have

$$y-6 = \tfrac{3}{4}(x-7) \Rightarrow y = \tfrac{3}{4}x + \tfrac{3}{4}$$

9. **a.** 1 **b.** 5 **c.** 1 **d.** undefined **e.** Domain: $(-2,\infty)$ Range: $(0,\infty)$
f. Increasing on $(4,\infty)$ Decreasing on (0,4) Constant on (-2,0)

11. Yes, f is one-to-one since

$$f(x)=f(y)\Rightarrow\sqrt{x-4}=\sqrt{y-4}$$
$$\Rightarrow x-4=y-4$$
$$\Rightarrow x=y$$

13.

$$\begin{aligned}f(x)&=-4x^2+8x-5\\&=-4\left(x^2-2x\right)-5\\&=-4\left(x^2-2x+1\right)-5+4\\&=-4(x-1)^2-1\end{aligned}$$

So, the vertex is (1,-1).

15. Rewrite as: $\left(-x^4-4x^3+3x^2+7x-20\right)\div(x-(-4))$

Synthetic division then gives

$$\begin{array}{r|rrrrr} -4 & -1 & -4 & 3 & 7 & -20\\ & & 4 & 0 & -12 & 20\\ \hline & -1 & 0 & 3 & -5 & 0\end{array}$$

So, $Q(x)=-x^3+3x-5,\ r(x)=0$.

17. Vertical asymptote: $x=3$ Horizontal asymptote: None
Slant asymptote: $y=x+3$ since

$$\begin{array}{r} x+3\\ x-3\overline{\big)\ x^2+0x+7}\\ -\left(x^2+0x-9\right)\\ \hline 16\end{array}$$

19. $\left(\frac{1}{25}\right)^{-3/2}=25^{3/2}=5^3=125$

21. $\log_3 243=\log_3\left(3^5\right)=5$

23.

$$10^{2\log(4x+9)}=10^{\log(4x+9)^2}=(4x+9)^2=121$$
$$4x+9=\pm 11$$
$$4x=-9\pm 11$$
$$x=\tfrac{1}{4}(-9\pm 11)=\cancel{-5},\ 0.5$$

25. Use $A=Pe^{rt}$.
We know that $P=8500$ and $r=0.04$.
Determine t such that $A=12,000$.

$$12,000=8500e^{0.04\,t}$$
$$t=\tfrac{1}{0.04}\ln\left(\tfrac{12,000}{8,500}\right)\approx 8.62 \text{ years}$$

27. **a.** Using $N=N_0e^{rt}$ with (0,6) and (28,3), we need to find r:

$$3=6e^{28r}\ \Rightarrow\ r=\tfrac{1}{28}\ln\left(\tfrac{1}{2}\right)\approx -0.247553$$

So, the equation of dosage is given by

$$N=6e^{-0.247553t}\approx 6(0.9755486421)^t.$$

b. $N(32)\approx 2.72$ grams

CHAPTER 6

Section 6.1 Solutions

1.

Solve the system: $\begin{cases} x - y = 1 & \textbf{(1)} \\ x + y = 1 & \textbf{(2)} \end{cases}$

Solve **(1)** for x: $x = 1 + y$ **(3)**

Substitute **(3)** into **(2)** and solve for y:

$$(1+y)+y=1$$
$$1+2y=1$$
$$y=0$$

Substitute $y = 0$ into **(1)** to find x: $x = 1$.

So, the solution is $\boxed{(1,0)}$.

3.

Solve the system: $\begin{cases} x + y = 7 & \textbf{(1)} \\ x - y = 9 & \textbf{(2)} \end{cases}$

Solve **(1)** for y: $y = 7 - x$ **(3)**

Substitute **(3)** into **(2)** and solve for x:

$$x-(7-x)=9$$
$$x-7+x=9$$
$$x=8$$

Substitute $x = 8$ into **(1)** to find y: $y = -1$.

So, the solution is $\boxed{(8,-1)}$.

5.

Solve the system: $\begin{cases} 2x - y = 3 & \textbf{(1)} \\ x - 3y = 4 & \textbf{(2)} \end{cases}$

Solve **(2)** for x: $x = 3y + 4$ **(3)**

Substitute **(3)** into **(1)** and solve for y:

$$2(3y+4)-y=3$$
$$5y=-5$$
$$y=-1$$

Substitute $y = -1$ into **(1)** to find x:

$$2x-(-1)=3$$
$$2x=2.$$
$$x=1$$

So, the solution is $\boxed{(1,-1)}$.

7.

Solve the system: $\begin{cases} 3x + y = 5 & \textbf{(1)} \\ 2x - 5y = -8 & \textbf{(2)} \end{cases}$

Solve **(1)** for y: $y = 5 - 3x$ **(3)**

Substitute **(3)** into **(2)** and solve for x:

$$2x-5(5-3x)=-8$$
$$2x-25+15x=-8$$
$$17x=17$$
$$x=1$$

Substitute $x = 1$ into **(3)** to find y: $y = 2$

So, the solution is $\boxed{(1,2)}$.

9.

Solve the system: $\begin{cases} 2u+5v=7 & \textbf{(1)} \\ 3u-v=5 & \textbf{(2)} \end{cases}$

Solve **(2)** for v: $v=3u-5$ **(3)**
Substitute **(3)** into **(1)** and solve for u:

$$2u+5(3u-5)=7$$
$$17u=32$$
$$u=\tfrac{32}{17}$$

Substitute $u=\frac{32}{17}$ into **(2)** to find v:

$$3(\tfrac{32}{17})-v=5 \Rightarrow v=\tfrac{96-85}{17}=\tfrac{11}{17}.$$

So, the solution is $\boxed{u=\frac{32}{17},\ v=\frac{11}{17}}$.

11.

Solve the system: $\begin{cases} 2x+y=7 & \textbf{(1)} \\ -2x-y=5 & \textbf{(2)} \end{cases}$

Solve **(1)** for y: $y=7-2x$ **(3)**
Substitute **(3)** into **(2)** and solve for x:

$$-2x-(7-2x)=5$$
$$-7=5$$

So, the system is inconsistent. Thus, there is $\boxed{\text{no solution}}$.

13.

Solve the system: $\begin{cases} 4r-s=1 & \textbf{(1)} \\ 8r-2s=2 & \textbf{(2)} \end{cases}$

Solve **(1)** for s: $s=4r-1$ **(3)**
Substitute **(3)** into **(2)** and solve for r:

$$8r-2(4r-1)=2$$
$$2=2$$

So, the system is consistent. There are $\boxed{\text{infinitely many solutions}}$ of the form $(r,\ 4r-1)$, where r is any real number.

15.

Solve the system: $\begin{cases} 5r-3s=15 & \textbf{(1)} \\ -10r+6s=-30 & \textbf{(2)} \end{cases}$

Solve **(1)** for r: $r=\frac{1}{5}(3s+15)$ **(3)**
Substitute **(3)** into **(2)** and solve for s:

$$-10\left[\tfrac{1}{5}(3s+15)\right]+6s=-30$$
$$-6s-30+6s=-30$$
$$-30=-30$$

So, the system is consistent. There are $\boxed{\text{infinitely many solutions}}$ of the form $(r,\frac{5r-15}{3})$, where r is any real number.

17. Solve the system: $\begin{cases} 2x-3y=-7 & \textbf{(1)} \\ 3x+7y=24 & \textbf{(2)} \end{cases}$

Solve **(1)** for x: $x=\frac{1}{2}(3y-7)$ **(3)**
Substitute **(3)** into **(2)** and solve for y:

$$3\cdot\tfrac{1}{2}(3y-7)+7y=24$$
$$\tfrac{9}{2}y-\tfrac{21}{2}+7y=24$$
$$\tfrac{23}{2}y=\tfrac{69}{2}$$
$$\boxed{y=3}$$

Substitute this value back into **(3)** to obtain that $\boxed{x=1}$.

19. Solve the system: $\begin{cases} \frac{1}{3}x-\frac{1}{4}y=0 & \textbf{(1)} \\ -\frac{2}{3}x+\frac{3}{4}y=2 & \textbf{(2)} \end{cases}$

First, clear the fractions by multiplying both equations by 12 to obtain the equivalent system:

$$\begin{cases} 4x-3y=0 & \textbf{(3)} \\ -8x+9y=24 & \textbf{(4)} \end{cases}$$

Solve **(3)** for x: $x=\frac{3}{4}y$ **(5)**
Substitute **(5)** into **(4)** and solve for y:

$$-8(\tfrac{3}{4}y)+9y=24$$
$$3y=24 \Rightarrow \boxed{y=8}$$

Substitute this back into **(5)** to obtain that $\boxed{x=6}$.

21. Solve the system:

$$\begin{cases} 7.2x - 4.1y = 7 & \textbf{(1)} \\ -3.5x + 16.5y = 2.4 & \textbf{(2)} \end{cases}$$

Solve **(1)** for x: $x = \frac{1}{7.2}(7 + 4.1y)$ **(3)**

Substitute **(3)** into **(2)** and solve for y:

$$-3.5\left(\tfrac{1}{7.2}(7 + 4.1y)\right) + 16.5y = 2.4$$

$$\boxed{y = 0.4}$$

Substitute this value back into **(3)** to obtain that $\boxed{x = 1.2}$.

23.

Solve the system: $\begin{cases} x - y = 2 & \textbf{(1)} \\ x + y = 4 & \textbf{(2)} \end{cases}$

Add **(1)** and **(2)** to eliminate y:

$$2x = 6$$

$$x = 3 \quad \textbf{(3)}$$

Substitute **(3)** into **(1)** and solve for y:

$$3 - y = 2$$

$$y = 1$$

So, the solution is $\boxed{(3,1)}$.

25.

Solve the system: $\begin{cases} x - y = -3 & \textbf{(1)} \\ x + y = 7 & \textbf{(2)} \end{cases}$

Add **(1)** and **(2)** to eliminate y:

$$2x = 4$$

$$x = 2 \quad \textbf{(3)}$$

Substitute **(3)** into **(2)** and solve for y:

$$2 + y = 7$$

$$y = 5$$

So, the solution is $\boxed{(2,5)}$.

27.

Solve the system: $\begin{cases} 5x + 3y = -3 & \textbf{(1)} \\ 3x - 3y = -21 & \textbf{(2)} \end{cases}$

Add **(1)** and **(2)** to eliminate y:

$$8x = -24$$

$$x = -3 \quad \textbf{(3)}$$

Substitute **(3)** into **(1)** and solve for y:

$$5(-3) + 3y = -3$$

$$3y = 12$$

$$y = 4$$

So, the solution is $\boxed{(-3,4)}$.

29.

Solve the system: $\begin{cases} 2x - 7y = 4 & \textbf{(1)} \\ 5x + 7y = 3 & \textbf{(2)} \end{cases}$

Add **(1)** and **(2)** to eliminate y:

$$7x = 7$$

$$x = 1 \quad \textbf{(3)}$$

Substitute **(3)** into **(2)** and solve for y:

$$5(1) + 7y = 3$$

$$7y = -2$$

$$y = -\tfrac{2}{7}$$

So, the solution is $\boxed{\left(1, -\tfrac{2}{7}\right)}$.

31.

Solve the system: $\begin{cases} 2x + 5y = 7 & \textbf{(1)} \\ 3x - 10y = 5 & \textbf{(2)} \end{cases}$

Multiply **(1)** by 2: $4x + 10y = 14$ **(3)**

Add **(2)** and **(3)** to eliminate y:

$$7x = 19$$

$$x = \tfrac{19}{7} \quad \textbf{(4)}$$

Substitute **(4)** into **(1)** and solve for y:

$$2\left(\tfrac{19}{7}\right) + 5y = 7$$

$$y = \tfrac{11}{35}$$

So, the solution is $\boxed{\left(\tfrac{19}{7}, \tfrac{11}{35}\right)}$.

33. Solve the system: $\begin{cases} 2x + 5y = 5 & \textbf{(1)} \\ -4x - 10y = -10 & \textbf{(2)} \end{cases}$

Multiply **(1)** by 2: $4x + 10y = 10$ **(3)**

Add **(2)** and **(3)** to eliminate y: $0 = 0$ **(4)**

So, the system is consistent. Thus, there are $\boxed{\text{infinitely many solutions}}$ of the form $(x, \frac{5-2x}{5})$, where x is any real number.

35.

Solve the system: $\begin{cases} 3x - 2y = 12 & \textbf{(1)} \\ 4x + 3y = 16 & \textbf{(2)} \end{cases}$

Multiply **(1)** by 4: $12x - 8y = 48$ **(3)**

Multiply **(2)** by -3: $-12x - 9y = -48$ **(4)**

Add **(3)** and **(4)** to eliminate x:

$$-17y = 0$$

$$y = 0 \quad \textbf{(5)}$$

Substitute **(5)** into **(1)** and solve for x:

$$3x - 2(0) = 12$$

$$x = 4$$

So, the solution is $\boxed{(4,0)}$.

37.

Solve the system: $\begin{cases} 6x - 3y = -15 & \textbf{(1)} \\ 7x + 2y = -12 & \textbf{(2)} \end{cases}$

Multiply **(1)** by 2: $12x - 6y = -30$ **(3)**

Multiply **(2)** by 3: $21x + 6y = -36$ **(4)**

Add **(3)** and **(4)** to eliminate y:

$$33x = -66 \Rightarrow x = -2 \quad \textbf{(5)}$$

Substitute **(5)** into **(1)** and solve for y:

$$6(-2) - 3y = -15$$

$$-3y = -3$$

$$y = 1$$

So, the solution is $\boxed{(-2,1)}$.

39.

Solve the system:

$$\begin{cases} 0.02x + 0.05y = 1.25 & \textbf{(1)} \\ -0.06x - 0.15y = -3.75 & \textbf{(2)} \end{cases}$$

Multiply **(1)** by 3:

$$0.06x + 0.15y = 3.75 \quad \textbf{(3)}$$

Add **(2)** and **(3)** to eliminate y:

$$0 = 0 \quad \textbf{(4)}$$

So, the system is consistent. Thus, there are $\boxed{\text{infinitely many solutions}}$ of the form $(x, \frac{125-2x}{5})$, where x is any real number.

41.

Solve the system: $\begin{cases} \frac{1}{3}x + \frac{1}{2}y = 1 & \textbf{(1)} \\ \frac{1}{5}x + \frac{7}{2}y = 2 & \textbf{(2)} \end{cases}$

Multiply **(1)** by -7: $-\frac{7}{3}x - \frac{7}{2}y = -7$ **(3)**

Add **(2)** and **(3)** to eliminate y:

$$-\tfrac{32}{15}x = -5$$

$$x = \tfrac{75}{32} \quad \textbf{(4)}$$

Substitute **(4)** into **(1)** and solve for y:

$$\tfrac{1}{3}(\tfrac{75}{32}) + \tfrac{1}{2}y = 1$$

$$\tfrac{1}{2}y = \tfrac{21}{96}$$

$$y = \tfrac{7}{16}$$

So, the solution is $\boxed{\left(\frac{75}{32}, \frac{7}{16}\right)}$.

43. c Adding the equations yields $6x = 6$, so that $x = 1$. Thus, $3(1) - y = 1$, so that $y = 2$.

45. d Subtracting the equations yields $0 = -4$. Thus, the system is inconsistent and so, the lines should be parallel.

47.

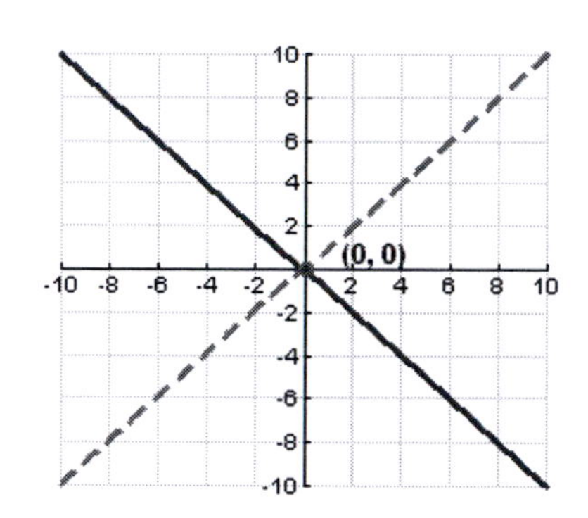

Notes on the graph:
Solid curve is $y = -x$
Dashed curve is $y = x$
So, the solution is (0,0).

49.

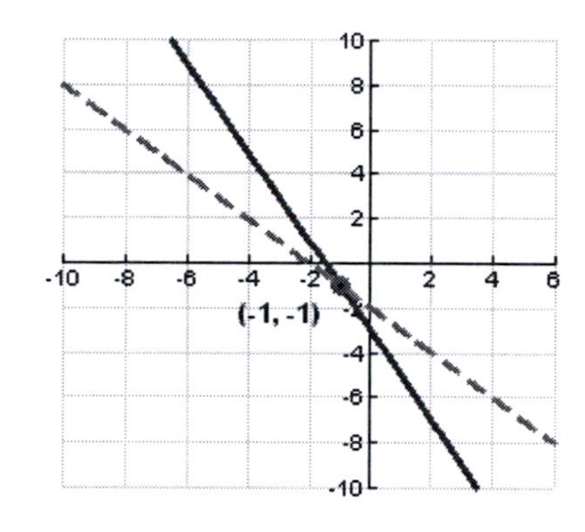

Notes on the graph:
Solid curve is $2x + y = -3$
Dashed curve is $x + y = -2$
So, the solution is $(-1,-1)$.

51.

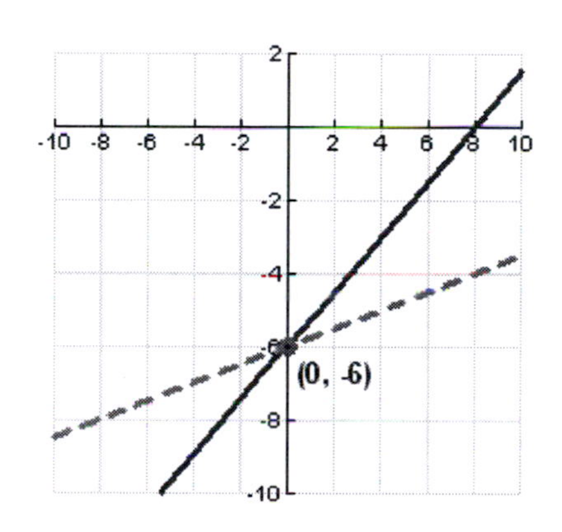

Notes on the graph:
Solid curve is $\frac{1}{2}x - \frac{2}{3}y = 4$
Dashed curve is $\frac{1}{4}x - y = 6$
So, the solution is $(0,-6)$.

53.

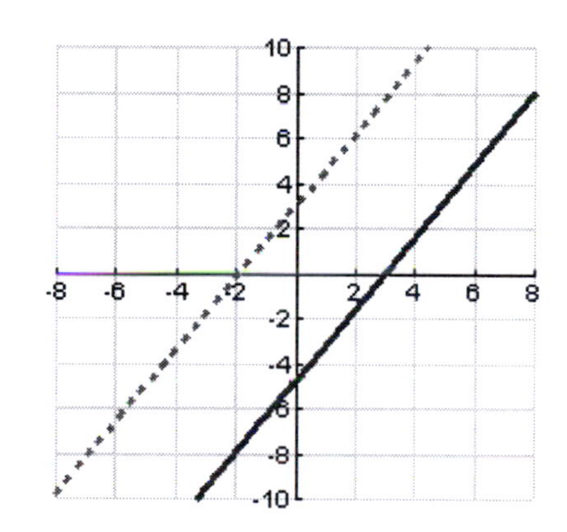

Notes on the graph:
Solid curve is $1.6x - y = 4.8$
Dashed curve is $-0.8x + 0.5y = 1.5$
So, there is no solution.

55. Let x = number of Auspens in a kit,
y = number of refill bottles in a kit.
Must solve the system:

$$\begin{cases} x + 40y = 246 & \textbf{(1)} \\ x = y & \textbf{(2)} \end{cases}$$

Substituting **(2)** into **(1)** yields

$$41x = 246, \text{ so that } x = 6.$$

So, there are 6 Auspens per kit and 6 refill bottles per kit.

57. Let x = # of Montblanc pens and y = # of Cross pens

Must solve the system: $\begin{cases} x+y=69 & \textbf{(1)} \\ 72x+10y=1000 & \textbf{(2)} \end{cases}$

Solve **(1)** for y: $y=69-x$ **(3)**

Substitute **(3)** into **(2)**: $72x+10(69-x)=1000 \Rightarrow x=5$

Substitute this value of x into **(3)**: $y=64$

So, have 5 Montblanc pens and 64 Cross pens.

59. Let x = number of ml of 8% HCl.
y = number of ml of 15% HCl.

Must solve the system:

$$\begin{cases} 0.08x+0.15y=(0.12)(37) & \textbf{(1)} \\ x+y=37 & \textbf{(2)} \end{cases}$$

Multiply **(2)** by -0.08:

$$-0.08x-0.08y=-2.96 \quad \textbf{(3)}$$

Add **(1)** and **(3)** to eliminate x:

$$0.07y=1.48 \Rightarrow y \cong 21.14 \quad \textbf{(4)}$$

Substitute **(4)** into **(2)** and solve for x:

$$x \cong 15.86$$

So, should use approximately 15.86 ml of 8% HCl and 21.14 ml of 15% HCl.

61. Let x = total annual sales.

Salary at Autocount: $15,000+0.10x$

Salary at Polk: $30,000+0.05x$

We need to find x such that

$$15,000+0.10x>30,000+0.05x.$$

Solving this yields:

$$0.05x>15,000 \Rightarrow x>300,000$$

So, he needs to make at least \$300,000 of sales in order to make more money at Autocount.

63. Let x= gallons used for highway miles and y = gallons used for city miles

Must solve the system:

$$\begin{cases} 26x+19y=349.5 & \textbf{(1)} \\ x+y=16 & \textbf{(2)} \end{cases}$$

Multiply **(2)** by -19: $-19x-19y=-304$ **(3)**

Add **(1)** and **(3)** to eliminate y: $7x=45.5 \Rightarrow x=6.5$ **(4)**

Substitute **(4)** into **(2)** and solve for y: $y=9.5$

Thus, there are 169 highway miles and 180.5 city miles.

65. Let x = speed of the plane and y = wind speed.

	Rate (mph)	**Time (hours)**	**Distance**
Atlanta to Paris	$x+y$	8	4000
Paris to Atlanta	$x-y$	10	4000

So, using Distance = Rate × Time, we see that we must solve the system:

$$\begin{cases} 8(x+y)=4000 \\ 10(x-y)=4000 \end{cases}, \text{ which is equivalent to } \begin{cases} x+y=500 & \textbf{(1)} \\ x-y=400 & \textbf{(2)} \end{cases}$$

Adding **(1)** and **(2)** yields: $2x=900$, so that $x=450$. Substituting this into **(1)** then yields $y=50$. So, the average ground speed of the plane is 450 mph, and the average wind speed is 50 mph.

67. Let x = amount invested in 10% stock
y = amount invested in 14% stock
We must solve the system:

$$\begin{cases} 0.10x+0.14y=1260 & \textbf{(1)} \\ x+y=10{,}000 & \textbf{(2)} \end{cases}$$

Multiply **(2)** by -0.10:

$$-0.10x-0.10y=-1000 \quad \textbf{(3)}$$

Add **(1)** and **(3)** to eliminate x:

$$0.04y=260$$

$$y=6500 \quad \textbf{(4)}$$

Substitute **(4)** into **(2)** and solve for y:

$$x=3500$$

So, should invest \$3500 in the 10% stock, and \$6500 in the 14% stock.

69. Let x = # CD players sold
Cost equation: $y=15x+120$ **(1)**
Revenue equation: $y=30x$ **(2)**
We want the intersection of these two equations to find the break even point. To this end, substitute **(2)** into **(1)** to find that $x=8$. So, must sell 8 CD players to break even.

71. Every term in the first equation is not multiplied by -1 correctly. The equation should be $-2x-y=3$, and the resulting solution should be $x=11,\ y=-25$.

73. Did not distribute -1 correctly. In Step 3, the calculation should be $-(-3y-4)=3y+4$ and the resulting answer should be (2, -2).

75. False. If the lines are coincident, then there are infinitely many solutions.

77. False. The lines could be coincident.

79. Substitute the pair $(2,-3)$ into both equations to get the system:

$$\begin{cases} 2A-3B=-29 & \textbf{(1)} \\ 2A+3B=13 & \textbf{(2)} \end{cases}$$

Add **(1)** and **(2)** to eliminate B:

$$4A=-16$$

$$A=-4 \quad \textbf{(3)}$$

Substitute **(3)** into **(1)** and solve for B:

$$2(-4)-3B=-29$$

$$-3B=-21$$

$$B=7$$

Hence, the solution is $A=-4,\ B=7$.

81. Let $x=$ # cups of pineapple juice for 2% drink, and $y=$ # cups of pomegranate juice for 2% drink.

Must solve the system:

$$\begin{cases} \frac{y}{x+y}=0.02 & \textbf{(1)} \\ \frac{4-y}{(100-x)+(4-y)}=0.04 & \textbf{(2)} \end{cases}$$

This system, after simplification, is equivalent to:

$$\begin{cases} 0.02x-0.98y=0 & \textbf{(3)} \\ 0.04x-0.96y=0.16 & \textbf{(4)} \end{cases}$$

Substitute **(3)** into **(4)**: $y=\frac{0.02}{0.98}x$ **(5)**

Substitute **(5)** into **(4)** and simplify:

$$x=7.84$$

Substitute this value of x back into **(5)**:

$$y=0.16$$

So, she can make 8 cups of the 2% drink and 96 cups of the 4% drink.

83.

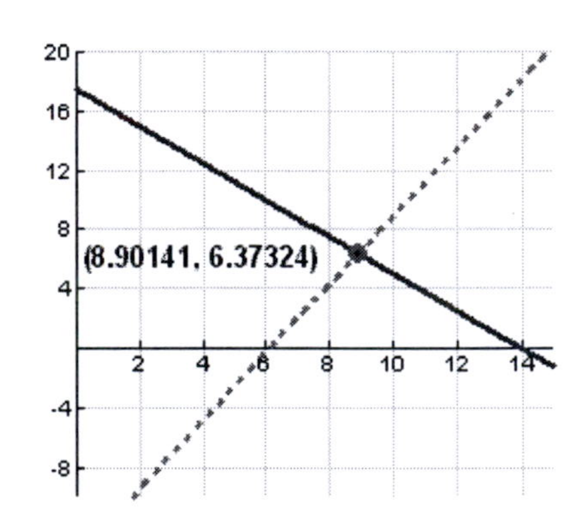

Notes on the graph:

Solid curve is $y=-1.25x+17.5$

Dashed curve is $y=2.3x-14.1$

The approximate solution is (8.9, 6.4).

85.

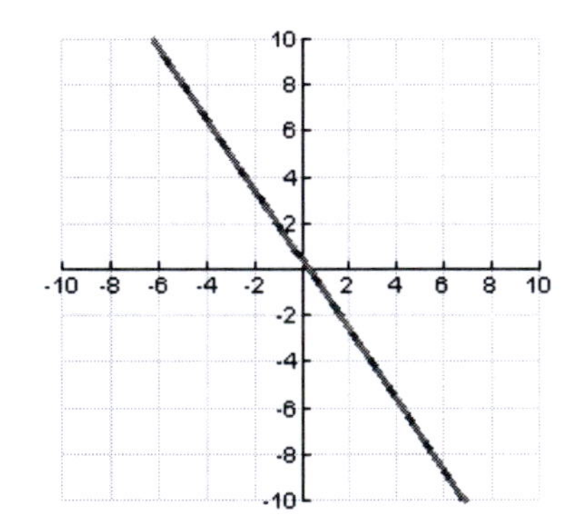

Notes on the graph:

Solid curve is $23x+15y=7$

Dashed curve is $46x+30y=14$

Note that the graphs are the same line, so there are infinitely many solutions to this system.

87. The graph of the system of equations is as follows:

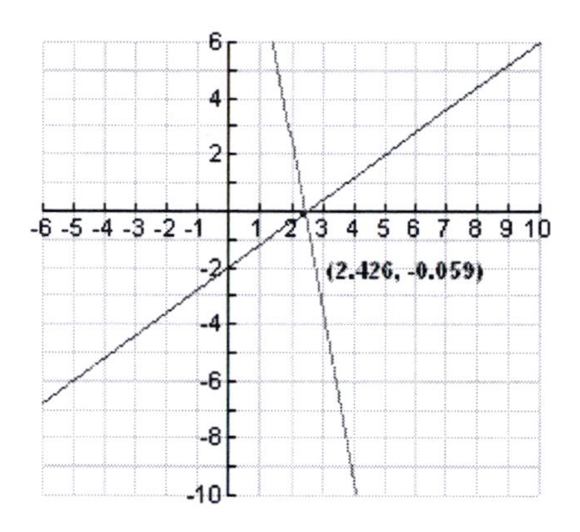

The solution is approximately (2.426, -0.059).

Section 6.2 Solutions ---

1. Solve the system:

$$\begin{cases} x - y + z = 6 & \textbf{(1)} \\ -x + y + z = 3 & \textbf{(2)} \\ -x - y - z = 0 & \textbf{(3)} \end{cases}$$

Step 1: Obtain a system of 2 equations in 2 unknowns by eliminating the same variable in two pairs of equations.

Add **(1)** and **(2)**:

$$2z = 9 \Rightarrow \boxed{z = \tfrac{9}{2}} \quad \textbf{(4)}$$

Add **(2)** and **(3)**:

$$-2x = 3 \Rightarrow \boxed{x = -\tfrac{3}{2}} \quad \textbf{(5)}$$

Step 2: Solve for the remaining variable using **(4)** and **(5)**.

Substitute **(4)** and **(5)** into **(1)**:

$$-\tfrac{3}{2} - y + \tfrac{9}{2} = 6 \Rightarrow \boxed{y = -3}$$

3. Solve the system:

$$\begin{cases} x + y - z = 2 & \textbf{(1)} \\ -x - y - z = -3 & \textbf{(2)} \\ -x + y - z = 6 & \textbf{(3)} \end{cases}$$

Step 1: Obtain a system of 2 equations in 2 unknowns by eliminating the same variable in two pairs of equations.

Add **(1)** and **(2)**: $-2z = -1 \Rightarrow \boxed{z = \tfrac{1}{2}}$ **(4)**

Add **(2)** and **(3)**, and sub. in **(4)**:

$$-2x - 2z = 3 \Rightarrow -2x - 2\left(\tfrac{1}{2}\right) = 3$$

$$\Rightarrow -2x = 4 \Rightarrow \boxed{x = -2} \quad \textbf{(5)}$$

Step 2: Solve for the remaining variable using **(4)** and **(5)**.

Substitute **(4)** and **(5)** into **(1)**:

$$-2 + y - \tfrac{1}{2} = 2 \Rightarrow \boxed{y = \tfrac{9}{2}}$$

5. Solve the system:

$$\begin{cases} -x + y - z = -1 & \textbf{(1)} \\ x - y - z = 3 & \textbf{(2)} \\ x + y - z = 9 & \textbf{(3)} \end{cases}$$

Step 1: Obtain a system of 2 equations in 2 unknowns by eliminating the same variable in two pairs of equations.

Add **(1)** and **(2)**: $-2z = 2 \Rightarrow \boxed{z = -1}$ **(4)**

Add **(2)** and **(3)**, and sub. in **(4)**:

$$2x - 2z = 12 \Rightarrow 2x - 2(-1) = 12$$

$$\Rightarrow 2x = 10 \Rightarrow \boxed{x = 5} \quad \textbf{(5)}$$

Step 2: Solve for the remaining variable using **(4)** and **(5)**.

Substitute **(4)** and **(5)** into **(1)**:

$$-5 + y - (-1) = -1 \Rightarrow \boxed{y = 3}$$

7. Solve the system:

$$\begin{cases} 2x - 3y + 4z = -3 & \textbf{(1)} \\ -x + y + 2z = 1 & \textbf{(2)} \\ 5x - 2y - 3z = 7 & \textbf{(3)} \end{cases}$$

Step 1: Obtain a system of 2 equations in 2 unknowns by eliminating the same variable in two pairs of equations.

Multiply **(2)** by 3:

$$-3x + 3y + 6z = 3 \quad \textbf{(4)}$$

Add **(4)** and **(1)** to eliminate y:

$$-x + 10z = 0 \quad \textbf{(5)}$$

Next, multiply **(2)** by 2:

$$-2x + 2y + 4z = 2 \quad \textbf{(6)}$$

Add **(6)** and **(3)** to eliminate y:

$$3x + z = 9 \quad \textbf{(7)}$$

These steps yield the following system:

$$(*) \begin{cases} -x + 10z = 0 & \textbf{(5)} \\ 3x + z = 9 & \textbf{(7)} \end{cases}$$

Step 2: Solve system **(*)** from Step 1.

Multiply **(5)** by 3:

$$-3x + 30z = 0 \quad \textbf{(8)}$$

Add **(8)** and **(7)** to eliminate x and solve for z:

$$31z = 9 \text{ so that } z = \tfrac{9}{31} \quad \textbf{(9)}$$

Substitute **(9)** into **(7)** to find x:

$$-x + 10(\tfrac{9}{31}) = 0$$

$$x = \tfrac{90}{31} \quad \textbf{(10)}$$

Step 3: Use the solution of the system in Step 2 to find the value of the third variable in the original system.

Substitute **(9)** and **(10)** into **(2)** to find y:

$$-(\tfrac{90}{31}) + y + 2(\tfrac{9}{31}) = 1$$

$$y - \tfrac{72}{31} = 1 \text{ so that } y = \tfrac{103}{31}$$

Thus, the solution is:

$$\boxed{x = \tfrac{90}{31},\ y = \tfrac{103}{31},\ z = \tfrac{9}{31}}.$$

9. Solve the system:

$$\begin{cases} -4x + 3y + 5z = 2 & \textbf{(1)} \\ 2x - 3y - 2z = -3 & \textbf{(2)} \\ -2x + 4y + 3z = 1 & \textbf{(3)} \end{cases}$$

Step 1: Obtain a system of 2 equations in 2 unknowns by eliminating the same variable in two pairs of equations.

Add **(2)** and **(3)** to eliminate x:

$$y + z = -2 \quad \textbf{(4)}$$

Next, multiply **(2)** by 2:

$$4x - 6y - 4z = -6 \quad \textbf{(5)}$$

Add **(5)** and **(1)** to eliminate x:

$$-3y + z = -4 \quad \textbf{(6)}$$

These steps yield the following system:

$$(*) \begin{cases} y + z = -2 & \textbf{(4)} \\ -3y + z = -4 & \textbf{(6)} \end{cases}$$

Step 2: Solve system **(*)** from Step 1.

Multiply **(6)** by -1: $3y - z = 4$ **(7)**

Add **(4)** and **(7)** to eliminate z and solve for y:

$$4y = 2 \Rightarrow y = \tfrac{1}{2} \quad \textbf{(8)}$$

Substitute **(8)** into **(4)** to find z:

$$\tfrac{1}{2} + z = -2$$

$$z = -\tfrac{5}{2} \quad \textbf{(9)}$$

Step 3: Use the solution of the system in Step 2 to find the value of the third variable in the original system.

Substitute **(8)** and **(9)** into **(2)** to find y:

$$2x - 3(\tfrac{1}{2}) - 2(-\tfrac{5}{2}) = -3$$

$$2x = -\tfrac{13}{2}$$

$$x = -\tfrac{13}{4} \quad \textbf{(10)}$$

Thus, the solution is:

$$\boxed{x = -\tfrac{13}{4},\ y = \tfrac{1}{2},\ z = -\tfrac{5}{2}}.$$

11. Solve the system:

$$\begin{cases} x - y + z = -1 & \textbf{(1)} \\ y - z = -1 & \textbf{(2)} \\ -x + y + z = 1 & \textbf{(3)} \end{cases}$$

Add **(1)** and **(3)** to eliminate both x and y:

$$2z = 0$$

$$z = 0 \quad \textbf{(4)}$$

Substitute **(4)** into **(2)** to find y:

$$y - 0 = -1 \quad \textbf{(5)}$$

Substitute **(4)** and **(5)** into **(1)** to find x:

$$x - (-1) + 0 = -1$$

$$x = -2 \quad \textbf{(6)}$$

Thus, the solution is:

$$\boxed{x = -2,\ y = -1,\ z = 0}.$$

13. Solve the system:

$$\begin{cases} 3x - 2y - 3z = -1 & \textbf{(1)} \\ x - y + z = -4 & \textbf{(2)} \\ 2x + 3y + 5z = 14 & \textbf{(3)} \end{cases}$$

Step 1: Obtain a system of 2 equations in 2 unknowns by eliminating the same variable in two pairs of equations.

Multiply **(2)** by 3:

$$3x - 3y + 3z = -12 \quad \textbf{(4)}$$

Add **(4)** and **(1)** to eliminate z:

$$6x - 5y = -13 \quad \textbf{(5)}$$

Next, multiply **(2)** by -5:

$$-5x + 5y - 5z = 20 \quad \textbf{(6)}$$

Add **(6)** and **(3)** to eliminate z:

$$-3x + 8y = 34 \quad \textbf{(7)}$$

These steps yield the following system:

$$(*) \begin{cases} 6x - 5y = -13 & \textbf{(5)} \\ -3x + 8y = 34 & \textbf{(7)} \end{cases}$$

Step 2: Solve system (*) from Step 1.

Multiply **(7)** by 2:

$$-6x + 16y = 68 \quad \textbf{(8)}$$

Add **(5)** and **(8)** to eliminate x and solve for y:

$$11y = 55$$

$$y = 5 \quad \textbf{(9)}$$

Substitute **(9)** into **(5)** to find x:

$$6x - 5(5) = -13$$

$$6x = 12$$

$$x = 2 \quad \textbf{(10)}$$

Step 3: Use the solution of the system in Step 2 to find the value of the third variable in the original system.

Substitute **(9)** and **(10)** into **(2)** to find z:

$$2 - 5 + z = -4$$

$$z = -1 \quad \textbf{(11)}$$

Thus, the solution is:

$$\boxed{x = 2,\ y = 5,\ z = -1}.$$

15. Solve the system:

$$\begin{cases} -3x - y - z = 2 & \textbf{(1)} \\ x + 2y - 3z = 4 & \textbf{(2)} \\ 2x - y + 4z = 6 & \textbf{(3)} \end{cases}$$

Step 1: Obtain a system of 2 equations in 2 unknowns by eliminating the same variable in two pairs of equations.

Multiply **(2)** by 3:

$$3x + 6y - 9z = 12 \quad \textbf{(4)}$$

Add **(1)** and **(4)** to eliminate x:

$$5y - 10z = 14 \quad \textbf{(5)}$$

Next, multiply **(2)** by -2:

$$-2x - 4y + 6z = -8 \quad \textbf{(6)}$$

Add **(3)** and **(6)** to eliminate x:

$$-5y + 10z = -2 \quad \textbf{(7)}$$

These steps yield the following system:

$$(*) \begin{cases} 5y - 10z = 14 & \textbf{(5)} \\ -5y + 10z = -2 & \textbf{(7)} \end{cases}$$

Step 2: Solve system (*) from Step 1.

Adding **(5)** and **(7)** yields the false statement 0 = 12. Hence, this system has $\boxed{\text{no solution}}$.

17. Solve the system:

$$\begin{cases} 3x + 2y + z = 4 & \textbf{(1)} \\ -4x - 3y - z = -15 & \textbf{(2)} \\ x - 2y + 3z = 12 & \textbf{(3)} \end{cases}$$

Step 1: Obtain a system of 2 equations in 2 unknowns by eliminating the same variable in two pairs of equations.

Multiply **(1)** by -3:

$$-9x - 6y - 3z = -12 \quad \textbf{(4)}$$

Add **(3)** and **(4)** to eliminate z:

$$-8x - 8y = 0 \;\Rightarrow\; x + y = 0 \; \textbf{(5)}$$

Next, multiply **(2)** by 3:

$$-12x - 9y - 3z = -45 \quad \textbf{(6)}$$

Add **(3)** and **(6)** to eliminate z:

$$-11x - 11y = -33 \;\Rightarrow\; x + y = 3 \quad \textbf{(7)}$$

These steps yield the following system:

$$(*) \begin{cases} x + y = 0 & \textbf{(5)} \\ x + y = 3 & \textbf{(7)} \end{cases}$$

Step 2: Solve system **(∗)** from Step 1. Subtracting **(5)** and **(7)** yields the false statement 0 = 3. Hence, this system has $\boxed{\text{no solution}}$.

19. Solve the system:

$$\begin{cases} -x + 2y + z = -2 & \textbf{(1)} \\ 3x - 2y + z = 4 & \textbf{(2)} \\ 2x - 4y - 2z = 4 & \textbf{(3)} \end{cases}$$

Step 1: Obtain a system of 2 equations in 2 unknowns by eliminating the same variable in two pairs of equations.

Multiply **(1)** by 3:

$$-3x + 6y + 3z = -6 \quad \textbf{(4)}$$

Add **(4)** and **(2)** to eliminate x:

$$4y + 4z = -2 \quad \textbf{(5)}$$

Next, multiply **(1)** by 2:

$$-2x + 4y + 2z = -4 \quad \textbf{(6)}$$

Add **(6)** and **(3)** to eliminate x:

$$0 = 0 \quad \textbf{(7)}$$

Hence, we know that the system has infinitely many solutions.

Let $z = a$. Then, substituting this value into **(5)**, we can find the value of y:

$$2y = -2a - 1$$
$$y = -(a + \tfrac{1}{2})$$

Now, substitute the values of z and y into **(1)** to find x:

$$-x - 2(a + \tfrac{1}{2}) + a = -2$$
$$-x - 2a - 1 + a = -2$$
$$x = 1 - a$$

Thus, the solution is $\boxed{x = 1 - a,\; y = -(a + \tfrac{1}{2}),\; z = a}$, where a is any real number.

21. Solve the system:

$$\begin{cases} x - y - z = 10 & \textbf{(1)} \\ 2x - 3y + z = -11 & \textbf{(2)} \\ -x + y + z = -10 & \textbf{(3)} \end{cases}$$

Step 1: Obtain a system of 2 equations in 2 unknowns by eliminating the same variable in two pairs of equations.

Add **(1)** and **(3)**:

$$0 = 0 \quad \textbf{(4)}$$

Add **(1)** and **(2)**:

$$3x - 4y = -1 \quad \textbf{(5)}$$

Hence, we know that the system has infinitely many solutions.

To determine them, let $x = a$. Substitute this into **(5)** to find y:

$$3a - 4y = -1$$

$$y = \tfrac{1}{4}(3a+1)$$

Now, substitute the values of x and y into **(1)** to find z:

$$a - \tfrac{1}{4}(3a+1) - z = 10$$

$$z = \tfrac{4a-3a-1-40}{4}$$

$$z = \tfrac{a-41}{4}$$

Thus, the solution is:

$$x = a,\ \ y = \tfrac{1}{4}(3a+1),\ \ z = \tfrac{a-41}{4}.$$

Equivalently,

$$\boxed{x = 41 + 4a,\ y = 31 + 3a,\ z = a}.$$

23. Solve the system:

$$\begin{cases} 3x_1 + x_2 - x_3 = 1 & \textbf{(1)} \\ x_1 - x_2 + x_3 = -3 & \textbf{(2)} \\ 2x_1 + x_2 + x_3 = 0 & \textbf{(3)} \end{cases}$$

Step 1: Obtain a system of 2 equations in 2 unknowns by eliminating the same variable in two pairs of equations.

Add **(1)** and **(2)** to eliminate x_2 and x_3:

$$4x_1 = -2$$

$$x_1 = -\tfrac{1}{2} \quad \textbf{(4)}$$

Add **(2)** and **(3)** to eliminate x_2:

$$3x_1 + 2x_3 = -3 \quad \textbf{(5)}$$

These steps yield the following system:

$$(*) \begin{cases} x_1 = -\tfrac{1}{2} & \textbf{(4)} \\ 3x_1 + 2x_3 = -3 & \textbf{(5)} \end{cases}$$

Step 2: Solve system **(*)** from Step 1.

Substitute **(4)** into **(5)**:

$$3(-\tfrac{1}{2}) + 2x_3 = -3$$

$$2x_3 = -\tfrac{3}{2}$$

$$x_3 = -\tfrac{3}{4} \quad \textbf{(6)}$$

Step 3: Use the solution of the system in Step 2 to find the value of the third variable in the original system.

Substitute **(4)** and **(6)** into **(1)** to find x_2:

$$3(-\tfrac{1}{2}) + x_2 - (-\tfrac{3}{4}) = 1$$

$$x_2 = \tfrac{7}{4} \quad \textbf{(7)}$$

Thus, the solution is:

$$\boxed{x_1 = -\tfrac{1}{2},\ x_2 = \tfrac{7}{4},\ x_3 = -\tfrac{3}{4}}.$$

25. Solve the system:

$$\begin{cases} 2x + 5y \qquad = 9 & \textbf{(1)} \\ x + 2y - z = 3 & \textbf{(2)} \\ -3x - 4y + 7z = 1 & \textbf{(3)} \end{cases}$$

Step 1: Obtain a system of 2 equations in 2 unknowns by eliminating the same variable in two pairs of equations.

Multiply **(2)** by 7:

$$7x + 14y - 7z = 21 \quad \textbf{(4)}$$

Add **(4)** and **(3)** to eliminate z:

$$4x + 10y = 22 \quad \textbf{(5)}$$

These steps yield the following system:

$$(*) \begin{cases} 2x + 5y = 9 & \textbf{(1)} \\ 4x + 10y = 22 & \textbf{(5)} \end{cases}$$

Step 2: Solve system $(*)$ from Step 1.

Multiply **(1)** by -2:

$$-4x - 10y = -18 \quad \textbf{(6)}$$

Add **(5)** and **(6)** to eliminate x and solve for y:

$$0 = 4 \quad \textbf{(7)}$$

Hence, we conclude from **(7)** that the system has $\boxed{\text{no solution}}$.

27. Solve the system:

$$\begin{cases} 2x_1 - x_2 + x_3 = 3 & \textbf{(1)} \\ x_1 - x_2 + x_3 = 2 & \textbf{(2)} \\ -2x_1 + 2x_2 - 2x_3 = -4 & \textbf{(3)} \end{cases}$$

Step 1: Obtain a system of 2 equations in 2 unknowns by eliminating the same variable in two pairs of equations.

Multiply **(1)** by -1:

$$-2x_1 + x_2 - x_3 = -3 \quad \textbf{(4)}$$

Add **(2)** and **(4)** to eliminate x_2 and x_3:

$$-x_1 = -1$$

$$x_1 = 1 \quad \textbf{(5)}$$

Substitute **(5)** into **(2)**:

$$1 - x_2 + x_3 = 2$$

$$-x_2 + x_3 = 1 \quad \textbf{(6)}$$

Substitute **(5)** into **(3)**:

$$-2 + 2x_2 - 2x_3 = -4$$

$$x_2 - x_3 = -1 \quad \textbf{(7)}$$

These steps yield the following system:

$$(*) \begin{cases} -x_2 + x_3 = 1 & \textbf{(6)} \\ x_2 - x_3 = -1 & \textbf{(7)} \end{cases}$$

Step 2: Solve system $(*)$ from Step 1.

Add **(6)** and **(7)**: $0 = 0$ **(8)**

Hence, we conclude from **(8)** that there are infinitely many solutions. To determine them, let $x_3 = a$. Substitute this value into **(7)** to see that $x_2 = a - 1$.

Thus, the solution is:

$$\boxed{x_1 = 1,\ x_2 = -1 + a,\ x_3 = a}.$$

29. Solve the system:

$$\begin{cases} 2x+y-z=2 & \textbf{(1)} \\ x-y-z=6 & \textbf{(2)} \end{cases}$$

Since there are two equations and three unknowns, we know there are infinitely many solutions (as long as neither statement is inconsistent). To this end, let $z=a$.

Add **(1)** and **(2)**: $3x-2z=8$ **(3)**

Substitute $z=a$ into **(3)** to find x: $x=\frac{2}{3}a+\frac{8}{3}$ **(4)**

Finally, substitute $z=a$ and **(4)** into **(2)** to find y: $y=-\frac{1}{3}a-\frac{10}{3}$

Thus, the solution is

$$\boxed{x=\tfrac{2}{3}a+\tfrac{8}{3},\ y=-\tfrac{1}{3}a-\tfrac{10}{3},\ z=a}.$$

31. Let x = number of basic widgets ($10 each)
y = number of mid-price widgets ($12 each)
z = number of top-of-the-line widgets ($15 each)

Solve the system:

$$\begin{cases} x+y+z=300 & \textbf{(1)} \\ 10x+12y+15z=3,700 & \textbf{(2)} \\ x=z & \textbf{(3)} \end{cases}$$

Step 1: Obtain a system of 2 equations in 2 unknowns by eliminating the same variable in two pairs of equations. To this end, substitute **(3)** into both **(1)** and **(2)** to obtain the following system:

$$(*)\quad \begin{cases} y+2z=300 & \textbf{(4)} \\ 12y+25z=3,700 & \textbf{(5)} \end{cases}$$

Step 2: Solve system **(*)** from Step 1.

Solve **(4)** for y: $y=300-2z$. **(6)**
Substitute **(6)** into **(5)** and solve for z: $z=100$ **(7)**
Substitute **(7)** into **(6)** to find y: $y=100$ **(8)**

Substitute **(7)** and **(8)** into **(1)** to find x: $x=100$.

So, 100 basic widgets, 100 mid-price widgets, and 100 top-of-the-line widgets are produced.

33. Let x = number of touchdowns
y = number of extra points
z = number of field goals

Solve the system:

$$\begin{cases} x + y + z = 18 & \textbf{(1)} \\ 6x + y + 3z = 66 & \textbf{(2)} \\ x = z + 4 & \textbf{(3)} \end{cases}$$

Step 1: Obtain a system of 2 equations in 2 unknowns by eliminating the same variable in two pairs of equations.

Substitute **(3)** into **(1)** and simplify:

$$y + 2z = 14 \quad \textbf{(4)}$$

Substitute **(3)** into **(2)** and simplify:

$$y + 9z = 42 \quad \textbf{(5)}$$

These steps yield the following system:

$$(*) \begin{cases} y + 2z = 14 & \textbf{(4)} \\ y + 9z = 42 & \textbf{(5)} \end{cases}$$

Step 2: Solve system **(*)** from Step 1.

Subtract **(5) – (4)**:

$$7z = 28 \Rightarrow z = 4 \quad \textbf{(6)}$$

Substitute **(6)** into **(3)** to solve for x:

$$x = 8 \quad \textbf{(7)}$$

Substitute **(6)** and **(7)** into **(1)** to find y:

$$y = 6 \quad \textbf{(8)}$$

Thus, there were
8 touchdowns, 6 extra points, and 4 field goals.

35. Let x = # Mediterranean chicken sand.
y = # Six-inch tuna sandwiches
z = # Six-inch roast beef sandwiches

We must solve the system:

$$(*) \begin{cases} x + y + z = 14 & \textbf{(1)} \\ 350x + 430y + 290z = 4840 & \textbf{(2)} \\ 18x + 19y + 5z = 190 & \textbf{(3)} \end{cases}$$

Step 1: Obtain a system of 2 equations in 2 unknowns by eliminating the same variable in two pairs of equations.

Solve **(1)** for x: $x = 14 - y - z$ **(4)**

Substitute **(4)** into **(2)** and simplify:

$$4y - 3z = -3 \quad \textbf{(5)}$$

Substitute **(4)** into **(3)** and simplify:

$$y - 13z = -62 \quad \textbf{(6)}$$

These steps yield the following system:

$$(*) \begin{cases} 4y - 3z = -3 & \textbf{(5)} \\ y - 13z = -62 & \textbf{(6)} \end{cases}$$

Step 2: Solve system **(*)** from Step 1.

Solve **(6)** for y:

$$y = 13z - 62 \quad \textbf{(7)}$$

Substitute **(7)** into **(5)** and simplify:

$$z = 5 \quad \textbf{(8)}$$

Substitute **(8)** into **(7)** and simplify:

$$y = 3 \quad \textbf{(8)}$$

Step 3: Use the solution of the system in Step 2 to find the value of the third variable in the original system.

Substitute **(8)** and **(9)** into **(1)** to find x:

$$x = 6 \quad \textbf{(10)}$$

Thus, there are:
6 Mediterranean chicken sandwiches,
3 Six-inch tuna sandwiches,
5 Six-inch roast beef sandwiches.

37. The system that must be solved is:

$$\begin{cases} 36 = \frac{1}{2}a(1)^2 + v_0(1) + h_0 \\ 40 = \frac{1}{2}a(2)^2 + v_0(2) + h_0 \\ 12 = \frac{1}{2}a(3)^2 + v_0(3) + h_0 \end{cases}$$

which is equivalent to

$$\begin{cases} 72 = a + 2v_0 + 2h_0 & \textbf{(1)} \\ 40 = 2a + 2v_0 + h_0 & \textbf{(2)} \\ 24 = 9a + 6v_0 + 2h_0 & \textbf{(3)} \end{cases}$$

Step 1: Obtain a 2×2 system:

Multiply **(1)** by -2, and add to **(2)**:

$$-2v_0 - 3h_0 = -104 \quad \textbf{(4)}$$

Multiply **(1)** by -9, and add to **(3)**:

$$-12v_0 - 16h_0 = -624 \quad \textbf{(5)}$$

These steps yield the following 2×2 system:

$$\begin{cases} -2v_0 - 3h_0 = -104 & \textbf{(4)} \\ -12v_0 - 16h_0 = -624 & \textbf{(5)} \end{cases}$$

Step 2: Solve the system in Step 1.

Multiply **(4)** by -6 and add to **(5)**:

$$2h_0 = 0$$

$$\boxed{h_0 = 0} \quad \textbf{(6)}$$

Substitute **(6)** into **(4)**:

$$-2v_0 - 3(0) = -104$$

$$-2v_0 = -104$$

$$\boxed{v_0 = 52} \quad \textbf{(7)}$$

Step 3: Find values of remaining variables.

Substitute **(6)** and **(7)** into **(1)**:

$$72 = a + 2(52) + 2(0)$$

$$\boxed{-32 = a}$$

Thus, the polynomial has the equation $h = -16t^2 + 52t$.

39. Since we are given that the points $(20,30)$, $(40,60)$, and $(60,40)$, the system that must be solved is:

$$\begin{cases} 30 = a(20)^2 + b(20) + c \\ 60 = a(40)^2 + b(40) + c \\ 40 = a(60)^2 + b(60) + c \end{cases}$$

which is equivalent to

$$\begin{cases} 30 = 400a + 20b + c & \textbf{(1)} \\ 60 = 1600a + 40b + c & \textbf{(2)} \\ 40 = 3600a + 60b + c & \textbf{(3)} \end{cases}$$

Step 1: Obtain a 2×2 system:

Multiply **(1)** by -4, and add to **(2)**:

$$-40b - 3c = -60 \quad \textbf{(4)}$$

Multiply **(1)** by -9, and add to **(3)**:

$$-120b - 8c = -230 \quad \textbf{(5)}$$

These steps yield the following 2×2 system:

$$\begin{cases} -40b - 3c = -60 & \textbf{(4)} \\ -120b - 8c = -230 & \textbf{(5)} \end{cases}$$

Step 2: Solve the system in Step 1.

Multiply **(4)** by -3 and add to **(5)**:

$$c = -50 \quad \textbf{(6)}$$

Substitute **(6)** into **(4)**:

$$-40b - 3(-50) = -60$$

$$-40b = -210$$

$$b = \tfrac{21}{4} = 5.25 \quad \textbf{(7)}$$

Step 3: Find values of remaining variables.

Substitute **(6)** and **(7)** into **(1)**:

$$30 = 400a + 20(5.25) + (-50)$$

$$-25 = 400a$$

$$0.0625 = a$$

Thus, the polynomial has the equation $\boxed{y = -0.0625x^2 + 5.25x - 50}$.

41. Let x = amount in money market
y = amount in mutual fund
z = amount in stock

The system we must solve is:

$$\begin{cases} x = y + 6000 & \textbf{(1)} \\ x + y + z = 20,000 & \textbf{(2)} \\ 0.03x + 0.07y + 0.10z = 1180 & \textbf{(3)} \end{cases}$$

Step 1: Obtain a 2×2 system:
Substitute **(1)** into both **(2)** and **(3)**:

$$(y+6000)+y+z=20,000$$

$$0.03(y+6000)+0.07y+0.10z=1180$$

This yields the following system:

$$\begin{cases} 2y+z=14,000 & \textbf{(4)} \\ 0.10y+0.10z=1000 & \textbf{(5)} \end{cases}$$

Step 2: Solve the system in Step 1.
Multiply **(4)** by -0.10 and add to **(5)** to find y:

$$-0.10y=-400$$

$$y=4000 \quad \textbf{(6)}$$

Substitute **(6)** into **(1)** to find x:

$$x=6000+4000=10,000 \quad \textbf{(7)}$$

Substitute **(6)** and **(7)** into **(2)** to find z:

$$10,000+4000+z=20,000$$

$$z=6000$$

Thus, the following allocation of funds should be made:

$10,000 in money market
$4000 in mutual fund
$6000 in stock

43. Let x = # regular models skis
y = # trick skis
z = # slalom skis
Solve the system:

$$\begin{cases} x+y+z=110 & \textbf{(1)} \\ 2x+3y+3z=297 & \textbf{(2)} \\ x+2y+5z=202 & \textbf{(3)} \end{cases}$$

Step 1: Obtain a 2×2 system:
Solve **(1)** for x: $x=110-y-z$ **(4)**
Substitute **(4)** into **(2)**, and simplify:

$$y+z=77 \quad \textbf{(5)}$$

Substitute **(4)** into **(3)**, and simplify:

$$y+4z=92 \quad \textbf{(6)}$$

This yields the following system:

$$(*)\begin{cases} y+z=77 & \textbf{(5)} \\ y+4z=92 & \textbf{(6)} \end{cases}$$

Step 2: Solve system (*) from Step 1.
Subtract **(6) – (5)**:

$$3z=15 \Rightarrow \boxed{z=5} \quad \textbf{(7)}$$

Substitute **(7)** into **(5)**:

$$\boxed{y=72} \quad \textbf{(8)}$$

Substitute **(7)** and **(8)** into **(1)**:

$$x+72+5=110 \Rightarrow \boxed{x=33}$$

So, need to sell 33 regular model skis, 72 trick skis, and 5 slalom skis.

45. Let x = # points scored in game 1
y = # points scored in game 2
z = # points scored in game 3
Solve the system:

$$\begin{cases} x+y+z=2{,}591 & \textbf{(1)} \\ x-y=62 & \textbf{(2)} \\ x-z=2 & \textbf{(3)} \end{cases}$$

Step 1: Obtain a 2×2 system:
Solve **(1)** for x: $x=2591-y-z$ **(4)**
Substitute **(4)** into **(2)**, and simplify:
$$2y+z=2529 \quad \textbf{(5)}$$
Substitute **(4)** into **(3)**, and simplify:
$$y+2z=2589 \quad \textbf{(6)}$$

This yields the following system:
$$(*)\begin{cases} 2y+z=2529 & \textbf{(5)} \\ y+2z=2589 & \textbf{(6)} \end{cases}$$

Step 2: Solve system (*) from Step 1.
Multiply **(6)** by -2:
$$-2y-4z=-5178 \quad \textbf{(7)}$$
Add **(7)** + **(5)**:
$$-3z=-2649 \Rightarrow \boxed{z=883} \quad \textbf{(8)}$$
Substitute **(8)** into **(6)**:
$$\boxed{y=823} \quad \textbf{(9)}$$
Substitute **(9)** and **(8)** into **(1)**:
$$x+823+883=2591 \Rightarrow \boxed{x=885}$$

So, 885 points scored in game 1, 823 points scored in game 2, and 883 points scored in game 3.

47. Equation (2) and Equation (3) must be added correctly – should be $2x-y+z=2$. Also, should begin by eliminating one variable from Equation (1).

49. True

51. Substitute the given points into the equation to obtain the following system:

$$\begin{cases} (-2)^2+4^2+a(-2)+b(4)+c=0 \\ 1^2+1^2+a(1)+b(1)+c=0 \\ (-2)^2+(-2)^2+a(-2)+b(-2)+c=0 \end{cases}$$

which is equivalent to (after simplification)

$$\begin{cases} -2a+4b+c=-20 & \textbf{(1)} \\ a+b+c=-2 & \textbf{(2)} \\ -2a-2b+c=-8 & \textbf{(3)} \end{cases}$$

Multiply **(2)** by 2 and then, add to **(3)**:
$$3c=-12 \Rightarrow c=-4 \quad \textbf{(4)}$$

Multiply **(2)** by 2 and then, add to **(1)**:
$$6b+3c=-24 \quad \textbf{(5)}$$
Substitute **(4)** into **(5)** to find b:
$$6b+3(-4)=-24$$
$$6b=-12$$
$$b=-2 \quad \textbf{(6)}$$
Finally, substitute **(4)** and **(6)** into **(2)** to find a:
$$a+(-2)+(-4)=-2$$
$$a=4$$
Thus, the equation is
$$\boxed{x^2+y^2+4x-2y-4=0}.$$

53. We deduce from the diagram that the following points are on the curve:

$(-2,46)$, $(-1,51)$, $(0,44)$, $(1,51)$, $(2,43)$

Substituting these points into the given equation gives rise to the system:

$$\begin{cases} 46 = a(-2)^4 + b(-2)^3 + c(-2)^2 + d(-2) + e \\ 51 = a(-1)^4 + b(-1)^3 + c(-1)^2 + d(-1) + e \\ 44 = a(0)^4 + b(0)^3 + c(0)^2 + d(0) + e \\ 51 = a(1)^4 + b(1)^3 + c(1)^2 + d(1) + e \\ 43 = a(2)^4 + b(2)^3 + c(2)^2 + d(2) + e \end{cases}$$

Observe that the third equation above simplifies to

$$\boxed{e = 44}.$$

Substitute this value into the remaining four equations to obtain the following system:

$$\begin{cases} 16a - 8b + 4c - 2d = 2 & \textbf{(1)} \\ a - b + c - d = 7 & \textbf{(2)} \\ a + b + c + d = 7 & \textbf{(3)} \\ 16a + 8b + 4c + 2d = -1 & \textbf{(4)} \end{cases}$$

Now, add **(2)** and **(3)**: $2a + 2c = 14$ **(5)**

Add **(1)** and **(4)**: $32a + 8c = 1$ **(6)**

Solve the system: $\begin{cases} 2a + 2c = 14 & \textbf{(5)} \\ 32a + 8c = 1 & \textbf{(6)} \end{cases}$

Multiply **(5)** by -4, and then add to **(6)**:

$$24a = -55$$

$$\boxed{a = -\tfrac{55}{24}} \quad \textbf{(7)}$$

Substitute **(7)** into **(5)** to find c:

$$2(-\tfrac{55}{24}) + 2c = 14$$

$$\boxed{c = 7 + \tfrac{55}{24} = \tfrac{223}{24}} \quad \textbf{(8)}$$

At this point, we have values for three of the five unknowns. We now use **(7)** and **(8)** to obtain a 2×2 system involving these two remaining unknowns:

Substitute **(7)** and **(8)** into **(1):**

$$2 = 16(-\tfrac{55}{24}) - 8b + 4(\tfrac{223}{24}) - 2d$$

$$36 = -192b - 48d \quad \textbf{(9)}$$

Substitute **(7)** and **(8)** into **(2):**

$$7 = -\tfrac{55}{24} - b + \tfrac{223}{24} - d$$

$$0 = b + d \quad \textbf{(10)}$$

Solve the system:

$$\begin{cases} 36 = -192b - 48d & \textbf{(9)} \\ 0 = b + d & \textbf{(10)} \end{cases}$$

Solve **(10)** for b: $b = -d$ **(11)**

Substitute **(11)** into **(9)** to find d:

$$36 = -192(-d) - 48d$$

$$36 = 144d$$

$$\boxed{\tfrac{1}{4} = d} \quad \textbf{(12)}$$

Substitute **(12)** into **(10)** to find b:

$$\boxed{b = -\tfrac{1}{4}}$$

Hence, the equation of the polynomial is $y = -\tfrac{55}{24}x^4 - \tfrac{1}{4}x^3 + \tfrac{223}{24}x^2 + \tfrac{1}{4}x + 44$.

55. Solve the system

$$\begin{cases} 2y+z=3 & \textbf{(1)} \\ 4x-z=-3 & \textbf{(2)} \\ 7x-3y-3z=2 & \textbf{(3)} \\ x-y-z=-2 & \textbf{(4)} \end{cases}$$

Step 1: Obtain a 3×3 system:

Solve **(2)** for z: $z=4x+3$ **(5)**

Substitute **(5)** into each of **(1)**, **(3)**, and **(4)**. Simplifying yields the following system:

$$\begin{cases} 2y+4x=0 & \textbf{(6)} \\ -3y-5x=11 & \textbf{(7)} \\ -y-3x=1 & \textbf{(8)} \end{cases}$$

Step 2: Solve the system in Step 1.

Solve **(6)** for y: $y=-2x$ **(9)**

Substitute **(9)** into both **(7)** and **(8)** to obtain the following two equations:

$$-3(-2x)-5x=11 \text{ so that } x=11 \quad \textbf{(10)}$$

$$-(-2x)-3x=1 \text{ so that } x=-1 \quad \textbf{(11)}$$

Observe that **(10)** and **(11)** yield different values of x. As such, there can be no solution to this system.

57. Solve the system

$$\begin{cases} 3x_1-2x_2+x_3+2x_4=-2 & \textbf{(1)} \\ -x_1+3x_2+4x_3+3x_4=4 & \textbf{(2)} \\ x_1+x_2+x_3+x_4=0 & \textbf{(3)} \\ 5x_1+3x_2+x_3+2x_4=-1 & \textbf{(4)} \end{cases}$$

Step 1: Obtain a 3×3 system:

Add **(2)** and **(3)**:

$$4x_2+5x_3+4x_4=4 \quad \textbf{(5)}$$

Multiply **(2)** by 3, and then add to **(1)**:

$$7x_2+13x_3+11x_4=10 \quad \textbf{(6)}$$

Multiply **(2)** by 5, and then add to **(4)**:

$$18x_2+21x_3+17x_4=19 \quad \textbf{(7)}$$

These steps yield the following system:

$$\begin{cases} 4x_2+5x_3+4x_4=4 & \textbf{(5)} \\ 7x_2+13x_3+11x_4=10 & \textbf{(6)} \\ 18x_2+21x_3+17x_4=19 & \textbf{(7)} \end{cases}$$

CONTINUED ONTO NEXT PAGE!!

These steps yield the following system:

$$\begin{cases} 17x_3+16x_4=12 & \textbf{(8)} \\ 87x_3+79x_4=47 & \textbf{(9)} \end{cases}$$

Solve this system:

Multiply **(8)** by -87, **(9)** by 17, and then add :

$$-49x_4=-245$$

$$x_4=5 \quad \textbf{(10)}$$

Substitute **(10)** into **(8)**:

$$17x_3+16(5)=12$$

$$x_3=-4 \quad \textbf{(11)}$$

Now, substitute **(10)** and **(11)** into **(5)**:

$$4x_2+5(-4)+4(5)=4$$

$$x_2=1 \quad \textbf{(12)}$$

Finally, substitute **(10)** – **(12)** into **(3)**:

$$x_1+1-4+5=0$$

$$x_1=-2$$

Step 2: Solve the system in Step 1.

Multiply **(5)** by -7, **(6)** by 4, and then add them: $17x_3 + 16x_4 = 12$ **(8)**

Multiply **(6)** by 18, **(7)** by -7, and then add them: $87x_3 + 79x_4 = 47$ **(9)**

Thus, the solution is:

$\boxed{x_1 = -2,\ x_2 = 1, x_3 = -4,\ x_4 = 5}$.

59. See the solution to #21.

61. Write the system in the form:

$$\begin{cases} z = x - y - 10 \\ z = -2x + 3y - 11 \\ z = x - y - 10 \end{cases}$$

A graphical solution is given to the right: Notice that there are only two planes since the first and third equations of the system are the same. Hence, the line of intersection of these two planes constitutes the infinitely many solutions of the system. This is precisely what was found to be the case in Exercises 21 and 59.

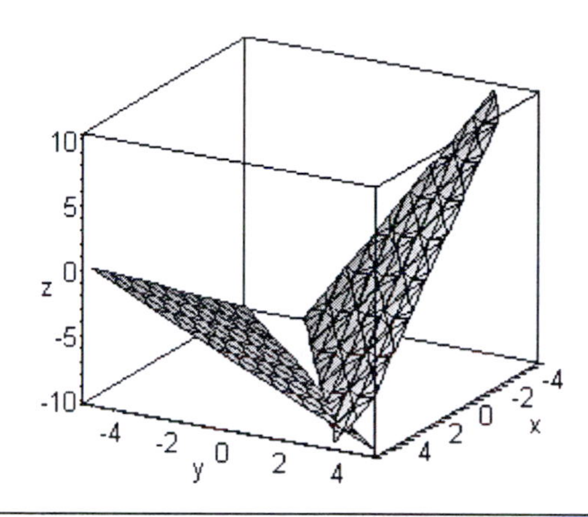

63. If your calculator has 3-dimensional graphing capabilities, then generate the graph of the system to obtain:

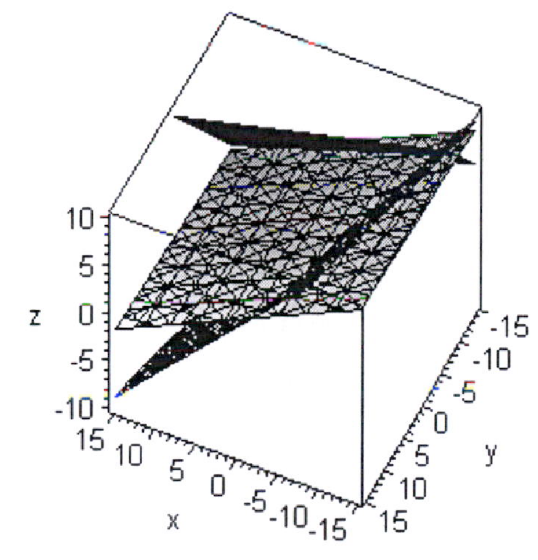

It's a bit difficult to tell from the graph, but the three planes intersect in a common point with coordinates $\left(-\frac{80}{7}, -\frac{80}{7}, \frac{48}{7}\right)$, which can be easily verified using your calculator.

Section 6.3 Solutions ---

1. d $x\left(x^2-25\right)=x(x-5)(x+5)$ has three distinct linear factors

3. a $x^2\left(x^2+25\right)$ has one repeated linear factor and one irreducible quadratic factor.

5. b $x\left(x^2+25\right)^2$ has one linear factor and one repeated irreducible quadratic factor.

7.

$$\frac{9}{x^2-x-20}=\frac{9}{(x-5)(x+4)}$$
$$=\frac{A}{x-5}+\frac{B}{x+4}$$

9.

$$\frac{2x+5}{x^3-4x^2}=\frac{2x+5}{x^2(x-4)}$$
$$=\frac{A}{x-4}+\frac{B}{x}+\frac{C}{x^2}$$

11. Must long divide in this case.

$$\begin{array}{r} 2x-6 \\ x^2+x+5\overline{\big)\ 2x^3-4x^2+7x+3} \\ \underline{-(2x^3+2x^2+10x)} \\ -6x^2-3x+3 \\ \underline{-(-6x^2-6x-30)} \\ 3x+33 \end{array}$$

So,

$$\frac{2x^3-4x^2+7x+3}{x^2+x+5}=2x-6+\frac{3x+33}{x^2+x+5}$$

13.

$$\frac{3x^3-x+9}{\left(x^2+10\right)^2}=\frac{Ax+B}{x^2+10}+\frac{Cx+D}{\left(x^2+10\right)^2}$$

15. The partial fraction decomposition has the form:

$$\frac{1}{x(x+1)}=\frac{A}{x}+\frac{B}{x+1} \quad \textbf{(1)}$$

To find the coefficients, multiply both sides of **(1)** by $x(x+1)$, and gather like terms:

$$1=A(x+1)+Bx$$
$$1=(A+B)x+A \quad \textbf{(2)}$$

Equate corresponding coefficients in **(2)** to obtain the following system:

$$(*)\begin{cases} A+B=0 & \textbf{(3)} \\ A=1 & \textbf{(4)} \end{cases}$$

Now, solve system $(*)$:

Substitute **(4)** into **(3)** to see that $B=-1$.

Thus, the partial fraction decomposition **(1)** becomes:

$$\boxed{\frac{1}{x(x+1)}=\frac{1}{x}-\frac{1}{x+1}}$$

17. Observe that simplifying the expression first yields

$$\frac{\cancel{x}}{\cancel{x}(x-1)}=\frac{1}{x-1}.$$

This IS the partial fraction decomposition.

19. The partial fraction decomposition has the form:

$$\frac{9x-11}{(x-3)(x+5)}=\frac{A}{x-3}+\frac{B}{x+5} \quad \textbf{(1)}$$

To find the coefficients, multiply both sides of **(1)** by $(x-3)(x+5)$, and gather like terms:

$$9x-11=A(x+5)+B(x-3)$$

$$9x-11=(A+B)x+(5A-3B) \quad \textbf{(2)}$$

Equate corresponding coefficients in **(2)** to obtain the following system:

$$\textbf{(*)} \begin{cases} A+B=9 & \textbf{(3)} \\ 5A-3B=-11 & \textbf{(4)} \end{cases}$$

Now, solve system **(*)**:

Multiply **(3)** by -5:

$$-5A-5B=-45 \quad \textbf{(5)}$$

Add **(5)** and **(3)** to solve for B:

$$-8B=-56$$

$$B=7 \quad \textbf{(6)}$$

Substitute **(6)** into **(3)** to find A:

$$A=2$$

Thus, the partial fraction decomposition **(1)** becomes:

$$\boxed{\frac{9x-11}{(x-3)(x+5)}=\frac{2}{x-3}+\frac{7}{x+5}}$$

21. The partial fraction decomposition has the form:

$$\frac{3x+1}{(x-1)^2}=\frac{A}{x-1}+\frac{B}{(x-1)^2} \quad \textbf{(1)}$$

To find the coefficients, multiply both sides of **(1)** by $(x-1)^2$, and gather like terms:

$$3x+1=A(x-1)+B$$

$$3x+1=Ax+(B-A) \quad \textbf{(2)}$$

Equate corresponding coefficients in **(2)** to obtain the following system:

$$\textbf{(*)} \begin{cases} A=3 & \textbf{(3)} \\ B-A=1 & \textbf{(4)} \end{cases}$$

Substitute **(3)** into **(4)** to find B:

$$B-3=1 \Rightarrow B=4$$

Thus, the partial fraction decomposition **(1)** becomes:

$$\boxed{\frac{3x+1}{(x-1)^2}=\frac{3}{x-1}+\frac{4}{(x-1)^2}}$$

23. The partial fraction decomposition has the form: $\frac{4x-3}{(x+3)^2}=\frac{A}{x+3}+\frac{B}{(x+3)^2}$ **(1)**

To find the coefficients, multiply both sides of **(1)** by $(x+3)^2$, gather like terms:

$$4x-3=A(x+3)+B \quad \Rightarrow \quad 4x-3=Ax+(3A+B) \ \textbf{(2)}$$

Equate corresponding coefficients in **(2)** to obtain the following system:

$$(*)\begin{cases} A=4 & \textbf{(3)} \\ 3A+B=-3 & \textbf{(4)} \end{cases}$$

Substitute **(3)** into **(4)** to find B: $3(4)+B=-3 \ \Rightarrow \ B=-15$

Thus, the partial fraction decomposition **(1)** becomes: $\boxed{\frac{4x-3}{(x+3)^2}=\frac{4}{x+3}+\frac{-15}{(x+3)^2}}$

25. The partial fraction decomposition has the form:

$$\frac{4x^2-32x+72}{(x+1)(x-5)^2}=\frac{A}{x+1}+\frac{B}{x-5}+\frac{C}{(x-5)^2} \ \textbf{(1)}$$

To find the coefficients, multiply both sides of **(1)** by $(x+1)(x-5)^2$, and gather like terms:

$$\begin{aligned} 4x^2-32x+72 &= A(x-5)^2+B(x-5)(x+1)+C(x+1) \\ &= A(x^2-10x+25)+B(x^2-4x-5)+C(x+1) \\ &= (A+B)x^2+(-10A-4B+C)x+(25A-5B+C) \ \textbf{(2)} \end{aligned}$$

Equate corresponding coefficients in **(2)** to obtain the following system:

$$(*)\begin{cases} A+B=4 & \textbf{(3)} \\ -10A-4B+C=-32 & \textbf{(4)} \\ 25A-5B+C=72 & \textbf{(5)} \end{cases}$$

Solve **(3)** for B: $B=4-A$ **(6)**

Substitute **(6)** into **(4)**:

$$-10A-4(4-A)+C=-32 \ \text{ which is equivalent to } \ -6A+C=-16 \ \textbf{(7)}$$

Substitute **(6)** into **(5)**:

$$25A-5(4-A)+C=72 \ \text{ which is equivalent to } \ 30A+C=92 \ \textbf{(8)}$$

Now, solve the 2×2 system: $\begin{cases} -6A+C=-16 & \textbf{(7)} \\ 30A+C=92 & \textbf{(8)} \end{cases}$

Multiply **(7)** by -1, and add to **(8)** to find A: $36A=108$ so that $A=3$.

Substitute this value for A into **(7)** to find C: $-6(3)+C=-16$ so that $C=2$.

Finally, substitute the value of A into **(3)** to find B: $3+B=4$ so that $B=1$.

Thus, the partial fraction decomposition **(1)** becomes:

$$\boxed{\frac{4x^2-32x+72}{(x+1)(x-5)^2}=\frac{3}{x+1}+\frac{1}{x-5}+\frac{2}{(x-5)^2}}$$

27. The partial fraction decomposition has the form:

$$\frac{5x^2+28x-6}{(x+4)(x^2+3)}=\frac{A}{x+4}+\frac{Bx+C}{x^2+3} \quad \textbf{(1)}$$

To find the coefficients, multiply both sides of **(1)** by $(x+4)(x^2+3)$, and gather like terms:

$$\begin{aligned}5x^2+28x-6&=A(x^2+3)+(Bx+C)(x+4)\\&=Ax^2+3A+Bx^2+Cx+4Bx+4C\\&=(A+B)x^2+(4B+C)x+(3A+4C) \quad \textbf{(2)}\end{aligned}$$

Equate corresponding coefficients in **(2)** to obtain the following system:

$$(*)\ \begin{cases}A+B=5 & \textbf{(3)}\\4B+C=28 & \textbf{(4)}\\3A+4C=-6 & \textbf{(5)}\end{cases}$$

Solve **(3)** for B: $B=5-A$ **(6)**

Substitute **(6)** into **(4)**:

$$4(5-A)+C=28 \text{ which is equivalent to } -4A+C=8 \quad \textbf{(7)}$$

Now, solve the 2×2 system:

$$\begin{cases}3A+4C=-6 & \textbf{(5)}\\-4A+C=8 & \textbf{(7)}\end{cases}$$

Multiply **(7)** by -4 and then add to **(5)** to find A: $19A=-38$ so that $A=-2$.

Substitute this value for A into **(7)** to find C: $-4(-2)+C=8$ so that $C=0$.

Finally, substitute the value of A into **(3)** to find B: $-2+B=5$ so that $B=7$.

Thus, the partial fraction decomposition **(1)** becomes:

$$\boxed{\frac{5x^2+28x-6}{(x+4)(x^2+3)}=\frac{-2}{x+4}+\frac{7x}{x^2+3}}$$

29. The partial fraction decomposition has the form:

$$\frac{-2x^2-17x+11}{(x-7)\underbrace{\left(3x^2-7x+5\right)}_{\text{Irreducible Quadratic Term}}}=\frac{A}{x-7}+\frac{Bx+C}{3x^2-7x+5}\quad\textbf{(1)}$$

To find the coefficients, multiply both sides of **(1)** by $(x-7)\left(3x^2-7x+5\right)$, and gather like terms:

$$\begin{aligned}-2x^2-17x+11&=A\left(3x^2-7x+5\right)+(Bx+C)(x-7)\\&=3Ax^2-7Ax+5A+Bx^2+Cx-7Bx-7C\\&=(3A+B)x^2+(-7A-7B+C)x+(5A-7C)\quad\textbf{(2)}\end{aligned}$$

Equate corresponding coefficients in **(2)** to obtain the following system:

$$(*)\begin{cases}3A+B=-2 & \textbf{(3)}\\-7A-7B+C=-17 & \textbf{(4)}\\5A-7C=11 & \textbf{(5)}\end{cases}$$

Solve **(3)** for B: $B=-2-3A$ **(6)**
Substitute **(6)** into **(4)**:

$-7A-7(-2-3A)+C=-17$ which is equivalent to $14A+C=-31$ **(7)**

Now, solve the 2×2 system:

$$\begin{cases}5A-7C=11 & \textbf{(5)}\\14A+C=-31 & \textbf{(7)}\end{cases}$$

Multiply **(7)** by 7 and then add to **(5)** to find A: $103A=-206$ so that $A=-2$.
Substitute this value for A into **(5)** to find C: $5(-2)-7C=11$ so that $C=-3$.
Finally, substitute the value of A into **(3)** to find B: $3(-2)+B=-2$ so that $B=4$.

Thus, the partial fraction decomposition **(1)** becomes:

$$\boxed{\frac{-2x^2-17x+11}{(x-7)\left(3x^2-7x+5\right)}=\frac{-2}{x-7}+\frac{4x-3}{3x^2-7x+5}}$$

31. The partial fraction decomposition has the form:

$$\frac{x^3}{\left(x^2+9\right)^2}=\frac{Ax+B}{x^2+9}+\frac{Cx+D}{\left(x^2+9\right)^2} \quad \textbf{(1)}$$

To find the coefficients, multiply both sides of **(1)** by $\left(x^2+9\right)^2$, and gather like terms:

$$\begin{aligned} x^3 &= (Ax+B)\left(x^2+9\right)+(Cx+D) \\ &= Ax^3+Bx^2+9Ax+9B+Cx+D \\ &= Ax^3+Bx^2+(9A+C)x+(9B+D) \quad \textbf{(2)} \end{aligned}$$

Equate corresponding coefficients in **(2)** to obtain the following system:

$$(*)\begin{cases} A=1 & \textbf{(3)} \\ B=0 & \textbf{(4)} \\ 9A+C=0 & \textbf{(5)} \\ 9B+D=0 & \textbf{(6)} \end{cases}$$

Substitute **(3)** into **(5)** to find *C*: $9(1)+C=0$ so that $C=-9$

Substitute **(4)** into **(6)** to find *D*: $9(0)+D=0$ so that $D=0$

Thus, the partial fraction decomposition **(1)** becomes:

$$\boxed{\frac{x^3}{\left(x^2+9\right)^2}=\frac{x}{x^2+9}-\frac{9x}{\left(x^2+9\right)^2}}$$

33. The partial fraction decomposition has the form:

$$\frac{2x^3-3x^2+7x-2}{\left(x^2+1\right)^2}=\frac{Ax+B}{x^2+1}+\frac{Cx+D}{\left(x^2+1\right)^2} \quad \textbf{(1)}$$

To find the coefficients, multiply both sides of **(1)** by $\left(x^2+1\right)^2$, and gather like terms:

$$\begin{aligned} 2x^3-3x^2+7x-2 &= (Ax+B)\left(x^2+1\right)+(Cx+D) \\ &= Ax^3+Bx^2+Ax+B+Cx+D \\ &= Ax^3+Bx^2+(A+C)x+(B+D) \quad \textbf{(2)} \end{aligned}$$

Equate corresponding coefficients in **(2)** to obtain the following system:

$$(*)\begin{cases} A=2 & \textbf{(3)} \\ B=-3 & \textbf{(4)} \\ A+C=7 & \textbf{(5)} \\ B+D=-2 & \textbf{(6)} \end{cases}$$

Substitute **(3)** into **(5)** to find C: $2+C=7$ so that $C=5$
Substitute **(4)** into **(6)** to find D: $-3+D=-2$ so that $D=1$

Thus, the partial fraction decomposition **(1)** becomes:

$$\boxed{\frac{2x^3-3x^2+7x-2}{\left(x^2+1\right)^2}=\frac{2x-3}{x^2+1}+\frac{5x+1}{\left(x^2+1\right)^2}}$$

35. The partial fraction decomposition has the form:

$$\frac{3x+1}{x^4-1}=\frac{3x+1}{\left(x^2-1\right)\left(x^2+1\right)}=\frac{3x+1}{(x-1)(x+1)\left(x^2+1\right)}=\frac{A}{x-1}+\frac{B}{x+1}+\frac{Cx+D}{x^2+1} \quad \textbf{(1)}$$

To find the coefficients, multiply both sides of **(1)** by $(x-1)(x+1)\left(x^2+1\right)$, and gather like terms:

$$\begin{aligned}3x+1 &= A(x+1)\left(x^2+1\right)+B(x-1)\left(x^2+1\right)+\left(Cx+D\right)(x-1)(x+1)\\ &= Ax^3+Ax^2+Ax+A+Bx^3-Bx^2+Bx-B+Cx^3+Dx^2-Cx-D\\ &= (A+B+C)x^3+(A-B+D)x^2+(A+B-C)x+(A-B-D) \quad \textbf{(2)}\end{aligned}$$

Equate corresponding coefficients in **(2)** to obtain the following system:

$$(*)\ \begin{cases} A+B+C=0 & \textbf{(3)}\\ A-B+D=0 & \textbf{(4)}\\ A+B-C=3 & \textbf{(5)}\\ A-B-D=1 & \textbf{(6)}\end{cases}$$

To solve this system, first obtain a 3×3 system:

Add **(4)** and **(6)**: $2A-2B=1$ **(7)**

This enables us to consider the following 3×3 system:

$$\begin{cases} A+B+C=0 & \textbf{(3)}\\ A+B-C=3 & \textbf{(5)}\\ 2A-2B=1 & \textbf{(7)}\end{cases}$$

Now, to solve this system, obtain a 2×2 system:

Add **(3)** and **(5)**: $2A+2B=3$ **(8)**

This enables us to consider the following 2×2 system:

$$\begin{cases} 2A-2B=1 & \textbf{(7)}\\ 2A+2B=3 & \textbf{(8)}\end{cases}$$

Add **(7)** and **(8)** to find A: $4A=4$ so that $A=1$

Substitute this value of A into **(7)** to find B: $2(1)-2B=1$ so that $B=\frac{1}{2}$.

Now, substitute these values of A and B into **(6)** to find D:

$$1-\tfrac{1}{2}-D=1 \text{ so that } D=-\tfrac{1}{2}$$

Finally, substitute these values of A and B into **(3)** to find C:

$$1+\tfrac{1}{2}+C=0 \text{ so that } C=-\tfrac{3}{2}.$$

Thus, the partial fraction decomposition **(1)** becomes:

$$\boxed{\frac{3x+1}{x^4-1}=\frac{1}{x-1}+\frac{1}{2(x+1)}+\frac{-3x-1}{2\left(x^2+1\right)}}$$

37. The partial fraction decomposition has the form:

$$\frac{5x^2+9x-8}{(x-1)\underbrace{\left(x^2+2x-1\right)}_{\text{Irreducible Quadratic Term}}}=\frac{A}{x-1}+\frac{Bx+C}{x^2+2x-1} \quad \textbf{(1)}$$

To find the coefficients, multiply both sides of **(1)** by $(x-1)\left(x^2+2x-1\right)$, and gather like terms:

$$\begin{aligned} 5x^2+9x-8 &= A\left(x^2+2x-1\right)+(Bx+C)(x-1) \\ &= Ax^2+2Ax-A+Bx^2+Cx-Bx-C \\ &= (A+B)x^2+(2A-B+C)x+(-A-C) \quad \textbf{(2)} \end{aligned}$$

Equate corresponding coefficients in **(2)** to obtain the following system:

$$(*)\begin{cases} A+B=5 & \textbf{(3)} \\ 2A-B+C=9 & \textbf{(4)} \\ -A-C=-8 & \textbf{(5)} \end{cases}$$

Solve **(3)** for B: $B=5-A$ **(6)**
Substitute **(6)** into **(4)**:

$$3A+C=14 \quad \textbf{(7)}$$

Now, solve the 2×2 system:

$$\begin{cases} -A-C=-8 & \textbf{(5)} \\ 3A+C=14 & \textbf{(7)} \end{cases}$$

Add **(5)** and **(7)** to find A: $2A=6$ so that $A=3$.
Substitute this value for A into **(7)** to find C: $3(3)+C=14$ so that $C=5$.
Finally, substitute the value of A into **(3)** to find B: $B=2$.

Thus, the partial fraction decomposition **(1)** becomes:

$$\boxed{\frac{5x^2+9x-8}{(x-1)\left(x^2+2x-1\right)}=\frac{3}{x-1}+\frac{2x+5}{x^2+2x-1}}$$

39. The partial fraction decomposition has the form:

$$\frac{3x}{x^3-1}=\frac{3x}{(x-1)\left(x^2+x+1\right)}=\frac{A}{x-1}+\frac{Bx+C}{x^2+x+1} \quad \textbf{(1)}$$

To find the coefficients, multiply both sides of **(1)** by $(x-1)\left(x^2+x+1\right)$, and gather like terms:

$$\begin{aligned}3x &= A\left(x^2+x+1\right)+\left(Bx+C\right)(x-1)\\ &= Ax^2+Ax+A+Bx^2+Cx-Bx-C\\ &= (A+B)x^2+\left(A-B+C\right)x+\left(A-C\right) \quad \textbf{(2)}\end{aligned}$$

Equate corresponding coefficients in **(2)** to obtain the following system:

$$\textbf{(*)}\begin{cases} A+B=0 & \textbf{(3)}\\ A-B+C=3 & \textbf{(4)}\\ A-C=0 & \textbf{(5)}\end{cases}$$

Solve **(3)** for B: $B=-A$ **(6)**
Solve **(5)** for C: $C=A$ **(7)**
Substitute **(6)** and **(7)** into **(4)**: $A-(-A)+(A)=3$ so that $A=1$.
Substitute this value of A into **(6)** to find B: $B=-1$
Finally, substitute this value of A into **(7)** to find C: $C=1$

Thus, the partial fraction decomposition **(1)** becomes:

$$\boxed{\frac{3x}{x^3-1}=\frac{1}{x-1}+\frac{1-x}{x^2+x+1}}$$

41. The partial fraction decomposition has the form:

$$\frac{f(d_i+d_0)}{d_i d_0}=\frac{A}{d_i}+\frac{B}{d_0} \quad \textbf{(1)}$$

To find the coefficients, multiply both sides of **(1)** by $d_i d_0$, and gather like terms:

$$f(d_i+d_0)=Ad_0+Bd_i$$
$$fd_i+fd_0=Ad_0+Bd_i$$

Equate corresponding coefficients in **(2)** to obtain the following system:

$$(*) \begin{cases} A=f & \textbf{(3)} \\ B=f & \textbf{(4)} \end{cases}$$

Thus, the partial fraction decomposition **(1)** becomes

$$\frac{f(d_i+d_0)}{d_i d_0}=\frac{f}{d_i}+\frac{f}{d_0} \quad \textbf{(5)}$$

Hence, using **(5)** enables us to write the lens law as $\frac{f}{d_i}+\frac{f}{d_0}=1$, or as $\frac{1}{d_0}+\frac{1}{d_i}=\frac{1}{f}$.

43. The form of the decomposition is incorrect. It should be $\frac{A}{x}+\frac{Bx+C}{x^2+1}$. Once this correction is made, the correct decomposition is $\frac{1}{x}+\frac{2x+3}{x^2+1}$.

45. False. The degree of the numerator must be less than or equal to the degree of the denominator in order to apply the partial fraction decomposition procedure.

47. The first step in forming the partial fraction decomposition of $\frac{x^2+4x-8}{x^3-x^2-4x+4}$ is to factor the denominator. To do so, we begin by applying the Rational Root Test:

Factors of 4: $\pm1, \pm2, \pm4$

Factors of 1: ±1

Possible Rational Zeros: $\pm1, \pm2, \pm4$

Applying synthetic division to the zeros, one can see that 1 is a rational zero:

$$\begin{array}{r|rrrr} 1 & 1 & -1 & -4 & 4 \\ & & 1 & 0 & -4 \\ \hline & 1 & 0 & -4 & 0 \end{array}$$

So, $x^3-x^2-4x+4=(x-1)(x^2-4)=(x-1)(x-2)(x+2)$.

Now, the partial fraction decomposition has the form:

$$\frac{x^2+4x-8}{x^3-x^2-4x+4}=\frac{x^2+4x-8}{(x-1)(x-2)(x+2)}=\frac{A}{x-1}+\frac{B}{x-2}+\frac{C}{x+2} \quad \textbf{(1)}$$

To find the coefficients, multiply both sides of **(1)** by $(x-1)(x-2)(x+2)$, and gather like terms:

$$\begin{aligned} x^2+4x-8 &= A(x-2)(x+2)+B(x-1)(x+2)+C(x-1)(x-2) \\ &= A(x^2-4)+B(x^2+x-2)+C(x^2-3x+2) \\ &= (A+B+C)x^2+(B-3C)x+(-4A-2B+2C) \quad \textbf{(2)} \end{aligned}$$

Equate corresponding coefficients in **(2)** to obtain the following system:

$$(*)\ \begin{cases} A+B+C=1 & \textbf{(3)} \\ B-3C=4 & \textbf{(4)} \\ -4A-2B+2C=-8 & \textbf{(5)} \end{cases}$$

Now, solve system **(*)**:

Multiply **(3)** by 4 and then, add to **(5)**: $2B+6C=-4$ **(6)**

This leads to the following 2×2 system:

$$\begin{cases} B-3C=4 & \textbf{(4)} \\ 2B+6C=-4 & \textbf{(6)} \end{cases}$$

Multiply **(4)** by -2 and then add to **(6)** to find C: $12C=-12$ so that $C=-1$

Substitute this value of C into **(4)** to find B: $B-3(-1)=4$ so that $B=1$.

Finally, substitute these values of B and C into **(3)** to find A: $A+1-1=1$ so that $A=1$.

Thus, the partial fraction decomposition **(1)** becomes:

$$\boxed{\frac{x^2+4x-8}{x^3-x^2-4x+4}=\frac{1}{x-1}-\frac{1}{x+2}+\frac{1}{x-2}}$$

49. The partial fraction decomposition has the form:

$$\frac{2x^3+x^2-x-1}{x^4+x^3}=\frac{2x^3+x^2-x-1}{x^3(x+1)}=\frac{A}{x}+\frac{B}{x^2}+\frac{C}{x^3}+\frac{D}{x+1} \quad \textbf{(1)}$$

To find the coefficients, multiply both sides of **(1)** by $x^3(x+1)$, and gather like terms:

$$\begin{aligned} 2x^3+x^2-x-1 &= Ax^2(x+1)+Bx(x+1)+C(x+1)+Dx^3 \\ &= Ax^3+Ax^2+Bx^2+Bx+Cx+C+Dx^3 \\ &= (A+D)x^3+(A+B)x^2+(B+C)x+C \quad \textbf{(2)} \end{aligned}$$

Equate corresponding coefficients in **(2)** to obtain the following system:

$$(*)\begin{cases} A+D=2 & \textbf{(3)} \\ A+B=1 & \textbf{(4)} \\ B+C=-1 & \textbf{(5)} \\ C=-1 & \textbf{(6)} \end{cases}$$

Substitute **(6)** into **(5)** to find *B*: $B-1=-1$ so that $B=0$.
Substitute this value of *B* into **(4)** to find *A*: $A+0=1$ so that $A=1$.
Substitute this value of *A* into **(3)** to find *D*: $1+D=2$ so that $D=1$.

Thus, the partial fraction decomposition **(1)** becomes:

$$\boxed{\frac{2x^3+x^2-x-1}{x^4+x^3}=\frac{1}{x}+\frac{1}{x+1}-\frac{1}{x^3}}$$

51. The partial fraction decomposition has the form:

$$\frac{x^5+2}{\left(x^2+1\right)^3}=\frac{Ax+B}{x^2+1}+\frac{Cx+D}{\left(x^2+1\right)^2}+\frac{Ex+F}{\left(x^2+1\right)^3} \quad \textbf{(1)}$$

To find the coefficients, multiply both sides of **(1)** by $\left(x^2+1\right)^3$, and gather like terms:

$$\begin{aligned} x^5+2 &= (Ax+B)\underbrace{\left(x^2+1\right)^2}_{x^4+2x^2+1}+(Cx+D)\left(x^2+1\right)+(Ex+F) \\ &= Ax^5+Bx^4+2Ax^3+2Bx^2+Ax+B+Cx^3+Dx^2+Cx+D+Ex+F \\ &= Ax^5+Bx^4+(2A+C)x^3+(2B+D)x^2+(A+C+E)x+(B+D+F) \quad \textbf{(2)} \end{aligned}$$

Equate corresponding coefficients in **(2)** to obtain the following system:

$$(*)\begin{cases} A=1 & \textbf{(3)} \\ B=0 & \textbf{(4)} \\ 2A+C=0 & \textbf{(5)} \\ 2B+D=0 & \textbf{(6)} \\ A+C+E=0 & \textbf{(7)} \\ B+D+F=2 & \textbf{(8)} \end{cases}$$

Substitute **(3)** into **(5)** to find *C*: $2+C=0$ so that $C=-2$.
Substitute **(4)** into **(6)** to find *D*: $0+D=0$ so that $D=0$.
Substitute the values of *A* and *C* into **(7)** to find *E*: $1+(-2)+E=0$ so that $E=1$.
Substitute the values of *B* and *D* into **(8)** to find *F*: $0+0+F=2$ so that $F=2$.

Thus, the partial fraction decomposition **(1)** becomes:

$$\boxed{\frac{x^5+2}{\left(x^2+1\right)^3}=\frac{x}{x^2+1}-\frac{2x}{\left(x^2+1\right)^2}+\frac{x+2}{\left(x^2+1\right)^3}}$$

53.

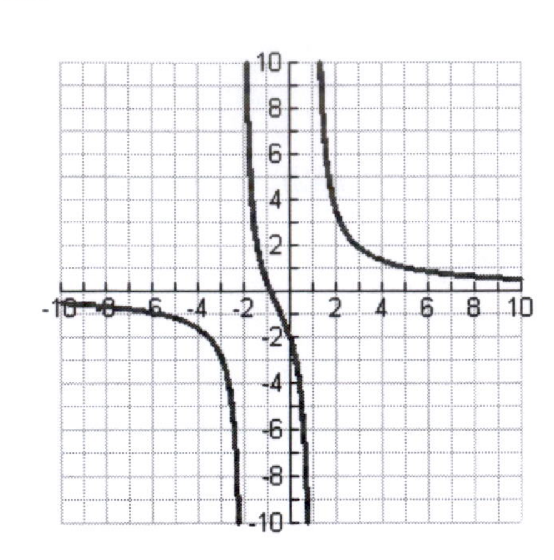

Notes on the graph:
Solid curve: Graph of y_1,
Dashed curve: Graph of y_2
In this case, since the graphs coincide, we know y_2 is, in fact, the partial fraction decomposition of y_1.

55.

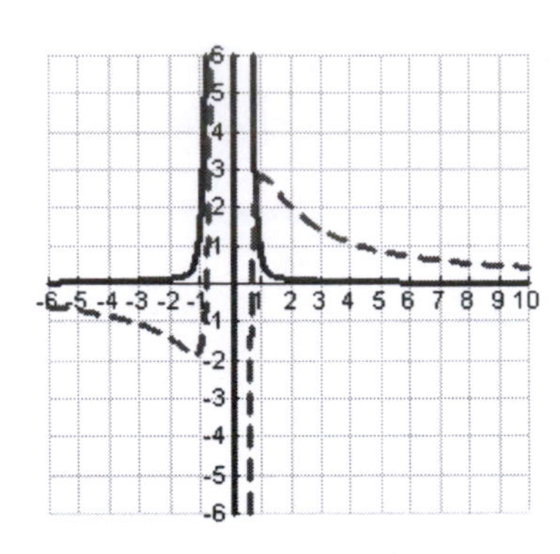

Notes on the graph:
Solid curve: Graph of y_1,
Dashed curve: Graph of y_2
In this case, since the graphs do not coincide, we know y_2 is not the partial fraction decomposition of y_1.

57. Since the graphs coincide, as seen below, y2 is the partial fraction decomposition of y1.

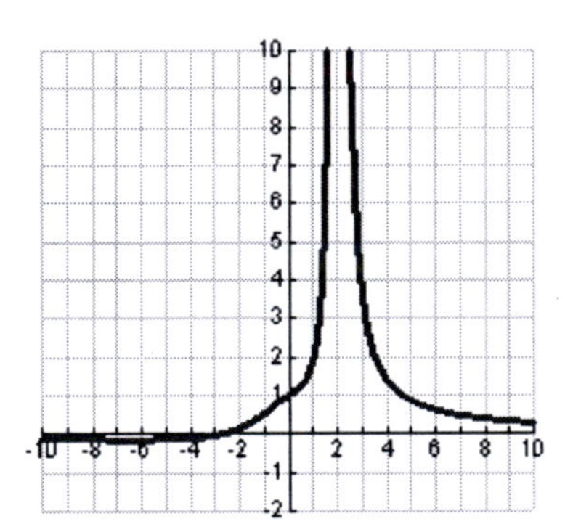

Section 6.4 Solutions ---

1. d	Above the line $y = x$, and do not include the line itself.
3. b	Below the line $y = x$, and do not include the line itself.

5.

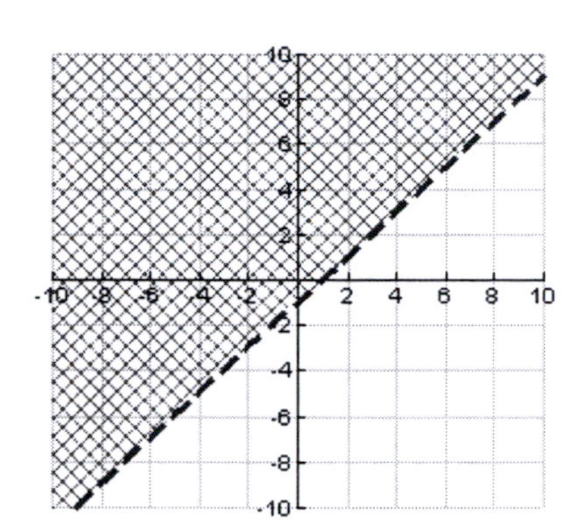

The equation of dashed curve is $y = x - 1$.

7.

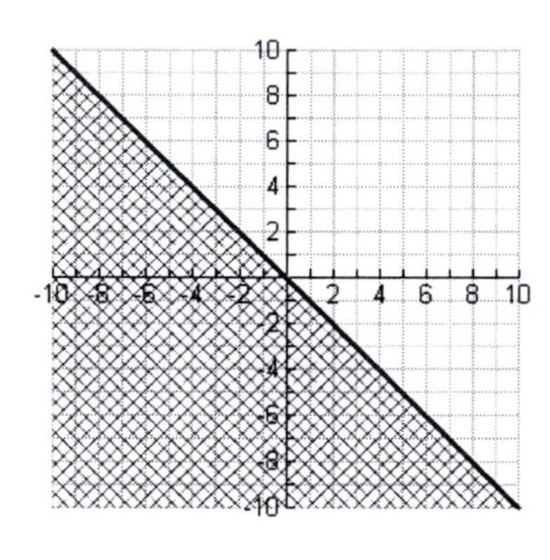

The equation of the solid curve is $y = -x$.

9.

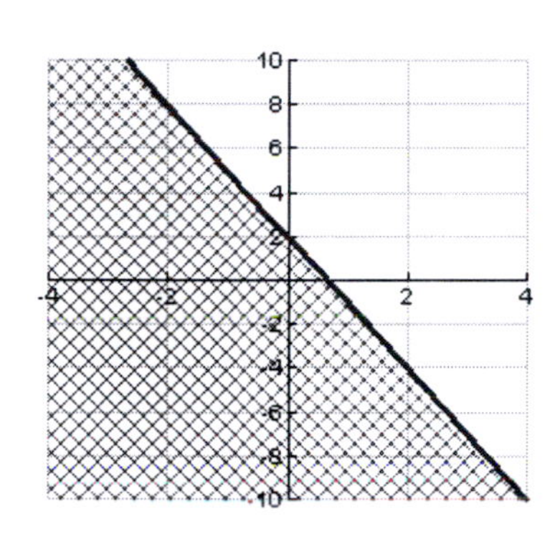

The equation of the solid curve is $y = -3x + 2$.

11.

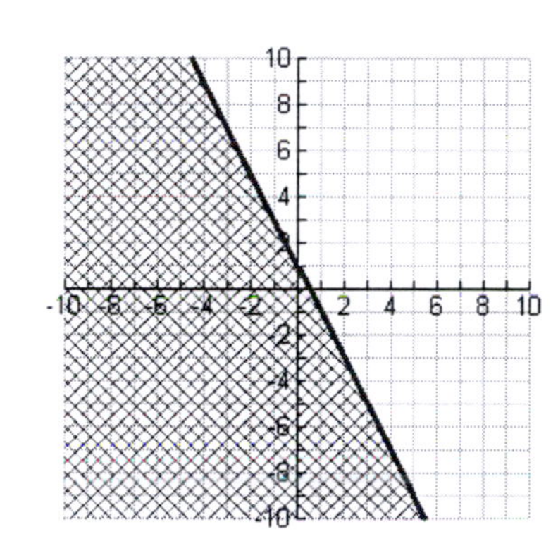

The equation of the solid curve is $y = -2x + 1$

13. Write the inequality as $y < \frac{1}{4}(2 - 3x)$.

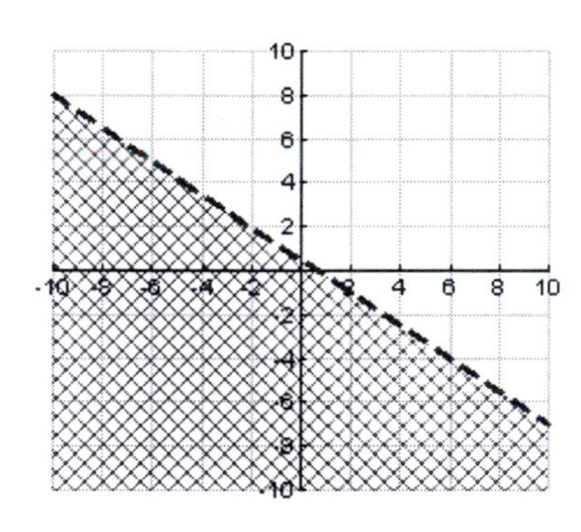

The equation of the dashed curve is $y = \frac{1}{4}(2 - 3x)$.

15.

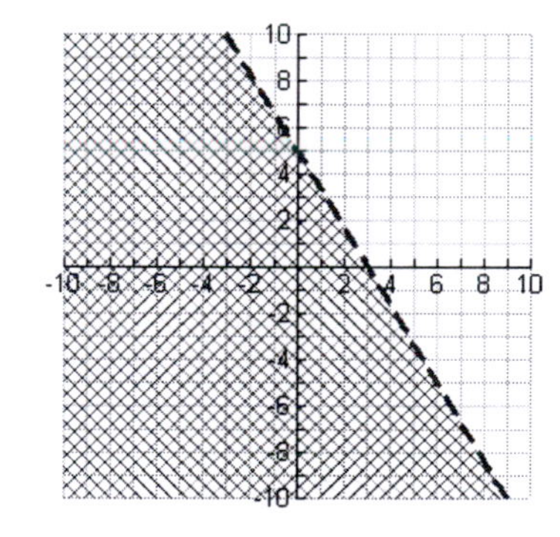

The equation of the dashed curve is $y = -\frac{5}{3}x + 5$.

17. Write the inequality as $y \le 2x - 3$.

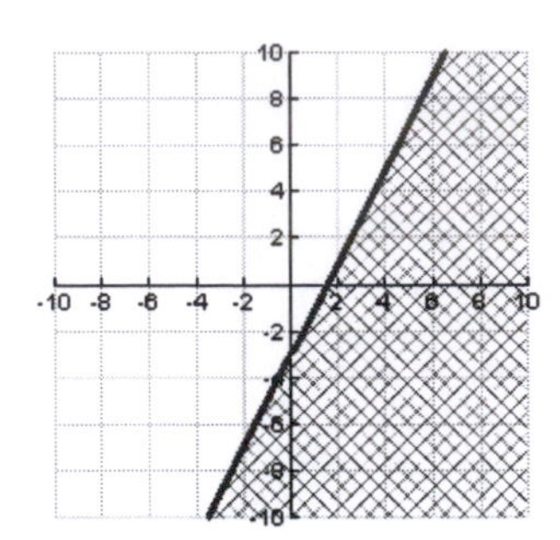

The equation of the solid curve is $y = 2x - 3$.

19.

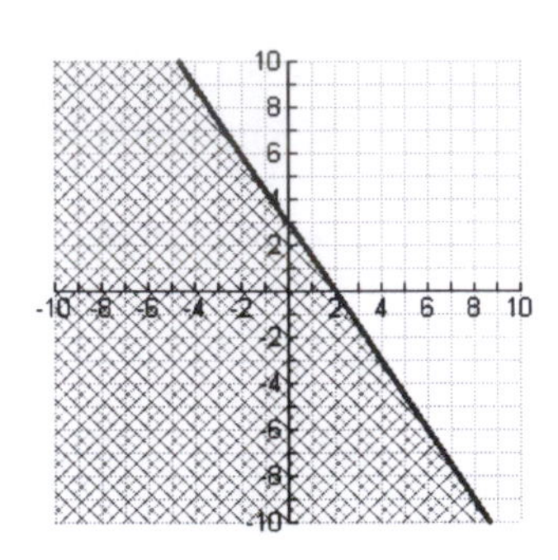

The equation of the solid curve is $y = -\frac{3}{2}x + 3$.

21.

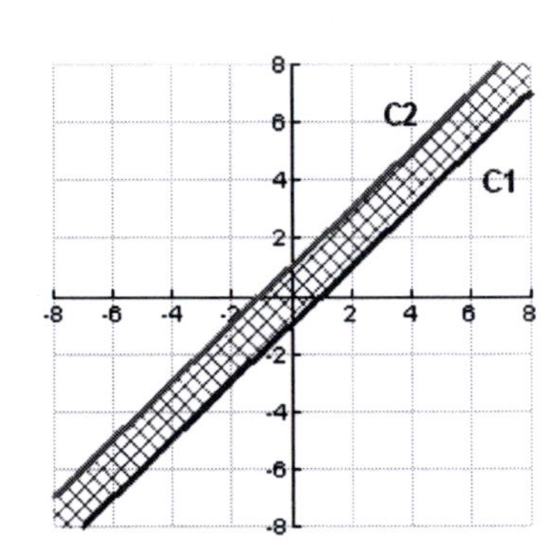

<u>Notes on the graph</u>:
C1: $y = x - 1$
C2: $y = x + 1$

23.

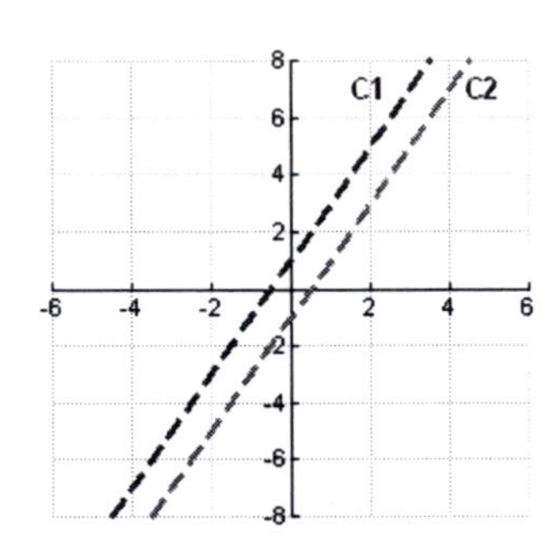

There is no common region, hence the system has no solution.
<u>Notes on the graph</u>:
C1: $y = 2x + 1$ **C2:** $y = 2x - 1$

25.

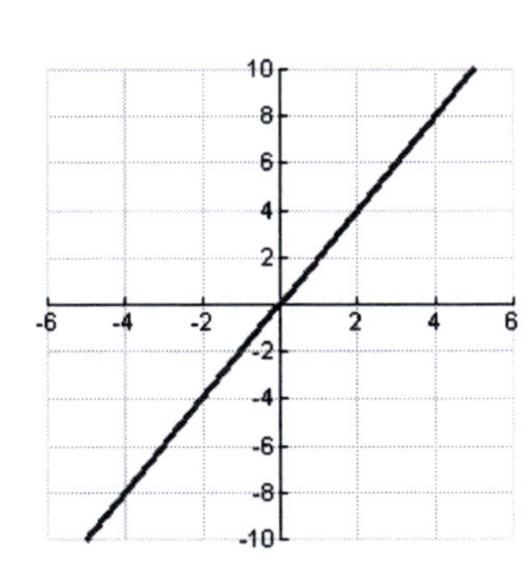

In this case, the common region is the line itself.
<u>Notes on the graph</u>:
C1 and **C2:** $y = 2x$

27.

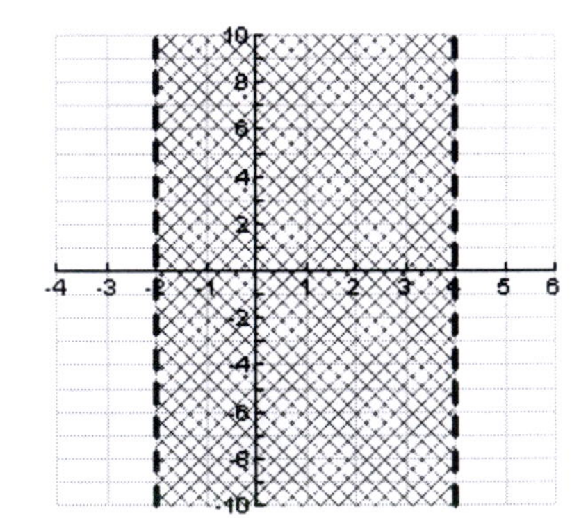

29.

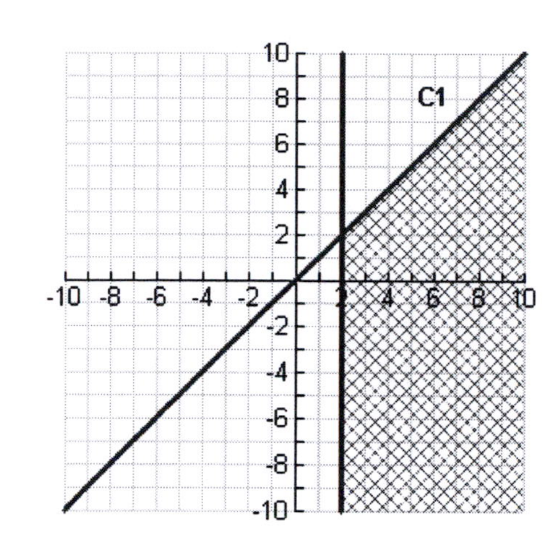

Notes on the graph:
C1: $y = x$

31.

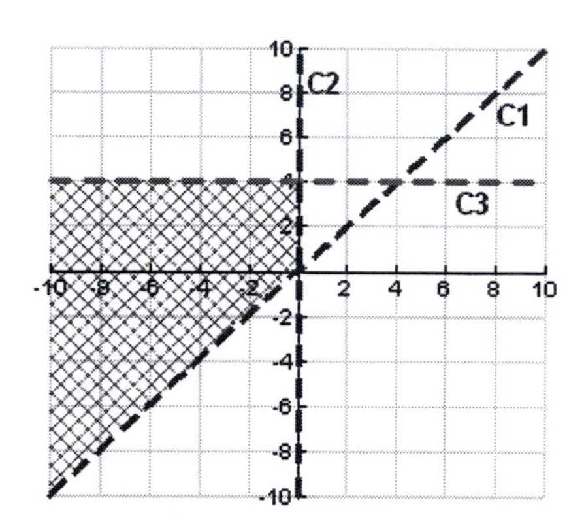

Notes on the graph:
C1: $y = x$
C2: $x = 0$
C3: $y = 4$

33.

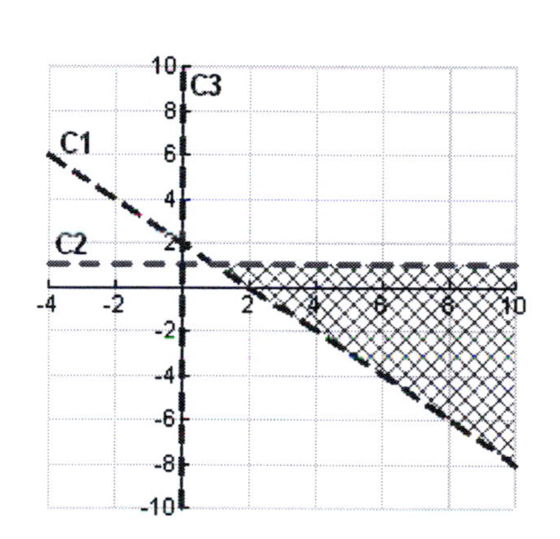

Notes on the graph:
C1: $y = -x + 2$
C2: $y = 1$
C3: $x = 0$

35.

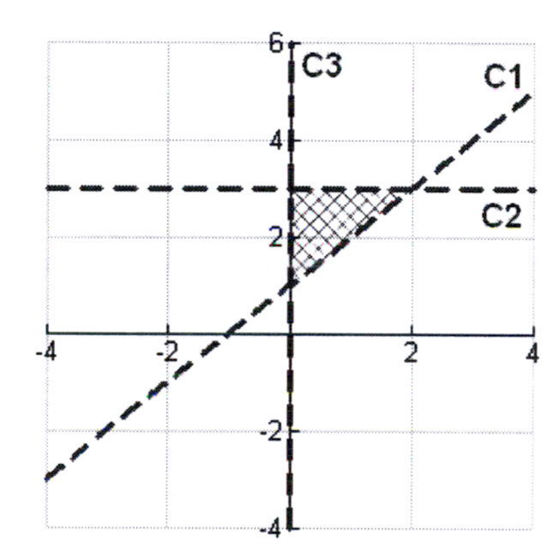

Notes on the graph:
C1: $y = x + 1$
C2: $y = 3$
C3: $x = 0$

37.

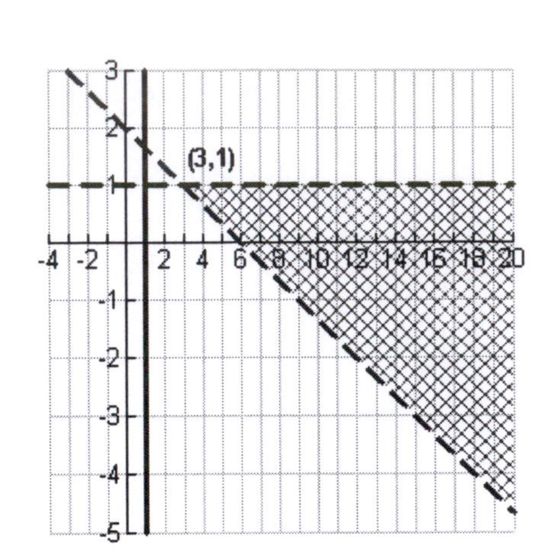

39.

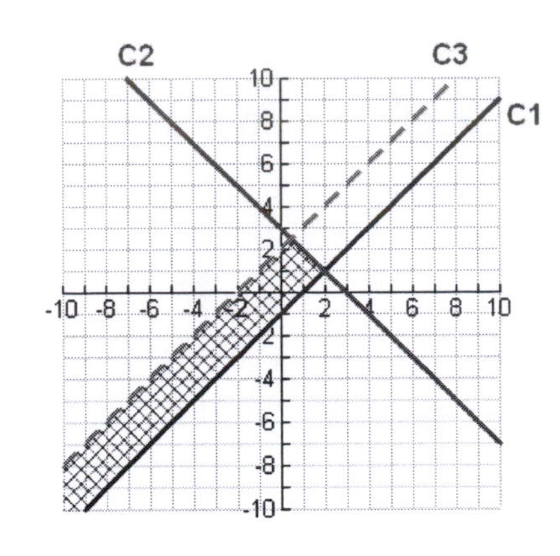

Notes on the graph:

C1: $y = x - 1$

C2: $y = -x + 3$

C3: $y = x + 2$

41.

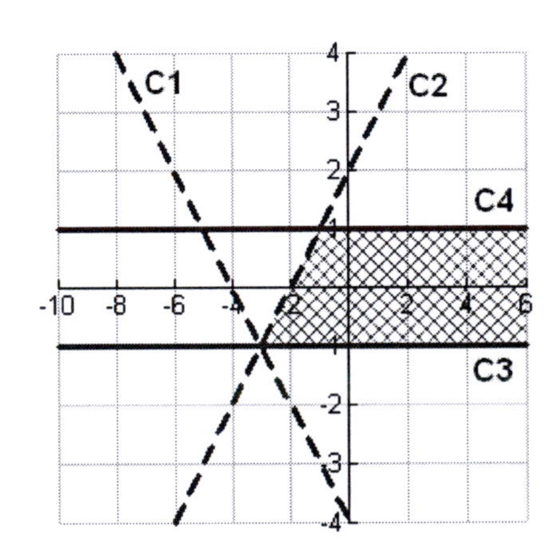

Notes on the graph:

C1: $y = -x - 4$

C2: $y = x + 2$

C3: $y = -1$

C4: $y = 1$

43.

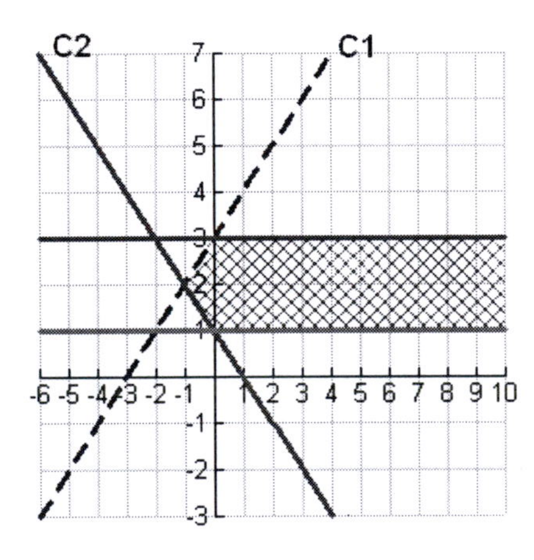

Notes on the graph:

C1: $y = x + 3$

C2: $y = -x + 1$

45. There is no solution since the regions do not overlap.

47.

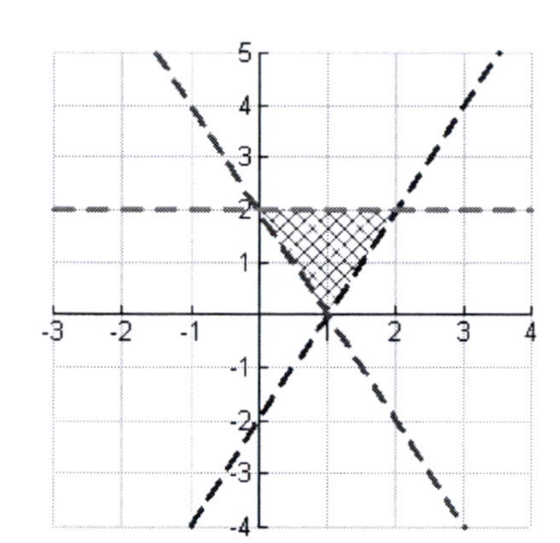

49. The shaded region is a triangle, seen below. Note that the base b has length 4, and the height h is 2. Hence, the area is:

$$\text{Area} = \tfrac{1}{2}(4)(2) = 4 \text{ units}^2$$

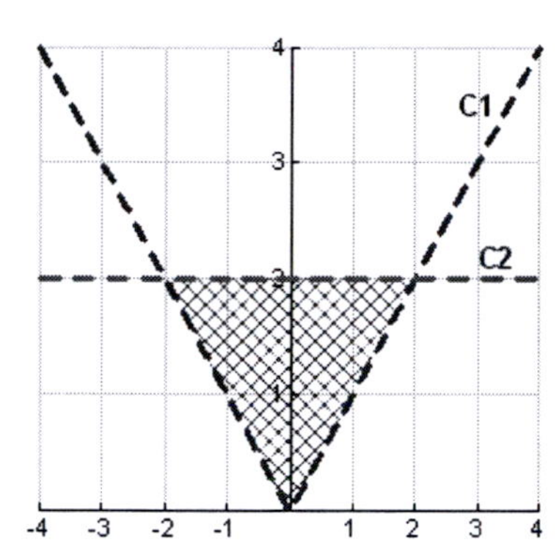

<u>Notes on the graph</u>:

C1: $y = |x|$

C2: $y = 2$

51.

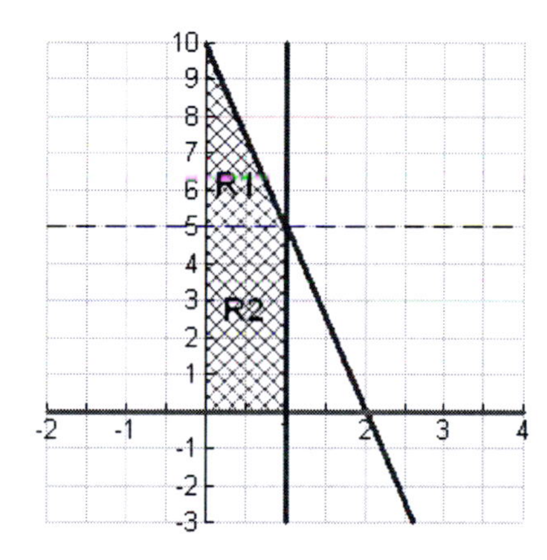

Area of **R1**: $\frac{1}{2}(1)(5) = \frac{5}{2}$ units^2

Area of **R2**: $1(5) = 5$ units^2

So, the area of the shaded region is 7.5 units^2.

53. Let x = number of cases of water and y = number of generators
Certainly, $x \geq 0,\ y \geq 0$. Then, we also have the following restrictions:
Cubic feet restriction: $x + 20y \leq 2400$
Weight restriction: $25x + 150y \leq 6000$

So, we obtain the following system of inequalities: $\begin{cases} x \geq 0,\ y \geq 0 \\ x + 20y \leq 2400 \\ 25x + 150y \leq 6000 \end{cases}$

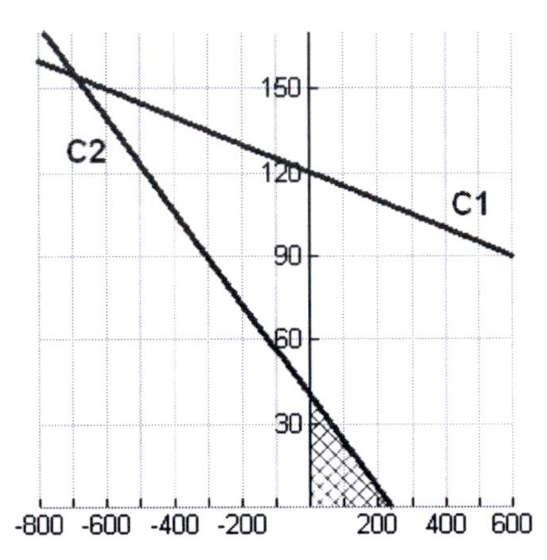

Notes on the graph: **C1:** $x + 20y = 2400$ **C2:** $25x + 150y = 6000$

55. a. Let x = number of ounces of food A, y = number of ounces of food B.

$275 \leq 10x + 20y, \quad x \geq 0,\ y \geq 0$

$125 \leq 15x + 10y$

$200 \leq 20x + 15y$

b.

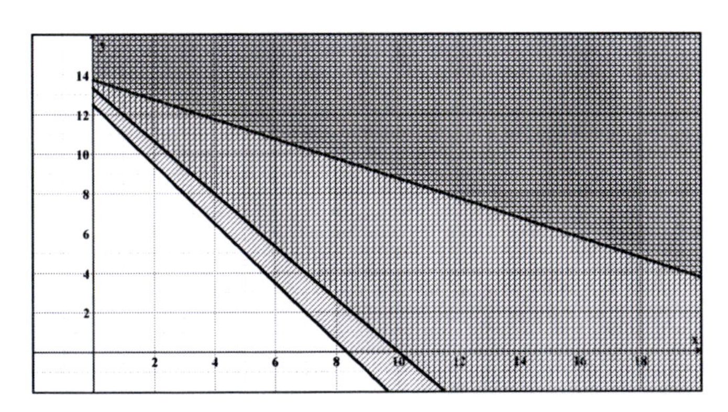

c. Two possible diet combinations are 2 ounces of food A and 14 ounces of food B or 10 ounces of food A and 10 ounces of food B.

57. Let x = number of USB wireless mice, y = number of Bluetooth mice.

a.

$$x \geq 2y, \qquad x \geq 0,\ y \geq 0$$
$$x + y \geq 1000$$

b.

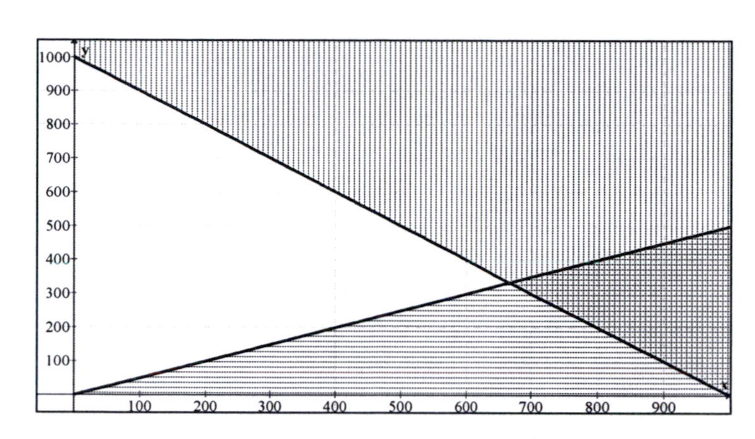

c. Two possible solutions would be for the manufacturer to produce 700 USB wireless mouse and 300 Bluetooth mouse or 800 USB wireless mouse and 300 Bluetooth mouse.

59. First, find the point of intersection of

$$\begin{cases} P = 80 - 0.01x \\ P = 20 + 0.02x \end{cases}$$

Equating these and solving for x yields $x = 2000$. Then, substituting this into the first equation yields $P = 60$. Hence, the system for consumer surplus is:

$$\begin{cases} P \leq 80 - 0.01x \\ P \geq 60 \\ x \geq 0 \end{cases}$$

61. The graph of the system in Exercise 55 is as follows:

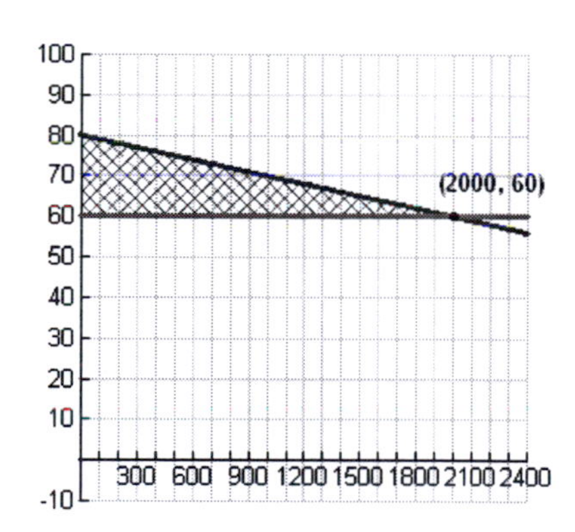

The consumer surplus is the area of the shaded region, which is

$$\tfrac{1}{2}(20)(2000) = 20{,}000 \text{ units}^2.$$

63. The shading should be <u>above</u> the line.

65. True. The line cuts the plane into two half planes, and one must either shade above or below the line.

67. False A dashed curve is used.

69. Given that $a < b$ and $c < d$, the solution region is a shaded rectangle which includes the upper and left sides – shown below:

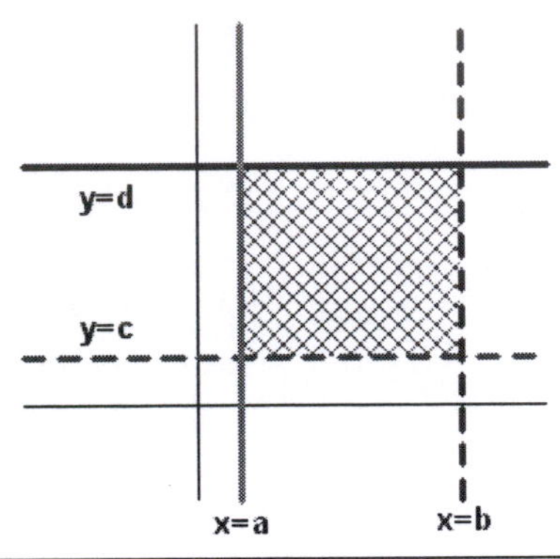

71. For any value of b, the following system has a solution:

$$\begin{cases} y \le ax + b \\ y \ge -ax + b \end{cases}$$

The solution region occurs in the first and fourth quadrants, and is shown graphically below:

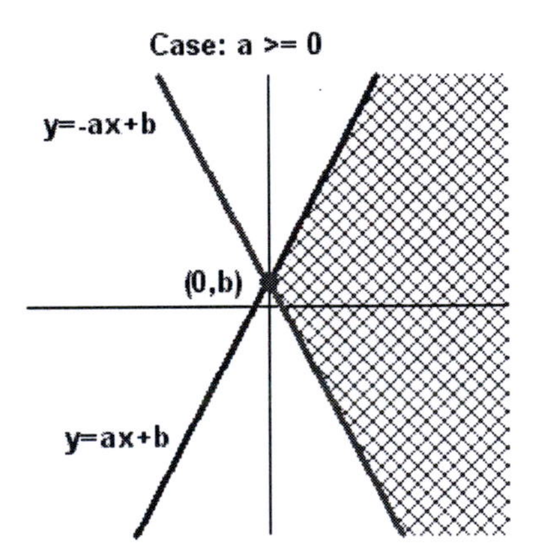

If $a = 0$, then the system becomes:

$$\begin{cases} y \le b \\ y \ge b \end{cases}$$

The solution to this system is the horizontal line $y = b$.

73.

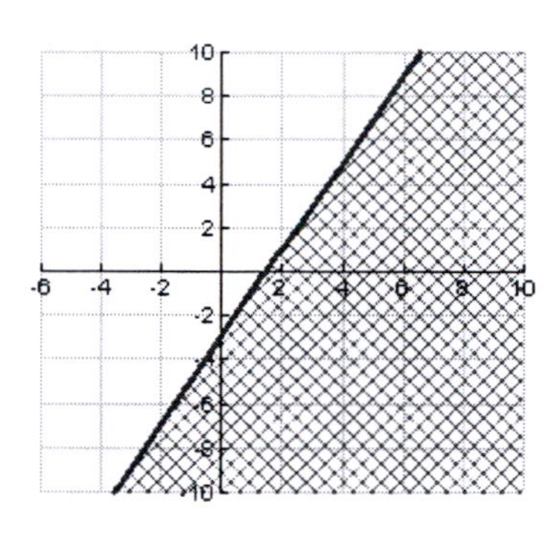

The solid curve is the graph of $y = 2x - 3$.

75.

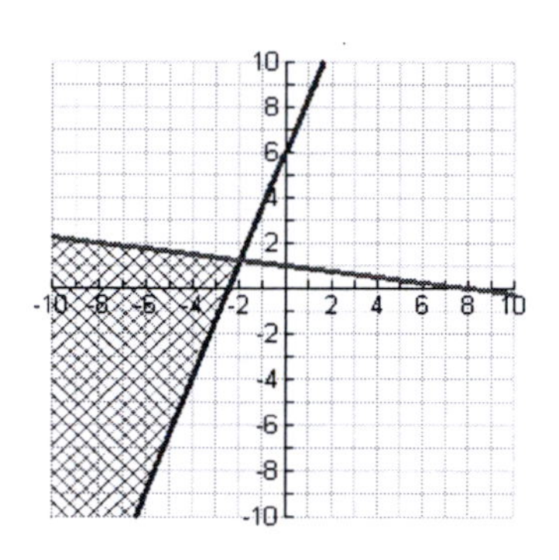

Section 6.5 Solutions --

1.

Vertex	Objective Function $z = 2x + 3y$
$(-1,4)$	$z = 2(-1) + 3(4) = 10$
$(2,4)$	$z = 2(2) + 3(4) = 16$
$(-2,-1)$	$z = 2(-2) + 3(-1) = -7$
$(1,-1)$	$z = 2(1) + 3(-1) = -1$

So, the maximum value of z is 16, and the minimum value of z is -7.

3.

Vertex	Objective Function $z = 1.5x + 4.5y$
$(-1,4)$	$z = 1.5(-1) + 4.5(4) = 16.5$
$(2,4)$	$z = 1.5(2) + 4.5(4) = 21$
$(-2,-1)$	$z = 1.5(-2) + 4.5(-1) = -7.5$
$(1,-1)$	$z = 1.5(1) + 4.5(-1) = -3$

So, the maximum value of z is 21, and the minimum value of z is -7.5.

5. The region in this case is:

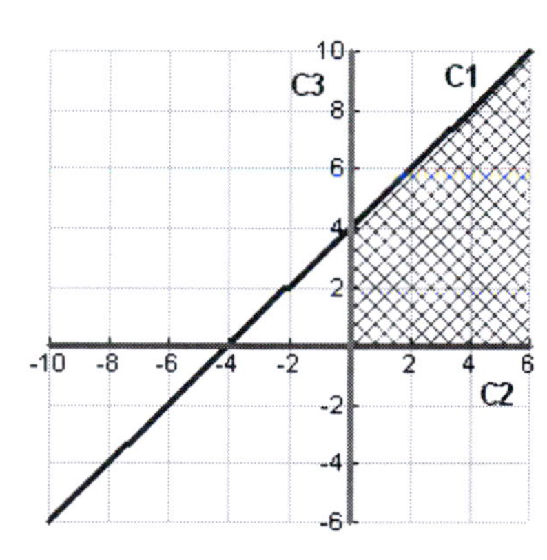

Notes on the graph:

C1: $y = 4 + x$

C2: $y = 0$

C3: $x = 0$

Vertex	Objective Function $z = 7x + 4y$
$(0,0)$	$z = 7(0) + 4(0) = 0$
$(0,4)$	$z = 7(0) + 4(4) = 16$

So, the minimum value of z is 0.

7. The region in this case is:

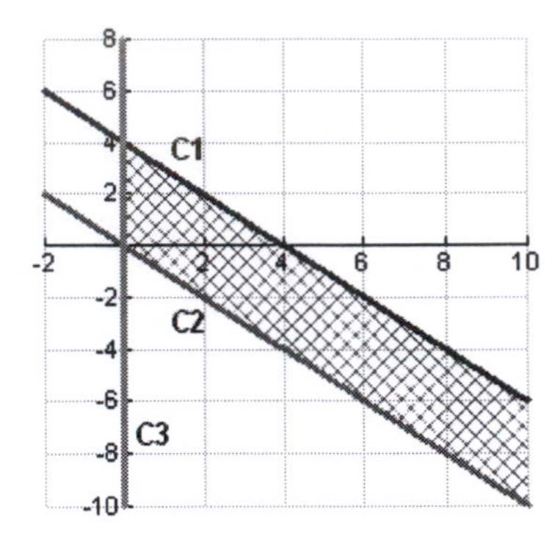

Notes on the graph:
C1: $y = 4 - x$ **C2:** $y = -x$ **C3:** $x = 0$

Vertex	Objective Function $z = 4x + 3y$
$(0,0)$	$z = 4(0) + 3(0) = 0$
$(0,4)$	$z = 4(0) + 3(4) = 12$
Additional point $(4,0)$	$z = 4(4) + 3(0) = 16$

Since the region is unbounded and the maximum must occur at a vertex, we conclude that the objective function does not have a maximum in this case.

9. The region in this case is:

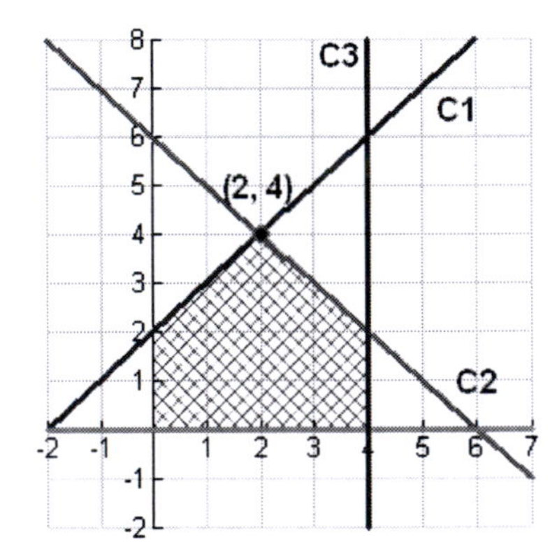

Notes on the graph:
C1: $y = x + 2$ **C2:** $y = -x + 6$ **C3:** $x = 4$

Vertex	Objective Function $z = 2.5x + 3.1y$
$(2,4)$	$z = 2.5(2) + 3.1(4) = 17.4$
$(0,0)$	$z = 2.5(0) + 3.1(0) = 0$
$(4,2)$	$z = 2.5(4) + 3.1(2) = 16.2$
$(0,2)$	$z = 2.5(0) + 3.1(2) = 6.2$
$(4, 0)$	$z = 2.5(4) + 3.1(0) = 10$

So, the minimum value of z is 0.

11. The region in this case is:

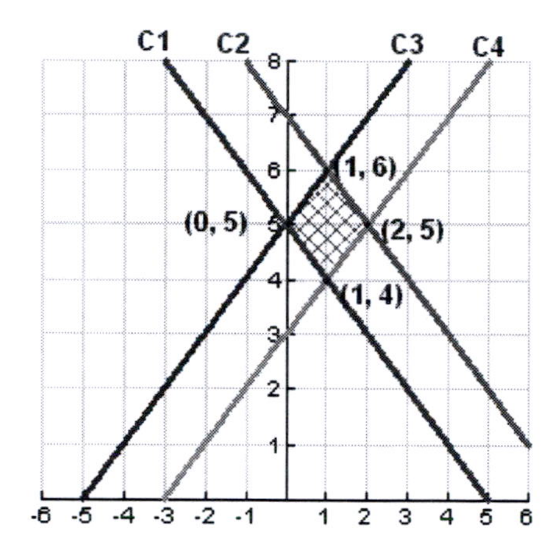

Notes on the graph:
C1: $y=-x+5$ **C2:** $y=-x+7$
C3: $y=x+5$ **C4:** $y=x+3$

We need all of the intersection points since they constitute the vertices:

Intersection of $y=x+5$ and $y=-x+7$:

$$x+5=-x+7$$
$$2x=2$$
$$x=1$$

So, the intersection point is (1,6).

Intersection of $y=x+3$ and $y=-x+7$:

$$x+3=-x+7$$
$$2x=4 \text{ so that } x=2$$

So, the intersection point is (2,5).

Intersection of $y=x+3$ and $y=-x+5$:

$$x+3=-x+5$$
$$2x=2 \text{ so that } x=1$$

So, the intersection point is (1,4).

Intersection of $y=x+5$ and $y=-x+5$:

$$x+5=-x+5$$
$$2x=0 \text{ so that } x=0$$

So, the intersection point is (0,5).

Now, compute the objective function at the vertices:

Vertex	Objective Function $z=\frac{1}{4}x+\frac{2}{5}y$
(1,6)	$z=\frac{1}{4}(1)+\frac{2}{5}(6)=\frac{53}{20}$
(2,5)	$z=\frac{1}{4}(2)+\frac{2}{5}(5)=\frac{5}{2}$
(1,4)	$z=\frac{1}{4}(1)+\frac{2}{5}(4)=\frac{37}{20}$
(0,5)	$z=\frac{1}{4}(0)+\frac{2}{5}(5)=2$

So, the maximum value of z is $\frac{53}{20}=2.65$.

13. Let x = number of Charley T-shirts
y = number of Francis T-shirts

Profit from Charley T-shirts:

Revenue – cost = $13x-7x=6x$

Profit from Francis T-shirts:

Revenue – cost = $10y-5y=5y$

So, to maximize profit, we must maximize the objective function $z=6x+5y$. We have the following constraints:

$$\begin{cases} 7x+5y\le 1000 \\ x+y\le 180 \\ x\ge 0,\ y\ge 0 \end{cases}$$

CONTINUED ONTO NEXT PAGE!

Notes on the graph:
C1: $y=-\frac{7}{5}x+200$
C2: $y=-x+180$
C3: $x=0$

We need all of the intersection points since they constitute the vertices:

Intersection of $7x+5y=1000$ and $x+y=180$:

$$-x+180=-\frac{7}{5}x+200$$
$$\frac{2}{5}x=20$$
$$x=50$$

So, the intersection point is (50, 130).

The region in this case is:

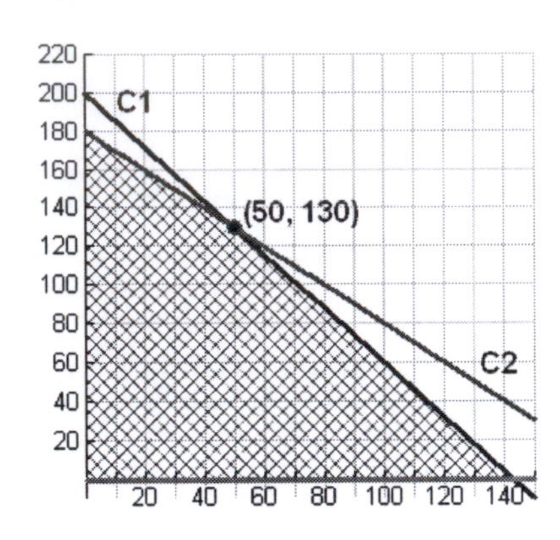

Now, compute the objective function at the vertices:

Vertex	Objective Function $z = 6x + 5y$
(0,180)	$z = 6(0) + 5(180) = 900$
$(\frac{1000}{7}, 0)$	$z = 6(\frac{1000}{7}) + 5(0) \cong 857.14$
(50, 130)	$z = 6(50) + 5(130) = 950$

So, to attain a maximum profit of \$950, he should sell 130 Francis T-shirts, and 50 Charley T-shirts.

15. Let x = # of desktops
y = # of laptops

We must maximize the objective function $z = 500x + 300y$ subject to the following constraints:

$$\begin{cases} 5x + 3y \le 90 \\ 700x + 400y \le 10{,}000 \\ y \ge 3x \\ x \ge 0,\ y \ge 0 \end{cases}$$

The region in this case is:

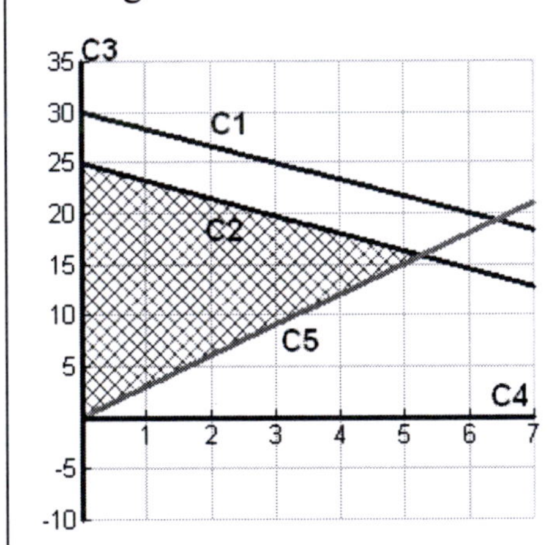

Notes on the graph:
C1: $y = 30 - \frac{5}{3}x$ **C2:** $y = 25 - \frac{7}{4}x$
C3: $x = 0$ **C4:** $y = 0$ **C5:** $y = 3x$

We need all of the intersection points since they constitute the vertices:

Intersection of $5x + 3y = 90$ and $700x + 400y = 10{,}000$:

$$30 - \tfrac{5}{3}x = 25 - \tfrac{7}{4}x$$
$$360 - 20x = 300 - 21x$$
$$x = -60$$

So, the intersection point is $(-60, 130)$.

Not a vertex, however, since x must be non-negative (see the region).

Intersection of **C5** and **C2**:

$$3x = 25 - \tfrac{7}{4}x$$
$$\tfrac{19}{4}x = 25 \text{ so that } x = \tfrac{100}{19}$$

So, the intersection point is $\left(\frac{100}{19}, \frac{300}{19}\right)$

Now, compute the objective function at the vertices:

Vertex	Objective Function $z = 500x + 300y$
(0,25)	$z = 500(0) + 300(25) = 7500$
(0,0)	$z = 500(0) + 300(0) = 0$
$\left(\frac{100}{19}, \frac{300}{19}\right)$	$z = 500(\frac{100}{19}) + 300(\frac{300}{19}) = \frac{140{,}000}{19}$

In order to attain a maximum profit of \$7500, he must sell 25 laptops and 0 desktops.

17. Let x = # first class cars
y = # second class cars
Let p denote the profit for each second class car. Then, the profit for each first class car is $2p$. Hence, we seek to maximize the objective function

$$z = py + 2px = p(y+2x).$$

We have the following constraints:

$$\begin{cases} x + y = 30 \\ 2 \le x \le 4 \\ y \ge 8x \end{cases}$$

The feasible region in this case is simply the bolded segment in the below diagram:

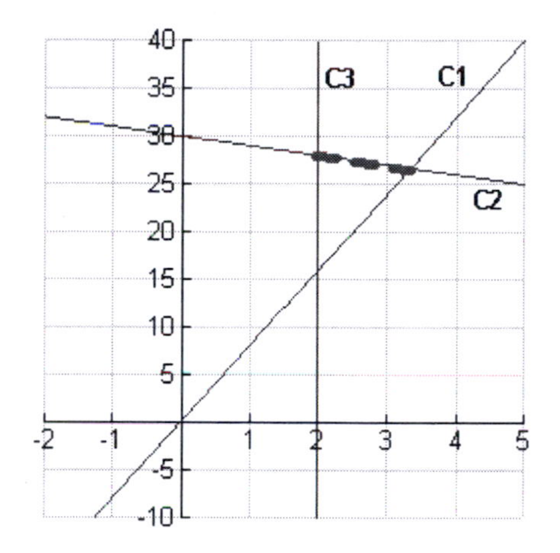

Notes on the graph:
C1: $y = 8x$
C2: $y = -x + 30$
C3: $x = 2$

We need all of the intersection points since they constitute the vertices:
Intersection of $x + y = 30$ and $y = 8x$:

$$-x + 30 = 8x$$
$$30 = 9x \text{ so that } \tfrac{10}{3} = x$$

So, the intersection point is $(\frac{10}{3}, \frac{80}{3})$.
Now, compute the objective function at the vertices:

Vertex	Objective Function $z = p(2x+y)$
(2, 16)	$z = p(2(2)+16) = 20p$
(2, 28)	$z = p(2(2)+28) = 32p$
$(\frac{10}{3}, \frac{80}{3})$	$z = p(2(\frac{10}{3}) + \frac{80}{3}) = \frac{100}{3}p$
(3, 27)	$z = p(2(3)+27) = 33p$

The maximum in this case would occur at $(\frac{10}{3}, \frac{80}{3})$. However, this is not tenable since one cannot have a fraction of a car. However, very near at the vertex (3, 27) the profit is very near to this one, as are the number of cars. Hence, to maximize profit, they should use:
3 first class cars and 27 second class cars.

19. Let $x = \#$ of regular skis
$y = \#$ of slalom skis

We must maximize the objective function $z = 25x + 50y$ subject to the following constraints:

$$\begin{cases} x + y \leq 400 \\ x \geq 200, \ \ y \geq 80 \\ x \geq 0, \ y \geq 0 \end{cases}$$

The region in this case is:

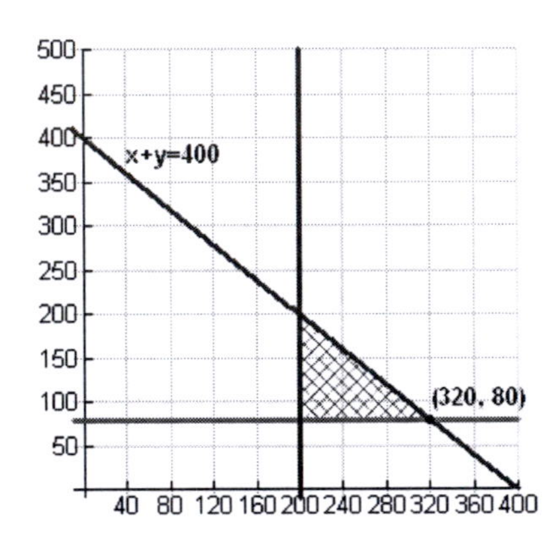

Now, compute the objective function at the vertices:

Vertex	Objective Function $z = 25x + 50y$
(200,80)	$z = 25(200) + 50(80) = 9000$
(200,200)	$z = 25(200) + 50(200) = 15,000$
(320,80)	$z = 25(320) + 50(80) = 12,000$

So, he should sell 200 of each type of ski.

21. Should compare the values of the objective function at the vertices rather than comparing the y-values of the vertices.

23. False. The region could be the entire plane (i.e., unconstrained).

25. Assume that $a > 2$. Then, the region looks like:

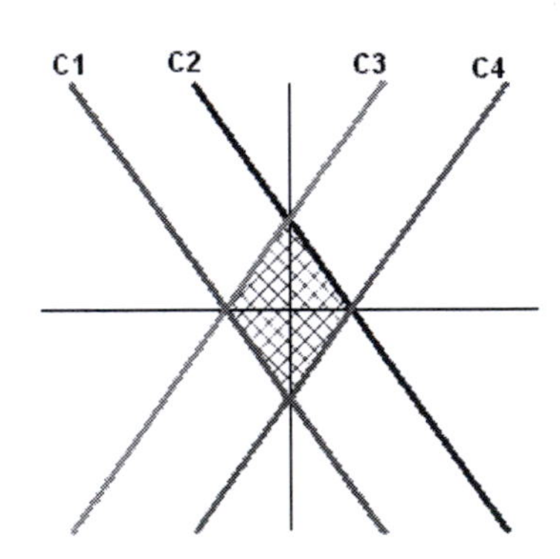

Notes on the graph:

C1: $y = -ax - a$ **C3:** $y = ax + a$

C2: $y = -ax + a$ **C4:** $y = ax - a$

Now, compute the objective function at the vertices:

Vertex	Objective Function $z = 2x + y$
$(0, a)$	$z = 2(0) + a = a$
$(-1, 0)$	$z = 2(-1) + 0 = -2$
$(0, -a)$	$z = 2(0) - a = -a$
$(1, 0)$	$z = 2(1) + 0 = 2$

Since $a > 2$, the maximum in this case would occur at $(0, a)$ and is a.

27. This is the same as #9 – the minimum occurs at (0, 0) and is 0.

29.

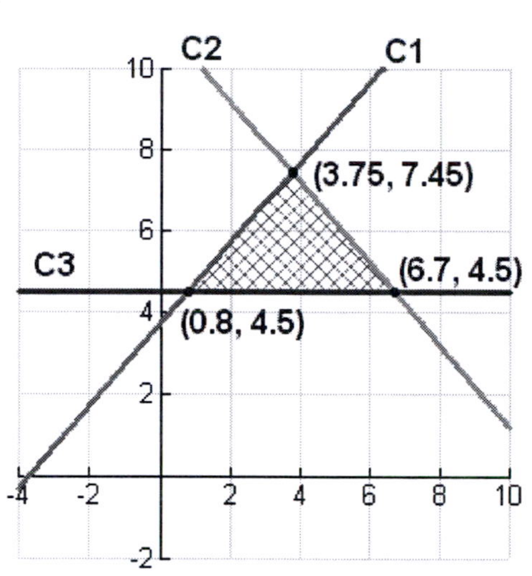

Notes on the graph:

C1: $y = x + 3.7$

C2: $y = -x + 11.2$

C3: $y = 4.5$

Vertex	Objective Function $z = 17x + 14y$
(0.8,4.5)	$z = 17(0.8) + 14(4.5) = 76.6$
(6.7,4.5)	$z = 17(6.7) + 14(4.5) = 176.9$
(3.75,7.45)	$z = 17(3.75) + 14(7.45)$ $= 168.05$

The maximum occurs at (6.7, 4.5) and is 176.9.

31.

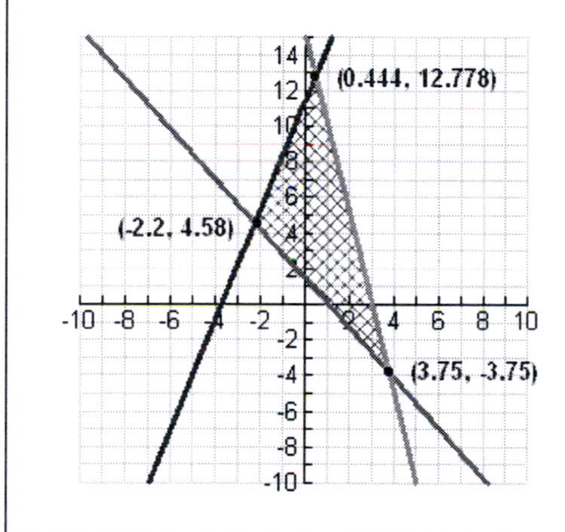

Vertex	Objective Function $z = 4.5x + 1.8y$
(-2.2, 4.58)	-1.656
(0.444, 12.778)	24.9984
(3.75, -3.75)	10.125

The maximum occurs at (0.444,12.778) and is approximately 25.

Chapter 6 Review Solutions

1. Solve the system: $\begin{cases} r-s=3 & \textbf{(1)} \\ r+s=3 & \textbf{(2)} \end{cases}$

Add **(1)** and **(2)**: $2r=6$ so that $r=3$
Substitute this value of r into **(1)** to find s:
$3-s=3$ so that $s=0$.

So, the solution is $\boxed{(3,0)}$.

3. Solve the system:

$$\begin{cases} -4x+2y=3 & \textbf{(1)} \\ 4x-y=5 & \textbf{(2)} \end{cases}$$

Add **(1)** and **(2)** to eliminate x:

$$y=8 \quad \textbf{(3)}$$

Substitute **(3)** into **(2)** and solve for x:

$$4x-8=5$$

$$4x=13$$

$$x=\tfrac{13}{4}$$

So, the solution is $\boxed{\left(\tfrac{13}{4},8\right)}$.

5. Solve the system: $\begin{cases} x+y=3 & \textbf{(1)} \\ x-y=1 & \textbf{(2)} \end{cases}$

Solve **(1)** for y: $y=3-x$ **(3)**
Substitute **(3)** into **(2)** and solve for x:

$$x-(3-x)=1$$

$$x-3+x=1$$

$$2x=4 \text{ so that } x=2$$

Substitute this value of x into **(3)** to find y:

$$y=1.$$

So, the solution is $\boxed{(2,1)}$.

7. Solve the system: $\begin{cases} 4c-4d=3 & \textbf{(1)} \\ c+d=4 & \textbf{(2)} \end{cases}$

Solve **(2)** for c: $c=4-d$ **(3)**
Substitute **(3)** into **(1)** and solve for d:

$$4(4-d)-4d=3$$

$$16-8d=3$$

$$8d=13$$

$$d=\tfrac{13}{8}$$

Now, substitute this value of d into **(3)** to find c: $c=4-\tfrac{13}{8}=\tfrac{19}{8}$

So, the solution is $\boxed{\left(\tfrac{19}{8},\tfrac{13}{8}\right)}$.

9.

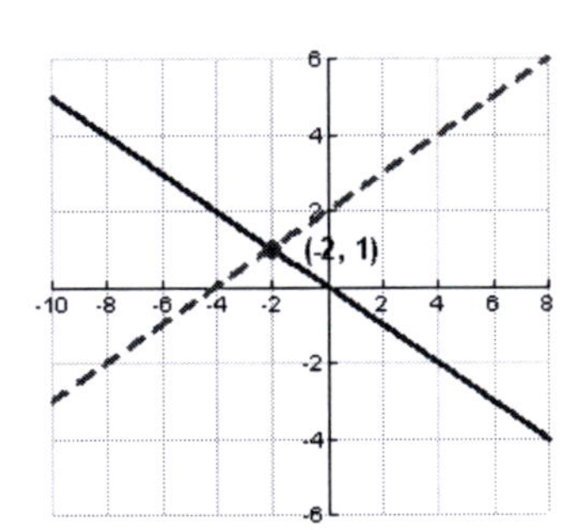

Notes on the graph:
Solid curve: $y=-\tfrac{1}{2}x$
Dashed curve: $y=\tfrac{1}{2}x+2$
So, the solution is $(-2,1)$.

11.

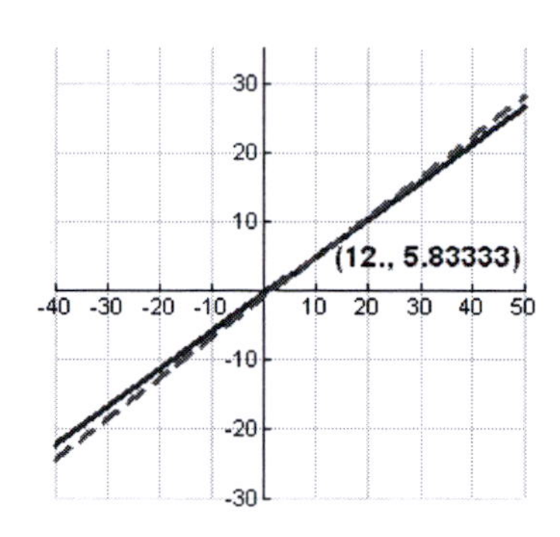

Notes on the graph: (Careful! The curves are very close together in a vicinity of the point of intersection.)

Solid curve: $1.3x - 2.4y = 1.6$

Dashed curve: $0.7x - 1.2y = 1.4$

So, the solution is $(12, 5.8\overline{3})$.

13. Solve the system:

$$\begin{cases} 5x - 3y = 21 & \textbf{(1)} \\ -2x + 7y = -20 & \textbf{(2)} \end{cases}$$

To eliminate x, multiply **(1)** by 2:

$$10x - 6y = 42 \quad \textbf{(3)}$$

Multiply **(2)** by 5: $-10x + 35y = -100$ **(4)**

Add **(3)** and **(4)**: $29y = -58 \Rightarrow y = -2$

Substitute this into **(1)** to find x:

$$5x = 15 \Rightarrow x = 3$$

So, the solution is $\boxed{(3, -2)}$.

15. Solve the system:

$$\begin{cases} 10x - 7y = -24 & \textbf{(1)} \\ 7x + 4y = 1 & \textbf{(2)} \end{cases}$$

To eliminate y, multiply **(1)** by 4:

$$40x - 28y = -96 \quad \textbf{(3)}$$

Multiply **(2)** by 7: $49x + 28y = 7$ **(4)**

Add **(3)** and **(4)**: $89x = -89 \Rightarrow x = -1$

Substitute this into **(1)** to find y:

$$-7y = -14 \Rightarrow y = 2$$

So, the solution is $\boxed{(-1, 2)}$.

17. c The intersection point is $(\frac{25}{11}, \frac{2}{11})$

19. d Multiplying the first equation by 2 reveals that the two equations are equivalent. Hence, the graphs are the same line.

21. Let x = number of ml of 6% NaCl.
y = number of ml of 18% NaCl.
Must solve the system:

$$\begin{cases} 0.06x + 0.18y = (0.15)(42) & \textbf{(1)} \\ x + y = 42 & \textbf{(2)} \end{cases}$$

First, for convenience, simplify **(1)** to get the equivalent system:

$$\begin{cases} x + 3y = 105 & \textbf{(3)} \\ x + y = 42 & \textbf{(4)} \end{cases}$$

Multiply **(1)** by -1, and then add to **(2)**:

$$-2y = -63$$

$$y = \tfrac{63}{2} = 31.5 \quad \textbf{(6)}$$

Substitute **(6)** into **(4)** to find x:

$$x + 31.5 = 42$$

$$x = 10.5$$

So, should use approximately 10.5 ml of 6% NaCl and 31.5 ml of 18% NaCl.

23. Solve the system:

$$\begin{cases} x + y + z = 1 & \textbf{(1)} \\ x - y - z = -3 & \textbf{(2)} \\ -x + y + z = 3 & \textbf{(3)} \end{cases}$$

Add **(2)** and **(3)**: $0 = 0$ **(4)**

Hence, we know that the system has infinitely many solutions.

Add **(1)** and **(2)**:
$2x = -2$ so that $x = -1$ **(5)**

Substitute this value of x into **(1)**:

$$-1 + y + z = 1$$

$$y + z = 2 \quad \textbf{(6)}$$

Let $z = a$. Substitute this into **(6)** to find z:

$$y = -a + 2$$

Thus, the solutions are:
$\boxed{x = -1,\ y = -a + 2,\ z = a}$.

25. Solve the system:

$$\begin{cases} x + y + z = 7 & \textbf{(1)} \\ x - y - z = 17 & \textbf{(2)} \\ y + z = 5 & \textbf{(3)} \end{cases}$$

Step 1: Obtain a system of 2 equations in 2 unknowns by eliminating the same variable in two pairs of equations.

Subtract **(1) - (2)** and simplify: $y + z = -5$ **(4)**

Solve the system:

$$(*) \begin{cases} y + z = 5 & \textbf{(3)} \\ y + z = -5 & \textbf{(4)} \end{cases}$$

Note that equating **(3)** and **(4)** yields the false statement 5 = -5. Hence, the system has $\boxed{\text{no solution}}$.

27. Since we are given that the points $(16,2)$, $(40,6)$, and $(65,4)$, the system that must be solved is:

$$\begin{cases} 2 = a(16)^2 + b(16) + c \\ 6 = a(40)^2 + b(40) + c \\ 4 = a(65)^2 + b(65) + c \end{cases}$$

which is equivalent to

$$\begin{cases} 2 = \ \ 256a + 16b + c & \textbf{(1)} \\ 6 = 1600a + 40b + c & \textbf{(2)} \\ 4 = 4225a + 65b + c & \textbf{(3)} \end{cases}$$

Step 1: Obtain a 2×2 system:

Multiply **(1)** by -1, and add to **(2)**:

$$1344a + 24b = 4 \quad \textbf{(4)}$$

Multiply **(1)** by -1, and add to **(3)**:

$$3969a + 49b = 2 \quad \textbf{(5)}$$

These steps yield the following 2×2 system:

Step 2: Solve the system in Step 1.

$$\begin{cases} 1344a + 24b = 4 & \textbf{(4)} \\ 3969a + 49b = 2 & \textbf{(5)} \end{cases}$$

Multiply **(4)** by -49 and **(5)** by 24. Then, add them together:

$$29,400a = -148$$

$$a \cong -0.0050 \quad \textbf{(6)}$$

Substitute **(6)** into **(4)**:

$$1344(-0.005) + 24b = 4$$

$$b \cong 0.4486 \quad \textbf{(7)}$$

Step 3: Find values of remaining variables.

Substitute **(6)** and **(7)** into **(1)**:

$$256(-0.005) + 16(0.447) + c = 2$$

$$c \cong -3.8884$$

Thus, the polynomial has the approximate equation $\boxed{y = -0.0050x^2 + 0.4486x - 3.8884}$.

29.

$$\frac{4}{(x-1)^2(x+3)(x-5)} = \frac{A}{x-1} + \frac{B}{(x-1)^2} + \frac{C}{x+3} + \frac{D}{x-5}$$

31.

$$\frac{12}{x(4x+5)(2x+1)^2} = \frac{A}{x} + \frac{B}{4x+5} + \frac{C}{2x+1} + \frac{D}{(2x+1)^2}$$

33.

$$\frac{3}{x^2 + x - 12} = \frac{3}{(x-3)(x+4)} = \frac{A}{x-3} + \frac{B}{x+4}$$

35.

$$\frac{3x^3 + 4x^2 + 56x + 62}{(x^2+17)^2} = \frac{Ax+B}{x^2+17} + \frac{Cx+D}{(x^2+17)^2}$$

37. The partial fraction decomposition has the form:

$$\frac{9x+23}{(x-1)(x+7)} = \frac{A}{x-1} + \frac{B}{x+7} \quad \textbf{(1)}$$

To find the coefficients, multiply both sides of **(1)** by $(x-1)(x+7)$, and gather like terms: $9x+23 = A(x+7)+B(x-1) = (A+B)x+(7A-B)$ **(2)**

Equate corresponding coefficients in **(2)** to obtain the following system:

$$\textbf{(*)} \begin{cases} A+B=9 & \textbf{(3)} \\ 7A-B=23 & \textbf{(4)} \end{cases}$$

Now, solve system **(*)**:

Add **(3)** and **(4)**: $8A=32 \Rightarrow A=4$

Substitute this value of A into **(3)** to see that $B=5$.

Thus, the partial fraction decomposition **(1)** becomes:

$$\boxed{\frac{9x+23}{(x-1)(x+7)} = \frac{4}{x-1} + \frac{5}{x+7}}$$

39. The partial fraction decomposition has the form:

$$\frac{13x^2+90x-25}{2x^3-50x} = \frac{13x^2+90x-25}{2x(x-5)(x+5)} = \frac{A}{2x} + \frac{B}{x-5} + \frac{C}{x+5} \quad \textbf{(1)}$$

To find the coefficients, multiply both sides of **(1)** by $2x(x-5)(x+5)$, and gather like terms:

$$\begin{aligned} 13x^2+90x-25 &= A(x^2-25)+B(2x)(x+5)+C(2x)(x-5) \\ &= (A+2B+2C)x^2+(10B-10C)x-25A \end{aligned} \quad \textbf{(2)}$$

Equate corresponding coefficients in **(2)** to obtain the following system:

$$\textbf{(*)} \begin{cases} A+2B+2C=13 & \textbf{(3)} \\ 10B-10C=90 & \textbf{(4)} \\ -25A=-25 & \textbf{(5)} \end{cases}$$

Now, solve system **(*)**:

Solve **(5)**: $A=1$

Substitute this value of A into **(3)**: $B+C=6$ **(6)**

Note that **(4)** is equivalent to $B-C=9$ **(7)**. So, solve the system:

$$\begin{cases} B+C=6 & \textbf{(6)} \\ B-C=9 & \textbf{(7)} \end{cases}$$

Add **(6)** and **(7)**: $2B=15 \Rightarrow B=\frac{15}{2}$

Substitute this value of B into **(6)**: $C=-\frac{3}{2}$

Thus, the partial fraction decomposition **(1)** becomes:

$$\boxed{\frac{13x^2+90x-25}{2x^3-50x} = \frac{1}{2x} + \frac{15}{2(x-5)} - \frac{3}{2(x+5)}}$$

41. The partial fraction decomposition has the form:

$$\frac{2}{x(x+1)} = \frac{A}{x} + \frac{B}{x+1} \quad \textbf{(1)}$$

To find the coefficients, multiply both sides of **(1)** by $x(x+1)$, and gather like terms:

$$2 = A(x+1) + Bx$$

$$2 = (A+B)x + A \quad \textbf{(2)}$$

Equate corresponding coefficients in **(2)** to obtain the following system:

$$(*) \begin{cases} A + B = 0 & \textbf{(3)} \\ A = 2 & \textbf{(4)} \end{cases}$$

Now, solve system $(*)$:

Substitute **(4)** into **(3)** to see that $B = -2$.

Thus, the partial fraction decomposition **(1)** becomes: $\boxed{\frac{2}{x(x+1)} = \frac{-2}{x+1} + \frac{2}{x}}$

43. The partial fraction decomposition has the form:

$$\frac{5x-17}{(x+2)^2} = \frac{A}{x+2} + \frac{B}{(x+2)^2} \quad \textbf{(1)}$$

To find the coefficients, multiply both sides of **(1)** by $(x+2)^2$, and gather like terms:

$$5x - 17 = A(x+2) + B$$

$$5x - 17 = Ax + (2A + B) \quad \textbf{(2)}$$

Equate corresponding coefficients in **(2)** to obtain the following system:

$$(*) \begin{cases} A = 5 & \textbf{(3)} \\ 2A + B = -17 & \textbf{(4)} \end{cases}$$

Substitute **(3)** into **(4)** to find *B*:

$$2(5) + B = -17 \text{ so that } B = -27$$

Thus, the partial fraction decomposition **(1)** becomes:

$$\boxed{\frac{5x-17}{(x+2)^2} = \frac{5}{x+2} - \frac{27}{(x+2)^2}}$$

45.

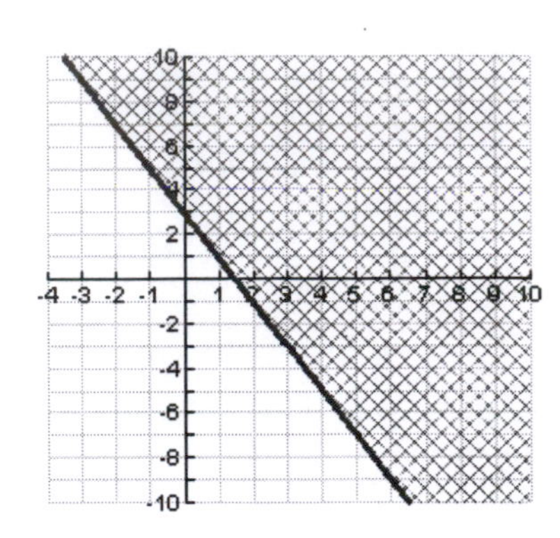

The solid curve is the graph of $y = -2x + 3$.

47. Shade above the dashed line $y = \frac{1}{4}(5 - 2x)$.

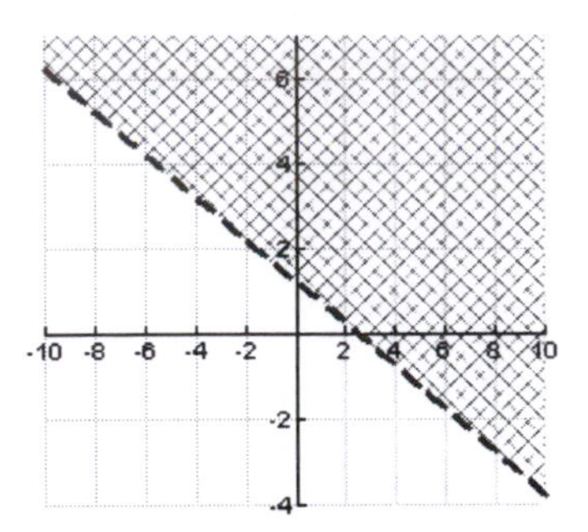

49. Shade above the solid line $y = -3x + 2$.

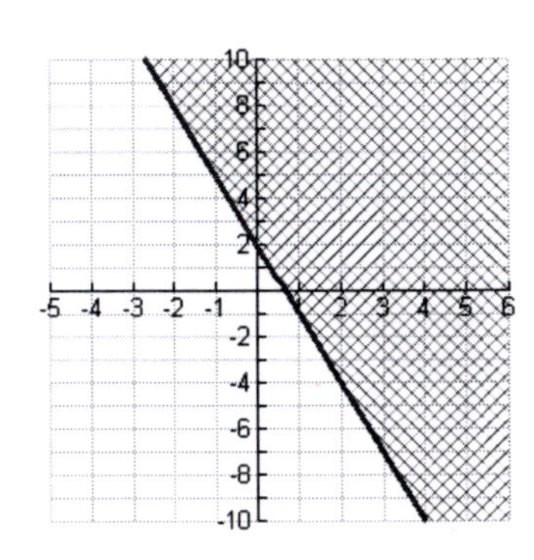

51. Shade below the solid line $y = -\frac{3}{8}x + 2$.

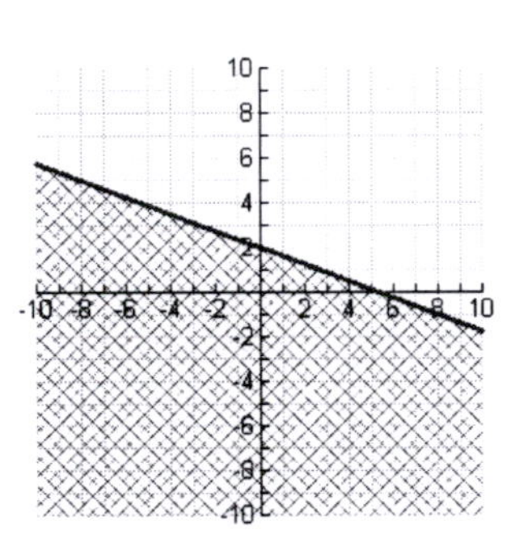

53.

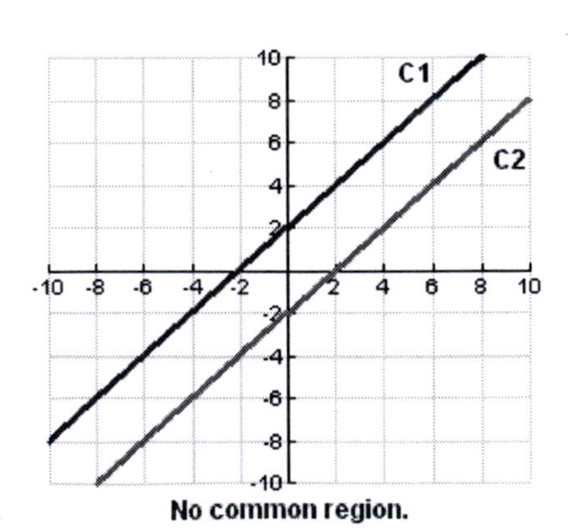

Notes on the graph:

C1: $y = x + 2$

C2: $y = x - 2$

Since there is no region in common with both inequalities, the system has no solution.

55.

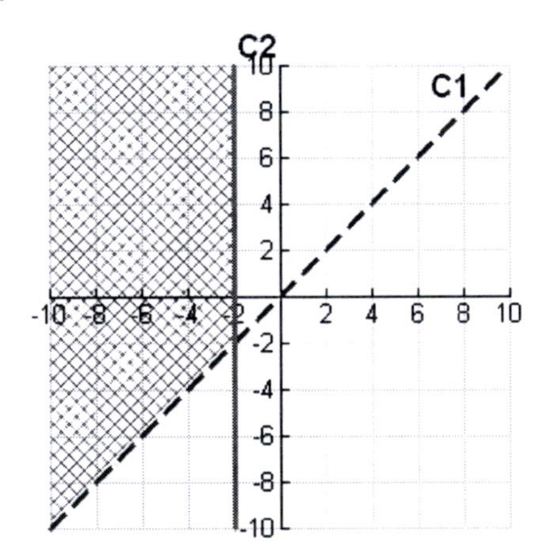

Notes on the graph:

C1: $y = x$

C2: $x = -2$

41. The partial fraction decomposition has the form:

$$\frac{2}{x(x+1)} = \frac{A}{x} + \frac{B}{x+1} \quad \textbf{(1)}$$

To find the coefficients, multiply both sides of **(1)** by $x(x+1)$, and gather like terms:

$$2 = A(x+1) + Bx$$

$$2 = (A+B)x + A \quad \textbf{(2)}$$

Equate corresponding coefficients in **(2)** to obtain the following system:

$$(*) \begin{cases} A+B=0 & \textbf{(3)} \\ A=2 & \textbf{(4)} \end{cases}$$

Now, solve system $(*)$:

Substitute **(4)** into **(3)** to see that $B=-2$.

Thus, the partial fraction decomposition **(1)** becomes: $\boxed{\frac{2}{x(x+1)} = \frac{-2}{x+1} + \frac{2}{x}}$

43. The partial fraction decomposition has the form:

$$\frac{5x-17}{(x+2)^2} = \frac{A}{x+2} + \frac{B}{(x+2)^2} \quad \textbf{(1)}$$

To find the coefficients, multiply both sides of **(1)** by $(x+2)^2$, and gather like terms:

$$5x-17 = A(x+2) + B$$

$$5x-17 = Ax + (2A+B) \quad \textbf{(2)}$$

Equate corresponding coefficients in **(2)** to obtain the following system:

$$(*) \begin{cases} A=5 & \textbf{(3)} \\ 2A+B=-17 & \textbf{(4)} \end{cases}$$

Substitute **(3)** into **(4)** to find *B*:

$$2(5) + B = -17 \text{ so that } B = -27$$

Thus, the partial fraction decomposition **(1)** becomes:

$$\boxed{\frac{5x-17}{(x+2)^2} = \frac{5}{x+2} - \frac{27}{(x+2)^2}}$$

45.

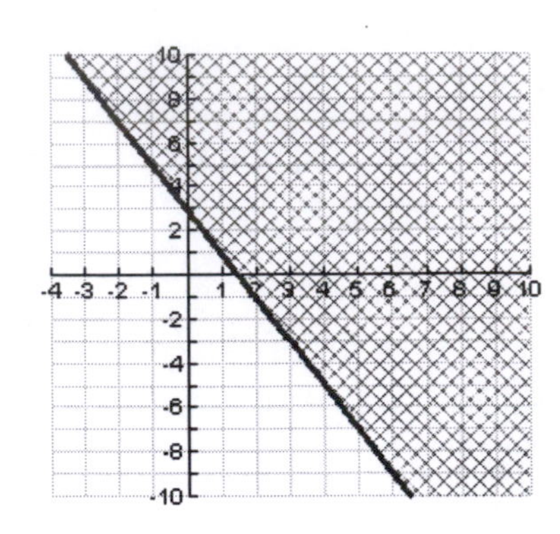

The solid curve is the graph of $y = -2x + 3$.

47. Shade above the dashed line $y = \frac{1}{4}(5 - 2x)$.

49. Shade above the solid line $y = -3x + 2$.

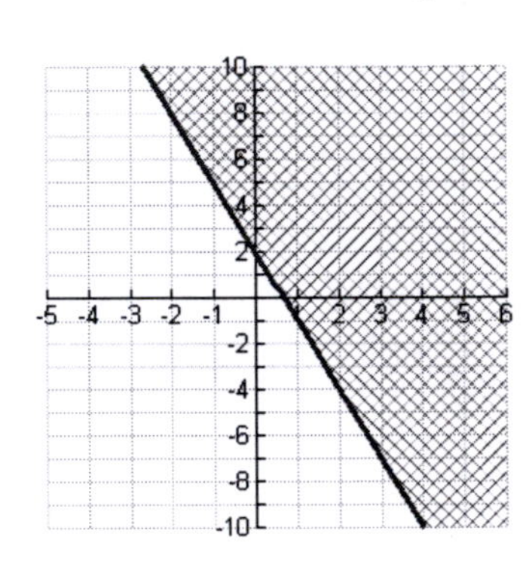

51. Shade below the solid line $y = -\frac{3}{8}x + 2$.

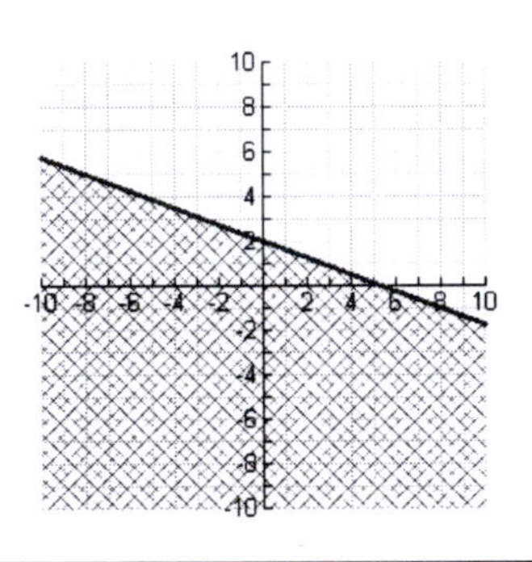

53.

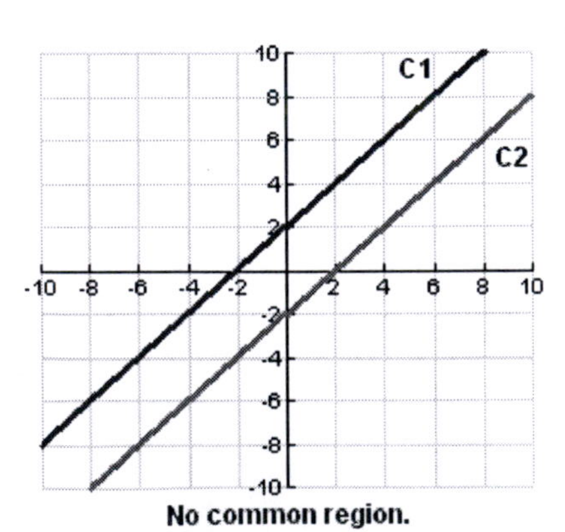

Notes on the graph:

C1: $y = x + 2$

C2: $y = x - 2$

Since there is no region in common with both inequalities, the system has no solution.

55.

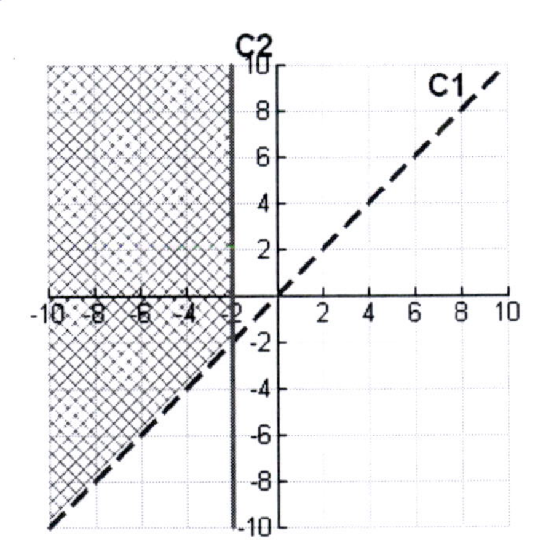

Notes on the graph:

C1: $y = x$

C2: $x = -2$

57.

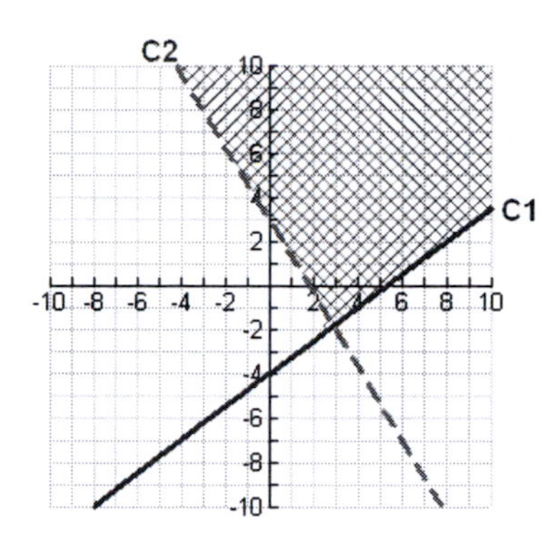

Notes on the graph:

C1: $y = \frac{3}{4}x - 4$

C2: $y = 3 - \frac{5}{3}x$

59. The region in this case is:

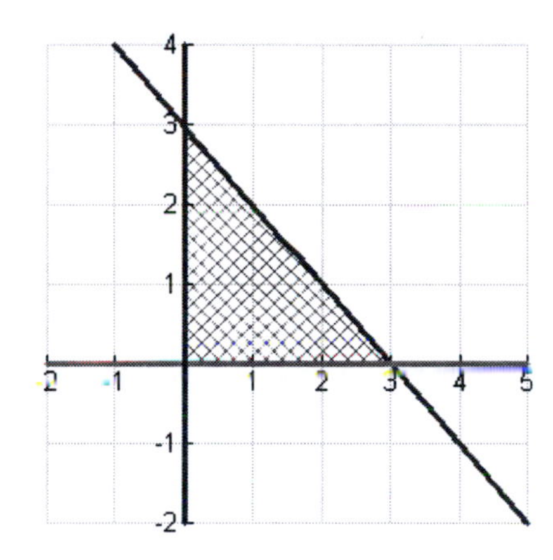

Vertex	Objective Function $z = 2x + y$
$(0,0)$	$z = 2(0) + (0) = 0$
$(3,0)$	$z = 2(3) + (0) = 6$
$(0,3)$	$z = 2(0) + (3) = 3$

So, the minimum value of z is 0.

61. The region in this case is:

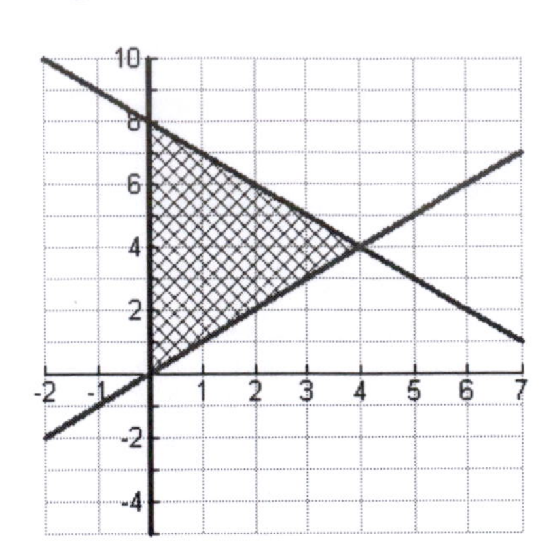

We need all of the intersection points since they constitute the vertices:

Intersection of $y = -x + 8$ and $y = x$:

$$-x + 8 = x$$
$$2x = 8$$
$$x = 4$$

So, the intersection point is (4,4).

Vertex	Objective Function $z = 2.5x + 3.2y$
$(0,0)$	$z = 2.5(0) + 3.2(0) = 0$
$(0,8)$	$z = 2.5(0) + 3.2(8) = 25.6$
$(4,4)$	$z = 2.5(4) + 3.2(4) = 22.8$

So, the maximum value of z is 25.6.

63. The region in this case is:

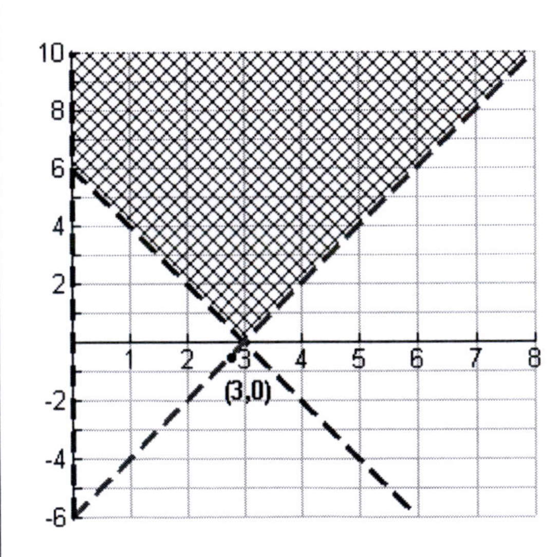

Vertex	Objective Function $z = 3x - 5y$
$(3,0)$	9
$(0,6)$	-30

The region is unbounded. There is no minimum.

65. Let x = number of ocean watercolor coaster sets
y = number of geometric shape coaster sets

Profit from ocean watercolor sets:
Revenue – cost = $15x$
Profit from geometric shape sets:
Revenue – cost = $8y$

So, to maximize profit, we must maximize the objective function $z = 15x + 8y$. We have the following constraints:

$$\begin{cases} 4x + 2y \le 100 \\ 3x + 2y \le 90 \\ x \ge 0,\ y \ge 0 \end{cases}$$

The region in this case is:

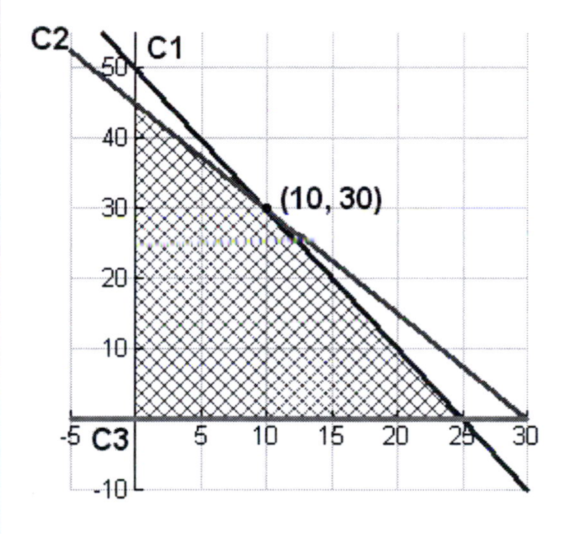

Notes on the graph:
C1: $y = 50 - 2x$
C2: $y = 45 - \frac{3}{2}x$
C3: $y = 0$

We need all of the intersection points since they constitute the vertices:
Intersection of $4x + 2y = 100$ and $3x + 2y = 90$:

$$50 - 2x = 45 - \tfrac{3}{2}x$$
$$5 = \tfrac{1}{2}x$$
$$x = 10$$

So, the intersection point is (10, 30).

Now, compute the objective function at the vertices:

Vertex	Objective Function $z = 15x + 8y$
(0,0)	$z = 15(0) + 8(0) = 0$
(0, 45)	$z = 15(0) + 8(45) = 360$
(10, 30)	$z = 15(10) + 8(30) = 390$
(25,0)	$z = 15(25) + 8(0) = 375$

So, to attain maximum profit, she should sell 10 ocean watercolor coaster sets and 30 geometric shape coaster sets.

67. The graph of this system of equation is as follows:

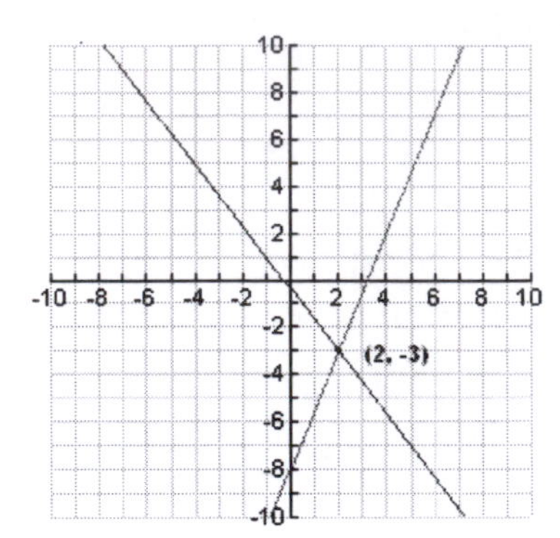

The solution is (2, -3).

69. If your calculator has 3-dimensional graphing capabilities, graph the system to obtain the following:

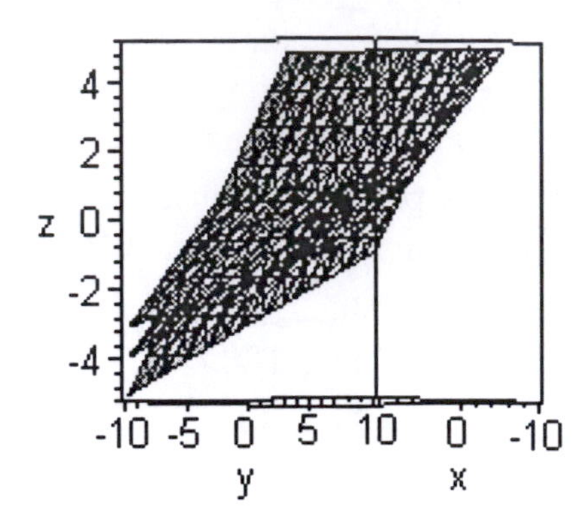

It is difficult to see from this graph, but zooming in will enable you to see that the solution is (3.6, 3, 0.8).

71. Since the graphs coincide, y2 is the partial fraction decomposition of y1.

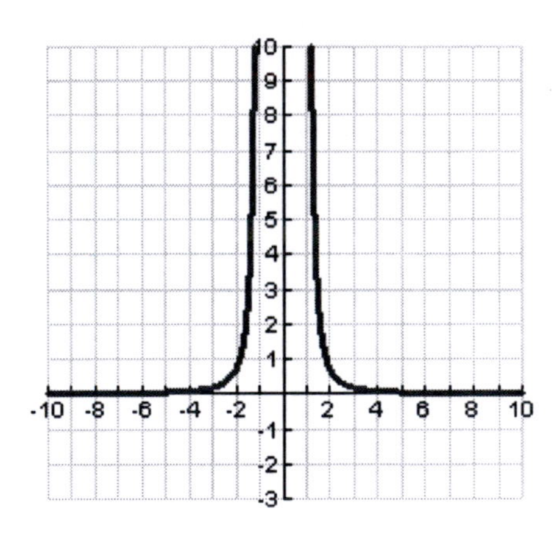

73. The region is as follows.

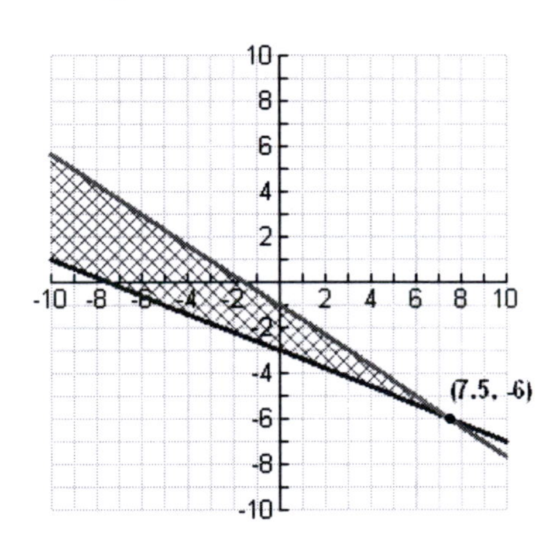

75.

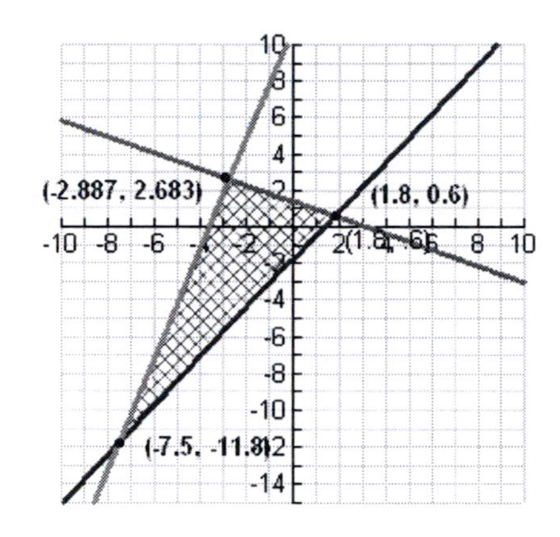

Vertex	Objective Function $z = 6.2x + 1.5y$
(-2.7, 2.6)	-12.84
(-7.5, -11.8)	-64.2
(1.8, 0.6)	12.06

The maximum is 12.06 and occurs at (1.8, 0.6).

Chapter 6 Practice Test

1. Solve the system:

$$\begin{cases} x - 2y = 1 & \textbf{(1)} \\ -x + 3y = 2 & \textbf{(2)} \end{cases}$$

Add **(1)** and **(2)**: $y = 3$

Substitute $y = 3$ into **(1)** to find x:

$$x - 2(3) = 1 \Rightarrow x = 7.$$

So, the solution is $\boxed{(7,3)}$.

3. Solve the system:

$$\begin{cases} x - y = 2 & \textbf{(1)} \\ -2x + 2y = -4 & \textbf{(2)} \end{cases}$$

Multiply **(1)** by 2, and then add to **(2)**:

$$0 = 0$$

So, the system is consistent. There are infinitely many solutions of the form $\boxed{x = a,\ y = a - 2}$.

5. Solve the system:

$$\begin{cases} x + y + z = -1 & \textbf{(1)} \\ 2x + y + z = 0 & \textbf{(2)} \\ -x + y + 2z = 0 & \textbf{(3)} \end{cases}$$

Step 1: Obtain a system of 2 equations in 2 unknowns by eliminating the same variable in two pairs of equations.

Multiply **(1)** by -2, and then add to **(2)** to eliminate x:

$$-y - z = 2 \quad \textbf{(4)}$$

Add **(1)** and **(3)** to eliminate x:

$$2y + 3z = -1 \quad \textbf{(5)}$$

These steps yield system:

$$(*) \begin{cases} -y - z = 2 & \textbf{(4)} \\ 2y + 3z = -1 & \textbf{(5)} \end{cases}$$

Step 2: Solve system **(*)** from Step 1.

Multiply **(4)** by 2, and then add to **(5)**:

$$z = 3 \quad \textbf{(6)}$$

Substitute **(6)** into **(4)** to find y:

$$-y - 3 = 2 \text{ so that } y = -5 \quad \textbf{(7)}$$

Step 3: Use the solution of the system in Step 2 to find the value of the third variable in the original system.

Substitute **(6)** and **(7)** into **(1)** to find x:

$$x - 5 + 3 = -1$$

$$x = 1$$

Thus, the solution is:

$\boxed{x = 1,\ y = -5,\ z = 3}$.

7. Solve the system:

$$\begin{cases} x-2y+3z=5 & \textbf{(1)} \\ -2x+y+4z=21 & \textbf{(2)} \\ 3x-5y+z=-14 & \textbf{(3)} \end{cases}$$

Step 1: Obtain a system of 2 equations in 2 unknowns by eliminating the same variable in two pairs of equations.

Multiply **(1)** by 2, and then add to **(2)** to eliminate x:

$$-3y+10z=31 \quad \textbf{(4)}$$

Multiply **(1)** by -3, and then add to **(3)** to eliminate x:

$$y-8z=-29 \quad \textbf{(5)}$$

These steps lead to the system:

$$\begin{cases} -3y+10z=31 & \textbf{(4)} \\ y-8z=-29 & \textbf{(5)} \end{cases}$$

Step 2: Solve the system in Step 1.

Multiply **(5)** by 3: $3y-24z=-87$ **(6)**

Add **(4)** and **(6)**: $-14z=-56 \Rightarrow \boxed{z=4}$

Substitute this value of z into **(6)** to find y:

$$y-8(4)=-29 \Rightarrow \boxed{y=3}$$

Substitute the values of y and z into **(1)** to find x: $\boxed{x=-1}$.

9. The partial fraction decomposition has the form:

$$\frac{2x+5}{x(x+1)}=\frac{A}{x}+\frac{B}{x+1} \quad \textbf{(1)}$$

To find the coefficients, multiply both sides of **(1)** by $x(x+1)$, and gather like terms:

$$2x+5=A(x+1)+Bx$$

$$2x+5=(A+B)x+A \quad \textbf{(2)}$$

Equate corresponding coefficients in **(2)** to obtain the following system:

$$\textbf{(*)} \begin{cases} A+B=2 & \textbf{(3)} \\ A=5 & \textbf{(4)} \end{cases}$$

Now, solve system **(*)**:

Substitute **(4)** into **(3)** to see that $B=-3$.

Thus, the partial fraction decomposition **(1)** becomes:

$$\boxed{\frac{2x+5}{x(x+1)}=\frac{5}{x}-\frac{3}{x+1}}$$

11. The partial fraction decomposition has the form:

$$\frac{7x+5}{(x+2)^2}=\frac{A}{x+2}+\frac{B}{(x+2)^2} \quad \textbf{(1)}$$

To find the coefficients, multiply both sides of **(1)** by $(x+2)^2$, and gather like terms:

$$7x+5=A(x+2)+B=Ax+(2A+B) \quad \textbf{(2)}$$

Equate corresponding coefficients in **(2)** to obtain the following system:

$$\textbf{(*)} \begin{cases} A=7 & \textbf{(3)} \\ 2A+B=5 & \textbf{(4)} \end{cases}$$

Now, solve system **(*)**:

Substitute **(3)** into **(4)** to see that $B=-9$.

Thus, the partial fraction decomposition **(1)** becomes: $\boxed{\frac{7x+5}{(x+2)^2}=\frac{7}{x+2}-\frac{9}{(x+2)^2}}$

13. The partial fraction decomposition has the form:

$$\frac{5x-3}{x(x-3)(x+3)}=\frac{A}{x}+\frac{B}{x-3}+\frac{C}{x+3} \quad \textbf{(1)}$$

To find the coefficients, multiply both sides of **(1)** by $x(x-3)(x+3)$, and gather like terms:

$$\begin{aligned}5x-3&=A(x-3)(x+3)+Bx(x+3)+Cx(x-3)\\&=A(x^2-9)+B(x^2+3x)+C(x^2-3x)\\&=(A+B+C)x^2+(3B-3C)x+(-9A) \quad \textbf{(2)}\end{aligned}$$

Equate corresponding coefficients in **(2)** to obtain the following system:

$$(*)\begin{cases}A+B+C=0 & \textbf{(3)}\\ 3B-3C=5 & \textbf{(4)}\\ -9A=-3 & \textbf{(5)}\end{cases}$$

Solve **(5)** for A: $A=\frac{1}{3}$ **(6)**

Substitute **(6)** into **(1)** to eliminate A: $B+C=-\frac{1}{3}$ **(7)**

Now, solve the 2×2 system: $\begin{cases}3B-3C=5 & \textbf{(4)}\\ B+C=-\frac{1}{3} & \textbf{(7)}\end{cases}$

Multiply **(7)** by 3, and then add to **(4)** to find B: $6B=4$ so that $B=\frac{2}{3}$.

Substitute this value for B into **(7)** to find C: $\frac{2}{3}+C=-\frac{1}{3}$ so that $C=-1$.

Finally, substitute the values of B and C into **(1)** to find A: $A+\frac{2}{3}-1=0$ so that $A=\frac{1}{3}$.

Thus, the partial fraction decomposition **(1)** becomes:

$$\boxed{\frac{5x-3}{x(x-3)(x+3)}=\frac{1}{3x}+\frac{2}{3(x-3)}-\frac{1}{x+3}}$$

15.

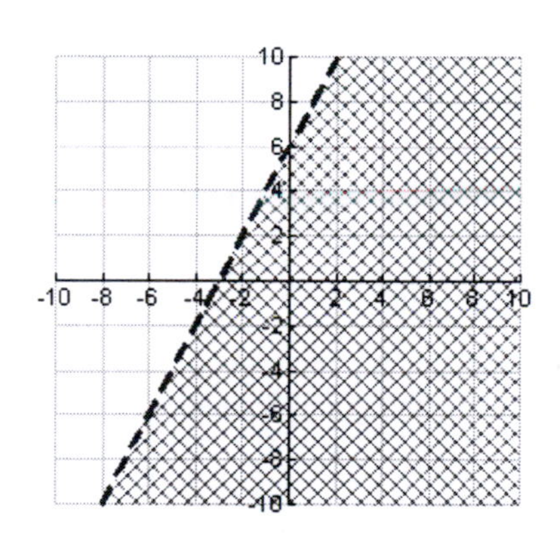

The dashed curve is the graph of $y=2x+6$.

17.

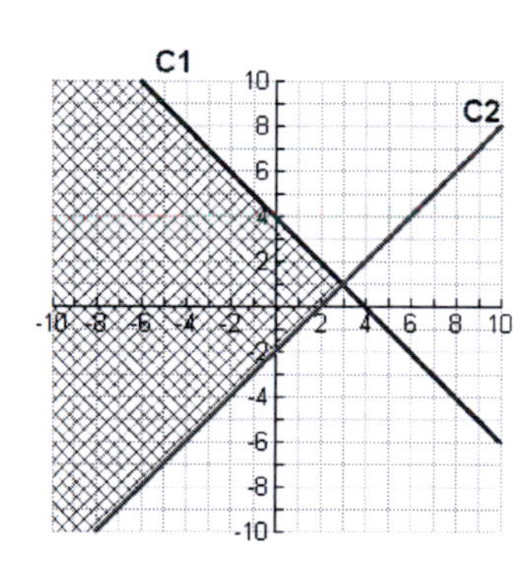

Notes on the graph:

C1: $y=4-x$

C2: $y=x-2$

19.

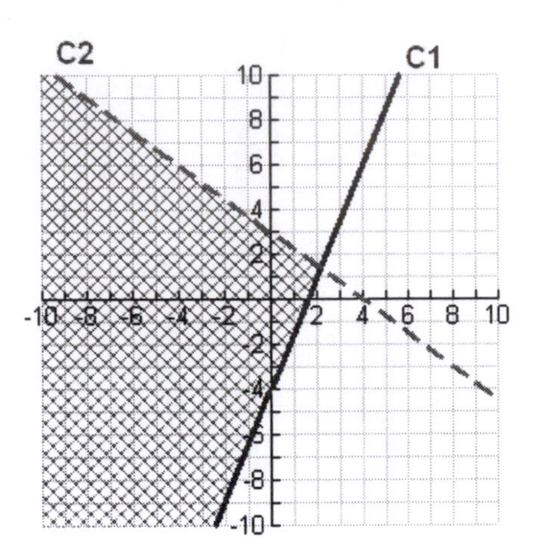

Notes on the graph:
C1: $y = \frac{5}{2}x - 4$
C2: $y = -\frac{3}{4}x + 3$

21. The region in this case is:

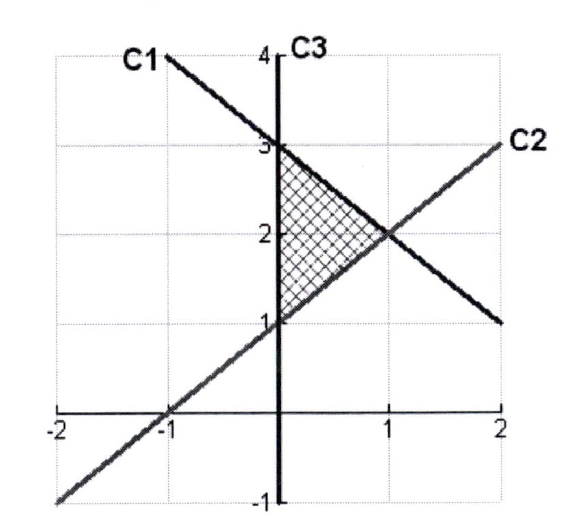

Notes on the graph:
C1: $y = -x + 3$
C2: $y = x + 1$
C3: $x = 0$

We need all of the intersection points since they constitute the vertices:

Intersection of $y = -x + 3$ and $y = x + 1$:

$$-x + 3 = x + 1$$
$$2x = 2$$
$$x = 1$$

So, the intersection point is (1,2).

Vertex	**Objective Function** $z = 5x + 7y$
(0,3)	$z = 5(0) + 7(3) = 21$
(0,1)	$z = 5(0) + 7(1) = 7$
(1,2)	$z = 5(1) + 7(2) = 19$

So, the minimum value of z is 7.

23. Let x = amount in money market
y = amount in aggressive stock
z = amount in conservative stock

The system we must solve is:

$$\begin{cases} y = z + 1000 \\ x + y + z = 30{,}000 \\ 0.03x + 0.12y + 0.06z = 1890 \end{cases}$$

First, simplify this system by multiplying the third equation by 100:

$$\begin{cases} y = z + 1000 & \textbf{(1)} \\ x + y + z = 30{,}000 & \textbf{(2)} \\ 3x + 12y + 6z = 189{,}000 & \textbf{(3)} \end{cases}$$

Step 1: Obtain a 2×2 system:
Substitute **(1)** into both **(2)** and **(3)**:

$$x + (z + 1000) + z = 30{,}000$$

$$3x + 12(z + 1000) + 6z = 189{,}000$$

which are equivalent to

$$\begin{cases} x + 2z = 29{,}000 & \textbf{(4)} \\ 3x + 18z = 177{,}000 & \textbf{(5)} \end{cases}$$

Step 2: Solve the system in Step 1.
Multiply **(4)** by -3 and add to **(5)** to find z:

$$12z = 90{,}000 \Rightarrow z = 7500 \quad \textbf{(6)}$$

Substitute **(6)** into **(1)** to find y:

$$y = 7500 + 1000 = 8500 \quad \textbf{(7)}$$

Substitute **(6)** and **(7)** into **(2)** to find x:

$$x + 8500 + 7500 = 30{,}000$$

$$x = 14{,}000$$

Thus, the following allocation of funds should be made:

\$14,000 in money market
\$8,500 in aggressive stock
\$7,500 in conservative stock

25. If your calculator has 3-dimensional graphing capabilities, use it to graph the system of equations to obtain:

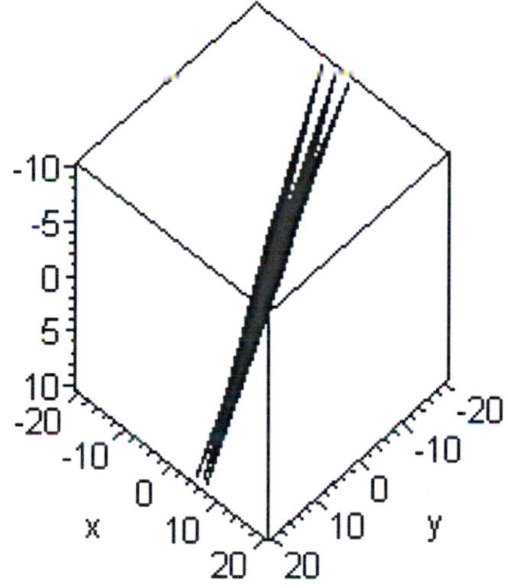

It is difficult to see from the graph, but if you zoom in, you will see that the solution is (11,19,1).

Chapter 6 Cumulative Review--

1. $\dfrac{6x^2-11x+5}{30x-25}=\dfrac{\cancel{(6x-5)}(x-1)}{5\cancel{(6x-5)}}=\boxed{\dfrac{x-1}{5}}$. Note that $x\neq\frac{5}{6}$.

3. The LCD is $x(x+1)$, $x\neq 0,-1$.
Multiply both sides by the LCD and solve for x:

$$5(x+1)-5x=-5 \Rightarrow 5=-5$$

Since this statement is false, the equation has $\boxed{\text{no solution}}$.

5.

$$\frac{2t^2}{t-1}-\frac{3t(t-1)}{t-1}\geq 0 \Rightarrow \frac{-t^2+3t}{t-1}\geq 0 \Rightarrow \frac{t(3-t)}{t-1}\geq 0$$

CPs: $t=0,1,3$ (sign chart: $+$ before 0, $-$ between 0 and 1, $+$ between 1 and 3, $-$ after 3)

The solution set is $(-\infty,0]\cup(1,3]$.

7. The slope is $m=\dfrac{6.2-5.0}{5.6-3.2}=0.5$. To find b, use the point (3.2, 5.0):

$$5.0=0.5(3.2)+b \Rightarrow b=3.4$$

So, the line has equation $y=0.5x+3.4$.

9. $\dfrac{f(4)-f(2)}{4-2}=\dfrac{\frac{5}{4}-\frac{5}{2}}{4-2}=-\frac{5}{8}$

11. $f(-1)=\sqrt[3]{-1-7}=-2$. So, $g(f(-1))=g(-2)=1$.

13. Since (0,7) is the vertex, we know that $f(x)=a(x-0)^2+7$. Use the point (2,-1) to find a:

$$-1=a(2-0)^2+7 \Rightarrow a=-2$$

Hence, $f(x)=-2x^2+7$.

15. Factors of 10: $\pm1, \pm2, \pm5, \pm10$ Factors of 2: $\pm1, \pm2$
So, possible rational zeros: $\pm1, \pm2, \pm5, \pm10, \pm\frac{1}{2}, \pm\frac{5}{2}$
Observe that

$$\begin{array}{r|rrrrr} -1 & 2 & 7 & -18 & -13 & 10 \\ & & -2 & -5 & 23 & -10 \\ \hline 2 & 2 & 5 & -23 & 10 & 0 \\ & & 4 & 18 & -10 & \\ \hline & 2 & 9 & -5 & 0 & \end{array}$$

Thus, $P(x)=(x+1)(x-2)\left(2x^2+9x-5\right)=(x+1)(x-2)(x+5)(2x-1)$.
So, the rational zeros are $-5,-1,\frac{1}{2},2$.

17. The graph is below.
Domain: $(-\infty,\infty)$ Range: $(-1,\infty)$
x-intercept: $0=5^{-x}-1 \Rightarrow x=0$. So, (0,0).
y-intercept: (0,0)
Horizontal asymptote: $y = -1$

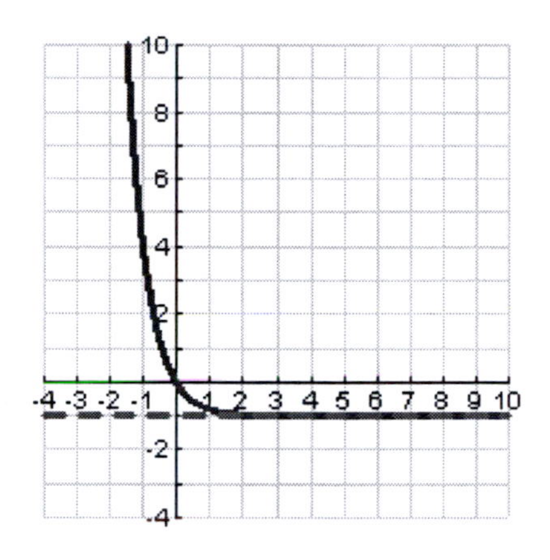

19. Shift the graph of $y=\ln x$ left 1 unit, and then down 3 units.
The graph is as follows:

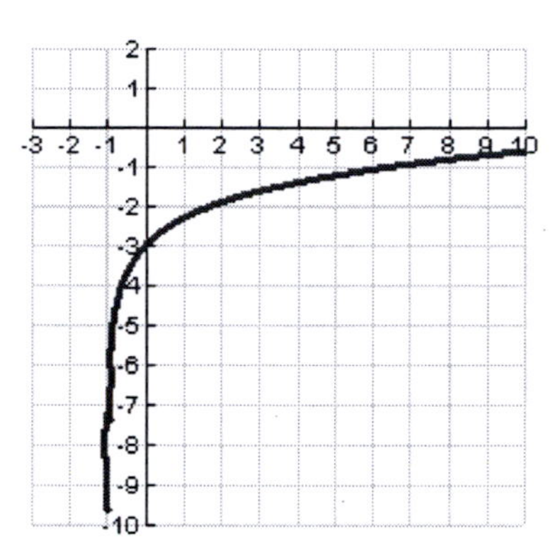

21.

$$5\left(10^{2x}\right)=37$$
$$10^{2x}=\tfrac{37}{5}$$
$$2x=\log\left(\tfrac{37}{5}\right)$$
$$x=\tfrac{1}{2}\log\left(\tfrac{37}{5}\right)\approx 0.435$$

23. Solve the system:

$$\begin{cases} x+6y-z=-3 & \textbf{(1)} \\ 2x-5y+z=9 & \textbf{(2)} \\ x+4y+2z=12 & \textbf{(3)} \end{cases}$$

To eliminate z, add **(1)** and **(2)**: $3x+y=6$ **(4)**
Multiply **(1)** by 2 and then add to **(3)**: $3x+16y=6$ **(5)**
Solve the system **(4) & (5)**:
Subtract **(4) – (5)**: $\boxed{y=0}$
Substitute this value of y into **(4)**: $\boxed{x=2}$
Finally, substitute these values of x and y into **(1)**: $\boxed{z=5}$.

25. The region is as follows:

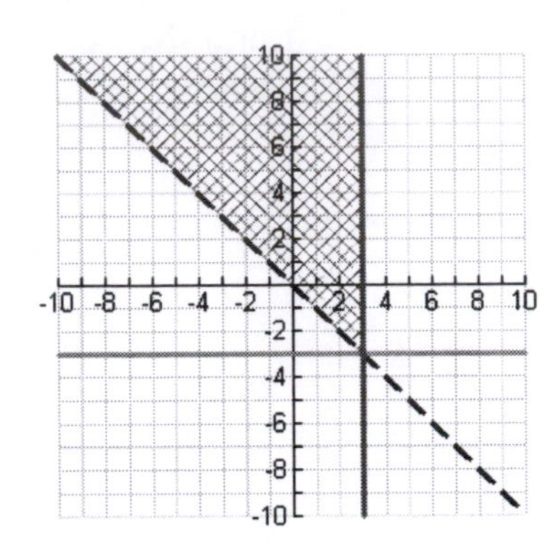

27. Since the graphs coincide, y2 is the partial fraction decomposition of y1.

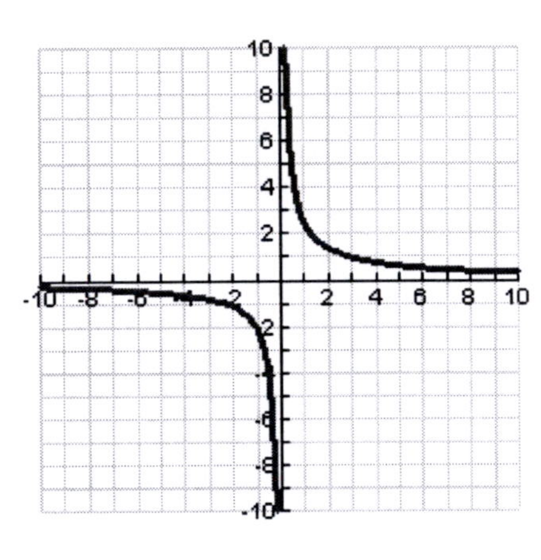

CHAPTER 7

Section 7.1 Solutions --

1. 2×3	**3.** 1×4
5. 1×1	**7.** $\left[\begin{array}{cc\|c} 3 & -2 & 7 \\ -4 & 6 & -3 \end{array}\right]$
9. $\left[\begin{array}{ccc\|c} 2 & -3 & 4 & -3 \\ -1 & 1 & 2 & 1 \\ 5 & -2 & -3 & 7 \end{array}\right]$	**11.** $\left[\begin{array}{ccc\|c} 1 & 1 & 0 & 3 \\ 1 & 0 & -1 & 2 \\ 0 & 1 & 1 & 5 \end{array}\right]$
13. $\left[\begin{array}{ccc\|c} -4 & 3 & 5 & 2 \\ 2 & -3 & -2 & -3 \\ -2 & 4 & 3 & 1 \end{array}\right]$	**15.** $\begin{cases} -3x+7y=2 \\ x+5y=8 \end{cases}$
17. $\begin{cases} -x=4 \\ 7x+9y+3z=-3 \\ 4x+6y-5z=8 \end{cases}$	**19.** $\begin{cases} x=a \\ y=b \end{cases}$
21. Not in row echelon form since Condition 3 is violated.	**23.** In reduced form.
25. Not in row echelon form since Condition 4 is violated.	**27.** In reduced form.
29. In row echelon form.	**31.** $\left[\begin{array}{cc\|c} 1 & -2 & -3 \\ 0 & 7 & 5 \end{array}\right]$
33. $\left[\begin{array}{ccc\|c} 1 & -2 & -1 & 3 \\ 0 & 5 & -1 & 0 \\ 3 & -2 & 5 & 8 \end{array}\right]$	**35.** $\left[\begin{array}{cccc\|c} 1 & -2 & 5 & -1 & 2 \\ 0 & 1 & 1 & -3 & 3 \\ 0 & -2 & 1 & -2 & 5 \\ 0 & 0 & 1 & -1 & -6 \end{array}\right]$
37. $\left[\begin{array}{cccc\|c} 1 & 0 & 5 & -10 & -5 \\ 0 & 1 & 2 & -3 & -2 \\ 0 & 0 & -7 & 6 & 3 \\ 0 & 0 & 8 & -10 & -9 \end{array}\right]$	**39.** $\left[\begin{array}{cccc\|c} 1 & 0 & 4 & 0 & 27 \\ 0 & 1 & 2 & 0 & -11 \\ 0 & 0 & 1 & 0 & 21 \\ 0 & 0 & 0 & 1 & -3 \end{array}\right]$

41.

$$\left[\begin{array}{cc|c}1&2&4\\2&3&2\end{array}\right]\xrightarrow{R_1\leftrightarrow R_2}\left[\begin{array}{cc|c}2&3&2\\1&2&4\end{array}\right]\xrightarrow{R_1-2R_2\to R_2}\left[\begin{array}{cc|c}2&3&2\\0&-1&-6\end{array}\right]\xrightarrow{-R_2\to R_2}\left[\begin{array}{cc|c}2&3&2\\0&1&6\end{array}\right]$$

$$\xrightarrow{\frac{1}{2}R_1\to R_1}\left[\begin{array}{cc|c}1&\frac{3}{2}&1\\0&1&6\end{array}\right]\xrightarrow{R_1-\frac{3}{2}R_2\to R_1}\left[\begin{array}{cc|c}1&0&-8\\0&1&6\end{array}\right]$$

43.

$$\left[\begin{array}{ccc|c}1&-1&1&-1\\0&1&-1&-1\\-1&1&1&1\end{array}\right]\xrightarrow{R_1+R_3\to R_3}\left[\begin{array}{ccc|c}1&-1&1&-1\\0&1&-1&-1\\0&0&2&0\end{array}\right]$$

$$\xrightarrow{\frac{1}{2}R_3\to R_3}\left[\begin{array}{ccc|c}1&-1&1&-1\\0&1&-1&-1\\0&0&1&0\end{array}\right]\xrightarrow[R_1+R_2\to R_1]{R_2+R_3\to R_2}\left[\begin{array}{ccc|c}1&0&0&-2\\0&1&0&-1\\0&0&1&0\end{array}\right]$$

45.

$$\left[\begin{array}{ccc|c}3&-2&-3&-1\\1&-1&1&-4\\2&3&5&14\end{array}\right]\xrightarrow[R_3-2R_2\to R_3]{R_1-3R_2\to R_1}\left[\begin{array}{ccc|c}0&1&-6&11\\1&-1&1&-4\\0&5&3&22\end{array}\right]\xrightarrow{R_1\leftrightarrow R_2}\left[\begin{array}{ccc|c}1&-1&1&-4\\0&1&-6&11\\0&5&3&22\end{array}\right]$$

$$\xrightarrow{R_3-5R_2\to R_3}\left[\begin{array}{ccc|c}1&-1&1&-4\\0&1&-6&11\\0&0&33&-33\end{array}\right]\xrightarrow{\frac{1}{33}R_3\to R_3}\left[\begin{array}{ccc|c}1&-1&1&-4\\0&1&-6&11\\0&0&1&-1\end{array}\right]$$

$$\xrightarrow{R_2+6R_3\to R_2}\left[\begin{array}{ccc|c}1&-1&1&-4\\0&1&0&5\\0&0&1&-1\end{array}\right]\xrightarrow{R_1+R_2-R_3\to R_1}\left[\begin{array}{ccc|c}1&0&0&2\\0&1&0&5\\0&0&1&-1\end{array}\right]$$

47.

$$\left[\begin{array}{ccc|c}2&1&-6&4\\1&-2&2&-3\end{array}\right]\xrightarrow{R_1-2R_2\to R_2}\left[\begin{array}{ccc|c}2&1&-6&4\\0&5&-10&10\end{array}\right]$$

$$\xrightarrow[\frac{1}{5}R_2\to R_2]{\frac{1}{2}R_1\to R_1}\left[\begin{array}{ccc|c}1&\frac{1}{2}&-3&2\\0&1&-2&2\end{array}\right]\xrightarrow{R_1-\frac{1}{2}R_2\to R_1}\left[\begin{array}{ccc|c}1&0&-2&1\\0&1&-2&2\end{array}\right]$$

49.

$$\left[\begin{array}{ccc|c} -1 & 2 & 1 & -2 \\ 3 & -2 & 1 & 4 \\ 2 & -4 & -2 & 4 \end{array}\right] \xrightarrow{R_2+3R_1\to R_2} \left[\begin{array}{ccc|c} -1 & 2 & 1 & -2 \\ 0 & 4 & 4 & -2 \\ 2 & -4 & -2 & 4 \end{array}\right] \xrightarrow{R_3+2R_1\to R_3} \left[\begin{array}{ccc|c} -1 & 2 & 1 & -2 \\ 0 & 4 & 4 & -2 \\ 0 & 0 & 0 & 0 \end{array}\right]$$

$$\xrightarrow[\frac{1}{4}R_2\to R_2]{-R_1\to R_1} \left[\begin{array}{ccc|c} 1 & -2 & -1 & 2 \\ 0 & 1 & 1 & -\frac{1}{2} \\ 0 & 0 & 0 & 0 \end{array}\right] \xrightarrow{R_1+2R_2\to R_1} \left[\begin{array}{ccc|c} 1 & 0 & 1 & 1 \\ 0 & 1 & 1 & -\frac{1}{2} \\ 0 & 0 & 0 & 0 \end{array}\right]$$

51. First, reduce the augmented matrix down to row echelon form:

$$\left[\begin{array}{cc|c} 2 & 3 & 1 \\ 1 & 1 & -2 \end{array}\right] \xrightarrow{R_1-2R_2\to R_2} \left[\begin{array}{cc|c} 2 & 3 & 1 \\ 0 & 1 & 5 \end{array}\right] \xrightarrow{\frac{1}{2}R_1\to R_1} \left[\begin{array}{cc|c} 1 & \frac{3}{2} & \frac{1}{2} \\ 0 & 1 & 5 \end{array}\right]$$

Now, from this we see that $y = 5$, and then substituting this value into the equation obtained from the first row yields:

$$x + \tfrac{3}{2}(5) = \tfrac{1}{2}$$
$$x = -7$$

Hence, the solution is $\boxed{x = -7,\ y = 5}$.

53. First, reduce the augmented matrix down to row echelon form:

$$\left[\begin{array}{cc|c} -1 & 2 & 3 \\ 2 & -4 & -6 \end{array}\right] \xrightarrow{2R_1+R_2\to R_2} \left[\begin{array}{cc|c} -1 & 2 & 3 \\ 0 & 0 & 0 \end{array}\right]$$

From the bottom row, we conclude that there are infinitely many solutions. To get the precise form of these solutions, let $y = a$. Then, substituting this value into the equation obtained from the first row yields:

$$-x + 2a = 3$$
$$x = 2a - 3$$

Hence, the solution is $\boxed{x = 2a-3,\ y = a}$.

55. First, reduce the augmented matrix down to row echelon form:

$$\left[\begin{array}{cc|c} \frac{2}{3} & \frac{1}{3} & \frac{8}{9} \\ \frac{1}{2} & \frac{1}{4} & \frac{3}{4} \end{array}\right] \xrightarrow[2R_2\to R_2]{3R_1\to R_1} \left[\begin{array}{cc|c} 2 & 1 & \frac{8}{3} \\ 1 & \frac{1}{2} & \frac{3}{2} \end{array}\right] \xrightarrow{R_1-2R_2\to R_2} \left[\begin{array}{cc|c} 2 & 1 & \frac{8}{3} \\ 0 & 0 & -\frac{1}{3} \end{array}\right]$$

From the second row, we see that $0 = -\frac{1}{3}$, which means that the system has $\boxed{\text{no solution}}$.

57. First, rewrite the system as

$$\begin{cases} x-y-z=10 \\ 2x-3y+z=-11 \\ -x+y+z=-10 \end{cases}$$

Now, reduce the augmented matrix down to row echelon form:

$$\left[\begin{array}{ccc|c} 1 & -1 & -1 & 10 \\ 2 & -3 & 1 & -11 \\ -1 & 1 & 1 & -10 \end{array}\right] \xrightarrow[R_2-2R_1 \to R_2]{R_1+R_3 \to R_3} \left[\begin{array}{ccc|c} 1 & -1 & -1 & 10 \\ 0 & -1 & 3 & -31 \\ 0 & 0 & 0 & 0 \end{array}\right] \xrightarrow{-R_2 \to R_2} \left[\begin{array}{ccc|c} 1 & -1 & -1 & 10 \\ 0 & 1 & -3 & 31 \\ 0 & 0 & 0 & 0 \end{array}\right]$$

From the last row, we conclude that the system has infinitely many solutions. To find them, let $z=a$. Then, the second row implies that $y-3a=31$ so that $y=31+3a$. So, substituting these values of y and z into the first row yields $x-(3a+31)-a=10$, so that $x=4a+41$. Hence, the solutions are $\boxed{x=4a+41,\ y=31+3a,\ z=a}$.

59. Reduce the augmented matrix down to row echelon form:

$$\left[\begin{array}{ccc|c} 3 & 1 & -1 & 1 \\ 1 & -1 & 1 & -3 \\ 2 & 1 & 1 & 0 \end{array}\right] \xrightarrow[R_1-3R_2 \to R_2]{R_3-2R_2 \to R_3} \left[\begin{array}{ccc|c} 3 & 1 & -1 & 1 \\ 0 & 4 & -4 & 10 \\ 0 & 3 & -1 & 6 \end{array}\right] \xrightarrow[\frac{1}{3}R_3 \to R_3]{\frac{1}{3}R_1 \to R_1,\ \frac{1}{4}R_2 \to R_2} \left[\begin{array}{ccc|c} 1 & \frac{1}{3} & -\frac{1}{3} & \frac{1}{3} \\ 0 & 1 & -1 & \frac{5}{2} \\ 0 & 1 & -\frac{1}{3} & 2 \end{array}\right]$$

$$\xrightarrow{R_2-R_3 \to R_3} \left[\begin{array}{ccc|c} 1 & \frac{1}{3} & -\frac{1}{3} & \frac{1}{3} \\ 0 & 1 & -1 & \frac{5}{2} \\ 0 & 0 & -\frac{2}{3} & \frac{1}{2} \end{array}\right] \xrightarrow{-\frac{3}{2}R_3 \to R_3} \left[\begin{array}{ccc|c} 1 & \frac{1}{3} & -\frac{1}{3} & \frac{1}{3} \\ 0 & 1 & -1 & \frac{5}{2} \\ 0 & 0 & 1 & -\frac{3}{4} \end{array}\right]$$

From the last row, we conclude that $x_3=-\frac{3}{4}$. Then, the second row implies that $x_2-(-\frac{3}{4})=\frac{5}{2}$ so that $x_2=\frac{7}{4}$. So, substituting these values of x_2 and x_3 into the first row yields $x_1+\frac{1}{3}(\frac{7}{4})-\frac{1}{3}(-\frac{3}{4})=\frac{1}{3}$, so that $x_1=-\frac{1}{2}$. Hence, the solution is

$\boxed{x_1=-\frac{1}{2},\ x_2=\frac{7}{4},\ x_3=-\frac{3}{4}}$.

61. Reduce the augmented matrix down to row echelon form:

$$\left[\begin{array}{ccc|c} 2 & 5 & 0 & 9 \\ 1 & 2 & -1 & 3 \\ -3 & -4 & 7 & 1 \end{array}\right] \xrightarrow[R_1-2R_2 \to R_2]{R_3+3R_2 \to R_3} \left[\begin{array}{ccc|c} 2 & 5 & 0 & 9 \\ 0 & 1 & 2 & 3 \\ 0 & 2 & 4 & 10 \end{array}\right] \xrightarrow{R_3-2R_2 \to R_3} \left[\begin{array}{ccc|c} 2 & 5 & 0 & 9 \\ 0 & 1 & 2 & 3 \\ 0 & 0 & 0 & 4 \end{array}\right]$$

$$\xrightarrow{\frac{1}{2}R_1 \to R_1} \left[\begin{array}{ccc|c} 1 & \frac{5}{2} & 0 & \frac{9}{2} \\ 0 & 1 & 2 & 3 \\ 0 & 0 & 0 & 4 \end{array}\right]$$

From the last row, we have the false statement 0 = 4, so that we can conclude the system has $\boxed{\text{no solution}}$.

63. Reduce the augmented matrix down to row echelon form:

$$\left[\begin{array}{ccc|c} 2 & -1 & 1 & 3 \\ 1 & -1 & 1 & 2 \\ -2 & 2 & -2 & -4 \end{array}\right] \xrightarrow[2R_2+R_3\to R_3]{\frac{1}{2}R_1\to R_1} \left[\begin{array}{ccc|c} 1 & -\frac{1}{2} & \frac{1}{2} & \frac{3}{2} \\ 0 & 1 & -1 & -1 \\ 0 & 0 & 0 & 0 \end{array}\right]$$

From the last row, we conclude that the system has infinitely many solutions. To find them, let $x_3 = a$. Then, the second row implies that $x_2 - a = -1$ so that $x_2 = a - 1$. So, substituting these values of x_2 and x_3 into the first row yields $x_1 - \frac{1}{2}(a-1) + \frac{1}{2}a = \frac{3}{2}$, so that $x_1 = 1$. Hence, the solutions are $\boxed{x_1 = 1,\ x_2 = a-1,\ x_3 = a}$.

65. Reduce the augmented matrix down to row echelon form:

$$\left[\begin{array}{ccc|c} 2 & 1 & -1 & 2 \\ 1 & -1 & -1 & 6 \end{array}\right] \xrightarrow{R_1-2R_2\to R_2} \left[\begin{array}{ccc|c} 2 & 1 & -1 & 2 \\ 0 & 3 & 1 & -10 \end{array}\right] \xrightarrow[\frac{1}{3}R_2\to R_2]{\frac{1}{2}R_1\to R_1} \left[\begin{array}{ccc|c} 1 & \frac{1}{2} & -\frac{1}{2} & 1 \\ 0 & 1 & \frac{1}{3} & -\frac{10}{3} \end{array}\right]$$

Since there are two rows and three unknowns, we know from the above calculation that there are infinitely many solutions to this system. (Note: The only other possibility in such case would be that there was no solution, which would occur if one of the rows yielded a false statement.) To find the solutions, let $z = a$. Then, from row 2, we observe that $y + \frac{1}{3}a = -\frac{10}{3}$ so that $y = -\frac{1}{3}(a+10)$. Then, substituting these values of y and z into the equation obtained from row 1, we see that

$$x + \tfrac{1}{2}\left(-\tfrac{1}{3}(a+10)\right) - \tfrac{1}{2}a = 1$$
$$x - \tfrac{2}{3}a - \tfrac{5}{3} = 1$$
$$x = \tfrac{2}{3}a + \tfrac{8}{3}$$

Hence, the solutions are $\boxed{x = \frac{2}{3}(a+4),\ y = -\frac{1}{3}(a+10),\ z = a}$.

67. Reduce the augmented matrix down to row echelon form:

$$\left[\begin{array}{ccc|c} 0 & 2 & 1 & 3 \\ 4 & 0 & -1 & -3 \\ 7 & -3 & -3 & 2 \\ 1 & -1 & -1 & -2 \end{array}\right] \xrightarrow{R_1\leftrightarrow R_2} \left[\begin{array}{ccc|c} 4 & 0 & -1 & -3 \\ 0 & 2 & 1 & 3 \\ 7 & -3 & -3 & 2 \\ 1 & -1 & -1 & -2 \end{array}\right] \xrightarrow[R_1-4R_4\to R_4]{R_3-7R_4\to R_3} \left[\begin{array}{ccc|c} 4 & 0 & -1 & -3 \\ 0 & 2 & 1 & 3 \\ 0 & 4 & 4 & 16 \\ 0 & 4 & 3 & 5 \end{array}\right]$$

$$\xrightarrow[R_4-2R_2\to R_4]{R_3-2R_2\to R_3} \left[\begin{array}{ccc|c} 4 & 0 & -1 & -3 \\ 0 & 2 & 1 & 3 \\ 0 & 0 & 2 & 10 \\ 0 & 0 & 1 & -1 \end{array}\right] \xrightarrow[\frac{1}{2}R_3\to R_3]{\frac{1}{4}R_1\to R_1,\ \frac{1}{2}R_2\to R_2} \left[\begin{array}{ccc|c} 1 & 0 & -\frac{1}{4} & -\frac{3}{4} \\ 0 & 1 & \frac{1}{2} & \frac{3}{2} \\ 0 & 0 & 1 & 5 \\ 0 & 0 & 1 & -1 \end{array}\right]$$

The last two rows require that $z = 5$ and $z = -1$ simultaneously. Since this is not possible, we conclude that the system has $\boxed{\text{no solution}}$.

69. Reduce the augmented matrix down to row echelon form:

$$\left[\begin{array}{cccc|c} 3 & -2 & 1 & 2 & -2 \\ -1 & 3 & 4 & 3 & 4 \\ 1 & 1 & 1 & 1 & 0 \\ 5 & 3 & 1 & 2 & -1 \end{array}\right] \xrightarrow[R_4-5R_3\to R_4]{\substack{R_1+3R_2\to R_2 \\ R_1-3R_3\to R_3}} \left[\begin{array}{cccc|c} 0 & -5 & -2 & -1 & -2 \\ 0 & 4 & 5 & 4 & 4 \\ 1 & 1 & 1 & 1 & 0 \\ 0 & -2 & -4 & -3 & -1 \end{array}\right] \xrightarrow{R_1\leftrightarrow R_3} \left[\begin{array}{cccc|c} 1 & 1 & 1 & 1 & 0 \\ 0 & 4 & 5 & 4 & 4 \\ 0 & -5 & -2 & -1 & -2 \\ 0 & -2 & -4 & -3 & -1 \end{array}\right]$$

$$\xrightarrow[-\frac{1}{2}R_4\to R_4]{\substack{\frac{1}{4}R_2\to R_2 \\ -\frac{1}{5}R_3\to R_3}} \left[\begin{array}{cccc|c} 1 & 1 & 1 & 1 & 0 \\ 0 & 1 & \frac{5}{4} & 1 & 1 \\ 0 & 1 & \frac{2}{5} & \frac{1}{5} & \frac{2}{5} \\ 0 & 1 & 2 & \frac{3}{2} & \frac{1}{2} \end{array}\right] \xrightarrow[R_4-R_3\to R_4]{R_2-R_3\to R_2} \left[\begin{array}{cccc|c} 1 & 1 & 1 & 1 & 0 \\ 0 & 0 & \frac{17}{20} & \frac{4}{5} & \frac{3}{5} \\ 0 & 1 & \frac{2}{5} & \frac{1}{5} & \frac{2}{5} \\ 0 & 1 & \frac{8}{5} & \frac{13}{10} & \frac{1}{10} \end{array}\right] \xrightarrow{R_2\leftrightarrow R_3} \left[\begin{array}{cccc|c} 1 & 1 & 1 & 1 & 0 \\ 0 & 1 & \frac{2}{5} & \frac{1}{5} & \frac{2}{5} \\ 0 & 0 & \frac{17}{20} & \frac{4}{5} & \frac{3}{5} \\ 0 & 1 & \frac{8}{5} & \frac{13}{10} & \frac{1}{10} \end{array}\right]$$

$$\xrightarrow[\frac{5}{8}R_4\to R_4]{\frac{20}{17}R_3\to R_3} \left[\begin{array}{cccc|c} 1 & 1 & 1 & 1 & 0 \\ 0 & 1 & \frac{2}{5} & \frac{1}{5} & \frac{2}{5} \\ 0 & 0 & 1 & \frac{16}{17} & \frac{12}{17} \\ 0 & 0 & 1 & \frac{13}{16} & \frac{1}{16} \end{array}\right] \xrightarrow{R_3-R_4\to R_4} \left[\begin{array}{cccc|c} 1 & 1 & 1 & 1 & 0 \\ 0 & 1 & \frac{2}{5} & \frac{1}{5} & \frac{2}{5} \\ 0 & 0 & 1 & \frac{16}{17} & \frac{12}{17} \\ 0 & 0 & 0 & \frac{35}{272} & \frac{175}{272} \end{array}\right] \xrightarrow{\frac{272}{35}R_4\to R_4} \left[\begin{array}{cccc|c} 1 & 1 & 1 & 1 & 0 \\ 0 & 1 & \frac{2}{5} & \frac{1}{5} & \frac{2}{5} \\ 0 & 0 & 1 & \frac{16}{17} & \frac{12}{17} \\ 0 & 0 & 0 & 1 & 5 \end{array}\right]$$

Thus, Row 4 now implies that $x_4 = 5$.

Then, Row 3 implies that $x_3 + \frac{16}{17}(5) = \frac{12}{17}$ so that $x_3 = -4$.

Then, Row 2 implies that $x_2 + \frac{2}{5}(-4) + \frac{1}{5}(5) = \frac{2}{5}$ so that $x_2 = 1$.

Finally, Row 1 implies that $x_1 + 1 - 4 + 5 = 0$ so that $x_1 = -2$.

Hence, the solution is $\boxed{x_1 = -2,\ x_2 = 1,\ x_3 = -4,\ x_4 = 5}$.

71. Reduce the matrix down to reduced row echelon form:

$$\left[\begin{array}{cc|c} 1 & 3 & -5 \\ -2 & -1 & 0 \end{array}\right] \xrightarrow{R_2+2R_1\to R_2} \left[\begin{array}{cc|c} 1 & 3 & -5 \\ 0 & 5 & -10 \end{array}\right] \xrightarrow{\frac{1}{5}R_2\to R_2} \left[\begin{array}{cc|c} 1 & 3 & -5 \\ 0 & 1 & -2 \end{array}\right]$$

$$\xrightarrow{R_1-3R_2\to R_1} \left[\begin{array}{cc|c} 1 & 0 & 1 \\ 0 & 1 & -2 \end{array}\right]$$

Hence, the solution is $\boxed{(1,-2)}$.

73. Reduce the matrix down to reduced row echelon form:

$$\left[\begin{array}{cc|c} 1 & 1 & 4 \\ -3 & -3 & 10 \end{array}\right] \xrightarrow{3R_1+R_2\to R_1} \left[\begin{array}{cc|c} 0 & 0 & 22 \\ -3 & -3 & 10 \end{array}\right] \xrightarrow{-\frac{1}{3}R_2\to R_2} \left[\begin{array}{cc|c} 0 & 0 & 22 \\ 1 & 1 & -\frac{10}{3} \end{array}\right]$$

Since the first row translates to $0 = 22$, which is false, the solution has $\boxed{\text{no solution}}$.

75. Reduce the matrix down to reduced row echelon form:

$$\left[\begin{array}{ccc|c} 1 & -2 & 3 & 5 \\ 3 & 6 & -4 & -12 \\ -1 & -4 & 6 & 16 \end{array}\right] \xrightarrow[-3R_1+R_2\to R_2]{R_1+R_3\to R_3} \left[\begin{array}{ccc|c} 1 & -2 & 3 & 5 \\ 0 & 12 & -13 & -27 \\ 0 & -6 & 9 & 21 \end{array}\right] \xrightarrow{R_2+2R_3\to R_3} \left[\begin{array}{ccc|c} 1 & -2 & 3 & 5 \\ 0 & 12 & -13 & -27 \\ 0 & 0 & 5 & 15 \end{array}\right]$$

$$\xrightarrow{\frac{1}{5}R_3\to R_3} \left[\begin{array}{ccc|c} 1 & -2 & 3 & 5 \\ 0 & 12 & -13 & -27 \\ 0 & 0 & 1 & 3 \end{array}\right] \xrightarrow{13R_3+R_2\to R_2} \left[\begin{array}{ccc|c} 1 & -2 & 3 & 5 \\ 0 & 12 & 0 & 12 \\ 0 & 0 & 1 & 3 \end{array}\right]$$

$$\xrightarrow{\frac{1}{12}R_2\to R_2} \left[\begin{array}{ccc|c} 1 & -2 & 3 & 5 \\ 0 & 1 & 0 & 1 \\ 0 & 0 & 1 & 3 \end{array}\right] \xrightarrow{R_1+2R_2-3R_3\to R_1} \left[\begin{array}{ccc|c} 1 & 0 & 0 & -2 \\ 0 & 1 & 0 & 1 \\ 0 & 0 & 1 & 3 \end{array}\right]$$

Hence, the solution is $\boxed{(-2,1,3)}$.

77. Reduce the matrix down to reduced row echelon form:

$$\left[\begin{array}{ccc|c} 1 & 1 & 1 & 3 \\ 1 & 0 & -1 & 1 \\ 0 & 1 & -1 & -4 \end{array}\right] \xrightarrow{R_1-R_2\to R_2} \left[\begin{array}{ccc|c} 1 & 1 & 1 & 3 \\ 0 & 1 & 2 & 2 \\ 0 & 1 & -1 & -4 \end{array}\right] \xrightarrow{R_2-R_3\to R_3} \left[\begin{array}{ccc|c} 1 & 1 & 1 & 3 \\ 0 & 1 & 2 & 2 \\ 0 & 0 & 3 & 6 \end{array}\right]$$

$$\xrightarrow{\frac{1}{3}R_3\to R_3} \left[\begin{array}{ccc|c} 1 & 1 & 1 & 3 \\ 0 & 1 & 2 & 2 \\ 0 & 0 & 1 & 2 \end{array}\right] \xrightarrow{R_2-2R_3\to R_2} \left[\begin{array}{ccc|c} 1 & 1 & 1 & 3 \\ 0 & 1 & 0 & -2 \\ 0 & 0 & 1 & 2 \end{array}\right]$$

$$\xrightarrow{R_1-R_2-R_3\to R_1} \left[\begin{array}{ccc|c} 1 & 0 & 0 & 3 \\ 0 & 1 & 0 & -2 \\ 0 & 0 & 1 & 2 \end{array}\right]$$

Hence, the solution is $\boxed{(3,-2,2)}$.

79. Reduce the matrix down to reduced row echelon form:

$$\left[\begin{array}{ccc|c} 1 & 2 & 1 & 3 \\ 2 & -1 & 3 & 7 \\ 3 & 1 & 4 & 5 \end{array}\right] \xrightarrow[R_3-3R_1\to R_3]{R_2-2R_1\to R_2} \left[\begin{array}{ccc|c} 1 & 2 & 1 & 3 \\ 0 & -5 & 1 & 1 \\ 0 & -5 & 1 & -4 \end{array}\right] \xrightarrow{R_2-R_3\to R_3} \left[\begin{array}{ccc|c} 1 & 2 & 1 & 3 \\ 0 & -5 & 1 & 1 \\ 0 & 0 & 0 & 5 \end{array}\right]$$

The last row is equivalent to the statement $0 = 5$, which is false. Hence, the system has $\boxed{\text{no solution}}$.

81. Reduce the matrix down to reduced row echelon form:

$$\left[\begin{array}{ccc|c} 3 & -1 & 1 & 8 \\ 1 & 1 & -2 & 4 \end{array}\right] \xrightarrow{R_1-3R_2\to R_2} \left[\begin{array}{ccc|c} 3 & -1 & 1 & 8 \\ 0 & -4 & 7 & -4 \end{array}\right] \xrightarrow[-\frac{1}{4}R_2\to R_2]{R_1-\frac{1}{4}R_2\to R_1} \left[\begin{array}{ccc|c} 3 & 0 & -\frac{3}{4} & 9 \\ 0 & 1 & -\frac{7}{4} & 1 \end{array}\right]$$

$$\xrightarrow{\frac{1}{3}R_1\to R_1} \left[\begin{array}{ccc|c} 1 & 0 & -\frac{1}{4} & 3 \\ 0 & 1 & -\frac{7}{4} & 1 \end{array}\right]$$

Now, must solve the dependent system:

$$\begin{cases} x-\frac{1}{4}z=3 & \textbf{(1)} \\ y-\frac{7}{4}z=1 & \textbf{(2)} \end{cases}$$

Let $z=a$. Then, $x=\frac{a}{4}+3$ and $y=\frac{7a}{4}+1$.

83. Reduce the matrix down to reduced row echelon form:

$$\left[\begin{array}{ccc|c} 4 & -2 & 5 & 20 \\ 1 & 3 & -2 & 6 \end{array}\right] \xrightarrow{R_1-4R_2\to R_2} \left[\begin{array}{ccc|c} 4 & -2 & 5 & 20 \\ 0 & -14 & 13 & -4 \end{array}\right] \xrightarrow{-7R_1+R_2\to R_1} \left[\begin{array}{ccc|c} -28 & 0 & -22 & -144 \\ 0 & -14 & 13 & -4 \end{array}\right]$$

$$\xrightarrow{-\frac{1}{2}R_1\to R_1} \left[\begin{array}{ccc|c} 14 & 0 & 11 & 72 \\ 0 & -14 & 13 & -4 \end{array}\right]$$

Now, must solve the dependent system:

$$\begin{cases} 14x+11y=72 & \textbf{(1)} \\ -14y+13z=-4 & \textbf{(2)} \end{cases}$$

Multiply **(1)** by 14 and **(2)** by 11 and add to obtain $196x+143z=964$. Let $z=a$. Then, $x=\frac{72-11a}{14}$ and $y=\frac{13a+4}{14}$.

85. Reduce the matrix down to reduced row echelon form:

$$\left[\begin{array}{cccc|c} 1 & -1 & -1 & -1 & 1 \\ 2 & 1 & 1 & 2 & 3 \\ 1 & -2 & -2 & -3 & 0 \\ 3 & -4 & 1 & 5 & -3 \end{array}\right] \xrightarrow[\substack{R_2-2R_3\to R_3 \\ R_4-3R_3\to R_4}]{R_2-2R_1\to R_2} \left[\begin{array}{cccc|c} 1 & -1 & -1 & -1 & 1 \\ 0 & 3 & 3 & 4 & 1 \\ 0 & 5 & 5 & 8 & 3 \\ 0 & 2 & 7 & 14 & -3 \end{array}\right] \xrightarrow{\frac{1}{5}R_3\to R_3} \left[\begin{array}{cccc|c} 1 & -1 & -1 & -1 & 1 \\ 0 & 3 & 3 & 4 & 1 \\ 0 & 1 & 1 & \frac{8}{5} & \frac{3}{5} \\ 0 & 2 & 7 & 14 & -3 \end{array}\right]$$

$$\xrightarrow[R_4-2R_3\to R_4]{R_2-3R_3\to R_2} \left[\begin{array}{cccc|c} 1 & -1 & -1 & -1 & 1 \\ 0 & 0 & 0 & -\frac{4}{5} & -\frac{4}{5} \\ 0 & 1 & 1 & \frac{8}{5} & \frac{3}{5} \\ 0 & 0 & 5 & \frac{54}{5} & -\frac{21}{5} \end{array}\right] \xrightarrow[R_3\leftrightarrow R_2]{R_1\leftrightarrow R_3} \left[\begin{array}{cccc|c} 1 & -1 & -1 & -1 & 1 \\ 0 & 1 & 1 & \frac{8}{5} & \frac{3}{5} \\ 0 & 0 & 5 & \frac{54}{5} & -\frac{21}{5} \\ 0 & 0 & 0 & -\frac{4}{5} & -\frac{4}{5} \end{array}\right] \xrightarrow[-\frac{5}{4}R_4\to R_4]{\frac{1}{5}R_3\to R_3} \left[\begin{array}{cccc|c} 1 & -1 & -1 & -1 & 1 \\ 0 & 1 & 1 & \frac{8}{5} & \frac{3}{5} \\ 0 & 0 & 1 & \frac{54}{25} & -\frac{21}{25} \\ 0 & 0 & 0 & 1 & 1 \end{array}\right]$$

$$\xrightarrow{R_3-\frac{54}{25}R_4\to R_3} \left[\begin{array}{cccc|c} 1 & -1 & -1 & -1 & 1 \\ 0 & 1 & 1 & \frac{8}{5} & \frac{3}{5} \\ 0 & 0 & 1 & 0 & -3 \\ 0 & 0 & 0 & 1 & 1 \end{array}\right] \xrightarrow{R_2-R_3-\frac{8}{5}R_4\to R_2} \left[\begin{array}{cccc|c} 1 & -1 & -1 & -1 & 1 \\ 0 & 1 & 0 & 0 & 2 \\ 0 & 0 & 1 & 0 & -3 \\ 0 & 0 & 0 & 1 & 1 \end{array}\right] \xrightarrow{R_1+R_2+R_3+R_4\to R_1} \left[\begin{array}{cccc|c} 1 & 0 & 0 & 0 & 1 \\ 0 & 1 & 0 & 0 & 2 \\ 0 & 0 & 1 & 0 & -3 \\ 0 & 0 & 0 & 1 & 1 \end{array}\right]$$

Hence, the solution is $\boxed{w=1, z=-3, y=2, x=1}$

87. Let w = number of touchdowns
x = number of extra points
y = number of two-point conversions
z = number of field goals

We must solve the system:

$$\begin{cases} w+x+y+z=16 \\ w=4z \\ x=5y \\ 6w+x+2y+3z=61 \end{cases} \text{ which is equivalent to } \begin{cases} w+x+y+z=16 \\ w-4z=0 \\ x-5y=0 \\ 6w+x+2y+3z=61 \end{cases}$$

Now, write down the augmented matrix, and reduce it down to row echelon form:

$$\left[\begin{array}{cccc|c} 1 & 1 & 1 & 1 & 16 \\ 1 & 0 & 0 & -4 & 0 \\ 0 & 1 & -5 & 0 & 0 \\ 6 & 1 & 2 & 3 & 61 \end{array}\right] \xrightarrow[R_4-6R_1\to R_4]{R_1-R_2\to R_2} \left[\begin{array}{cccc|c} 1 & 1 & 1 & 1 & 16 \\ 0 & 1 & 1 & 5 & 16 \\ 0 & 1 & -5 & 0 & 0 \\ 0 & -5 & -4 & -3 & -35 \end{array}\right] \xrightarrow[5R_2+R_4\to R_4]{R_2-R_3\to R_3} \left[\begin{array}{cccc|c} 1 & 1 & 1 & 1 & 16 \\ 0 & 1 & 1 & 5 & 16 \\ 0 & 0 & 6 & 5 & 16 \\ 0 & 0 & 1 & 22 & 45 \end{array}\right]$$

$$\xrightarrow{R_3-6R_4\to R_4} \left[\begin{array}{cccc|c} 1 & 1 & 1 & 1 & 16 \\ 0 & 1 & 1 & 5 & 16 \\ 0 & 0 & 6 & 5 & 16 \\ 0 & 0 & 0 & -127 & -254 \end{array}\right] \xrightarrow[-\frac{1}{127}R_4\to R_4]{\frac{1}{6}R_3\to R_3} \left[\begin{array}{cccc|c} 1 & 1 & 1 & 1 & 16 \\ 0 & 1 & 1 & 5 & 16 \\ 0 & 0 & 1 & \frac{5}{6} & \frac{16}{6} \\ 0 & 0 & 0 & 1 & 2 \end{array}\right]$$

Thus, Row 4 now implies that $z=2$.
Then, Row 3 implies that $y+\frac{5}{6}(2)=\frac{16}{6}$ so that $y=1$.
Then, Row 2 implies that $x+1+5(2)=16$ so that $x=5$.
Finally, Row 1 implies that $w+5+1+2=16$ so that $w=8$.
So, there were **8 touchdowns, 5 extra points, 1 two-point conversion, and 2 field goals**.

89. Let w = number of Mediterranean chicken sandwiches
x = number of 6-Inch Tuna sandwiches
y = number of 6-Inch Roast Beef sandwiches
z = number of Turkey Bacon wraps

We must solve the system:

$$\begin{cases} w + x + y + z = 14 \\ 17w + 46x + 45y + 20z = 526 \\ 18w + 19x + 5y + 27z = 168 \\ 36w + 20x + 19y + 34z = 332 \end{cases}$$

Now, write down the augmented matrix, and reduce it down to row echelon form:

$$\left[\begin{array}{cccc|c} 1 & 1 & 1 & 1 & 14 \\ 17 & 46 & 45 & 20 & 526 \\ 18 & 19 & 5 & 27 & 168 \\ 36 & 20 & 19 & 34 & 332 \end{array}\right] \xrightarrow[\substack{R_4-36R_1\to R_4 \\ R_3-18R_1\to R_3 \\ R_2-17R_1\to R_2}]{} \left[\begin{array}{cccc|c} 1 & 1 & 1 & 1 & 14 \\ 0 & 29 & 28 & 3 & 288 \\ 0 & 1 & -13 & 9 & -84 \\ 0 & -16 & -17 & -2 & -172 \end{array}\right]$$

$$\xrightarrow[\substack{R_2-29R_3\to R_2 \\ R_4+16R_3\to R_4}]{} \left[\begin{array}{cccc|c} 1 & 1 & 1 & 1 & 14 \\ 0 & 0 & 405 & -258 & 2724 \\ 0 & 1 & -13 & 9 & -84 \\ 0 & 0 & -225 & 142 & -1516 \end{array}\right] \xrightarrow{R_2 \leftrightarrow R_3} \left[\begin{array}{cccc|c} 1 & 1 & 1 & 1 & 14 \\ 0 & 1 & -13 & 9 & -84 \\ 0 & 0 & 405 & -258 & 2724 \\ 0 & 0 & -225 & 142 & -1516 \end{array}\right]$$

$$\xrightarrow{\frac{1}{405}R_3\to R_3} \left[\begin{array}{cccc|c} 1 & 1 & 1 & 1 & 14 \\ 0 & 1 & -13 & 9 & -84 \\ 0 & 0 & 1 & -\frac{258}{405} & \frac{2724}{405} \\ 0 & 0 & -225 & 142 & -1516 \end{array}\right] \xrightarrow{R_4+225R_3\to R_4} \left[\begin{array}{cccc|c} 1 & 1 & 1 & 1 & 14 \\ 0 & 1 & -13 & 9 & -84 \\ 0 & 0 & 1 & -\frac{258}{405} & \frac{2724}{405} \\ 0 & 0 & 0 & -\frac{540}{405} & -\frac{1080}{405} \end{array}\right]$$

$$\xrightarrow{-\frac{405}{540}R_4\to R_4} \left[\begin{array}{cccc|c} 1 & 1 & 1 & 1 & 14 \\ 0 & 1 & -13 & 9 & -84 \\ 0 & 0 & 1 & -\frac{258}{405} & \frac{2724}{405} \\ 0 & 0 & 0 & 1 & 2 \end{array}\right]$$

Thus, Row 4 now implies that $z = 2$.
Then, Row 3 implies that $y - \frac{258}{405}(2) = \frac{2724}{405}$ so that $y = 8$.
Then, Row 2 implies that $x - 13(8) + 9(2) = -84$ so that $x = 2$.
Then, Row 1 implies that $w + 2 + 8 + 2 = 14$ so that $w = 2$.
So, there were 2 Mediterranean chicken sandwiches, 2 6-Inch Tuna sandwich, 8 6-Inch Roast Beef sandwiches, and 2 Turkey Bacon wraps.

91. From the given information, we must solve the following system:

$$\begin{cases} 34 = \frac{1}{2}a(1)^2 + v_0(1) + h_0 \\ 36 = \frac{1}{2}a(2)^2 + v_0(2) + h_0 \\ 6 = \frac{1}{2}a(3)^2 + v_0(3) + h_0 \end{cases} \text{ is equivalent to } \begin{cases} 68 = a + 2v_0 + 2h_0 \\ 72 = 4a + 4v_0 + 2h_0 \\ 12 = 9a + 6v_0 + 2h_0 \end{cases}$$

Now, write down the augmented matrix, and reduce it down to row echelon form:

$$\left[\begin{array}{ccc|c} 1 & 2 & 2 & 68 \\ 4 & 4 & 2 & 72 \\ 9 & 6 & 2 & 12 \end{array}\right] \xrightarrow[R_3-9R_1\to R_3]{R_2-4R_1\to R_2} \left[\begin{array}{ccc|c} 1 & 2 & 2 & 68 \\ 0 & -4 & -6 & -200 \\ 0 & -12 & -16 & -600 \end{array}\right] \xrightarrow{R_3-3R_2\to R_3} \left[\begin{array}{ccc|c} 1 & 2 & 2 & 68 \\ 0 & -4 & -6 & -200 \\ 0 & 0 & 2 & 0 \end{array}\right]$$

Thus, Row 3 now implies that $h_0 = 0 \text{ ft.} = \text{initial height}$.

Then, Row 2 implies that $-4v_0 - 6(0) = -200$ so that $v_0 = 50 \text{ ft./sec.} = \text{initial velocity}$.

Then, Row 1 implies that $a + 2(50) + 2(0) = 68$ so that $a = -32 \text{ ft./sec}^2 = \text{acceleration.}$.

Thus, the equation of the curve is $y = -\frac{1}{2}(32)t^2 + 50t + 0 = -16t^2 + 50t$.

93. From the given information, we must solve the following system:

$$\begin{cases} 25 = a(16)^2 + b(16) + c \\ 64 = a(40)^2 + b(40) + c \\ 40 = a(65)^2 + b(65) + c \end{cases} \text{ is equivalent to } \begin{cases} 25 = 256a + 16b + c \\ 64 = 1600a + 40b + c \\ 40 = 4225a + 65b + c \end{cases}$$

Now, write down the augmented matrix, and reduce it down to row echelon form:

$$\left[\begin{array}{ccc|c} 256 & 16 & 1 & 25 \\ 1600 & 40 & 1 & 64 \\ 4225 & 65 & 1 & 40 \end{array}\right] \xrightarrow{\frac{1}{256}R_1\to R_1} \left[\begin{array}{ccc|c} 1 & \frac{1}{16} & \frac{1}{256} & \frac{25}{256} \\ 1600 & 40 & 1 & 64 \\ 4225 & 65 & 1 & 40 \end{array}\right] \xrightarrow[R_3-4225R_1\to R_3]{R_2-1600R_1\to R_1} \left[\begin{array}{ccc|c} 1 & \frac{1}{16} & \frac{1}{256} & \frac{25}{256} \\ 0 & -60 & -\frac{1344}{256} & -\frac{23616}{256} \\ 0 & -\frac{3185}{16} & -\frac{3969}{256} & -\frac{95385}{256} \end{array}\right]$$

$$\xrightarrow[-\frac{16}{3185}R_3\to R_3]{-\frac{1}{60}R_2\to R_2} \left[\begin{array}{ccc|c} 1 & \frac{1}{16} & \frac{1}{256} & \frac{25}{256} \\ 0 & 1 & \frac{1344}{15360} & \frac{23616}{15360} \\ 0 & 1 & \frac{63504}{815360} & \frac{1526160}{815360} \end{array}\right] \xrightarrow{R_3-R_2\to R_3} \left[\begin{array}{ccc|c} 1 & \frac{1}{16} & \frac{1}{256} & \frac{25}{256} \\ 0 & 1 & \frac{1344}{15360} & \frac{23616}{15360} \\ 0 & 0 & \frac{63504}{815360}-\frac{1344}{15360} & \frac{1526160}{815360}-\frac{23616}{15360} \end{array}\right]$$

$$\xrightarrow{\frac{1}{\frac{1344}{15360}-\frac{63504}{815360}}R_3\to R_3} \left[\begin{array}{ccc|c} 1 & \frac{1}{16} & \frac{1}{256} & \frac{25}{256} \\ 0 & 1 & \frac{1344}{15360} & \frac{23616}{15360} \\ 0 & 0 & 1 & -34.76326 \end{array}\right] \cong \left[\begin{array}{ccc|c} 1 & 0.0625 & 0.0039 & 0.0977 \\ 0 & 1 & 0.0875 & 1.5375 \\ 0 & 0 & 1 & -34.76326 \end{array}\right]$$

Thus, Row 3 now implies that $c \cong -34.76326$.
Then, Row 2 implies that $b + 0.0875(-34.76326) = 1.5375$ so that $b \cong 4.57928$.
Then, Row 1 implies that
$a + 0.0625(4.57928) + 0.0039(-34.76326) = 0.0977$ so that $a \cong -0.052755$.

Thus, the approximate equation of the curve is $\boxed{y = -0.053x^2 + 4.58x - 34.76}$.

95. Let x = number of mL of 1.5% solution,
y = number of mL of 30% solution.

We must solve the system:

$$\begin{cases} x+y=100 \\ 0.015x+0.30y=0.05(100) \end{cases}$$

Write down the augmented matrix, and reduce it down to row echelon form:

$$\left[\begin{array}{cc|c} 1 & 1 & 100 \\ 0.015 & 0.30 & 5 \end{array}\right] \xrightarrow{1000R_2 \to R_2} \left[\begin{array}{cc|c} 1 & 1 & 100 \\ 15 & 300 & 5000 \end{array}\right]$$

$$\xrightarrow{R_1 \leftrightarrow R_2} \left[\begin{array}{cc|c} 15 & 300 & 5000 \\ 1 & 1 & 100 \end{array}\right] \xrightarrow{R_1-15R_2 \to R_2} \left[\begin{array}{cc|c} 15 & 300 & 5000 \\ 0 & 285 & 3500 \end{array}\right]$$

$$\xrightarrow{\frac{1}{285}R_1 \to R_2} \left[\begin{array}{cc|c} 15 & 300 & 5000 \\ 0 & 1 & 12.28 \end{array}\right] \xrightarrow{\frac{1}{15}R_1 \to R_1} \left[\begin{array}{cc|c} 1 & 20 & 333.33 \\ 0 & 1 & 12.28 \end{array}\right]$$

$$\xrightarrow{R_1-20R_2 \to R_1} \left[\begin{array}{cc|c} 1 & 0 & 87.73 \\ 0 & 1 & 12.28 \end{array}\right]$$

So, about 88mL of 1.5% solution and 12 mL of 30% solution.

97. Let x = number of basic widgets ($12 each),
y = number of mid-price widgets ($15 each),
z = number of top-of-the-line widgets ($18 each).

We must solve the system:

$$\begin{cases} x+y+z=375 \\ 12x+15y+18z=5250 \\ x=2y \end{cases} \text{ which is equivalent to } \begin{cases} x+y+z=375 \\ 12x+15y+18z=5250 \\ x-2y=0 \end{cases}$$

Write down the augmented matrix, and reduce it down to row echelon form:

$$\left[\begin{array}{ccc|c} 1 & 1 & 1 & 375 \\ 1 & -2 & 0 & 0 \\ 12 & 15 & 18 & 5250 \end{array}\right] \xrightarrow{R_3-12R_2 \to R_3} \left[\begin{array}{ccc|c} 1 & 1 & 1 & 375 \\ 1 & -2 & 0 & 0 \\ 0 & 39 & 18 & 5250 \end{array}\right] \xrightarrow{R_1-R_2 \to R_1} \left[\begin{array}{ccc|c} 0 & 3 & 1 & 375 \\ 1 & -2 & 0 & 0 \\ 0 & 39 & 18 & 5250 \end{array}\right]$$

$$\xrightarrow{R_1 \leftrightarrow R_2} \left[\begin{array}{ccc|c} 1 & -2 & 0 & 0 \\ 0 & 3 & 1 & 375 \\ 0 & 39 & 18 & 5250 \end{array}\right] \xrightarrow{\frac{1}{3}R_2 \to R_2} \left[\begin{array}{ccc|c} 1 & -2 & 0 & 0 \\ 0 & 1 & \frac{1}{3} & 125 \\ 0 & 39 & 18 & 5250 \end{array}\right] \xrightarrow{R_3-29R_2 \to R_3} \left[\begin{array}{ccc|c} 1 & -2 & 0 & 0 \\ 0 & 1 & \frac{1}{3} & 125 \\ 0 & 0 & 5 & 375 \end{array}\right]$$

$$\xrightarrow{\frac{1}{5}R_3 \to R_3} \left[\begin{array}{ccc|c} 1 & -2 & 0 & 0 \\ 0 & 1 & \frac{1}{3} & 125 \\ 0 & 0 & 1 & 75 \end{array}\right] \xrightarrow{R_2-\frac{1}{3}R_3 \to R_2} \left[\begin{array}{ccc|c} 1 & -2 & 0 & 0 \\ 0 & 1 & 0 & 100 \\ 0 & 0 & 1 & 75 \end{array}\right] \xrightarrow{R_1+2R_2 \to R_1} \left[\begin{array}{ccc|c} 1 & 0 & 0 & 200 \\ 0 & 1 & 0 & 100 \\ 0 & 0 & 1 & 75 \end{array}\right]$$

So, 200 basic widgets, 100 mid-price widgets, and 75 top-of-the-line widgets will be produced.

99. Let x = amount in money market (3%)
y = amount in mutual fund (7%)
z = amount in stock (10%)

We must solve the system:

$$\begin{cases} x+y+z=10,000 \\ x=y+3000 \\ 0.03x+0.07y+0.10z=540 \end{cases} \text{ which is equivalent to } \begin{cases} x+y+z=10,000 \\ x-y=3000 \\ 3x+7y+10z=54,000 \end{cases}$$

Write down the augmented matrix, and reduce it down to row echelon form:

$$\left[\begin{array}{ccc|c} 1 & 1 & 1 & 10000 \\ 1 & -1 & 0 & 3000 \\ 3 & 7 & 10 & 54000 \end{array}\right] \xrightarrow[R_2-R_1\to R_2]{R_3-3R_1\to R_3} \left[\begin{array}{ccc|c} 1 & 1 & 1 & 10000 \\ 0 & -2 & -1 & -7000 \\ 0 & 4 & 7 & 24000 \end{array}\right] \xrightarrow{R_3+2R_2\to R_3} \left[\begin{array}{ccc|c} 1 & 1 & 1 & 10000 \\ 0 & -2 & -1 & -7000 \\ 0 & 0 & 5 & 10000 \end{array}\right]$$

$$\xrightarrow[-\frac{1}{2}R_2\to R_2]{\frac{1}{5}R_3\to R_3} \left[\begin{array}{ccc|c} 1 & 1 & 1 & 10000 \\ 0 & 1 & \frac{1}{2} & 3500 \\ 0 & 0 & 1 & 2000 \end{array}\right] \xrightarrow{R_2-\frac{1}{2}R_3\to R_2} \left[\begin{array}{ccc|c} 1 & 1 & 1 & 10000 \\ 0 & 1 & 0 & 2500 \\ 0 & 0 & 1 & 2000 \end{array}\right] \xrightarrow{R_1-R_2-R_3\to R_1} \left[\begin{array}{ccc|c} 1 & 0 & 0 & 5500 \\ 0 & 1 & 0 & 2500 \\ 0 & 0 & 1 & 2000 \end{array}\right]$$

So, they should invest \$5500 in money market, \$2500 in mutual fund, and \$2000 in stock.

101. Let a = units of product x, b = units of product y, and c = units of product z. We employ matrix methods to solve the system:

$$\begin{cases} 20a+25b+150c=2400 \\ 2a+5b+10c=310 \\ a+0b+\frac{1}{2}c=28 \end{cases}$$

$$\left[\begin{array}{ccc|c} 20 & 25 & 150 & 2400 \\ 2 & 5 & 10 & 310 \\ 1 & 0 & \frac{1}{2} & 28 \end{array}\right] \xrightarrow[R_2-2R_3\to R_3]{\frac{1}{5}R_1\to R_1} \left[\begin{array}{ccc|c} 4 & 5 & 30 & 480 \\ 2 & 5 & 10 & 310 \\ 0 & 5 & 9 & 254 \end{array}\right] \xrightarrow{R_1-2R_2\to R_2} \left[\begin{array}{ccc|c} 4 & 5 & 30 & 480 \\ 0 & -5 & 10 & -140 \\ 0 & 5 & 9 & 254 \end{array}\right]$$

$$\xrightarrow{R_2+R_3\to R_3} \left[\begin{array}{ccc|c} 4 & 5 & 30 & 480 \\ 0 & -5 & 10 & -140 \\ 0 & 0 & 19 & 114 \end{array}\right] \xrightarrow[\frac{1}{19}R_3\to R_3]{-\frac{1}{5}R_2\to R_2} \left[\begin{array}{ccc|c} 4 & 5 & 30 & 480 \\ 0 & 1 & -2 & 28 \\ 0 & 0 & 1 & 6 \end{array}\right]$$

$$\xrightarrow[R_2+2R_3\to R_2]{\frac{1}{4}R_1\to R_1} \left[\begin{array}{ccc|c} 1 & \frac{5}{4} & \frac{15}{2} & 120 \\ 0 & 1 & 0 & 40 \\ 0 & 0 & 1 & 6 \end{array}\right] \xrightarrow{R_1-\frac{5}{4}R_2-\frac{15}{2}R_3\to R_1} \left[\begin{array}{ccc|c} 1 & 0 & 0 & 25 \\ 0 & 1 & 0 & 40 \\ 0 & 0 & 1 & 6 \end{array}\right]$$

Hence, there are 25 units of product x, 40 units of product y, and 6 units of product z.

103. Let x = # of general admission tickets, y = # of reserved tickets, and z = # of end zone tickets.

We solve the following system using matrix methods:

$$\begin{cases} y = x+5 \\ z = x+20 \\ 20x+40y+15z = 2375 \end{cases}$$

$$\left[\begin{array}{ccc|c} 1 & -1 & 0 & -5 \\ 1 & 0 & -1 & -20 \\ 20 & 40 & 15 & 2375 \end{array}\right] \xrightarrow[\frac{1}{5}R_3 \to R_3]{R_1 - R_2 \to R_2} \left[\begin{array}{ccc|c} 1 & -1 & 0 & -5 \\ 0 & -1 & 1 & 15 \\ 4 & 8 & 3 & 475 \end{array}\right] \xrightarrow{R_3 - 4R_1 \to R_3} \left[\begin{array}{ccc|c} 1 & -1 & 0 & -5 \\ 0 & -1 & 1 & 15 \\ 0 & 12 & 3 & 495 \end{array}\right]$$

$$\xrightarrow[-R_2 \to R_2]{R_3 + 12R_2 \to R_3} \left[\begin{array}{ccc|c} 1 & -1 & 0 & -5 \\ 0 & 1 & -1 & -15 \\ 0 & 0 & 15 & 675 \end{array}\right] \xrightarrow{\frac{1}{15}R_3 \to R_3} \left[\begin{array}{ccc|c} 1 & -1 & 0 & -5 \\ 0 & 1 & -1 & -15 \\ 0 & 0 & 1 & 45 \end{array}\right]$$

$$\xrightarrow{R_2 + R_3 \to R_2} \left[\begin{array}{ccc|c} 1 & -1 & 0 & -5 \\ 0 & 1 & 0 & 30 \\ 0 & 0 & 1 & 45 \end{array}\right] \xrightarrow{R_1 + R_2 \to R_1} \left[\begin{array}{ccc|c} 1 & 0 & 0 & 25 \\ 0 & 1 & 0 & 30 \\ 0 & 0 & 1 & 45 \end{array}\right]$$

Hence, there are 25 general admission tickets, 30 reserved tickets, and 45 end zone tickets.

105. Since the points $(4,4)$, $(-3,-1)$, and $(1,-3)$ are assumed to lie on the circle, they must each satisfy the equation. This generates the following system:

$$\begin{cases} (4)^2+(4)^2+a(4)+b(4)+c=0 \\ (-3)^2+(-1)^2+a(-3)+b(-1)+c=0 \\ (1)^2+(-3)^2+a(1)+b(-3)+c=0 \end{cases} \text{ which is equivalent to } \begin{cases} 4a+4b+c=-32 \\ -3a-b+c=-10 \\ a-3b+c=-10 \end{cases}$$

Write down the augmented matrix, and reduce it down to row echelon form:

$$\left[\begin{array}{ccc|c} 4 & 4 & 1 & -32 \\ -3 & -1 & 1 & -10 \\ 1 & -3 & 1 & -10 \end{array}\right] \xrightarrow[R_2+3R_3 \to R_2]{R_1 - 4R_3 \to R_3} \left[\begin{array}{ccc|c} 4 & 4 & 1 & -32 \\ 0 & -10 & 4 & -40 \\ 0 & 16 & -3 & 8 \end{array}\right] \xrightarrow[-\frac{1}{10}R_2 \to R_2]{\frac{1}{4}R_1 \to R_1} \left[\begin{array}{ccc|c} 1 & 1 & \frac{1}{4} & -8 \\ 0 & 1 & -\frac{2}{5} & 4 \\ 0 & 16 & -3 & 8 \end{array}\right]$$

$$\xrightarrow{R_3 - 16R_2 \to R_3} \left[\begin{array}{ccc|c} 1 & 1 & \frac{1}{4} & -8 \\ 0 & 1 & -\frac{2}{5} & 4 \\ 0 & 0 & \frac{17}{5} & -56 \end{array}\right] \xrightarrow{\frac{5}{17}R_3 \to R_3} \left[\begin{array}{ccc|c} 1 & 1 & \frac{1}{4} & -8 \\ 0 & 1 & -\frac{2}{5} & 4 \\ 0 & 0 & 1 & -\frac{280}{17} \end{array}\right]$$

Row 3 implies that $\boxed{c = -\frac{280}{17}}$.

Then, Row 2 implies that $b - \frac{2}{5}(-\frac{280}{17}) = 4$ so that $\boxed{b = -\frac{44}{17}}$.

Then, Row 1 implies that $a + (-\frac{44}{17}) + \frac{1}{4}(-\frac{280}{17}) = -8$ so that $\boxed{a = -\frac{22}{17}}$.

Hence, the equation of the circle is $x^2 + y^2 - \frac{22}{7}x - \frac{44}{17}y - \frac{280}{17} = 0$.

107. Need to line up a single variable in a given column before forming the augmented matrix. Specifically, write the system as follows:

$$\begin{cases} -x + y + z = 2 \\ x + y - 2z = -3 \\ x + y + z = 6 \end{cases}$$

So, the correct matrix is $\left[\begin{array}{ccc|c} -1 & 1 & 1 & 2 \\ 1 & 1 & -2 & -3 \\ 1 & 1 & 1 & 6 \end{array}\right]$, and after reducing, $\left[\begin{array}{ccc|c} 1 & 0 & 0 & 2 \\ 0 & 1 & 0 & 1 \\ 0 & 0 & 1 & 3 \end{array}\right]$.

109. Row 3 implies that $1z = 0$, which is valid. Also, Rows 2 and 1 imply $y = 2,\ x = 1$, respectively. So, the solution is $x = 1,\ y = 2,\ z = 0$.

111. False

113. True

115. Assume year 1999 corresponds to 0, 2000 to 1, 2001 to 2, 2002 to 3, and 2003 to 4. Then, from the given data, we can infer that the following four points are on the graph of $f(x) = ax^4 + bx^3 + cx^2 + dx + e$:

$$(0,44),\ (1,51),\ (2,46),\ (3,51),\ (4,44)$$

Hence, they must all satisfy the equation. This leads to the following 5×5 system:

$$\begin{cases} 44 = a(0)^4 + b(0)^3 + c(0)^2 + d(0) + e \\ 51 = a(1)^4 + b(1)^3 + c(1)^2 + d(1) + e \\ 46 = a(2)^4 + b(2)^3 + c(2)^2 + d(2) + e \\ 51 = a(3)^4 + b(3)^3 + c(3)^2 + d(3) + e \\ 44 = a(4)^4 + b(4)^3 + c(4)^2 + d(4) + e \end{cases}$$

which is equivalent to:

$$\begin{cases} 44 = e \\ 51 = a + b + c + d + e \\ 46 = 16a + 8b + 4c + 2d + e \\ 51 = 81a + 27b + 9c + 3d + e \\ 44 = 256a + 64b + 16c + 4d + e \end{cases}$$

The corresponding augmented matrix is:

$$\left[\begin{array}{ccccc|c} 0 & 0 & 0 & 0 & 1 & 44 \\ 1 & 1 & 1 & 1 & 1 & 51 \\ 16 & 8 & 4 & 2 & 1 & 46 \\ 81 & 27 & 9 & 3 & 1 & 51 \\ 256 & 64 & 16 & 4 & 1 & 44 \end{array}\right]$$

One can perform Gaussian elimination as in the previous problems, but this would be a fine example as to when using technology can be useful. Indeed, using a calculator with matrix capabilities can be used to verify the following choices for a, b, c, d, and e: $a = -\frac{11}{6}$, $b = \frac{44}{3}$, $c = -\frac{223}{6}$, $d = \frac{94}{3}$, $e = 44$. Hence, the equation of the polynomial is

$\boxed{y = -\frac{11}{6}x^4 + \frac{44}{3}x^3 - \frac{223}{6}x^2 + \frac{94}{3}x + 44}$.

117. The answers coincide.

119.

a.

```
rref([A])
[[1 0 0 -.23653…
 [0 1 0 .928846…
 [0 0 1 6.08846…
```

$y = -0.24x^2 + 0.93x + 6.09$

b.

```
QuadReg
 y=ax²+bx+c
 a=-.2365384615
 b=.9288461538
 c=6.088461538
```

$y = -0.24x^2 + 0.93x + 6.09$

Section 7.2 Solutions ---

1. 2×3

3. 2×2

5. 3×3

7. 1×1

9. 4×4

11. Since corresponding entries of equal matrices must themselves be equal, we have $x=-5, y=1$.

13. Since corresponding entries of equal matrices must themselves be equal, we have

$$\begin{cases} x+y=-5 & \textbf{(1)} \\ x-y=-1 & \textbf{(2)} \\ z=3 \end{cases}$$

Solve the system:

Add **(1)** and **(2)**: $2x=-6 \Rightarrow x=-3$

Substitute this into **(1)** to obtain $y=-2$.

15. Since corresponding entries of equal matrices must themselves be equal, we have

$$\begin{cases} x-y=3 & \textbf{(1)} \\ x+2y=12 & \textbf{(2)} \end{cases}$$

Solve the system:

Multiply **(1)** by -1: $-x+y=-3$ **(3)**

Add **(2)** and **(3)**: $3y=9 \Rightarrow y=3$

Substitute this into **(1)** to obtain $x=6$.

17.

$$\begin{bmatrix} -1 & 3 & 0 \\ 2 & 4 & 1 \end{bmatrix} + \begin{bmatrix} 0 & 2 & 1 \\ 3 & -2 & 4 \end{bmatrix} = \begin{bmatrix} -1 & 5 & 1 \\ 5 & 2 & 5 \end{bmatrix}$$

19.

$$\begin{bmatrix} -2 & 4 \\ 2 & -2 \\ -1 & 3 \end{bmatrix}$$

21. Not defined since B is 2×3 and C is 3×2

23. This is not defined since the matrices have different orders.

25.

$$\begin{bmatrix} -2 & 6 & 0 \\ 4 & 8 & 2 \end{bmatrix}$$

27.

$$\begin{bmatrix} 0 & -5 \\ -10 & 5 \\ -15 & -5 \end{bmatrix}$$

29.

$$2\begin{bmatrix} -1 & 3 & 0 \\ 2 & 4 & 1 \end{bmatrix} + 3\begin{bmatrix} 0 & 2 & 1 \\ 3 & -2 & 4 \end{bmatrix} = \begin{bmatrix} -2 & 6 & 0 \\ 4 & 8 & 2 \end{bmatrix} + \begin{bmatrix} 0 & 6 & 3 \\ 9 & -6 & 12 \end{bmatrix} = \begin{bmatrix} -2 & 12 & 3 \\ 13 & 2 & 14 \end{bmatrix}$$

31. $\begin{bmatrix} 8 & 3 \\ 11 & 5 \end{bmatrix}$

33. $\begin{bmatrix} -3 & 21 & 6 \\ -4 & 7 & 1 \\ 13 & 14 & 9 \end{bmatrix}$

35. $\begin{bmatrix} 3 & 6 \\ -2 & -2 \\ 17 & 24 \end{bmatrix}$

37. Not possible you cannot multiply a 2×2 matrix by a 3×2 matrix.

39. $-4\begin{bmatrix} 0 & -15 \end{bmatrix} = \begin{bmatrix} 0 & 60 \end{bmatrix}$

41. $\begin{bmatrix} 2 & 0 & -3 \end{bmatrix}\begin{bmatrix} 0 & 2 & 0 \\ 2 & 4 & 5 \\ 2 & 1 & 3 \end{bmatrix} = \begin{bmatrix} -6 & 1 & -9 \end{bmatrix}$

43.

$$\begin{bmatrix} 2 & 0 & -3 \\ 0 & 0 & 0 \\ -2 & 0 & 3 \end{bmatrix} + \begin{bmatrix} 5 & 10 & -5 \\ 0 & 15 & 5 \\ 25 & 0 & -10 \end{bmatrix} = \begin{bmatrix} 7 & 10 & -8 \\ 0 & 15 & 5 \\ 23 & 0 & -7 \end{bmatrix}$$

45. $\begin{bmatrix} 7 & 10 \\ 15 & 22 \end{bmatrix} + \begin{bmatrix} 5 & 10 \\ 15 & 20 \end{bmatrix} = \begin{bmatrix} 12 & 20 \\ 30 & 42 \end{bmatrix}$

47. $\begin{bmatrix} -2 & 0 & 2 \\ 4 & 2 & 8 \\ -6 & 2 & 10 \end{bmatrix}\begin{bmatrix} 1 \\ 0 \\ -1 \end{bmatrix} = \begin{bmatrix} -4 \\ -4 \\ -16 \end{bmatrix}$

49. Not possible since the inner dimensions are different.

51. $A = \begin{bmatrix} \textit{Yes} \text{ response} \\ \textit{No} \text{ response} \end{bmatrix} = \begin{bmatrix} 0.70 \\ 0.30 \end{bmatrix}$ $\quad B = \begin{bmatrix} \textit{Yes} \text{ to "increase lung cancer"} \\ \textit{Yes} \text{ to "would shorten lives"} \end{bmatrix} = \begin{bmatrix} 0.89 \\ 0.84 \end{bmatrix}$

a. $46A = 46\begin{bmatrix} 0.70 \\ 0.30 \end{bmatrix} = \begin{bmatrix} 32.2 \\ 13.8 \end{bmatrix}$. This matrix tells us that out of 46 million people, 32.2 million said that they had tried to quit smoking, while 13.8 million said that they had not.

b. $46B = 46\begin{bmatrix} 0.89 \\ 0.84 \end{bmatrix} = \begin{bmatrix} 40.94 \\ 38.64 \end{bmatrix}$. This matrix tells us that out of 46 million people, 40.94 million believed that smoking would increase the chance of getting lung cancer, and that 38.64 million believed that smoking would shorten their lives.

53. Since $A = \begin{bmatrix} 0.589 & 0.628 \\ 0.414 & 0.430 \end{bmatrix}$ and $B = \begin{bmatrix} 100M \\ 110M \end{bmatrix}$, it follows that $AB = \begin{bmatrix} 127.98M \\ 88.7M \end{bmatrix}$.

This matrix tells us the number of registered voters and the number of actual voters. In particular, there are 127.98 million registered voters, and of those 88.7 million actually vote.

55.

$$A = \begin{bmatrix} 0.45 & 0.50 & 1.00 \end{bmatrix} \quad B = \begin{bmatrix} 7,523 \\ 2,700 \\ 15,200 \end{bmatrix} \quad AB = \begin{bmatrix} 19,935.35 \end{bmatrix}$$

57.

$$A = \begin{bmatrix} 230 & 3 & 44 & 9 \\ 430 & 19 & 46 & 20 \\ 290 & 5 & 45 & 19 \\ 330 & 5 & 47 & 24 \end{bmatrix} \quad 2A = \begin{bmatrix} 460 & 6 & 88 & 18 \\ 860 & 38 & 92 & 40 \\ 580 & 10 & 90 & 38 \\ 660 & 10 & 94 & 48 \end{bmatrix}$$

This matrix represents the nutritional information corresponding to 2 sandwiches.

$$0.5A = \begin{bmatrix} 115 & 1.5 & 22 & 4.5 \\ 215 & 9.5 & 23 & 10 \\ 145 & 2.5 & 22.5 & 9.5 \\ 165 & 2.5 & 23.5 & 12 \end{bmatrix}$$

This matrix represents the nutritional information corresponding to one-half of a sandwich.

59.

$$A = \begin{bmatrix} 0.06 & 0 \\ 0.02 & 0.1 \\ 0 & 0.3 \end{bmatrix} \quad B = \begin{bmatrix} 3.80 \\ 0.05 \end{bmatrix} \quad AB = \begin{bmatrix} 0.228 \\ 0.081 \\ 0.015 \end{bmatrix}$$

The entries in AB represent the total cost to run each type of automobile per mile.

61. $N = \begin{bmatrix} 2 \\ 1 \\ 0 \end{bmatrix} \quad XN = \begin{bmatrix} 10 \\ 16 \\ 20 \end{bmatrix}$

The nutritional content of the meal is 10 g of carbohydrates, 16 g of protein and 20 g of fat.

63. $N = \begin{bmatrix} 200 \\ 25 \\ 0 \end{bmatrix} \quad XN = \begin{bmatrix} 9.25 \\ 13.25 \\ 15.75 \end{bmatrix}$

Company 1 would charge \$9.25, Company 2 would charge \$13.25 and Company 3 would charge \$15.75 respectively for 200 minutes of talking and 25 text messages. The better cell phone provider for this employee would be Company 1.

65. Not multiplying correctly. It should be:

$$\begin{bmatrix} 3 & 2 \\ 1 & 4 \end{bmatrix} \cdot \begin{bmatrix} -1 & 3 \\ -2 & 5 \end{bmatrix} = \begin{bmatrix} 3(-1)+2(-2) & 3(3)+2(5) \\ 1(-1)+4(-2) & 1(3)+4(5) \end{bmatrix} = \begin{bmatrix} -7 & 19 \\ -9 & 23 \end{bmatrix}$$

67. False. In general,

$$AB = \begin{bmatrix} a_{11} & a_{12} \\ a_{21} & a_{22} \end{bmatrix} \cdot \begin{bmatrix} b_{11} & b_{12} \\ b_{21} & b_{22} \end{bmatrix} = \begin{bmatrix} a_{11}b_{11} + a_{12}b_{21} & a_{11}b_{12} + a_{12}b_{22} \\ a_{21}b_{11} + a_{22}b_{21} & a_{21}b_{12} + a_{22}b_{22} \end{bmatrix}$$

69. True

71. $A^2 = AA = \begin{bmatrix} a_{11}^2 + a_{12}a_{21} & a_{11}a_{12} + a_{12}a_{22} \\ a_{21}a_{11} + a_{22}a_{21} & a_{22}^2 + a_{21}a_{12} \end{bmatrix}$

73. Observe that

$$A = \begin{bmatrix} 1 & 1 \\ 1 & 1 \end{bmatrix},$$

$$A^2 = AA = \begin{bmatrix} 2 & 2 \\ 2 & 2 \end{bmatrix},$$

$$A^3 = A^2A = \begin{bmatrix} 4 & 4 \\ 4 & 4 \end{bmatrix} = \begin{bmatrix} 2^2 & 2^2 \\ 2^2 & 2^2 \end{bmatrix} = 2^2 \begin{bmatrix} 1 & 1 \\ 1 & 1 \end{bmatrix} = 2^2 A$$

$$\vdots$$

$$A^n = \begin{bmatrix} 2^{n-1} & 2^{n-1} \\ 2^{n-1} & 2^{n-1} \end{bmatrix} = 2^{n-1} \begin{bmatrix} 1 & 1 \\ 1 & 1 \end{bmatrix} = 2^{n-1} A, \quad n \geq 1$$

75. $A_{m\times n}B_{n\times p}$ is a matrix of order $m \times p$. Since a product CD is defined if and only if the number of columns of C equals the number of rows of D, if $C = D$, then C has the same number of columns and rows. Applying this to the given product, we see that in order for $\left(A_{m\times n}B_{n\times p}\right)^2$ to be defined, we need $m = p$.

77. $AB = \begin{bmatrix} 33 & 35 \\ -96 & -82 \\ 31 & 19 \\ 146 & 138 \end{bmatrix}$

79. BB is not defined.

81. $\begin{bmatrix} 5 & -4 & 4 \\ 2 & -15 & -3 \\ 26 & 4 & -8 \end{bmatrix}$

Section 7.3 Solutions--

1. $\begin{bmatrix} -2 & 5 \\ 7 & -2 \end{bmatrix} \begin{bmatrix} x \\ y \end{bmatrix} = \begin{bmatrix} 10 \\ -4 \end{bmatrix}$

3. $\begin{bmatrix} 1 & -2 \\ -3 & 1 \end{bmatrix} \begin{bmatrix} x \\ y \end{bmatrix} = \begin{bmatrix} 8 \\ 6 \end{bmatrix}$

5. $\begin{bmatrix} 3 & 5 & -1 \\ 1 & 0 & 2 \\ -1 & 1 & -1 \end{bmatrix} \begin{bmatrix} x \\ y \\ z \end{bmatrix} = \begin{bmatrix} 2 \\ 17 \\ 4 \end{bmatrix}$

7. $\begin{bmatrix} 3 & 0 & 1 \\ 0 & 1 & -2 \\ 1 & 2 & 0 \end{bmatrix} \begin{bmatrix} x \\ y \\ z \end{bmatrix} = \begin{bmatrix} 10 \\ 4 \\ 6 \end{bmatrix}$

9. Since $AB = \begin{bmatrix} 8 & -11 \\ -5 & 7 \end{bmatrix} \cdot \begin{bmatrix} 7 & 11 \\ 5 & 8 \end{bmatrix} = \begin{bmatrix} 1 & 0 \\ 0 & 1 \end{bmatrix} = I$, B must be the inverse of A.

11. Since $AB = \begin{bmatrix} 3 & 1 \\ 1 & -2 \end{bmatrix} \cdot \begin{bmatrix} \frac{2}{7} & \frac{1}{7} \\ \frac{1}{7} & -\frac{3}{7} \end{bmatrix} = \begin{bmatrix} 1 & 0 \\ 0 & 1 \end{bmatrix} = I$, B must be the inverse of A.

13. Since $AB = \begin{bmatrix} 1 & 2 \\ 3 & 4 \end{bmatrix} \cdot \begin{bmatrix} 4 & -2 \\ -3 & 1 \end{bmatrix} = \begin{bmatrix} -2 & 0 \\ 0 & -2 \end{bmatrix} \neq I$, B cannot be the inverse of A.

15. Since $AB = \begin{bmatrix} 1 & -1 & 1 \\ 1 & 0 & -1 \\ 0 & 1 & -1 \end{bmatrix} \cdot \begin{bmatrix} 1 & 0 & 1 \\ 1 & -1 & 2 \\ 1 & -1 & 1 \end{bmatrix} = \begin{bmatrix} 1 & 0 & 0 \\ 0 & 1 & 0 \\ 0 & 0 & 1 \end{bmatrix} = I$, B must be the inverse of A.

17. Since $AB = \begin{bmatrix} 2 & 0 & 1 \\ 0 & 3 & 1 \\ 0 & 2 & -1 \end{bmatrix} \cdot \begin{bmatrix} 0 & 2 & 1 \\ 0 & 3 & 0 \\ 2 & 0 & 2 \end{bmatrix} = \begin{bmatrix} 2 & 4 & 4 \\ 2 & 9 & 2 \\ -2 & 6 & -2 \end{bmatrix} \neq I$, B cannot be the inverse of A.

19. $A^{-1} = \frac{1}{1}\begin{bmatrix} 0 & -1 \\ 1 & 2 \end{bmatrix} = \begin{bmatrix} 0 & -1 \\ 1 & 2 \end{bmatrix}$.

21. $A^{-1} = \frac{1}{-\frac{39}{4}}\begin{bmatrix} \frac{3}{4} & -2 \\ -5 & \frac{1}{3} \end{bmatrix} = \begin{bmatrix} -\frac{4}{39}(\frac{3}{4}) & -\frac{4}{39}(-2) \\ -\frac{4}{39}(-5) & -\frac{4}{39}(\frac{1}{3}) \end{bmatrix} = \begin{bmatrix} -\frac{1}{13} & \frac{8}{39} \\ \frac{20}{39} & -\frac{4}{117} \end{bmatrix}$.

23. $A^{-1} = \frac{1}{-10.51}\begin{bmatrix} 1.7 & -2.4 \\ -5.3 & 1.3 \end{bmatrix} \cong \begin{bmatrix} -0.1618 & 0.2284 \\ 0.5043 & -0.1237 \end{bmatrix}$.

25. Using Gaussian elimination, we obtain the following:

$$\left[\begin{array}{ccc|ccc} 1 & 1 & 1 & 1 & 0 & 0 \\ 1 & -1 & -1 & 0 & 1 & 0 \\ -1 & 1 & -1 & 0 & 0 & 1 \end{array}\right] \xrightarrow{R_2+R_3 \to R_3} \left[\begin{array}{ccc|ccc} 1 & 1 & 1 & 1 & 0 & 0 \\ 1 & -1 & -1 & 0 & 1 & 0 \\ 0 & 0 & -2 & 0 & 1 & 1 \end{array}\right]$$

$$\xrightarrow{R_1+R_2 \to R_1} \left[\begin{array}{ccc|ccc} 2 & 0 & 0 & 1 & 1 & 0 \\ 1 & -1 & -1 & 0 & 1 & 0 \\ 0 & 0 & -2 & 0 & 1 & 1 \end{array}\right] \xrightarrow[-\frac{1}{2}R_3 \to R_3]{\frac{1}{2}R_1 \to R_1} \left[\begin{array}{ccc|ccc} 1 & 0 & 0 & \frac{1}{2} & \frac{1}{2} & 0 \\ 1 & -1 & -1 & 0 & 1 & 0 \\ 0 & 0 & 1 & 0 & -\frac{1}{2} & -\frac{1}{2} \end{array}\right]$$

$$\xrightarrow{R_1-R_2 \to R_2} \left[\begin{array}{ccc|ccc} 1 & 0 & 0 & \frac{1}{2} & \frac{1}{2} & 0 \\ 0 & 1 & 1 & \frac{1}{2} & -\frac{1}{2} & 0 \\ 0 & 0 & 1 & 0 & -\frac{1}{2} & -\frac{1}{2} \end{array}\right] \xrightarrow{R_2-R_3 \to R_2} \left[\begin{array}{ccc|ccc} 1 & 0 & 0 & \frac{1}{2} & \frac{1}{2} & 0 \\ 0 & 1 & 0 & \frac{1}{2} & 0 & \frac{1}{2} \\ 0 & 0 & 1 & 0 & -\frac{1}{2} & -\frac{1}{2} \end{array}\right]$$

So, $A^{-1} = \begin{bmatrix} \frac{1}{2} & \frac{1}{2} & 0 \\ \frac{1}{2} & 0 & \frac{1}{2} \\ 0 & -\frac{1}{2} & -\frac{1}{2} \end{bmatrix}$.

27. Using Gaussian elimination, we obtain the following:

$$\left[\begin{array}{ccc|ccc} 1 & 0 & 1 & 1 & 0 & 0 \\ 0 & 1 & 1 & 0 & 1 & 0 \\ 1 & -1 & 0 & 0 & 0 & 1 \end{array}\right] \xrightarrow{R_2+R_3\to R_3} \left[\begin{array}{ccc|ccc} 1 & 0 & 1 & 1 & 0 & 0 \\ 0 & 1 & 1 & 0 & 1 & 0 \\ 1 & 0 & 1 & 0 & 1 & 1 \end{array}\right] \xrightarrow{R_1-R_3\to R_3} \left[\begin{array}{ccc|ccc} 1 & 0 & 1 & 1 & 0 & 0 \\ 0 & 1 & 1 & 0 & 1 & 1 \\ 0 & 0 & 0 & 1 & -1 & -1 \end{array}\right]$$

From the bottom row, we deduce that A is not invertible.

29. Using Gaussian elimination, we obtain the following:

$$\left[\begin{array}{ccc|ccc} 2 & 4 & 1 & 1 & 0 & 0 \\ 1 & 1 & -1 & 0 & 1 & 0 \\ 1 & 1 & 0 & 0 & 0 & 1 \end{array}\right] \xrightarrow[R_2\leftrightarrow R_3]{\frac{1}{2}R_1\to R_1} \left[\begin{array}{ccc|ccc} 1 & 2 & \frac{1}{2} & \frac{1}{2} & 0 & 0 \\ 1 & 1 & 0 & 0 & 0 & 1 \\ 1 & 1 & -1 & 0 & 1 & 0 \end{array}\right]$$

$$\xrightarrow[R_1-R_3\to R_3]{R_2-R_3\to R_2} \left[\begin{array}{ccc|ccc} 1 & 2 & \frac{1}{2} & \frac{1}{2} & 0 & 0 \\ 0 & 0 & 1 & 0 & -1 & 1 \\ 0 & 1 & \frac{3}{2} & \frac{1}{2} & -1 & 0 \end{array}\right] \xrightarrow{R_2\leftrightarrow R_3} \left[\begin{array}{ccc|ccc} 1 & 2 & \frac{1}{2} & \frac{1}{2} & 0 & 0 \\ 0 & 1 & \frac{3}{2} & \frac{1}{2} & -1 & 0 \\ 0 & 0 & 1 & 0 & -1 & 1 \end{array}\right]$$

$$\xrightarrow{R_2-\frac{3}{2}R_3\to R_2} \left[\begin{array}{ccc|ccc} 1 & 2 & \frac{1}{2} & \frac{1}{2} & 0 & 0 \\ 0 & 1 & 0 & \frac{1}{2} & \frac{1}{2} & -\frac{3}{2} \\ 0 & 0 & 1 & 0 & -1 & 1 \end{array}\right] \xrightarrow{R_1-2R_2-\frac{1}{2}R_3\to R_1} \left[\begin{array}{ccc|ccc} 1 & 0 & 0 & -\frac{1}{2} & -\frac{1}{2} & \frac{5}{2} \\ 0 & 1 & 0 & \frac{1}{2} & \frac{1}{2} & -\frac{3}{2} \\ 0 & 0 & 1 & 0 & -1 & 1 \end{array}\right]$$

So, $A^{-1} = \begin{bmatrix} -\frac{1}{2} & -\frac{1}{2} & \frac{5}{2} \\ \frac{1}{2} & \frac{1}{2} & -\frac{3}{2} \\ 0 & -1 & 1 \end{bmatrix}$.

31. Using Gaussian elimination, we obtain the following:

$$\left[\begin{array}{ccc|ccc} 1 & 1 & -1 & 1 & 0 & 0 \\ 1 & -1 & 1 & 0 & 1 & 0 \\ 2 & -1 & -1 & 0 & 0 & 1 \end{array}\right] \xrightarrow[R_2-R_1\to R_2]{R_3-2R_1\to R_3} \left[\begin{array}{ccc|ccc} 1 & 1 & -1 & 1 & 0 & 0 \\ 0 & -2 & 2 & -1 & 1 & 0 \\ 0 & -3 & 1 & -2 & 0 & 1 \end{array}\right]$$

$$\xrightarrow[-\frac{1}{3}R_3\to R_3]{-\frac{1}{2}R_2\to R_2} \left[\begin{array}{ccc|ccc} 1 & 1 & -1 & 1 & 0 & 0 \\ 0 & 1 & -1 & \frac{1}{2} & -\frac{1}{2} & 0 \\ 0 & 1 & -\frac{1}{3} & \frac{2}{3} & 0 & -\frac{1}{3} \end{array}\right] \xrightarrow{R_3-R_2\to R_3} \left[\begin{array}{ccc|ccc} 1 & 1 & -1 & 1 & 0 & 0 \\ 0 & 1 & -1 & \frac{1}{2} & -\frac{1}{2} & 0 \\ 0 & 0 & \frac{2}{3} & \frac{1}{6} & \frac{1}{2} & -\frac{1}{3} \end{array}\right]$$

$$\xrightarrow{\frac{3}{2}R_3\to R_3} \left[\begin{array}{ccc|ccc} 1 & 1 & -1 & 1 & 0 & 0 \\ 0 & 1 & -1 & \frac{1}{2} & -\frac{1}{2} & 0 \\ 0 & 0 & 1 & \frac{1}{4} & \frac{3}{4} & -\frac{1}{2} \end{array}\right] \xrightarrow{R_2+R_3\to R_2} \left[\begin{array}{ccc|ccc} 1 & 1 & -1 & 1 & 0 & 0 \\ 0 & 1 & 0 & \frac{3}{4} & \frac{1}{4} & -\frac{1}{2} \\ 0 & 0 & 1 & \frac{1}{4} & \frac{3}{4} & -\frac{1}{2} \end{array}\right]$$

$$\xrightarrow{R_1-R_2+R_3\to R_1} \left[\begin{array}{ccc|ccc} 1 & 0 & 0 & \frac{1}{2} & \frac{1}{2} & 0 \\ 0 & 1 & 0 & \frac{3}{4} & \frac{1}{4} & -\frac{1}{2} \\ 0 & 0 & 1 & \frac{1}{4} & \frac{3}{4} & -\frac{1}{2} \end{array}\right].$$

So, $A^{-1} = \begin{bmatrix} \frac{1}{2} & \frac{1}{2} & 0 \\ \frac{3}{4} & \frac{1}{4} & -\frac{1}{2} \\ \frac{1}{4} & \frac{3}{4} & -\frac{1}{2} \end{bmatrix}$

33. First, write the system in the form $AX = B$:

$$\begin{bmatrix} 2 & -1 \\ 1 & 1 \end{bmatrix}\begin{bmatrix} x \\ y \end{bmatrix} = \begin{bmatrix} 5 \\ 1 \end{bmatrix}$$

The solution is $\begin{bmatrix} x \\ y \end{bmatrix} = \begin{bmatrix} 2 & -1 \\ 1 & 1 \end{bmatrix}^{-1}\begin{bmatrix} 5 \\ 1 \end{bmatrix}$. Since $\begin{bmatrix} 2 & -1 \\ 1 & 1 \end{bmatrix}^{-1} = \frac{1}{3}\begin{bmatrix} 1 & 1 \\ -1 & 2 \end{bmatrix}$, the solution is

$\begin{bmatrix} x \\ y \end{bmatrix} = \frac{1}{3}\begin{bmatrix} 1 & 1 \\ -1 & 2 \end{bmatrix}\begin{bmatrix} 5 \\ 1 \end{bmatrix} = \begin{bmatrix} 2 \\ -1 \end{bmatrix}$. So, $\boxed{x = 2\ , y = -1}$.

35. First, write the system in the form $AX = B$:

$$\begin{bmatrix} 4 & -9 \\ 7 & -3 \end{bmatrix}\begin{bmatrix} x \\ y \end{bmatrix} = \begin{bmatrix} -1 \\ \frac{5}{2} \end{bmatrix}$$

The solution is $\begin{bmatrix} x \\ y \end{bmatrix} = \begin{bmatrix} 4 & -9 \\ 7 & -3 \end{bmatrix}^{-1}\begin{bmatrix} -1 \\ \frac{5}{2} \end{bmatrix}$. Since $\begin{bmatrix} 4 & -9 \\ 7 & -3 \end{bmatrix}^{-1} = \frac{1}{51}\begin{bmatrix} -3 & 9 \\ -7 & 4 \end{bmatrix}$, the solution is

$\begin{bmatrix} x \\ y \end{bmatrix} = \frac{1}{51}\begin{bmatrix} -3 & 9 \\ -7 & 4 \end{bmatrix}\begin{bmatrix} -1 \\ \frac{5}{2} \end{bmatrix} = \begin{bmatrix} \frac{1}{2} \\ \frac{1}{3} \end{bmatrix}$. So, $\boxed{x = \frac{1}{2}, y = \frac{1}{3}}$.

37. First, write the system in the form $AX = B$:

$$\begin{bmatrix} \frac{3}{4} & -\frac{2}{3} \\ -\frac{1}{2} & -\frac{5}{3} \end{bmatrix}\begin{bmatrix} x \\ y \end{bmatrix} = \begin{bmatrix} 5 \\ 3 \end{bmatrix}$$

The solution is $\begin{bmatrix} x \\ y \end{bmatrix} = \begin{bmatrix} \frac{3}{4} & -\frac{2}{3} \\ -\frac{1}{2} & -\frac{5}{3} \end{bmatrix}^{-1}\begin{bmatrix} 5 \\ 3 \end{bmatrix}$. Since $\begin{bmatrix} \frac{3}{4} & -\frac{2}{3} \\ -\frac{1}{2} & -\frac{5}{3} \end{bmatrix}^{-1} = -\frac{12}{19}\begin{bmatrix} -\frac{5}{3} & \frac{2}{3} \\ \frac{1}{2} & \frac{3}{4} \end{bmatrix}$, the solution

is $\begin{bmatrix} x \\ y \end{bmatrix} = -\frac{12}{19}\begin{bmatrix} -\frac{5}{3} & \frac{2}{3} \\ \frac{1}{2} & \frac{3}{4} \end{bmatrix}\begin{bmatrix} 5 \\ 3 \end{bmatrix} = \begin{bmatrix} 4 \\ -3 \end{bmatrix}$. So, $\boxed{x = 4, y = -3}$.

39. From #25, we know that $A^{-1} = \begin{bmatrix} \frac{1}{2} & \frac{1}{2} & 0 \\ \frac{1}{2} & 0 & \frac{1}{2} \\ 0 & -\frac{1}{2} & -\frac{1}{2} \end{bmatrix}$. Further, in the current problem, we

identify $B = \begin{bmatrix} 1 \\ -1 \\ -1 \end{bmatrix}$. Since $A\begin{bmatrix} x \\ y \\ z \end{bmatrix} = B$ is equivalent to $\begin{bmatrix} x \\ y \\ z \end{bmatrix} = A^{-1}B = \begin{bmatrix} 0 \\ 0 \\ 1 \end{bmatrix}$, we conclude that

the solution of this system is $\boxed{x = 0,\ y = 0,\ z = 1}$.

41. From #27, we know that A^{-1} does not exist.

43. From #29, we know that $A^{-1} = \begin{bmatrix} -\frac{1}{2} & -\frac{1}{2} & \frac{5}{2} \\ \frac{1}{2} & \frac{1}{2} & -\frac{3}{2} \\ 0 & -1 & 1 \end{bmatrix}$. Further, in the current problem, we identify $B = \begin{bmatrix} -5 \\ 7 \\ 0 \end{bmatrix}$. Since $A\begin{bmatrix} x \\ y \\ z \end{bmatrix} = B$ is equivalent to $\begin{bmatrix} x \\ y \\ z \end{bmatrix} = A^{-1}B = \begin{bmatrix} -1 \\ 1 \\ -7 \end{bmatrix}$, we conclude that the solution of this system is $\boxed{x=-1,\ y=1,\ z=-7}$.

45. From #31, we know that $A^{-1} = \begin{bmatrix} \frac{1}{2} & \frac{1}{2} & 0 \\ \frac{3}{4} & \frac{1}{4} & -\frac{1}{2} \\ \frac{1}{4} & \frac{3}{4} & -\frac{1}{2} \end{bmatrix}$. Further, in the current problem, we identify $B = \begin{bmatrix} 4 \\ 2 \\ -3 \end{bmatrix}$. Since $A\begin{bmatrix} x \\ y \\ z \end{bmatrix} = B$ is equivalent to $\begin{bmatrix} x \\ y \\ z \end{bmatrix} = A^{-1}B = \begin{bmatrix} 3 \\ 5 \\ 4 \end{bmatrix}$, we conclude that the solution of this system is $\boxed{x=3,\ y=5,\ z=4}$.

47. a. $A^{-1}B = -\frac{1}{9\times 10^9}\begin{bmatrix} -450,000,000,000 \\ -180,000,000,000 \end{bmatrix} = \begin{bmatrix} 50 \\ 20 \end{bmatrix}$.

b. The price of a sweatshirt is \$50 and the price of a t-shirt is \$20.

49. For Problems 49 – 52, we need to compute K^{-1}:

$$\left[\begin{array}{ccc|ccc} 1 & 1 & 0 & 1 & 0 & 0 \\ -1 & 0 & 1 & 0 & 1 & 0 \\ 2 & 0 & -1 & 0 & 0 & 1 \end{array}\right] \xrightarrow[R_3-2R_1\to R_3]{R_1+R_2\to R_2} \left[\begin{array}{ccc|ccc} 1 & 1 & 0 & 1 & 0 & 0 \\ 0 & 1 & 1 & 1 & 1 & 0 \\ 0 & -2 & -1 & -2 & 0 & 1 \end{array}\right] \xrightarrow{R_3+2R_2\to R_3}$$

$$\left[\begin{array}{ccc|ccc} 1 & 1 & 0 & 1 & 0 & 0 \\ 0 & 1 & 1 & 1 & 1 & 0 \\ 0 & 0 & 1 & 0 & 2 & 1 \end{array}\right] \xrightarrow{R_2-R_3\to R_2} \left[\begin{array}{ccc|ccc} 1 & 1 & 0 & 1 & 0 & 0 \\ 0 & 1 & 0 & 1 & -1 & -1 \\ 0 & 0 & 1 & 0 & 2 & 1 \end{array}\right] \xrightarrow{R_1-R_2\to R_1}$$

$$\left[\begin{array}{ccc|ccc} 1 & 0 & 0 & 0 & 1 & 1 \\ 0 & 1 & 0 & 1 & -1 & -1 \\ 0 & 0 & 1 & 0 & 2 & 1 \end{array}\right].\ \text{So, } K^{-1} = \begin{bmatrix} 0 & 1 & 1 \\ 1 & -1 & -1 \\ 0 & 2 & 1 \end{bmatrix}.$$

As such, $\begin{bmatrix} 55 & 10 & -22 \end{bmatrix}\begin{bmatrix} 0 & 1 & 1 \\ 1 & -1 & -1 \\ 0 & 2 & 1 \end{bmatrix} = \begin{bmatrix} 10 & 1 & 23 \end{bmatrix}$, which corresponds to JAW.

51. Since $K^{-1} = \begin{bmatrix} 0 & 1 & 1 \\ 1 & -1 & -1 \\ 0 & 2 & 1 \end{bmatrix}$ (see Problem #49), we see that

$[21 \quad 12 \quad -2]\begin{bmatrix} 0 & 1 & 1 \\ 1 & -1 & -1 \\ 0 & 2 & 1 \end{bmatrix} = [12 \quad 5 \quad 7]$, which corresponds to LEG.

53. Since $K^{-1} = \begin{bmatrix} 0 & 1 & 1 \\ 1 & -1 & -1 \\ 0 & 2 & 1 \end{bmatrix}$ (see Problem #49), we see that

$[-10 \quad 5 \quad 20]\begin{bmatrix} 0 & 1 & 1 \\ 1 & -1 & -1 \\ 0 & 2 & 1 \end{bmatrix} = [5 \quad 25 \quad 5]$, which corresponds to EYE.

55. Using technology to compute, we see that

$$X = \begin{bmatrix} 8 & 4 & 6 \\ 6 & 10 & 5 \\ 10 & 4 & 8 \end{bmatrix}^{-1} \begin{bmatrix} 18 \\ 21 \\ 22 \end{bmatrix} = \begin{bmatrix} 1 \\ 1 \\ 1 \end{bmatrix}$$

So, the combination of 1 serving each of food A, B and C will create a meal of 18 g carbohydrates, 21 g of protein and 22 g of fat.

57. Using technology to compute, we see that

$$X = \begin{bmatrix} 0.03 & 0.06 & 0.15 \\ 0.04 & 0.05 & 0.18 \\ 0.05 & 0.07 & 0.13 \end{bmatrix}^{-1} \begin{bmatrix} 49.50 \\ 52.00 \\ 58.50 \end{bmatrix} = \begin{bmatrix} 350 \\ 400 \\ 100 \end{bmatrix}$$

The employee's normal monthly usage is 350 minutes talking, 400 text messages and 100 megabytes of data usage.

59. The third row of A is comprised of all zeros. Hence, $\det(A) = 0$, which implies A is not invertible. Also, note that the identity matrix doesn't occur on the left as it should once you have computed the inverse.

61. False. In general,

$A^{-1} = \frac{1}{a_{22}a_{11} - a_{21}a_{12}} \begin{bmatrix} a_{22} & -a_{12} \\ -a_{21} & a_{11} \end{bmatrix}$, provided that $a_{22}a_{11} - a_{21}a_{12} \neq 0$

63. By definition of inverse, the value of x for which A^{-1} does not exist would be the one for which $2x$ - 18 is 0, namely $\boxed{x = 9}$, since we need to divide by the quantity 2(6) – 3(4) in order to form the inverse.

65. Observe that $$A\cdot A^{-1}=\begin{bmatrix} a & b \\ c & d \end{bmatrix}\cdot\left(\frac{1}{ad-bc}\begin{bmatrix} d & -b \\ -c & a \end{bmatrix}\right)=\frac{1}{ad-bc}\left(\begin{bmatrix} a & b \\ c & d \end{bmatrix}\cdot\begin{bmatrix} d & -b \\ -c & a \end{bmatrix}\right)$$ $$=\frac{1}{ad-bc}\begin{bmatrix} ad-bc & 0 \\ 0 & ad-bc \end{bmatrix}=\begin{bmatrix} \frac{ad-bc}{ad-bc} & 0 \\ 0 & \frac{ad-bc}{ad-bc} \end{bmatrix}=\begin{bmatrix} 1 & 0 \\ 0 & 1 \end{bmatrix}=I$$
67. Since $da-bc=0$ the inverse does not exist.
69. $A^{-1}=\begin{bmatrix} -\frac{115}{6008} & \frac{431}{6008} & \frac{-1067}{6008} & \frac{103}{751} \\ \frac{411}{6008} & \frac{-391}{6008} & \frac{731}{6008} & \frac{-22}{751} \\ \frac{57}{751} & \frac{28}{751} & \frac{-85}{751} & \frac{3}{751} \\ \frac{-429}{6008} & \frac{145}{6008} & \frac{1035}{6008} & \frac{12}{751} \end{bmatrix}$
71. The system in matrix form is given by $$\begin{bmatrix} 2.7 & -3.1 \\ 1.5 & 2.7 \end{bmatrix}\begin{bmatrix} x \\ y \end{bmatrix}=\begin{bmatrix} 9.82 \\ -1.62 \end{bmatrix}$$ and its solution is $\begin{bmatrix} x \\ y \end{bmatrix}=\begin{bmatrix} 2.7 & -3.1 \\ 1.5 & 2.7 \end{bmatrix}^{-1}\begin{bmatrix} 9.82 \\ -1.62 \end{bmatrix}$. Since $\begin{bmatrix} 2.7 & -3.1 \\ 1.5 & 2.7 \end{bmatrix}^{-1}=\frac{1}{11.94}\begin{bmatrix} 2.7 & 3.1 \\ -1.5 & 2.7 \end{bmatrix}$, we see that the solution is $\begin{bmatrix} x \\ y \end{bmatrix}=\frac{1}{11.94}\begin{bmatrix} 2.7 & 3.1 \\ -1.5 & 2.7 \end{bmatrix}\begin{bmatrix} 9.82 \\ -1.62 \end{bmatrix}=\begin{bmatrix} 1.8 \\ -1.6 \end{bmatrix}$. So, $\boxed{x=1.8\,,y=-1.6}$.
73. The augmented matrix to enter into the calculator is: $$\left[\begin{array}{ccc\|c} 5.1 & 7.3 & 1.2 & 12.51 \\ 2.3 & -1.5 & 4.5 & 53.96 \\ -8.1 & 5.4 & -9.4 & -130.35 \end{array}\right]$$ Solving then yields the solution of the system as $(3.7,-2.4,9.3)$.

Section 7.4 Solutions ---

1. $(1)(4)-(3)(2)=-2$	**3.** $(7)(-2)-(-5)(9)=31$	
5. $(0)(-1)-(4)7=-28$	**7.** $(-1.2)(1.5)-(-0.5)(2.4)$ $=-0.6$	**9.** $\left(\frac{3}{4}\right)\left(\frac{8}{9}\right)-(2)\left(\frac{1}{3}\right)=0$

11.

$$D=\begin{vmatrix}1 & 1\\1 & -1\end{vmatrix}=(1)(-1)-(1)(1)=-2$$

$$D_x=\begin{vmatrix}-1 & 1\\11 & -1\end{vmatrix}=(-1)(-1)-(11)(1)=-10$$

$$D_y=\begin{vmatrix}1 & -1\\1 & 11\end{vmatrix}=(1)(11)-(1)(-1)=12$$

So, from Cramer's Rule, we have:

$$x=\frac{D_x}{D}=\frac{-10}{-2}=5$$

$$y=\frac{D_y}{D}=\frac{12}{-2}=-6$$

Thus, the solution is $x=5,\ y=-6$.

13.

$$D=\begin{vmatrix}3 & 2\\-2 & 1\end{vmatrix}=(3)(1)-(-2)(2)=7$$

$$D_x=\begin{vmatrix}-4 & 2\\5 & 1\end{vmatrix}=(-4)(1)-(5)(2)=-14$$

$$D_y=\begin{vmatrix}3 & -4\\-2 & 5\end{vmatrix}=(3)(5)-(-2)(-4)=7$$

So, from Cramer's Rule, we have:

$$x=\frac{D_x}{D}=\frac{-14}{7}=-2$$

$$y=\frac{D_y}{D}=\frac{7}{7}=1$$

Thus, the solution is $x=-2,\ y=1$.

15.

$$D=\begin{vmatrix}3 & -2\\5 & 4\end{vmatrix}=(3)(4)-(5)(-2)=22$$

$$D_x=\begin{vmatrix}-1 & -2\\-31 & 4\end{vmatrix}=(-1)(4)-(-31)(-2)=-66$$

$$D_y=\begin{vmatrix}3 & -1\\5 & -31\end{vmatrix}=(3)(-31)-(5)(-1)=-88$$

So, from Cramer's Rule, we have:

$$x=\frac{D_x}{D}=\frac{-66}{22}=-3$$

$$y=\frac{D_y}{D}=\frac{-88}{22}=-4$$

Thus, the solution is $x=-3,\ y=-4$.

17.

$$D=\begin{vmatrix}7 & -3\\5 & 2\end{vmatrix}=(7)(2)-(5)(-3)=29$$

$$D_x=\begin{vmatrix}-29 & -3\\0 & 2\end{vmatrix}=(-29)(2)-(0)(-3)=-58$$

$$D_y=\begin{vmatrix}7 & -29\\5 & 0\end{vmatrix}=(7)(0)-(5)(-29)=145$$

So, from Cramer's Rule, we have:

$$x=\frac{D_x}{D}=\frac{-58}{29}=-2$$

$$y=\frac{D_y}{D}=\frac{145}{29}=5$$

Thus, the solution is $x=-2,\ y=5$.

19.

$$D=\begin{vmatrix}3 & 5\\ -1 & 1\end{vmatrix}=(3)(1)-(-1)(5)=8$$

$$D_x=\begin{vmatrix}16 & 5\\ 0 & 1\end{vmatrix}=(16)(1)-(0)(5)=16$$

$$D_y=\begin{vmatrix}3 & 16\\ -1 & 0\end{vmatrix}=(3)(0)-(-1)(16)=16$$

So, from Cramer's Rule, we have:

$$x=\frac{D_x}{D}=\frac{16}{8}=2$$

$$y=\frac{D_y}{D}=\frac{16}{8}=2$$

Thus, the solution is $x=2$, $y=2$.

21.

$$D=\begin{vmatrix}3 & -5\\ -6 & 10\end{vmatrix}=(3)(10)-(-6)(-5)=0$$

$$D_x=\begin{vmatrix}7 & -5\\ -21 & 10\end{vmatrix}=(7)(10)-(-21)(-5)$$
$$=-35$$

From this information alone, we can conclude that the system is inconsistent or dependent.

23.

$$D=\begin{vmatrix}2 & -3\\ -10 & 15\end{vmatrix}=(2)(15)-(-10)(-3)=0$$

$$D_x=\begin{vmatrix}4 & -3\\ -20 & 15\end{vmatrix}=(4)(15)-(-20)(-3)=0$$

$$D_y=\begin{vmatrix}2 & 4\\ -10 & -20\end{vmatrix}=(2)(-20)-(-10)(4)=0$$

Hence, the system is inconsistent or dependent.

25. First, rewrite the system as:

$$\begin{cases}6x+y=2\\ 12x+y=5\end{cases}.$$

Now, compute the determinants from Cramer's Rule:

$$D=\begin{vmatrix}6 & 1\\ 12 & 1\end{vmatrix}=(6)(1)-(12)(1)=-6$$

$$D_x=\begin{vmatrix}2 & 1\\ 5 & 1\end{vmatrix}=(2)(1)-(5)(1)=-3$$

$$D_y=\begin{vmatrix}6 & 2\\ 12 & 5\end{vmatrix}=(6)(5)-(12)(2)=6$$

So, from Cramer's Rule, we have:

$$x=\frac{D_x}{D}=\frac{-3}{-6}=\frac{1}{2}$$

$$y=\frac{D_y}{D}=\frac{6}{-6}=-1$$

Thus, the solution is $x=\frac{1}{2}$, $y=-1$.

27.

$$D=\begin{vmatrix}0.3 & -0.5\\ 0.2 & 1\end{vmatrix}=(0.3)(1)-(0.2)(-0.5)$$
$$=0.4$$

$$D_x=\begin{vmatrix}-0.6 & -0.5\\ 2.4 & 1\end{vmatrix}=(-0.6)(1)-(2.4)(-0.5)$$
$$=-0.6$$

$$D_y=\begin{vmatrix}0.3 & -0.6\\ 0.2 & 2.4\end{vmatrix}=(0.3)(2.4)-(0.2)(-0.6)$$
$$=0.84$$

So, from Cramer's Rule, we have:

$$x=\frac{D_x}{D}=\frac{0.6}{0.4}=1.5$$
$$y=\frac{D_y}{D}=\frac{0.84}{0.4}=2.1$$

Thus, the solution is $x=1.5,\ y=2.1$.

29. First, rewrite the system as:

$$\begin{cases}17x-y=-7\\ -15x-y=-7\end{cases}.$$

Now, compute the determinants from Cramer's Rule:

$$D=\begin{vmatrix}17 & -1\\ -15 & -1\end{vmatrix}=(17)(-1)-(-15)(-1)$$
$$=-32$$

$$D_x=\begin{vmatrix}-7 & -1\\ -7 & -1\end{vmatrix}=(-7)(-1)-(-7)(-1)$$
$$=0$$

$$D_y=\begin{vmatrix}17 & -7\\ -15 & -7\end{vmatrix}=(17)(-7)-(-15)(-7)$$
$$=-224$$

So, from Cramer's Rule, we have:

$$x=\frac{D_x}{D}=\frac{0}{-32}=0$$
$$y=\frac{D_y}{D}=\frac{-224}{-32}=7$$

Thus, the solution is $x=0,\ y=7$.

31. First, rewrite the system as:

$$\begin{cases}2\left(\frac{1}{x}\right)-3\left(\frac{1}{y}\right)=2\\ 5\left(\frac{1}{x}\right)-6\left(\frac{1}{y}\right)=7\end{cases}.$$

Now, compute the determinants from Cramer's Rule:

$$D=\begin{vmatrix}2 & -3\\ 5 & -6\end{vmatrix}=(2)(-6)-(5)(-3)=3$$

$$D_{\frac{1}{x}}=\begin{vmatrix}2 & -3\\ 7 & -6\end{vmatrix}=(2)(-6)-(7)(-3)=9$$

$$D_{\frac{1}{y}}=\begin{vmatrix}2 & 2\\ 5 & 7\end{vmatrix}=(2)(7)-(5)(2)=4$$

So, from Cramer's Rule, we have:

$$\frac{1}{x}=\frac{D_{\frac{1}{x}}}{D}=\frac{9}{3}=3$$
$$\frac{1}{y}=\frac{D_{\frac{1}{y}}}{D}=\frac{4}{3}$$

Thus, the solution is $x=\frac{1}{3},\ y=\frac{3}{4}$.

33.

$$\begin{vmatrix} 3 & 1 & 0 \\ 2 & 0 & -1 \\ -4 & 1 & 0 \end{vmatrix} = 3\begin{vmatrix} 0 & -1 \\ 1 & 0 \end{vmatrix} - 1\begin{vmatrix} 2 & -1 \\ -4 & 0 \end{vmatrix} + 0\begin{vmatrix} 2 & 0 \\ -4 & 1 \end{vmatrix}$$

$$= 3\underbrace{\left[(0)(0)-(1)(-1)\right]}_{=1} - 1\underbrace{\left[(2)(0)-(-4)(-1)\right]}_{=-4} + 0\underbrace{\left[(2)(1)-(-4)(0)\right]}_{=2} = \boxed{7}$$

35.

$$\begin{vmatrix} 2 & 1 & -5 \\ 3 & 0 & -1 \\ 4 & 0 & 7 \end{vmatrix} = 2\begin{vmatrix} 0 & -1 \\ 0 & 7 \end{vmatrix} - \begin{vmatrix} 3 & -1 \\ 4 & 7 \end{vmatrix} + (-5)\begin{vmatrix} 3 & 0 \\ 4 & 0 \end{vmatrix}$$

$$= 2\underbrace{\left[(0)(7)-(0)(-1)\right]}_{=0} - \underbrace{\left[(3)(7)-(4)(-1)\right]}_{=25} - 5\underbrace{\left[(3)(0)-(4)(0)\right]}_{=0} = \boxed{-25}$$

37.

$$\begin{vmatrix} 1 & 1 & -5 \\ 3 & -7 & -4 \\ 4 & -6 & 9 \end{vmatrix} = \begin{vmatrix} -7 & -4 \\ -6 & 9 \end{vmatrix} - \begin{vmatrix} 3 & -4 \\ 4 & 9 \end{vmatrix} + (-5)\begin{vmatrix} 3 & -7 \\ 4 & -6 \end{vmatrix}$$

$$= \underbrace{\left[(-7)(9)-(-6)(-4)\right]}_{=-87} - \underbrace{\left[(3)(9)-(4)(-4)\right]}_{=43} - 5\underbrace{\left[(3)(-6)-(4)(-7)\right]}_{=10}$$

$$= \boxed{-180}$$

39.

$$\begin{vmatrix} 1 & 3 & 4 \\ 2 & -1 & 1 \\ 3 & -2 & 1 \end{vmatrix} = \begin{vmatrix} -1 & 1 \\ -2 & 1 \end{vmatrix} - 3\begin{vmatrix} 2 & 1 \\ 3 & 1 \end{vmatrix} + 4\begin{vmatrix} 2 & -1 \\ 3 & -2 \end{vmatrix}$$

$$= \underbrace{\left[(-1)(1)-(-2)(1)\right]}_{=1} - 3\underbrace{\left[(2)(1)-(3)(1)\right]}_{=-1} + 4\underbrace{\left[(2)(-2)-(3)(-1)\right]}_{=-1}$$

$$= \boxed{0}$$

41.

$$\begin{vmatrix} -3 & 1 & 5 \\ 2 & 0 & 6 \\ 4 & 7 & -9 \end{vmatrix} = -3\begin{vmatrix} 0 & 6 \\ 7 & -9 \end{vmatrix} - 1\begin{vmatrix} 2 & 6 \\ 4 & -9 \end{vmatrix} + 5\begin{vmatrix} 2 & 0 \\ 4 & 7 \end{vmatrix}$$

$$= -3\underbrace{\left[(0)(-9)-(6)(7)\right]}_{-42} - \underbrace{\left[(2)(-9)-(6)(4)\right]}_{-42} + 5\underbrace{\left[(2)(7)-(4)(0)\right]}_{14}$$

$$= \boxed{238}$$

43.

$$\begin{vmatrix} -2 & 1 & -7 \\ 4 & -2 & 14 \\ 0 & 1 & 8 \end{vmatrix} = -2\begin{vmatrix} -2 & 14 \\ 1 & 8 \end{vmatrix} - 1\begin{vmatrix} 4 & 14 \\ 0 & 8 \end{vmatrix} + (-7)\begin{vmatrix} 4 & -2 \\ 0 & 1 \end{vmatrix}$$

$$= -2\underbrace{\left[(-2)(8)-(1)(14)\right]}_{-30} - 1\underbrace{\left[(4)(8)-(0)(14)\right]}_{32} - 7\underbrace{\left[(4)(1)-(0)(-2)\right]}_{4}$$

$$= \boxed{0}$$

45.

$$\begin{vmatrix} \frac{3}{4} & -1 & 0 \\ 0 & \frac{1}{5} & -12 \\ 8 & 0 & -2 \end{vmatrix} = \frac{3}{4}\begin{vmatrix} \frac{1}{5} & -12 \\ 0 & -2 \end{vmatrix} - (-1)\begin{vmatrix} 0 & -12 \\ 8 & -2 \end{vmatrix} + 0\begin{vmatrix} 0 & \frac{1}{5} \\ 8 & 0 \end{vmatrix}$$

$$= \frac{3}{4}\underbrace{\left[\left(\tfrac{1}{5}\right)(-2)-(0)(-12)\right]}_{-\frac{2}{5}} + 1\underbrace{\left[(0)(-2)-(8)(-12)\right]}_{96} + 0\underbrace{\left[(0)(0)-(8)\left(\tfrac{1}{5}\right)\right]}_{-\frac{8}{5}}$$

$$= \boxed{95.7}$$

47.

$$D=\begin{vmatrix}1 & 1 & -1\\ 1 & -1 & 1\\ 1 & 1 & 1\end{vmatrix}=1\begin{vmatrix}-1 & 1\\ 1 & 1\end{vmatrix}-1\begin{vmatrix}1 & 1\\ 1 & 1\end{vmatrix}+(-1)\begin{vmatrix}1 & -1\\ 1 & 1\end{vmatrix}$$

$$=1\underbrace{\left[(-1)(1)-(1)(1)\right]}_{=-2}-1\underbrace{\left[(1)(1)-(1)(1)\right]}_{=0}+(-1)\underbrace{\left[(1)(1)-(1)(-1)\right]}_{=2}=-4$$

$$D_x=\begin{vmatrix}0 & 1 & -1\\ 4 & -1 & 1\\ 10 & 1 & 1\end{vmatrix}=0\begin{vmatrix}-1 & 1\\ 1 & 1\end{vmatrix}-1\begin{vmatrix}4 & 1\\ 10 & 1\end{vmatrix}+(-1)\begin{vmatrix}4 & -1\\ 10 & 1\end{vmatrix}$$

$$=0\underbrace{\left[(-1)(1)-(1)(1)\right]}_{=-2}-1\underbrace{\left[(4)(1)-(10)(1)\right]}_{=-6}+(-1)\underbrace{\left[(4)(1)-(10)(-1)\right]}_{=14}=-8$$

$$D_y=\begin{vmatrix}1 & 0 & -1\\ 1 & 4 & 1\\ 1 & 10 & 1\end{vmatrix}=1\begin{vmatrix}4 & 1\\ 10 & 1\end{vmatrix}-0\begin{vmatrix}1 & 1\\ 1 & 1\end{vmatrix}+(-1)\begin{vmatrix}1 & 4\\ 1 & 10\end{vmatrix}$$

$$=1\underbrace{\left[(4)(1)-(10)(1)\right]}_{=-6}-0\underbrace{\left[(1)(1)-(1)(1)\right]}_{=0}+(-1)\underbrace{\left[(1)(10)-(1)(4)\right]}_{=6}=-12$$

$$D_z=\begin{vmatrix}1 & 1 & 0\\ 1 & -1 & 4\\ 1 & 1 & 10\end{vmatrix}=1\begin{vmatrix}-1 & 4\\ 1 & 10\end{vmatrix}-1\begin{vmatrix}1 & 4\\ 1 & 10\end{vmatrix}+0\begin{vmatrix}1 & -1\\ 1 & 1\end{vmatrix}$$

$$=1\underbrace{\left[(-1)(10)-(1)(4)\right]}_{=-14}-1\underbrace{\left[(1)(10)-(1)(4)\right]}_{=6}+0\underbrace{\left[(1)(1)-(1)(-1)\right]}_{=2}=-20$$

So, by Cramer's Rule,

$$\boxed{x=\frac{D_x}{D}=\frac{-8}{-4}=2 \quad y=\frac{D_y}{D}=\frac{-12}{-4}=3 \quad z=\frac{D_z}{D}=\frac{-20}{-4}=5}$$

49.

$$D=\begin{vmatrix}3&8&2\\-2&5&3\\4&9&2\end{vmatrix}=3\begin{vmatrix}5&3\\9&2\end{vmatrix}-8\begin{vmatrix}-2&3\\4&2\end{vmatrix}+2\begin{vmatrix}-2&5\\4&9\end{vmatrix}$$

$$=3\underbrace{\left[(5)(2)-(9)(3)\right]}_{=-17}-8\underbrace{\left[(-2)(2)-(4)(3)\right]}_{=-16}+2\underbrace{\left[(-2)(9)-(4)(5)\right]}_{=-38}=1$$

$$D_x=\begin{vmatrix}28&8&2\\34&5&3\\29&9&2\end{vmatrix}=28\begin{vmatrix}5&3\\9&2\end{vmatrix}-8\begin{vmatrix}34&3\\29&2\end{vmatrix}+2\begin{vmatrix}34&5\\29&9\end{vmatrix}$$

$$=28\underbrace{\left[(5)(2)-(9)(3)\right]}_{=-17}-8\underbrace{\left[(34)(2)-(29)(3)\right]}_{=-19}+2\underbrace{\left[(34)(9)-(29)(5)\right]}_{=161}=-2$$

$$D_y=\begin{vmatrix}3&28&2\\-2&34&3\\4&29&2\end{vmatrix}=3\begin{vmatrix}34&3\\29&2\end{vmatrix}-28\begin{vmatrix}-2&3\\4&2\end{vmatrix}+2\begin{vmatrix}-2&34\\4&29\end{vmatrix}$$

$$=3\underbrace{\left[(34)(2)-(29)(3)\right]}_{=-19}-28\underbrace{\left[(-2)(2)-(4)(3)\right]}_{=-16}+2\underbrace{\left[(-2)(29)-(4)(34)\right]}_{=-194}=3$$

$$D_z=\begin{vmatrix}3&8&28\\-2&5&34\\4&9&29\end{vmatrix}=3\begin{vmatrix}5&34\\9&29\end{vmatrix}-8\begin{vmatrix}-2&34\\4&29\end{vmatrix}+28\begin{vmatrix}-2&5\\4&9\end{vmatrix}$$

$$=3\underbrace{\left[(5)(29)-(9)(34)\right]}_{=-161}-8\underbrace{\left[(-2)(29)-(4)(34)\right]}_{=-194}+28\underbrace{\left[(-2)(9)-(4)(5)\right]}_{=-38}=5$$

So, by Cramer's Rule,

$$\boxed{x=\frac{D_x}{D}=-2 \qquad y=\frac{D_y}{D}=3 \qquad z=\frac{D_z}{D}=5}$$

51.

$$D=\begin{vmatrix}3&0&5\\0&4&3\\2&-1&0\end{vmatrix}=3\begin{vmatrix}4&3\\-1&0\end{vmatrix}-0\begin{vmatrix}0&3\\2&0\end{vmatrix}+5\begin{vmatrix}0&4\\2&-1\end{vmatrix}$$

$$=3\underbrace{\left[(4)(0)-(-1)(3)\right]}_{=3}-0\underbrace{\left[(0)(0)-(2)(3)\right]}_{=-6}+5\underbrace{\left[(0)(-1)-(2)(4)\right]}_{=-8}=-31$$

$$D_x=\begin{vmatrix}11&0&5\\-9&4&3\\7&-1&0\end{vmatrix}=11\begin{vmatrix}4&3\\-1&0\end{vmatrix}-0\begin{vmatrix}-9&3\\7&0\end{vmatrix}+5\begin{vmatrix}-9&4\\7&-1\end{vmatrix}$$

$$=11\underbrace{\left[(4)(0)-(-1)(3)\right]}_{=3}-0\underbrace{\left[(-9)(0)-(7)(3)\right]}_{=-21}+5\underbrace{\left[(-9)(-1)-(7)(4)\right]}_{=-19}=-62$$

$$D_y=\begin{vmatrix}3&11&5\\0&-9&3\\2&7&0\end{vmatrix}=3\begin{vmatrix}-9&3\\7&0\end{vmatrix}-11\begin{vmatrix}0&3\\2&0\end{vmatrix}+5\begin{vmatrix}0&-9\\2&7\end{vmatrix}$$

$$=3\underbrace{\left[(-9)(0)-(7)(3)\right]}_{=-21}-11\underbrace{\left[(0)(0)-(2)(3)\right]}_{=-6}+5\underbrace{\left[(0)(7)-(2)(-9)\right]}_{=18}=93$$

$$D_z=\begin{vmatrix}3&0&11\\0&4&-9\\2&-1&7\end{vmatrix}=3\begin{vmatrix}4&-9\\-1&7\end{vmatrix}-0\begin{vmatrix}0&-9\\2&7\end{vmatrix}+11\begin{vmatrix}0&4\\2&-1\end{vmatrix}$$

$$=3\underbrace{\left[(4)(7)-(-1)(-9)\right]}_{=19}-0\underbrace{\left[(0)(7)-(2)(-9)\right]}_{=18}+11\underbrace{\left[(0)(-1)-(2)(4)\right]}_{=-8}=-31$$

So, by Cramer's Rule, $\boxed{x=\frac{D_x}{D}=\frac{-62}{-31}=2 \quad y=\frac{D_y}{D}=\frac{93}{-31}=-3 \quad z=\frac{D_z}{D}=\frac{-31}{-31}=1}$

53.

$$D=\begin{vmatrix}1&1&-1\\1&-1&1\\-2&-2&2\end{vmatrix}=1\begin{vmatrix}-1&1\\-2&2\end{vmatrix}-1\begin{vmatrix}1&1\\-2&2\end{vmatrix}+(-1)\begin{vmatrix}1&-1\\-2&-2\end{vmatrix}$$

$$=1\underbrace{\left[(-1)(2)-(-2)(1)\right]}_{=0}-1\underbrace{\left[(1)(2)-(-2)(1)\right]}_{=4}+(-1)\underbrace{\left[(1)(-2)-(-2)(-1)\right]}_{=-4}=0$$

So, Cramer's Rule implies that the system is either inconsistent or dependent.

55.

$$D=\begin{vmatrix}1&1&1\\1&-1&1\\-1&1&-1\end{vmatrix}=1\begin{vmatrix}-1&1\\1&-1\end{vmatrix}-1\begin{vmatrix}1&1\\-1&-1\end{vmatrix}+1\begin{vmatrix}1&-1\\-1&1\end{vmatrix}$$

$$=1\underbrace{\left[(-1)(-1)-(1)(1)\right]}_{=0}-1\underbrace{\left[(1)(-1)-(-1)(1)\right]}_{=0}+1\underbrace{\left[(1)(1)-(-1)(-1)\right]}_{=0}=0$$

So, Cramer's Rule implies that the system is either inconsistent or dependent.

57. First, write the system in matrix form as:

$$\begin{bmatrix} 1 & 2 & 3 \\ -2 & 3 & 5 \\ 4 & -1 & 8 \end{bmatrix}\begin{bmatrix} x \\ y \\ z \end{bmatrix} = \begin{bmatrix} 11 \\ 29 \\ 19 \end{bmatrix}$$

Now, apply Cramer's rule to solve the system:

$$D = \begin{vmatrix} 1 & 2 & 3 \\ -2 & 3 & 5 \\ 4 & -1 & 8 \end{vmatrix} = 1\begin{vmatrix} 3 & 5 \\ -1 & 8 \end{vmatrix} - 2\begin{vmatrix} -2 & 5 \\ 4 & 8 \end{vmatrix} + 3\begin{vmatrix} -2 & 3 \\ 4 & -1 \end{vmatrix} = 71$$

$$D_x = \begin{vmatrix} 11 & 2 & 3 \\ 29 & 3 & 5 \\ 19 & -1 & 8 \end{vmatrix} = 11\begin{vmatrix} 3 & 5 \\ -1 & 8 \end{vmatrix} - 2\begin{vmatrix} 29 & 5 \\ 19 & 8 \end{vmatrix} + 3\begin{vmatrix} 29 & 3 \\ 19 & -1 \end{vmatrix} = -213$$

$$D_y = \begin{vmatrix} 1 & 11 & 3 \\ -2 & 29 & 5 \\ 4 & 19 & 8 \end{vmatrix} = 1\begin{vmatrix} 29 & 5 \\ 19 & 8 \end{vmatrix} - 11\begin{vmatrix} -2 & 5 \\ 4 & 8 \end{vmatrix} + 3\begin{vmatrix} -2 & 29 \\ 4 & 19 \end{vmatrix} = 71$$

$$D_z = \begin{vmatrix} 1 & 2 & 11 \\ -2 & 3 & 29 \\ 4 & -1 & 19 \end{vmatrix} = 1\begin{vmatrix} 3 & 29 \\ -1 & 19 \end{vmatrix} - 2\begin{vmatrix} -2 & 29 \\ 4 & 19 \end{vmatrix} + 11\begin{vmatrix} -2 & 3 \\ 4 & -1 \end{vmatrix} = 284$$

So, by Cramer's rule:

$$x = \frac{D_x}{D} = \frac{-213}{71} = -3 \quad y = \frac{D_y}{D} = \frac{71}{71} = 1 \quad z = \frac{D_z}{D} = \frac{284}{71} = 4$$

59. First, write the system in matrix form as:

$$\begin{bmatrix} 1 & -4 & 7 \\ -3 & 2 & -1 \\ 5 & 8 & -2 \end{bmatrix} \begin{bmatrix} x \\ y \\ z \end{bmatrix} = \begin{bmatrix} 49 \\ -17 \\ -24 \end{bmatrix}$$

Now, apply Cramer's rule to solve the system:

$$D = \begin{vmatrix} 1 & -4 & 7 \\ -3 & 2 & -1 \\ 5 & 8 & -2 \end{vmatrix} = 1\begin{vmatrix} 2 & -1 \\ 8 & -2 \end{vmatrix} + 4\begin{vmatrix} -3 & -1 \\ 5 & -2 \end{vmatrix} + 7\begin{vmatrix} -3 & 2 \\ 5 & 8 \end{vmatrix} = -190$$

$$D_x = \begin{vmatrix} 49 & -4 & 7 \\ -17 & 2 & -1 \\ -24 & 8 & -2 \end{vmatrix} = 49\begin{vmatrix} 2 & -1 \\ 8 & -2 \end{vmatrix} + 4\begin{vmatrix} -17 & -1 \\ -24 & -2 \end{vmatrix} + 7\begin{vmatrix} -17 & 2 \\ -24 & 8 \end{vmatrix} = -380$$

$$D_y = \begin{vmatrix} 1 & 49 & 7 \\ -3 & -17 & -1 \\ 5 & -24 & -2 \end{vmatrix} = 1\begin{vmatrix} -17 & -1 \\ -24 & -2 \end{vmatrix} - 49\begin{vmatrix} -3 & -1 \\ 5 & -2 \end{vmatrix} + 7\begin{vmatrix} -3 & -17 \\ 5 & -24 \end{vmatrix} = 570$$

$$D_z = \begin{vmatrix} 1 & -4 & 49 \\ -3 & 2 & -17 \\ 5 & 8 & -24 \end{vmatrix} = 1\begin{vmatrix} 2 & -17 \\ 8 & -24 \end{vmatrix} + 4\begin{vmatrix} -3 & -17 \\ 5 & -24 \end{vmatrix} + 49\begin{vmatrix} -3 & 2 \\ 5 & 8 \end{vmatrix} = -950$$

So, by Cramer's rule:

$$x = \frac{D_x}{D} = \frac{-380}{-190} = 2 \quad y = \frac{D_y}{D} = \frac{570}{-190} = -3 \quad z = \frac{D_z}{D} = \frac{-950}{-190} = 5$$

61. Let $x_1 = 3,\ x_2 = 5,\ x_3 = 3,\ y_1 = 2,\ y_2 = 2,\ y_3 = -4$. Then, we have

$$\text{Area} = \pm\tfrac{1}{2}\begin{vmatrix} 3 & 2 & 1 \\ 5 & 2 & 1 \\ 3 & -4 & 1 \end{vmatrix} = \pm\tfrac{1}{2}\left[3\begin{vmatrix} 2 & 1 \\ -4 & 1 \end{vmatrix} - 2\begin{vmatrix} 5 & 1 \\ 3 & 1 \end{vmatrix} + 1\begin{vmatrix} 5 & 2 \\ 3 & -4 \end{vmatrix}\right]$$

$$= \pm\tfrac{1}{2}\left(3\underbrace{\left[(2)(1)-(-4)(1)\right]}_{=6} - 2\underbrace{\left[(5)(1)-(3)(1)\right]}_{=2} + 1\underbrace{\left[(5)(-4)-(3)(2)\right]}_{=-26}\right) = \pm\tfrac{1}{2}(-12)$$

Hence, choosing the positive value above, we conclude that the area is 6 units^2. Alternatively, we could compute the area by identifying the height and base of the triangle. Indeed, consider the following diagram:

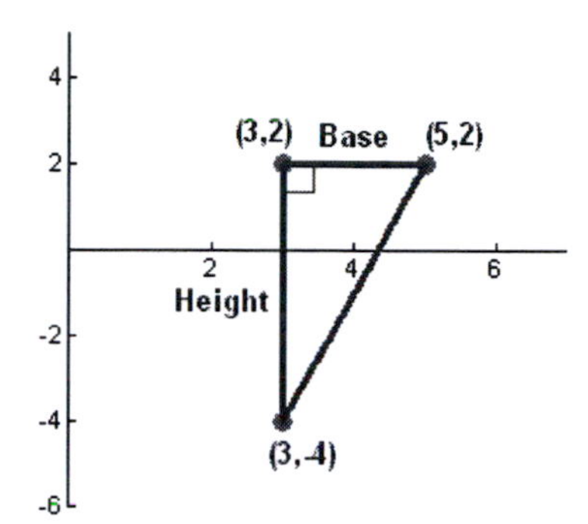

Base = 2
Height = 6

Hence, Area = $\tfrac{1}{2}(\text{Base})(\text{Height}) = \tfrac{1}{2}(2)(6) = 6$ units^2.

63. Let $x_1 = 1,\ x_2 = 3,\ x_3 = -2,\ y_1 = 2,\ y_2 = 4,\ y_3 = 5$. Then, we have

$$\text{Area} = \pm\tfrac{1}{2}\begin{vmatrix} 1 & 2 & 1 \\ 3 & 4 & 1 \\ -2 & 5 & 1 \end{vmatrix} = \pm\tfrac{1}{2}\left[1\begin{vmatrix} 4 & 1 \\ 5 & 1 \end{vmatrix} - 2\begin{vmatrix} 3 & 1 \\ -2 & 1 \end{vmatrix} + 1\begin{vmatrix} 3 & 4 \\ -2 & 5 \end{vmatrix}\right]$$

$$= \pm\tfrac{1}{2}\left(1\underbrace{\left[(4)(1)-(5)(1)\right]}_{=-1} - 2\underbrace{\left[(3)(1)-(-2)(1)\right]}_{=5} + 1\underbrace{\left[(3)(5)-(-2)(4)\right]}_{=23}\right) = \pm\tfrac{1}{2}(12)$$

Hence, choosing the positive value above, we conclude that the area is 6 units^2.

65. Let $x_1 = 1,\ x_2 = 2,\ y_1 = 2,\ y_2 = 4$. The equation of the line through the two points (x_1, y_1) and (x_2, y_2) is:

$$0 = \begin{vmatrix} x & y & 1 \\ 1 & 2 & 1 \\ 2 & 4 & 1 \end{vmatrix} \text{ which simplifies to}$$

$$0 = x\begin{vmatrix} 2 & 1 \\ 4 & 1 \end{vmatrix} - y\begin{vmatrix} 1 & 1 \\ 2 & 1 \end{vmatrix} + 1\begin{vmatrix} 1 & 2 \\ 2 & 4 \end{vmatrix}$$

$$= x\underbrace{\left[(2)(1)-(4)(1)\right]}_{=-2} - y\underbrace{\left[(1)(1)-(2)(1)\right]}_{=-1} + 1\underbrace{\left[(1)(4)-(2)(2)\right]}_{=0} = -2x + y + 0$$

Hence, the equation of this line is $\boxed{y = 2x}$.

67. The system we must solve is

$$\begin{cases} I_1 - I_2 - I_3 = 0 \\ 4I_1 + 0I_2 + 2I_3 = 16 \\ 4I_1 + 4I_2 + 0I_3 = 24 \end{cases}, \text{ which is equivalent to } \begin{bmatrix} 1 & -1 & -1 \\ 4 & 0 & 2 \\ 4 & 4 & 0 \end{bmatrix}\begin{bmatrix} I_1 \\ I_2 \\ I_3 \end{bmatrix} = \begin{bmatrix} 0 \\ 16 \\ 24 \end{bmatrix}.$$

The solution is

$$\begin{bmatrix} I_1 \\ I_2 \\ I_3 \end{bmatrix} = \begin{bmatrix} 1 & -1 & -1 \\ 4 & 0 & 2 \\ 4 & 4 & 0 \end{bmatrix}^{-1}\begin{bmatrix} 0 \\ 16 \\ 24 \end{bmatrix}.$$

We now compute the inverse:

$$\left[\begin{array}{ccc|ccc} & -1 & -1 & 1 & 0 & 0 \\ 4 & 0 & 2 & 0 & 1 & 0 \\ 4 & 4 & 0 & 0 & 0 & 1 \end{array}\right] \xrightarrow[R_2 - 4R_1 \to R_2]{R_3 - 4R_1 \to R_3} \left[\begin{array}{ccc|ccc} 1 & -1 & -1 & -1 & 0 & 0 \\ 0 & 4 & 6 & -4 & 1 & 0 \\ 0 & 8 & 4 & -4 & 0 & 1 \end{array}\right] \xrightarrow{R_3 - 2R_2 \to R_3}$$

$$\left[\begin{array}{ccc|ccc} 1 & -1 & -1 & 1 & 0 & 0 \\ 0 & 4 & 6 & -4 & 1 & 0 \\ 0 & 0 & -8 & 4 & -2 & 1 \end{array}\right] \xrightarrow{-\frac{1}{8}R_3 \to R_3} \left[\begin{array}{ccc|ccc} 1 & -1 & -1 & 1 & 0 & 0 \\ 0 & 4 & 6 & -4 & 1 & 0 \\ 0 & 0 & 1 & -\frac{1}{2} & \frac{1}{4} & -\frac{1}{8} \end{array}\right] \xrightarrow{R_2 - 6R_3 \to R_2}$$

$$\left[\begin{array}{ccc|ccc} 1 & -1 & -1 & 1 & 0 & 0 \\ 0 & 4 & 0 & -1 & -\frac{1}{2} & \frac{3}{4} \\ 0 & 0 & 1 & -\frac{1}{2} & \frac{1}{4} & -\frac{1}{8} \end{array}\right] \xrightarrow{\frac{1}{4}R_2 \to R_2} \left[\begin{array}{ccc|ccc} 1 & -1 & -1 & 1 & 0 & 0 \\ 0 & 1 & 0 & -\frac{1}{4} & -\frac{1}{8} & \frac{3}{16} \\ 0 & 0 & 1 & -\frac{1}{2} & \frac{1}{4} & -\frac{1}{8} \end{array}\right] \xrightarrow{R_1 + R_2 + R_3 \to R_1}$$

$$\left[\begin{array}{ccc|ccc} 1 & 0 & 0 & \frac{1}{4} & \frac{1}{8} & \frac{1}{16} \\ 0 & 1 & 0 & -\frac{1}{4} & -\frac{1}{8} & \frac{3}{16} \\ 0 & 0 & 1 & -\frac{1}{2} & \frac{1}{4} & -\frac{1}{8} \end{array}\right]$$

So, the solution is

$$\begin{bmatrix} I_1 \\ I_2 \\ I_3 \end{bmatrix} = \begin{bmatrix} \frac{1}{4} & \frac{1}{8} & \frac{1}{16} \\ -\frac{1}{4} & -\frac{1}{8} & \frac{3}{16} \\ -\frac{1}{2} & \frac{1}{4} & -\frac{1}{8} \end{bmatrix}\begin{bmatrix} 0 \\ 16 \\ 24 \end{bmatrix} = \begin{bmatrix} \frac{7}{2} \\ \frac{5}{2} \\ 1 \end{bmatrix}.$$

So, $I_1 = \frac{7}{2}$, $I_2 = \frac{5}{2}$, $I_3 = 1$.

69. The second determinant should be subtracted; that is, it should be

$$-1\begin{vmatrix} -3 & 2 \\ 1 & -1 \end{vmatrix}.$$

71. In D_x and D_y, the column $\begin{bmatrix} 6 \\ -3 \end{bmatrix}$ should replace the column corresponding to the variable that is being solved for in each case. Precisely, D_x should be $\begin{vmatrix} 6 & 3 \\ -3 & -1 \end{vmatrix}$ and D_y should be $\begin{vmatrix} 2 & 6 \\ -1 & -3 \end{vmatrix}$.

73 True

75. False. Observe that

$$\begin{vmatrix} 2 & 6 & 4 \\ 0 & 2 & 8 \\ 4 & 0 & 10 \end{vmatrix} = 2\begin{vmatrix} 2 & 8 \\ 0 & 10 \end{vmatrix} - 6\begin{vmatrix} 0 & 8 \\ 4 & 10 \end{vmatrix} + 4\begin{vmatrix} 0 & 2 \\ 4 & 0 \end{vmatrix}$$

$$= 2\underbrace{\left[(2)(10)-(0)(8)\right]}_{=20} - 6\underbrace{\left[(0)(10)-(4)(8)\right]}_{=-32} + 4\underbrace{\left[(0)(0)-(4)(2)\right]}_{=-8} = 200$$

whereas

$$2\begin{vmatrix} 1 & 3 & 2 \\ 0 & 1 & 4 \\ 2 & 0 & 5 \end{vmatrix} = 2\left(1\begin{vmatrix} 1 & 4 \\ 0 & 5 \end{vmatrix} - 3\begin{vmatrix} 0 & 4 \\ 2 & 5 \end{vmatrix} + 2\begin{vmatrix} 0 & 1 \\ 2 & 0 \end{vmatrix}\right)$$

$$= 2\left(1\underbrace{\left[(1)(5)-(0)(4)\right]}_{=5} - 3\underbrace{\left[(0)(5)-(2)(4)\right]}_{=-8} + 2\underbrace{\left[(0)(0)-(2)(1)\right]}_{=-2}\right) = 2(25) = 50$$

77.

$$\begin{vmatrix} a & 0 & 0 \\ 0 & b & 0 \\ 0 & 0 & c \end{vmatrix} = a\begin{vmatrix} b & 0 \\ 0 & c \end{vmatrix} - 0\begin{vmatrix} 0 & 0 \\ 0 & c \end{vmatrix} + 0\begin{vmatrix} 0 & b \\ 0 & 0 \end{vmatrix}$$

$$= a\underbrace{\left[(b)(c)-(0)(0)\right]}_{=bc} - 0\underbrace{\left[(0)(c)-(0)(0)\right]}_{=0} + 0\underbrace{\left[(0)(0)-(0)(b)\right]}_{=0} = \boxed{abc}$$

79.

$$\begin{vmatrix} 1 & -2 & -1 & 3 \\ 4 & 0 & 1 & 2 \\ 0 & 3 & 2 & 4 \\ 1 & -3 & 5 & -4 \end{vmatrix} = 1\begin{vmatrix} 0 & 1 & 2 \\ 3 & 2 & 4 \\ -3 & 5 & -4 \end{vmatrix} - (-2)\begin{vmatrix} 4 & 1 & 2 \\ 0 & 2 & 4 \\ 1 & 5 & -4 \end{vmatrix} - 1\begin{vmatrix} 4 & 0 & 2 \\ 0 & 3 & 4 \\ 1 & -3 & -4 \end{vmatrix} - 3\begin{vmatrix} 4 & 0 & 1 \\ 0 & 3 & 2 \\ 1 & -3 & 5 \end{vmatrix} \quad \textbf{(1)}$$

where

$$\begin{vmatrix} 0 & 1 & 2 \\ 3 & 2 & 4 \\ -3 & 5 & -4 \end{vmatrix} = 0\begin{vmatrix} 2 & 4 \\ 5 & -4 \end{vmatrix} - 1\begin{vmatrix} 3 & 4 \\ -3 & -4 \end{vmatrix} + 2\begin{vmatrix} 3 & 2 \\ -3 & 5 \end{vmatrix}$$

$$= 0\underbrace{\left[(2)(-4)-(5)(4)\right]}_{=-28} - 1\underbrace{\left[(3)(-4)-(-3)(4)\right]}_{=0} + 2\underbrace{\left[(3)(5)-(-3)(2)\right]}_{=21} = 42$$

$$\begin{vmatrix} 4 & 1 & 2 \\ 0 & 2 & 4 \\ 1 & 5 & -4 \end{vmatrix} = 4\begin{vmatrix} 2 & 4 \\ 5 & -4 \end{vmatrix} - 1\begin{vmatrix} 0 & 4 \\ 1 & -4 \end{vmatrix} + 2\begin{vmatrix} 0 & 2 \\ 1 & 5 \end{vmatrix}$$

$$= 4\underbrace{\left[(2)(-4)-(5)(4)\right]}_{=-28} - 1\underbrace{\left[(0)(-4)-(1)(4)\right]}_{=-4} + 2\underbrace{\left[(0)(5)-(1)(2)\right]}_{=-2} = -112$$

$$\begin{vmatrix} 4 & 0 & 2 \\ 0 & 3 & 4 \\ 1 & -3 & -4 \end{vmatrix} = 4\begin{vmatrix} 3 & 4 \\ -3 & -4 \end{vmatrix} - 0\begin{vmatrix} 0 & 4 \\ 1 & -4 \end{vmatrix} + 2\begin{vmatrix} 0 & 3 \\ 1 & -3 \end{vmatrix}$$

$$= 4\underbrace{\left[(3)(-4)-(-3)(4)\right]}_{=0} - 0\underbrace{\left[(0)(-4)-(1)(4)\right]}_{=-4} + 2\underbrace{\left[(0)(-3)-(1)(3)\right]}_{=-3} = -6$$

$$\begin{vmatrix} 4 & 0 & 1 \\ 0 & 3 & 2 \\ 1 & -3 & 5 \end{vmatrix} = 4\begin{vmatrix} 3 & 2 \\ -3 & 5 \end{vmatrix} - 0\begin{vmatrix} 0 & 2 \\ 1 & 5 \end{vmatrix} + 1\begin{vmatrix} 0 & 3 \\ 1 & -3 \end{vmatrix}$$

$$= 4\underbrace{\left[(3)(5)-(-3)(2)\right]}_{=21} - 0\underbrace{\left[(0)(5)-(1)(2)\right]}_{=-2} + 1\underbrace{\left[(0)(-3)-(1)(3)\right]}_{=-3} = 81$$

So, substituting the values of the determinants of the four individual 3×3 matrices, we conclude that the given determinant is $1(42)+2(-112)-(-6)-3(81)=\boxed{-419}$

81.

$$\begin{vmatrix} a_1 & b_1 & c_1 \\ a_2 & b_2 & c_2 \\ a_3 & b_3 & c_3 \end{vmatrix} = -b_1\begin{vmatrix} a_2 & c_2 \\ a_3 & c_3 \end{vmatrix} + b_2\begin{vmatrix} a_1 & c_1 \\ a_3 & c_3 \end{vmatrix} - b_3\begin{vmatrix} a_1 & c_1 \\ a_2 & c_2 \end{vmatrix}$$

$$= -b_1\left[(a_2)(c_3)-(a_3)(c_2)\right] + b_2\left[(a_1)(c_3)-(a_3)(c_1)\right] - b_3\left[(a_1)(c_2)-(a_2)(c_1)\right]$$

$$= -a_2b_1c_3 + a_3b_1c_2 + a_1b_2c_3 - a_3b_2c_1 - a_1b_3c_2 + a_2b_3c_1$$

The last line above is equal to the given right-side since addition and multiplication of real numbers is commutative.

83. -180	**85.** -1019

87. First, write the system in matrix form as:

$$\begin{bmatrix} 3.1 & 1.6 & -4.8 \\ 5.2 & -3.4 & 0.5 \\ 0.5 & -6.4 & 11.4 \end{bmatrix}\begin{bmatrix} x \\ y \\ z \end{bmatrix} = \begin{bmatrix} -33.76 \\ -36.68 \\ 25.96 \end{bmatrix}$$

Now, apply Cramer's rule to solve the system:

$$D = \begin{vmatrix} 3.1 & 1.6 & -4.8 \\ 5.2 & -3.4 & 0.5 \\ 0.5 & -6.4 & 11.4 \end{vmatrix}, \quad D_x = \begin{vmatrix} -33.76 & 1.6 & -4.8 \\ -36.68 & -3.4 & 0.5 \\ 25.96 & -6.4 & 11.4 \end{vmatrix},$$

$$D_y = \begin{vmatrix} 3.1 & -33.76 & -4.8 \\ 5.2 & -36.68 & 0.5 \\ 0.5 & 25.96 & 11.4 \end{vmatrix}, \quad D_z = \begin{vmatrix} 3.1 & 1.6 & -33.76 \\ 5.2 & -3.4 & -36.68 \\ 0.5 & -6.4 & 25.96 \end{vmatrix}$$

Using a calculator to compute the determinants, we obtain from Cramer's rule that:

$$\boxed{x = \frac{D_x}{D} = -6.4 \quad y = \frac{D_y}{D} = 1.5 \quad z = \frac{D_z}{D} = 3.4}$$

Chapter 7 Review Solutions

1. $\left[\begin{array}{cc|c} 5 & 7 & 2 \\ 3 & -4 & -2 \end{array}\right]$

3. $\left[\begin{array}{ccc|c} 2 & 0 & -1 & 3 \\ 0 & 1 & -3 & -2 \\ 1 & 0 & 4 & -3 \end{array}\right]$

5. Row echelon form. Not reduced row echelon form.

7. Not in row echelon form.

9. $\left[\begin{array}{cc|c} 1 & -2 & 1 \\ 0 & 1 & -1 \end{array}\right]$

11. $\left[\begin{array}{ccc|c} 1 & -4 & 3 & -1 \\ 0 & -2 & 3 & -2 \\ 0 & 1 & -4 & 8 \end{array}\right]$

13.

$$\left[\begin{array}{cc|c} 1 & 3 & 0 \\ 3 & 4 & 1 \end{array}\right] \xrightarrow{R_1 \leftrightarrow R_2} \left[\begin{array}{cc|c} 3 & 4 & 1 \\ 1 & 3 & 0 \end{array}\right] \xrightarrow{R_1 - 3R_2 \to R_2} \left[\begin{array}{cc|c} 3 & 4 & 1 \\ 0 & -5 & 1 \end{array}\right]$$

$$\xrightarrow[-\frac{1}{5}R_2 \to R_2]{\frac{1}{3}R_1 \to R_1} \left[\begin{array}{cc|c} 1 & \frac{4}{3} & \frac{1}{3} \\ 0 & 1 & -\frac{1}{5} \end{array}\right] \xrightarrow{R_1 - \frac{4}{3}R_2 \to R_1} \left[\begin{array}{cc|c} 1 & 0 & \frac{3}{5} \\ 0 & 1 & -\frac{1}{5} \end{array}\right]$$

15.

$$\left[\begin{array}{ccc|c} 4 & 1 & -2 & 0 \\ 1 & 0 & -1 & 0 \\ -2 & 1 & 1 & 12 \end{array}\right] \xrightarrow[R_1+2R_3 \to R_3]{R_1-4R_2 \to R_2} \left[\begin{array}{ccc|c} 4 & 1 & -2 & 0 \\ 0 & 1 & 2 & 0 \\ 0 & 3 & 0 & 24 \end{array}\right] \xrightarrow{R_3-3R_2 \to R_2} \left[\begin{array}{ccc|c} 4 & 1 & -2 & 0 \\ 0 & 0 & -6 & 24 \\ 0 & 3 & 0 & 24 \end{array}\right]$$

$$\xrightarrow{R_2 \leftrightarrow R_3} \left[\begin{array}{ccc|c} 4 & 1 & -2 & 0 \\ 0 & 3 & 0 & 24 \\ 0 & 0 & -6 & 24 \end{array}\right] \xrightarrow[-\frac{1}{6}R_3 \to R_3]{\frac{1}{4}R_1 \to R_1,\ \frac{1}{3}R_2 \to R_2} \left[\begin{array}{ccc|c} 1 & \frac{1}{4} & -\frac{1}{2} & 0 \\ 0 & 1 & 0 & 8 \\ 0 & 0 & 1 & -4 \end{array}\right] \xrightarrow{R_1-\frac{1}{4}R_2+\frac{1}{2}R_3 \to R_1} \left[\begin{array}{ccc|c} 1 & 0 & 0 & -4 \\ 0 & 1 & 0 & 8 \\ 0 & 0 & 1 & -4 \end{array}\right]$$

17. First, reduce the augmented matrix down to row echelon form:

$$\left[\begin{array}{cc|c} 3 & -2 & 2 \\ -2 & 4 & 1 \end{array}\right] \xrightarrow{-\frac{1}{2}R_2 \to R_2} \left[\begin{array}{cc|c} 3 & -2 & 2 \\ 1 & -2 & -\frac{1}{2} \end{array}\right] \xrightarrow{R_1-3R_2 \to R_2} \left[\begin{array}{cc|c} 3 & -2 & 2 \\ 0 & 4 & \frac{7}{2} \end{array}\right]$$

$$\xrightarrow[\frac{1}{4}R_2 \to R_2]{\frac{1}{3}R_1 \to R_1} \left[\begin{array}{cc|c} 1 & -\frac{2}{3} & \frac{2}{3} \\ 0 & 1 & \frac{7}{8} \end{array}\right]$$

Now, from this we see that $y = \frac{7}{8}$, and then substituting this value into the equation obtained from the first row yields:

$$x - \frac{2}{3}\left(\frac{7}{8}\right) = \frac{2}{3}$$
$$x = \frac{5}{4}$$

Hence, the solution is $\boxed{x = \frac{5}{4},\ y = \frac{7}{8}}$.

19. First, reduce the augmented matrix down to reduced row echelon form:

$$\left[\begin{array}{cc|c} 5 & -1 & 9 \\ 1 & 4 & 6 \end{array}\right] \xrightarrow[\frac{1}{5}R_1 \to R_1]{R_1-5R_2 \to R_2} \left[\begin{array}{cc|c} 1 & -\frac{1}{5} & \frac{9}{5} \\ 0 & -21 & -21 \end{array}\right] \xrightarrow{-\frac{1}{21}R_2 \to R_2} \left[\begin{array}{cc|c} 1 & -\frac{1}{5} & \frac{9}{5} \\ 0 & 1 & 1 \end{array}\right]$$

$$\xrightarrow{R_1+\frac{1}{5}R_2 \to R_1} \left[\begin{array}{cc|c} 1 & 0 & 2 \\ 0 & 1 & 1 \end{array}\right]$$

Hence, the solution is $\boxed{x = 2, y = 1}$.

21. First, reduce the augmented matrix down to row echelon form:

$$\left[\begin{array}{ccc|c} 1 & -2 & 1 & 3 \\ 2 & -1 & 1 & -4 \\ 3 & -3 & -5 & 2 \end{array}\right] \xrightarrow[R_2-2R_1 \to R_2]{R_3-3R_1 \to R_3} \left[\begin{array}{ccc|c} 1 & -2 & 1 & 3 \\ 0 & 3 & -1 & -10 \\ 0 & 3 & -8 & -7 \end{array}\right] \xrightarrow{R_2-R_3 \to R_3} \left[\begin{array}{ccc|c} 1 & -2 & 1 & 3 \\ 0 & 3 & -1 & -10 \\ 0 & 0 & 7 & -3 \end{array}\right]$$

$$\xrightarrow[\frac{1}{3}R_2 \to R_2]{\frac{1}{7}R_3 \to R_3} \left[\begin{array}{ccc|c} 1 & -2 & 1 & 3 \\ 0 & 1 & -\frac{1}{3} & -\frac{10}{3} \\ 0 & 0 & 1 & -\frac{3}{7} \end{array}\right]$$

Now, from this we see that $z = -\frac{3}{7}$. Then, substituting this value into the equation obtained from the second row yields:

$$y - \tfrac{1}{3}(-\tfrac{3}{7}) = -\tfrac{10}{3} \text{ so that } y = -\tfrac{73}{21}$$

Finally, substituting these values of y and z into the equation obtained from the first row subsequently yields:

$$x - 2(-\tfrac{73}{21}) + (-\tfrac{3}{7}) = 3 \text{ so that } x = -\tfrac{74}{21}.$$

Hence, the solution is $\boxed{x = -\frac{74}{21},\ y = -\frac{73}{21},\ z = -\frac{3}{7}}$.

23. We reduce the matrix down to reduced row echelon form:

$$\left[\begin{array}{ccc|c} 1 & -4 & 10 & -61 \\ 3 & -5 & 8 & -52 \\ -5 & 1 & -2 & 8 \end{array}\right] \xrightarrow[R_3+5R_1 \to R_3]{R_2-3R_1 \to R_2} \left[\begin{array}{ccc|c} 1 & -4 & 10 & -61 \\ 0 & 7 & -22 & 131 \\ 0 & -19 & 48 & -297 \end{array}\right] \xrightarrow[-\frac{1}{19}R_3 \to R_3]{\frac{1}{7}R_2 \to R_2} \left[\begin{array}{ccc|c} 1 & -4 & 10 & -61 \\ 0 & 1 & -\frac{22}{7} & \frac{131}{7} \\ 0 & 1 & -\frac{48}{19} & \frac{297}{19} \end{array}\right]$$

$$\xrightarrow{R_2-R_3 \to R_3} \left[\begin{array}{ccc|c} 1 & 4 & 10 & 61 \\ 0 & 1 & -\frac{22}{7} & \frac{131}{7} \\ 0 & 0 & -\frac{82}{19(7)} & \frac{410}{19(7)} \end{array}\right] \xrightarrow{-\frac{19(7)}{82}R_3 \to R_3} \left[\begin{array}{ccc|c} 1 & -4 & 10 & -61 \\ 0 & 1 & -\frac{22}{7} & \frac{131}{7} \\ 0 & 0 & 1 & -5 \end{array}\right] \xrightarrow{R_2+\frac{22}{7}R_3 \to R_2}$$

$$\left[\begin{array}{ccc|c} 1 & -4 & 10 & -61 \\ 0 & 1 & 0 & 3 \\ 0 & 0 & 1 & -5 \end{array}\right] \xrightarrow{R_1+4R_2-10R_3 \to R_1} \left[\begin{array}{ccc|c} 1 & 0 & 0 & 1 \\ 0 & 1 & 0 & 3 \\ 0 & 0 & 1 & -5 \end{array}\right]$$

Hence, the solution is $\boxed{x = 1, y = 3, z = -5}$.

25. First, reduce the augmented matrix down to row echelon form:

$$\left[\begin{array}{ccc|c} 3 & 1 & 1 & -4 \\ 1 & -2 & 1 & -6 \end{array}\right] \xrightarrow{R_1 - 3R_2 \to R_2} \left[\begin{array}{ccc|c} 3 & 1 & 1 & -4 \\ 0 & 7 & -2 & 14 \end{array}\right] \xrightarrow[\frac{1}{7}R_2 \to R_2]{\frac{1}{3}R_1 \to R_1} \left[\begin{array}{ccc|c} 1 & \frac{1}{3} & \frac{1}{3} & -\frac{4}{3} \\ 0 & 1 & -\frac{2}{7} & 2 \end{array}\right]$$

Since there are two rows and three unknowns, we know from the above calculation that there are infinitely many solutions to this system. (Note: The only other possibility in such case would be that there was no solution, which would occur if one of the rows yielded a false statement.) To find the solutions, let $z = a$. Then, from row 2, we observe that $y - \frac{2}{7}a = 2$ so that $y = \frac{2}{7}a + 2$. Then, substituting these values of y and z into the equation obtained from row 1, we see that

$$x + \tfrac{1}{3}\left(\tfrac{2}{7}a + 2\right) + \tfrac{1}{3}a = -\tfrac{4}{3}$$
$$x = -\tfrac{3}{7}a - 2$$

Hence, the solutions are $\boxed{x = -\frac{3}{7}a - 2,\ y = \frac{2}{7}a + 2,\ z = a}$.

27. From the given information, we must solve the following system:

$$\begin{cases} 2 = a(16)^2 + b(16) + c \\ 6 = a(40)^2 + b(40) + c \\ 4 = a(65)^2 + b(65) + c \end{cases} \text{ is equivalent to } \begin{cases} 2 = 256a + 16b + c \\ 6 = 1600a + 40b + c \\ 4 = 4225a + 65b + c \end{cases}$$

Identify $A = \begin{bmatrix} 256 & 16 & 1 \\ 1600 & 40 & 1 \\ 4225 & 65 & 1 \end{bmatrix}$ and $B = \begin{bmatrix} 2 \\ 6 \\ 4 \end{bmatrix}$, and rewrite this system in matrix form:

$$A\begin{bmatrix} a \\ b \\ c \end{bmatrix} = B$$

The solution of this system is $\begin{bmatrix} a \\ b \\ c \end{bmatrix} = A^{-1}B$. In this case, rather than using Gaussian elimination to compute the inverse, using technology would be beneficial. Indeed, in this case, the above simplifies to:

$$\begin{bmatrix} a \\ b \\ c \end{bmatrix} = A^{-1}B = \begin{bmatrix} -0.00503 \\ 0.44857 \\ -3.888 \end{bmatrix}$$

Thus, the equation of the curve is $\boxed{y = -0.005x^2 + 0.45x - 3.89}$.

29. Not defined since the matrices have different orders.

31. $\begin{bmatrix} 3 & 5 & 2 \\ 7 & 8 & 1 \end{bmatrix}$

33. $\begin{bmatrix} 9 & -4 \\ 9 & 9 \end{bmatrix}$

35. $\begin{bmatrix} 4 & 13 \\ 18 & 11 \end{bmatrix}$

37. $\begin{bmatrix} 0 & -19 \\ -18 & -9 \end{bmatrix}$

39. $\begin{bmatrix} -7 & -11 & -8 \\ 3 & 7 & 2 \end{bmatrix}$

41. $\begin{bmatrix} 10 & -13 \\ 18 & -20 \end{bmatrix}$

43. $\begin{bmatrix} 17 & -8 & 18 \\ 33 & 0 & 42 \end{bmatrix}$

45. $\begin{bmatrix} 10 & 9 & 20 \\ 22 & -4 & 2 \end{bmatrix}$

47. Since $AB = \begin{bmatrix} 6 & 4 \\ 4 & 2 \end{bmatrix} \cdot \begin{bmatrix} -0.5 & 1 \\ 1 & -1.5 \end{bmatrix} = \begin{bmatrix} 1 & 0 \\ 0 & 1 \end{bmatrix} = I$, B must be the inverse of A.

49. Since $AB = \begin{bmatrix} 1 & -2 & 6 \\ 2 & 3 & -2 \\ 0 & -1 & 1 \end{bmatrix} \cdot \begin{bmatrix} -\frac{1}{7} & \frac{4}{7} & 2 \\ \frac{2}{7} & -\frac{1}{7} & -2 \\ \frac{2}{7} & -\frac{1}{7} & -1 \end{bmatrix} = \begin{bmatrix} 1 & 0 & 0 \\ 0 & 1 & 0 \\ 0 & 0 & 1 \end{bmatrix} = I$, B must be the inverse of A.

51. $A^{-1} = \frac{1}{4(1)-(-3)(2)} \begin{bmatrix} 4 & -2 \\ 3 & 1 \end{bmatrix} = \frac{1}{10} \begin{bmatrix} 4 & -2 \\ 3 & 1 \end{bmatrix} = \begin{bmatrix} 0.4 & -0.2 \\ 0.3 & 0.1 \end{bmatrix}$

53. $A^{-1} = \frac{1}{(0)(0)-(-2)(1)} \begin{bmatrix} 0 & -1 \\ 2 & 0 \end{bmatrix} = \frac{1}{2} \begin{bmatrix} 0 & -1 \\ 2 & 0 \end{bmatrix} = \begin{bmatrix} 0 & -\frac{1}{2} \\ 1 & 0 \end{bmatrix}$

55. Using Gaussian elimination, we obtain the following:

$$\left[\begin{array}{ccc|ccc} 1 & 3 & -2 & 1 & 0 & 0 \\ 2 & 1 & -1 & 0 & 1 & 0 \\ 0 & 1 & -3 & 0 & 0 & 1 \end{array}\right] \xrightarrow{R_2 - 2R_1 \to R_2} \left[\begin{array}{ccc|ccc} 1 & 3 & -2 & 1 & 0 & 0 \\ 0 & -5 & 3 & -2 & 1 & 0 \\ 0 & 1 & -3 & 0 & 0 & 1 \end{array}\right]$$

$$\xrightarrow{R_2 + 5R_3 \to R_3} \left[\begin{array}{ccc|ccc} 1 & 3 & -2 & 1 & 0 & 0 \\ 0 & -5 & 3 & -2 & 1 & 0 \\ 0 & 0 & -12 & -2 & 1 & 5 \end{array}\right] \xrightarrow[-\frac{1}{12}R_3 \to R_3]{-\frac{1}{5}R_2 \to R_2} \left[\begin{array}{ccc|ccc} 1 & 3 & -2 & 1 & 0 & 0 \\ 0 & 1 & -\frac{3}{5} & \frac{2}{5} & -\frac{1}{5} & 0 \\ 0 & 0 & 1 & \frac{1}{6} & -\frac{1}{12} & -\frac{5}{12} \end{array}\right]$$

$$\xrightarrow{R_2 + \frac{3}{5}R_3 \to R_2} \left[\begin{array}{ccc|ccc} 1 & 3 & -2 & 1 & 0 & 0 \\ 0 & 1 & 0 & \frac{1}{2} & -\frac{1}{4} & -\frac{1}{4} \\ 0 & 0 & 1 & \frac{1}{6} & -\frac{1}{12} & -\frac{5}{12} \end{array}\right] \xrightarrow{R_1 - 3R_2 + 2R_3 \to R_1} \left[\begin{array}{ccc|ccc} 1 & 0 & 0 & -\frac{1}{6} & \frac{7}{12} & -\frac{1}{12} \\ 0 & 1 & 0 & \frac{1}{2} & -\frac{1}{4} & -\frac{1}{4} \\ 0 & 0 & 1 & \frac{1}{6} & -\frac{1}{12} & -\frac{5}{12} \end{array}\right]$$

So, $A^{-1} = \begin{bmatrix} -\frac{1}{6} & \frac{7}{12} & -\frac{1}{12} \\ \frac{1}{2} & -\frac{1}{4} & -\frac{1}{4} \\ \frac{1}{6} & -\frac{1}{12} & -\frac{5}{12} \end{bmatrix}$.

57. Using Gaussian elimination, we obtain the following:

$$\left[\begin{array}{ccc|ccc} -1 & 1 & 0 & 1 & 0 & 0 \\ -2 & 1 & 2 & 0 & 1 & 0 \\ 1 & 2 & 4 & 0 & 0 & 1 \end{array}\right] \xrightarrow[R_1+R_3 \to R_3]{R_2-2R_1 \to R_2} \left[\begin{array}{ccc|ccc} -1 & 1 & 0 & 1 & 0 & 0 \\ 0 & -1 & 2 & -2 & 1 & 0 \\ 0 & 3 & 4 & 1 & 0 & 1 \end{array}\right]$$

$$\xrightarrow[-R_1 \to R_1]{R_3+3R_2 \to R_3} \left[\begin{array}{ccc|ccc} 1 & -1 & 0 & -1 & 0 & 0 \\ 0 & -1 & 2 & -2 & 1 & 0 \\ 0 & 0 & 10 & -5 & 3 & 1 \end{array}\right] \xrightarrow[\frac{1}{10}R_3 \to R_3]{R_3-5R_2 \to R_2} \left[\begin{array}{ccc|ccc} 1 & -1 & 0 & -1 & 0 & 0 \\ 0 & 5 & 0 & 5 & -2 & 1 \\ 0 & 0 & 1 & -\frac{1}{2} & \frac{3}{10} & \frac{1}{10} \end{array}\right]$$

$$\xrightarrow{\frac{1}{5}R_2 \to R_2} \left[\begin{array}{ccc|ccc} 1 & -1 & 0 & -1 & 0 & 0 \\ 0 & 1 & 0 & 1 & -\frac{2}{5} & \frac{1}{5} \\ 0 & 0 & 1 & -\frac{1}{2} & \frac{3}{10} & \frac{1}{10} \end{array}\right] \xrightarrow{R_1+R_2 \to R_1} \left[\begin{array}{ccc|ccc} 1 & 0 & 0 & 0 & -\frac{2}{5} & \frac{1}{5} \\ 0 & 1 & 0 & 1 & -\frac{2}{5} & \frac{1}{5} \\ 0 & 0 & 1 & -\frac{1}{2} & \frac{3}{10} & \frac{1}{10} \end{array}\right]$$

So, $A^{-1} = \begin{bmatrix} 0 & -\frac{2}{5} & \frac{1}{5} \\ 1 & -\frac{2}{5} & \frac{1}{5} \\ -\frac{1}{2} & \frac{3}{10} & \frac{1}{10} \end{bmatrix}$.

59. The system in matrix form is: $\begin{bmatrix} 3 & -1 \\ 5 & 2 \end{bmatrix}\begin{bmatrix} x \\ y \end{bmatrix} = \begin{bmatrix} 11 \\ 33 \end{bmatrix}$.

The solution is

$$\begin{bmatrix} x \\ y \end{bmatrix} = \begin{bmatrix} 3 & -1 \\ 5 & 2 \end{bmatrix}^{-1}\begin{bmatrix} 11 \\ 33 \end{bmatrix} = \frac{1}{11}\begin{bmatrix} 2 & 1 \\ -5 & 3 \end{bmatrix}\begin{bmatrix} 11 \\ 33 \end{bmatrix} = \begin{bmatrix} 5 \\ 4 \end{bmatrix}.$$

Hence, $\boxed{x = 5, y = 4}$.

61. The system in matrix form is: $\begin{bmatrix} \frac{5}{8} & -\frac{2}{3} \\ \frac{3}{4} & \frac{5}{6} \end{bmatrix}\begin{bmatrix} x \\ y \end{bmatrix} = \begin{bmatrix} -3 \\ 16 \end{bmatrix}$

The solution is:

$$\begin{bmatrix} x \\ y \end{bmatrix} = \begin{bmatrix} \frac{5}{8} & -\frac{2}{3} \\ \frac{3}{4} & \frac{5}{6} \end{bmatrix}^{-1}\begin{bmatrix} -3 \\ 16 \end{bmatrix} = \frac{48}{49}\begin{bmatrix} \frac{5}{6} & \frac{2}{3} \\ -\frac{3}{4} & \frac{5}{8} \end{bmatrix}\begin{bmatrix} -3 \\ 16 \end{bmatrix} = \begin{bmatrix} 8 \\ 12 \end{bmatrix}.$$

Hence, $\boxed{x = 8, y = 12}$.

63. We reduce down to reduced row echelon form:

$$\left[\begin{array}{ccc|c} 3 & -2 & 4 & 11 \\ 6 & 3 & -2 & 6 \\ 1 & -1 & 7 & 20 \end{array}\right] \xrightarrow[R_1-3R_3 \to R_3]{R_2-6R_3 \to R_2} \left[\begin{array}{ccc|c} 3 & -2 & 4 & 11 \\ 0 & 9 & -44 & -114 \\ 0 & 1 & -17 & -49 \end{array}\right] \xrightarrow{R_2-9R_3 \to R_3} \left[\begin{array}{ccc|c} 3 & -2 & 4 & 11 \\ 0 & 9 & -44 & -114 \\ 0 & 0 & 109 & 327 \end{array}\right]$$

$$\xrightarrow[\frac{1}{109}R_3 \to R_3]{\frac{1}{9}R_2 \to R_2} \left[\begin{array}{ccc|c} 3 & -2 & 4 & 11 \\ 0 & 1 & -\frac{44}{9} & -114 \\ 0 & 0 & 109 & 327 \end{array}\right] \xrightarrow[R_2+\frac{44}{9}R_3 \to R_2]{\frac{1}{3}R_1 \to R_1} \left[\begin{array}{ccc|c} 1 & -\frac{2}{3} & \frac{4}{3} & \frac{11}{3} \\ 0 & 1 & 0 & 2 \\ 0 & 0 & 1 & 3 \end{array}\right] \xrightarrow{R_1+\frac{2}{3}R_2-\frac{4}{3}R_3 \to R_1} \left[\begin{array}{ccc|c} 1 & 0 & 0 & 1 \\ 0 & 1 & 0 & 2 \\ 0 & 0 & 1 & 3 \end{array}\right]$$

Hence, the solution is $\boxed{x = 1, y = 2, z = 3}$.

65. $\begin{vmatrix} 2 & 4 \\ 3 & 2 \end{vmatrix} = (2)(2)-(3)(4) = \boxed{-8}$

67. $\begin{vmatrix} 2.4 & -2.3 \\ 3.6 & -1.2 \end{vmatrix} = (2.4)(-1.2)-(3.6)(-2.3) = \boxed{5.4}$

69.

$$D = \begin{vmatrix} 1 & -1 \\ 1 & 1 \end{vmatrix} = (1)(1)-(1)(-1) = 2$$

$$D_x = \begin{vmatrix} 2 & -1 \\ 4 & 1 \end{vmatrix} = (2)(1)-(4)(-1) = 6$$

$$D_y = \begin{vmatrix} 1 & 2 \\ 1 & 4 \end{vmatrix} = (1)(4)-(1)(2) = 2$$

So, from Cramer's Rule, we have:

$$x = \frac{D_x}{D} = \frac{6}{2} = 3$$

$$y = \frac{D_y}{D} = \frac{2}{2} = 1$$

Thus, the solution is $\boxed{x=3,\ y=1}$.

71. Divide the first equation by 2. Then, we have:

$$D = \begin{vmatrix} 1 & 2 \\ 1 & -2 \end{vmatrix} = (1)(-2)-(1)(2) = -4$$

$$D_x = \begin{vmatrix} 6 & 2 \\ 6 & -2 \end{vmatrix} = (6)(-2)-(6)(2) = -24$$

$$D_y = \begin{vmatrix} 1 & 6 \\ 1 & 6 \end{vmatrix} = (1)(6)-(1)(6) = 0$$

So, from Cramer's Rule, we have:

$$x = \frac{D_x}{D} = \frac{-24}{-4} = 6$$

$$y = \frac{D_y}{D} = \frac{0}{-4} = 0$$

Thus, the solution is $\boxed{x=6,\ y=0}$.

73. First, write the system as:

$$\begin{cases} -3x+2y=40 \\ 2x-y=25 \end{cases}$$

$$D = \begin{vmatrix} -3 & 2 \\ 2 & -1 \end{vmatrix} = (-3)(-1)-(2)(2) = -1$$

$$D_x = \begin{vmatrix} 40 & 2 \\ 25 & -1 \end{vmatrix} = (40)(-1)-(25)(2) = -90$$

$$D_y = \begin{vmatrix} -3 & 40 \\ 2 & 25 \end{vmatrix} = (-3)(25)-(2)(40) = -155$$

So, from Cramer's Rule, we have:

$$x = \frac{D_x}{D} = \frac{-90}{-1} = 90$$

$$y = \frac{D_y}{D} = \frac{-155}{-1} = 155$$

Thus, the solution is $\boxed{x=90,\ y=155}$.

75.

$$\begin{vmatrix} 1 & 2 & 2 \\ 0 & 1 & 3 \\ 2 & -1 & 0 \end{vmatrix} = 1\begin{vmatrix} 1 & 3 \\ -1 & 0 \end{vmatrix} - 2\begin{vmatrix} 0 & 3 \\ 2 & 0 \end{vmatrix} + 2\begin{vmatrix} 0 & 1 \\ 2 & -1 \end{vmatrix}$$

$$= 1\underbrace{\left[(1)(0)-(-1)(3)\right]}_{=3} - 2\underbrace{\left[(0)(0)-(2)(3)\right]}_{=-6} + 2\underbrace{\left[(0)(-1)-(2)(1)\right]}_{=-2}$$

$$= \boxed{11}$$

77.

$$\begin{vmatrix} a & 0 & -b \\ -a & b & c \\ 0 & 0 & -d \end{vmatrix} = a\begin{vmatrix} b & c \\ 0 & -d \end{vmatrix} - 0\begin{vmatrix} -a & c \\ 0 & -d \end{vmatrix} + (-b)\begin{vmatrix} -a & b \\ 0 & 0 \end{vmatrix}$$

$$= a\underbrace{\left[(b)(-d)-(0)(c)\right]}_{=-bd} - 0\underbrace{\left[(-a)(-d)-(0)(c)\right]}_{=ad} + (-b)\underbrace{\left[(-a)(0)-(0)(b)\right]}_{=0}$$

$$= \boxed{-abd}$$

79.

$$D = \begin{vmatrix} 1 & 1 & -2 \\ 2 & -1 & 1 \\ 1 & 1 & 1 \end{vmatrix} = 1\begin{vmatrix} -1 & 1 \\ 1 & 1 \end{vmatrix} - 1\begin{vmatrix} 2 & 1 \\ 1 & 1 \end{vmatrix} + (-2)\begin{vmatrix} 2 & -1 \\ 1 & 1 \end{vmatrix}$$

$$= 1\underbrace{\left[(-1)(1)-(1)(1)\right]}_{=-2} - 1\underbrace{\left[(2)(1)-(1)(1)\right]}_{=1} + (-2)\underbrace{\left[(2)(1)-(1)(-1)\right]}_{=3} = -9$$

$$D_x = \begin{vmatrix} -2 & 1 & -2 \\ 3 & -1 & 1 \\ 4 & 1 & 1 \end{vmatrix} = (-2)\begin{vmatrix} -1 & 1 \\ 1 & 1 \end{vmatrix} - 1\begin{vmatrix} 3 & 1 \\ 4 & 1 \end{vmatrix} + (-2)\begin{vmatrix} 3 & -1 \\ 4 & 1 \end{vmatrix}$$

$$= (-2)\underbrace{\left[(-1)(1)-(1)(1)\right]}_{=-2} - 1\underbrace{\left[(3)(1)-(4)(1)\right]}_{=-1} + (-2)\underbrace{\left[(3)(1)-(4)(-1)\right]}_{=7} = -9$$

$$D_y = \begin{vmatrix} 1 & -2 & -2 \\ 2 & 3 & 1 \\ 1 & 4 & 1 \end{vmatrix} = 1\begin{vmatrix} 3 & 1 \\ 4 & 1 \end{vmatrix} - (-2)\begin{vmatrix} 2 & 1 \\ 1 & 1 \end{vmatrix} + (-2)\begin{vmatrix} 2 & 3 \\ 1 & 4 \end{vmatrix}$$

$$= 1\underbrace{\left[(3)(1)-(4)(1)\right]}_{=-1} - (-2)\underbrace{\left[(2)(1)-(1)(1)\right]}_{=1} + (-2)\underbrace{\left[(2)(4)-(1)(3)\right]}_{=5} = -9$$

$$D_z = \begin{vmatrix} 1 & 1 & -2 \\ 2 & -1 & 3 \\ 1 & 1 & 4 \end{vmatrix} = 1\begin{vmatrix} -1 & 3 \\ 1 & 4 \end{vmatrix} - 1\begin{vmatrix} 2 & 3 \\ 1 & 4 \end{vmatrix} + (-2)\begin{vmatrix} 2 & -1 \\ 1 & 1 \end{vmatrix}$$

$$= 1\underbrace{\left[(-1)(4)-(1)(3)\right]}_{=-7} - 1\underbrace{\left[(2)(4)-(1)(3)\right]}_{=5} + (-2)\underbrace{\left[(2)(1)-(1)(-1)\right]}_{=3} = -18$$

So, by Cramer's Rule,

$$\boxed{x = \frac{D_x}{D} = \frac{-9}{-9} = 1 \qquad y = \frac{D_y}{D} = \frac{-9}{-9} = 1 \qquad z = \frac{D_z}{D} = \frac{-18}{-9} = 2}$$

81.

$$D=\begin{vmatrix}3&0&4\\1&1&2\\0&1&-4\end{vmatrix}=3\begin{vmatrix}1&2\\1&-4\end{vmatrix}-0\begin{vmatrix}1&2\\0&-4\end{vmatrix}+4\begin{vmatrix}1&1\\0&1\end{vmatrix}$$

$$=3\underbrace{\left[(1)(-4)-(1)(2)\right]}_{=-6}-0\underbrace{\left[(1)(-4)-(0)(2)\right]}_{=-4}+4\underbrace{\left[(1)(1)-(0)(1)\right]}_{=1}=-14$$

$$D_x=\begin{vmatrix}-1&0&4\\-3&1&2\\-9&1&-4\end{vmatrix}=(-1)\begin{vmatrix}1&2\\1&-4\end{vmatrix}-0\begin{vmatrix}-3&2\\-9&-4\end{vmatrix}+4\begin{vmatrix}-3&1\\-9&1\end{vmatrix}$$

$$=(-1)\underbrace{\left[(1)(-4)-(1)(2)\right]}_{=-6}-0\underbrace{\left[(-3)(-4)-(-9)(2)\right]}_{=30}+4\underbrace{\left[(-3)(1)-(-9)(1)\right]}_{=6}=30$$

$$D_y=\begin{vmatrix}3&-1&4\\1&-3&2\\0&-9&-4\end{vmatrix}=3\begin{vmatrix}-3&2\\-9&-4\end{vmatrix}-(-1)\begin{vmatrix}1&2\\0&-4\end{vmatrix}+4\begin{vmatrix}1&-3\\0&-9\end{vmatrix}$$

$$=3\underbrace{\left[(-3)(-4)-(-9)(2)\right]}_{=30}-(-1)\underbrace{\left[(1)(-4)-(0)(2)\right]}_{=-4}+4\underbrace{\left[(1)(-9)-(0)(-3)\right]}_{=-9}=50$$

$$D_z=\begin{vmatrix}3&0&-1\\1&1&-3\\0&1&-9\end{vmatrix}=3\begin{vmatrix}1&-3\\1&-9\end{vmatrix}-0\begin{vmatrix}1&-3\\0&-9\end{vmatrix}+(-1)\begin{vmatrix}1&1\\0&1\end{vmatrix}$$

$$=3\underbrace{\left[(1)(-9)-(1)(-3)\right]}_{=-6}-0\underbrace{\left[(1)(-9)-(0)(-3)\right]}_{=-9}+(-1)\underbrace{\left[(1)(1)-(0)(1)\right]}_{=1}=-19$$

So, by Cramer's Rule,

$$\boxed{x=\frac{D_x}{D}=\frac{30}{-14}=-\frac{15}{7}\quad y=\frac{D_y}{D}=\frac{50}{-14}=-\frac{25}{7}\quad z=\frac{D_z}{D}=\frac{-19}{-14}=\frac{19}{14}}$$

83. Let $x_1=2,\ x_2=4,\ x_3=-4,\ y_1=4,\ y_2=4,\ y_3=3$. Then, we have

$$\text{Area}=\pm\tfrac{1}{2}\begin{vmatrix}2&4&1\\4&4&1\\-4&3&1\end{vmatrix}=\pm\tfrac{1}{2}\left[2\begin{vmatrix}4&1\\3&1\end{vmatrix}-4\begin{vmatrix}4&1\\-4&1\end{vmatrix}+1\begin{vmatrix}4&4\\-4&3\end{vmatrix}\right]$$

$$=\pm\tfrac{1}{2}\left(2\underbrace{\left[(4)(1)-(3)(1)\right]}_{=1}-4\underbrace{\left[(4)(1)-(-4)(1)\right]}_{=8}+1\underbrace{\left[(4)(3)-(-4)(4)\right]}_{=28}\right)$$

$$=\pm\tfrac{1}{2}(-2)$$

Hence, choosing the positive value above, we conclude that the area is 1 unit^2.

85.

a.

```
rref([A])
[[1 0 0 .161066...
 [0 1 0 -.04946...
 [0 0 1 -4.1012...
```

$y = 0.16x^2 - 0.05x - 4.10$

b.

```
QuadReg
 y=ax²+bx+c
 a=.1610661269
 b=-.0494601889
 c=-4.101214575
```

$y = 0.16x^2 - 0.05x - 4.10$

87. $\begin{bmatrix} -238 & 206 & 50 \\ -113 & 159 & 135 \\ 40 & -30 & 0 \end{bmatrix}$

89. The augmented matrix to enter into the calculator is

$$\left[\begin{array}{cc|c} 6.1 & -14.2 & 75.495 \\ -2.3 & 7.2 & -36.495 \end{array}\right].$$

Solving using the calculator then yields the solution of the system as (2.25, -4.35).

91. First, write the system in matrix form as:

$$\begin{bmatrix} 4.5 & -8.7 \\ -1.4 & 5.3 \end{bmatrix}\begin{bmatrix} x \\ y \end{bmatrix} = \begin{bmatrix} -72.33 \\ 31.32 \end{bmatrix}$$

Now, apply Cramer's rule to solve the system:

$$D = \begin{vmatrix} 4.5 & -8.7 \\ -1.4 & 5.3 \end{vmatrix},\quad D_x = \begin{vmatrix} -72.33 & -8.7 \\ 31.32 & 5.3 \end{vmatrix},\quad D_y = \begin{vmatrix} 4.5 & -72.33 \\ -1.4 & 31.32 \end{vmatrix}$$

Using a calculator to compute the determinants, we obtain from Cramer's rule that:

$$\boxed{x = \frac{D_x}{D} = -9.5 \quad y = \frac{D_y}{D} = 3.4}$$

Chapter 7 Practice Test Solutions

1. Augmented matrix is $\left[\begin{array}{cc|c} 1 & -2 & 1 \\ -1 & 3 & 2 \end{array}\right]$ Matrix equation is $\begin{bmatrix} 1 & -2 \\ -1 & 3 \end{bmatrix}\begin{bmatrix} x \\ y \end{bmatrix} = \begin{bmatrix} 1 \\ 2 \end{bmatrix}$

3. Augmented matrix is $\left[\begin{array}{ccc|c} 6 & 9 & 1 & 5 \\ 2 & -3 & 1 & 3 \\ 10 & 12 & 2 & 9 \end{array}\right]$ Matrix equation is $\begin{bmatrix} 6 & 9 & 1 \\ 2 & -3 & 1 \\ 10 & 12 & 2 \end{bmatrix}\begin{bmatrix} x \\ y \\ z \end{bmatrix} = \begin{bmatrix} 5 \\ 3 \\ 9 \end{bmatrix}$

5.

$$\begin{bmatrix} 1 & 3 & 5 \\ 2 & 7 & -1 \\ -3 & -2 & 0 \end{bmatrix} \xrightarrow{R_2 - 2R_1 \to R_2} \begin{bmatrix} 1 & 3 & 5 \\ 0 & 1 & -11 \\ -3 & -2 & 0 \end{bmatrix} \xrightarrow{R_3 + 3R_1 \to R_3} \begin{bmatrix} 1 & 3 & 5 \\ 0 & 1 & -11 \\ 0 & 7 & 15 \end{bmatrix}$$

7. Reduce the augmented matrix down to row echelon form:

$$\left[\begin{array}{ccc|c} 6 & 9 & 1 & 5 \\ 2 & -3 & 1 & 3 \\ 10 & 12 & 2 & 9 \end{array}\right] \xrightarrow[R_3-5R_2\to R_3]{R_1-3R_2\to R_2} \left[\begin{array}{ccc|c} 6 & 9 & 1 & 5 \\ 0 & 18 & -2 & -4 \\ 0 & 27 & -3 & -6 \end{array}\right] \xrightarrow{\substack{\frac{1}{18}R_2\to R_2 \\ \frac{1}{27}R_3\to R_3 \\ \frac{1}{6}R_1\to R_1}} \left[\begin{array}{ccc|c} 1 & \frac{3}{2} & \frac{1}{6} & \frac{5}{6} \\ 0 & 1 & -\frac{1}{9} & -\frac{2}{9} \\ 0 & 1 & -\frac{1}{9} & -\frac{2}{9} \end{array}\right]$$

$$\xrightarrow{R_2-R_3\to R_3} \left[\begin{array}{ccc|c} 1 & \frac{3}{2} & \frac{1}{6} & \frac{5}{6} \\ 0 & 1 & -\frac{1}{9} & -\frac{2}{9} \\ 0 & 0 & 0 & 0 \end{array}\right] \xrightarrow{R_1-\frac{3}{2}R_2\to R_1} \left[\begin{array}{ccc|c} 1 & 0 & \frac{1}{3} & \frac{7}{6} \\ 0 & 1 & -\frac{1}{9} & -\frac{2}{9} \\ 0 & 0 & 0 & 0 \end{array}\right]$$

From the last row, we conclude that the system has infinitely many solutions. To find them, let $z = a$. Then, the second row implies that $y - \frac{1}{9}a = -\frac{2}{9}$ so that $y = \frac{1}{9}a - \frac{2}{9}$. So, substituting these values of y and z into the first row yields $x + \frac{3}{2}(\frac{1}{9}a - \frac{2}{9}) + \frac{1}{6}a = \frac{5}{6}$, so that $x = -\frac{1}{3}a + \frac{7}{6}$. Hence, the solutions are $\boxed{x = -\frac{1}{3}a + \frac{7}{6},\ y = \frac{1}{9}a - \frac{2}{9},\ z = a}$.

9. $\begin{bmatrix} 1 & -2 & 5 \\ 0 & -1 & 3 \end{bmatrix} \cdot \begin{bmatrix} 0 & 4 \\ 3 & -5 \\ -1 & 1 \end{bmatrix} = \begin{bmatrix} -11 & 19 \\ -6 & 8 \end{bmatrix}$

11. $\begin{bmatrix} 4 & 3 \\ 5 & -1 \end{bmatrix}^{-1} = \frac{1}{4(-1)-5(3)} \begin{bmatrix} -1 & -3 \\ -5 & 4 \end{bmatrix} = \frac{1}{-19} \begin{bmatrix} -1 & -3 \\ -5 & 4 \end{bmatrix} = \begin{bmatrix} \frac{1}{19} & \frac{3}{19} \\ \frac{5}{19} & -\frac{4}{19} \end{bmatrix}$

13. No inverse – only square matrices can have inverses.

15. The system in matrix form is:

$$\begin{bmatrix} 3 & -1 & 4 \\ 1 & 2 & 3 \\ -4 & 6 & -1 \end{bmatrix} \begin{bmatrix} x \\ y \\ z \end{bmatrix} = \begin{bmatrix} 18 \\ 20 \\ 11 \end{bmatrix}.$$

The solution is:

$$\begin{bmatrix} x \\ y \\ z \end{bmatrix} = \begin{bmatrix} 3 & -1 & 4 \\ 1 & 2 & 3 \\ -4 & 6 & -1 \end{bmatrix}^{-1} \begin{bmatrix} 18 \\ 20 \\ 11 \end{bmatrix}$$

We calculate the inverse below:

$$\left[\begin{array}{ccc|ccc} 3 & -1 & 4 & 1 & 0 & 0 \\ 1 & 2 & 3 & 0 & 1 & 0 \\ -4 & 6 & -1 & 0 & 0 & 1 \end{array}\right] \xrightarrow{R_1 \leftrightarrow R_2} \left[\begin{array}{ccc|ccc} 1 & 2 & 3 & 0 & 1 & 0 \\ 3 & -1 & 4 & 1 & 0 & 0 \\ -4 & 6 & -1 & 0 & 0 & 1 \end{array}\right] \xrightarrow[R_3+4R_1 \to R_3]{R_2-3R_1 \to R_2} \left[\begin{array}{ccc|ccc} 1 & 2 & 3 & 0 & 1 & 0 \\ 0 & -7 & -5 & 1 & -3 & 0 \\ 0 & 14 & 11 & 0 & 4 & 1 \end{array}\right]$$

$$\xrightarrow{R_3+2R_2 \to R_3} \left[\begin{array}{ccc|ccc} 1 & 2 & 3 & 0 & 1 & 0 \\ 0 & -7 & -5 & 1 & -3 & 0 \\ 0 & 0 & 1 & 2 & -2 & 1 \end{array}\right] \xrightarrow{-\frac{1}{7}R_2 \to R_2} \left[\begin{array}{ccc|ccc} 1 & 2 & 3 & 0 & 1 & 0 \\ 0 & 1 & \frac{5}{7} & -\frac{1}{7} & \frac{3}{7} & 0 \\ 0 & 0 & 1 & 2 & -2 & 1 \end{array}\right]$$

$$\xrightarrow{R_2-\frac{5}{7}R_3 \to R_2} \left[\begin{array}{ccc|ccc} 1 & 2 & 3 & 0 & 1 & 0 \\ 0 & 1 & 0 & -\frac{11}{7} & \frac{13}{7} & -\frac{5}{7} \\ 0 & 0 & 1 & 2 & -2 & 1 \end{array}\right] \xrightarrow{R_1-2R_2-3R_3 \to R_1} \left[\begin{array}{ccc|ccc} 1 & 0 & 0 & -\frac{20}{7} & \frac{23}{7} & -\frac{11}{7} \\ 0 & 1 & 0 & -\frac{11}{7} & \frac{13}{7} & -\frac{5}{7} \\ 0 & 0 & 1 & 2 & -2 & 1 \end{array}\right]$$

Hence, the solution is

$$\begin{bmatrix} x \\ y \\ z \end{bmatrix} = \begin{bmatrix} -\frac{20}{7} & \frac{23}{7} & -\frac{11}{7} \\ -\frac{11}{7} & \frac{13}{7} & -\frac{5}{7} \\ 2 & -2 & 1 \end{bmatrix} \begin{bmatrix} 18 \\ 20 \\ 11 \end{bmatrix} = \begin{bmatrix} -3 \\ 1 \\ 7 \end{bmatrix}.$$

Thus, $\boxed{x=-3, y=1, z=7}$.

17.

$$\begin{vmatrix} 1 & -2 & -1 \\ 3 & -5 & 2 \\ 4 & -1 & 0 \end{vmatrix} = 1\begin{vmatrix} -5 & 2 \\ -1 & 0 \end{vmatrix} - (-2)\begin{vmatrix} 3 & 2 \\ 4 & 0 \end{vmatrix} + (-1)\begin{vmatrix} 3 & -5 \\ 4 & -1 \end{vmatrix}$$

$$= 1\underbrace{\left[(-5)(0)-(-1)(2)\right]}_{=2} - (-2)\underbrace{\left[(3)(0)-(4)(2)\right]}_{=-8} + (-1)\underbrace{\left[(3)(-1)-(4)(-5)\right]}_{=17} = \boxed{-31}$$

19.

$$D = \begin{vmatrix} 3 & 5 & -2 \\ 7 & 11 & 3 \\ 1 & -1 & 1 \end{vmatrix} = 3\begin{vmatrix} 11 & 3 \\ -1 & 1 \end{vmatrix} - 5\begin{vmatrix} 7 & 3 \\ 1 & 1 \end{vmatrix} - 2\begin{vmatrix} 7 & 11 \\ 1 & -1 \end{vmatrix}$$

$$= 3\underbrace{\left[(11)(1)-(-1)(3)\right]}_{=14} - 5\underbrace{\left[(7)(1)-(1)(3)\right]}_{=4} - 2\underbrace{\left[(7)(-1)-(1)(11)\right]}_{=-18} = 58$$

$$D_x = \begin{vmatrix} -6 & 5 & -2 \\ 2 & 11 & 3 \\ 4 & -1 & 1 \end{vmatrix} = -6\begin{vmatrix} 11 & 3 \\ -1 & 1 \end{vmatrix} - 5\begin{vmatrix} 2 & 3 \\ 4 & 1 \end{vmatrix} - 2\begin{vmatrix} 2 & 11 \\ 4 & -1 \end{vmatrix}$$

$$= -6\underbrace{\left[(11)(1)-(-1)(3)\right]}_{=14} - 5\underbrace{\left[(2)(1)-(4)(3)\right]}_{=-10} - 2\underbrace{\left[(2)(-1)-(4)(11)\right]}_{=-46} = 58$$

$$D_y = \begin{vmatrix} 3 & -6 & -2 \\ 7 & 2 & 3 \\ 1 & 4 & 1 \end{vmatrix} = 3\begin{vmatrix} 2 & 3 \\ 4 & 1 \end{vmatrix} + 6\begin{vmatrix} 7 & 3 \\ 1 & 1 \end{vmatrix} - 2\begin{vmatrix} 7 & 2 \\ 1 & 4 \end{vmatrix}$$

$$= 3\underbrace{\left[(2)(1)-(4)(3)\right]}_{=-10} + 6\underbrace{\left[(7)(1)-(1)(3)\right]}_{=4} - 2\underbrace{\left[(7)(4)-(1)(2)\right]}_{=26} = -58$$

$$D_z = \begin{vmatrix} 3 & 5 & -6 \\ 7 & 11 & 2 \\ 1 & -1 & 4 \end{vmatrix} = 3\begin{vmatrix} 11 & 2 \\ -1 & 4 \end{vmatrix} - 5\begin{vmatrix} 7 & 2 \\ 1 & 4 \end{vmatrix} - 6\begin{vmatrix} 7 & 11 \\ 1 & -1 \end{vmatrix}$$

$$= 3\underbrace{\left[(11)(4)-(-1)(2)\right]}_{=46} - 5\underbrace{\left[(7)(4)-(1)(2)\right]}_{=26} - 6\underbrace{\left[(7)(-1)-(1)(11)\right]}_{=-18} = 116$$

So, by Cramer's Rule,

$$\boxed{x = \frac{D_x}{D} = 1 \quad y = \frac{D_y}{D} = -1 \quad z = \frac{D_z}{D} = 2}$$

21. Let x = amount invested in money market (2%)
y = amount invested in conservative stock (4%)
z = amount invested in aggressive stock (22%)

Solve the system:

$$\begin{cases} y = x + 1000 & \textbf{(1)} \\ z = 2x & \textbf{(2)} \\ 0.02x + 0.04y + 0.22z = 1790 & \textbf{(3)} \end{cases}$$

First, observe that substituting **(2)** into **(3)** yields $0.46x + 0.04y = 1790$. Thus, we can consider this equation, together with **(1)** to get the following 2×2 system:

$$\begin{cases} x - y = -1000 \\ 0.46x + 0.04y = 1790 \end{cases}$$

Now, write this system in its equivalent matrix form:

$$\begin{bmatrix} 1 & -1 \\ 0.46 & 0.04 \end{bmatrix} \begin{bmatrix} x \\ y \end{bmatrix} = \begin{bmatrix} -1000 \\ 1790 \end{bmatrix}$$

Solving then yields:

$$\begin{bmatrix} x \\ y \end{bmatrix} = \begin{bmatrix} 1 & -1 \\ 0.46 & 0.04 \end{bmatrix}^{-1} \begin{bmatrix} -1000 \\ 1790 \end{bmatrix}$$

$$= \left(\frac{1}{(1)(0.04)-(0.46)(-1)} \begin{bmatrix} 0.04 & 1 \\ -0.46 & 1 \end{bmatrix} \right) \begin{bmatrix} -1000 \\ 1790 \end{bmatrix}$$

$$= \left(\frac{1}{0.5} \begin{bmatrix} 0.04 & 1 \\ -0.46 & 1 \end{bmatrix} \right) \begin{bmatrix} -1000 \\ 1790 \end{bmatrix} = \begin{bmatrix} 0.08 & 2 \\ -0.92 & 2 \end{bmatrix} \begin{bmatrix} -1000 \\ 1790 \end{bmatrix} = \begin{bmatrix} 3500 \\ 4500 \end{bmatrix}$$

So, $x = 3500$, $y = 4500$. Substituting this value of x into **(2)** yields $z = 2(3500) = 7000$.
Therefore, he should invest
$3500 in money market, $4500 in conservative stock, and $7000 in aggressive stock.

23. The augmented matrix that should be entered into the calculator is

$$\left[\begin{array}{cc|c} 5.6 & -2.7 & 87.28 \\ -4.2 & 8.4 & -106.26 \end{array}\right]$$

Solving yields the solution (12.5, -6.4).

Chapter 7 Cumulative Review---

1.

$$\begin{aligned} x^2-6x+9&=11+9\\ (x-3)^2&=20\\ x-3&=\pm\sqrt{20}=\pm 2\sqrt{5}\\ x&=3\pm 2\sqrt{5} \end{aligned}$$

3. The radius is $\sqrt{(-3-1)^2+(-1-2)^2}=5$. So, the equation of the circle is

$$(x+3)^2+(y+1)^2=25.$$

5. $g(-x)=\sqrt{2-(-x)^2}=\sqrt{2-x^2}=g(x)$, so g is even. As such, since $-g(-x)\neq g(x)$, g is not odd.

7. $f(g(x))=\left(\frac{1}{x}\right)^3-1=\frac{1}{x^3}-1$.
Domain: $(-\infty,0)\cup(0,\infty)$.

9. Completing the square yields

$$\begin{aligned} f(x)&=\tfrac{1}{4}x^2+\tfrac{3}{5}x-\tfrac{6}{25}\\ &=\tfrac{1}{4}\left(x^2+\tfrac{12}{5}x\right)-\tfrac{6}{25}\\ &=\tfrac{1}{4}\left(x^2+\tfrac{12}{5}x+\tfrac{36}{25}\right)-\tfrac{6}{25}-\tfrac{9}{25}\\ &=\tfrac{1}{4}\left(x+\tfrac{6}{5}\right)^2-\tfrac{3}{5} \end{aligned}$$

So, the vertex is $\left(-\frac{6}{5},-\frac{3}{5}\right)$.

11.

$$\begin{array}{r} 5x-4\\ x^2+0x+1\overline{\big)\,5x^3-4x^2+0x+3}\\ \underline{-\left(5x^3+0x^2+5x\right)}\\ -4x^2-5x+3\\ \underline{-\left(-4x^2+0x-4\right)}\\ -5x+7 \end{array}$$

Hence, $Q(x)=5x-4,\ \ r(x)=-5x+7$.

13. $f(x)=\dfrac{0.7x^2-5x+11}{(x-3)(x+2)}$

Vertical asymptotes: $x=3, x=-2$ Horizontal asymptote: $y=0.7$

15. Must have $x+3>0$, so the domain is $(-3,\infty)$.

17.

$$\log_5(x+2)+\log_5(6-x)=\log_5(3x)$$
$$\log_5(x+2)(6-x)=\log_5(3x)$$
$$(x+2)(6-x)=3x$$
$$x^2-x-12=0$$
$$(x-4)(x+3)=0$$
$$x=4, \cancel{-3}$$

19. Write the system in matrix form as

$$\begin{bmatrix} 2 & 3 \\ 1 & 1.5 \end{bmatrix}\begin{bmatrix} x \\ y \end{bmatrix}=\begin{bmatrix} 6 \\ -5.5 \end{bmatrix}.$$

The solution would be

$$\begin{bmatrix} x \\ y \end{bmatrix}=\begin{bmatrix} 2 & 3 \\ 1 & 1.5 \end{bmatrix}^{-1}\begin{bmatrix} 6 \\ -5.5 \end{bmatrix},$$

but since $\begin{bmatrix} 2 & 3 \\ 1 & 1.5 \end{bmatrix}^{-1}$ does not exist since $\begin{vmatrix} 2 & 3 \\ 1 & 1.5 \end{vmatrix}=0$, the system has no solution.

21. The region is as follows:

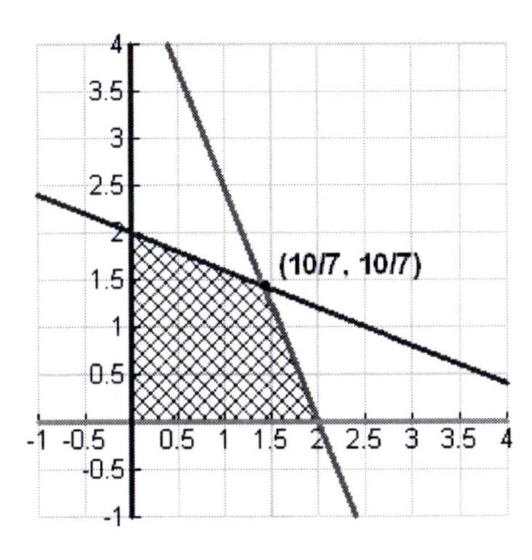

Vertex	$z=5x+7y$
(0,0)	0
(0,2)	14
(2,0)	10
$\left(\frac{10}{7},\frac{10}{7}\right)$	$\frac{120}{7}$ MAX

So, the maximum value of z is $\frac{120}{7}$ and occurs at $\left(\frac{10}{7},\frac{10}{7}\right)$

23. $\begin{bmatrix} 78 & -10 & 40 \\ 26 & 0 & 14 \end{bmatrix}$

25. First, write the system as

$$\begin{bmatrix} 7 & 5 \\ -1 & 4 \end{bmatrix}\begin{bmatrix} x \\ y \end{bmatrix} = \begin{bmatrix} 1 \\ -1 \end{bmatrix}.$$

$$D = \begin{vmatrix} 7 & 5 \\ -1 & 4 \end{vmatrix} = (7)(4) - (-1)(5) = 33$$

$$D_x = \begin{vmatrix} 1 & 5 \\ -1 & 4 \end{vmatrix} = (1)(4) - (-1)(5) = 9$$

$$D_y = \begin{vmatrix} 7 & 1 \\ -1 & -1 \end{vmatrix} = (7)(-1) - (-1)(1) = -6$$

So, by Cramer's Rule, we have

$$x = \frac{D_x}{D} = \frac{3}{11}$$

$$y = \frac{D_y}{D} = -\frac{2}{11}$$

27. Using the calculator, we see that $AB = \begin{bmatrix} 35 & 12 \\ -19 & -14 \end{bmatrix}$ and $(AB)^{-1} = \begin{bmatrix} \frac{7}{131} & \frac{6}{131} \\ -\frac{19}{262} & -\frac{35}{262} \end{bmatrix}$.

CHAPTER 8

Section 8.1 Solutions --

1. Since $A=1,\ B=1,\ C=-1$, $B^2-4AC=1+4=5>0$. Hence, this is the equation of a $\boxed{\text{hyperbola}}$.	**3.** Since $A=2,\ B=0,\ C=2$, $B^2-4AC=0-4(2)(2)=-16<0$. Hence, this is the equation of an $\boxed{\text{ellipse}}$. Note this is a circle.
5. Hyperbola	**7.** Ellipse
9. Parabola	**11.** Circle

Section 8.2 Solutions --

1. c Opens to the right	**3. d** Opens down
5. c Vertex is $(1,1)$, opens to the right	**7. a** Vertex is $(-1,-1)$, opens down

9. Vertex is $(0,0)$ and focus is $(0,3)$. So, the parabola opens up. As such, the general form is $x^2=4py$. In this case, $p=3$, so the equation is $\boxed{x^2=12y}$.

11. Vertex is $(0,0)$ and focus is $(-5,0)$. So, the parabola opens to the left. As such, the general form is $y^2=4px$. In this case, $p=-5$, so the equation is $\boxed{y^2=-20x}$.

13. Vertex is $(3,5)$ and focus is $(3,7)$. So, the parabola opens up. As such, the general form is $(x-3)^2=4p(y-5)$. In this case, $p=2$, so the equation is $\boxed{(x-3)^2=8(y-5)}$.

15. Vertex is $(2,4)$ and focus is $(0,4)$. So, the parabola opens to the left. As such, the general form is $(y-4)^2=4p(x-2)$. In this case, $p=-2$, so the equation is $\boxed{(y-4)^2=-8(x-2)}$.

17. Focus is $(2,4)$ and the directrix is $y=-2$. So, the parabola opens up. Since the distance between the focus and directrix is 6 and the vertex must occur halfway between them, we know $p=3$ and the vertex is $(2,1)$. Hence, the equation is $\boxed{(x-2)^2=4(3)(y-1)=12(y-1)}$.

19. Focus is $(3,-1)$ and the directrix is $x=1$. So, the parabola opens to the right. Since the distance between the focus and directrix is 2 and the vertex must occur halfway between them, we know $p=1$ and the vertex is $(2,-1)$. Hence, the equation is $\boxed{(y+1)^2=4(1)(x-2)=4(x-2)}$.

21. Since the vertex is $(-1,2)$ and the parabola opens to the right, the general form of its equation is $(y-2)^2 = 4p(x+1)$, for some $p > 0$. The fact that $(1,6)$ is on the graph implies:

$$\begin{aligned}(6-2)^2 &= 4p(2)\\ 16 &= 8p\\ 2 &= p\end{aligned}$$

Hence, the equation is $\boxed{(y-2)^2 = 8(x+1)}$.

23. Since the vertex is $(2,-1)$ and the parabola opens down, the general form of its equation is $(x-2)^2 = 4p(y+1)$, for some $p < 0$. The fact that $(6,-3)$ is on the graph implies:

$$\begin{aligned}(6-2)^2 &= 4p(-3+1)\\ 16 &= -8p\\ -2 &= p\end{aligned}$$

Hence, the equation is $\boxed{(x-2)^2 = -8(y+1)}$.

25.

Equation: $x^2 = 8y = 4(2)y$ **(1)**

So, $p = 2$ and opens up.

Vertex: $(0,0)$

Focus: $(0,2)$

Directrix: $y = -2$

Latus Rectum: Connects x-values corresponding to $y = 2$. Substituting this into **(1)** yields: $x^2 = 16$ so that $x = \pm 4$

So, the length of the latus rectum is 8.

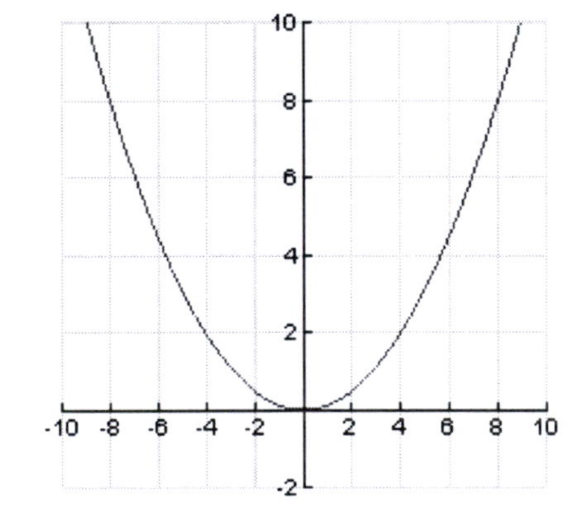

27.

Equation: $y^2 = -2x = 4(-\frac{1}{2})x$ **(1)**

So, $p = -\frac{1}{2}$ and opens to the left.

Vertex: $(0,0)$

Focus: $(-\frac{1}{2},0)$

Directrix: $x = \frac{1}{2}$

Latus Rectum: Connects y-values corresponding to $x = -\frac{1}{2}$. Substituting this into **(1)** yields: $y^2 = 1$ so that $y = \pm 1$

So, the length of the latus rectum is 2.

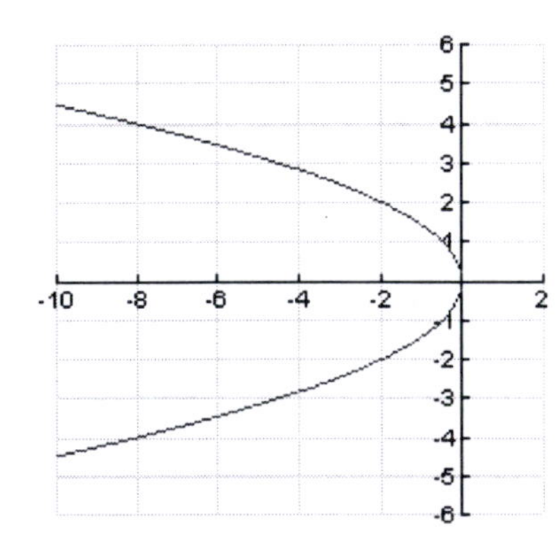

29.

Equation: $x^2 = 16y = 4(4)y$ **(1)**

So, $p = 4$ and opens up.

Vertex: $(0,0)$

Focus: $(0,4)$

Directrix: $y = -4$

Latus Rectum: Connects x-values corresponding to $y = 4$. Substituting this into **(1)** yields:

$$x^2 = 64 \text{ so that } x = \pm 8$$

So, the length of the latus rectum is 16.

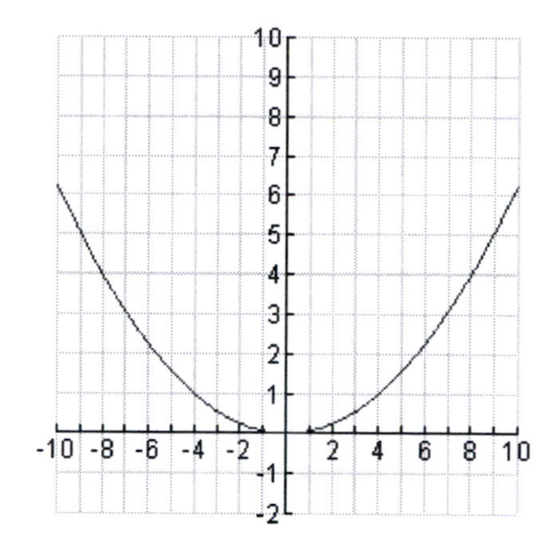

31.

Equation: $y^2 = 4x = 4(1)x$ **(1)**

So, $p = 1$ and opens to the right.

Vertex: $(0,0)$

Focus: $(1,0)$

Directrix: $x = -1$

Latus Rectum: Connects y-values corresponding to $x = 1$. Substituting this into **(1)** yields:

$$y^2 = 4 \text{ so that } y = \pm 2$$

So, the length of the latus rectum is 4.

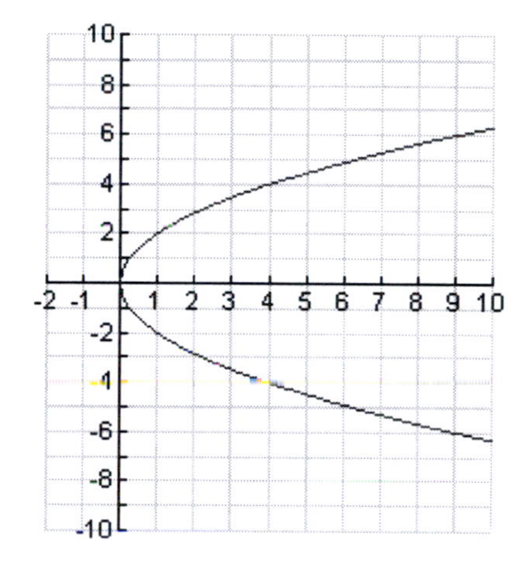

33.

Equation: $(y-2)^2 = 4(1)(x+3)$ **(1)**

So, $p = 1$ and opens to the right.

Vertex: $(-3,2)$

Focus: $(-2,2)$

Directrix: $x = -4$

Latus Rectum: Connects y-values corresponding to $x = -2$. Substituting this into **(1)** yields:

$$(y-2)^2 = 4 \text{ so that } y = 2 \pm 2 = 4,\ 0$$

So, the length of the latus rectum is 4.

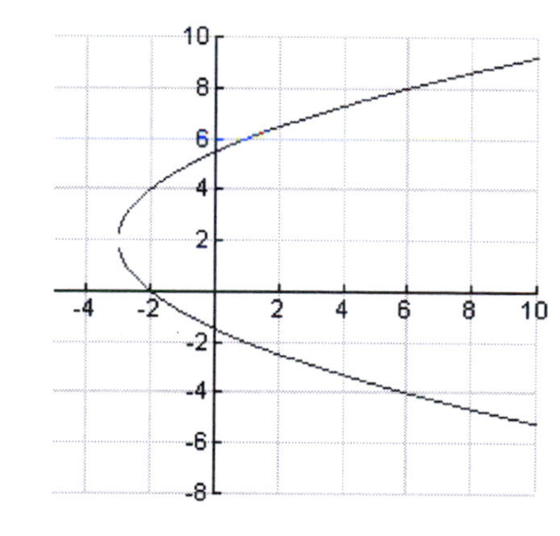

35.

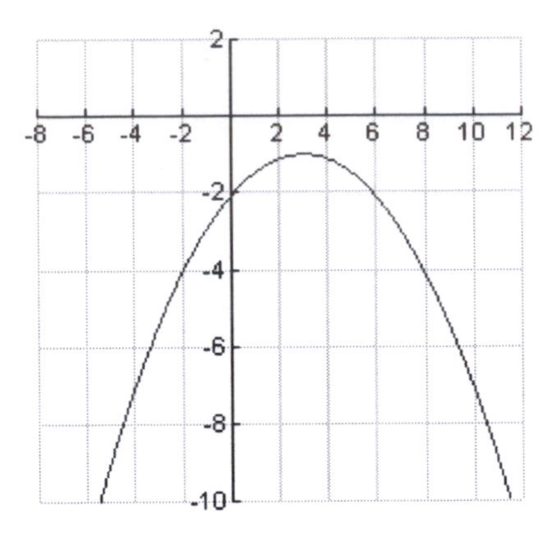

Equation: $(x-3)^2 = 4(-2)(y+1)$ **(1)**

So, $p = -2$ and opens down.

Vertex: $(3,-1)$

Focus: $(3,-3)$

Directrix: $y = 1$

Latus Rectum: Connects x-values corresponding to $y = -3$. Substituting this into **(1)** yields:

$$(x-3)^2 = 16 \text{ so that } x = 3 \pm 4 = -1,\ 7$$

So, the length of the latus rectum is 8.

37.

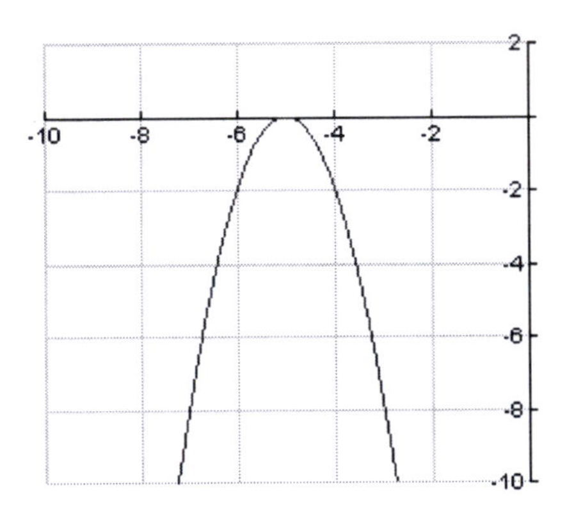

Equation: $(x+5)^2 = 4(-\frac{1}{2})(y-0)$ **(1)**

So, $p = -\frac{1}{2}$ and opens down.

Vertex: $(-5,0)$

Focus: $(-5,-\frac{1}{2})$

Directrix: $y = \frac{1}{2}$

Latus Rectum: Connects x-values corresponding to $y = -\frac{1}{2}$. Substituting this into **(1)** yields:

$$(x+5)^2 = 1 \text{ so that } x = -5 \pm 1 = -6,\ -4$$

So, the length of the latus rectum is 2.

39.

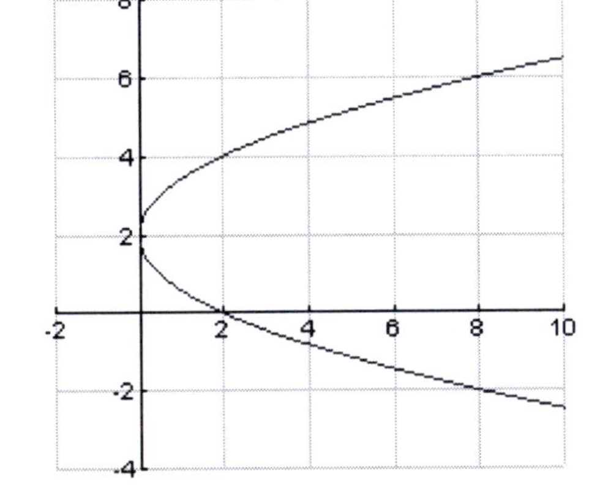

Equation: Completing the square yields:

$$y^2 - 4y - 2x + 4 = 0$$

$$(y^2 - 4y + 4) - 4 - 2x + 4 = 0$$

$$(y-2)^2 = 2x = 4(\tfrac{1}{2})x \quad \textbf{(1)}$$

So, $p = \frac{1}{2}$ and opens to the right.

Vertex: $(0,2)$

Focus: $(\frac{1}{2},2)$

Directrix: $x = -\frac{1}{2}$

Latus Rectum: Connects y-values corresponding to $x = \frac{1}{2}$. Substituting this into **(1)** yields:

$$(y-2)^2 = 1 \text{ so that } y = 2 \pm 1 = 1,\ 3$$

So, the length of the latus rectum is 2.

41.

Equation: Completing the square yields:

$$y^2+2y-8x-23=0$$

$$(y^2+2y+1)-1-8x-23=0$$

$$(y+1)^2=8x+24$$

$$(y+1)^2=8(x+3)$$

$$(y+1)^2=4(2)(x+3) \quad \textbf{(1)}$$

So, $p=2$ and opens to the right.

Vertex: $(-3,-1)$

Focus: $(-1,-1)$

Directrix: $x=-5$

Latus Rectum: Connects y-values corresponding to $x=-1$. Substituting this into **(1)** yields:

$(y+1)^2=16$ so that $y=-1\pm 4=3,-5$

So, the length of the latus rectum is 8.

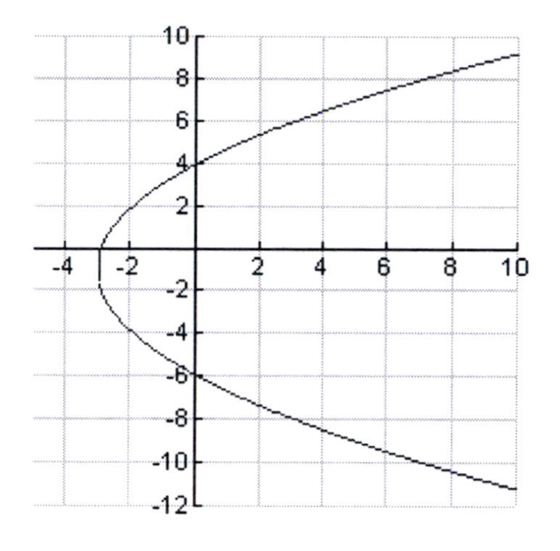

43.

Equation: Completing the square yields:

$$x^2-x+y-1=0$$

$$(x^2-x+\tfrac{1}{4})-\tfrac{1}{4}+y-1=0$$

$$(x-\tfrac{1}{2})^2=-y+\tfrac{5}{4}$$

$$(x-\tfrac{1}{2})^2=-(y-\tfrac{5}{4})$$

$$(x-\tfrac{1}{2})^2=4(-\tfrac{1}{4})(y-\tfrac{5}{4}) \quad \textbf{(1)}$$

So, $p=-\frac{1}{4}$ and opens down.

Vertex: $(\frac{1}{2},\frac{5}{4})$

Focus: $(\frac{1}{2},1)$

Directrix: $y=\frac{3}{2}$

Latus Rectum: Connects x-values corresponding to $y=1$. Substituting this into **(1)** yields:

$(x-\frac{1}{2})^2=\frac{1}{4}$ so that $x=\frac{1}{2}\pm\frac{1}{2}=1,0$

So, the length of the latus rectum is 1.

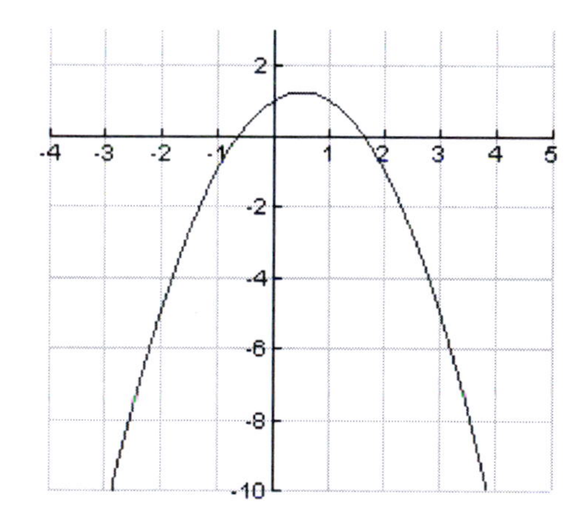

45. Assume the vertex is at (0,0). The distance from (0,0) to the opening of the dish is 2 feet. Identifying the opening as the latus rectum, we know that the focus will be at (0,2). So, the receiver should be placed 2 feet from the vertex.

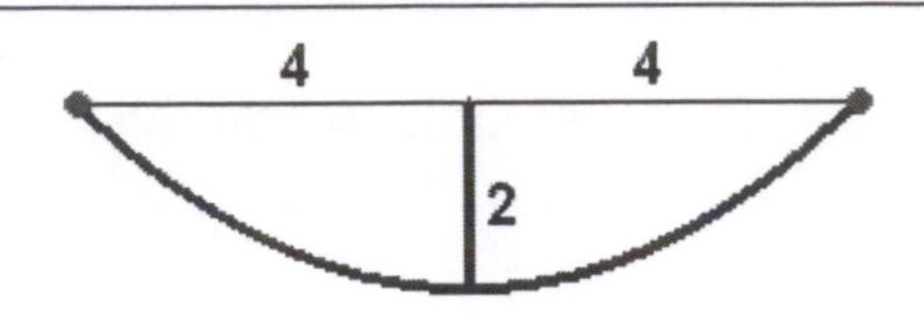

47. Assume the vertex is at the origin and that the parabola opens up. The general form of the equation, therefore, is $x^2 = 4py$. Since the focus is at (0,2), $p = 2$. Hence, the equation governing the shape of the lens is $x^2 = 8y$, or equivalently, $y = \frac{1}{8}x^2$. Since the latus rectum has length 5, the values of x are restricted to the interval $[-2.5, 2.5]$ for this model. Thus, the equation (with the restricted domain) is:

$$y = \tfrac{1}{8}x^2, \text{ for any } x \text{ in } [-2.5, 2.5].$$

Alternatively, assuming the dish opens to the right, we obtain a parabola whose equation is the same as above, but with the roles of x and y switched, namely,

$$x = \tfrac{1}{8}y^2, \text{ for any } y \text{ in } [-2.5, 2.5]$$

49. Assume the vertex is at (0,0) and that the parabola opens up. Since the rays are focused at (0, 40), we know that $p = 40$. Hence, the equation is $x^2 = 4(40)y = 160y$.

51. From the diagram, we can assume that the vertex is at (0, 20) and that the parabola opens down. Hence, the general form of the equation is $(x-0)^2 = 4p(y-20)$ **(1)**, for some $p<0$. Since the point $(-40,0)$ is on the graph, we plug it into **(1)** to find *p*: $$(-40-0)^2 = 4p(0-20)$$ $$1600 = -80p$$ $$-20 = p$$ So, the equation is $(x-0)^2 = 4(-20)(y-20)$ which simplifies to $x^2 = -80(y-20)$. Now, if the boat passes under the bridge 10 feet from the center, we need to compare the height of the bridge at $x=10$ (or $=-10$) to the height of the boat (which is 17 feet). Indeed, from the equation, the height of the bridge at $x=10$ is: $$(10)^2 = -80(y-20)$$ $$100 = -80y+1600$$ $$-1500 = -80y$$ $$18.75 = y$$ So, yes, the boat will pass under the bridge without scraping the mast.
53. The focal length is 374.25 feet. Since the standard form of the equation is $x^2 = 4py$, we see that $p = 374.25$. So, the equation is $x^2 = 1497y$.
55. Completing the square yields $$P(x) = -x^2 + 60x - 500$$ $$= -\left(x^2 - 60x\right) - 500$$ $$= -\left(x^2 - 60x + 900\right) - 500 + 900$$ $$= -(x-30)^2 + 400$$ So, the maximum profit of $400,000 is achieved when 3,000 units are produced.
57. If the vertex is at the origin and the focus is at (3,0), then the parabola must open to the right. So, the general equation is $y^2 = 4px$, for some $p>0$.
59. True
61. False. It is the same distance the focus is from the vertex, but on the opposite side.

63. Derive the equation for $x^2 = 4py$, where $p > 0$. Let (x, y) be any point on this parabola. Consult the diagram to the right. Using the distance formula, we find that

$$d_1 = \sqrt{(x-0)^2 + (y-p)^2}$$

$$d_2 = \sqrt{(x-x)^2 + (y-(-p))^2} = |y+p|$$

Equate d_1 and d_2 and simplify:

$$\sqrt{(x-0)^2 + (y-p)^2} = |y+p|$$

$$x^2 + (y-p)^2 = (y+p)^2$$

$$x^2 + y^2 - 2py + p^2 = y^2 + 2py + p^2$$

$$x^2 = 4py,$$

as desired.

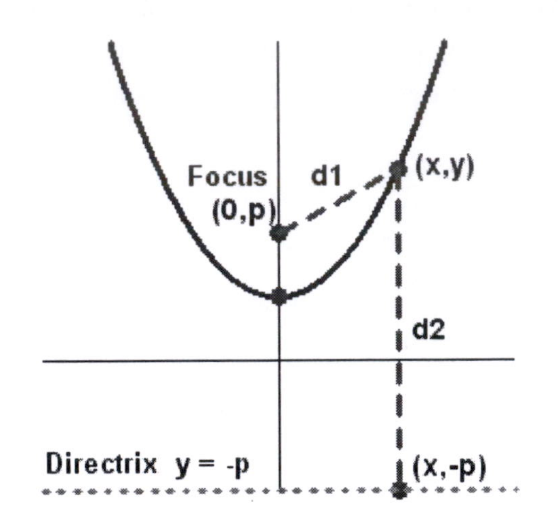

65.

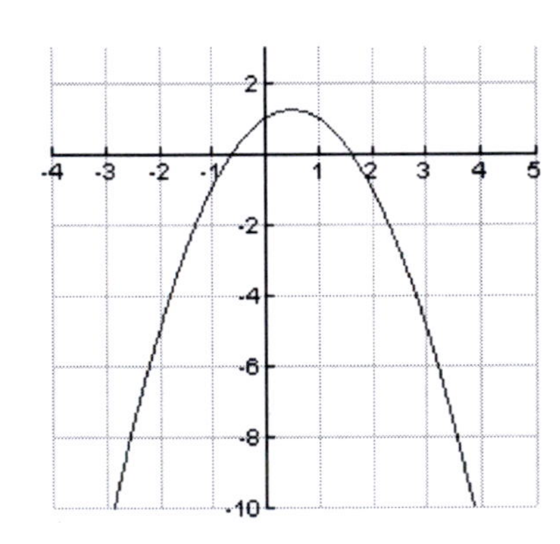

67. The vertex is $(2.5, -3.5)$. The parabola opens to the right since the square term is y^2 and $p > 0$. The graph of $(y+3.5)^2 = 10(x-2.5)$ is:

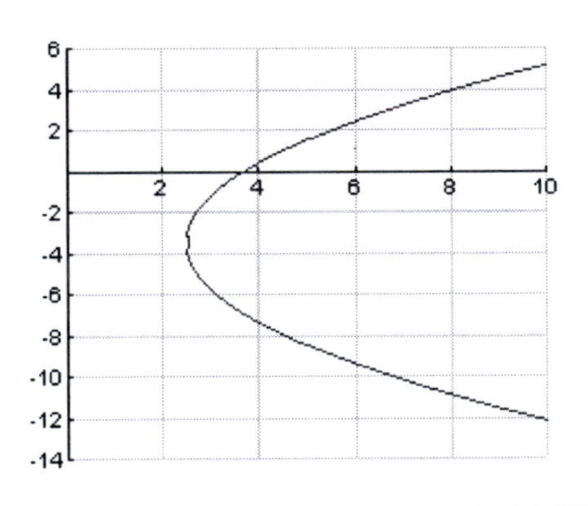

69. The vertex is at (1.8, 1.5) and it should open to the left. Graphing this parabola on the calculator confirms this:

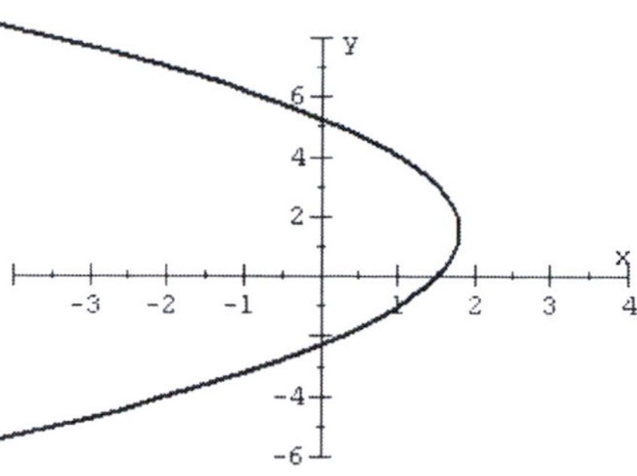

Section 8.3 Solutions ---

1. d x-axis is the major axis. The intercepts are $(6,0), (-6,0), (0,4), (0,-4)$.
3. a y-axis is the major axis. The intercepts are $(2\sqrt{2},0), (-2\sqrt{2},0), (0,6\sqrt{2}), (0,-6\sqrt{2})$.

5.	**7.**
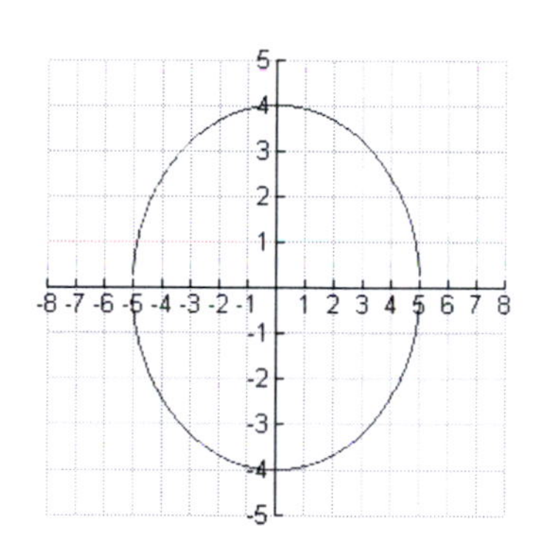	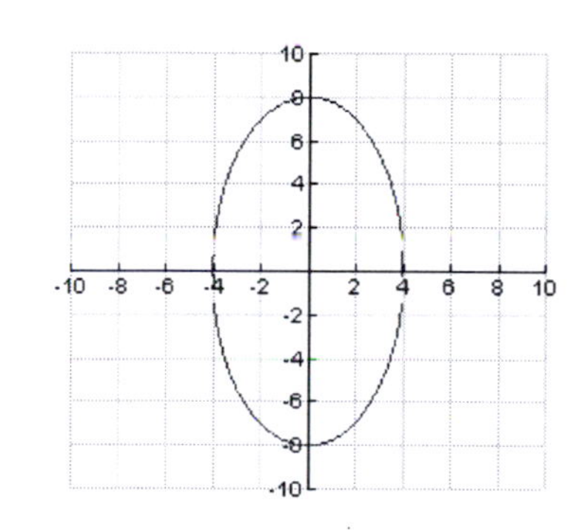
Center: (0,0) Vertices: $(\pm 5,0), (0,\pm 4)$	Center: (0,0) Vertices: $(\pm 4,0), (0,\pm 8)$

9.

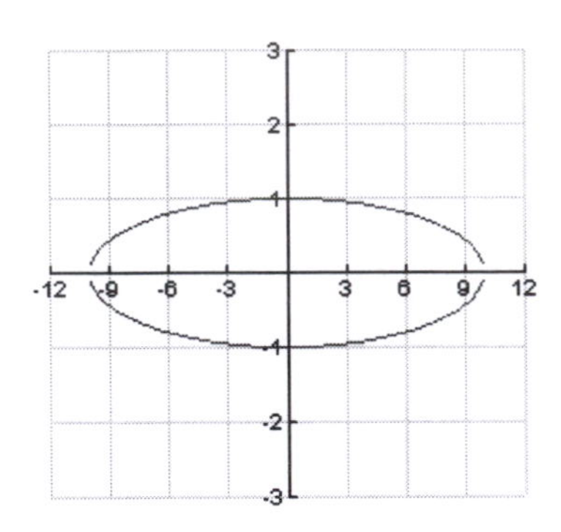

Center: (0,0) Vertices: $(\pm 10,0),(0,\pm 1)$

11.

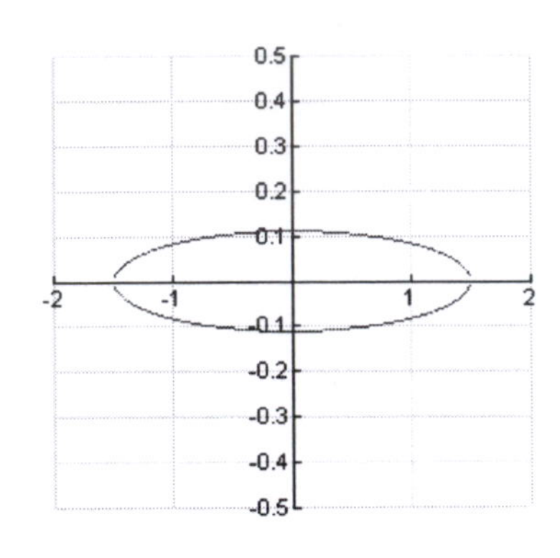

Center: (0,0) Vertices: $(\pm\frac{3}{2},0),(0,\pm\frac{1}{9})$

13. Write the equation in standard form as $\frac{x^2}{4}+\frac{y^2}{16}=1$.

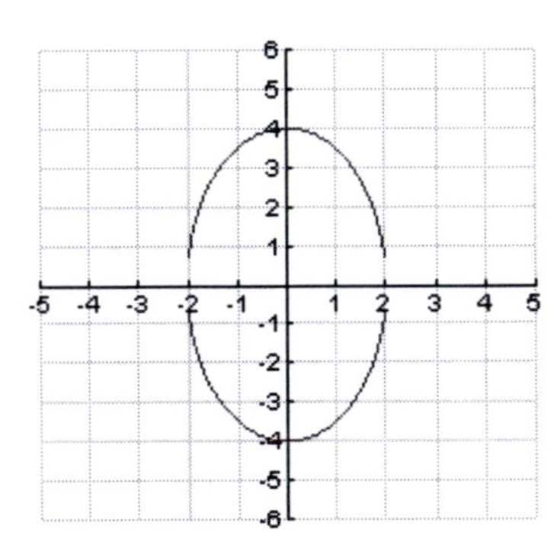

Center: (0,0) Vertices: $(\pm 2,0),(0,\pm 4)$

15. Write the equation in standard form as $\frac{x^2}{4}+\frac{y^2}{2}=1$.

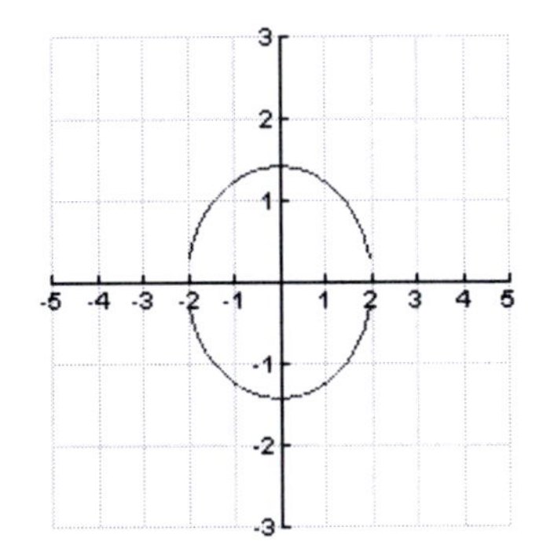

Center: (0,0) Vertices: $(\pm 2,0),\left(0,\pm\sqrt{2}\right)$

17. Since the foci are $(-4,0)$ and $(4,0)$, we know that $c=4$ and the center of the ellipse is $(0,0)$. Further, since the vertices are $(-6,0)$ and $(6,0)$ and they lie on the x-axis, the ellipse is horizontal and $a=6$. Hence,

$$c^2=a^2-b^2$$
$$16=36-b^2$$
$$b^2=20$$

Therefore, the equation of the ellipse is $\boxed{\frac{x^2}{36}+\frac{y^2}{20}=1}$.

19. Since the foci are $(0,-3)$ and $(0,3)$, we know that $c=3$ and the center of the ellipse is $(0,0)$. Further, since the vertices are $(0,-4)$ and $(0,4)$ and they lie on the y-axis, the ellipse is vertical and $a=4$. Hence,

$$c^2=a^2-b^2$$
$$9=16-b^2$$
$$b^2=7$$

Therefore, the equation of the ellipse is $\boxed{\frac{x^2}{7}+\frac{y^2}{16}=1}$.

21. Since the ellipse is centered at $(0,0)$, the length of the vertical major axis being 8 implies that $a=4$, and the length of the horizontal minor axis being 4 implies that $b=2$. Since the major axis is vertical, the equation of the ellipse must be $\boxed{\frac{x^2}{4}+\frac{y^2}{16}=1}$.

23. Since the vertices are $(0,-7)$, $(0,7)$, we know that the ellipse is centered at $(0,0)$ (since it is halfway between the vertices), the major axis is vertical, and $a=7$. Further, since the endpoints of the minor axis are $(-3,0)$, $(3,0)$, we know that $b=3$. Thus, the equation of the ellipse is $\boxed{\frac{x^2}{9}+\frac{y^2}{49}=1}$.

25. c Center is $(3,-2)$ and the major axis is vertical.

27. b Center is $(3,-2)$ and the major axis is horizontal.

29.

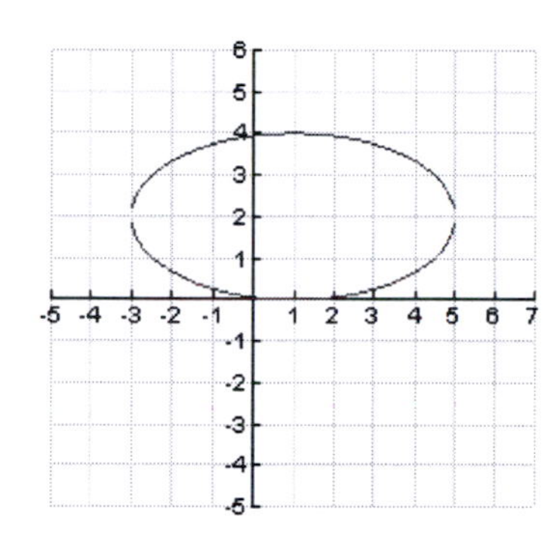

Center: (1,2) Vertices: (-3,2), (5,2), (1,0), (1,4)

31. First, write the equation in standard form: $\frac{(x+3)^2}{8}+\frac{(y-4)^2}{80}=1$

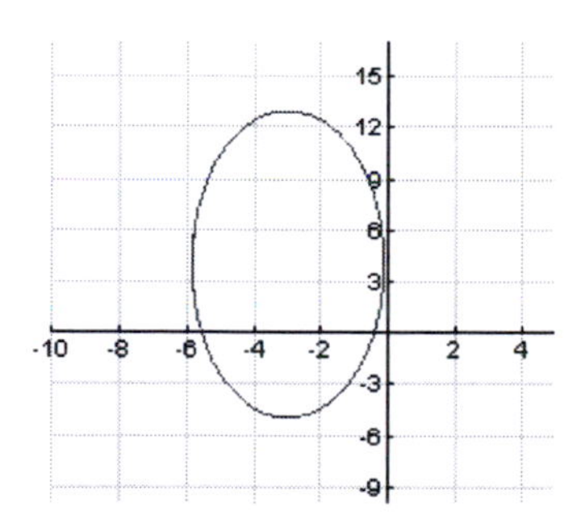

Center: (-3,4)

Vertices: $\left(-2\sqrt{2}-3,4\right)$, $\left(2\sqrt{2}-3,4\right)$, $\left(-3,4+4\sqrt{5}\right)$, $\left(-3,4-4\sqrt{5}\right)$

33. First, write the equation in standard form by completing the square:

$$x^2+4(y^2-6y)=-32$$

$$x^2+4(y^2-6y+9)=-32+36$$

$$x^2+4(y-3)^2=4 \Rightarrow \frac{x^2}{4}+(y-3)^2=1$$

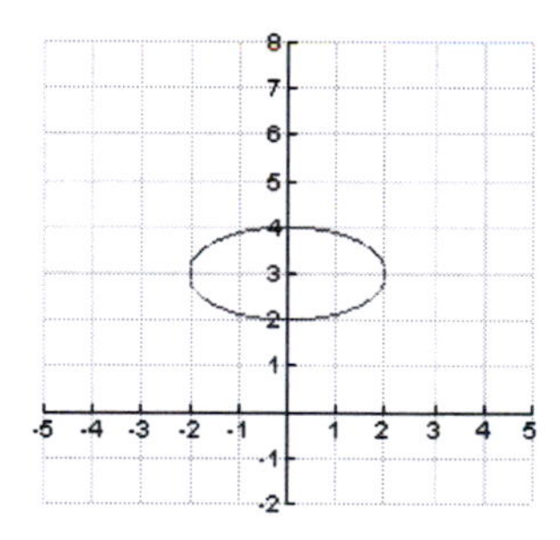

Center: (0,3)

Vertices: (-2,3), (2,3), (0,2), (0,4)

35. First, write the equation in standard form by completing the square:

$$(x^2-2x)+2(y^2-2y)=5$$

$$(x^2-2x+1)+2(y^2-2y+1)=5+1+2$$

$$(x-1)^2+2(y-1)^2=8$$

$$\frac{(x-1)^2}{8}+\frac{(y-1)^2}{4}=1$$

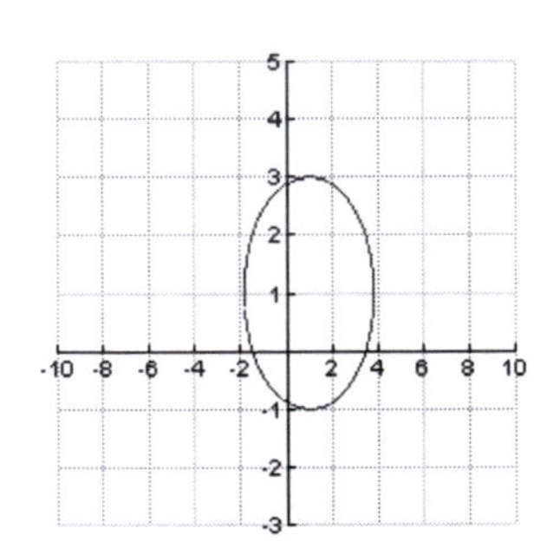

Center: (1,1)

Vertices: $\left(1\pm 2\sqrt{2},1\right)$, (1,3), (1,−1)

37. First, write the equation in standard form by completing the square:

$$5(x^2+4x)+(y^2+6y)=21$$

$$5(x^2+4x+4)+(y^2+6y+9)=21+20+9$$

$$5(x+2)^2+(y+3)^2=50$$

$$\frac{(x+2)^2}{10}+\frac{(y+3)^2}{50}=1$$

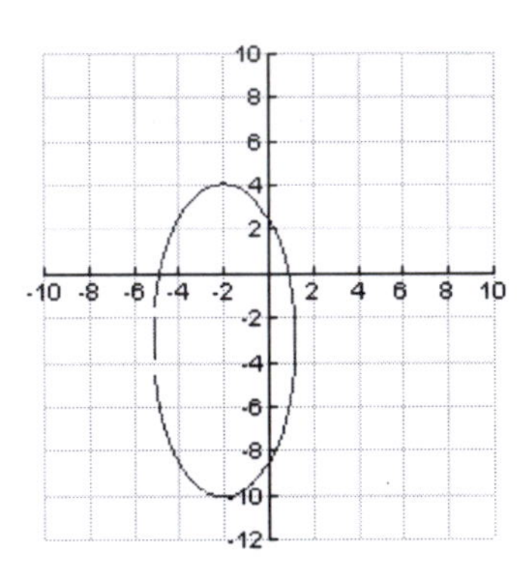

Center: (-2,-3)

Vertices: $\left(-2\pm\sqrt{10},-3\right)$, $\left(-2,-3\pm 5\sqrt{2}\right)$

39. Since the foci are $(-2,5)$, $(6,5)$ and the vertices are $(-3,5)$, $(7,5)$, we know:

i) the ellipse is horizontal with center at $(\frac{-3+7}{2},5)=(2,5)$,

ii) $2-c=-2$ so that $c=4$,

iii) the length of the major axis is $7-(-3)=10$. Hence, $a=5$.

Now, since $c^2=a^2-b^2$, $16=25-b^2$ so that $b^2=9$. Hence, the equation of the ellipse is $\boxed{\frac{(x-2)^2}{25}+\frac{(y-5)^2}{9}=1}$.

41. Since the foci are $(4,-7)$, $(4,-1)$ and the vertices are $(4,-8)$, $(4,0)$, we know:

i) the ellipse is vertical with center at $(4,\frac{-8+0}{2})=(4,-4)$,

ii) $-4-c=-7$ so that $c=3$,

iii) the length of the major axis is $0-(-8)=8$. Hence, $a=4$.

Now, since $c^2=a^2-b^2$, $9=16-b^2$ so that $b^2=7$. Hence, the equation of the ellipse is $\boxed{\frac{(x-4)^2}{7}+\frac{(y+4)^2}{16}=1}$.

43. Since the length of the major axis is 8, we know that $a=4$; since the length of the minor axis is 4, we know that $b=2$. As such, since the center is $(3,2)$ and the major axis is vertical, the equation of the ellipse must be $\boxed{\frac{(x-3)^2}{4}+\frac{(y-2)^2}{16}=1}$.

45. Since the vertices are $(-1,-9), (-1,1)$, we know that:
i) the ellipse is vertical with center at $(-1, \frac{-9+1}{2}) = (-1,-4)$,
ii) $a = \frac{1-(-9)}{2} = 5$.
Now, since the endpoints of the minor axis are $(-4,-4), (2,-4)$, we know that
$b = \frac{2-(-4)}{2} = 3$. Hence, the equation of the ellipse is $\boxed{\frac{(x+1)^2}{9} + \frac{(y+4)^2}{25} = 1}$.

47. Since the major axis has length 150 feet, we know that $a = 75$; since the minor axis has length 30 feet, we know that $b = 15$. From the diagram we know that the *y*-axis is the major axis. So, since the ellipse is centered at the origin, its equation must be
$\frac{x^2}{225} + \frac{y^2}{5625} = 1$

49. a. Since the major axis has length 150 yards, we know that $a = 75$; since the minor axis has length 40 yards, we know that $b = 20$. So, since the ellipse is centered at the origin, its equation must be $\frac{x^2}{5625} + \frac{y^2}{400} = 1$.
b. Since the football field is 120 yards long and it is completely surrounded by an elliptical track, we can compute the equation in part **a.** at $x = 60$ to find the corresponding *y*-values, and then subtract them to find the width at that particular position in the field:

$$y^2 = 400(1 - \tfrac{3600}{5625}) = \tfrac{400(2025)}{5625} = 144 \text{ so that } y = \pm 12.$$

Hence, the width of the track at the end of the football field is 24 yards. Since the field itself is 30 yards wide, the track will NOT encompass the field.

51. Assume that the sun is at the origin. Then, the vertices of Pluto's horizontal elliptical trajectory are: $(-4.447\times10^8, 0), (7.38\times10^8, 0)$
From this, we know that:
i) The length of the major axis is $7.38\times10^8 - (-4.447\times10^8) = 11.827\times10^8$, and so the value of a is half of this, namely 5.9135×10^8;
ii) The center of the ellipse is $\left(\frac{7.38\times10^8 + (-4.447\times10^8)}{2}, 0\right) = (1.4665\times10^8, 0)$;
iii) The value of c is 1.4665×10^8.
Now, since $c^2 = a^2 - b^2$, we have $\left(1.4665\times10^8\right)^2 = \left(5.9135\times10^8\right)^2 - b^2$, so that $b^2 = 3.2818\times10^{17}$. Hence, the equation of the ellipse must be

$$\boxed{\frac{\left(x-(1.4665\times10^8)\right)^2}{3.4969\times10^{17}} + \frac{(y-0)^2}{3.2818\times10^{17}} = 1}.$$

(Note: There is more than one correct way to set up this problem, and it begins with choosing the center of the ellipse. If, alternatively, you choose the center to be at (0,0), then the actual coordinates of the vertices and foci will change, and will result in the following slightly different equation for the trajectory: $\frac{x^2}{5{,}914{,}000{,}000^2} + \frac{y^2}{5{,}720{,}000{,}000^2} = 1$
Both are equally correct!.)

53. The length of the semi-major axis = $a = 150\times10^6$, so that $a^2 = 2.25\times10^{16}$.
Observe that the formula for eccentricity can be rewritten as follows:

$$e^2 = 1-\tfrac{b^2}{a^2}$$
$$\tfrac{b^2}{a^2} = 1-e^2$$
$$b^2 = a^2(1-e^2)$$
$$b = a\sqrt{1-e^2}$$

Since e = 0.223 and (from above) a = 150,000,000, we find that b = 146,000,000.

Hence, the equation of the ellipse is $\boxed{\frac{x^2}{150{,}000{,}000^2}+\frac{y^2}{146{,}000{,}000^2}=1}$.

55. It approaches the graph of a straight line. Indeed, note that

$$1=\sqrt{1-\frac{b^2}{a^2}} \Rightarrow 1-\frac{b^2}{a^2}=1 \Rightarrow \frac{b^2}{a^2}=0 \Rightarrow b=0.$$

Hence, $a^2x^2+b^2y^2=a^2b^2$ becomes $a^2x^2=0$, and since $a\neq 0$, we conclude that $x=0$.

57. a. Assuming the center of the trainer is located at (0,0), we must identify a and b such that the ellipse traced by the pedals is of the form $\frac{x^2}{a^2}+\frac{y^2}{b^2}=1$.

The x-intercepts are $\pm a$, and the distance between them (namely $a - (-a) = 2a$) should be 16. So, $2a = 16$, so that $a = 8$.

The minimum and maximum step heights occur at 2.5 and 12.5, respectively. So, the distance between them is 12.5 – 2.5 = 10. Hence, $2b = 10$, so that $b = 5$.

Thus, the equation of the ellipse is $\frac{x^2}{64}+\frac{y^2}{25}=1$.

b. $p=\pi\sqrt{2(64+25)}\approx 42$ inches

c. 1 step = 42 inches, and 1 mile = 5280 feet = 5280(12) inches = 63,360 inches.
So, the number of steps for 1 mile is $\frac{63{,}360}{42}\approx 1{,}509$.

59. It should be $a^2=6,\ b^2=4$, so that $a=\pm\sqrt{6},\ b=\pm 2$.

61. False. Since you could not deduce the coordinates of the endpoints of the minor axis from this information alone. Hence, you couldn't determine the value of b.

63. True. The equation would have the general form $\frac{x^2}{a^2}+\frac{y^2}{b^2}=1$, so that substituting in $-x$ for x and $-y$ for y yields the exact same equation.

65. Pluto: $e=\frac{1.4665\times10^8}{5.9135\times10^8}\cong 0.25$ (from Problem 51)

Earth: $e=\frac{2.75\times10^6}{1.4985\times10^8}\cong 0.02$ (from Problem 52)

67. Note that from the graphs, we see that as c increases, the graph of the ellipse described by $x^2+cy^2=1$ narrows in the y-direction and becomes more elongated.

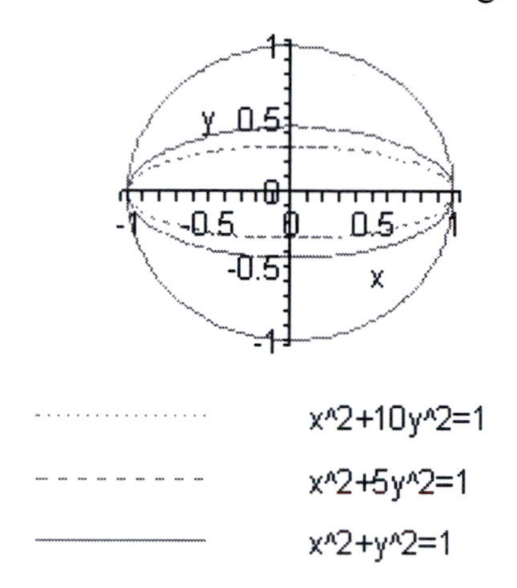

69. Note that from the graphs, we see that as c increases, the graph of the ellipse described by $cx^2+cy^2=1$, which is equivalent to $x^2+y^2=\frac{1}{c}$ (a circle) shrinks towards the origin. (This makes sense since the radius is $\frac{1}{c}$, and as c gets larger, the value of $\frac{1}{c}$ goes to 0.)

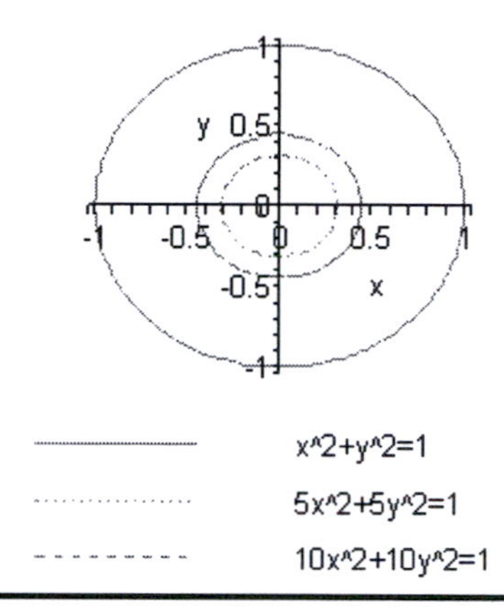

71. Note that from the graphs, we see that as c decreases, the major axis (along the x-axis) of the ellipse described by $cx^2+y^2=1$ becomes longer.

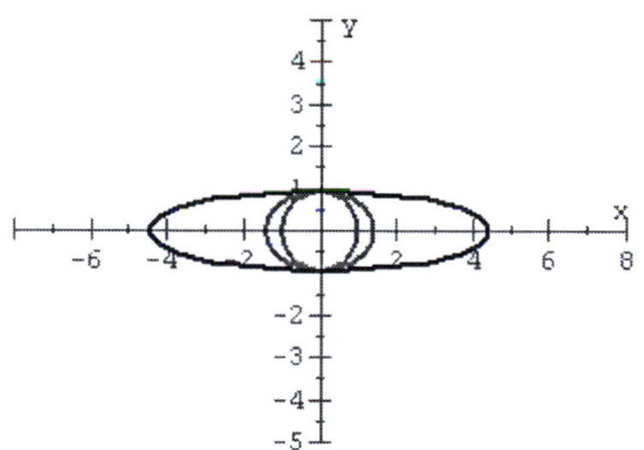

Section 8.4 Solutions --

1. b The transverse axis is the x-axis, so that the hyperbola will open left/right. The vertices are $(-6,0)$, $(6,0)$.

3. d The transverse axis is the x-axis, so that the hyperbola will open left/right. The vertices are $(-2\sqrt{2},0)$, $(2\sqrt{2},0)$.

5.

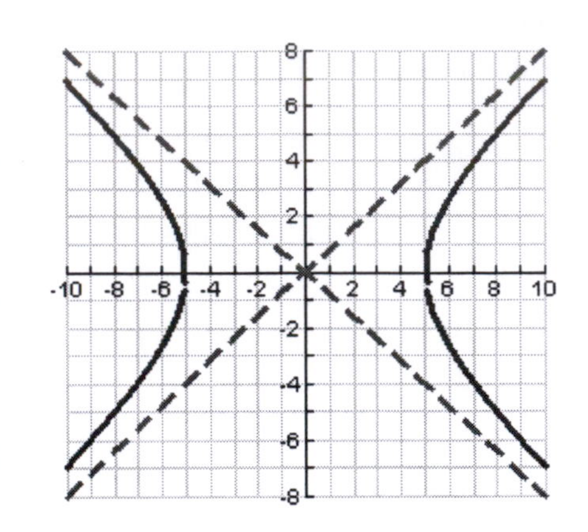

Notes on the Graph:
Equations of the asymptotes are $y = \pm\frac{4}{5}x$.

7.

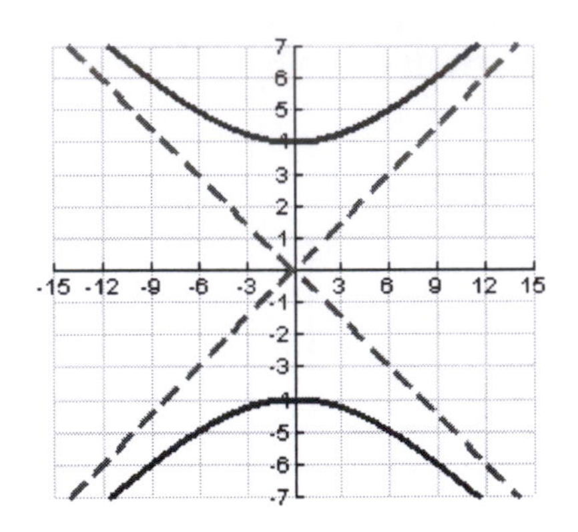

Notes on the Graph:
Equations of the asymptotes are $y = \pm\frac{1}{2}x$.

9.

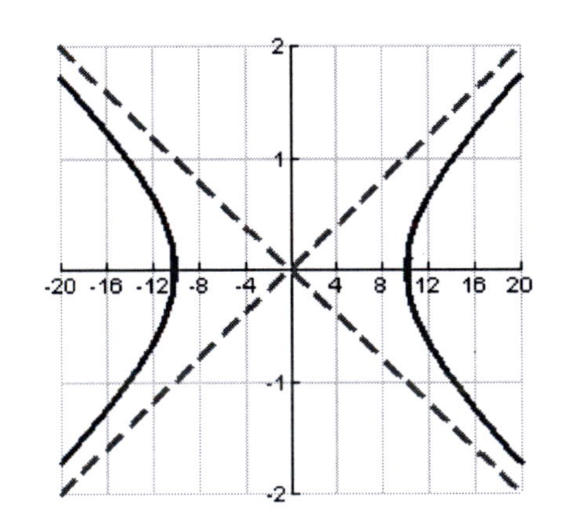

Notes on the Graph:
Equations of the asymptotes are $y = \pm\frac{1}{10}x$

11.

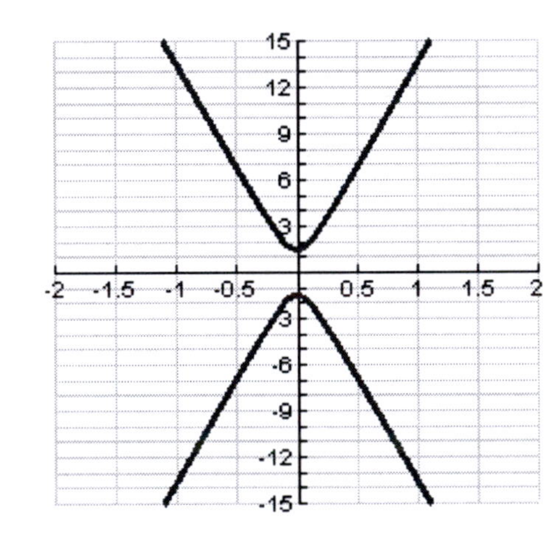

Notes on the Graph:
First, write the equation in standard form as $\frac{y^2}{\frac{9}{4}} - \frac{x^2}{\frac{1}{81}} = 1$. The equations of the asymptotes are $y = \pm\frac{27}{2}x$.

13.

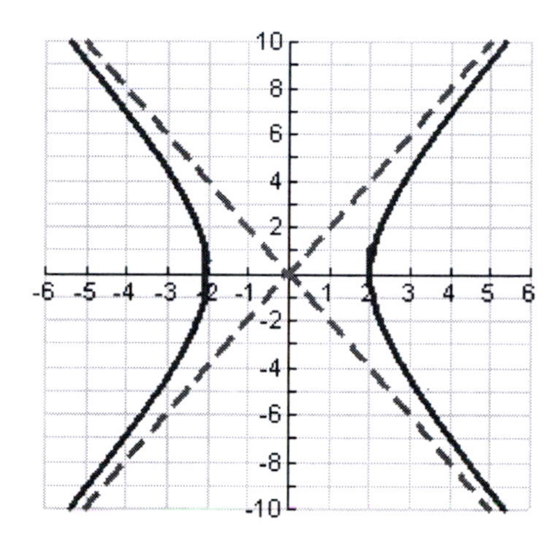

Notes on the Graph:
First, write the equation in standard form as $\frac{x^2}{4} - \frac{y^2}{16} = 1$. The equations of the asymptotes are $y = \pm 2x$.

15.

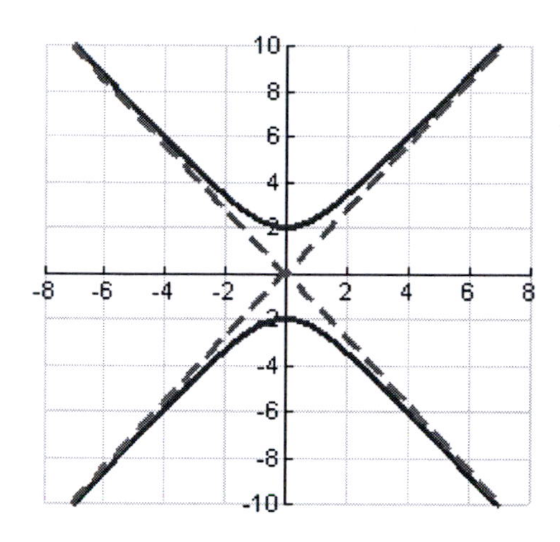

Notes on the Graph:
First, write the equation in standard form as $\frac{y^2}{4} - \frac{x^2}{2} = 1$. The equations of the asymptotes are $y = \pm \frac{2}{\sqrt{2}} x$.

17. Since the foci are $(-6,0), (6,0)$ and the vertices are $(-4,0), (4,0)$, we know that:
i) the hyperbola opens right/left with center at $(0,0)$, and ii) $a = 4, c = 6$,
Now, since $c^2 = a^2 + b^2$, $36 = 16 + b^2$ so that $b^2 = 20$. Hence, the equation of the hyperbola is $\boxed{\frac{x^2}{16} - \frac{y^2}{20} = 1}$.

19. Since the foci are $(0,-4), (0,4)$ and the vertices are $(0,-3), (0,3)$, we know that:
i) the hyperbola opens up/down with center at $(0,0)$, and ii) $a = 3, c = 4$,
Now, since $c^2 = a^2 + b^2$, $16 = 9 + b^2$ so that $b^2 = 7$. Hence, the equation of the hyperbola is $\boxed{\frac{y^2}{9} - \frac{x^2}{7} = 1}$.

21. Since the center is $(0,0)$ and the transverse axis is the x-axis, the general form of the equation of this hyperbola is $\frac{x^2}{a^2} - \frac{y^2}{b^2} = 1$ with asymptotes $y = \pm \frac{b}{a} x$. In this case, $\frac{b}{a} = 1$ so that $a = b$. Thus, the equation simplifies to $\frac{x^2}{a^2} - \frac{y^2}{a^2} = 1$, or $\boxed{x^2 - y^2 = a^2}$.

23. Since the center is $(0,0)$ and the transverse axis is the y-axis, the general form of the equation of this hyperbola is $\frac{y^2}{a^2} - \frac{x^2}{b^2} = 1$ with asymptotes $y = \pm \frac{a}{b} x$. In this case, $\frac{a}{b} = 2$ so that $a = 2b$. Thus, the equation simplifies to $\frac{y^2}{(2b)^2} - \frac{x^2}{b^2} = 1$, or equivalently $\boxed{\frac{y^2}{4} - x^2 = b^2}$.

25. c The transverse axis is parallel to the x-axis, so that the hyperbola will open left/right. The vertices are $(1,-2)$, $(5,-2)$ and the center is $(3,-2)$.
27. b The transverse axis is parallel to the y-axis, so that the hyperbola will open up/down. The vertices are $(-2,8)$, $(-2,-2)$ and the center is $(-2,3)$.

29.

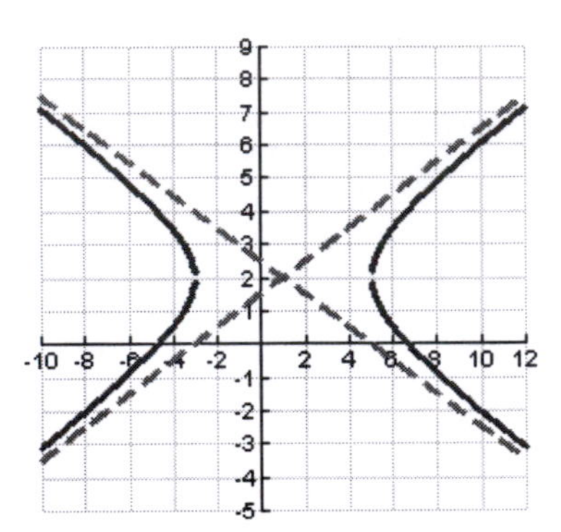

Notes on the Graph: The equations of the asymptotes are $y = \pm\frac{1}{2}(x-1)+2$.

31.

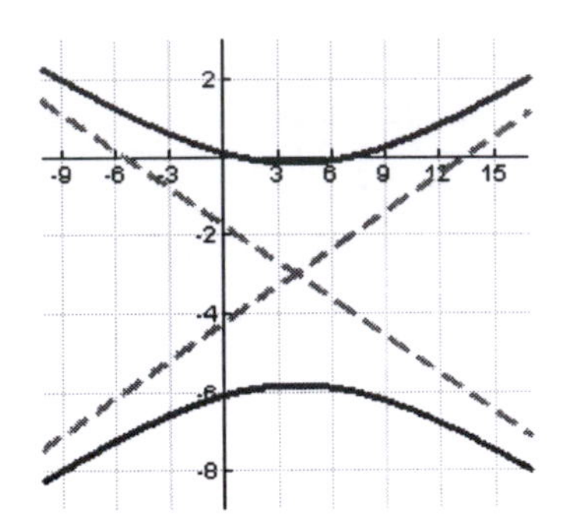

Notes on the Graph: The equations of the asymptotes are $y = \pm\frac{\sqrt{10}}{10}(x-4)-3$.

33.

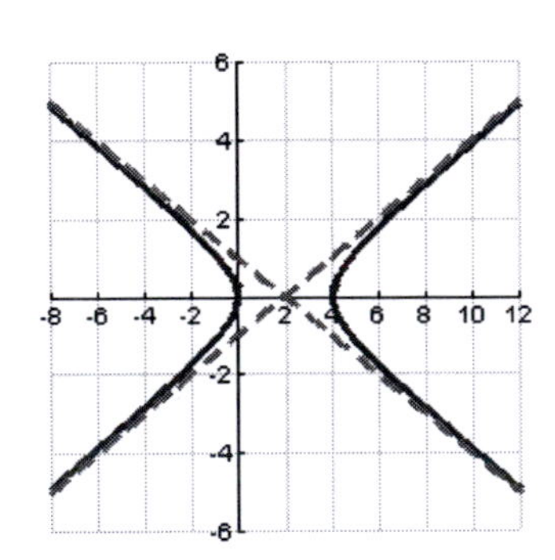

Notes on the Graph:
First, write the equation in standard form by completing the square:

$$x^2 - 4x - 4y^2 = 0$$

$$\left(x^2 - 4x + 4\right) - 4y^2 = 0 + 4$$

$$(x-2)^2 - 4y^2 = 4$$

$$\frac{(x-2)^2}{4} - y^2 = 1$$

The equations of the asymptotes are $y = \pm\frac{1}{2}(x-2)$.

35.

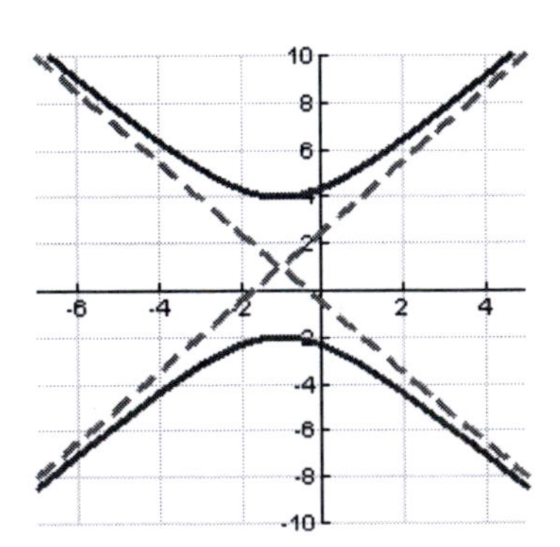

Notes on the Graph:
First, write the equation in standard form by completing the square:

$$-9\left(x^2 + 2x\right) + 4\left(y^2 - 2y\right) = 41$$

$$-9\left(x^2 + 2x + 1\right) + 4\left(y^2 - 2y + 1\right) = 41 - 9 + 4$$

$$-9(x+1)^2 + 4(y-1)^2 = 36$$

$$-\frac{(x+1)^2}{4} + \frac{(y-1)^2}{9} = 1$$

$$\frac{(y-1)^2}{9} - \frac{(x+1)^2}{4} = 1$$

The equations of the asymptotes are $y = \pm\frac{3}{2}(x+1)+1$.

37.

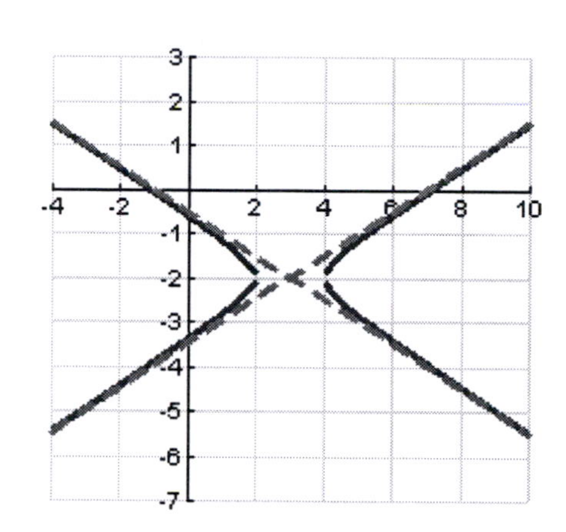

Notes on the Graph:
First, write the equation in standard form by completing the square:

$$\left(x^2-6x\right)-4\left(y^2+4y\right)=8$$

$$\left(x^2-6x+9\right)-4\left(y^2+4y+4\right)=8+9-16$$

$$(x-3)^2-4(y+2)^2=1$$

$$(x-3)^2-\frac{(y+2)^2}{\frac{1}{4}}=1$$

The equations of the asymptotes are $y=\pm\frac{1}{2}(x-3)-2$.

39. Since the vertices are $(-2,5),\ (6,5)$, we know that the center is $(2,5)$ and the transverse axis is parallel to the x-axis. Hence, $2-a=-2$ so that $a=4$. Also, since the foci are $(-3,5),\ (7,5)$, we know that $2+c=-3$ so that $c=-5$. Now, to find b, we substitute the values of c and a obtained above into $c^2=a^2+b^2$ to see that $25=16+b^2$ so that $b^2=9$. Hence, the equation of the hyperbola is $\boxed{\frac{(x-2)^2}{16}-\frac{(y-5)^2}{9}=1}$.

41. Since the vertices are $(4,-7),\ (4,-1)$, we know that the center is $(4,-4)$ and the transverse axis is parallel to the y-axis. Hence, $-4-a=-7$ so that $a=3$. Also, since the foci are $(4,-8),\ (4,0)$, we know that $-4-c=-8$ so that $c=4$. Now, to find b, we substitute the values of c and a obtained above into $c^2=a^2+b^2$ to see that $16=9+b^2$ so that $b^2=7$. Hence, the equation of the hyperbola is $\boxed{\frac{(y+4)^2}{9}-\frac{(x-4)^2}{7}=1}$.

43. Assume that the stations coincide with the foci and are located at $(-75,0),\ (75,0)$. The difference in distance between the ship and each of the two stations must remain constantly $2a$, where $(a,0)$ is the vertex. Assume that the radio signal speed is 186,000 $^{mi}/_{sec}$ and the time difference is 0.0005 sec. Then, using distance = rate × time, we obtain:

$$2a=(186,000)(0.0005)=93 \text{ so that } a=46.5$$

So, the ship will come ashore between the two stations 28.5 miles from one and 121.5 miles from the other.

45. Here, we want $a=45$. So, $2a=90$. Using the same radio speed, observe that $90=186,000(t)$, so that $t=0.000484$ sec.

47. Assume that the vertices are $(0,a), (0,-a)$. Then, $2a=2$ so that $a=1$, and the center is at (0,0). Further, the transverse axis is parallel to the *y*-axis. Since the foci are $(0,1.5), (0,-1.5)$, we know that $c=1.5$. Thus, to find *b*, we substitute these values of *a* and *c* into $c^2=a^2+b^2$ to see that $(1.5)^2=1^2+b^2$, and so $b^2=1.25$. Hence, the equation of the hyperbola is $y^2-\frac{x^2}{1.25}=1$, which is equivalent to $\boxed{y^2-\frac{4}{5}x^2=1}$.
49. Find the *x* values when *y* = 150 (which correspond to the top of the cooling tower): $$\frac{x^2}{8{,}100}-\frac{150^2}{16{,}900}=1$$ $$\frac{x^2}{8{,}100}=1+\frac{22{,}500}{16{,}900}=\frac{39{,}400}{16{,}900}$$ $$x^2=8{,}100\left(\frac{39{,}400}{16{,}900}\right)\approx 18{,}884.02367$$ $$x=\pm\sqrt{18{,}884.02367}\approx\pm 137.4$$ So, the diameter is 137.4 – (-137.4) = 275 feet.
51. The transverse axis should be vertical. The points are $(3,0), (-3,0)$ and the vertices are $(0,2), (0,-2)$.
53. False. You won't be able to find the value of *b* in such case without more information.
55. True. Since the general forms of the equations are $\frac{x^2}{a^2}-\frac{y^2}{b^2}=1$ and $\frac{y^2}{a^2}-\frac{x^2}{b^2}=1$, substituting $-x$ for *x* and $-y$ for *y* doesn't change the equation – hence, the symmetry follows.
57. Assume the center of the hyperbola is (0,0). Assume that the two asymptotes are perpendicular. We separate our discussion into two cases, depending on which axis is the transverse axis. <u>Case 1</u>: Here, the equations of the asymptotes are $y=\pm\frac{a}{b}x$. Hence, in order for them to be perpendicular, the products of their slopes must be -1. So, we have $\left(\frac{a}{b}\right)\left(-\frac{a}{b}\right)=-1$, which simplifies to $-\frac{a^2}{b^2}=-1$ so that $a^2=b^2$. Hence, in such case, the equation is $\frac{y^2}{a^2}-\frac{x^2}{a^2}=1$, which is equivalent to $y^2-x^2=a^2$. <u>Case 2</u>: Here, the equations of the asymptotes are $y=\pm\frac{b}{a}x$. Hence, in order for them to be perpendicular, the products of their slopes must be -1. So, we have $\left(\frac{b}{a}\right)\left(-\frac{b}{a}\right)=-1$, which simplifies to $-\frac{b^2}{a^2}=-1$ so that $a^2=b^2$. Hence, in such case, the equation is $\frac{x^2}{a^2}-\frac{y^2}{a^2}=1$, which is equivalent to $x^2-y^2=a^2$.

59. Observe from the graphs that as c increases, the graphs of the hyperbolas described by $x^2 - cy^2 = 1$ become more and more squeezed down towards the x-axis.

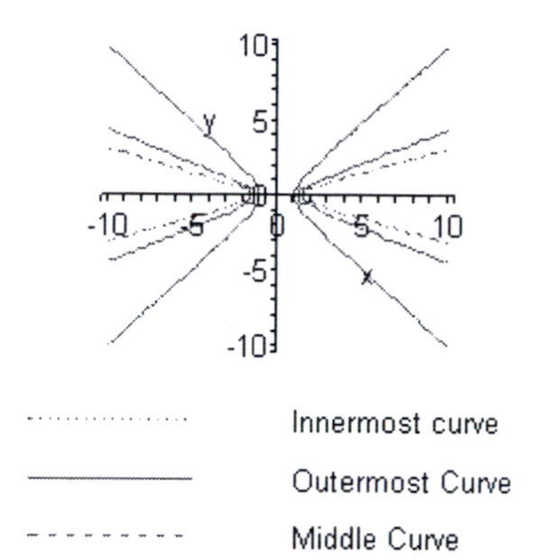

61. Note from the graphs that as c decreases, the vertices of the hyperbola whose equation is given by $cx^2 - y^2 = 1$ are located at $\left(\pm\frac{1}{c}, 0\right)$, and are moving away from the origin:

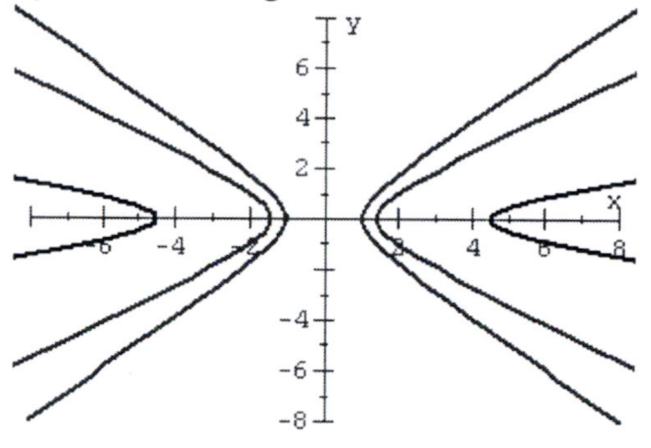

Section 8.5 Solutions ---

1. Solve the system $\begin{cases} x^2 - y = -2 & \textbf{(1)} \\ -x + y = 4 & \textbf{(2)} \end{cases}$

Add **(1)** and **(2)** to get an equation in terms of x:

$$x^2 - x - 2 = 0$$
$$(x-2)(x+1) = 0$$
$$x = 2, -1$$

Now, substitute each value of x back into **(2)** to find the corresponding values of y:

$$x = 2: \quad -2 + y = 4 \text{ so that } y = 6$$
$$x = -1: \quad -(-1) + y = 4 \text{ so that } y = 3$$

So, the solutions are $\boxed{(2,6) \text{ and } (-1,3)}$.

3. Solve the system $\begin{cases} x^2 + y = 1 & \textbf{(1)} \\ 2x + y = 2 & \textbf{(2)} \end{cases}$

Multiply **(2)** by -1, and then add to **(1)** to get an equation in terms of x:

$$x^2 - 2x + 1 = 0$$
$$(x-1)^2 = 0$$
$$x = 1$$

Now, substitute this value of x back into **(1)** to find the corresponding value of y:

$$x = 1: \quad 1^2 + y = 1 \text{ so that } y = 0$$

So, the solution is $\boxed{(1,0)}$.

5. Solve the system $\begin{cases} x^2+y=-5 & \textbf{(1)} \\ -x+y=3 & \textbf{(2)} \end{cases}$

Multiply **(2)** by -1 and then add to **(1)** to get an equation in terms of x:

$$x^2+x=-8$$
$$x^2+x+8=0$$
$$x=\tfrac{-1\pm\sqrt{1-4(8)}}{2}$$

Since the values of x are not real numbers, there is $\boxed{\text{no solution}}$ to this system.

7. Solve the system $\begin{cases} x^2+y^2=1 & \textbf{(1)} \\ x^2-y=-1 & \textbf{(2)} \end{cases}$

Multiply **(2)** by -1, and then add to **(1)** to get an equation in terms of y:

$$y^2+y=2$$
$$y^2+y-2=0$$
$$(y+2)(y-1)=0$$
$$y=-2,\ 1$$

Now, substitute each value of y back into **(2)** to find the corresponding values of x:

$y=-2:\quad x^2-(-2)=-1$ so that $x^2=-3$ (No real solutions)

$y=1:\quad x^2-1=-1$

$x^2=0$ so that $x=0$

So, the solution is $\boxed{(0,1)}$.

9. Solve the system $\begin{cases} x^2+y^2=3 & \textbf{(1)} \\ 4x^2+y=0 & \textbf{(2)} \end{cases}$

Multiply **(1)** by -4, and then add to **(2)** to get an equation in terms of y:

$$-4y^2+y=-12$$
$$4y^2-y-12=0$$
$$y=\tfrac{1\pm\sqrt{1-4(4)(-12)}}{2(4)}=\tfrac{1\pm\sqrt{193}}{8}$$
$$\cong 1.862,\ -1.612$$

Now, substitute each value of y back into **(2)** to find the corresponding values of x:

$y=1.862:\quad 4x^2+1.862=0$ so that $x^2=-0.466$ (No real solutions)

$y=-1.612:\quad 4x^2-1.612=0$

$x^2=0.403$ so that $x\cong\pm0.635$

So, the solutions are $\boxed{(0.63,-1.61) \text{ and } (-0.63,-1.61)}$.

11. Solve the system $\begin{cases} x^2 + y^2 = -6 & \textbf{(1)} \\ -2x^2 + y = 7 & \textbf{(2)} \end{cases}$

Multiply **(1)** by 2, and then add to **(2)** to get an equation in terms of y:

$$2y^2 + y = -5$$

$$2y^2 + y + 5 = 0$$

$$y = \frac{-1 \pm \sqrt{1-4(2)(5)}}{2(2)}$$

Since the values of y are not real numbers, there is $\boxed{\text{no solution}}$ to this system.

13. Solve the system $\begin{cases} x + y = 2 & \textbf{(1)} \\ x^2 + y^2 = 2 & \textbf{(2)} \end{cases}$

Solve **(1)** for y: $y = 2 - x$ **(3)**

Substitute **(3)** into **(2)** to get an equation in terms of x:

$$x^2 + 4 - 4x + x^2 = 2$$

$$2x^2 - 4x + 2 = 0$$

$$2(x-1)^2 = 0$$

$$x = 1$$

Now, substitute this value of x back into **(3)** to find the corresponding value of y:

$$y = 2 - (1) = 1$$

So, the solution is $\boxed{(1,1)}$.

15. Solve the system $\begin{cases} xy = 4 & \textbf{(1)} \\ x^2 + y^2 = 10 & \textbf{(2)} \end{cases}$

Solve **(1)** for y: $y = \frac{4}{x}$ **(3)**

Substitute **(3)** into **(2)** to get an equation in terms of x:

$$x^2 + \left(\tfrac{4}{x}\right)^2 = 10$$

$$x^2 + \tfrac{16}{x^2} = 10$$

$$\frac{x^4 + 16}{x^2} = 10$$

$$\frac{x^4 + 16}{x^2} - \frac{10x^2}{x^2} = 0$$

$$\frac{x^4 - 10x^2 + 16}{x^2} = 0$$

$$\frac{\left(x^2 - 8\right)\left(x^2 - 2\right)}{x^2} = 0$$

$$x^2 - 8 = 0 \quad \text{or} \quad x^2 - 2 = 0$$

Hence, $x = \pm 2\sqrt{2}$ or $x = \pm\sqrt{2}$.

Now, substitute each of these values of x back into **(3)** to find the corresponding value of y:

$$x = 2\sqrt{2}: \quad y = \tfrac{4}{2\sqrt{2}} = \tfrac{2}{\sqrt{2}}$$

$$x = -2\sqrt{2}: \quad y = \tfrac{4}{-2\sqrt{2}} = -\tfrac{2}{\sqrt{2}}$$

$$x = \sqrt{2}: \quad y = \tfrac{4}{\sqrt{2}} = \tfrac{4}{\sqrt{2}}$$

$$x = -\sqrt{2}: \quad y = \tfrac{4}{-\sqrt{2}} = -\tfrac{4}{\sqrt{2}}$$

So, the solutions are (after rationalizing)

$$\boxed{\left(2\sqrt{2}, \sqrt{2}\right), \left(-2\sqrt{2}, -\sqrt{2}\right), \left(\sqrt{2}, 2\sqrt{2}\right), \text{ and } \left(-\sqrt{2}, -2\sqrt{2}\right)}.$$

17. Solve the system $\begin{cases} y = x^2 - 3 & \textbf{(1)} \\ y = -4x + 9 & \textbf{(2)} \end{cases}$

Substitute **(1)** into **(2)** to get an equation in terms of x:

$$x^2 - 3 = -4x + 9$$

$$x^2 + 4x - 12 = 0$$

$$(x + 6)(x - 2) = 0$$

$$x = -6,\ 2$$

Now, substitute this value of y back into **(1)** to find the corresponding value of y:

$$x = -6: \quad y = (-6)^2 - 3 = 33$$

$$x = 2: \quad y = (2)^2 - 3 = 1$$

So, the solutions are $\boxed{(2,1) \text{ and } (-6,33)}$.

19. Solve the system $\begin{cases} x^2 + xy - y^2 = 5 & \textbf{(1)} \\ x - y = -1 & \textbf{(2)} \end{cases}$

Solve **(2)** for x: $x = y - 1$ **(3)**

Substitute **(3)** into **(1)** to get an equation in terms of y:

$$\begin{aligned} (y-1)^2 + (y-1)y - y^2 &= 5 \\ y^2 - 2y + 1 + y^2 - y - y^2 &= 5 \\ y^2 - 3y - 4 &= 0 \\ (y-4)(y+1) &= 0 \\ y &= 4, \ -1 \end{aligned}$$

Now, substitute these values of y back into **(3)** to find the corresponding values of x:

$$y = 4: \quad x = 4 - 1 = 3$$

$$y = -1: \quad x = -1 - 1 = -2$$

So, the solutions are $\boxed{(3,4) \text{ and } (-2,-1)}$.

21. Solve the system $\begin{cases} 2x - y = 3 & \textbf{(1)} \\ x^2 + y^2 - 2x + 6y = -9 & \textbf{(2)} \end{cases}$

Solve **(1)** for y: $y = 2x - 3$ **(3)**

Substitute **(3)** into **(2)** to get an equation in terms of x:

$$\begin{aligned} x^2 + (2x-3)^2 - 2x + 6(2x-3) &= -9 \\ x^2 + 4x^2 - 12x + 9 - 2x + 12x - 18 + 9 &= 0 \\ 5x^2 - 2x &= 0 \\ x(5x-2) &= 0 \\ x &= 0, \ \tfrac{2}{5} \end{aligned}$$

Now, substitute these values of x back into **(3)** to find the corresponding values of y:

$$x = 0: \quad y = 2(0) - 3 = -3$$

$$x = \tfrac{2}{5}: \quad y = 2(\tfrac{2}{5}) - 3 = -\tfrac{11}{5}$$

So, the solutions are $\boxed{(0,-3) \text{ and } (\tfrac{2}{5}, -\tfrac{11}{5})}$.

23. Solve the system $\begin{cases} 4x^2+12xy+9y^2=25 & \textbf{(1)} \\ -2x+y=1 & \textbf{(2)} \end{cases}$

Solve **(2)** for y: $y=2x+1$ **(3)**

Substitute **(3)** into **(1)** to get an equation in terms of x:

$$4x^2+12x(2x+1)+9(2x+1)^2=25$$

$$4x^2+24x^2+12x+\underbrace{9(4x^2+4x+1)}_{36x^2+36x+9}-25=0$$

$$64x^2+48x-16=0$$

$$4x^2+3x-1=0$$

$$(x+1)(4x-1)=0$$

$$x=-1,\ \tfrac{1}{4}$$

Now, substitute these values of x back into **(3)** to find the corresponding values of y:

$$x=-1:\ \ y=2(-1)+1=-1$$

$$x=\tfrac{1}{4}:\ \ y=2(\tfrac{1}{4})+1=\tfrac{3}{2}$$

So, the solutions are $\boxed{(-1,-1) \text{ and } (\tfrac{1}{4},\tfrac{3}{2})}$.

25. Solve the system $\begin{cases} x^3-y^3=63 & \textbf{(1)} \\ x-y=3 & \textbf{(2)} \end{cases}$

Solve **(2)** for x: $x=y+3$ **(3)**

Substitute **(3)** into **(1)** to get an equation in terms of x:

$$\underbrace{(y+3)^3}_{(y+3)(y^2+6y+9)}-y^3=63$$

$$(y^3+6y^2+9y+3y^2+18y+27)-y^3-63=0$$

$$9y^2+27y-36=0$$

$$y^2+3y-4=0$$

$$(y+4)(y-1)=0$$

$$y=-4,\ 1$$

Now, substitute these values of y back into **(3)** to find the corresponding values of x:

$$y=-4:\ x=-4+3=-1$$

$$y=1:\ \ x=1+3=4$$

So, the solutions are $\boxed{(4,1) \text{ and } (-1,-4)}$.

27. Solve the system $\begin{cases} 4x^2 - 3xy = -5 & \textbf{(1)} \\ -x^2 + 3xy = 8 & \textbf{(2)} \end{cases}$

Add **(1)** and **(2)** to get an equation in terms of x:

$$3x^2 = 3$$
$$x^2 = 1$$
$$x = -1, 1$$

Now, substitute these values of x back into **(2)** to find the corresponding values of y:

$$x = 1: \; -(1)^2 + 3(1)(y) = 8$$
$$3y = 9 \text{ so that } y = 3$$
$$x = -1: \; -(-1)^2 + 3(-1)(y) = 8$$
$$-3y = 9 \text{ so that } y = -3$$

So, the solutions are $\boxed{(1,3) \text{ and } (-1,-3)}$.

29. Solve the system $\begin{cases} \log_x(2y) = 3 & \textbf{(1)} \\ \log_x(y) = 2 & \textbf{(2)} \end{cases}$

This system is equivalent to: $\begin{cases} x^3 = 2y & \textbf{(3)} \\ x^2 = y & \textbf{(4)} \end{cases}$

Substitute **(4)** into **(3)** to get an equation in terms of x:

$$x^3 = 2(x^2)$$
$$x^3 - 2x^2 = 0$$
$$x^2(x-2) = 0$$
$$x = \not{0},\ 2$$

Recall that 0 is not an allowable base for a logarithm. So, although $x = 0$ would give rise to a corresponding y (namely $y = 0$) such that the pair would solve system **(3) – (4)**, the equations **(1) – (2)** would not be defined for such x and y values. So, we only substitute the value $x = 2$ back into **(4)** to find the corresponding value of y that will yield a solution to the original system:

$$x = 2: \; y = 2(2) = 4$$

So, the solution is $\boxed{(2,4)}$.

31. Solve the system $\begin{cases} \frac{1}{x^3}+\frac{1}{y^2}=17 & \textbf{(1)} \\ \frac{1}{x^3}-\frac{1}{y^2}=-1 & \textbf{(2)} \end{cases}$

Add **(1)** and **(2)** to get an equation in terms of x:

$$\frac{2}{x^3}=16$$

$$2=16x^3$$

$$x^3=\tfrac{1}{8} \text{ so that the only real solution is } x=\tfrac{1}{2}$$

Now, substitute this value of x back into **(2)** to find the corresponding value of y:

$$\frac{1}{(\frac{1}{2})^3}-\frac{1}{y^2}=-1$$

$$8-\frac{1}{y^2}=-1$$

$$9=\frac{1}{y^2} \text{ so that } y^2=\tfrac{1}{9}. \text{ Hence, } y=\pm\tfrac{1}{3}.$$

So, the solutions are $\boxed{(\frac{1}{2},\frac{1}{3}) \text{ and } (\frac{1}{2},-\frac{1}{3})}$.

33. Solve the system $\begin{cases} 2x^2+4y^4=-2 & \textbf{(1)} \\ 6x^2+3y^4=-1 & \textbf{(2)} \end{cases}$

Note that the left side of **(1)** (and **(2)**) is always ≥ 0, while the right side is negative. Hence, there are no real values of x and y that can satisfy either equation, not to mention both simultaneously. Hence, there is $\boxed{\text{no solution}}$ of this system.

35.

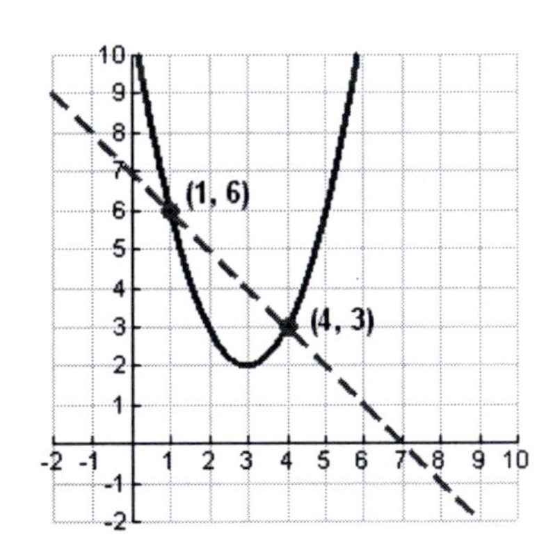

Notes on the graph:

Solid curve: $y=x^2-6x+11$, Dashed curve: $y=-x+7$

The solutions of the system are $(1,6)$ and $(4,3)$.

37. The graphs of the two equations are as follows:

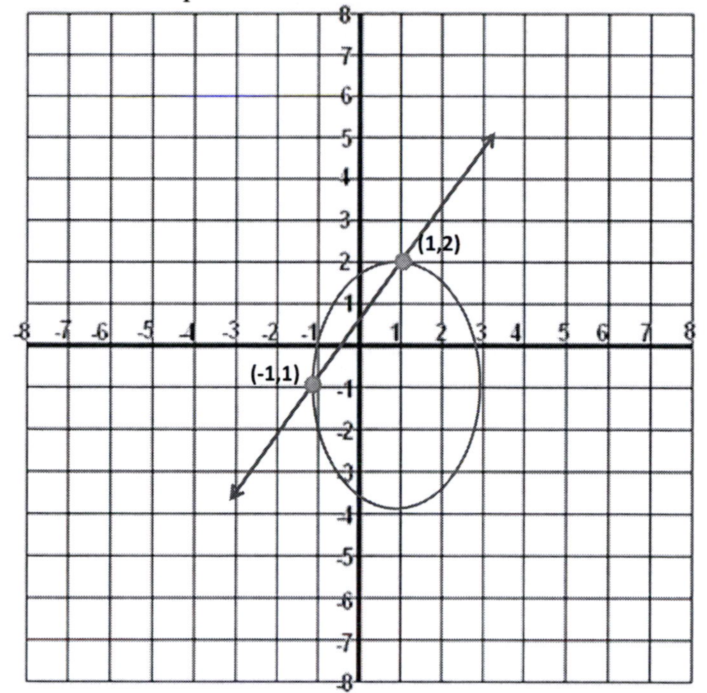

The two solutions are (-1, -1) and (1, 2).

39. Let x and y be two numbers.
The system that must be solved is:

$$\begin{cases} x + y = 10 & \textbf{(1)} \\ x^2 - y^2 = 40 & \textbf{(2)} \end{cases}$$

Solve **(1)** for y: $y = 10 - x$ **(3)**
Substitute **(3)** into **(2)** to get an equation in x:

$$x^2 - (10 - x)^2 = 40$$

$$\cancel{x^2} - 100 + 20x - \cancel{x^2} = 40$$

$$20x = 140$$

$$x = 7$$

Substitute this value of x into **(3)** to find the corresponding value of y: $y = 10 - 7 = 3$
So, the two numbers are $\boxed{3 \text{ and } 7}$.

41. Let x and y be two numbers.
The system that must be solved is:

$$\begin{cases} xy = \dfrac{1}{\frac{1}{x} - \frac{1}{y}} & \textbf{(1)} \\ xy = 72 & \textbf{(2)} \end{cases}$$

First, simplify **(1)**:

$$xy = \frac{1}{\frac{1}{x} - \frac{1}{y}} = \frac{1}{\frac{y-x}{xy}} = \frac{xy}{y-x} \quad \Rightarrow \quad 1 = y - x \quad \textbf{(3)}$$

So, we now solve the following simplified system: $\begin{cases} xy = 72 & \textbf{(2)} \\ y - x = 1 & \textbf{(3)} \end{cases}$

To do so, solve **(3)** for y: $y = x + 1$ **(4)**
Substitute **(4)** into **(2)** to get an equation in x:

$$x(x+1) = 72$$

$$x^2 + x - 72 = 0$$

$$(x+9)(x-8) = 0 \text{ so that } x = -9, 8$$

Substitute these values of x into **(4)** to find the corresponding values of y:

$$x = -9: \quad y = (-9) + 1 = -8$$

$$x = 8: \quad y = (8) + 1 = 9$$

So, there are two pairs of numbers that will work, namely the pair 8 and 9, and the pair -8 and -9.

43. Let x = width of rectangular pen (in cm)
y = length of rectangular pen.
The system that must be solved is:

$$\begin{cases} 2x + 2y = 36 & \textbf{(1)} \text{ (perimeter)} \\ xy = 80 & \textbf{(2)} \text{ (area)} \end{cases}$$

Solve **(1)** for y: $y = 18 - x$ **(3)**
Substitute **(3)** into **(2)** to get an equation in x:

$$x(18 - x) = 80$$

$$18x - x^2 = 80$$

$$x^2 - 18x + 80 = 0$$

$$(x-10)(x-8) = 0 \text{ so that } x = 10, 8$$

Substitute these values of x into **(3)** to find the corresponding values of y:

$$x = 10: \quad y = 18 - 10 = 8$$

$$x = 8: \quad y = 18 - 8 = 10$$

Hence, in terms of the context of the problem, both x values result in the same solution, namely that the dimensions of the rectangular pen should be $\boxed{8\text{ cm} \times 10\text{ cm}}$.

45. Let x = width of one of the two congruent rectangular pens (in ft)
y = length of one of the two congruent rectangular pens.

The system that must be solved is:

$$\begin{cases} 3x+4y=2200 & \textbf{(1)} \text{ (total amount of fence needed)} \\ x\cdot 2y = 200,000 & \textbf{(2)} \text{ (combined area of the two pens)} \end{cases}$$

Solve **(1)** for y: $y=\frac{1}{4}(2200-3x)$ **(3)**

Substitute **(3)** into **(2)** to get an equation in x:

$$x\cdot 2\left[\tfrac{1}{4}(2200-3x)\right]=200,000$$

$$2200x-3x^2=400,000$$

$$3x^2-2200x+400,000=0$$

$$x=\frac{2200\pm\sqrt{4,840,000-4,800,000}}{2(3)}=\frac{2200\pm\sqrt{40,000}}{6}$$

$$=\frac{2200\pm 200}{6}=400,\ \frac{1000}{3}$$

Substitute these values of x into **(3)** to find the corresponding values of y:

$$x=400:\ \tfrac{1}{4}(2200-3(400))=250$$

$$x=\tfrac{1000}{3}:\ \tfrac{1}{4}(2200-3(\tfrac{1000}{3}))=300$$

Hence, in terms of the context of the problem, there are two distinct solutions.

Solution 1: Each of the two congruent rectangular pens should be $400\text{ ft}\times 250\text{ ft}$, so that the two combined have dimensions $\boxed{400\text{ ft}\times 500\text{ ft}}$.

Solution 2: Each of the two congruent rectangular pens should be $\frac{1000}{3}\text{ ft}\times 300\text{ ft}$, so that the two combined have dimensions $\boxed{\frac{1000}{3}\text{ ft}\times 600\text{ ft}}$.

47. Let r = speed of the professor (in $^{m}/_{\text{min}}$)

t = number of minutes it took Jeremy to complete the 400m race

We have the following information:

	Distance (in m)	Rate (in $^{m}/_{\text{min}}$)	Time (in min)
Professor	400	r	$t + \underbrace{1}_{\text{Head start}} + \underbrace{\tfrac{5}{3}}_{\text{Finished 1 min 40 sec after Jerermy}} = t + \tfrac{8}{3}$
Jeremy	400	$5r$	t

Since Distance = Rate × Time, the following system must be solved:

$$\begin{cases} 400 = r\left(t + \frac{8}{3}\right) & \textbf{(1)} \\ 400 = 5rt & \textbf{(2)} \end{cases}$$

Solve **(2)** for t: $t = \frac{80}{r}$ **(3)**

Substitute **(3)** into **(1)** to get an equation in x:

$$400 = r\left(\frac{80}{r} + \frac{8}{3}\right)$$

$$400 = \cancel{r}\left(\frac{240+8x}{3\cancel{r}}\right)$$

$$1200 = 240 + 8r$$

$$960 = 8r \quad \text{so that} \quad 120 = r$$

Hence, the professor's average speed was 120 $^{m}/_{\text{min}}$ = $\boxed{2\ ^{m}/_{\text{sec}}}$, while Jeremy's average speed was 600 $^{m}/_{\text{min}}$ = $\boxed{10\ ^{m}/_{\text{sec}}}$. (Note: Though the value of t was not needed to answer the question posed in the problem, one could substitute this value of r into **(3)** to find that the corresponding value of t is $\frac{2}{3}$, meaning that it took Jeremy 40 sec to complete the race).

49. In general, $y^2 - y \neq 0$. Must solve this system using substitution.

51. False.

There are at most two intersection points. Solving a system of the form

$$\begin{cases} y = ax^2 + bx + c \\ y = dx + e \end{cases}$$

amounts to solving the quadratic equation $ax^2 + bx + c = dx + e$. Such an equation can have at most two real solutions. So, the graphs can intersect in at most two points.

53. False.

For example, the system

$$\begin{cases} x - y = 1 \\ x^2 + y^2 = 5 \end{cases}$$

cannot be solved using elimination since there is no common term in both equations.

55. Let $y = a_n x^n + a_{n-1}x^{n-1} + \cdots + a_1 x + a_0$ be a given nth degree polynomial. Then, consider a system of the form:

$$\begin{cases} y = a_n x^n + a_{n-1}x^{n-1} + \cdots + a_1 x + a_0 & \textbf{(1)} \\ (x-h)^2 + (y-k)^2 = r^2 & \textbf{(2)} \end{cases}$$

Substitute **(1)** into **(2)** to obtain:

$$(x-h)^2 + \underbrace{(a_n x^n + a_{n-1}x^{n-1} + \cdots + a_1 x + a_0 - k)^2}_{\text{Has degree } 2n} = r^2 \quad \textbf{(3)}$$

The degree of the left side of **(3)** is $2n$, so there can be at most $2n$ real solutions. Consequently, there are at most $2n$ intersection points of the graphs of the equations of **(1)** and **(2)**.

57. Consider $\begin{cases} y = x^2 + 1 \\ y = 1 \end{cases}$.

Any system in which the linear equation is the tangent line to the parabola at its vertex will have only one solution.

59.

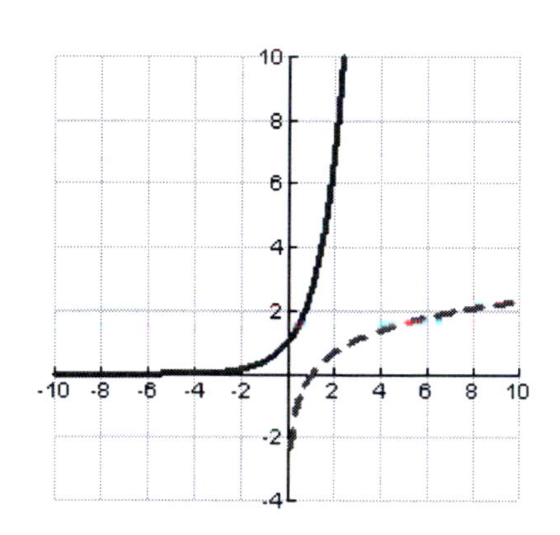

Notes on the graph:

Solid curve: $y = e^x$

Dashed curve: $y = \ln x$

Since the two graphs never intersect, the system has no solution.

61.

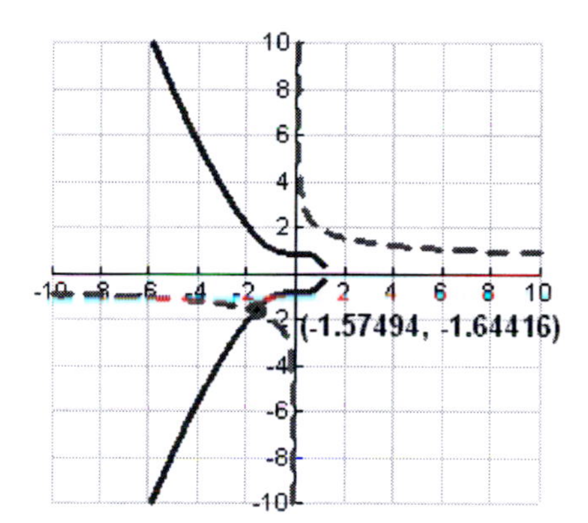

Notes on the graph:

Solid curve: $2x^3 + 4y^2 = 3$

Dashed curve: $xy^3 = 7$

The approximate solution of the system is $(-1.57, -1.64)$.

63. The graph of the nonlinear system is given by:

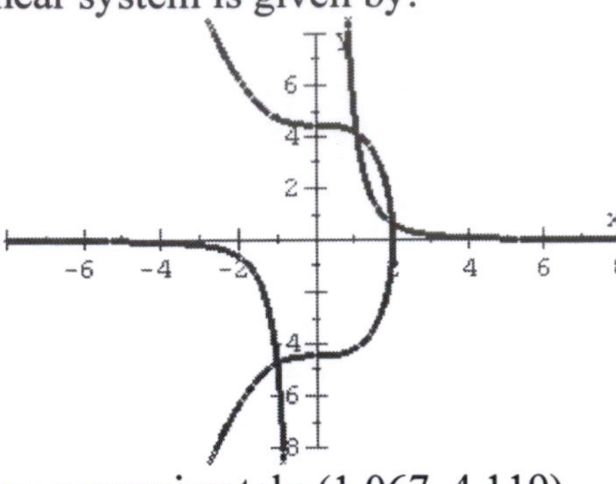

The points of intersection are approximately (1.067, 4.119), (1.986, 0.638), and (–1.017, –4.757).

Section 8.6 Solutions

1. b	**3. j**
5. h	**7. c**
9. d	**11. k**

13.

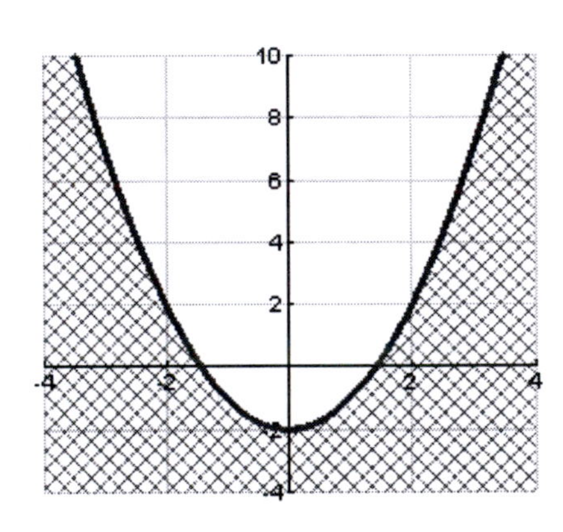

The equation of the solid curve is $y = x^2 - 2$.

15.

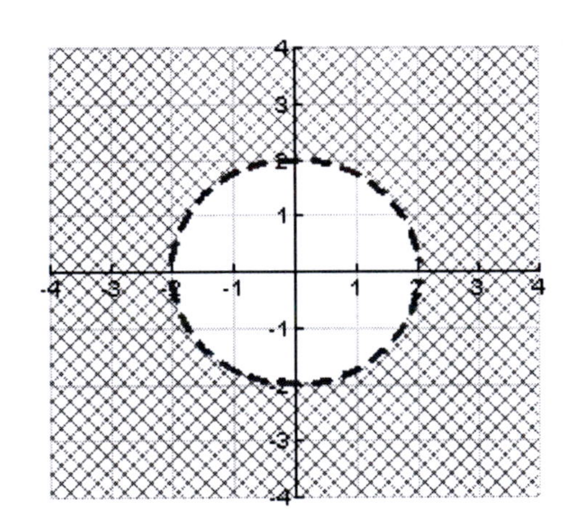

The equation of the dashed curve is $x^2 + y^2 = 4$.

17. Begin by writing the inequality in standard form (so the radius and center are easily identified):

$$\left(x^2-2x+\ \right)+\left(y^2+4y+\ \right)\geq -4$$

$$\left(x^2-2x+1\right)+\left(y^2+4y+4\right)\geq -4+1+4$$

$$(x-1)^2+(y+2)^2\geq 1$$

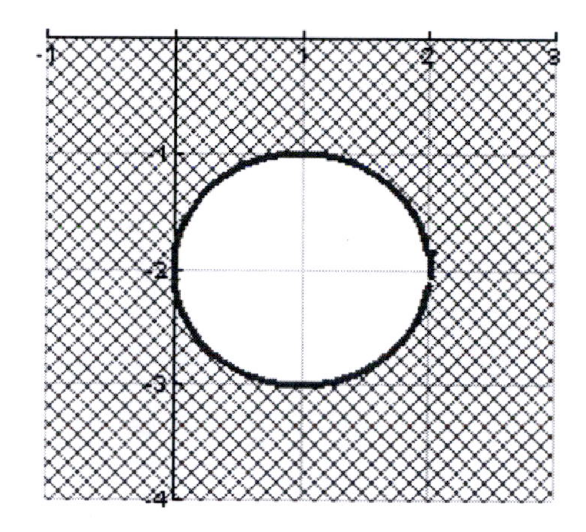

The equation of the solid curve is $(x-1)^2+(y+2)^2=1$.

19.

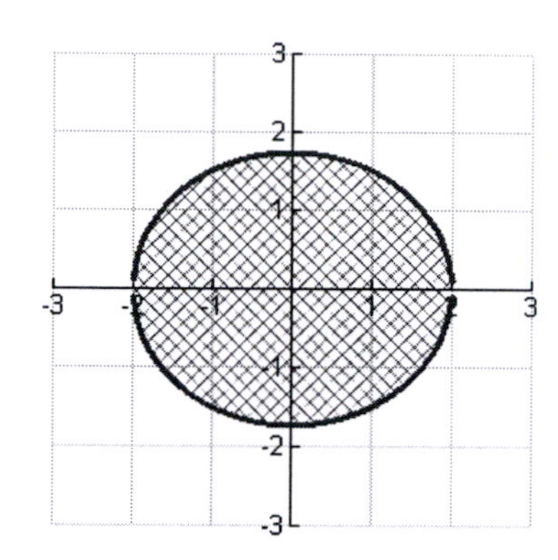

The equation of the solid curve is $\frac{x^2}{4}+\frac{y^2}{3}=1$.

21. Upon completing the square, the inequality becomes

$$9(x-1)^2+16(y+3)^2>144$$

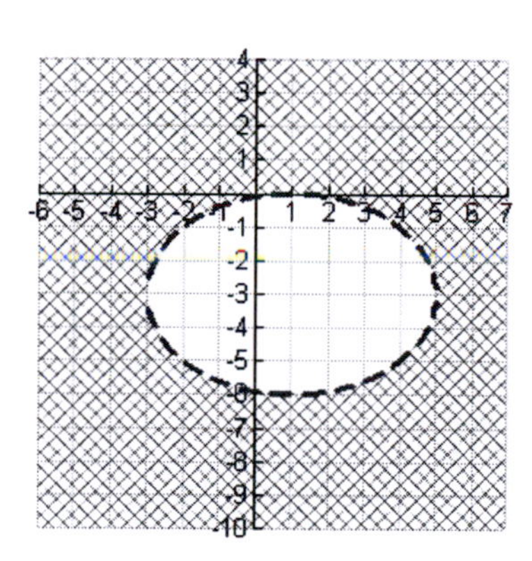

23.

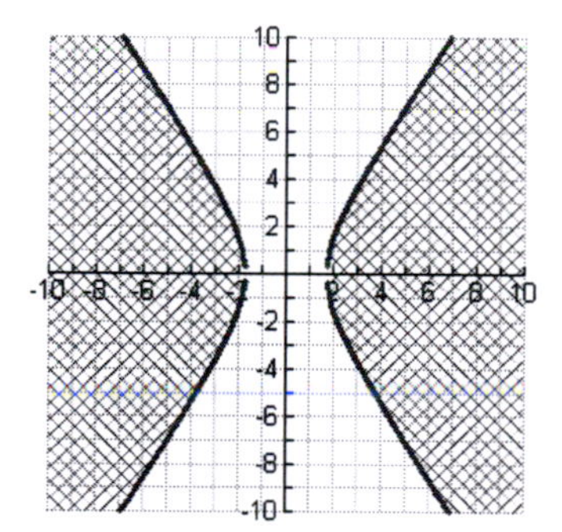

25.

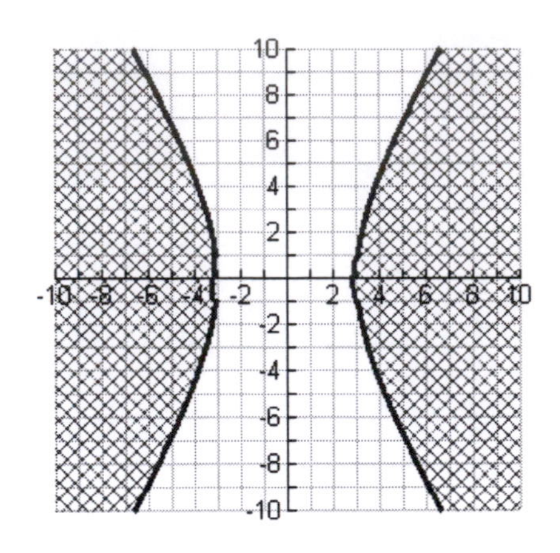

27.

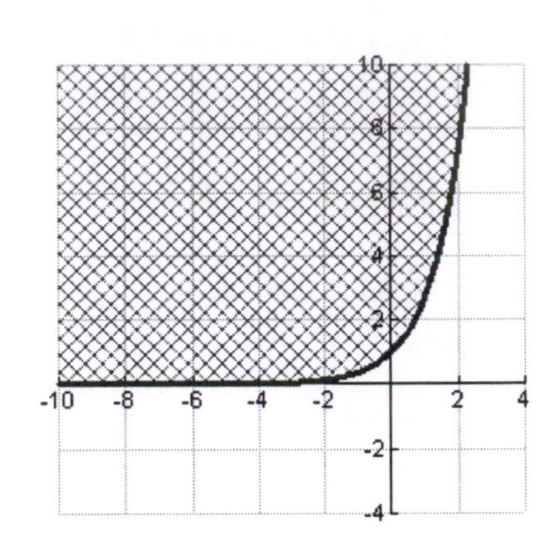

The equation of the solid curve is $y = e^x$.

29.

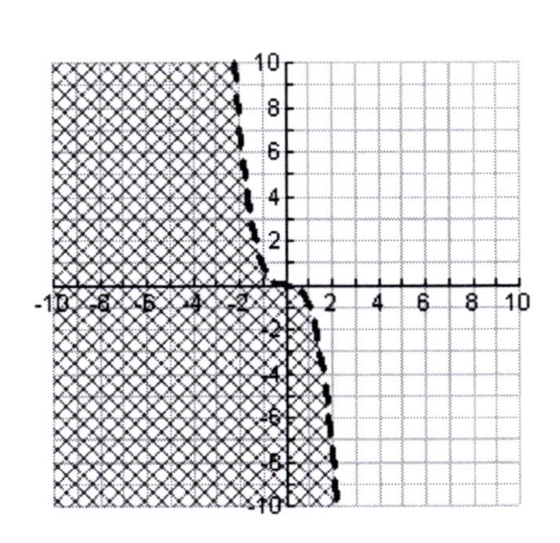

The equation of the dashed curve is $y = -x^3$.

31.

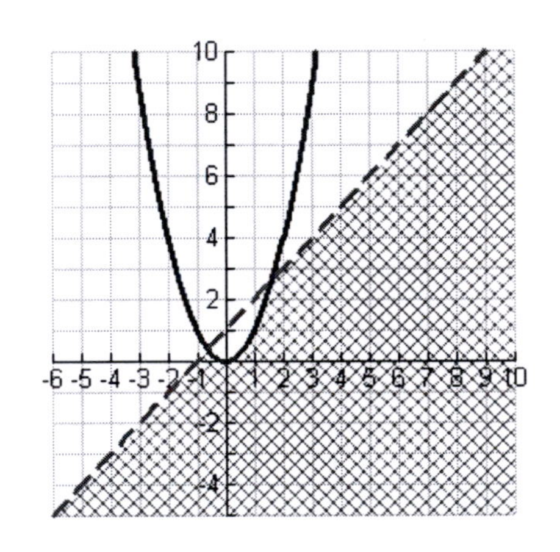

33.

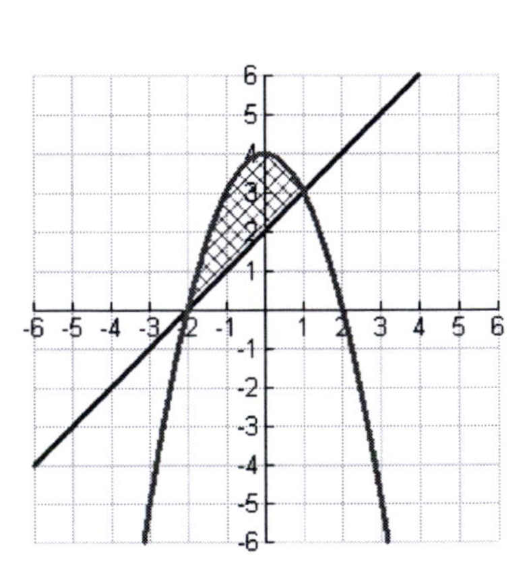

35.

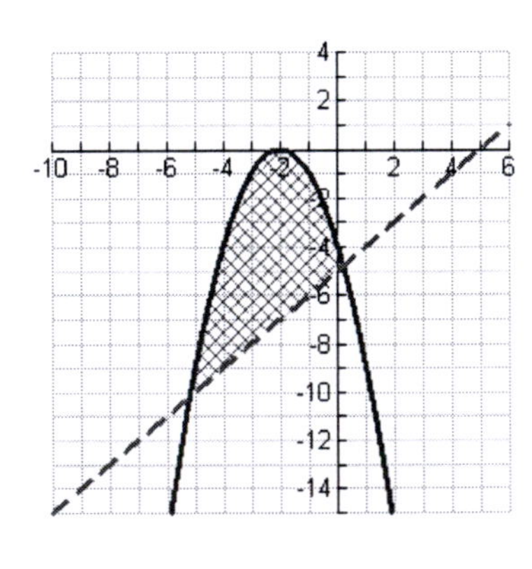

37.

First write the inequalities as:

$$y > x^2 - 1$$
$$y < 1 - x^2$$

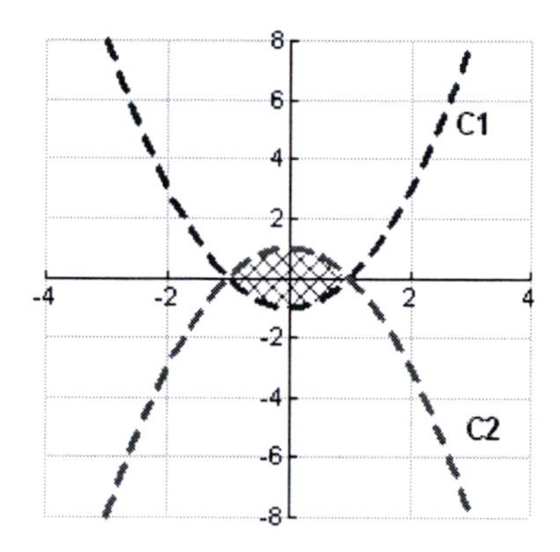

Notes on the graph:

C1: $y = x^2 - 1$ **C2:** $y = -x^2 + 1$

39.

First write the inequalities as:

$$y \geq x^2$$
$$x \geq y^2$$

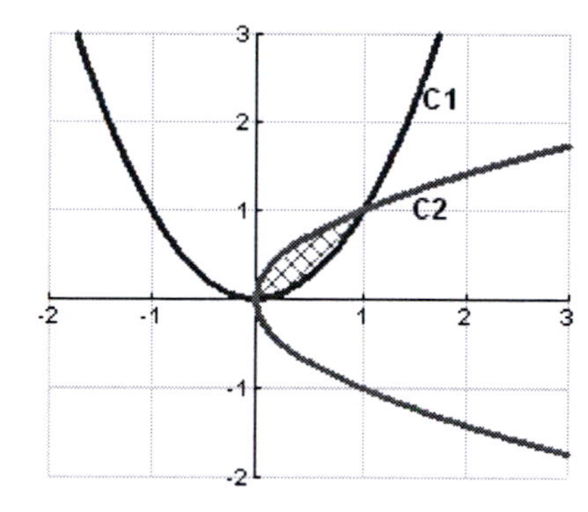

Notes on the graph:

C1: $y = x^2$ **C2:** $x = y^2$

41.

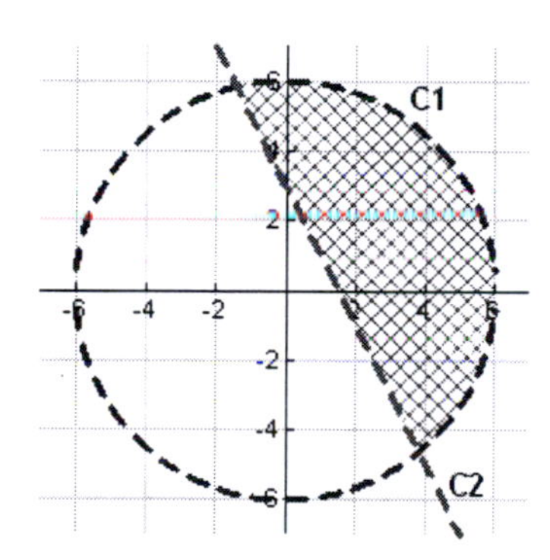

Notes on the graph:

C1: $x^2 + y^2 = 36$ **C2:** $2x + y = 3$

43.

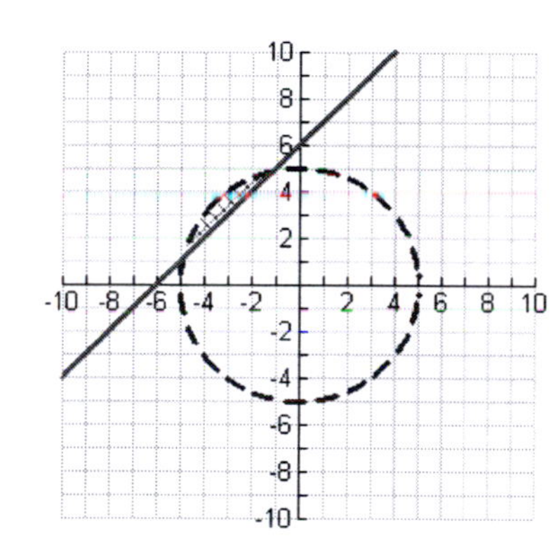

45.

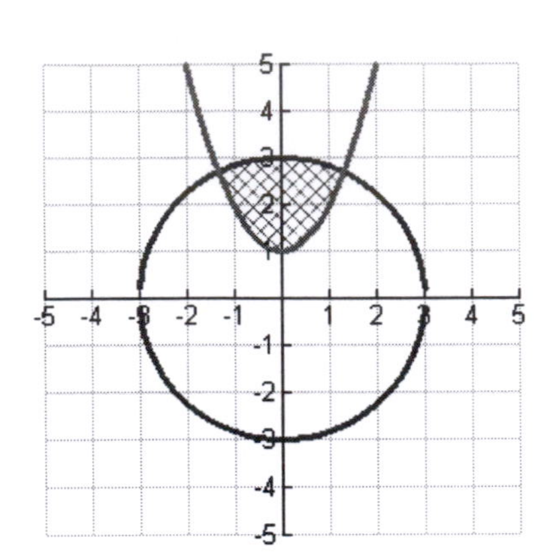

47.

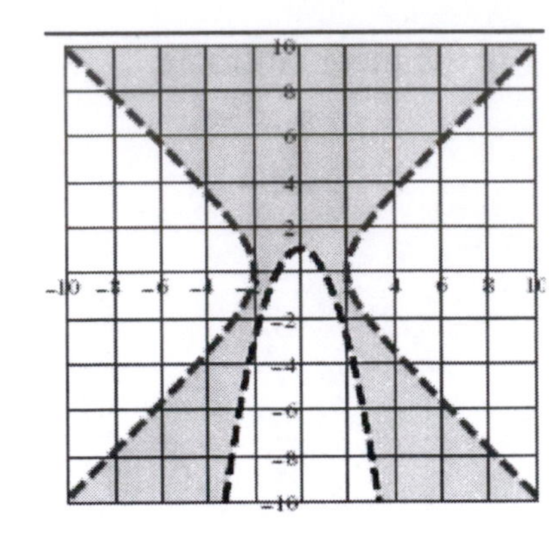

49.

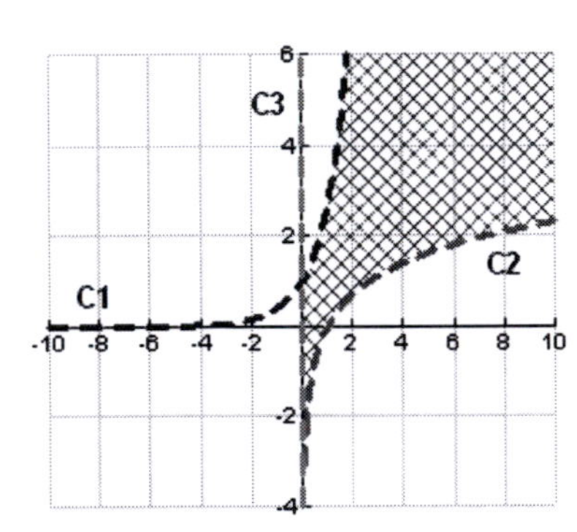

Notes on the graph:
C1: $y = \exp(x)$ **C2:** $y = \ln(x)$ **C3:** $x = 0$

51. The shaded region is a semi-circle of radius 3, seen below. Hence, the area is:

$$\text{Area} = \tfrac{1}{2}\left(\pi (3)^2\right) = \tfrac{9}{2}\pi \text{ units}^2$$

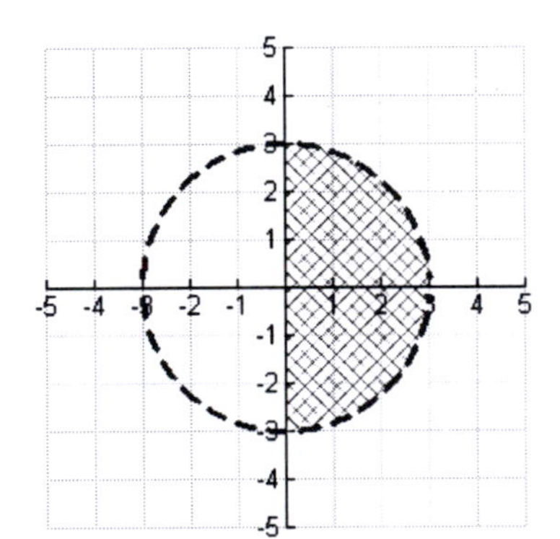

53. There is no common region here – it is empty, as is seen in the graph below:

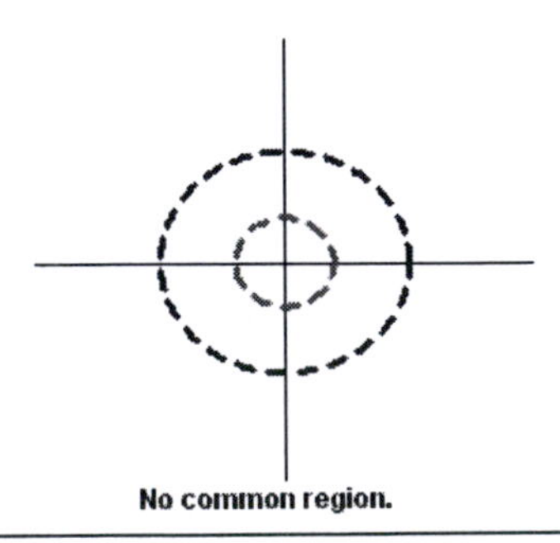

55. False Shade below the parabola, and include the parabola itself in the region; the inequality is $y \le x^2$

57. Assume that a and b are both positive. Must have $a^2 \le b^2$, which is equivalent to:

$$a^2 - b^2 = (a-b)(a+b) \le 0$$

This requires that

$$a - b \le 0 \text{ and } \underbrace{a+b \ge 0}_{\substack{\text{Automatically holds since} \\ a \text{ and } b \text{ are assumed to be} \\ \text{positive.}}}.$$

This implies that $a \le b$ and $a \ge -b$, and so $-b \le a \le b$. Since a is assumed to be positive, this simplifies to $0 \le a \le b$.

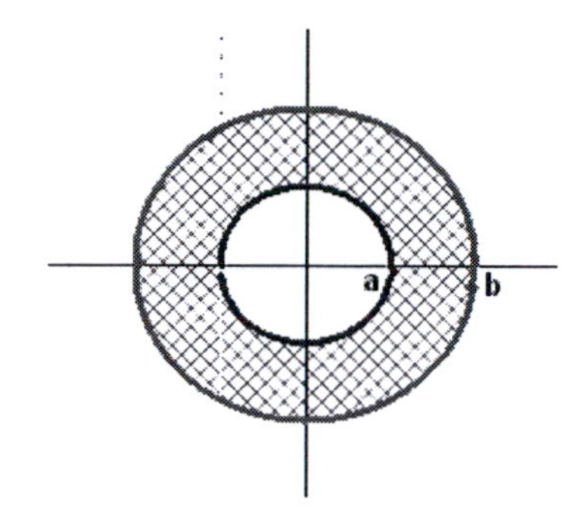

59. Begin by writing the inequality in standard form (so the radius and center are easily identified):

$$\left(x^2 - 2x + \right) + \left(y^2 + 4y + \right) \ge -4$$

$$\left(x^2 - 2x + 1\right) + \left(y^2 + 4y + 4\right) \ge -4 + 1 + 4$$

$$(x-1)^2 + (y+2)^2 \ge 1$$

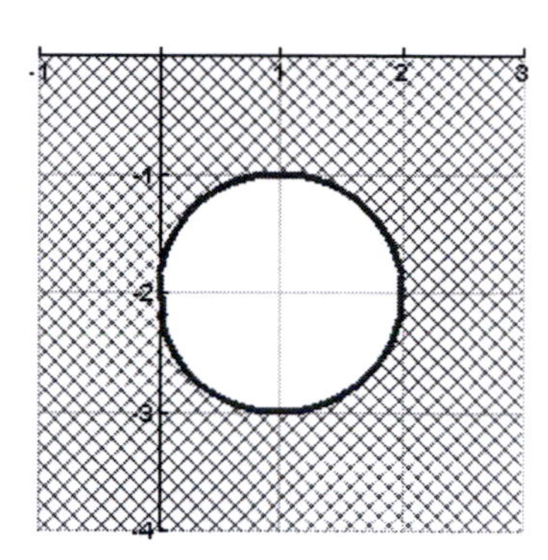

The solid curve is the graph of $(x-1)^2 + (y+2)^2 = 1$.

61.

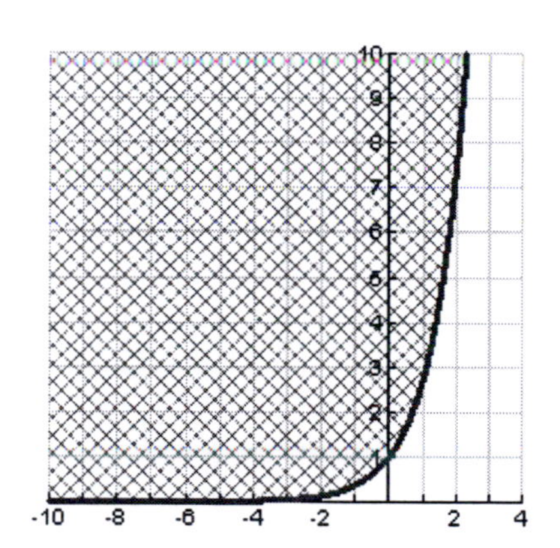

The solid curve is the graph of $y = e^x$.

63.

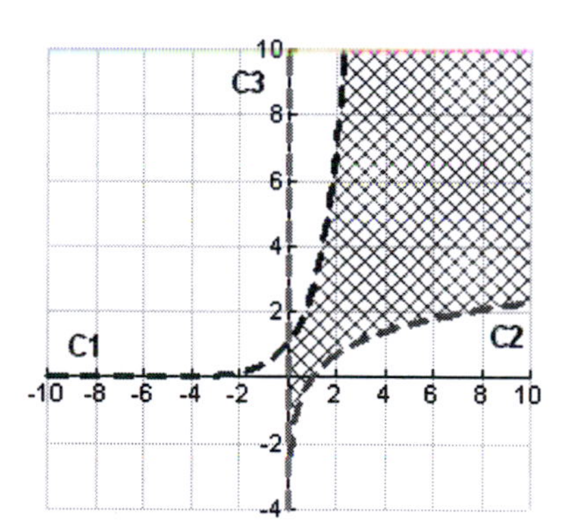

Notes on the graph:
C1: $y = \exp(x)$ **C2:** $y = \ln(x)$ **C3:** $x = 0$

65. From the calculator, the dark shaded region is the one desired:

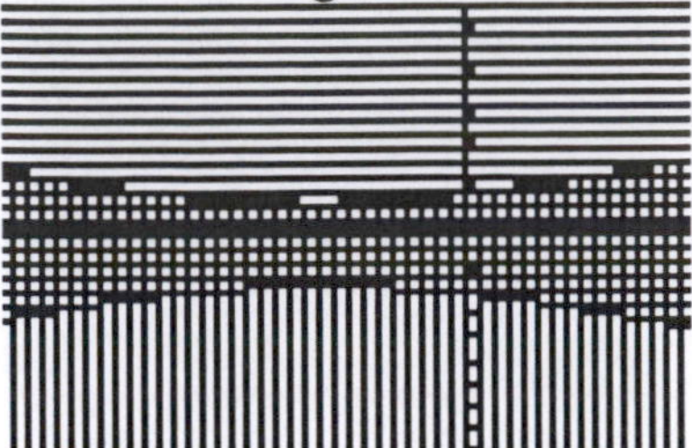

Chapter 8 Review Solutions ---

1. False. The focus is always in the region "inside" the parabola.

3. True.

5. Vertex is $(0,0)$ and focus is $(3,0)$. So, the parabola opens to the right. As such, the general form is $y^2 = 4px$. In this case, $p = 3$, so the equation is $\boxed{y^2 = 12x}$.

7. Vertex is $(0,0)$ and the directrix is $x = 5$. So, the parabola opens to the left. Since the distance between the vertex and the directrix is 5, and the focus must be equidistant from the vertex, we know $p = -5$. Hence, the equation is

$\boxed{(y-0)^2 = 4(-5)(x-0) = -20x \;\Rightarrow\; y^2 = -20x}$

9. Vertex is $(2,3)$ and focus is $(2,5)$. So, the parabola opens up. As such, the general form is $(x-2)^2 = 4p(y-3)$. In this case, $p = 2$, so the equation is $\boxed{(x-2)^2 = 8(y-3)}$.

11. Focus is $(1,5)$ and the directrix is $y = 7$. So, the parabola opens down. Since the distance between the focus and directrix is 2 and the vertex must occur halfway between them, we know $p = -1$ and the vertex is $(1,6)$. Hence, the equation is

$\boxed{(x-1)^2 = 4(-1)(y-6) = -4(y-6)}$.

13. Equation: $x^2 = -12y = 4(-3)y$ **(1)**

So, $p = -3$ and opens down.

Vertex: $(0,0)$

Focus: $(0,-3)$

Directrix: $y = 3$

Latus Rectum: Connects x-values corresponding to $y = -3$. Substituting this into **(1)** yields:

$$x^2 = 36 \text{ so that } x = \pm 6$$

So, the length of the latus rectum is 12.

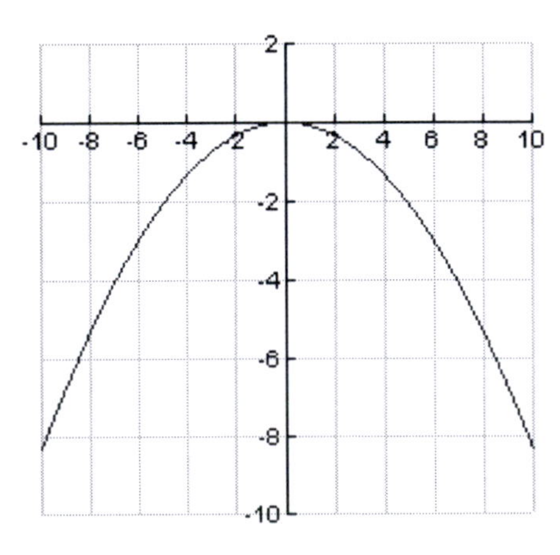

15. Equation: $y^2 = x = 4(\frac{1}{4})x$ **(1)**

So, $p = \frac{1}{4}$ and opens to the right.

Vertex: $(0,0)$

Focus: $(\frac{1}{4},0)$

Directrix: $x = -\frac{1}{4}$

Latus Rectum: Connects y-values corresponding to $x = \frac{1}{4}$. Substituting this into **(1)** yields:

$$y^2 = \frac{1}{4} \text{ so that } y = \pm\frac{1}{2}$$

So, the length of the latus rectum is 1.

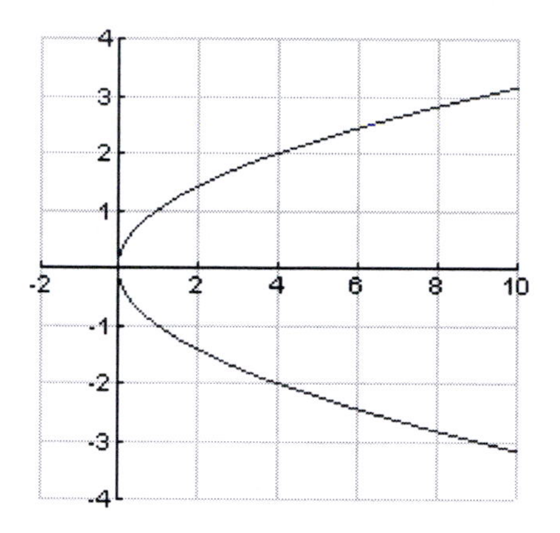

17. Equation: $(y+2)^2 = 4(1)(x-2)$ **(1)**

So, $p = 1$ and opens to the right.

Vertex: $(2,-2)$

Focus: $(3,-2)$

Directrix: $x = 1$

Latus Rectum: Connects y-values corresponding to $x = 3$. Substituting this into **(1)** yields:

$$(y+2)^2 = 4 \text{ so that } y = -2 \pm 2 = -4,\ 0$$

So, the length of the latus rectum is 4.

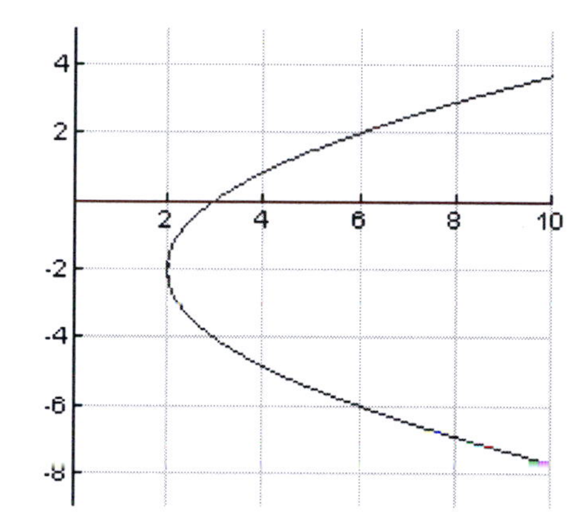

19. Equation: $(x+3)^2 = 4(-2)(y-1)$ **(1)**

So, $p = -2$ and opens down.

Vertex: $(-3,1)$ Focus: $(-3,-1)$

Directrix: $y = 3$

Latus Rectum: Connects x-values corresponding to $y = -1$. Substituting this into **(1)** yields:

$$(x+3)^2 = 16 \text{ so that } x = -3 \pm 4 = 1,\ -7$$

So, the length of the latus rectum is 8.

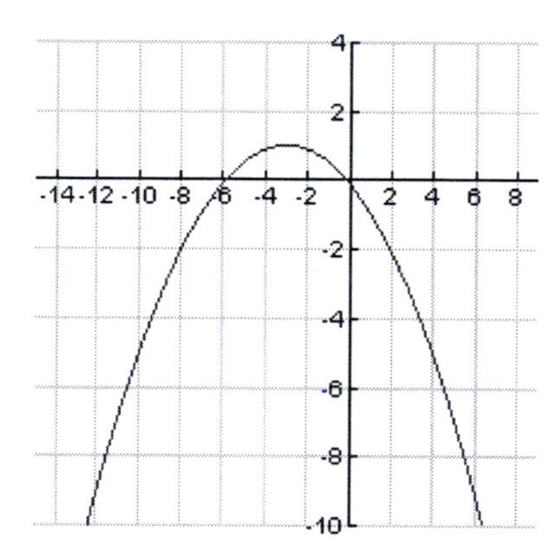

21. Equation: Completing the square yields:

$$x^2+5x+2y+25=0$$
$$\left(x^2+5x+\tfrac{25}{4}\right)=-2y-25+\tfrac{25}{4}$$
$$\left(x+\tfrac{5}{2}\right)^2=-2y-\tfrac{75}{4}$$
$$\left(x+\tfrac{5}{2}\right)^2=-2\left(y+\tfrac{75}{8}\right)$$
$$\left(x+\tfrac{5}{2}\right)^2=4\left(-\tfrac{1}{2}\right)\left(y+\tfrac{75}{8}\right) \quad \textbf{(1)}$$

So, $p=-\frac{1}{2}$ and opens down.

Vertex: $\left(-\frac{5}{2},-\frac{75}{8}\right)$ Focus: $\left(-\frac{5}{2},-\frac{79}{8}\right)$

Directrix: $y=-\frac{71}{8}$

Latus Rectum: Connects x-values corresponding to $y=-\frac{79}{8}$. Substituting this into **(1)** yields:

$$\left(x+\tfrac{5}{2}\right)^2=1 \text{ so that } x=-\tfrac{5}{2}\pm1=-\tfrac{3}{2},-\tfrac{7}{2}$$

So, the length of the latus rectum is 2.

23. Assume the vertex is at $(0,-2)$. Then, the general equation of this parabola is $x^2=4p(y+2)$ **(1)**. Since $(-5,0)$ is on the graph, we can substitute it into **(1)** to see:

$$25=4p(2)=8p \text{ so that } p=\tfrac{25}{8}$$

So, the focus is at $\left(0,\frac{9}{8}\right)$, and the receiver should be placed $\frac{25}{8}=3.125$ feet from the vertex.

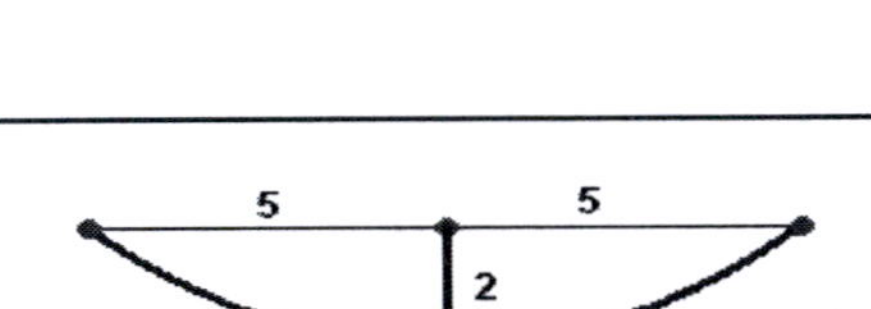

25.

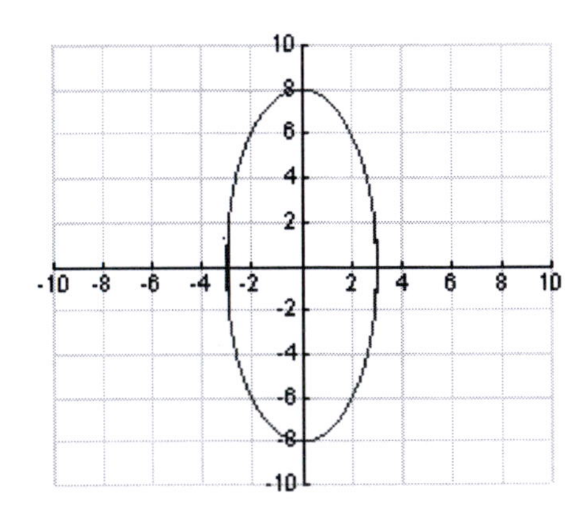

27. Write the equation in standard form as $x^2+\frac{y^2}{25}=1$.

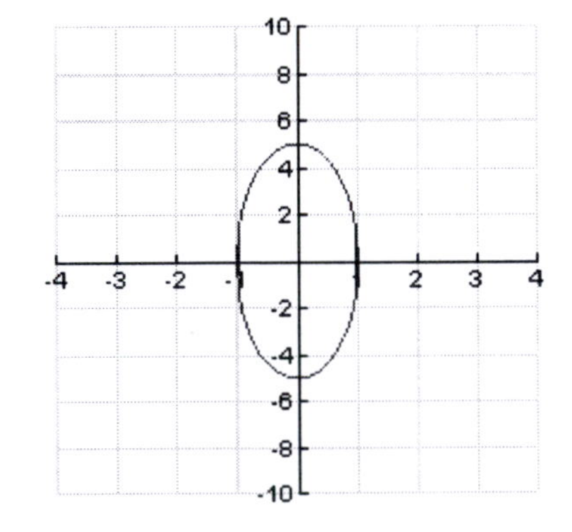

29. Since the foci are $(-3,0)$ and $(3,0)$, we know that $c=3$ and the center of the ellipse is $(0,0)$. Further, since the vertices are $(-5,0)$ and $(5,0)$ and they lie on the x-axis, the ellipse is horizontal and $a=5$. Hence,

$$c^2 = a^2 - b^2$$
$$9 = 25 - b^2$$
$$b^2 = 16$$

Therefore, the equation of the ellipse is $\boxed{\frac{x^2}{25}+\frac{y^2}{16}=1}$.

31. Since the ellipse is centered at $(0,0)$, the length of the horizontal major axis being 16 implies that $a=8$, and the length of the horizontal minor axis being 6 implies that $b=3$. Since the major axis is vertical, the equation of the ellipse is $\boxed{\frac{x^2}{9}+\frac{y^2}{64}=1}$.

33.

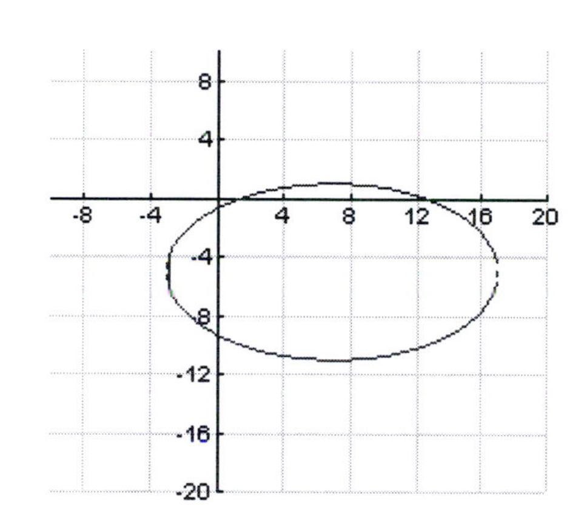

35. First, write the equation in standard form by completing the square:

$$4(x^2-4x)+12(y^2+6y)=-123$$
$$4(x^2-4x+4)+12(y^2+6y+9)=-123+16+108$$
$$4(x-2)^2+12(y+3)^2=1$$
$$\frac{(x-2)^2}{\frac{1}{4}}+\frac{(y+3)^2}{\frac{1}{12}}=1$$

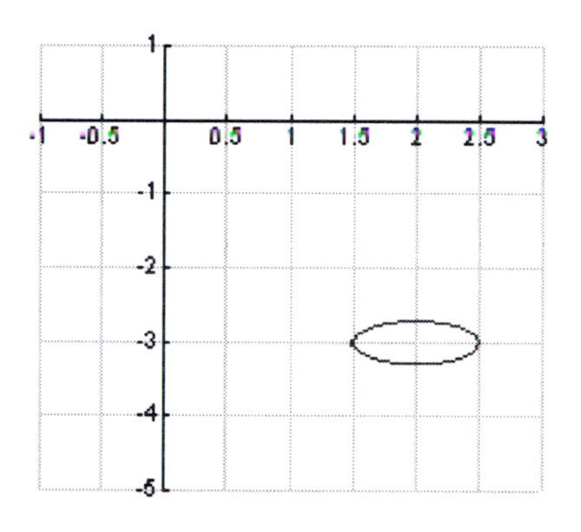

37. Since the foci are $(-1,3)$, $(7,3)$ and the vertices are $(-2,3)$, $(8,3)$, we know:

i) the ellipse is horizontal with center at $(\frac{-2+8}{2},3)=(3,3)$,

ii) $3-c=-1$ so that $c=4$,

iii) the length of the major axis is $8-(-2)=10$. Hence, $a=5$.

Now, since $c^2=a^2-b^2$, $16=25-b^2$ so that $b^2=9$. Hence, the equation of the ellipse is $\boxed{\frac{(x-3)^2}{25}+\frac{(y-3)^2}{9}=1}$.

39. Assume that the sun is at the origin. Then, the vertices of Jupiter's horizontal elliptical trajectory are: $(-7.409\times10^8, 0)$, $(8.157\times10^8, 0)$

From this, we know that:

i) The length of the major axis is $8.157\times10^8 - (-7.409\times10^8) = 1.5566\times10^9$, and so the value of a is half of this, namely 7.783×10^8;

ii) The center of the ellipse is $\left(\frac{-7.409\times10^8 + (8.157\times10^8)}{2}, 0\right) = (3.74\times10^7, 0)$;

iii) The value of c is 3.74×10^7.

Now, since $c^2 = a^2 - b^2$, we have $\left(3.74\times10^7\right)^2 = \left(7.783\times10^8\right)^2 - b^2$, so that $b^2 = 6.044\times10^{17}$. Hence, the equation of the ellipse must be

$$\boxed{\frac{\left(x-(3.74\times10^7)\right)^2}{6.058\times10^{17}} + \frac{(y-0)^2}{6.044\times10^{17}} = 1}.$$

(Note: There is more than one correct way to set up this problem, and it begins with choosing the center of the ellipse. If, alternatively, you choose the center to be at (0,0), then the actual coordinates of the vertices and foci will change, and will result in the following slightly different equation for the trajectory: $\frac{x^2}{778{,}300{,}000^2} + \frac{y^2}{777{,}400{,}000^2} = 1$

Both are equally correct!.)

41.

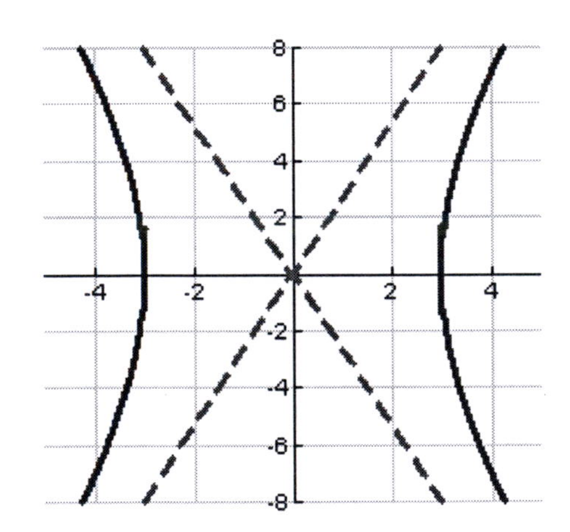

Notes on the Graph:

The equations of the asymptotes are $y = \pm\frac{8}{3}x$.

43.

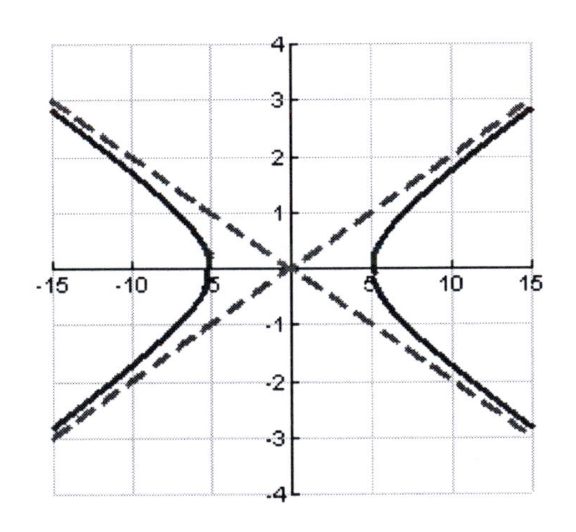

Notes on the Graph:

First, write the equation in standard form as $\frac{x^2}{25} - y^2 = 1$. The equations of the asymptotes are $y = \pm\frac{1}{5}x$.

45. Since the foci are $(-5,0)$, $(5,0)$ and the vertices are $(-3,0)$, $(3,0)$, we know that:
i) the hyperbola opens right/left with center at $(0,0)$,
ii) $a=3$, $c=5$,
Now, since $c^2=a^2+b^2$, $25=9+b^2$ so that $b^2=16$. Hence, the equation of the hyperbola is $\boxed{\frac{x^2}{9}-\frac{y^2}{16}=1}$.

47. Since the center is $(0,0)$ and the transverse axis is the y-axis, the general form of the equation of this hyperbola is $\frac{y^2}{a^2}-\frac{x^2}{b^2}=1$ with asymptotes $y=\pm\frac{a}{b}x$. In this case, $\frac{a}{b}=3$ so that $a=3b$. Thus, the equation simplifies to $\frac{y^2}{(3b)^2}-\frac{x^2}{b^2}=1$, or equivalently $\frac{y^2}{9}-x^2=b^2$. If we took $b=1$, then this simplifies to $\boxed{\frac{y^2}{9}-x^2=1}$.

49.

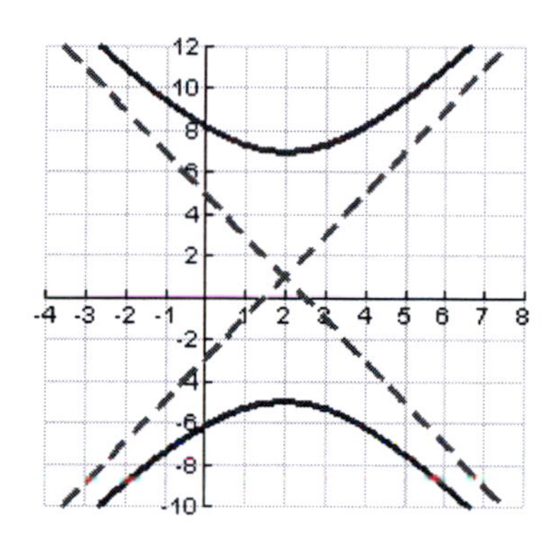

Notes on the Graph:
The equations of the asymptotes are $y=\pm 2(x-2)+1$.

51.

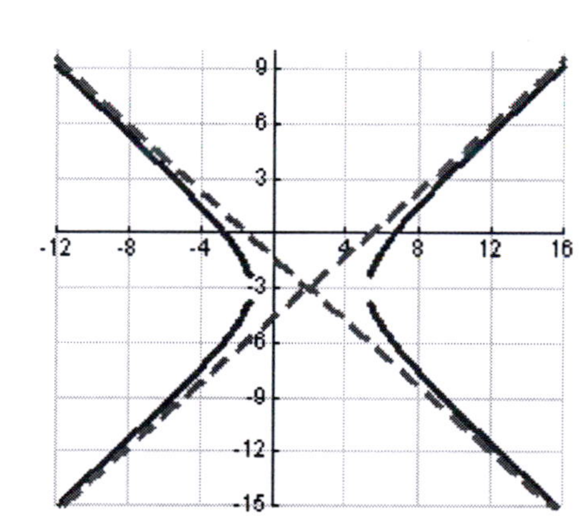

Notes on the Graph:
First, write the equation in standard form by completing the square:

$$8\left(x^2-4x\right)-10\left(y^2+6y\right)=138$$

$$8\left(x^2-4x+4\right)-10\left(y^2+6y+9\right)=138+32-90$$

$$8(x-2)^2-10(y+3)^2=80$$

$$\frac{(x-2)^2}{10}-\frac{(y+3)^2}{8}=1$$

The equations of the asymptotes are $y=\pm\frac{2}{\sqrt{5}}(x-2)-3$.

53. Since the vertices are $(0,3)$, $(8,3)$, we know that the center is $(4,3)$ and the transverse axis is parallel to the x-axis. Hence, $4-a=0$ so that $a=4$. Also, since the foci are $(-1,3)$, $(9,3)$, we know that $4+c=-1$ so that $c=-5$. Now, to find b, we substitute the values of c and a obtained above into $c^2=a^2+b^2$ to see that $25=16+b^2$ so that $b^2=9$. Hence, the equation of the hyperbola is $\boxed{\frac{(x-4)^2}{16}-\frac{(y-3)^2}{9}=1}$.

55. Assume that the stations coincide with the foci and are located at $(-110,0)$, $(110,0)$. The difference in distance between the ship and each of the two stations must remain constantly $2a$, where $(a,0)$ is the vertex. Assume that the radio signal speed is 186,000 $^{mi}/_{sec}$ and the time difference is 0.00048 sec. Then, using distance = rate × time, we obtain:

$$2a=(186,000)(0.00048)=89.28 \text{ so that } a=44.64$$

So, the ship will come ashore between the two stations 65.36 miles from one and 154.64 miles from the other.

57. Solve the system $\begin{cases} x^2+y=-3 & \textbf{(1)} \\ x-y=5 & \textbf{(2)} \end{cases}$

Add **(1)** and **(2)** to get an equation in terms of x:

$$x^2+x-2=0$$
$$(x+2)(x-1)=0$$
$$x=-2,\ 1$$

Now, substitute each value of x back into **(2)** to find the corresponding values of y:

$$x=-2: \quad -2-y=5 \text{ so that } y=-7$$
$$x=1: \quad 1-y=5 \text{ so that } y=-4$$

So, the solutions are $\boxed{(-2,-7) \text{ and } (1,-4)}$.

59. Solve the system $\begin{cases} x^2+y^2=5 & \textbf{(1)} \\ 2x^2-y=0 & \textbf{(2)} \end{cases}$

Solve **(2)** for y: $y=2x^2$ **(3)**

Substitute **(3)** into **(1)** to get an equation in x:

$$x^2+(2x^2)^2=5$$
$$x^2+4x^4=5$$
$$4x^4+x^2-5=0$$
$$\left(4x^2+5\right)\left(x^2-1\right)=0$$
$$x=\pm 1$$

Now, substitute each value of x back into **(2)** to find the corresponding values of y:

$$x=1:\quad y=2(1)^2=2$$
$$x=-1:\quad y=2(-1)^2=2$$

So, the solutions are $\boxed{(1,2) \text{ and } (-1,2)}$.

61. Solve the system $\begin{cases} x+y=3 & \textbf{(1)} \\ x^2+y^2=4 & \textbf{(2)} \end{cases}$

Solve **(1)** for y: $y=3-x$ **(3)**

Substitute **(3)** into **(2)** to get an equation in terms of y:

$$x^2+(3-x)^2=4$$
$$x^2+9-6x+x^2=4$$
$$2x^2-6x+5=0$$
$$x=\tfrac{6\pm\sqrt{36-40}}{2(2)} \text{ which are not real numbers}$$

Hence, the system has $\boxed{\text{no solution}}$.

63. Solve the system $\begin{cases} x^2+xy+y^2=-12 & \textbf{(1)} \\ x-y=2 & \textbf{(2)} \end{cases}$

Solve **(2)** for x: $x=y+2$ **(3)**

Substitute **(3)** into **(1)** to get an equation in terms of y:

$$(y+2)^2+(y+2)y+y^2=-12$$
$$y^2+4y+4+y^2+2y+y^2=-12$$
$$3y^2+6y+16=0 \text{ so that } y=\tfrac{-6\pm\sqrt{36-4(3)(16)}}{2} \text{ which are not real numbers}$$

Hence, the system has $\boxed{\text{no solution}}$.

65. Solve the system $\begin{cases} x^3 - y^3 = -19 & \textbf{(1)} \\ x - y = -1 & \textbf{(2)} \end{cases}$

Solve **(2)** for x: $x = y - 1$ **(3)**

Substitute **(3)** into **(1)** to get an equation in terms of x:

$$\underbrace{(y-1)^3}_{(y-1)(y^2-2y+1)} - y^3 = -19$$

$$(y^3 - 2y^2 + y - y^2 + 2y - 1) - y^3 + 19 = 0$$

$$-3y^2 + 3y + 18 = 0$$

$$y^2 - y - 6 = 0$$

$$(y+2)(y-3) = 0 \text{ so that } y = -2,\ 3$$

Now, substitute these values of y back into **(3)** to find the corresponding values of x:

$$y = -2:\ x = -2 - 1 = -3, \qquad y = 3: \quad x = 3 - 1 = 2$$

So, the solutions are $\boxed{(2,3) \text{ and } (-3,-2)}$.

67. Solve the system $\begin{cases} \dfrac{2}{x^2} + \dfrac{1}{y^2} = 15 & \textbf{(1)} \\ \dfrac{1}{x^2} - \dfrac{1}{y^2} = -3 & \textbf{(2)} \end{cases}$

Add **(1)** and **(2)** to get an equation in terms of x:

$$\frac{3}{x^2} = 12$$

$$12x^2 = 3$$

$$x^2 = \tfrac{1}{4} \text{ so that } x = \pm\tfrac{1}{2}.$$

Now, substitute these values of x back into **(2)** to find the corresponding value of y:

$$x = \tfrac{1}{2}:\ \frac{1}{(\frac{1}{2})^2} - \frac{1}{y^2} = -3 \qquad\qquad x = -\tfrac{1}{2}:\ \frac{1}{(-\frac{1}{2})^2} - \frac{1}{y^2} = -3$$

$$4 - \frac{1}{y^2} = -3 \qquad\qquad 4 - \frac{1}{y^2} = -3$$

$$7 = \frac{1}{y^2} \qquad\qquad 7 = \frac{1}{y^2}$$

$$7y^2 = 1 \qquad\qquad 7y^2 = 1$$

$$y^2 = \tfrac{1}{7} \qquad\qquad y^2 = \tfrac{1}{7}$$

$$y = \pm\sqrt{\tfrac{1}{7}} \qquad\qquad y = \pm\sqrt{\tfrac{1}{7}}$$

So, the solutions are $\boxed{(\tfrac{1}{2}, \tfrac{1}{\sqrt{7}}),\ (-\tfrac{1}{2}, \tfrac{1}{\sqrt{7}}),\ (\tfrac{1}{2}, -\tfrac{1}{\sqrt{7}}), \text{ and } (-\tfrac{1}{2}, -\tfrac{1}{\sqrt{7}})}$.

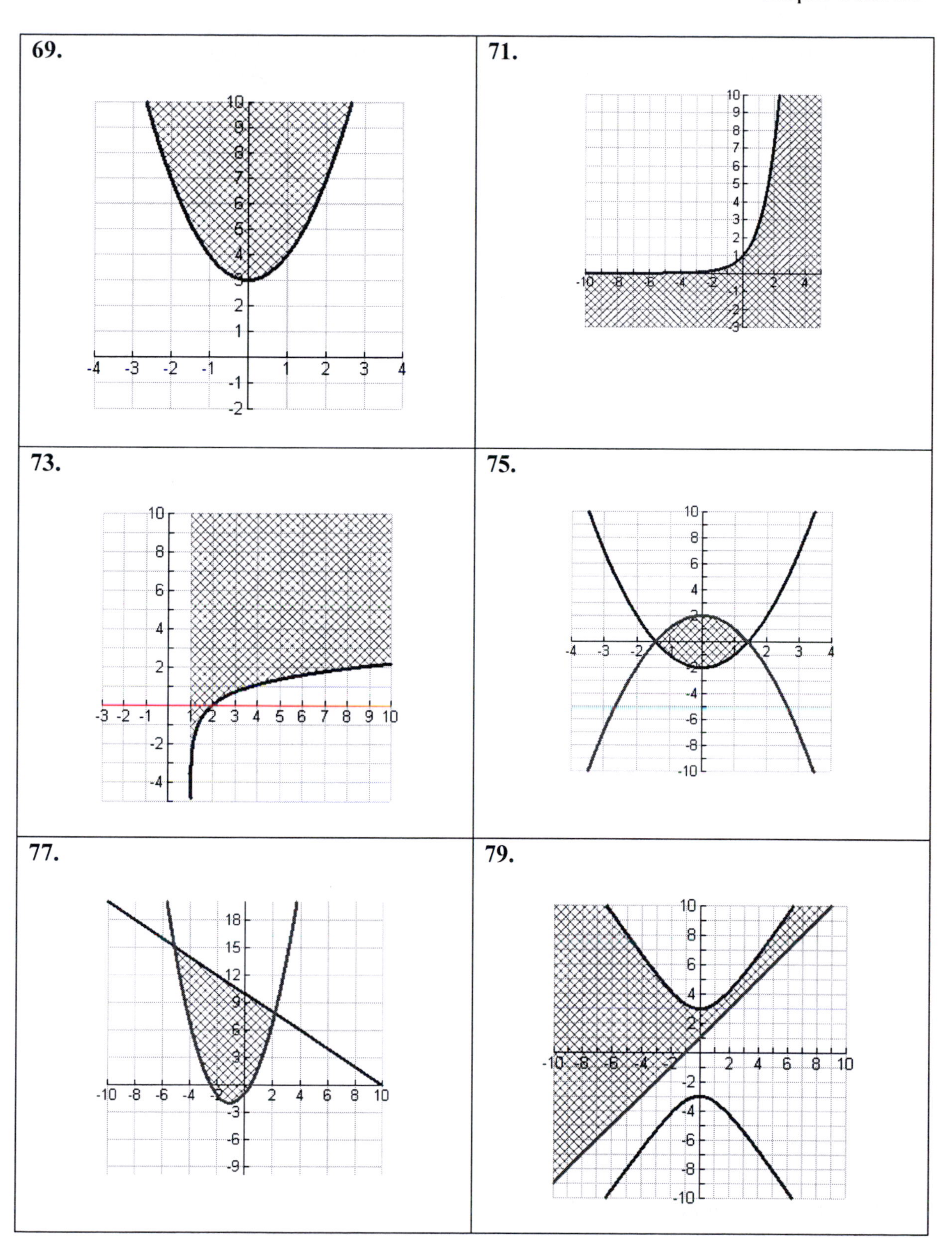

69.
71.
73.
75.
77.
79.

81. The vertex is at (0.6, -1.2), and it should open down. The graph below confirms this:

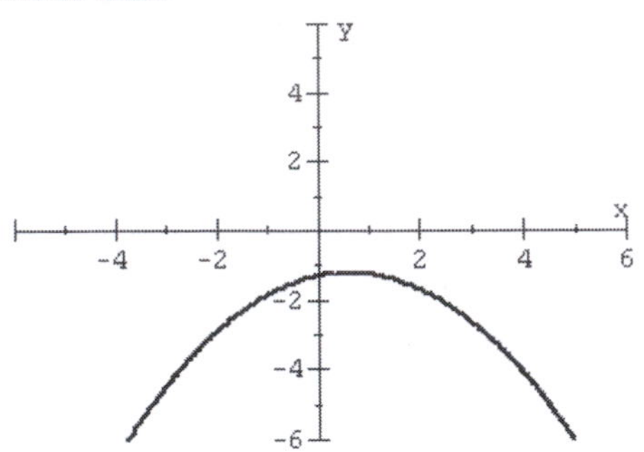

83. From the graphs, we see that as c increases the minor axis (along the x-axis) of the ellipse whose equation is given by $4(cx)^2 + y^2 = 1$ decreases.

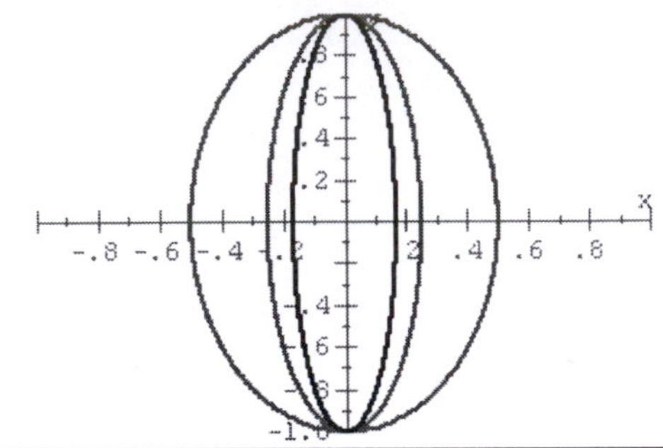

85. From the graphs, we see that as c increases, the vertices of the hyperbola whose equation is given by $4(cx)^2 - y^2 = 1$ are at $\left(\pm\frac{1}{2c}, 0\right)$ are moving towards the origin:

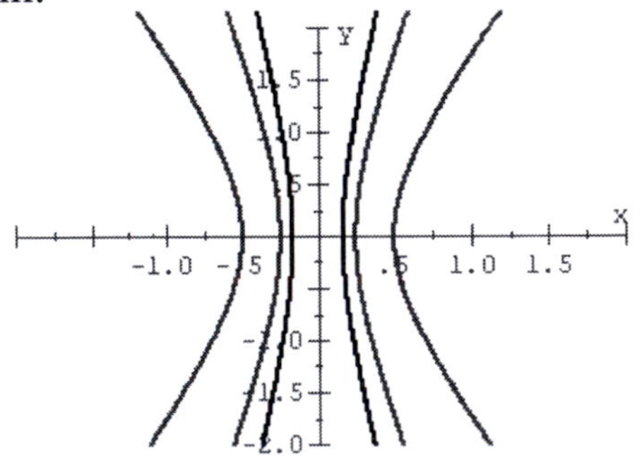

87. The graph of this nonlinear system is as follows:

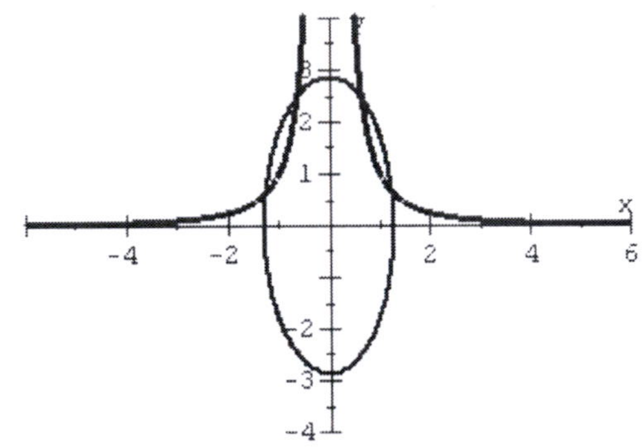

The points of intersection are approximately (0.635, 2.480), (–0.635, 2.480), (–1.245, 0.645), and (1.245, 0.645).

89. The dark region below is the one desired:

Chapter 8 Practice Test Solutions---

1. c	Parabola opens to the right
3. d	Ellipse is horizontal
5. f	Ellipse is vertical

7. Vertex is $(0,0)$ and focus is $(-4,0)$. So, the parabola opens to the left. As such, the general form is $y^2 = 4px$. In this case, $p = -4$, so the equation is $\boxed{y^2 = -16x}$.
9. Vertex is $(-1,5)$ and focus is $(-1,2)$. So, the parabola opens down. As such, the general form is $(x+1)^2 = 4p(y-5)$. In this case, $p = -3$, so the equation is $\boxed{(x+1)^2 = -12(y-5)}$.
11. Since the foci are $(0,-3), (0,3)$ and the vertices are $(0,-4), (0,4)$, we know: i) the ellipse is vertical with center at $(0,0)$, ii) $c = 3$, iii) the length of the major axis is $4-(-4) = 8$. Hence, $a = 4$. Now, since $c^2 = a^2 - b^2$, $9 = 16 - b^2$ so that $b^2 = 7$. Hence, the equation of the ellipse is $\boxed{\frac{x^2}{7} + \frac{y^2}{16} = 1}$.
13. Since the foci are $(2,-4), (2,4)$ and the vertices are $(2,-6), (2,6)$, we know: i) the ellipse is vertical with center at $(2, \frac{-6+6}{2}) = (2,0)$, ii) $c = 4$, iii) the length of the major axis is $6-(-6) = 12$. Hence, $a = 6$. Now, since $c^2 = a^2 - b^2$, $16 = 36 - b^2$ so that $b^2 = 20$. Hence, the equation of the ellipse is $\boxed{\frac{(x-2)^2}{20} + \frac{y^2}{36} = 1}$.
15. Since the vertices are $(-1,0), (1,0)$, and the center is the midpoint of the segment connecting them, the center must be (0,0). Moreover, since the form of the vertices is $(h \pm a, k)$ (which in this case is $(0 \pm a, 0)$) we see that $a = 1$. It remains to find *b*. At this point, we need to use the fact that we are given that the equations of the asymptotes are $y = \pm 2x$. Since the transverse axis is parallel to the *x*-axis, we know that the slopes of the asymptotes for such a hyperbola are $\pm \frac{b}{a}$. Thus, $\pm \frac{b}{1} = \pm 2$, which implies that $b = \pm 2$. Hence, the equation of the hyperbola must be $\boxed{x^2 - \frac{y^2}{4} = 1}$.
17. Since the vertices are $(2,-4), (2,4)$, we know that the center is $(2,0)$ and the transverse axis is parallel to the *y*-axis. Hence, $0 - a = -4$ so that $a = 4$. Also, since the foci are $(2,-6), (2,6)$, we know that $0 - c = -6$ so that $c = 6$. Now, to find *b*, we substitute the values of *c* and *a* obtained above into $c^2 = a^2 + b^2$ to see that $36 = 16 + b^2$ so that $b^2 = 20$. Hence, the equation of the hyperbola is $\boxed{\frac{y^2}{16} - \frac{(x-2)^2}{20} = 1}$.

19.

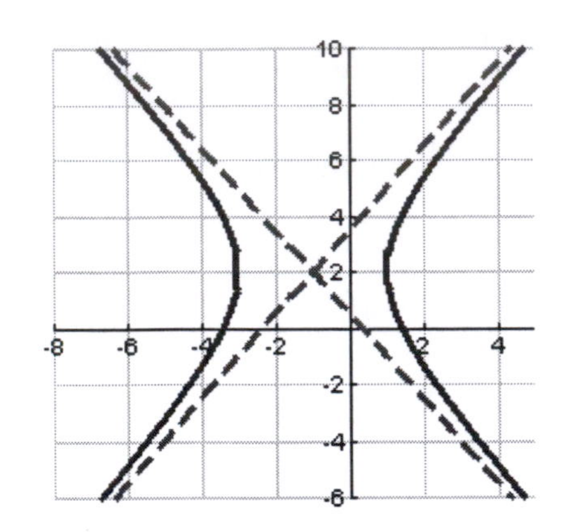

Notes on the Graph:
First, write the equation in standard form by completing the square:

$$9\left(x^2+2x\right)-4\left(y^2-4y\right)=43$$

$$9\left(x^2+2x+1\right)-4\left(y^2-4y+4\right)=43+9-16$$

$$9(x+1)^2-4(y-2)^2=36$$

$$\tfrac{(x+1)^2}{4}-\tfrac{(y-2)^2}{9}=1$$

The equations of the asymptotes are $y=\pm\frac{3}{2}(x+1)+2$.

21.

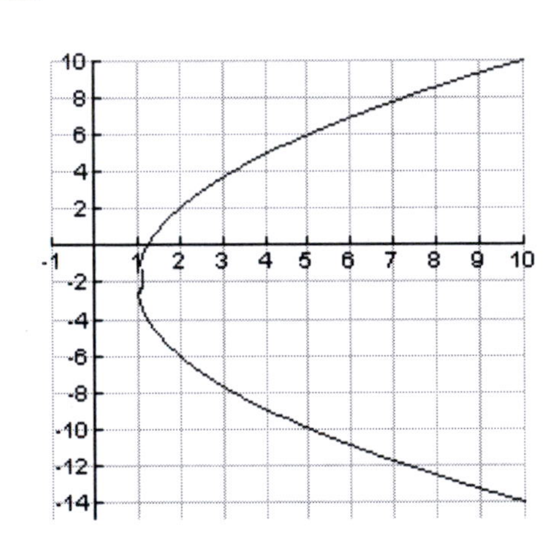

Equation: Completing the square yields:

$$y^2+4y=16x-20$$

$$(y^2+4y+4)=16x-20+4$$

$$(y+2)^2=16x-16$$

$$(y+2)^2=16(x-1)$$

$$(y+2)^2=4(4)(x-1) \quad \textbf{(1)}$$

So, $p=4$ and opens to the right and its vertex is $(1,-2)$.

23. Assume that the vertex is at the origin and that the parabola opens up. We are given that $p=1.5$. Hence, the equation is $x^2=4(1.5)y=6y,\quad -2\le x\le 2$.

25. Solve the system: $\begin{cases} x^2+y=4 & \textbf{(1)} \\ -x^2+y=-4 & \textbf{(2)} \end{cases}$

Solve **(2)** for y: $y=x^2-4$ **(3)**
Substitute **(3)** into **(1)** and solve for x:

$$x^2+x^2-4=4$$

$$2x^2=8$$

$$x=\pm 2$$

Substitute these back into **(3)** to find y, and conclude that the solutions of the system are $(\pm 2,0)$.

27.

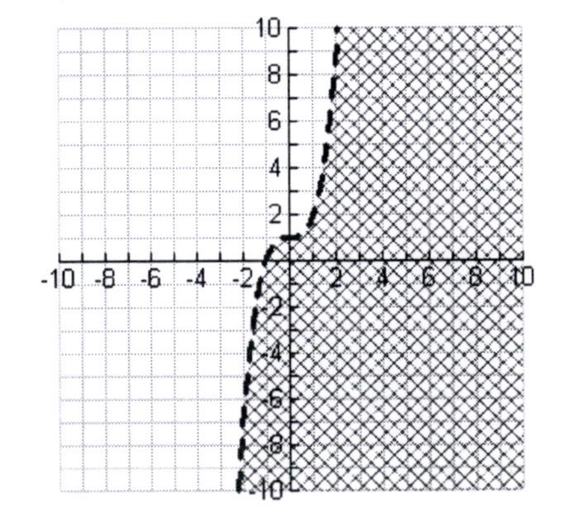

29.

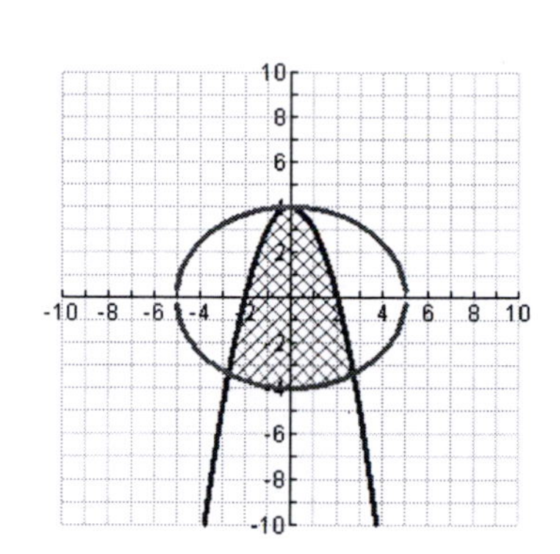

31. The graph using the calculator is as follows:

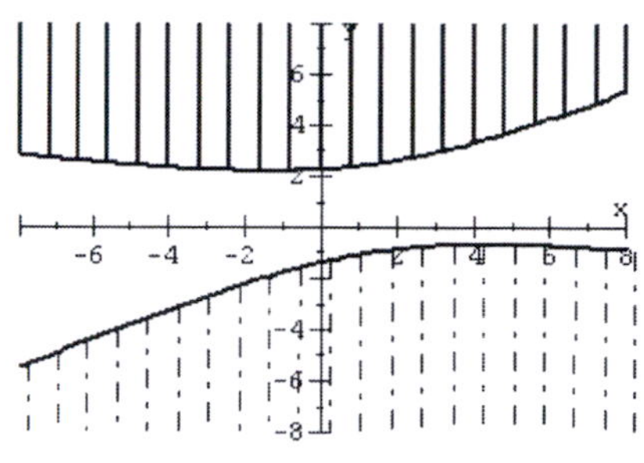

Chapter 8 Cumulative Review

1.

$$(2-x)^2 = -16$$
$$2-x = \pm 4i$$
$$\boxed{x = 2 \pm 4i}$$

3.

$$m = \frac{\frac{3}{4}-\left(-\frac{1}{4}\right)}{\frac{1}{3}-\left(-\frac{1}{6}\right)} = \boxed{2}$$

5. $\frac{f(x+h)-f(x)}{h} = \frac{\left(5-(x+h)^2\right)-\left(5-x^2\right)}{h} = \frac{-2hx-h^2}{h} = \boxed{-2x-h}$

7. This is undefined since $f(-1)$ is undefined.

9. 0 (multiplicity 2), -2 (multiplicity 3)

11.

Vertical asymptote: $x = 2$
Horizontal asymptote: none
Slant asymptote: $y = x+2$, as seen from the synthetic divison below:

$$\begin{array}{r|rrr} 2 & 1 & 0 & 3 \\ & & 2 & 4 \\ \hline & 1 & 2 & 7 \end{array}$$

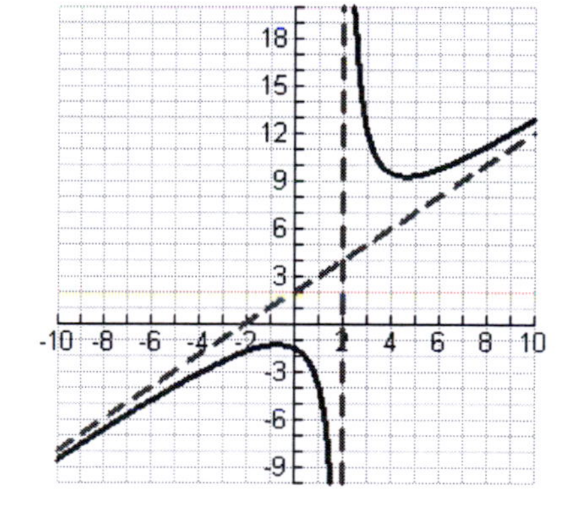

13. Use $A = Pe^{rt}$.

$$85{,}000 = Pe^{0.055(15)}$$
$$P = \frac{85{,}000}{e^{0.055(15)}} \approx \boxed{\$37{,}250}$$

15.

$$10^{-3\log 10} = 10^{\log 10^{-3}} = \boxed{0.001}$$

17. Let x = cost of a soda and
y = cost of a soft pretzel.

Solve the system:

$$\begin{cases} 3x+2y=6.77 & \textbf{(1)} \\ 5x+4y=12.25 & \textbf{(2)} \end{cases}$$

Multiply **(1)** by -5 and **(2)** by 3, and add:

$$\begin{aligned} -15x-10y &= -33.85 \\ +\ 15x+12y &= 36.75 \\ \hline 2y &= 2.90 \\ y &= 1.45 \end{aligned}$$

Substitute this value of y into **(1)** to solve for x:

$$3x+2(1.45)=6.77 \Rightarrow x=1.29$$

So, a soda costs \$1.29 and a soft pretzel costs \$1.45.

19.

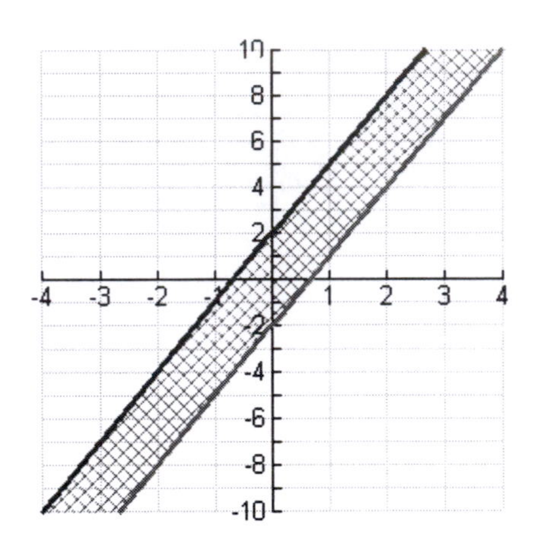

21.

$$2B-3A=\begin{bmatrix} 16 & -4 & 12 \\ 18 & 0 & -2 \end{bmatrix}-\begin{bmatrix} 9 & 12 & -21 \\ 0 & 3 & 15 \end{bmatrix}$$
$$=\begin{bmatrix} 7 & -16 & 33 \\ 18 & -3 & -17 \end{bmatrix}$$

23. Since the vertices are (6,3) and (6,-7), the center is $\left(6,\frac{3-7}{2}\right)=(6,-2)$. Using the foci, we see that

$$-2+c=2 \text{ and } -2-c=-6,$$

so that $c=4$. Also, we have

$$-2+a=3 \text{ and } -2-a=-7,$$

so that $a=5$.
As such, $c^2=a^2-b^2 \Rightarrow b=3$.
Thus, the equation of the ellipse is

$$\frac{(x-6)^2}{9}+\frac{(y+2)^2}{25}=1.$$

25. Solve the system:

$$\begin{cases} x + y = 6 & \textbf{(1)} \\ x^2 + y^2 = 20 & \textbf{(2)} \end{cases}$$

Solve **(1)** for y: $y = 6 - x$ **(3)**

Substitute **(3)** into **(2)** and solve for x:

$$x^2 + (6-x)^2 = 20$$
$$2x^2 - 12x + 36 = 20$$
$$x^2 - 6x + 8 = 0$$
$$(x-4)(x-2) = 0$$
$$x = 2, 4$$

Substitute each value into **(1)** to find the corresponding y-value. This yields two solutions, namely (2,4) and (4,2).

27. The graph from the calculator is as follows:

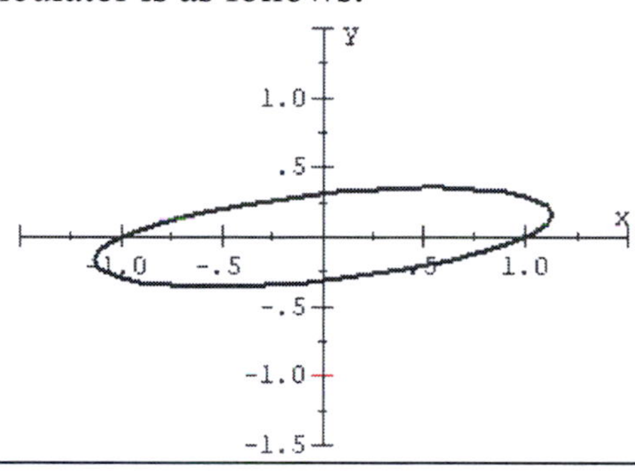

CHAPTER 9

Section 9.1 Solutions--

1. $1,2,3,4$	**3.** $1,3,5,7$
5. $\frac{1}{2},\frac{2}{3},\frac{3}{4},\frac{4}{5}$	**7.** $2, \underbrace{\frac{4}{2!}}_{=\frac{4}{2\cdot 1}}, \underbrace{\frac{8}{3!}}_{=\frac{8}{3\cdot 2\cdot 1}}, \underbrace{\frac{16}{4!}}_{=\frac{16}{4\cdot 3\cdot 2\cdot 1}},$ which is equivalent to $2,\ 2,\frac{4}{3},\frac{2}{3}$
9. $-x^2,x^3,-x^4,x^5$	**11.** $\frac{-1}{2\cdot 3},\frac{1}{3\cdot 4},\frac{-1}{4\cdot 5},\frac{1}{5\cdot 6}$ which is equivalent to $\frac{-1}{6},\frac{1}{12},\frac{-1}{20},\frac{1}{30}$
13. $\left(\frac{1}{2}\right)^9=\frac{1}{512}$	**15.** $\frac{(-1)^{19}19!}{(21)!}=\frac{-1}{21\cdot 20}=\frac{-1}{420}$
17. $\left(1+\frac{1}{100}\right)^2=\left(\frac{101}{100}\right)^2=\frac{10,201}{10,000}$	**19.** $\log 10^{23}=23\log 10=23$
21. $a_n=2n, n\geq 1$	**23.** $a_n=\frac{1}{(n+1)n}, n\geq 1$
25. $a_n=(-1)^n\left(\frac{2}{3}\right)^n=\frac{(-1)^n 2^n}{3^n},\ n\geq 1$	**27.** $a_n=(-1)^{n+1}, n\geq 1$
29. $\frac{9!}{7!}=\frac{9\cdot 8\cdot \cancel{7!}}{\cancel{7!}}=72$	**31.** $\frac{29!}{27!}=\frac{29\cdot 28\cdot \cancel{27!}}{\cancel{27!}}=812$
33. $\frac{75!}{77!}=\frac{\cancel{75!}}{77\cdot 76\cdot \cancel{75!}}=\frac{1}{5852}$	**35.** $\frac{97!}{93!}=\frac{97\cdot 96\cdot 95\cdot 94\cdot \cancel{93!}}{\cancel{93!}}=83,156,160$
37. $\frac{(n-1)!}{(n+1)!}=\frac{\cancel{(n-1)!}}{(n+1)n\cancel{(n-1)!}}=\frac{1}{(n+1)n}$	**39.** $\frac{(2n+3)!}{(2n+1)!}=\frac{(2n+3)(2n+2)\cancel{(2n+1)!}}{\cancel{(2n+1)!}}$ $=(2n+3)(2n+2)$
41. $7,10,13,16$	**43.** $1,2,6,24$

45.

$$100,\ \underbrace{\frac{100}{2!}}_{=50},\ \underbrace{\frac{50}{3!}}_{=\frac{50}{3\cdot 2}=\frac{25}{3}},\ \underbrace{\frac{\frac{25}{3}}{4!}}_{=\frac{25}{3\cdot 4\cdot 3\cdot 2}=\frac{25}{72}}$$ which is equivalent to $100, 50, \frac{25}{3}, \frac{25}{72}$.

47. $1,2,2,4$ (Note that $a_3 = a_2 \cdot a_1,\ a_4 = a_3 \cdot a_2$)

49. $$1, -1, -\underbrace{\left[(1)^2 + (-1)^2\right]}_{=-2},\ \underbrace{\left[(-1)^2 + (-2)^2\right]}_{=5}$$

which is equivalent to $1, -1, -2, 5$

51. $2 \cdot 5 = 10$

53. $0^2 + 1^2 + 2^2 + 3^2 + 4^2 = 30$

55. $$\sum_{n=1}^{6} (2n-1) = 1+3+5+7+9+11 = 36$$

57. $1^n = 1$, for all n. So, $\sum_{n=0}^{4} 1^n = \sum_{n=0}^{4} 1 = 5(1) = 5$.

59. $1 - x + x^2 - x^3$

61. $$\frac{2^0}{0!} + \frac{2^1}{1!} + \underbrace{\frac{2^2}{2!}}_{=\frac{4}{2\cdot 1}} + \underbrace{\frac{2^3}{3!}}_{=\frac{8}{3\cdot 2}} + \underbrace{\frac{2^4}{4!}}_{=\frac{16}{4\cdot 3\cdot 2}} + \underbrace{\frac{2^5}{5!}}_{=\frac{32}{5\cdot 4\cdot 3\cdot 2}} = 1 + 2 + 2 + \frac{4}{3} + \frac{2}{3} + \frac{4}{15} = 7 + \frac{4}{15} = \frac{109}{15}$$

63. $$\frac{1}{0!} + \frac{x}{1!} + \frac{x^2}{2!} + \frac{x^3}{3!} + \frac{x^4}{4!} = 1 + x + \frac{x^2}{2} + \frac{x^3}{6} + \frac{x^4}{24}$$

65. $2\frac{(0.1)^0}{1-0.1} = \frac{2}{0.9} = 2.\overline{2} = \frac{20}{9}$

67. Not possible. (The result is infinite.)

69. $\sum_{n=0}^{6} (-1)^n \frac{1}{2^n}$

71. $\sum_{n=1}^{\infty} (-1)^{n-1} n$

73. $\sum_{n=1}^{6} \frac{(n+1)!}{(n-1)!} = \sum_{n=1}^{5} n(n+1)$

(Remember that $0! = 1! = 1$.)

75. $\sum_{n=1}^{\infty} (-1)^{n-1} \frac{x^{n-1}}{(n-1)!} = \sum_{n=0}^{\infty} (-1)^n \frac{x^n}{n!}$

77. $A_{72} = 20{,}000\left(1 + \frac{0.06}{12}\right)^{72} \approx \$28{,}640.89$.

So, she has approximately \$28,640.89 in her account after 6 years (or 72 months).

79. Let $n =$ number of years experience. Then, the salary per hour is given by

$$a_n = 20 + 2n,\ n \geq 0.$$

So, $a_{20} = 20 + 2(20) = 60$. Thus, a paralegal with 20 years experience would make $60 per hour.

81. Let $n =$ number of years on the job. Then, the salary is given recursively by

$$a_0 = 30{,}000$$

$$a_n = \underbrace{a_{n-1}}_{\text{previous year}} + \underbrace{0.03a_{n-1}}_{\text{raise}} = 1.03a_{n-1}$$

83. Let $n =$ number of years.
Then, the number of T-cells in body is given by:

$$a_1 = 1000$$

$$a_{n+1} = 1000 - 75n,\ n \geq 1.$$

We must find n such that $a_{n+1} \leq 200$.
To do so, note that the above formula can be expressed explicitly as $a_{n+1} = 1000 - 75n$.
So, we must solve $1000 - 75n = 200$:

$$75n = 800 \Rightarrow n \approx 10.7$$

As such, after approximately 10.7 years, the person would have full blown AIDS.

85. Observe that

$$A_1 = 100{,}000(1.001 - 1) = 100$$

$$A_2 = 100{,}000\left((1.001)^2 - 1\right) = 200.10$$

$$A_3 = 100{,}000\left((1.001)^3 - 1\right) \approx 300.30$$

$$A_4 = 100{,}000\left((1.001)^4 - 1\right) \approx 400.60$$

$$A_{36} = 100{,}000\left((1.001)^{36} - 1\right) \approx 3663.72$$

87.

	At $x = 2$
1	1
$1 + x$	3
$1 + x + \frac{x^2}{2}$	5
$1 + x + \frac{x^2}{2} + \frac{x^3}{3!}$	≈ 6.33
$1 + x + \frac{x^2}{2} + \frac{x^3}{3!} + \frac{x^4}{4!}$	≈ 7.0
	Approx. of $e^2 \approx 7.38906$

89.

	At $x = 1.1$
$(x-1)$	0.1
$(x-1) - \frac{(x-1)^2}{2}$	0.0950
$(x-1) - \frac{(x-1)^2}{2} + \frac{(x-1)^3}{3}$	0.0953
$(x-1) - \frac{(x-1)^2}{2} + \frac{(x-1)^3}{3} - \frac{(x-1)^4}{4}$	0.09531
$(x-1) - \frac{(x-1)^2}{2} + \frac{(x-1)^3}{3} - \frac{(x-1)^4}{4} + \frac{(x-1)^5}{5}$	0.095310 Approx of ln(1.1)

91. The mistake is that $6! \neq 3!2!$, but rather $6! = 6\cdot5\cdot4\cdot3\cdot2\cdot1$.

93.

$$(-1)^{n+1} = \begin{cases} 1, & n = 1,3,5,\ldots \\ -1, & n = 2,4,6,\ldots \end{cases}$$

So, the terms should all be the opposite sign.

95. True.

97. False. Let $a_k = 1 = b_k$ and $n = 5$.
Then, observe that

$$\sum_{k=1}^{n} a_k b_k = \sum_{k=1}^{5} 1 = 1\cdot5 = 5$$

whereas

$$\left(\sum_{k=1}^{n} a_k\right)\cdot\left(\sum_{k=1}^{n} b_k\right) = \left(\sum_{k=1}^{5} 1\right)\cdot\left(\sum_{k=1}^{5} 1\right) = 25.$$

99.

$$a_1 = C$$
$$a_2 = (C) + D$$
$$a_3 = (C + D) + D = C + 2D$$
$$a_4 = (C + 2D) + D = C + 3D$$

101.

$$F_1 = \frac{(1+\sqrt{5})-(1-\sqrt{5})}{2\sqrt{5}} = \frac{2\sqrt{5}}{2\sqrt{5}} = 1$$

$$F_2 = \frac{(1+\sqrt{5})^2-(1-\sqrt{5})^2}{2^2\sqrt{5}} = \frac{1+2\sqrt{5}+5-1+2\sqrt{5}-5}{4\sqrt{5}} = 1$$

103. Using a calculator yields the following table of values:

n	$\left(1+\frac{1}{n}\right)^n$
100	$2.704813 \approx 2.705$
1000	$2.716923 \approx 2.717$
10000	$2.718146 \approx 2.718$
100000	2.718268
1000000	2.718280
10000000	2.718281693
$\downarrow$	$\downarrow$
∞	e

105. The calculator gives $\frac{109}{15}$, which agrees with Exercise 61.

Section 9.2 Solutions---

1. Yes, $d=3$.	**3.** No.
5. Yes, $d=-0.03$.	**7.** Yes, $d=2/3$.
9. No.	**11.** First four terms: $3,1,-1,-3$ So, yes and $d=-2$.
13. First four terms: $1,4,9,16$ So, no.	**15.** First four terms: $2,7,12,17$ So, yes and $d=5$.
17. First four terms: $0,10,20,30$ So, yes and $d=10$.	**19.** First four terms: $-1,2,-3,4$ So, no.
21. $a_n=11+(n-1)5=5n+6$	**23.** $a_n=-4+(n-1)(2)=-6+2n$
25. $a_n=0+(n-1)\frac{2}{3}=\frac{2}{3}n-\frac{2}{3}$	**27.** $a_n=0+(n-1)e=en-e$

29. Here, $a_1=7,\ d=13$. So, $a_n=7+(n-1)13=13n-6$. Thus, $a_{10}=130-6=124$.

31. Here, $a_1=9,\ d=-7$. So, $a_n=9+(n-1)(-7)=-7n+16$ Thus, $a_{100}=-700+16=-684$.

33. Here, $a_1=\frac{1}{3},\ d=\frac{1}{4}$. So, $a_n=\frac{1}{3}+(n-1)(\frac{1}{4})$. Thus, $a_{21}=\frac{1}{3}+(20)(\frac{1}{4})=\frac{16}{3}$.

35. $a_5=44$, $a_{17}=152$

Find d: $a_{17}=a_5+12d$
$$152=44+12d$$
$$9=d$$
Find a_1: $a_5=a_1+4(9)$
$$44=a_1+36$$
$$8=a_1$$
So, $a_n=8+(n-1)9=9n-1$

37. $a_7=-1$, $a_{17}=-41$

Find d: $a_{17}=a_7+10d$
$$-41=-1+10d$$
$$-40=10d$$
$$-4=d$$
Find a_1: $a_7=a_1+6(-4)$
$$-1=a_1-24$$
$$23=a_1$$
So, $a_n=23+(n-1)(-4)=-4n+27$

39. $a_4=3$, $a_{22}=15$

Find d: $a_{22}=a_4+18d$
$$15=3+18d$$
$$12=18d$$
$$\tfrac{2}{3}=d$$
Find a_1: $a_4=a_1+3\left(\frac{2}{3}\right)$
$$3=a_1+2$$
$$1=a_1$$
So, $a_n=1+(n-1)\frac{2}{3}=\frac{2}{3}n+\frac{1}{3}$

41. $\sum_{k=1}^{23}2k=2\sum_{k=1}^{23}k=\frac{23}{2}[2+46]=552$

43. $\sum_{n=1}^{30}(-2n+5)=-2\sum_{n=1}^{30}n+\sum_{n=1}^{30}5=\frac{30}{2}[3-55]=-930+150=-780$

45. $\sum_{j=3}^{14} 0.5j = \left[\sum_{j=1}^{14} 0.5j - \sum_{j=1}^{2} 0.5j\right] = \frac{14}{2}[0.5+7] - \frac{3}{2} = 0.5[102] = 51$

47. $\underbrace{2+7+12+\ldots+62}_{\text{arithmetic}}$. We need to find n such that $a_n = 62$. To this end, observe that $a_1 = 2$ and $d = 5$. Thus, since the sequence is arithmetic, $a_n = 2+(n-1)5 = 5n-3$. So, we need to solve $a_n = 5n-3 = 62$: $5n = 65$, so that $n = 13$. Therefore, this sum equals $\frac{13}{2}[2+62] = 416$.

49. $\underbrace{4+7+10+\ldots+151}_{\text{arithmetic}}$. We need to find n such that $a_n = 151$. To this end, observe that $a_1 = 4$ and $d = 3$. Thus, since the sequence is arithmetic, $a_n = 4+(n-1)3 = 3n+1$. So, we need to solve $a_n = 3n+1 = 151$: $3n = 150$, so that $n = 50$. Therefore, this sum equals $= \frac{50}{2}(4+151) = 3875$.

51. $\underbrace{\frac{1}{6} - \frac{1}{6} - \frac{1}{2} - \ldots - \frac{13}{2}}_{\text{arithmetic}}$. We need to find n such that $a_n = -\frac{13}{2}$. To this end, observe that $a_1 = \frac{1}{6}$ and $d = -\frac{1}{3}$. Since the sequence is arithmetic, $a_n = \frac{1}{6} + (n-1)\left(\frac{-1}{3}\right) = -\frac{1}{3}n + \frac{1}{2}$.
So, we need to solve $a_n = -\frac{1}{3}n + \frac{1}{2} = -\frac{13}{2}$: $\frac{-1}{3}n = -7$, so that $n = 21$. Therefore, this sum equals $\frac{21}{2}\left[\frac{1}{6} - \frac{13}{2}\right] = \frac{21}{2}\left(\frac{1-39}{6}\right) = -\frac{133}{2}$.

53. Here, $d = 4$ and $a_1 = 1$. Hence, $a_n = 1 + (n-1)(4) = 4n-3$. So,
$S_n = \frac{n}{2}(a_1 + a_n) = \frac{n}{2}(1+4n-3) = n(2n-1)$. As such, $S_{18} = 18(35) = 630$.

55. Here, $d = -\frac{1}{2}$ and $a_1 = 1$. Hence, $a_n = 1 + (n-1)\left(-\frac{1}{2}\right) = -\frac{1}{2}n + \frac{3}{2}$. So,
$S_n = \frac{n}{2}(a_1 + a_n) = \frac{n}{2}\left(1 - \frac{1}{2}n + \frac{3}{2}\right) = \frac{n}{2}\left(-\frac{1}{2}n + \frac{5}{2}\right)$. As such, $S_{43} = \frac{43}{4}(5-43) = -\frac{817}{2}$.

57. Here, $d = 10$ and $a_1 = -9$. Hence, $a_n = -9 + (n-1)(10) = 10n - 19$. So,
$S_n = \frac{n}{2}(a_1 + a_n) = \frac{n}{2}(-9 + 10n - 19) = n(5n-14)$. As such, $S_{18} = 18(5(18) - 14) = 1368$.

59.

Colin	Camden
$a_1 = 28{,}000$	$a_1 = 25{,}000$
$d = 1500$	$d = 2000$
So $a_n = 28{,}000 + (n-1)(1500)$	So, $a_n = 25{,}000 + (n-1)2000$

$=26,500+1500n$ Thus, $a_{10}=41,500$. After 10 years, Colin will have accumulated $\frac{10}{2}(28,000+41,500)$ = \$347,500.	$=23,000+2000n$ So, $a_{10}=43,000$ After 10 years, Camden will have accumulated $\frac{10}{2}(25,000+43,000)$ = \$340,000.

61. We are given that $a_1=22$. We seek $\sum_{i=1}^{25} a_i$. Observe that $d=1$. So, $a_n=22+(n-1)(1)=21+n$. Thus, $a_{25}=21+25=46$. As such, using the formula $S_n=\frac{n}{2}(a_1+a_n)$, we conclude that $\sum_{i=1}^{25} a_i = =\frac{25}{2}[22+46]=850$.

63. We are given that $a_1=1$ and $n=56$. We need to find a_{56} and d. Well, note that $\sum_{i=1}^{56} a_i=30,856=\frac{56}{2}(a_1+a_{56})=\underbrace{\frac{56}{2}}_{28}(1+a_{56})$. Hence,

$$30,856=28(1+a_{56}), \text{ so that } 1101=\frac{30,828}{28}=a_{56}.$$

Now, since $a_n=a_1+(n-1)d$, we can substitute $n=56,\ a_1=1$ in to find d:

$$1101=1+(56-1)d, \text{ so that } 1100=55d \text{ and hence, } d=20.$$

So, there are 1101 glasses on the bottom row and each row had 20 fewer glasses than the one before.

65. We are given that $a_1=16, d=32, n=10$. Hence,

$$a_n=16+(n-1)(32)=-16+32n.$$

So, $a_{10}=-16+320=304$. Now, observe that $\sum_{i=1}^{10} a_i=\frac{10}{2}[16+304]$
$=1600$ feet in 10 seconds

67. The sum is 20 + 19 + 18 + . . . + 1. This is arithmetic with $a_1=20,\ d=-1$. So, $a_n=20+(n-1)(-1)=21-n$. Thus, $S_n=\frac{n}{2}(20+21-n)=\frac{n}{2}(41-n)$. So, $S_{20}=10(21)=210$. There are 210 oranges in the display.

69. $a_1+a_2+a_3+26+27+28+...+a_n$ Here, $d=1$, so that $a_1=23$.

a. There are 23 seats in the first row.

b. $a_n=23+(n-1)(1)=n+22$. So, $S_n=\frac{n}{2}(23+n+22)$, and hence, $S_{30}=1125$ seats in the theater.

71. $a_n=a_1+(n-1)d$, not a_1+nd

73. There are 11 terms, not 10. So, $n=11$, and thus, $S_{11}=\frac{11}{2}(1+21)=121$.

75. False. In a series you are adding terms, while in a sequence you are not.

77. True, since d must be constant. If a sequence alternates, the difference between consecutive terms would need to change sign.
79. $a+(a+b)+\ldots+(a+nb)=\sum_{k=1}^{n+1}(a+(k-1)b)=\frac{n+1}{2}(a+a+nb)=\frac{(n+1)(2a+nb)}{2}$
81. As n gets larger, $\frac{1}{n^2}$ goes to zero. So, $v=R\left(\frac{1}{k^2}-\frac{1}{n^2}\right)$ gets closer to $\frac{R}{k^2}$, which in this case is 27,419.5.
83. Compute $\sum_{i=1}^{100} i$ - you should get 5050.
85. Compute $\sum_{i=1}^{50}(2i-1)$ - you should get 2500.
87. 18,850

Section 9.3 Solutions---

1. Yes, $r=3$	**3.** No, $\frac{9}{4}\neq\frac{16}{9}$ for instance
5. Yes, $r=\frac{1}{2}$	**7.** Yes, $r=1.7$
9. $6, 18, 54, 162, 486$	**11.** $1, -4, 16, -64, 256$
13. $10{,}000,\ 10{,}600,\ 11{,}236,\ 11{,}910.16,\ 12{,}624.77$	**15.** $\frac{2}{3}, \frac{1}{3}, \frac{1}{6}, \frac{1}{12}, \frac{1}{24}$
17. $a_n=5(2)^{n-1}$	**19.** $a_n=1(-3)^{n-1}$
21. $a_n=1000(1.07)^{n-1}$	**23.** $a_n=\frac{16}{3}\left(-\frac{1}{4}\right)^{n-1}$
25. Since $a_1=-2,\ r=-2$, we have $a_n=-2(-2)^{n-1}$. In particular, $a_7=-2(-2)^{7-1}=-128$.	
27. Since $a_1=\frac{1}{3},\ r=2$, we have $a_n=\frac{1}{3}(2)^{n-1}$ In particular, $a_{13}=\frac{1}{3}(2)^{13-1}=\frac{4096}{3}$.	**29.** Since $a_1=1000,\ r=\frac{1}{20}$, we have $a_n=1000\left(\frac{1}{20}\right)^{n-1}$ In particular, $a_{15}=1000\left(\frac{1}{20}\right)^{15-1}\approx 6.10\times10^{-16}$.

31. Use the formula $S_n = a_1 \frac{(1-r^n)}{1-r}$.

Since $a_1 = \frac{1}{3}$, $n = 13$, $r = 2$, the given sum is $S_{13} = \frac{1}{3}\left(\frac{1-2^{13}}{1-2}\right) = \frac{8191}{3}$.

33. Use the formula $S_n = a_1 \frac{(1-r^n)}{1-r}$.

Since $a_1 = 2$, $n = 10$, $r = 3$, the given sum is

$$S_{10} = 2\left(\frac{1-3^{10}}{1-3}\right) = 59{,}048.$$

35. Use the formula $S_n = a_1 \frac{(1-r^n)}{1-r}$.

Since $a_1 = 2$, $r = 0.1$, $n = 11$, the given sum is

$$S_{11} = 2\left(\frac{1-(0.1)^{11}}{1-0.1}\right) = 2.\overline{2}.$$

(Note: The sum starts with $n = 0$, so there are 11 terms.)

37. Use the formula $S_n = a_1 \frac{(1-r^n)}{1-r}$.

Since $a_1 = 2$, $r = 3$, $n = 8$, the given sum is

$$S_8 = 2\left(\frac{1-3^8}{1-3}\right) = 6560.$$

39. Use the formula $S_n = a_1 \frac{(1-r^n)}{1-r}$.

Since $a_1 = 1$, $r = 2$, $n = 14$, the given sum is $S_{14} = 1\left(\frac{1-2^{14}}{1-2}\right) = 16{,}383$.

41. Use the formula $S_\infty = \frac{a_1}{1-r}$.

Since $a_1 = 1$, $r = \frac{1}{2}$, the given sum is

$$S_\infty = \frac{1}{1-\frac{1}{2}} = 2.$$

43. Use the formula $S_\infty = \frac{a_1}{1-r}$.

Since $a_1 = -\frac{1}{3}$, $r = -\frac{1}{3}$, the given sum is

$$S_\infty = \frac{-\frac{1}{3}}{1-\left(-\frac{1}{3}\right)} = -\frac{1}{4}.$$

45. Not possible - diverges.

47. Use the formula $S_\infty = \frac{a_1}{1-r}$.

Since $a_1 = -9$, $r = \frac{1}{3}$, the given sum is

$$S_\infty = \frac{-9}{1-\frac{1}{3}} = -\frac{27}{2}.$$

49. Use the formula $S_\infty = \frac{a_1}{1-r}$.

Since $a_1 = 10{,}000$, $r = 0.05$, the given sum is $S_\infty = \frac{10{,}000}{1-0.05} \approx 10{,}526$.

51. The sum is $\frac{0.4}{1-0.4}=\frac{2}{3}$.	**53.** The sum is $\frac{(0.99)^0}{1-0.99}=100$.
55. Observe that $a_1=34{,}000,\ r=1.025$. We need to find a_{12}. Observe that the nth term is given by $a_n=a_1r^{n-1}=34{,}000(1.025)^{n-1}$. Hence, $$a_{12}=34000(1.025)^{11} \approx 44{,}610.95$$ So, the salary after 12 years is approximately \$44,610.95.	**57.** Since $a_1=2000,\ r=0.5$, we see that $$a_n=\underbrace{2000(0.5)^n}_{\text{Value of laptop after } n \text{ years}}.$$ In particular, we have $$a_4=\underbrace{2000(0.5)^4=125}_{\text{Worth when graduating from college.}}$$ $$a_7=\underbrace{2000(0.5)^7=16}_{\text{Worth when graduating from graduate school.}}$$
59. Since $a_1=100,\ r=0.7$, we see that $$a_n=100(0.7)^n.$$ We need to find a_5. We see that $$a_5=100(0.7)^5\approx 17 \text{ feet}.$$	**61.** Since $$a_1=36{,}000,\ r=\frac{37{,}800}{36{,}000}=1.05,$$ we see that $a_n=36{,}000(1.05)^{n-1}$. So, in particular, $a_{11}=36{,}000(1.05)^{10}\approx 58{,}640$. As such, in the year 2010, there will be approximately 58,640 students.

63. Since $a_1=1000,\ r=0.90$, we know that $a_n=1000(0.9)^{n-1}$.
First, find the value of n_0 such that $a_{n_0}\le 1$.
Since
$$1000(0.9)^{n-1}<1 \quad\Rightarrow\quad (0.9)^{n-1}<0.001 \quad\Rightarrow\quad n>\frac{\ln(0.001)}{\ln(0.9)}\approx 67$$
We see that after 67 days, he will pay less than \$1 per day. Next, the total amount of money he paid in January is given by: $\sum_{n=1}^{31}1000(0.9)^{n-1}=1000\frac{\left(1-0.9^{31}\right)}{1-0.9}\approx 9618$

65. Use the formula $A = P\left(1+\frac{r}{n}\right)^{nt}$. Note that in the current problem, $r = 0.05,\ n = 12,$ so that the formula becomes:

$$A = P\left(1+\frac{0.05}{12}\right)^{12t} = P(1.0042)^{12t}$$

Let $t = \frac{n}{12}$, where n is the number of months of investment. Also, let $A_n = P(1.0042)^n$.

First deposit of \$100 gains interest for 36 months: $A_{36} = \$100(1.0042)^{36}$

Second deposit of \$100 gains interest for 35 months: $A_{35} = \$100(1.0042)^{35}$

$\vdots$

Last deposit of \$100 gains interest for 1 month: $A_1 = \$100(1.0042)^1$

Now, sum the amount accrued for 36 deposits:

$$A_1 + \ldots + A_{36} = \sum_{n=1}^{36} 100(1.0042)^{n-1}$$

Note that here $a_1 = 100, r = 1.0042$. So,

$$\text{So, } S_{36} = 100\left(\frac{1-(1.0042)^{36}}{1-1.0042}\right)$$

$$\approx 3877.64$$

So, they saved \$3,877.64 in 3 years.

67. Use the formula $A = P\left(1+\frac{r}{n}\right)^{nt}$. Note that in the current problem, $r = 0.06,\ n = 52,$ so that the formula becomes: $A = P\left(1+\frac{0.06}{52}\right)^{52t} = P(1.0012)^{52t}$

Let $t = \frac{n}{52}$, where n is the number of weeks of investment. Also, let $A_n = P(1.0012)^n$.

<u>**26 week investment**</u>:

First deposit of \$500 gains interest for 26 weeks: $A_{26} = 500(1.0012)^{26}$

Second deposit of \$500 gains interest for 25 weeks: $A_{25} = 500(1.0012)^{25}$

$\vdots$

Last deposit of \$500 gains interest for 1 week: $A_1 = 500(1.0012)^1$

Now, the amount accrued for 26 weeks:

$$A_1 + \ldots + A_{26} = \sum_{n=1}^{26} 500(1.0012)^{n-1}$$

Note that here $a_1 = 500,\ r = 1.0012$. So,

$$S_{26} = 500\left(\frac{1-(1.0012)^{26}}{1-1.0012}\right) \approx 13{,}196.88.$$

<u>**52 week investment**</u>:

First deposit of \$500 gains interest for 52 weeks: $A_{52} = 500(1.0012)^{52}$

Second deposit of \$500 gains interest for 51 weeks: $A_{51} = 500(1.0012)^{51}$

$\vdots$

Last deposit of \$500 gains interest for 1 week: $A_1 = 500(1.0012)^1$

Now, the amount accrued for 52 weeks:

$$A_1 + \ldots + A_{52} = \sum_{n=1}^{52} 500(1.0012)^{n-1}$$

Note that here $a_1 = 500,\ r = 1.0012$. So,

$$S_{52} = 500\left(\frac{1-(1.0012)^{52}}{1-1.0012}\right) \approx 26{,}811.75.$$

69. Here, $a_1 = 195{,}000$ and $r = 1 + 0.065 = 1.065$. So, $a_n = 195{,}000(1.065)^{n-1}$, $n \geq 1$. So, its worth after 15 years is $a_{15} = \$501{,}509$.

71. The sum is $\frac{\frac{1}{2}}{1-\frac{1}{2}} = 1$.

73. Should be $r = -\frac{1}{3}$

75. Should use $r = -3$ all the way through the calculation. Also, $a_1 = -12$(not 4).

77. False. $1, -\frac{1}{2}, \frac{1}{4}, -\frac{1}{8}$ is geometric with $a_1 = 1, r = -\frac{1}{2}$

79. True. We could have either of the following:

$1, \frac{1}{2}, \frac{1}{4}, \cdots$ (Here $r = \frac{1}{2}$.)

$1, -\frac{1}{2}, \frac{1}{4}, -\frac{1}{8}, \cdots$ (Here $r = -\frac{1}{2}$.)

81.

$$a + a \cdot b + \ldots + a \cdot b^n + \ldots = \sum_{n=1}^{\infty} ab^{n-1}$$

This sum exists for $|b| < 1$. In such case, the sum is $\frac{a}{1-b}$. (Here, $a_1 = a, r = b, S_\infty = \frac{a_1}{1-r}$)

83. This decimal can be represented as $\sum_{n=1}^{\infty} 47\left(\frac{1}{100}\right)^n = \frac{\frac{47}{100}}{1-\frac{1}{100}} = \frac{47}{99}$.

85. Since $a_1 = 1$, $r = -2$, $n = 50$, we have

$$\sum_{k=1}^{50} 1(-2)^{k-1} = 1\left(\frac{1-(-2)^{50}}{1-(-2)}\right) = -375{,}299{,}968{,}947{,}541$$

87. We expect that $\sum_{n=0}^{\infty} x^n = \frac{1}{1-x}$ since as $n \to \infty$, the graph of $y_n = 1 + \ldots + x^n$ gets closer to the graph of $\frac{1}{1-x}$.

Notes on the graph:

Solid curve: $y_1 = 1 + x + x^2 + x^3 + x^4$

Dashed curve: $y_2 = \frac{1}{1-x}$, assuming $|x| < 1$.

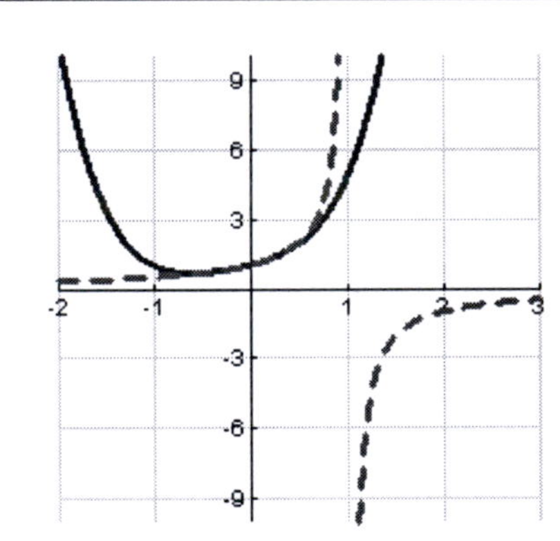

89. The plot of the two functions is:

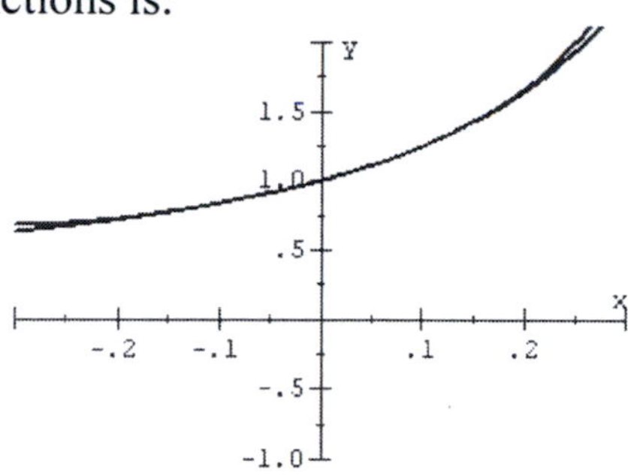

The series will sum to $\frac{1}{1-2x}$, assuming that $|x| < ½$

Section 9.4 Solutions --

1. **Claim:** $n^2 \leq n^3$, for all $n \geq 1$.

Proof.

Step 1: Show the statement is true for $n = 1$.

$$1^2 \leq 1^3 \text{ is true since } 1^2 = 1^3 = 1.$$

Step 2: Assume the statement is true for $n = k$: $k^2 \leq k^3$

Show the statement is true for $n = k+1$: $(k+1)^2 \leq (k+1)^3$

$$\begin{aligned}(k+1)^2 &= k^2 + 2k + 1 \\ &\leq k^3 + 2k + 1 \text{ (by assumption)} \\ &\leq k^3 + 3k + 1 \text{ (since } 2k \leq 3k, \text{ for } k > 0) \\ &\leq k^3 + 3k^2 + 3k + 1 \text{ (since } 3k^2 > 0) \\ &= (k+1)^3\end{aligned}$$

This completes the proof. ■

3. **Claim:** $2n \leq 2^n$, for all $n \geq 1$.

Proof.

Step 1: Show the statement is true for $n = 1$.

$$2(1) \leq 2^1 \text{ is true since both terms equal 2.}$$

Step 2: Assume the statement is true for $n = k$: $2k \leq 2^k$

Show the statement is true for $n = k+1$: $2(k+1) \leq 2^{k+1}$

$$\begin{aligned}2(k+1) = 2k + 2 &\leq 2^k + 2^1 \text{ (by assumption)} \\ &\leq 2^k + 2^k \text{ (since } 2 \leq 2^k, \text{ for } k > 0) \\ &= 2(2^k) = 2^{k+1}\end{aligned}$$

This completes the proof. ■

5. **Claim:** $n! > 2^n$, for all $n \geq 4$.

Proof.

Step 1: Show the statement is true for $n = 4$.

$$4! > 2^4 \text{ since } 4! = 4 \cdot 3 \cdot 2 \cdot 1 = 24 \text{ and } 2^4 = 16.$$

Step 2: Assume the statement is true for $n = k$: $k! > 2^k$

Show the statement is true for $n = k+1$: $(k+1)! > 2^{k+1}$

$$\begin{aligned} (k+1)! &= (k+1) \cdot k! \\ &> (k+1) \cdot 2^k \quad \text{(by assumption)} \\ &> 2 \cdot 2^k \quad \text{(since } k+1 > 2 \text{ for } k \geq 4) \\ &= 2^{k+1} \end{aligned}$$

This completes the proof. ■

7. **Claim:** $n(n+1)(n-1)$ is divisible by 3, for all $n \geq 1$.

Proof.

Step 1: Show the statement is true for $n = 1$.

$(1)(1+1)(1-1) = 1(2)(0) = 0$, which is clearly divisible by 3.

Step 2: Assume the statement is true for $n = k$: $k(k+1)(k-1)$ is divisible by 3

Show the statement is true for $n = k+1$: $(k+1)(k+2)(k)$ is divisible by 3

First, note that since by assumption $k(k+1)(k-1)$ is divisible by 3, we know that there exists an integer m such that $k(k+1)(k-1) = 3m$ **(1)**.

Now, observe that

$$(k+1)(k+2)(k) = (k^2 + 3k + 2)(k) = k^3 + 3k^2 + 2k$$

(At this point, write $2k = 3k - k$ and group the terms as shown.)

$$\begin{aligned} &= (k^3 - k) + (3k^2 + 3k) \\ &= 3m + 3(k^2 + k) \quad \text{(by (1))} \\ &= 3(m + k^2 + k) \end{aligned}$$

Now, choose $p = m + k^2 + k$ to see that you have expressed $(k+1)(k+2)(k)$ as $3p$, thereby showing $(k+1)(k+2)(k)$ is divisible by 3.

This completes the proof. ■

9. **Claim:** n^2+3n is divisible by 2, for all $n \geq 1$.

Proof.

Step 1: Show the statement is true for $n=1$.

$$1^2+3(1)=4=2(2)\text{, which is clearly divisible by 2.}$$

Step 2: Assume the statement is true for $n=k$: k^2+3k is divisible by 2

Show the statement is true for $n=k+1$: $(k+1)^2+3(k+1)$ is divisible by 2

First, note that since by assumption k^2+3k is divisible by 2, we know that there exists an integer m such that $k^2+3k=2m$ **(1)**.

Now, observe that

$$(k+1)^2+3(k+1)=\left(k^2+2k+1\right)+3\left(k+1\right)=k^2+5k+4$$

(At this point, write $5k=3k+2k$ and group the terms as shown.)

$$=(k^2+3k)+(2k+4)$$
$$=2m+2(k+2) \quad \text{(by (1))}$$
$$=2(m+k+2)$$

Now, choose $p=m+k+2$ to see that you have expressed $(k+1)^2+3(k+1)$ as $2p$, thereby showing $(k+1)^2+3(k+1)$ is divisible by 2.

This completes the proof. ■

11. **Claim:** $2+4+...+2n=n(n+1)$, for all $n \geq 1$.

Proof.

Step 1: Show the statement is true for $n=1$.

$$2=1(2)\text{, which is clearly true.}$$

Step 2: Assume the statement is true for $n=k$: $2+4+...+2k=k(k+1)$

Show the statement is true for $n=k+1$: $2+4+...+2(k+1)=(k+1)(k+2)$

Observe that

$$2+4+...+2k+2(k+1)=\left(2+4+...+2k\right)+2(k+1)$$
$$=k(k+1)+2(k+1) \quad \text{(by assumption)}$$
$$=(k+2)(k+1)$$

This completes the proof. ■

13. Claim: $1+3+...+3^n = \frac{3^{n+1}-1}{2}$, for all $n \geq 1$

Proof.

Step 1: Show the statement is true for $n=1$.

$$1+3^1 = \frac{3^2-1}{2} = 4 \text{ is clearly true.}$$

Step 2: Assume the statement is true for $n=k$: $1+3+...+3^k = \frac{3^{k+1}-1}{2}$

Show the statement is true for $n=k+1$: $1+3+...+3^{k+1} = \frac{3^{k+2}-1}{2}$

$$\begin{aligned} 1+3+...+3^k+3^{k+1} &= \left(1+3+...+3^k\right)+3^{k+1} \\ &= \frac{3^{k+1}-1}{2}+3^{k+1} \quad \text{(by assumption)} \\ &= \frac{3^{k+1}-1+2(3^{k+1})}{2} \\ &= \frac{3(3^{k+1})-1}{2} = \frac{3^{k+2}-1}{2} \end{aligned}$$

This completes the proof. ■

15. Claim: $1^2+...+n^2 = \frac{n(n+1)(2n+1)}{6}$, for all $n \geq 1$

Proof.

Step 1: Show the statement is true for $n=1$.

$$1^2 = \frac{1(1+1)(2(1)+1)}{6} = \frac{6}{6} = 1 \text{ is clearly true.}$$

Step 2: Assume the statement is true for $n=k$: $1^2+...+k^2 = \frac{k(k+1)(2k+1)}{6}$

Show the statement is true for $n=k+1$: $1^2+...+(k+1)^2 = \frac{(k+1)(k+2)\overbrace{(2(k+1)+1)}^{2k+3}}{6}$

$$\begin{aligned} 1^2+...+k^2+(k+1)^2 &= \left(1^2+...+k^2\right)+(k+1)^2 \\ &= \frac{k(k+1)(2k+1)}{6}+(k+1)^2 \quad \text{(by assumption)} \\ &= \frac{k(k+1)(2k+1)+6(k+1)^2}{6} \\ &= \frac{2k^3+3k^2+k+6k^2+12k+6}{6} \\ &= \frac{2k^3+9k^2+13k+6}{6} \\ &= \frac{(k+1)(k+2)(2k+3)}{6} \end{aligned}$$

This completes the proof. ■

17. Claim: $\frac{1}{1\cdot 2} + \frac{1}{2\cdot 3} + \ldots + \frac{1}{n(n+1)} = \frac{n}{n+1}$, for all $n \geq 1$

Proof.

Step 1: Show the statement is true for $n = 1$.

$$\frac{1}{1\cdot 2} = \frac{1}{1+1} = \frac{1}{2}\text{, which is clearly true.}$$

Step 2: Assume the statement is true for $n = k$: $\frac{1}{1\cdot 2} + \frac{1}{2\cdot 3} + \ldots + \frac{1}{k(k+1)} = \frac{k}{k+1}$

Show the statement is true for $n = k+1$: $\frac{1}{1\cdot 2} + \frac{1}{2\cdot 3} + \ldots + \frac{1}{(k+1)(k+2)} = \frac{k+1}{k+2}$

$$\begin{aligned}
\frac{1}{1\cdot 2} + \frac{1}{2\cdot 3} + \ldots + \frac{1}{k(k+1)} + \frac{1}{(k+1)(k+2)} &= \left(\frac{1}{1\cdot 2} + \frac{1}{2\cdot 3} + \ldots + \frac{1}{k(k+1)}\right) + \frac{1}{(k+1)(k+2)} \\
&= \frac{k}{k+1} + \frac{1}{(k+1)(k+2)} \quad \text{(by assumption)} \\
&= \frac{k(k+2)+1}{(k+1)(k+2)} \\
&= \frac{k^2+2k+1}{(k+1)(k+2)} \\
&= \frac{(k+1)^2}{(k+1)(k+2)} \\
&= \frac{k+1}{(k+2)}
\end{aligned}$$

This completes the proof. ■

19. Claim: $(1\cdot 2) + (2\cdot 3) + \ldots + n(n+1) = \frac{n(n+1)(n+2)}{3}$, for all $n \geq 1$.

Proof.

Step 1: Show the statement is true for $n = 1$.

$$(1\cdot 2) = \frac{1(1+1)(1+2)}{3} = \frac{1(2)(3)}{3} = 2\text{, which is clearly true.}$$

Step 2: Assume the statement is true for $n = k$:

$$(1\cdot 2) + (2\cdot 3) + \ldots + k(k+1) = \frac{k(k+1)(k+2)}{3}$$

Show the statement is true for $n = k+1$:

$(1\cdot 2) + (2\cdot 3) + \ldots + (k+1)(k+2) = \frac{(k+1)(k+2)(k+3)}{3}$

$$\begin{aligned}
(1\cdot 2) + (2\cdot 3) + \ldots + k(k+1) + (k+1)(k+2) &= \left((1\cdot 2) + (2\cdot 3) + \ldots + k(k+1)\right) + (k+1)(k+2) \\
&= \frac{k(k+1)(k+2)}{3} + (k+1)(k+2) \quad \text{(by assumption)} \\
&= \frac{k(k+1)(k+2)+3(k+1)(k+2)}{3} \\
&= \frac{(k+1)(k+2)(k+3)}{3} \quad \text{(factor out } (k+1)(k+2)\text{)}
\end{aligned}$$

This completes the proof. ■

21. Claim: $1+x+...+x^{n-1}=\frac{1-x^n}{1-x}$, for all $n \geq 1$.

Proof.

Step 1: Show the statement is true for $n=1$.

$$1=\frac{1-x^1}{1-x}, \text{ which is clearly true.}$$

Step 2: Assume the statement is true for $n=k$: $1+x+...+x^{k-1}=\frac{1-x^k}{1-x}$

Show the statement is true for $n=k+1$: $1+x+...+x^k=\frac{1-x^{k+1}}{1-x}$

$$\begin{aligned} 1+...+x^{k-1}+x^k &= \left(1+...+x^{k-1}\right)+x^k \\ &= \frac{1-x^k}{1-x}+x^k \\ &= \frac{1-x^k+x^k(1-x)}{1-x} \\ &= \frac{1-\cancel{x^k}+\cancel{x^k}-x^k x}{1-x} \\ &= \frac{1-x^{k+1}}{1-x} \end{aligned}$$

This completes the proof. ■

23. Claim: $a_1+(a_1+d)+...+(a_1+(n-1)d)=\frac{n}{2}[2a_1+(n-1)d]$, for all $n \geq 1$.

Proof.

Step 1: Show the statement is true for $n=1$.

$$a_1=\frac{1}{2}[2a_1+(1-1)d]=\frac{1}{2}[2a_1], \text{ which is clearly true.}$$

Step 2: Assume the statement is true for $n=k$:

$$a_1+(a_1+d)+...+(a_1+(k-1)d)=\frac{k}{2}[2a_1+(k-1)d]$$

Show the statement is true for $n=k+1$:

$$a_1+(a_1+d)+...+(a_1+((k+1)-1)d)=\frac{k+1}{2}\left[2a_1+(\overbrace{k+1-1}^{k})d\right]$$

Observe that

$$\begin{aligned} & a_1+(a_1+d)+...+(a_1+(k-1)d)+(a_1+kd) \\ &= \left(a_1+(a_1+d)+...+(a_1+(k-1)d)\right)+(a_1+kd) \\ &= \frac{k}{2}[2a_1+(k-1)d]+(a_1+kd) \\ &= ka_1+\frac{k(k-1)}{2}d+a_1+kd \\ &= (k+1)a_1+kd(\frac{k-1}{2}+1) \\ &= (k+1)a_1+kd(\frac{k+1}{2}) \\ &= \frac{(k+1)}{2}2a_1+kd(\frac{k+1}{2}) \\ &= \frac{(k+1)}{2}[2a_1+kd] \end{aligned}$$

This completes the proof. ■

25. Label the disks 1, 2, and 3 (smallest = 1 and largest = 3), and label the posts A, B, and C. The following are the moves on would take to solve the problem in the fewest number of step. (Note: The manner in which the disks are stacked (from top to bottom) on each peg form the contents of each cell; a blank cell means that no disk is on that peg in that particular move.)

	Post A	**Post B**	**Post C**
Initial placement	1 2 3		
Move 1	2 3	1	
Move 2	3	1	2
Move 3	3		1 2
Move 4		3	1 2
Move 5	1	3	2
Move 6	1	2 3	
Move 7		1 2 3	

So, the puzzle can be solved in as few as 7 steps. An argument as to why this is actually the fewest number of steps is beyond the scope of the text. (Note: Alternatively, we could have initially placed disk 1 on post C, and proceeded in a similar manner.)

27. Using the strategy of Problem 25 (and 26), this puzzle can be solved in as few as 31 steps. Have fun trying it!! There are many classical references that discuss this problem, as well as several internet sites.

29. If $n = 2$, then the number of wires needed is $\frac{2(2-1)}{2} = 1$.

Assume the formula holds for k cities. Then, if k cities are to be connected directly to each other, the number of wires needed is $\frac{k(k-1)}{2}$. Now, if one more city is added, then you must have k additional wires to connect the telephone to this additional city. So, the total wires in such case is $k + \frac{k(k-1)}{2} = \frac{2k+k^2-k}{2} = \frac{k(k+1)}{2}$. Thus, the statement holds for $k+1$ cities. Hence, we have proven the statement by induction.

31. False. You first need to show that S_1 is true.

33. Claim: $\sum_{k=1}^{n} k^4 = \frac{n(n+1)(2n+1)(3n^2+3n-1)}{30}$, for all $n \geq 1$.

Proof.

Step 1: Show the statement is true for $n=1$.

$$1^4 = \frac{1(1+1)(2(1)+1)(3(1)^2+3(1)-1)}{30} = 1 \text{, which is clearly true.}$$

Step 2: Assume the statement is true for $n = p$: $\sum_{k=1}^{p} k^4 = \frac{p(p+1)(2p+1)(3p^2+3p-1)}{30}$

Show the statement is true for $n = p+1$: $\sum_{k=1}^{p+1} k^4 = \frac{(p+1)(p+2)(2(p+1)+1)(3(p+1)^2+3(p+1)-1)}{30}$

Observe that

$$\begin{aligned}
\sum_{k=1}^{p+1} k^4 &= \sum_{k=1}^{p} k^4 + (p+1)^4 \\
&= \frac{p(p+1)(2p+1)(3p^2+3p-1)}{30} + (p+1)^4 \quad \text{(by assumption)} \\
&= \frac{p(p+1)(2p+1)(3p^2+3p-1)+30(p+1)^4}{30} \\
&= \frac{(p+1)\left[p(2p+1)(3p^2+3p-1)+30(p+1)^3\right]}{30} \\
&= \frac{(p+1)\left[6p^4+39p^3+91p^2+89p+30\right]}{30} \quad \textbf{(1)}
\end{aligned}$$

Next, note that multiplying out these terms yields

$$\begin{aligned}
\frac{(p+1)(p+2)(2(p+1)+1)(3(p+1)^2+3(p+1)-1)}{30} &= \frac{(p^2+3p+2)(2p+3)(3p^2+6p+3+3p+3-1)}{30} \\
&= \frac{(2p^3+9p^2+13p+6)(3p^2+9p+5)}{30} \\
&= \frac{6p^5+45p^4+130p^3+180p^2+119p+30}{30} \quad \textbf{(2)}
\end{aligned}$$

Comparing **(1)** and **(2)**, we see they are, in fact, equal. This is precisely what we needed to show to establish the claim.

This completes the proof. ■

35. Claim: $(1+\frac{1}{1})\cdot(1+\frac{1}{2})\cdot\ldots\cdot(1+\frac{1}{n})=n+1$, for all $n\geq 1$.

Proof.

Step 1: Show the statement is true for $n=1$.

$$(1+\tfrac{1}{1}) = 1+1=2, \text{ which is clearly true.}$$

Step 2: Assume the statement is true for $n=k$: $(1+\frac{1}{1})\cdot(1+\frac{1}{2})\cdot\ldots\cdot(1+\frac{1}{k})=k+1$

Show the statement is true for $n=k+1$: $(1+\frac{1}{1})\cdot(1+\frac{1}{2})\cdot\ldots\cdot(1+\frac{1}{k+1})=\underbrace{(k+1)+1}_{=k+2}$

Observe that

$$\begin{aligned}(1+\tfrac{1}{1})\cdot(1+\tfrac{1}{2})\cdot\ldots\cdot(1+\tfrac{1}{k})\cdot(1+\tfrac{1}{k+1}) &= \left[(1+\tfrac{1}{1})\cdot(1+\tfrac{1}{2})\cdot\ldots\cdot(1+\tfrac{1}{k})\right]\cdot(1+\tfrac{1}{k+1})\\ &= (k+1)\cdot(1+\tfrac{1}{k+1}) \quad \text{(by assumption)}\\ &= \cancel{(k+1)}\left(\tfrac{k+1+1}{\cancel{k+1}}\right)\\ &= k+2\end{aligned}$$

This completes the proof. ■

37. Claim: $\ln(c_1\cdot c_2\cdot\ldots\cdot c_n)=\ln(c_1)+\ldots+\ln(c_n)$, for all $n\geq 1$.

Proof.

Step 1: Show the statement is true for $n=1$.

$$\ln(c_1)=\ln(c_1) \text{ is clearly true.}$$

Step 2: Assume the statement is true for $n=k$:

$$\ln(c_1\cdot c_2\cdot\ldots\cdot c_k)=\ln(c_1)+\ldots+\ln(c_k)$$

Show the statement is true for $n=k+1$:

$$\ln(c_1\cdot c_2\cdot\ldots\cdot c_{k+1})=\ln(c_1)+\ldots+\ln(c_{k+1})$$

Observe that

$$\begin{aligned}\ln\left(\underbrace{c_1\cdot c_2\cdot\ldots\cdot c_k}_{\text{Treat as a single quantity}}\cdot c_{k+1}\right) &= \ln(c_1\cdot\ldots\cdot c_k)+\ln(c_{k+1})\\ &= \ln(c_1)+\ldots+\ln(c_k)+\ln(c_{k+1}) \quad \text{(by assumption)}\end{aligned}$$

This completes the proof. ■

39. $\frac{255}{256}$. Yes.

Section 9.5 Solutions ---

1. $\binom{7}{3}=\dfrac{7!}{(7-3)!3!}=\dfrac{7!}{4!3!}=\dfrac{7\cdot6\cdot5\cdot\cancel{4!}}{\cancel{4!}(3\cdot2\cdot1)}=35$

3. $\binom{10}{8} = \frac{10!}{(10-8)!8!} = \frac{10!}{2!8!} = \frac{10\cdot 9\cdot \cancel{8!}}{\cancel{8!}(2\cdot 1)} = 45$

5. $\binom{17}{0} = \frac{17!}{(17-0)!0!} = \frac{\cancel{17!}}{\cancel{17!}0!} = \frac{1}{0!} = 1$

7. $\binom{99}{99} = \frac{99!}{(99-99)!99!} = \frac{\cancel{99!}}{0!\cancel{99!}} = \frac{1}{0!} = 1$

9. $\binom{48}{45} = \frac{48!}{(48-45)!45!} = \frac{48!}{3!45!} = \frac{48\cdot 47\cdot 46\cdot \cancel{45!}}{\cancel{45!}(3\cdot 2\cdot 1)} = 17{,}296$

11.

$$
\begin{aligned}
(x+2)^4 &= \sum_{k=0}^{4}\binom{4}{k}x^{4-k}2^k \\
&= \binom{4}{0}x^4 2^0 + \binom{4}{1}x^3 2^1 + \binom{4}{2}x^2 2^2 + \binom{4}{3}x^1 2^3 + \binom{4}{4}x^0 2^4 \\
&= \underbrace{\tfrac{4!}{4!0!}}_{=1}x^4 2^0 + \underbrace{\tfrac{4!}{3!1!}}_{=4}x^3 2^1 + \underbrace{\tfrac{4!}{2!2!}}_{=6}x^2 2^2 + \underbrace{\tfrac{4!}{1!3!}}_{=4}x^1 2^3 + \underbrace{\tfrac{4!}{0!4!}}_{=1}x^0 2^4 \\
&= x^4 + 8x^3 + 24x^2 + 32x + 16
\end{aligned}
$$

13.

$$
\begin{aligned}
(y-3)^5 &= \sum_{k=0}^{5}\binom{5}{k}y^{5-k}(-3)^k \\
&= \binom{5}{0}y^5(-3)^0 + \binom{5}{1}y^4(-3)^1 + \binom{5}{2}y^3(-3)^2 + \binom{5}{3}y^2(-3)^3 + \binom{5}{4}y^1(-3)^4 + \binom{5}{5}y^0(-3)^5 \\
&= \underbrace{\tfrac{5!}{5!0!}}_{=1}y^5(-3)^0 + \underbrace{\tfrac{5!}{4!1!}}_{=5}y^4(-3)^1 + \underbrace{\tfrac{5!}{3!2!}}_{=10}y^3(-3)^2 + \underbrace{\tfrac{5!}{2!3!}}_{=10}y^2(-3)^3 + \underbrace{\tfrac{5!}{1!4!}}_{=5}y^1(-3)^4 + \underbrace{\tfrac{5!}{0!5!}}_{=1}y^0(-3)^5 \\
&= y^5 - 15y^4 + 90y^3 - 270y^2 + 405y - 243
\end{aligned}
$$

15.

$$
\begin{aligned}
(x+y)^5 &= \sum_{k=0}^{5}\binom{5}{k}x^{5-k}y^k \\
&= \binom{5}{0}x^5y^0 + \binom{5}{1}x^4y^1 + \binom{5}{2}x^3y^2 + \binom{5}{3}x^2y^3 + \binom{5}{4}x^1y^4 + \binom{5}{5}x^0y^5 \\
&= \underbrace{\tfrac{5!}{5!0!}}_{=1}x^5y^0 + \underbrace{\tfrac{5!}{4!1!}}_{=5}x^4y^1 + \underbrace{\tfrac{5!}{3!2!}}_{=10}x^3y^2 + \underbrace{\tfrac{5!}{2!3!}}_{=10}x^2y^3 + \underbrace{\tfrac{5!}{1!4!}}_{=5}x^1y^4 + \underbrace{\tfrac{5!}{0!5!}}_{=1}x^0y^5 \\
&= x^5 + 5x^4y + 10x^3y^2 + 10x^2y^3 + 5xy^4 + y^5
\end{aligned}
$$

17.

$$(x+3y)^3 = \sum_{k=0}^{3}\binom{3}{k}x^{3-k}(3y)^k$$

$$= \binom{3}{0}x^3(3y)^0 + \binom{3}{1}x^2(3y)^1 + \binom{3}{2}x^1(3y)^2 + \binom{3}{3}x^0(3y)^3$$

$$= \underbrace{\tfrac{3!}{3!0!}}_{=1}x^3(3y)^0 + \underbrace{\tfrac{3!}{2!1!}}_{=3}x^2(3y)^1 + \underbrace{\tfrac{3!}{1!2!}}_{=3}x^1(3y)^2 + \underbrace{\tfrac{3!}{0!3!}}_{=1}x^0(3y)^3$$

$$= x^3 + 9x^2y + 27xy^2 + 27y^3$$

19.

$$(5x-2)^3 = \sum_{k=0}^{3}\binom{3}{k}(5x)^{3-k}(-2)^k$$

$$= \binom{3}{0}(5x)^3(-2)^0 + \binom{3}{1}(5x)^2(-2)^1 + \binom{3}{2}(5x)^1(-2)^2 + \binom{3}{3}(5x)^0(-2)^3$$

$$= \underbrace{\tfrac{3!}{3!0!}}_{=1}(5x)^3(-2)^0 + \underbrace{\tfrac{3!}{2!1!}}_{=3}(5x)^2(-2)^1 + \underbrace{\tfrac{3!}{1!2!}}_{=3}(5x)^1(-2)^2 + \underbrace{\tfrac{3!}{0!3!}}_{=1}(5x)^0(-2)^3$$

$$= 125x^3 - 150x^2 + 60x - 8$$

21.

$$(\tfrac{1}{x}+5y)^4 = \sum_{k=0}^{4}\binom{4}{k}(\tfrac{1}{x})^{4-k}(5y)^k, \quad x \neq 0$$

$$= \binom{4}{0}(\tfrac{1}{x})^4(5y)^0 + \binom{4}{1}(\tfrac{1}{x})^3(5y)^1 + \binom{4}{2}(\tfrac{1}{x})^2(5y)^2 + \binom{4}{3}(\tfrac{1}{x})^1(5y)^3 + \binom{4}{4}(\tfrac{1}{x})^0(5y)^4$$

$$= \underbrace{\tfrac{4!}{4!0!}}_{=1}(\tfrac{1}{x})^4(5y)^0 + \underbrace{\tfrac{4!}{3!1!}}_{=4}(\tfrac{1}{x})^3(5y)^1 + \underbrace{\tfrac{4!}{2!2!}}_{=6}(\tfrac{1}{x})^2(5y)^2 + \underbrace{\tfrac{4!}{1!3!}}_{=4}(\tfrac{1}{x})^1(5y)^3 + \underbrace{\tfrac{4!}{0!4!}}_{=1}(\tfrac{1}{x})^0(5y)^4$$

$$= \tfrac{1}{x^4} + 20\tfrac{y}{x^3} + 150\tfrac{y^2}{x^2} + 500\tfrac{y^3}{x} + 625y^4$$

23.

$$(x^2+y^2)^4 = \sum_{k=0}^{4}\binom{4}{k}(x^2)^{4-k}(y^2)^k$$

$$= \binom{4}{0}(x^2)^4(y^2)^0 + \binom{4}{1}(x^2)^3(y^2)^1 + \binom{4}{2}(x^2)^2(y^2)^2 + \binom{4}{3}(x^2)^1(y^2)^3 + \binom{4}{4}(x^2)^0(y^2)^4$$

$$= \underbrace{\tfrac{4!}{4!0!}}_{=1}(x^2)^4(y^2)^0 + \underbrace{\tfrac{4!}{3!1!}}_{=4}(x^2)^3(y^2)^1 + \underbrace{\tfrac{4!}{2!2!}}_{=6}(x^2)^2(y^2)^2 + \underbrace{\tfrac{4!}{1!3!}}_{=4}(x^2)^1(y^2)^3 + \underbrace{\tfrac{4!}{0!4!}}_{=1}(x^2)^0(y^2)^4$$

$$= x^8 + 4x^6y^2 + 6x^4y^4 + 4x^2y^6 + y^8$$

25.

$$(ax+by)^5 = \sum_{k=0}^{5}\binom{5}{k}(ax)^{5-k}(by)^k$$

$$= \binom{5}{0}(ax)^5(by)^0 + \binom{5}{1}(ax)^4(by)^1 + \binom{5}{2}(ax)^3(by)^2 + \binom{5}{3}(ax)^2(by)^3 + \binom{5}{4}(ax)^1(by)^4 + \binom{5}{5}(ax)^0(by)^5$$

$$= \underbrace{\tfrac{5!}{5!0!}}_{=1}(ax)^5(by)^0 + \underbrace{\tfrac{5!}{4!1!}}_{=5}(ax)^4(by)^1 + \underbrace{\tfrac{5!}{3!2!}}_{=10}(ax)^3(by)^2 + \underbrace{\tfrac{5!}{2!3!}}_{=10}(ax)^2(by)^3 + \underbrace{\tfrac{5!}{1!4!}}_{=5}(ax)^1(by)^4 + \underbrace{\tfrac{5!}{0!5!}}_{=1}(ax)^0(by)^5$$

$$= a^5x^5 + 5a^4bx^4y + 10a^3b^2x^3y^2 + 10a^2b^3x^2y^3 + 5ab^4xy^4 + b^5y^5$$

27.

$$(\sqrt{x}+2)^6 = \sum_{k=0}^{6}\binom{6}{k}\left(\sqrt{x}\right)^{6-k}2^k,\ x \geq 0$$

$$= \binom{6}{0}\left(\sqrt{x}\right)^6 2^0 + \binom{6}{1}\left(\sqrt{x}\right)^5 2^1 + \binom{6}{2}\left(\sqrt{x}\right)^4 2^2 + \binom{6}{3}\left(\sqrt{x}\right)^3 2^3 + \binom{6}{4}\left(\sqrt{x}\right)^2 2^4 + \binom{6}{5}\left(\sqrt{x}\right)^1 2^5 + \binom{6}{6}\left(\sqrt{x}\right)^0 2^6$$

$$= \underbrace{\tfrac{6!}{6!0!}}_{=1}\left(\sqrt{x}\right)^6 2^0 + \underbrace{\tfrac{6!}{5!1!}}_{=6}\left(\sqrt{x}\right)^5 2^1 + \underbrace{\tfrac{6!}{4!2!}}_{=15}\left(\sqrt{x}\right)^4 2^2 + \underbrace{\tfrac{6!}{3!3!}}_{=20}\left(\sqrt{x}\right)^3 2^3 + \underbrace{\tfrac{6!}{2!4!}}_{=15}\left(\sqrt{x}\right)^2 2^4 + \underbrace{\tfrac{6!}{1!5!}}_{=6}\left(\sqrt{x}\right)^1 2^5 + \underbrace{\tfrac{6!}{0!6!}}_{=1}\left(\sqrt{x}\right)^0 2^6$$

$$= x^3 + 12x^{5/2} + 60x^2 + 160x^{3/2} + 240x + 192x^{1/2} + 64$$

29.

$$(a^{3/4}+b^{1/4})^4=\sum_{k=0}^{4}\binom{4}{k}\left(a^{3/4}\right)^{4-k}\left(b^{1/4}\right)^{k},\quad a,b\geq 0$$

$$=\binom{4}{0}\left(a^{3/4}\right)^4\left(b^{1/4}\right)^0+\binom{4}{1}\left(a^{3/4}\right)^3\left(b^{1/4}\right)^1+\binom{4}{2}\left(a^{3/4}\right)^2\left(b^{1/4}\right)^2$$

$$+\binom{4}{3}\left(a^{3/4}\right)^1\left(b^{1/4}\right)^3+\binom{4}{4}\left(a^{3/4}\right)^0\left(b^{1/4}\right)^4$$

$$=\underbrace{\tfrac{4!}{4!0!}}_{=1}\left(a^{3/4}\right)^4\left(b^{1/4}\right)^0+\underbrace{\tfrac{4!}{3!1!}}_{=4}\left(a^{3/4}\right)^3\left(b^{1/4}\right)^1+\underbrace{\tfrac{4!}{2!2!}}_{=6}\left(a^{3/4}\right)^2\left(b^{1/4}\right)^2$$

$$+\underbrace{\tfrac{4!}{1!3!}}_{=4}\left(a^{3/4}\right)^1\left(b^{1/4}\right)^3+\underbrace{\tfrac{4!}{0!4!}}_{=1}\left(a^{3/4}\right)^0\left(b^{1/4}\right)^4$$

$$=a^3+4a^{9/4}b^{1/4}+6a^{3/2}b^{1/2}+4a^{3/4}b^{3/4}+b$$

31.

$$(x^{1/4}+2\sqrt{y})^4=\sum_{k=0}^{4}\binom{4}{k}\left(x^{1/4}\right)^{4-k}\left(2\sqrt{y}\right)^{k},\quad x,y\geq 0$$

$$=\binom{4}{0}\left(x^{1/4}\right)^4\left(2\sqrt{y}\right)^0+\binom{4}{1}\left(x^{1/4}\right)^3\left(2\sqrt{y}\right)^1+\binom{4}{2}\left(x^{1/4}\right)^2\left(2\sqrt{y}\right)^2$$

$$+\binom{4}{3}\left(x^{1/4}\right)^1\left(2\sqrt{y}\right)^3+\binom{4}{4}\left(x^{1/4}\right)^0\left(2\sqrt{y}\right)^4$$

$$=\underbrace{\tfrac{4!}{4!0!}}_{=1}\left(x^{1/4}\right)^4\left(2\sqrt{y}\right)^0+\underbrace{\tfrac{4!}{3!1!}}_{=4}\left(x^{1/4}\right)^3\left(2\sqrt{y}\right)^1+\underbrace{\tfrac{4!}{2!2!}}_{=6}\left(x^{1/4}\right)^2\left(2\sqrt{y}\right)^2$$

$$+\underbrace{\tfrac{4!}{1!3!}}_{=4}\left(x^{1/4}\right)^1\left(2\sqrt{y}\right)^3+\underbrace{\tfrac{4!}{0!4!}}_{=1}\left(x^{1/4}\right)^0\left(2\sqrt{y}\right)^4$$

$$=x+8x^{3/4}y^{1/2}+24x^{1/2}y+32x^{1/4}y^{3/2}+16y^2$$

33.

$$(r-s)^4=r^4+4r^3(-s)+6r^2(-s)^2+4r(-s)^3+(-s)^4$$
$$=r^4-4r^3s+6r^2s^2-4rs^3+s^4$$

35.

$$\left(ax+by\right)^6=(ax)^6+6(ax)^5(by)+15(ax)^4(by)^2+20(ax)^3(by)^3$$
$$+15(ax)^2(by)^4+6(ax)^1(by)^5+(by)^6$$
$$=a^6x^6+6a^5bx^5y+15a^4b^2x^4y^2+20a^3b^3x^3y^3+15a^2b^4x^2y^4+6ab^5xy^5+b^6y^6$$

37. This term is $\binom{10}{4}x^{10-4}(2)^4 = \frac{10!}{4!6!}(16)x^6 = \frac{10\cdot9\cdot8\cdot7\cdot\cancel{6!}}{(4\cdot3\cdot2\cdot1)\cancel{6!}}x^6 = 3360x^6$

So, the desired coefficient is 3360.

39. This term is $\binom{8}{4}y^{8-4}(-3)^4 = \frac{8!}{4!4!}(81)y^4 = \frac{8\cdot7\cdot6\cdot5\cdot\cancel{4!}}{(4\cdot3\cdot2\cdot1)\cancel{4!}}y^4 = 5670y^4$

So, the desired coefficient is 5670.

41. This term is

$\binom{7}{4}(2x)^{7-4}(3y)^4 = \frac{7!}{4!3!}(8\cdot81)x^3y^4 = \frac{7\cdot6\cdot5\cdot\cancel{4!}}{(3\cdot2\cdot1)\cancel{4!}}(648)x^3y^4 = 22{,}680x^3y^4$

So, the desired coefficient is 22,680.

43. This term is $\binom{8}{4}(x^2)^{8-4}(y)^4 = \frac{8!}{4!4!}x^8y^4 = \frac{8\cdot7\cdot6\cdot5\cdot\cancel{4!}}{(4\cdot3\cdot2\cdot1)\cancel{4!}}x^8y^4 = 70x^8y^4$

So, the desired coefficient is 70.

45. Since $\binom{40}{6} = \frac{40!}{6!34!} = \frac{40\cdot39\cdot38\cdot37\cdot36\cdot35\cdot\cancel{34!}}{6!\cancel{34!}} = 3{,}838{,}380$, there are 3,838,380 such 6-number lottery numbers.

47. Since $\binom{52}{5} = \frac{52!}{5!47!} = \frac{52\cdot51\cdot50\cdot49\cdot48\cdot\cancel{47!}}{5!\cancel{47!}} = 2{,}598{,}960$, there are 2,598,960 possible 5-card poker hands.

49. $\binom{7}{5} \neq \frac{7!}{5!}$, $\binom{7}{5} = \frac{7!}{5!2!}$

51. False, there are 11 terms.

53. True

55. Claim: $\binom{n}{k} = \binom{n}{n-k}$, $0 \le k \le n$, for any $n \ge 1$.

Proof. Observe that $\binom{n}{k} = \frac{n!}{k!(n-k)!} = \frac{n!}{(n-k)!k!} = \frac{n!}{(n-k)!\underbrace{(n-(n-k))}_{\text{This is } k \text{ written in a different form.}}!} = \binom{n}{n-k}$ ■

57. The binomial expansion of $(1-x)^3$ is

$$1-3x+3x^2-x^3.$$

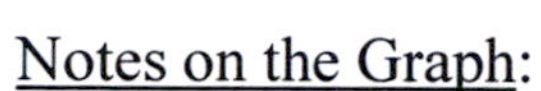

Notes on the Graph:

Heavy Solid Curve: $y_1=1-3x+3x^2-x^3$

Heavy Dashed Curve: $y_2=-1+3x-3x^2+x^3$

Heavy Dotted Curve: $y_3=(1-x)^3$

The graphs of y_1 and y_3 are the same, so you only see two graphs.

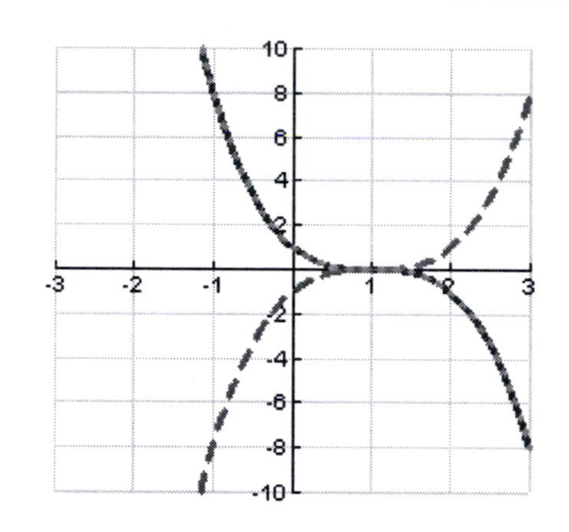

59. We see from the graph that as each term is added, the graphs of the respective functions get closer to the graph of $y_4=(1-x)^3$ when $1<x<2$. However, when $x>1$, this is no longer true.

Notes on the graph:

Heavy solid curve: y_4

Heavy dashed curve: y_1

Heavy dotted curve: y_2

Thin dashed curve: y_3

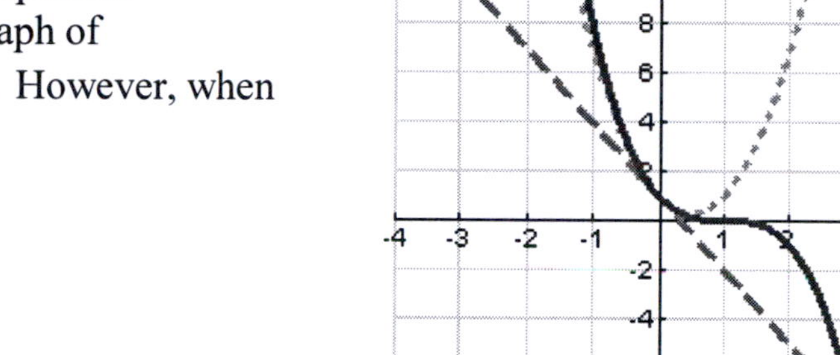

61. As each new term is added, the corresponding graph of the curve is a better approximation to the graph of $y=\left(1+\frac{1}{x}\right)^3$, for $1<x<2$. The series of functions does not get closer to this graph if $0<x<1$.

Section 9.6 Solutions--

1.

$${}_6P_4=\frac{6!}{(6-4)!}=\frac{6!}{2!}=\frac{6\cdot5\cdot4\cdot3\cdot\cancel{2!}}{\cancel{2!}}$$
$$=360$$

3.

$${}_9P_5=\frac{9!}{(9-5)!}=\frac{9!}{4!}=\frac{9\cdot8\cdot7\cdot6\cdot5\cdot\cancel{4!}}{\cancel{4!}}$$
$$=15{,}120$$

5. ${}_8P_8=\dfrac{8!}{(8-8)!}=\dfrac{8!}{0!}=8!=40{,}320$

7. ${}_{13}P_3=\dfrac{13!}{(13-3)!}=\dfrac{13!}{10!}=\dfrac{13\cdot12\cdot11\cdot\cancel{10!}}{\cancel{10!}}=1716$

9. $_{10}C_5 = \frac{10!}{(10-5)!5!} = \frac{10!}{5!5!} = \frac{10 \cdot 9 \cdot 8 \cdot 7 \cdot 6 \cdot \cancel{5!}}{\cancel{5!}(5 \cdot 4 \cdot 3 \cdot 2 \cdot 1)} = 252$

11. $_{50}C_6 = \frac{50!}{(50-6)!6!} = \frac{50 \cdot 49 \cdot 48 \cdot 47 \cdot 46 \cdot 45 \cdot \cancel{44!}}{\cancel{44!}(6 \cdot 5 \cdot 4 \cdot 3 \cdot 2 \cdot 1)} = 15,890,700$

13. $_7C_7 = \frac{7!}{(7-7)!7!} = \frac{7!}{0!7!} = 1$

15. $_{30}C_4 = \frac{30!}{(30-4)!4!} = \frac{30 \cdot 29 \cdot 28 \cdot 27 \cdot \cancel{26!}}{\cancel{26!}(4 \ldots 1)} = 27,405$

17. $_{45}C_8 = \frac{45!}{(45-8)!8!} = \frac{45 \cdot 44 \cdot 43 \cdot 42 \cdot 41 \cdot 40 \cdot 39 \cdot 38 \cdot \cancel{37!}}{\cancel{37!}(8 \cdot 7 \ldots 1)} = 215,553,195$

19. $4 \cdot 3 \cdot 2 = 24$ different system configurations

21. $\underbrace{2}_{color} \cdot \underbrace{3}_{writing} \cdot \underbrace{2}_{envelopes} = 12$ different types of invitations

23. Each slot can have $0,1,\ldots,9$ (10 choices)

Can allow repetition of digits and order is important. So, there are $10 \cdot 10 \cdot 10 \cdot 10 = 10^4$ $= 10,000$ different pin numbers

25. $\frac{15!}{11!} = \frac{15 \cdot 14 \cdot 13 \cdot 12 \cdot \cancel{11!}}{\cancel{11!}} = 32,760$ different leadership teams

27. Each of 20 questions has 4 answer choices. So, there are $4^{20} \approx 1.1 \times 10^{12}$ possible ways to answer the questions on the exam.

29. The number of 5-digit zip codes where any of 10 digits can be used in each slot is 10^5=100,000. The number of 5-digit zip codes where 0 cannot be used in 1st and last slot:

1st—9 digits
2nd—10 digits
3rd—10 digits
4th—10 digits
5th—9 digits

$9^2 \times 10^3 = 81,000$ such zip codes.

31. $30! \approx 2.65 \times 10^{32}$

33. $40 \cdot 39 \cdot 38 = 59{,}280$

40 ↑ 1st #

39 ↑ 2nd # (1 less digit to choose from)

38 ↑ 3rd # (2 less digits to choose from)

(Note: Here, order is important.)

So, there are 59,280 different locker combinations.

35.

$${}_{1000}P_3 = \frac{1000!}{(1000-3)!} = 1000 \cdot 999 \cdot 998 =$$

$997{,}002{,}000$ possible winning scenarios

37.

$$\binom{53}{6} = \frac{53!}{6!(53-6)!} = \frac{53 \cdot 52 \cdot 51 \cdot 50 \cdot 49 \cdot 48 \cdot \cancel{47!}}{6!\cancel{47!}}$$

$= 22{,}957{,}480$ different 6-number combinations

39.

$$\binom{52}{5} = \frac{52!}{5!(52-5)!} = \frac{52 \cdot 51 \cdot 50 \cdot 49 \cdot 48 \cdot \cancel{47!}}{(5 \ldots 1)\cancel{47!}}$$

$= 2{,}598{,}960$ different 5-card hands (Note: Here, order is not important.)

41. $\binom{52}{2} = \frac{52!}{(52-2)!2!} = \frac{52 \cdot 51 \cdot \cancel{50!}}{\cancel{50!}2} = 1326$ different blackjack hands

43.

$$\underbrace{\binom{64}{16}}_{\substack{\text{\# possible} \\ \text{scenarios for} \\ \text{Sweet Sixteen}}} = \frac{64!}{16!48!} \approx 4.9 \times 10^{14}$$

45. $\underset{\substack{1st\ slot \\ AFC}}{\underbrace{16}} \cdot \underset{\substack{2nd\ slot \\ NFC}}{\underbrace{16}} = 256$

possible scenarios for Super Bowl

47. $\underset{player1}{\underbrace{5}} \quad \underset{player2}{\underbrace{5}} \quad \underset{player3}{\underbrace{5}} \quad \ldots \quad \underset{player6}{\underbrace{5}}$ (Nobody votes themselves off)

$= 5^6 = 15{,}625$ voting combinations for Survivor when down to 6 contestants

49.

$$\binom{6}{3} \cdot \binom{6}{3} = \left(\frac{6!}{3!3!}\right)^2 = \left(\frac{\cancel{6} \cdot 5 \cdot 4}{\cancel{3!}}\right)^2$$

$$= 20^2 = 400$$

51. The combination formula ${}_nC_r$ should be used instead.

53. True

55. False. Since there will be repetition when permuting the letters in ABBA that are not distinguishable while no such repetition occurs in permuting ABCD.

57. Observe that

$$\frac{{}_nC_r}{{}_nC_{r+1}} = \frac{\cancel{n!}}{(n-r)!r!} \cdot \frac{(n-r-1)!(r+1)!}{\cancel{n!}} = \frac{\cancel{(n-r-1)!}\,(r+1)\cancel{r!}}{(n-r)\,\cancel{(n-r-1)!}\,\cancel{r!}} = \frac{r+1}{n-r}$$

59. $C(n,r)\cdot r! = \dfrac{n!}{(n-r)!\cdot \cancel{r!}}\cdot \cancel{r!} = \dfrac{n!}{(n-r)!}$
61. Use a calculator to do so, if you have one.
63. a. 5,040 **b.** 5,040 **c.** Yes. **d.** This is true because ${}_nP_r = r!\,{}_nC_r$.

Section 9.7---

1. Sum the two dice. So, the possible rolls constitute the sample space: $S = \{2,3\ldots12\}$

3. $S = \{BBBB, BBBG, BBGB, BBGG, BGBB, BGBG, BGGB, BGGG, GBBB, GBBG,$
$GBGB, GBGG, GGBB, GGBG, GGGB, GGGG\}$

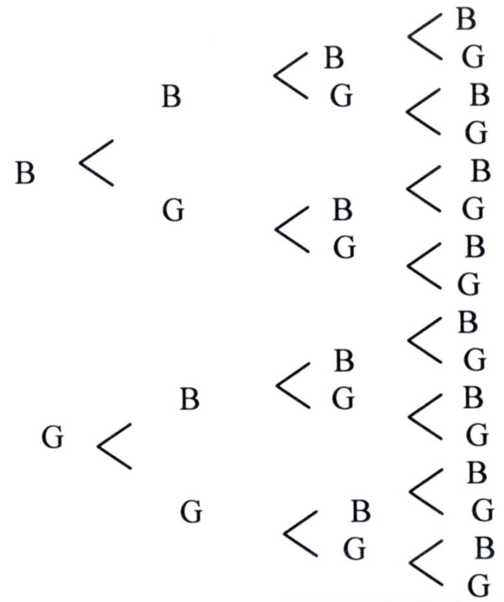

5. $S = \{RR, RB, RW, BB, BR, BW, WR, WB\}$

(Can't have WW since there is only one white ball in the container.)

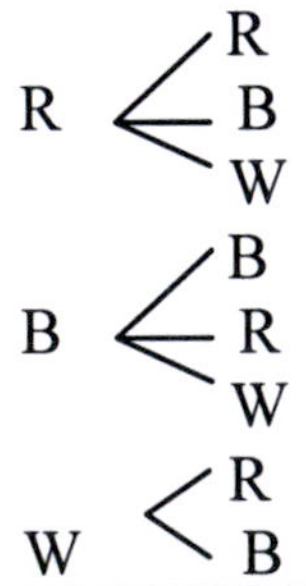

Note: For problems 7-10, the sample space is given in Problem 2. (Note that there are 8 distinct elements all with the same probability of occurring.)

7. $P(\text{all heads}) = P(HHH) = \frac{1}{8}$

9. $P(\text{at least one head}) = 1 - P(TTT) = \frac{7}{8}$

Note: For Problems 11-16, there are 36 different possible outcomes of throwing 2 dice, assuming that the two dice are distinguishable from each other. The sample space is given in Example 4 on page 841 of the text.

11. $P(\underbrace{sum = 3}_{(1,2)\text{or}(2,1)}) = \frac{2}{36} = \frac{1}{18}$

13. $P(\text{sum is even}) = 1 - P(\text{sum is odd})$
$= 1 - \frac{1}{2} = \frac{1}{2}$

15. Note that the sum of two dice can be greater than 7 only when the sum is 8, 9, 10, 11, or 12. Consulting sample space on page 662 of the text, we see that this can occur in 15 distinct ways. So, $P(sum > 7) = \frac{15}{36} = \frac{5}{12}$

Note: For Problems 17-20, the sample space is $S = \text{set of 52 distinct cards}$.

17. There are 12 face cards (4 kings, 4 queens, 4 jacks). So, $P(\text{drawing a nonface card}) = \frac{40}{52} = \frac{10}{13}$.

19. There are 4 twos, 4 fours, 4 sixes, 4 eights. So, $P(\text{drawing a 2, 4, 6, or 8}) = \frac{16}{52} = \frac{4}{13}$.

21. $P(\text{not } E_1) = 1 - P(E_1) = 1 - \frac{1}{4} = \frac{3}{4}$

23. If E_1, E_2 are mutually exclusive, then
$P(E_1 \cup E_2) = P(E_1) + P(E_2) = \frac{3}{4}$

25. If E_1, E_2 are mutually exclusive, then $P(E_1 \text{ and } E_2) = 0$.

27. a.
$$\binom{52}{4} = \frac{52!}{4!48!} = \frac{52 \cdot 51 \cdot 50 \cdot 49 \cdot \cancel{48!}}{4!\cancel{48!}} = 270,725$$

b. Number of ways to get 4 spades: $\binom{13}{4} = 715$

So, $P(\text{getting 4 spades}) = \frac{715}{270,725} \cong 0.0026$

c. Since there are 13 ways of getting 4 of a kind,
we see
that $P(\text{4 of a kind}) = \frac{13}{270,725} \approx 0.00005$

So, there is about a 0.005% chance of getting four of a kind.

29. Since drawing a 7 and drawing an 8 are mutually exclusive events,
P (drawing 7 or 8) = P(draw 7) + P(draw 8)
$= \frac{4}{52} + \frac{4}{52} = \frac{8}{52} = \frac{2}{13} \approx 0.154$

So, there is about a 15.4% chance of drawing a 7 or an 8.

31. Since these two events are independent, we see that $$P\,(7 \text{ on } 1^{st} \text{ draw AND } 8 \text{ on } 2^{nd} \text{ draw}) = \frac{4}{52}\cdot\frac{4}{51} = \frac{16}{2652} = \frac{4}{663} \approx 0.006.$$ So, there is about a 0.6% chance of getting a 7 on the first draw and an 8 on the second draw.
33. Sample space S has 2^5 elements, all elements are presumed equally likely. So, P (5 daughters) $=\frac{1}{32}=0.03125$. Thus, there is a 3.1% chance of getting 5 daughters.
35. Sample space has 2^5 elements, all elements presumed equally likely. So, $P(\geq 1 \text{ boy}) = 1 - P(0 \text{ boys}) = 1 - P(5 \text{ daughters}) = 1 - \frac{1}{32} = \frac{31}{32} = 0.96875$ So, there is a 96.9% chance of having at least one boy.
37. Assume independence of spins. Then, $$P(\text{red on } 1^{st} \text{ spin AND } 2^{nd} \text{ AND } 3^{rd} \text{ AND } 4^{th})$$ $$= P(\text{red on } 1^{st}) \cdot P(\text{red on } 2^{nd}) \ldots P(\text{red on } 4^{th}) = \left(\frac{18}{38}\right)^4 \approx 0.0503.$$ So, about a 5.03% chance.
39. P(not defective) = 9/10. Since you are drawing the 8 DVD players from different batches, the 8 choices are independent. So, the probability that none of the 8 are defective is $\left(\frac{9}{10}\right)^8 \approx 0.43$. So, about 43% chance.

41. The events are independent since they are chosen at random. So, P(1st number even AND 2nd number even) = $P(\text{1st number even}) \cdot P(\text{2nd number even}) =$ $\frac{1}{2}\cdot\frac{1}{2}=\frac{1}{4}$ So, 25% chance.	**43.** Number of 2 card hands is given by: $\binom{52}{2} = \frac{52!}{2!52!} = \frac{52\cdot 51\cdot \cancel{50!}}{2\cdot \cancel{50!}} = 1326$ Number of possible blackjack hands is $\underbrace{4}_{\text{ace}} \cdot \underbrace{16}_{\text{face or ten}} = 64$ So, $P(\text{blackjack}) = \frac{64}{1326} \approx 0.0483$ So, about a 4.8% chance.

45. Assume independence of games from week to week. Then, the probability of going $16-0$ is $\left(\frac{1}{2}\right)^{16} \approx 0.00001526$. So, about a 0.001526% chance.
47. a. An outcome is of the form: (*gene from mother*, *gene from father*) So, the sample space is given by S = { (Brown, Blue), (Brown, Brown), (Blue, Brown), (Blue, Blue) } **b.** ¼ since outcomes are equally likely **c.** ¾

49.
$$\binom{52}{5} = \frac{52!}{47! \cdot 5!} = \frac{52 \cdot 51 \cdot 50 \cdot 49 \cdot 48}{5 \cdot 4 \cdot 3 \cdot 2 \cdot 1} = 2,598,960$$

51. Using #49 for the total number of 5-card hands, we obtain
$$\frac{\binom{13}{5}}{2,598,960} = \frac{\frac{13!}{8! \cdot 5!}}{2,598,960} = \frac{1287}{2,598,960}$$

53. The events aren't mutually exclusive. So, probability $= \frac{4}{52} + \frac{13}{52} - \underbrace{\frac{1}{52}}_{2 \text{ of spades}} = \frac{16}{52} = \frac{4}{13}$

55. True. Since all events are mutually exclusive,
$$1 = P(E_1 \cup E_2 \cup E_3) = P(E_1) + P(E_2) + P(E_3).$$

57. False

59. The probability that Person 1 has birthday on Day A is simply $\frac{1}{365}$. Now, to compute the probability that Persons 1 and 2 have the same birthday, we note, first and foremost, that these events are independent since the people are being chosen at random. Since there are 365 days in a year, we can see that the probability of two people having the same birthday is $365\left(\frac{1}{365} \cdot \frac{1}{365}\right) = \frac{1}{365} = 0.0027$.

61. Sample space
$S = \{31, 32, 33, 34, 35, 36, 41, 4243, 44, 45, 46\}$

So, $P(\text{Sum is } 2, 5, \text{or } 6)$
$= P(\text{Sum is } 2) + P(\text{Sum is } 5) + P(\text{Sum is } 6)$
$= 0 + \frac{2}{12} + \frac{2}{12}$
$= \frac{1}{3} \approx 0.333$

63. Do so if you have access to a computer.

65. Use $n = 10$ and $k = 2$ in the formula provided to obtain the approximate probability of 0.2907.

Chapter 9 Review Solutions --

1. $1, 8, 27, 64$

3. $5, 8, 11, 14$

5. $a_5 = \left(\frac{2}{3}\right)^5 = \frac{32}{243} = 0.13$

7.
$$a_{15} = \frac{(-1)^{15}(15-1)!}{15(15+1)!} = -\frac{(14)!}{15(16)!}$$
$$= -\frac{\cancel{14!}}{15 \cdot 16 \cdot 15 \cdot \cancel{14!}} = -\frac{1}{3600}$$

9. $a_n = (-1)^{n+1} 3n, n \geq 1$

11. $a_n = (-1)^n, n \geq 1$

13. $\frac{8!}{6!} = \frac{8 \cdot 7 \cdot \cancel{6!}}{\cancel{6!}} = 56$	**15.** $\frac{n(n-1)!}{(n+1)!} = \frac{n\cancel{(n-1)!}}{(n+1)\,n\cancel{(n-1)!}} = \frac{1}{n+1}$
17. $5, 3, 1, -1$	**19.** $1, 2, \underbrace{(2)^2 \cdot (1)}_{=4}, \underbrace{4^2 \cdot 2}_{=32}$ which is equivalent to $1, 2, 4, 32$.
21. $\sum_{n=1}^{5} 3 = 3(5) = 15$	**23.** $\sum_{n=1}^{6}(3n+1) = \frac{6}{2}(4+19) = 69$
25. $\sum_{n=1}^{7} \frac{(-1)^n}{2^{n-1}}$	**27.** $\sum_{n=0}^{\infty} \frac{x^n}{n!}$ (Note: $2! = 2,\ 3! = 6,\ 4! = 24$)

29. $A_{60} = 30{,}000\left(1 + \frac{0.04}{12}\right)^{60} = \$36{,}639.90$

So, the amount in the account after 5 years is \$36,639.90.

31. Yes, it is arithmetic with $d = -2$.	**33.** Yes, it is arithmetic with $d = \frac{1}{2}$.

35. $a_n = \frac{(n+1)!}{n!} = \frac{(n+1)\cancel{n!}}{\cancel{n!}} = n+1$ Yes, it is arithmetic with $d = 1$.

37. $a_n = a_1 + (n-1)d = -4 + (n-1)(5) = 5n - 9$

39. $a_n = 1 + (n-1)\left(-\frac{2}{3}\right) = -\frac{2}{3}n + \frac{5}{3}$

41. We are given that

$$a_5 = a_i + (5-1)d = 13,$$
$$a_{17} = a_i + (17-1)d = 37.$$

In order to find a_n, we must first find *d*. To do so, subtract the above formulae to eliminate a_1:

$$\begin{array}{r} a_1 + 4d = 13 \\ -\quad a_1 + 16d = 37 \\ \hline -12d = -24 \end{array}$$

Hence, $d = 2$. So, $a_n = a_1 + (n-1)(2)$.

To find a_1, observe that

$$a_5 - a_1 = 2(4) = 8$$

So, $13 - a_1 = 8$ so that $a_1 = 5$.

Thus, $a_n = 5 + (n-1)(2) = 2n + 3, n \geq 1$

43. We are given that

$$a_8 = 52 = a_1 + (8-1)d,$$
$$a_{21} = 130 = a_1 + (21-1)d.$$

In order to find a_n, we must first find *d*. To do so, subtract the above formulae to eliminate a_1:

$$\begin{array}{r} a_1 + 7d = 52 \\ -\quad a_1 + 20d = 130 \\ \hline -13d = -78 \end{array}$$

Hence, $d = 6$. So, $a_n = a_1 + (n-1)(6)$.

To find a_1, observe that

$$a_8 - a_1 = 6(7) = 42$$

So, $52 - a_1 = 42$, so that $a_1 = 10$.

Thus, $a_n = 10 + (n-1)6 = 6n + 4,\ n \geq 1$.

45. $\sum_{k=1}^{20} 3k = \frac{20}{2}(3+60) = 630$

47. The sequence is arithmetic with $n=12$, $a_1 = 2$, and $a_{12} = 68$. So, the given sum is $\frac{12}{2}(2+68) = 420$.

49.

Bob	Tania
$a_n = 45{,}000 + (n-1)2000$	$a_n = 38{,}000 + 4000(n-1)$
$\sum_{n=1}^{15} a_n = \frac{15}{2}\left[45{,}000 + \underbrace{73{,}000}_{=a_{15}}\right] = 885{,}000$	$\sum_{n=1}^{15} a_n = \frac{15}{2}\left[38{,}000 + \underbrace{94{,}000}_{=a_{15}}\right] = 990;000$

So, Bob earns $885,000 in 15 years, while Tania earns $990,000 in 15 years.

51. Yes, it is geometric with $r = -2$.

53. Yes, it is geometric with $r = \frac{1}{2}$.

55. $3, 6, 12, 24, 48$

57. $100, -400, 1600, -6400, 25{,}600$

59. $a_n = a_1 r^{n-1} = 7 \cdot 2^{n-1}$, $n \geq 1$

61. $a_n = 1(-2)^{n-1}$, $n \geq 1$

63. $a_n = 2(2)^{n-1}, n \geq 1$ So, $a_{25} = 2(2)^{24} = 33{,}554{,}432$

65. $a_n = 100\left(\frac{-1}{5}\right)^{n-1}$, $n \geq 1$. So, in particular, $a_{12} = 100\left(\frac{-1}{5}\right)^{11} = -2.048 \times 10^{-6}$

67. $\sum_{n=1}^{9}\left(\frac{1}{2}\right)3^{n-1} = \frac{1}{2}\left[\frac{1-3^9}{1-3}\right] = \frac{19{,}682}{4} = 4920.50$

69. $\sum_{n=1}^{8} 5(3)^{n-1} = 5\left[\frac{1-3^8}{1-3}\right] = 16{,}400$

71. $\frac{1}{1-\frac{2}{3}} = 3$

73. $a_n = 48{,}000(1.02)^{n-1}$, $n \geq 1$. So, in particular, $a_{12} = 48000(1.02)^{13-1} = 60{,}875.61$. So, the salary after 12 years is $60,875.61.

75. Claim: $3n \le 3^n$, for all $n \ge 1$.

Proof.

Step 1: Show the statement is true for $n = 1$.

$3(1) \le 3^1$ is true since both terms equal 3.

Step 2: Assume the statement is true for $n = k$: $3k \le 3^k$

Show the statement is true for $n = k+1$: $3(k+1) \le 3^{k+1}$

$$\begin{aligned} 3(k+1) &= 3k+3 \\ &\le 3^k + 3^1 \quad \text{(by assumption)} \\ &\le 3^k + 3^k \quad (\text{since } 3 \le 3^k, \text{ for } k \ge 1) \\ &= 2(3^k) < 3(3^k) = 3^{k+1} \end{aligned}$$

This completes the proof. ■

77. Claim: $2 + 7 + \ldots + (5n-3) = \frac{n}{2}(5n-1)$, for all $n \ge 1$.

Proof.

Step 1: Show the statement is true for $n = 1$.

$2 = \frac{1}{2}(5(1)-1) = \frac{1}{2}(4)$ is clearly true.

Step 2: Assume the statement is true for $n = k$: $2 + 7 + \ldots + (5k-3) = \frac{k}{2}(5k-1)$

Show the statement is true for $n = k+1$:

$$2 + 7 + \ldots + (5(k+1)-3) = \tfrac{k+1}{2}(5(k+1)-1)$$

Observe that

$$\begin{aligned} 2 + 7 + \ldots + (5k-3) + (5(k+1)-3) &= \big(2 + 7 + \ldots + (5k-3)\big) + (5(k+1)-3) \\ &= \tfrac{k}{2}(5k-1) + (5(k+1)-3) \quad \text{(by assumption)} \\ &= \tfrac{k(5k-1)+2(5k+2)}{2} \\ &= \tfrac{5k^2+9k+4}{2} = \tfrac{(k+1)(5k+4)}{2} = \tfrac{k+1}{2}(5(k+1)-1) \end{aligned}$$

This completes the proof. ■

79.

$$\binom{11}{8} = \frac{11!}{(11-8)!8!} = \frac{11!}{3!8!} = \frac{11 \cdot 10 \cdot 9 \cdot \cancel{8!}}{\cancel{8!}(3 \cdot 2 \cdot 1)} = 165$$

81.

$$\binom{22}{22} = \frac{22!}{(22-22)!22!} = \frac{\cancel{22!}}{\cancel{22!}0!} = \frac{1}{0!} = 1$$

83.

$$(x-5)^4 = \sum_{k=0}^{4} \binom{4}{k} x^{4-k}(-5)^k$$

$$= \binom{4}{0} x^4(-5)^0 + \binom{4}{1} x^3(-5)^1 + \binom{4}{2} x^2(-5)^2 + \binom{4}{3} x^1(-5)^3 + \binom{4}{4} x^0(-5)^4$$

$$= \underbrace{\tfrac{4!}{4!0!}}_{=1} x^4(-5)^0 + \underbrace{\tfrac{4!}{3!1!}}_{=4} x^3(-5)^1 + \underbrace{\tfrac{4!}{2!2!}}_{=6} x^2(-5)^2 + \underbrace{\tfrac{4!}{1!3!}}_{=4} x^1(-5)^3 + \underbrace{\tfrac{4!}{0!4!}}_{=1} x^0(-5)^4$$

$$= x^4 - 20x^3 + 150x^2 - 500x + 625$$

85.

$$(2x-5)^3 = \sum_{k=0}^{3} \binom{3}{k} (2x)^{3-k}(-5)^k$$

$$= \binom{3}{0}(2x)^3(-5)^0 + \binom{3}{1}(2x)^2(-5)^1 + \binom{3}{2}(2x)^1(-5)^2 + \binom{3}{3}(2x)^0(-5)^3$$

$$= \underbrace{\tfrac{3!}{3!0!}}_{=1}(2x)^3(-5)^0 + \underbrace{\tfrac{3!}{2!1!}}_{=3}(2x)^2(-5)^1 + \underbrace{\tfrac{3!}{1!2!}}_{=3}(2x)^1(-5)^2 + \underbrace{\tfrac{3!}{0!3!}}_{=1}(2x)^0(-5)^3$$

$$= 8x^3 - 60x^2 + 150x - 125$$

87.

$$(\sqrt{x}+1)^5 = \sum_{k=0}^{5} \binom{5}{k} (\sqrt{x})^{5-k}(1)^k, \quad x \ge 0$$

$$= \binom{5}{0}(\sqrt{x})^5(1)^0 + \binom{5}{1}(\sqrt{x})^4(1)^1 + \binom{5}{2}(\sqrt{x})^3(1)^2 + \binom{5}{3}(\sqrt{x})^2(1)^3$$

$$+ \binom{5}{4}(\sqrt{x})^1(1)^4 + \binom{5}{5}(\sqrt{x})^0(1)^5$$

$$= \underbrace{\tfrac{5!}{5!0!}}_{=1}(\sqrt{x})^5(1)^0 + \underbrace{\tfrac{5!}{4!1!}}_{=5}(\sqrt{x})^4(1)^1 + \underbrace{\tfrac{5!}{3!2!}}_{=10}(\sqrt{x})^3(1)^2 + \underbrace{\tfrac{5!}{2!3!}}_{=10}(\sqrt{x})^2(1)^3$$

$$+ \underbrace{\tfrac{5!}{1!4!}}_{=5}(\sqrt{x})^1(1)^4 + \underbrace{\tfrac{5!}{0!5!}}_{=1}(\sqrt{x})^0(1)^5$$

$$= x^{5/2} + 5x^2 + 10x^{3/2} + 10x + 5x^{1/2} + 1$$

89.

$$(r-s)^5 = r^5s^0 + 5r^4(-s) + 10r^3(-s)^2 + 10r^2(-s)^3 + 5r(-s)^4 + (-s)^5r^0$$

$$= r^5 - 5r^4s + 10r^3s^2 - 10r^2s^3 + 5rs^4 - s^5$$

91. This term is $\binom{8}{2} x^{8-2}(-2)^2 = \frac{8!}{2!6!}(4)x^6 = \frac{8\cdot 7\cdot \cancel{6!}}{(2\cdot 1)\cancel{6!}}(4)x^6 = 112x^6$

So, the desired coefficient is 112.

93. This term is

$$\binom{6}{4}(2x)^{6-4}(5y)^4 = \frac{6!}{4!2!}(4\cdot 625)x^2y^4 = \frac{6\cdot 5\cdot \cancel{4!}}{(2\cdot 1)\cancel{4!}}(2500)x^2y^4 = 37{,}500x^2y^4$$

So, the desired coefficient is 37,500.

95. Since $\binom{53}{6} = \frac{53!}{6!47!} = \frac{53\cdot 52\cdot 51\cdot 50\cdot 49\cdot 48\cdot \cancel{47!}}{6!\cancel{47!}} = 22{,}957{,}480$, there are 22,957,480 different 6-number combinations.

97. ${}_7P_4 = \frac{7!}{(7-4)!} = \frac{7!}{3!} = \frac{7\cdot 6\cdot 5\cdot 4\cdot \cancel{3!}}{\cancel{3!}} = 840$

99. ${}_{12}P_5 = \frac{12!}{(12-5)!} = \frac{12!}{7!} = \frac{12\cdot 11\cdot 10\cdot 9\cdot 8\cdot \cancel{7!}}{\cancel{7!}} = 95{,}040$

101. ${}_{12}C_7 = \frac{12!}{(12-7)!7!} = \frac{12!}{7!5!} = \frac{12\cdot 11\cdot 10\cdot 9\cdot 8\cdot \cancel{7!}}{\cancel{7!}(5\cdot 4\cdot 3\cdot 2\cdot 1)} = 792$

103. ${}_9C_9 = \frac{9!}{(9-9)!9!} = \frac{9!}{0!9!} = 1$

105. There are $\underbrace{3}_{\text{Models}} \cdot \underbrace{2}_{\text{Interior}} \cdot \underbrace{5}_{\text{Exterior}} = 30$ different cars to choose from.

107. Since all 4 jobs are different with different responsibilities, we know order is important in how we fill the slots. As such, there are

${}_{10}P_4 = \frac{10!}{(10-4)!} = \frac{10!}{6!} = \frac{10\cdot 9\cdot 8\cdot 7\cdot \cancel{6!}}{\cancel{6!}} = 5040$ different leadership teams.

109. There are $\underbrace{5}_{\text{Seat 1}} \cdot \underbrace{4}_{\text{Seat 2}} \cdot \underbrace{3}_{\text{Seat 3}} \cdot \underbrace{2}_{\text{Seat 4}} \cdot \underbrace{1}_{\text{Seat 5}} = 120$ different seating arrangements.

Since there are 8 home games in a season, it would take $\frac{120}{8} = 15$ years to go through all possible arrangements.

111. Since the prizes are different, we know order is important. So, there are

${}_{100}P_4 = \frac{100!}{(100-4)!} = \frac{100!}{96!} = \frac{100\cdot 99\cdot 98\cdot 97\cdot \cancel{96!}}{\cancel{96!}} = 94{,}109{,}400$ different ways to distribute the prizes.

113. Since $\binom{52}{6} = \frac{52!}{6!46!} = \frac{52\cdot 51\cdot 50\cdot 49\cdot 48\cdot 47\cdot \cancel{46!}}{6!\cancel{46!}} = 20{,}358{,}520$, there are 20,358,520 different 6-card hands.

115. The sample space has $2^4 = 16$ elements and there is only one way to get all heads. So, $P(HHHH) = \frac{1}{16} = 0.0625$. So, there is a 6.25% chance of getting all heads.

117. (The sample space for this problem can be found in Example 4 of Section 9.7 of the text.)

$$P(\text{not getting } 7) = 1 - P(\text{getting } 7) = 1 - \tfrac{6}{36} = \tfrac{30}{36} = 0.83\overline{3}$$

So, there is about an 83.3% chance of not getting 7.

119.
$P(\text{not } E_1) = 1 - P(E_1) = 1 - \tfrac{1}{3} = \tfrac{2}{3} \approx 66.7\%$

121. Since E_1 and E_2 are not mutually exclusive,

$$P(E_1 \cup E_2) = P(E_1) + P(E_2) - P(E_1 \cap E_2)$$
$$= \tfrac{1}{3} + \tfrac{1}{2} - \tfrac{1}{4} = \tfrac{7}{12} \approx 58.3\%$$

123. Since these two events are mutually exclusive, we see that

$$P(\text{drawing ace or drawing 2}) = P(\text{drawing ace}) + P(\text{drawing 2})$$
$$= \tfrac{4}{52} + \tfrac{4}{52} = \tfrac{8}{52} = \tfrac{2}{13} \approx 15.4\%.$$

125. The sample space in this problem has $2^5 = 32$ elements. So, we see that

$$P(\text{having at least one girl}) = 1 - P(\text{having no girls})$$
$$= 1 - P(\text{having all boys})$$
$$= 1 - \tfrac{1}{32} = \tfrac{31}{32} = 96.88\%$$

127. $\frac{5369}{3600}$

129. $\frac{34{,}875}{14}$

131. The graphs of the two curves are given by:

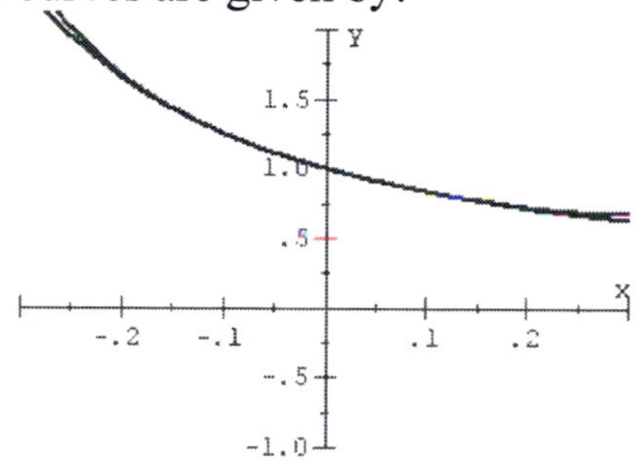

The series will sum to $\dfrac{1}{1+2x}$.

133. The sum is 99,900. Yes, it agrees with Exercise 77.

135. As each new term is added, the graphs become better approximations of the graph of $y = (1+2x)^4$ for $-0.1 < x < 0.1$. The series does not get closer to this graph for $0.1 < x < 1$.

137. a. 11,440 **b.** 11,440 **c.** Yes

d. This is true because $\frac{{}_nP_r}{r!} = {}_nC_r$.

139. Using the formula with $n = 10$ and $k = 9$ (and then cubing the result since the events are independent) yields the approximate probability of 0.0722.

Chapter 9 Practice Test

1. x^{n-1}	**3.** $S_n = \dfrac{1-x^n}{1-x}$
5. We must have $\lvert x \rvert < 1$ in order for the sum in Problem 4 to exist. In such case, the sum is $\frac{1}{1-x}$.	
7. $\sum_{n=1}^{10} 3\cdot\left(\frac{1}{4}\right)^n = 3\cdot\left(\frac{1}{4}\right)\cdot\dfrac{1-\left(\frac{1}{4}\right)^{10}}{1-\frac{1}{4}} = 1-\left(\frac{1}{4}\right)^{10} \approx 1$	**9.** $\sum_{n=1}^{100}(5n-3) = \dfrac{100}{2}(2+497) = 24{,}950$
11. $\dfrac{7!}{2!} = \dfrac{7\cdot6\cdot5\cdot4\cdot3\cdot\cancel{2!}}{\cancel{2!}} = 2520$	**13.** $\binom{15}{12} = \dfrac{15!}{12!3!} = \dfrac{15\cdot14\cdot13\cdot\cancel{12!}}{3!\cancel{12!}} = 455$
15. ${}_{14}P_3 = \dfrac{14!}{(14-3)!} = \dfrac{14!}{11!} = \dfrac{14\cdot13\cdot12\cdot\cancel{11!}}{\cancel{11!}} = 2184$	

17.

$$\left(x^2+\tfrac{1}{x}\right)^5 = \sum_{k=0}^{5}\binom{5}{k}\left(x^2\right)^{5-k}\left(\tfrac{1}{x}\right)^k,\ x\neq 0$$

$$=\binom{5}{0}\left(x^2\right)^5\left(\tfrac{1}{x}\right)^0+\binom{5}{1}\left(x^2\right)^4\left(\tfrac{1}{x}\right)^1+\binom{5}{2}\left(x^2\right)^3\left(\tfrac{1}{x}\right)^2$$

$$+\binom{5}{3}\left(x^2\right)^2\left(\tfrac{1}{x}\right)^3+\binom{5}{4}\left(x^2\right)^1\left(\tfrac{1}{x}\right)^4+\binom{5}{5}\left(x^2\right)^0\left(\tfrac{1}{x}\right)^5$$

$$=\underbrace{\tfrac{5!}{5!0!}}_{=1}\left(x^2\right)^5\left(\tfrac{1}{x}\right)^0+\underbrace{\tfrac{5!}{4!1!}}_{=5}\left(x^2\right)^4\left(\tfrac{1}{x}\right)^1+\underbrace{\tfrac{5!}{3!2!}}_{=10}\left(x^2\right)^3\left(\tfrac{1}{x}\right)^2$$

$$+\underbrace{\tfrac{5!}{2!3!}}_{=10}\left(x^2\right)^2\left(\tfrac{1}{x}\right)^3+\underbrace{\tfrac{5!}{1!4!}}_{=5}\left(x^2\right)^1\left(\tfrac{1}{x}\right)^4+\underbrace{\tfrac{5!}{0!5!}}_{=1}\left(x^2\right)^0\left(\tfrac{1}{x}\right)^5$$

$$=x^{10}+5x^7+10x^4+10x+\tfrac{5}{x^2}+\tfrac{1}{x^5}$$

19. Intuitively, there are more permutations than combinations since order is taken into account when determining the number of permutations, while order does not increase the number of combinations. More precisely, observe that

$${}_nC_r = \frac{n!}{r!(n-r)!} = \frac{1}{r!}\left(\frac{n!}{(n-r)!}\right) = \frac{1}{r!}\,{}_nP_r.$$

Since $r!\geq 1$, it follows that $\frac{1}{r!}\leq 1$, so that ${}_nP_r \geq {}_nC_r$.

21. $P(\text{red}) = \frac{18}{38} \cong 0.47$. So, about 47% chance.

23. Since spins are independent, the previous history has no effect on the probability of a particular spin. So, this probability is again $\frac{18}{38} \cong 0.47$.

25. Since these two events are not mutually exclusive (since there is an ace of diamonds), the probability is:

$$P(\text{ace or diamond}) = P(\text{ace}) + P(\text{diamond}) - P(\text{ace AND diamond})$$
$$= \tfrac{4}{52} + \tfrac{13}{52} - \tfrac{1}{52} = \tfrac{16}{52} = \tfrac{4}{13} \approx 0.308$$

27. Expanding this using a calculator reveals that the constant term is 184,756.

Chapter 9 Cumulative Review

1.

$$\frac{10x+1}{3x-2} + \frac{5+4x}{2-3x} = \frac{10x+1}{3x-2} - \frac{5+4x}{3x-2} = \frac{10x+1-(5+4x)}{3x-2} = \frac{6x-4}{3x-2} = \frac{2(3x-2)}{3x-2} = 2$$

3.

$$x = \frac{5 \pm \sqrt{(-5)^2 - 4(2)(11)}}{2(2)} = \frac{5 \pm \sqrt{-63}}{4} = \frac{5 \pm 3i\sqrt{7}}{4}$$

5. Since the slope is undefined, the line is vertical and hence, is described by an equation of the form $x = a$. Since the point (-8,0) is on the line, the equation is $x = -8$.

7.

$$\frac{f(x+h)-f(x)}{h} = \frac{(x+h)^2 - 3(x+h) + 2 - (x^2 - 3x + 2)}{h} = \frac{x^2 + 2hx + h^2 - 3x - 3h + 2 - x^2 + 3x - 2}{h} = \frac{h(2x+h-3)}{h} = 2x + h - 3$$

9.

$$f(x) = -0.04x^2 + 1.2x - 3$$
$$= -0.04(x^2 - 30x) - 3$$
$$= -0.04(x^2 - 30x + 225) - 3 + 9$$
$$= -0.04(x-15)^2 + 6$$

So, the vertex is (15,6).

11. Vertical asymptote: $x = 3$
Horizontal asymptote: $y = -5$

13. $\log_2 6 = \frac{\ln 6}{\ln 2} \approx 2.585$

15. Solve the system:

$$\begin{cases} 8x - 5y = 15 & \textbf{(1)} \\ y = \frac{8}{5}x + 10 & \textbf{(2)} \end{cases}$$

Substitute **(2)** into **(1)**:

$$8x - 5\left(\tfrac{8}{5}x + 10\right) = 15 \Rightarrow 0 = 65$$

Since this results in a false statement, the system has no solution.

17. The region is as follows:

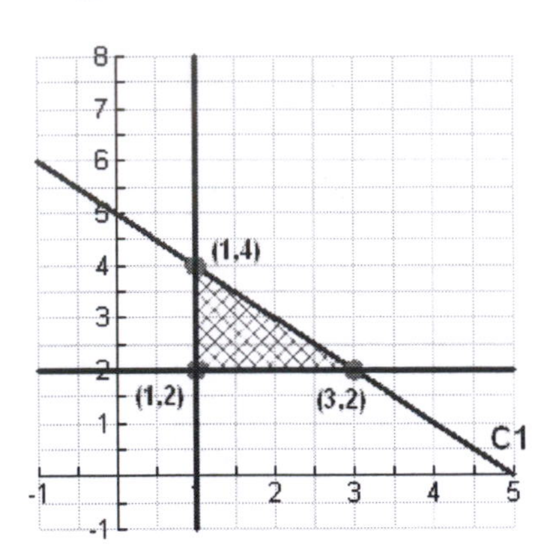

Vertex	$z = 4x + 5y$
(1,4)	24 MAX
(1,2)	14
(3,2)	22

So, the maximum is $z(1,4) = 24$.

Note on the graph:
C1: $y = 5 - x$

19.

$$\begin{bmatrix} 9 & 0 \\ 1 & 2 \end{bmatrix} \begin{bmatrix} 11 & 2 & -1 \\ 9 & 1 & 4 \end{bmatrix} = \begin{bmatrix} 99 & 18 & -9 \\ 29 & 4 & 7 \end{bmatrix}$$

21. Since (3,5) is the vertex and the directrix is $x = 7$, the parabola opens to the left and has general equation

$$-4p(x-3) = (y-5)^2$$

To find p, note the distance from vertex to directrix is 4 units, so $p = 4$. Hence, the equation is

$$-16(x-3) = (y-5)^2$$

$$x = -\frac{1}{16}(y-5)^2 + 3$$

23.

$$\frac{2^0}{1!} + \frac{2^1}{2!} + \frac{2^2}{3!} + \frac{2^3}{4!} = 1 + 1 + \tfrac{2}{3} + \tfrac{1}{3} = 3$$

25. $2^{10} = 1024$

27. Using a calculator to expand this quantity then reveals that the constant term is -3,432.